Recent Developments of DIAZO COMPOUNDS IN ORGANIC SYNTHESIS

Recent Developments of
DIAZO COMPOUNDS IN ORGANIC SYNTHESIS

Di Qiu
Tianjin Normal University, China

Jianbo Wang
Peking University, China

World Scientific

NEW JERSEY • LONDON • SINGAPORE • BEIJING • SHANGHAI • HONG KONG • TAIPEI • CHENNAI • TOKYO

Published by

World Scientific Publishing Europe Ltd.
57 Shelton Street, Covent Garden, London WC2H 9HE
Head office: 5 Toh Tuck Link, Singapore 596224
USA office: 27 Warren Street, Suite 401-402, Hackensack, NJ 07601

Library of Congress Cataloging-in-Publication Data
Names: Qiu, Di, 1967– author. | Wang, Jianbo, 1962– author.
Title: Recent developments of diazo compounds in organic synthesis /
Di Qiu, Tianjin Normal University, China; Jianbo Wang, Peking University, China.
Description: New Jersey : World Scientific, [2021] | Includes bibliographical references and index.
Identifiers: LCCN 2020032983 | ISBN 9781786348890 (hardcover) |
ISBN 9781786348906 (ebook for institutions) | ISBN 9781786348913 (ebook for individuals)
Subjects: LCSH: Diazo compounds--Synthesis. | Catalysts. | Chemical reactions.
Classification: LCC QD305.A9 Q58 2021 | DDC 547/.043--dc23
LC record available at https://lccn.loc.gov/2020032983

British Library Cataloguing-in-Publication Data
A catalogue record for this book is available from the British Library.

For any available supplementary material, please visit
https://www.worldscientific.com/worldscibooks/10.1142/Q0261#t=suppl

Desk Editors: Ramya Gangadharan/Michael Beale/Shi Ying Koe

Typeset by Stallion Press
Email: enquiries@stallionpress.com

Preface

Since the first preparation of ethyl diazoacetate by Curtius in 1883, diazo compounds have remained as attractive research topics over the past 137 years. Throughout the long history of research, various transformations based on diazo compounds have been established and some of them have been widely applied both in research laboratories and in industry. The 1998 classic book by Doyle, McKervey and Ye, *Modern Catalytic Methods for Organic Synthesis with Diazo Compounds*, which comprehensively summarized the chemistry of diazo compounds, has significantly promoted the related research. Subsequently, the chemistry related to diazo compounds remains as a fast-growing research area and new reports on these types of compounds appear in the literature on an almost day-to-day basis. Significant progress has been made in the asymmetric catalytic transformations with diazo compounds in the past 20 years. In particular, the reactions involving donor/acceptor carbene have shown remarkable regio- and stereoselectivity in asymmetric catalysis. On the other hand, various novel transformations of diazo compounds have been developed. For examples, diazo compounds have found applications in multi-component reactions, cross-coupling reactions, polymerizations, continuous-flow chemistry and chemical biology. Besides, diazo compounds have been frequently applied in the synthesis of natural products, drug molecules and other functional compounds. In view of the rapid developments in the field, it becomes appropriate to have a new book that can cover the recent developments of diazo compounds. This is the background for writing the current book, which summarizes the major developments made in the past 20 years in the field.

The book is divided into 16 chapters with the focus on reaction developments and synthetic applications with diazo compounds. The new developments in the preparative methods of diazo compounds and the transition metal catalysts are introduced in Chapters 2 and 3. Diazo compounds have been most widely used as the precursors for generating metal carbenes in transition-metal-catalyzed carbene transfer reactions. Thus, Chapters 4–6 cover the major carbene transfer reactions, namely the cyclopropanations, insertions and ylide generation and subsequent transformations. Other transition-metal-catalyzed reactions, such as Wolff rearrangement, olefinations, vinyl diazoacetate transformations, 1,3-dipole generation and the subsequent [3+2] cycloadditions are collected in Chapters 7–10. In addition to serving as carbene precursors, diazo compounds have also been extensively explored as either nucleophiles or electrophiles. These transformations are summarized in Chapter 11. In recent years, a significant development in transition-metal-catalyzed reaction of diazo compounds is their application in cross-coupling reactions, which is concisely introduced in Chapter 12. In addition, the polymerization with diazo compounds has attracted significant attention in recent years. The progress in this area is summarized in Chapter 13. The application of flow chemistry in the transformations of diazo compounds will without doubt enhance their usefulness, particularly in scale-up synthesis. The rapid growth along this line is shown in Chapter 14. The continuous and wide applications of diazo compounds in natural product synthesis have been demonstrated in Chapter 15. Finally, in Chapter 16 we have presented some of the reactions that have not been categorized in the aforementioned chapters.

The literature related to diazo compounds are enormous; however, due to the limited space of the book, unfortunately only a small portion of the representative work could be presented here. Nevertheless, from the limited work shown in this book, it is not difficult to conclude that the research on diazo compounds is still very active regardless of its long history. Thus, it can be expected that innovative applications of diazo compounds in organic synthesis will emerge continuously in the coming years.

We wish to thank the team of World Scientific, in particular Mary Simpson, Jennifer Brough and Koe Shi Ying for their support and patience in the past few years. We would also like to greatly appreciate Ramya Gangadharan for editing our manuscript and Michael Beale and his team for designing the cover art. Finally, we are indebted to the members of our research group at Peking University for their support with proof reading.

About the Authors

Di Qiu was born in Tianjin, China. He received his B.S. degree from Peking University in 2010. He was a graduate student in Prof. Jianbo Wang's laboratory from 2010 to 2015. Di Qiu obtained his Ph.D. degree in 2015. His research projects are focused on the development of basic organic functional group transformations, in particular the gold-catalyzed C–H bond functionalization and the transformation of aromatic C–N bonds based on diazonium salts. Di Qiu has published 20 papers in renowned international journals. Since 2015, Di Qiu has been an Assistant Professor at the College of Chemistry, Tianjin Normal University.

Jianbo Wang is a Professor of Chemistry at the College of Chemistry, Peking University, China. He received his B.S. degree from Nanjing University of Science and Technology in 1983, and his Ph.D. from Hokkaido University (under the supervision of Prof. Hiroshi Suginome) in 1990. He was a postdoctoral associate at the University of Geneva from 1990 to 1993 with Prof. Charles W. Jefford, and at the University of Wisconsin-Madison from 1993 to

1995 with Prof. Howard E. Zimmerman and Prof. Leonard A. Fahien. He began his independent academic career at Peking University in 1995 as an Associate Professor and was promoted to full Professor in 1999.

His research interests have focused on catalytic metal carbene transformations. In the past few years, his group has extensively studied the combination of cross-coupling reaction and metal carbene reaction. Another recent research topic in his group is the development of basic organic functional group transformations that are practically useful in organic synthesis. Jianbo Wang serves as associate editor for the *Journal of Physical Organic Chemistry* (Wiley) and *Chinese Journal of Organic Chemistry*. He is on the editorial advisory board of *Organic Letters*, *Tetrahedron* and *Tetrahedron Letters*.

Abbreviations

Abbreviation	**Chemical Name**
Ac	acetyl
acac	acetylacetonyl
ADMC	2-azido-1,3-dimethylimidazoliniumchloride
AIBN	2,2′-azo bisisobutyronitrile
aq	aqueous
B_2pin_2	bis(pinacolato)diboron
Bdan	1,8-naphthalenediaminatoboryl
BINAP	2,2′-bis(diphenylphosphino)-1,1′-binaphthyl
BINOL	1,1′-bi-2,2′-naphthol
Bn	benzyl
Boc	*tert*-butoxycarbonyl
BOR	Bestmann–Ohira reagent
Bpin	pinacolboryl
BPR	back pressure regular
BQ	Benzoquinone
Bz	Benzoyl
Cbz	Benzyloxycarbonyl
CFC	convection flow coil
CFL	compact fluorescent lamp
CHCR	C–H functionalization of donor–acceptor carbenoids/Cope Rearrangement
cod	1,5-cyclooctadiene
Cp	cyclopentadienyl

CPME	cyclopentyl methyl ether
$Cu(TBS)_2$	bis(*N-tert*-butylsalicylaldiminato) copper(II)
CuAAC	Cu-catalyzed azide-alkyne cycloaddition
DABCO	1,4-diazabicyclo[2.2.2]octane
dba	dibenzylideneacetone
DBU	1,8-diazabicycloundec-7-ene
DCE	1,2-dichloroethane
DCM	dichloromethane
DDQ	2,3-dichloro-5,6-dicyano-1,4-benzoquinone
de	diastereomeric excess
DEAD	diethyl azodicarboxylate
DIPEA(Hünig's base)	diisopropylethylamine
DMAP	*N,N*-4-dimethylaminopyridine
DMB	*m*-dimethoxybenzene
DMDO	dimethyl dioxirane
DMF	*N,N*-dimethylformamide
DMP	Dess–Martin periodinane
DMSO	dimethyl sulphoxide
DNA	deoxyribonucleic acid
dppf	1,1′-bis(diphenylphosphino)ferrocene
dppp	1,3-bis(diphenylphosphino)propane
dr	diastereomeric ratio
DTBP	2,6-di-*tert*-butylpyridine
EC	electrocyclization
EDA	ethyl diazoacetate
EDCl	1-ethyl-3-(3-dimethylaminopropyl) carbodiimide hydrochloride
EDG	electron-donating group
ee	enantiomeric excess
EPR	electron paramagnetic sesonance
er	enantiomeric ratio
EWG	electron-withdrawing group
Fmoc	9-fluorenylmethoxycarbonyl
hfacac	hexafluoroacetylacetone
HFIP	1,1,1,3,3,3-hexafluoro-2-propanol (hexafluoroisopropanol)
HMPA	hexamethylphosphoric acid triamide
HOMO	highest occupied molecular orbital
HWE	Horner–Wadsworth–Emmons reaction

*h*υ	irradiation with light
IR	infrared spectroscopy
KHMDS	potassium bis(trimethylsilyl)amide
LAH	lithium aluminum hydride
LDA	lithium diisopropylamide
LED	light emitting diode
LHMDS (LiHMDS)	lithium bis(trimethylsilyl)amide
L-selectride	lithium tri-*sec*-butylborohydride
LUMO	lowest unoccupied molecular orbital
MDA	methyl diazoacetate
MEM	methoxyethoxymethyl
Mes	mesityl
MOF	metal-organic framework
MOM	methoxymethyl
MS	molecular sieves
Ms	mesyl
MTBE	methyl *tert*-butyl ether
MW	microwave
NBS	*N*-bromosuccinimide
NCS	*N*-chlorosuccinimide
NHC	*N*-heterocyclic carbene
NMM	*N*-methylmorpholine
NMP	*N*-methyl-2-pyrrolidinone
NMR	nuclear magnetic resonance
oxone	potassium peroxymonosulfate
p-ABSA	*p*-acetamidobenzenesulfonyl azide
pfb	perfluorobutyrate
phen	phenanthroline
Piv	pivaloyl
PMP	*p*-methoxyphenyl
PPA	phosphoric acid
psi	pounds per square inch
PTC	phase-transfer catalyst
rac	racemic
RCM	ring-closing metathesis
$Rh_2(cap)_4$	dirhodium(II) caprolactamate
$Rh_2(esp)_2$	bis[rhodium(α,α,α′,α′-tetramethyl-1,3-benzenedipropionic acid)]
$Rh_2(oct)_4$	rhodium(II) octanoate

RNA	ribonucleic acid
Rose Bengal	2,4,5,7-tetraiodo-3′,4′,5′,6′-tetrachlorofluoresceindisodium salt
rt	room temperature
Salen	*N,N′*-ethylenebis(salicylideneiminato) bis(salicylidene)ethylenediamine
SET	single electron transfer
SGR	Seyferth–Gilbert reagent
TBAB	tetra-*n*-butylammonium bromide
TBAC	tetra-*n*-butylammonium chloride
TBAF	tetra-*n*-butylammonium fluoride
TBAI	tetrabromocyclohexadienone
TBDPS	*tert*-butyldiphenylsilyl
TBHP	*tert*-butylhydroperoxide
TBME	*tert*-butyl methyl ether
TBS	*tert*-butyldimethylsilyl
TCP	5,10,15,20-tetrachloroporphyrin
TEBAC	benzyl trimethylammonium chloride
TEMPO	2,2,6,6-tetramethyl-1-piperidinyloxy
TES	triethylsilyl
Tf	trifluoromethanesulfonyl
TFA	trifluoroacetamide
TFDA	trimethylsilyl 2-(fluorosulfonyl)-2,2-difluoroacetate
TFE	2,2,2-trifluoroethanol
THF	tetrahydrofuran
THP	2-tetrahydropyran
TMEDA	*N,N,N′,N′*-tetramethylethylenediamine
TMS	trimethylsilyl
TON	turnover number
TPP	5,10,15,20-tetraphenylporphyrin
Tr	trityl (triphenylmethyl)

Contents

Chapter 1

Introduction

Diazo compounds are versatile organic compounds and have become increasingly important in organic synthesis and many other fields. In particular, α-diazocarbonyl compounds are relatively stable and can be easily handled experimentally. While these compounds can be conveniently synthesized with a series of classic methods, novel and practical preparation/generation methods for diazo compounds have been continuously developed. For example, the *in situ* generation of non-stabilized diazo compounds from *N*-tosylhydrazones in the presence of a base, namely Bamford–Stevens reaction, has been recently extensively explored in transition-metal-catalyzed carbene transfer reactions. These advances have made diazo compounds easily accessible and versatile starting materials or key intermediates in various transformations.

Diazo compounds have shown diverse reactivity in organic synthesis. They can directly serve as electrophiles or nucleophiles. Particularly, diazo compounds have been commonly utilized as free carbene or metal carbene precursors. The catalytic carbene transfer reactions have found wide applications in organic synthesis. Traditionally, these transformations include insertion, cyclopropanation, ylide formation, Wolff rearrangement, 1,2-migration and olefination. Although these reactions have been well established in the past decades, there have been significant new developments in recent years. In this context, the following challenges have been particularly in focus: (1) establishment of highly chemo-, regio- and enantioselective synthetic methodologies based on the above-mentioned classic reactions, in most cases in a catalytic manner; (2) exploring novel types of reactions of diazo compounds, through which unprecedented transformations are achieved; (3) mechanistic understanding based on both theoretical and experimental studies. Among these significant developments, the evolution of chiral catalysts for asymmetric carbene transfers provides highly efficient methodologies for the construction of carbon–carbon and carbon–heteroatom bonds with high level of enantiocontrol.

In addition to the traditional transformations, in recent years diazo compounds have been established as general cross-coupling partners in transition-metal-catalyzed reactions. The coupling reactions with diazo compounds, though still in their infancy, have drawn significant attention as a new type of C–C bond forming transformations. These reactions, which can be viewed as the combination of traditional cross-coupling with carbene migratory insertion, have significantly expanded the classic chemistry of diazo compounds. Besides, it is noteworthy that diazo chemistry has also found remarkably wide applications in other areas such as chemical biology, drug discovery, materials science, polymer synthesis and total synthesis of natural products and functional molecules.

Other new developments in the transformations of diazo compounds include carbene metathesis, C(sp^2)–H bond functionalization, cyclization/cycloaddition involving metallocarbene species and so on. The recent development of continuous processing technology (flow chemistry) is expected to open up new possibilities for the application of diazo compounds, especially those that are non-stabilized, in both laboratory research and industrial-scale production.

The book published in 1998 by Doyle, McKervery and Ye, *Modern Catalytic Methods for Organic Synthesis with Diazo Compounds: From Cyclopropanes to Ylides*, has comprehensively summarized the important aspects of diazo compounds reported up to the last decade of the 20th century. The book has had an enormous impact on the community and has been cited extensively. The chemistry of diazo compounds has been evolving rapidly since 1998, as demonstrated by more than 80 related reviews, accounts, perspectives and feature articles published in recent years. We thus deemed it would be an appropriate time to write a new book that could comprehensively summarize the up-to-date achievements of the field. This book is divided into 16 chapters and is primarily focused on the developments of diazo chemistry in the past nearly two decades. For the historical backgrounds and the literature prior to 1999, the authors referred to the book by Doyle, McKervery and Ye, as well as the review articles published before 1999.

References

For selected reviews of diazo compounds before 1999, see:

1. Meier, H.; Zeller, K.-P. *Angew. Chem., Int. Ed. Engl.* **1975**, *14*, 32–43.
2. Regitz, M.; Maas, G. *Diazo Compounds — Properties and Synthesis*, Academic Press, Orlando, 1986.
3. Taber, D. F.; Hennessy, M. J.; Hoerrner, R. S.; Raman, K.; Ruckle, R. E., Jr.; Schuchardt, J. S. *Cyclopentane Construction by Rh-Catalyzed Intramolecular C–H Insertion: Scope and Selectivity in Catalysis of Organic Reactions*, Blackburn, D.W., Ed., Marcel Dekker, Inc., New York, 1990, pp. 43–60.
4. Taber, D. F. *Carbon–Carbon Bond Formation by C–H Insertion. Comprehensive Organic Synthesis*, V. 3, Pattenden, G. Ed., Pergamon Press, Oxford, 1991.
5. Adams, J.; Spero, D. M. *Tetrahedron* **1991**, *47*, 1765–1808.
6. Padwa, A.; Krumpe, K. E. *Tetrahedron* **1992**, *48*, 5385–5453.
7. Ye, T.; McKervey, M. A. *Chem. Rev.* **1994**, *94*, 1091–1160.
8. Padwa, A.; Austin, D. J. *Angew. Chem. Int. Ed.* **1994**, *33*, 1797–1815.
9. Taber, D. F. *Formation of C–C Bonds by C–H Insertion. Houben-Weyl, Methods of Organic Chemistry*, V. E21, Helmchen, G. Ed., Georg Thiem Verlag, Stuttgart, 1995.
10. Padwa, A.; Weingarten, M. D. *Chem. Rev.* **1996**, *96*, 223–269.
11. Doyle, M. P.; McKervey, M. A. *J. Chem. Soc., Chem. Commun.* **1997**, 983–989.
12. Doyle, M. P.; Forbes, D. C. *Chem. Rev.* **1998**, *98*, 911–935.
13. Doyle, M. P.; McKervey, M. A.; Ye, T. *Modern Catalytic Methods for Organic Synthesis with Diazo Compounds*, Wiley, New York, 1998.

For reviews, accounts, perspective and feature articles published since 1998, see:

14. Padwa, A. *J. Organomet. Chem.* **2001**, *617–618*, 3–16.
15. Davies, H. M. L.; Antoulinakis, E. G. *J. Organomet. Chem.* **2001**, *617–618*, 47–55.
16. Timmons, D. J.; Doyle, M. P. *J. Organomet. Chem.* **2001**, *617–618*, 98–104.
17. Hodgson, D. M.; Pierard, F. Y. T. M.; Stupple, P. A. *Chem. Soc. Rev.* **2001**, *30*, 50–61.
18. Doyle, M. P.; Hu, W. *Synlett* **2001**, 1364–1370.
19. Kirmse, W. *Eur. J. Org. Chem.* **2002**, 2193–2256.
20. Davies, H. M. L.; Beckwith, R. E. J. *Chem. Rev.* **2003**, *103*, 2861–2903.
21. Singh, G. S.; Mdee, L. K. *Curr. Org. Chem.* **2003**, *7*, 1821–1839.
22. Heydt, H. *Sci. Synth.* **2004**, *27*, 843–935.

23. Gois, P. M. P.; Afonso, C. A. M. *Eur. J. Org. Chem.* **2004**, 3773–3788.
24. Maas, G. *Chem. Soc. Rev.* **2004**, *33*, 183–190.
25. Díaz-Requejo, M. M.; Pérez, P. J. *J. Organomet. Chem.* **2005**, *690*, 5441–5450.
26. Fulton, J. R.; Aggarwal, V. K.; de Vicente, J. *Eur. J. Org. Chem.* **2005**, 1479–1492.
27. Davies, H. M. L.; Nikolai, J. *Org. Biomol. Chem.* **2005**, *3*, 4176–4187.
28. Singh, G. S. *Curr. Org. Synth.* **2005**, *2*, 377–391.
29. Zhao, Y.; Wang, J. *Synlett* **2005**, 2886–2892.
30. Wee, A. G. H. *Curr. Org. Synth.* **2006**, *3*, 499–555.
31. Noels, A. F. *Angew. Chem. Int. Ed.* **2007**, *46*, 1208–1210.
32. Davies, H. M. L.; Manning, J. R. *Nature* **2008**, *451*, 417–424.
33. Hansen, J.; Davies, H. M. L. *Coord. Chem. Rev.* **2008**, *252*, 545–555.
34. Zhang, Z.; Wang, J. *Tetrahedron* **2008**, *64*, 6577–6605.
35. Zhang, Y.; Lu, Z.; Wulff, W. D. *Synlett* **2009**, *17*, 2715–2739.
36. Zhang, Y.; Wang, J. *Chem. Commun.* **2009**, 5350–5361.
37. Maas, G. *Angew. Chem. Int. Ed.* **2009**, *48*, 8186–8195.
38. Honma, M.; Takeda, H.; Takano, M.; Nakada, M. *Synlett* **2009**, 1695–1712.
39. Davies, H. M. L.; Denton, J. R. *Chem. Soc. Rev.* **2009**, *38*, 3061–3071.
40. Padwa, A. *Chem. Soc. Rev.* **2009**, *38*, 3072–3081.
41. Sweeney, J. B. *Chem. Soc. Rev.* **2009**, *38*, 1027–1038.
42. Padwa, A. *J. Org. Chem.* **2009**, *74*, 6421–6441.
43. Zhou, C.-Y.; Huang, J.-S.; Che, C.-M. *Synlett* **2010**, 2681–2700.
44. Zhang, Y.; Wang, J. *Coord. Chem. Rev.* **2010**, *254*, 941–953.
45. Slattery, C. N.; Ford, A.; Maguire, A. R. *Tetrahedron* **2010**, *66*, 6681–6705.
46. Doyle, M. P.; Duffy, R.; Ratnikov, M.; Zhou, L. *Chem. Rev.* **2010**, *110*, 704–724.
47. Jellema, E.; Jongerius, A. L.; Reek, J. N. H.; de Bruin, B. *Chem. Soc. Rev.* **2010**, *39*, 1706–1723.
48. Johnston, J. N.; Muchalski, H.; Troyer, T. L. *Angew. Chem. Int. Ed.* **2010**, *49*, 2290–2298.
49. Zhang, Y.; Wang, J. *Eur. J. Org. Chem.* **2011**, 1015–1026.
50. Barluenga, J.; Valdés, C. *Angew. Chem. Int. Ed.* **2011**, *50*, 7486–7500.
51. Nawrat, C. C.; Moody, C. J. *Nat. Prod. Rep.* **2011**, *28*, 1426–1444.
52. Doyle, M. P.; Ratnikov, M.; Liu, Y. *Org. Biomol. Chem.* **2011**, *9*, 4007–4016.
53. Franssen, N. M. G.; Walters, A. J. C.; Reek, J. N. H.; de Bruin, B. *Catal. Sci. Technol.* **2011**, *1*, 153–165.
54. Zhang, Z.; Zhang, Y.; Wang, J. *ACS Catal.* **2011**, *1*, 1621–1630.
55. Davies, H. M. L.; Morton, D. *Chem. Soc. Rev.* **2011**, *40*, 1857–1869.
56. Zhao, X.; Zhang, Y.; Wang, J. *Chem. Commun.* **2012**, *48*, 10162–10173.
57. Shao, Z.; Zhang, H. *Chem. Soc. Rev.* **2012**, *41*, 560–572.

58. Davies, H. M. L.; Lian, Y. *Acc. Chem. Res.* **2012**, *45*, 923–935.
59. Zhu, S.-F.; Zhou, Q.-L. *Acc. Chem. Res.* **2012**, *45*, 1365–1377.
60. Chen, D. Y.-K.; Pouwer, R. H.; Richard, J. *Chem. Soc. Rev.* **2012**, *41*, 4631–4642.
61. Xiao, Q.; Zhang, Y.; Wang, J. *Acc. Chem. Res.* **2013**, *46*, 236–247.
62. Liu, Z.; Wang, J. *J. Org. Chem.* **2013**, *78*, 10024–10030.
63. Xia, Y.; Zhang, Y.; Wang, J. *ACS Catal.* **2013**, *3*, 2586–2598.
64. Li, H.; Zhang, Y.; Wang, J. *Synthesis* **2013**, *45*, 3090–3098.
65. Gillingham, D.; Fei, N. *Chem. Soc. Rev.* **2013**, *42*, 4918–4931.
66. Guo, X.; Hu, W. *Acc. Chem. Res.* **2013**, *46*, 2427–2440.
67. Murphy, G. K.; Stewart, C.; West, F. G. *Tetrahedron* **2013**, *69*, 2667–2686.
68. Padwa, A.; Cheng, B.; Zou, Y. *Aust. J. Chem.* **2014**, *67*, 343–353.
69. Xu, X.; Doyle, M. P. *Aust. J. Chem.* **2014**, *67*, 365–373.
70. Zheng, C.; You, S.-L. *RSC Adv.* **2014**, *4*, 6173–6214.
71. McNamara, O. A.; Buckley, N. R.; O'Leary, P.; Harrington, F.; Kelly, N.; O'Keeffe, S.; Stack, A.; O'Neill, S.; Lawrence, S. E.; Slattery, C. N.; Maguire, A. R. *Tetrahedron* **2014**, *70*, 6870–6878.
72. Hu, F.; Xia, Y.; Ma, C.; Zhang, Y.; Wang, J. *Chem. Commun.* **2015**, *51*, 7986–7995.

For the recent comprehensive review of diazo compounds, see:

73. Ford, A.; Miel, H.; Ring, A.; Slattery, C. N.; Maguire, A. R.; McKervey, M. A. *Chem. Rev.* **2015**, *115*, 9981–10080.
74. DeAngelis, A.; Panish, R.; Fox, J. M. *Acc. Chem. Res.* **2016**, *49*, 115–127.
75. Candeias, N. R.; Paterna, R.; Gois, P. M. P. *Chem. Rev.* **2016**, *116*, 2937–2981.

Chapter 2

New Developments in the Synthesis of Diazo Compounds

Since the first diazo compound, ethyl diazoacetate (EDA), was synthesized more than 130 years ago,[1,2] the preparative methods for these synthetically valuable and versatile substrates have been continuously explored. Scheme 2.1 summarizes the representative methods that have been established and widely practiced.[3]

The approaches toward preparation of diazo compounds are summarized as follows: (a) diazo-transfer reactions; (b) diazotization of primary aliphatic amines bearing α-electron-withdrawing substituent; (c) dehydrogenation or oxidation of hydrazones; (d) from sulfonylhydrazones in the presence of base; (e) alkaline cleavage of *N*-alkyl-*N*-nitroso sulfonamides, carboxamides, ureas and urethanes; (f) triazene fragmentation or 1,2,3-triazoles; (g) diazo compounds as the nucleophiles in addition to aldehydes, ketones or imines; (h) substituent modification of the diazo compounds bearing diazo functionality. Although these preparative methods have been well developed and some of them are widely used, new methods and the improvement of the existing methods have also been pursued in the past decades. In the following sections, some selected new developments are presented.

2.1 Synthesis of diazo compounds from azides

Raines and co-workers have developed a mild method for the conversion of azides into the corresponding diazo compound derivatives by using

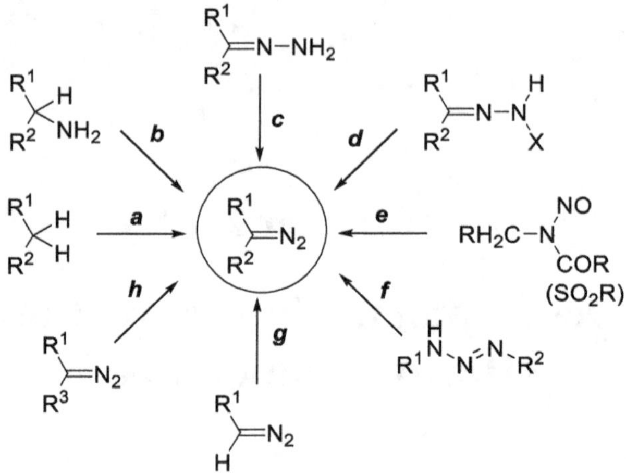

Scheme 2.1 General methods for the preparation of diazo compounds.

phosphine reagents.[4] First, azidoglycine and a series of phosphines bearing ester substituents were subjected to this reaction. The reaction of phosphine ester with azide only affords the Staudinger ligation product **1**. Initial success was achieved with phosphinothioester, which afforded a mixture of amide and α-diazoacetamide. Further study indicated that the reaction with succinimidyl ester cleanly gave acyltriazene **2**, which then underwent fragmentation leading to carboxamides **3** and α-diazoacetamide **4** (pathway *f*, Scheme 2.1). Based on these results, the (diphenylphosphanyl)propionate **5** was then chosen as the phosphine reagent, allowing the preparation of various functionalized diazo compounds in good yields (Scheme 2.2).

Subsequently, the authors have designed and optimized a series of phosphinoesters that mediate the efficient conversion of azides into diazo compounds in phosphate buffer at neutral pH and room temperature.[5] Thus, a water-soluble phosphinoester **6** was subjected to this transformation, which converted the azides into the corresponding diazo compounds with good tolerance of functional groups that are relevant to chemical biology (Scheme 2.3).

2.2 Generation of diazo compounds from hydrazones

When the diazo group is not stabilized by an adjacent electron-withdrawing group, the corresponding diazo compounds are in general not easy to

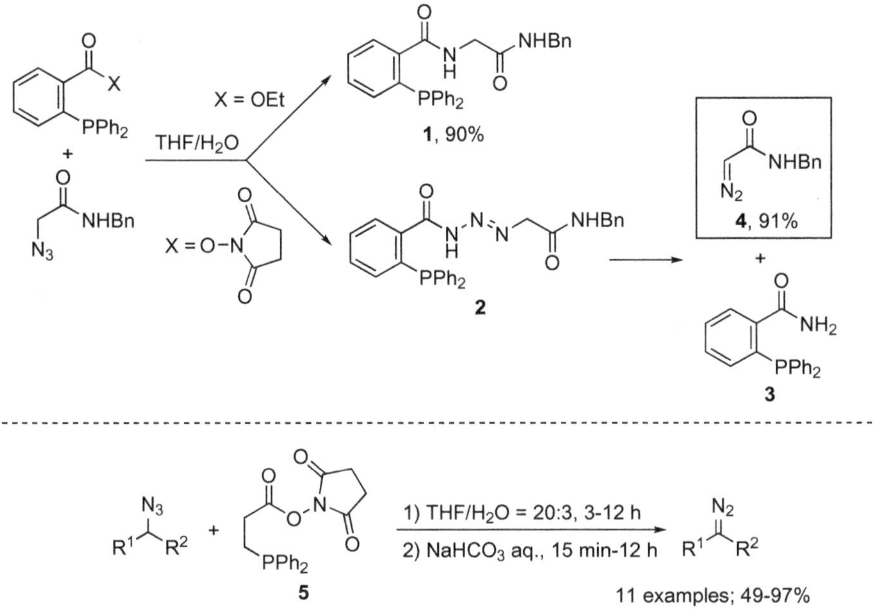

Scheme 2.2 Phosphine-mediated conversion of azides into diazo compounds.

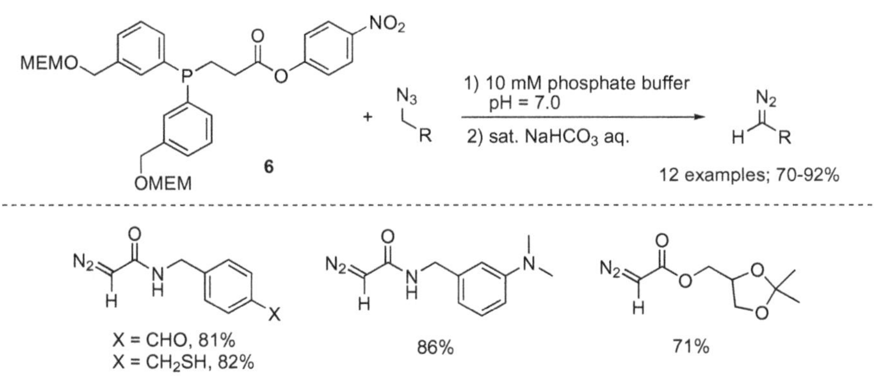

Scheme 2.3 Phosphine-mediated generation of diazo compounds in aqueous media.

handle because of their instability. In such cases, an *in situ* diazo generation strategy can be applied. Based on the Bamford–Stevens reaction, *N*-tosylhydrazone salts have been most successfully explored for the generation of diazo compounds under mild conditions (pathway ***d***, Scheme 2.1). The diazo compounds thus generated *in situ* can be coupled

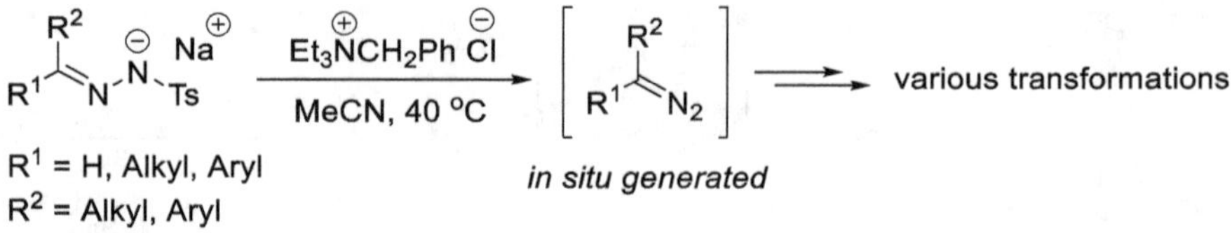

Scheme 2.4 Diazo compounds generated *in situ* from the salts of tosylhydrazones.

with various subsequent transformations, including transition-metal-catalyzed reactions (Scheme 2.4).[6]

The *in situ* diazo generation methods based on *N*-tosylhydrazones have been widely applied in transition-metal-catalyzed carbene transfer and carbene coupling reactions, as illustrated in the following chapters. For example, the *N*-alkylation of 1*H*-imidazole by a $Cu(acac)_2$-mediated reaction of diazoalkanes generated *in situ* from the corresponding *p*-toluenesulfonate hydrazones is reported by Cuevas-Yañez and co-workers (Eq. (2.1)).[7]

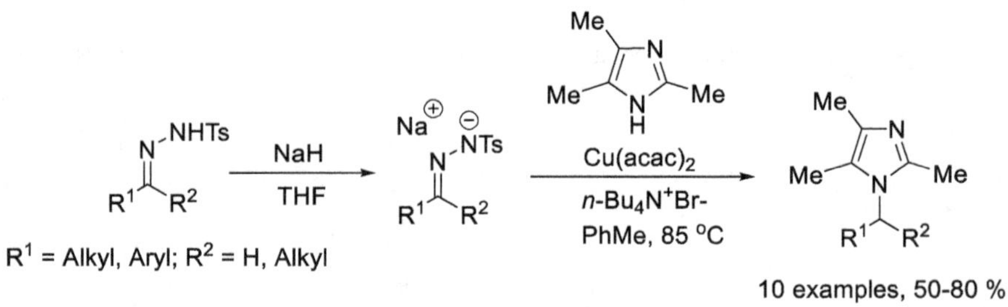

(2.1)

Besides, *N*-tosylhydrazones have been extensively used as reliable sources of unstable diazo compounds of theoretical interest. One of the examples is that reported by the group of McMahon, who has generated and characterized dialkynyl carbene species.[8] The triplet carbenes *tert*-butylpentadiynylidene and dimethylpentadiynylidene are produced photochemically from the corresponding diazo compound precursors, which are generated from the corresponding *N*-tosylhydrazones (Scheme 2.5).

The same group has also reported the synthesis of dialkynyl hydrazones from pentadiynone **7**. These compounds have been served as the precursors of dialkynyl-substituted diazomethane **8**, which are utilized for the generation of dialkynyl carbenes (Scheme 2.6).[9]

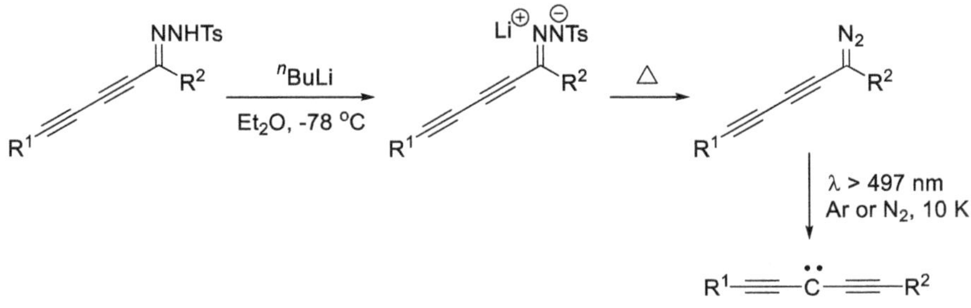

Scheme 2.5 Generation of dialkynyl carbene.

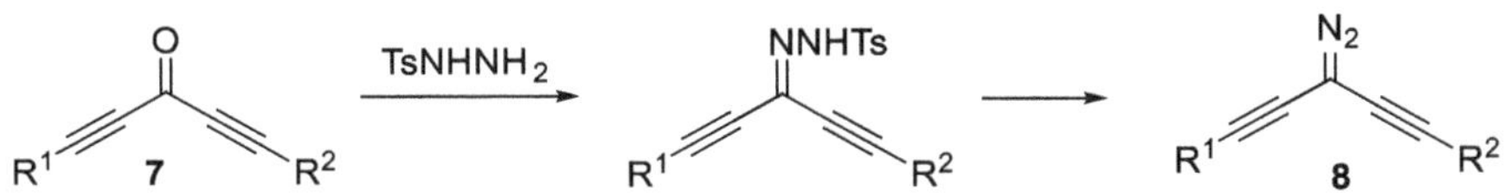

Scheme 2.6 Preparation of dialkynyl diazomethane.

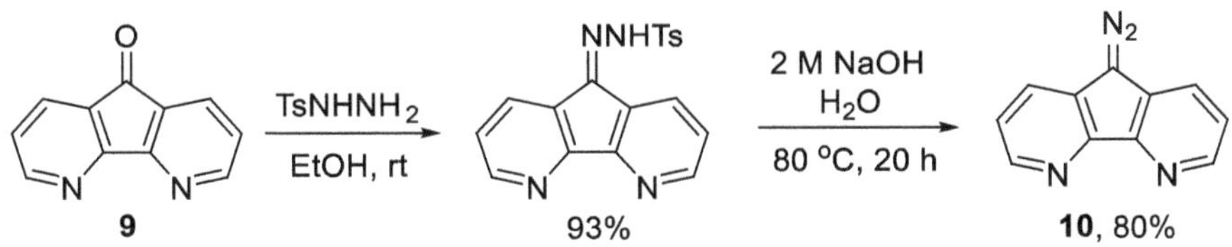

Scheme 2.7 Synthesis of diazofluorene derivatives.

Another example of the preparation of diazo compounds from *N*-tosylhydrazones is the synthesis of diazofluorene derivative **10** from the corresponding ketone **9** (Scheme 2.7).[10]

Diazo compounds can also be generated *in situ* from the precursors other than *N*-tosylhydrazone salts. Stoltz and May reported a stereoselective tandem rhodium-catalyzed Bamford–Stevens/Claisen reaction of *N*-aziridinylimines (Scheme 2.8).[11] The diazo compound generated *in situ* by thermal reaction of *N*-aziridinylimine was catalyzed by $Rh_2(OAc)_4$ to generate Rh(II) carbene intermediate **11**, which was followed by a 1,2-hydrogen shift to afford *Z*-enol ethers **12**. Further thermal Claisen rearrangement occurred from allylic enol ethers to deliver the final product, aldehyde.

Notably, the generation of diazo compounds from *N*-tosylhydrazone usually requires elevated reaction temperature, typically over 60°C. Thus,

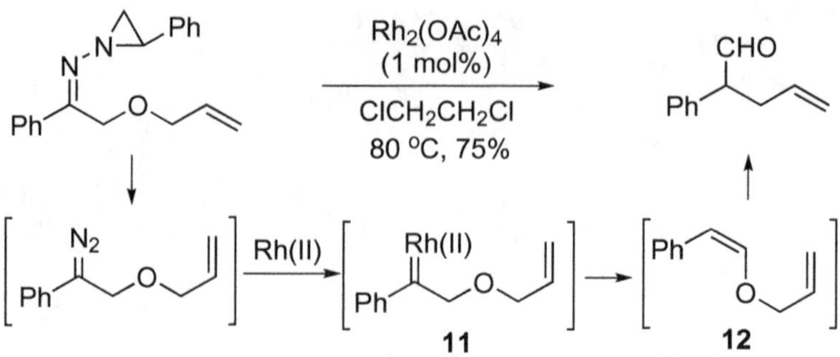

Scheme 2.8 Rh(II)-catalyzed reaction of diazo compound generated *in situ* by thermal reaction of *N*-aziridinylimines.

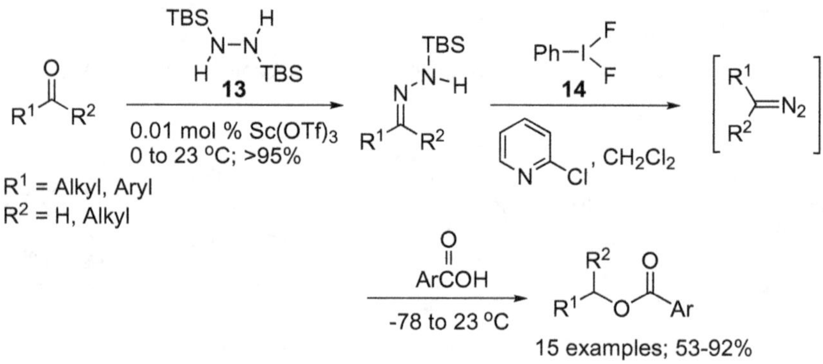

Scheme 2.9 Esterification of carboxylic acids with *in situ* generated diazoalkanes.

the protocol to generate diazo substrates at low temperature from readily available starting materials is highly desirable. In this context, Myers and Furrow have developed a general procedure for the esterification of carboxylic acids with diazoalkanes which are generated *in situ* by the oxidation of *N*-*tert*-butyldimethylsilyl hydrazones **13**. *N*-*tert*-butyldimethylsilyl hydrazones **13** can be easily prepared from aldehydes or ketones by the formation of hydrazones and subsequent reaction with (difluoroiodo) benzene **14** (Scheme 2.9).[12] The procedure can be applied to a wide range of diazo precursors and carboxylic substrates.

Fukuyama and co-workers described another *in situ* generation of diazo compounds at low temperature. As illustrated in Eq. (2.2), the diazoacetates are formed from the corresponding bromoacetates by the treatment with *N,N'*-ditosylhydrazine and DBU at 0°C (Eq. (2.2)).[13]

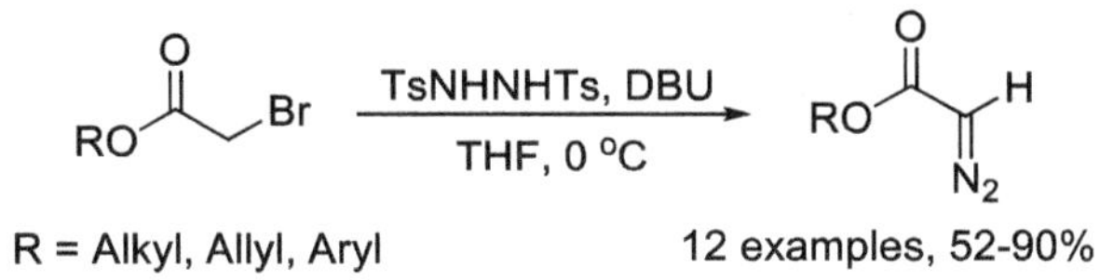

(2.2)

An alternative approach to diazo compounds is through oxidation of hydrazones, which has been known for a long time. For example, diazofluorenes can also be generated from hydrazones by mercury(II) oxidation.[10] Another example is illustrated in Eq. (2.3). A series of hydrazones are oxidized to the corresponding diazoalkanes using sodium hypochlorite (NaOCl) in the presence of catalytic amounts of TEMPO, followed by the esterification process with carboxylic acids.[14]

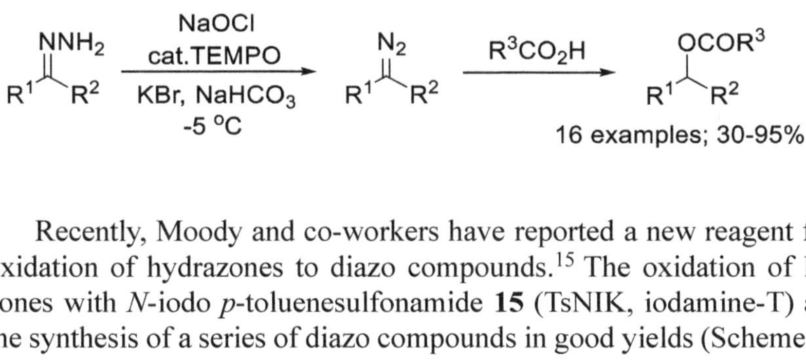

(2.3)

Recently, Moody and co-workers have reported a new reagent for the oxidation of hydrazones to diazo compounds.[15] The oxidation of hydrazones with *N*-iodo *p*-toluenesulfonamide **15** (TsNIK, iodamine-T) allows the synthesis of a series of diazo compounds in good yields (Scheme 2.10).

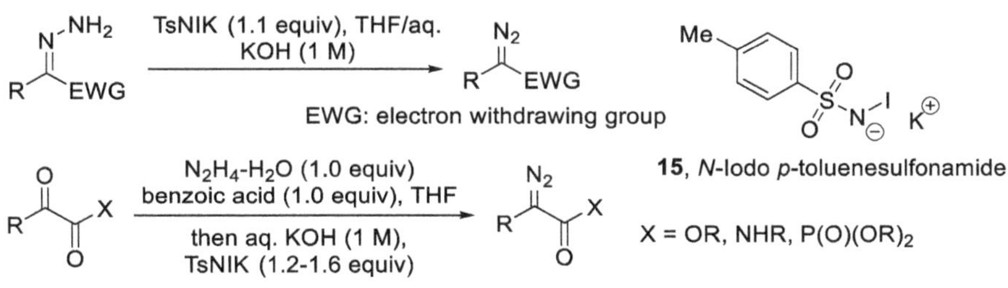

Scheme 2.10 Preparation of diazo compounds through the oxidation of hydrazones by TsNIK.

The method can also be performed in one-pot manner, affording the diazo compounds from the corresponding ketones in good yields.

Brewer reported an efficient dehydrogenating strategy of hydrazones by using "activated" dimethyl sulfoxide, affording the corresponding diazoalkanes under low temperature (Eq. (2.4)).[16] The diazo products can be either isolated in high yields, or be treated with carboxylic acids to afford the corresponding esters.

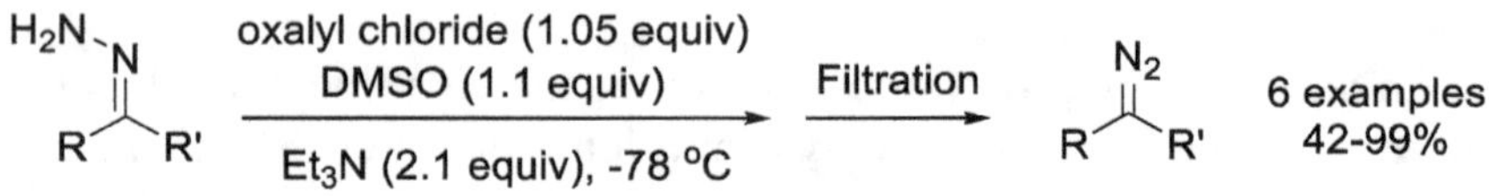

(2.4)

2.3 Synthesis of diazo compounds with diazomethane

Acylation of diazomethane by reaction of diazomethane with acyl halides or anhydrides, and the diazo transfer of an azide reagent to the carbon adjacent to a carbonyl group, represent two basic methods to synthesize diazocarbonyl compounds. Despite the hazardous nature of diazomethane, the former remains the single most important route to acyclic terminal α-diazoketones. Although these methods have been well established, improvements have been continuously pursued over the years. For example, efforts have been made to activate the acids under milder reaction conditions.[17–19] Addition of NBS to triphenylphosphine and carboxylic acids can generate an acylphosphonium salt **16**, which reacts with

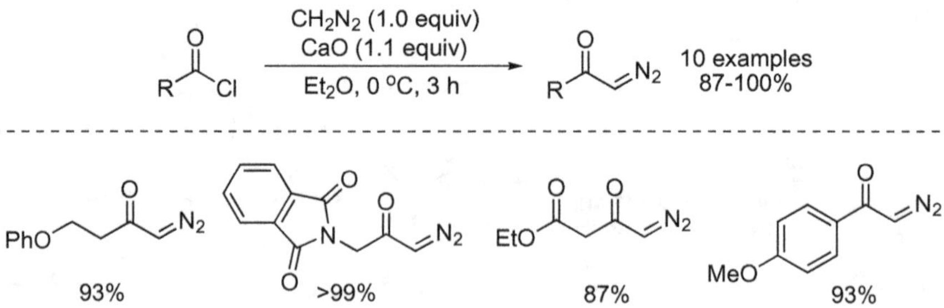

Scheme 2.11 Acylation of acyl chlorides with diazomethane in the presence of CaO.

diazomethane to afford α-diazoketones (Eq. (2.5)).[20] The reaction conditions are mild and various functional groups are tolerated.

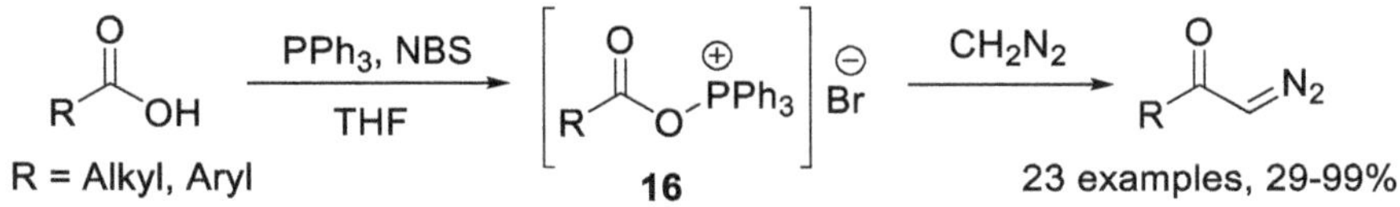

(2.5)

Historically, the synthesis of acyclic terminal α-diazocarbonyl substrates was closely related to classic Arndt–Eistert synthesis. This homologation approach, which is based on Wolff rearrangement, is a good preparative method for the chain elongation of carboxylic acids, and has found wide applications.[22] An improved Arndt–Eistert homologation was reported in 2010, in which a simple protocol based on the use of calcium oxide as acid scavenger was described.[21] An advantage of this method is the use of only stoichiometric amount of diazomethane, while in most of the transformations with diazomethane it is typically used in large excess (Scheme 2.11).

In 2012, Siciliano and Liguori reported a one-pot synthesis of *N*-9-fluorenylmethyloxycarbonyl-α-amino diazoketones with α-amino acids as the substrates.[23] The procedure involves *N*-protection with Fmoc-Cl in the presence of *N,N*-diisopropylethyl amine (DIEA), then *C*-activation with Fmoc-Cl, and finally Arndt–Eistert homologation, affording

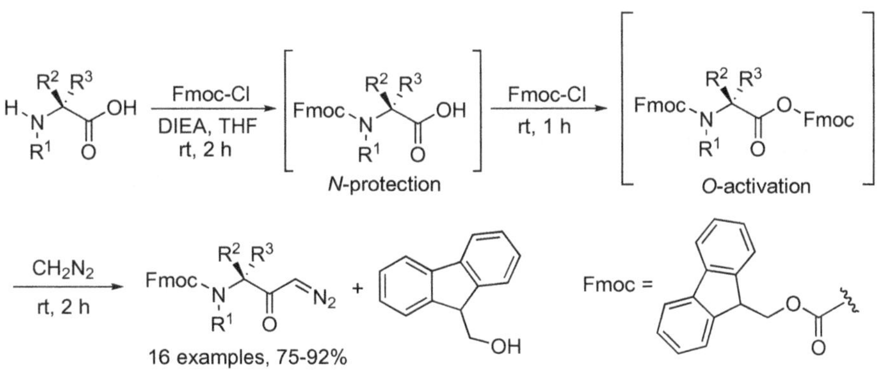

α-amino acid: Leu, Gly, Ala, Val, Ile, Phe, Pro, Gln, Ser(*t*-Bu), Thr(*t*-Bu), Lys(Boc), Asp(O*t*-Bu), Tyr(*t*-Bu), Trp, 2-aminoisobutyric acid.

Scheme 2.12 *N*-Fmoc-α-amino diazoketones from α-amino acids.

N-Fmoc-α-amino diazoketones in good to excellent isolated yields. A series of natural α-amino acids derivatives were subjected to this transformation (Scheme 2.12).

2.4 Synthesis of diazo compounds through diazo transfer reaction

The diazo transfer to an active methylene carbon represents the most effective method for directly introducing a diazo functionality into the carbon bearing electron-withdrawing substituent. Diazo group transfer reagents play a key role in these types of methods. Diazo-transfer reaction onto activated methylene compounds by using sulfonyl azides derivatives is a traditional method for the synthesis of α-diazo compounds (pathway ***a***, Scheme 2.1). A series of organoazide reagents, such as *p*-tosyl azide (TsN_3), mesyl azide (MsN_3) and *p*-acetamidobenzenesulfonyl azide (*p*-ABSA), have been well established as diazo transfer agents in this transformation in most of the cases. Nevertheless, new diazo group transfer reagents and procedures have been disclosed at times. For example, Charette and co-workers reported the highly electron-deficient trifluoromethanesulfonyl azide in the preparation of α-nitro-α-diazocarbonyl compounds, α-cyano-α-diazocarbonyl compounds and α-sulfonyl-α-diazocarbonyl compounds.[24,25]

Since diazo compounds are in general relatively unstable, it is desirable to improve the work-up procedure upon the completion of diazo transfer. For this purpose, polymer-bound sulfonyl azides have been employed as diazo transfer reagents. Flynn and Hanson have reported the development of a high-load, soluble oligomeric benzenesulfonyl azide.[26] The diazo transfer is efficient and the insoluble sulfonamide by-product can be completely removed by a single filtration through a silica gel SPE

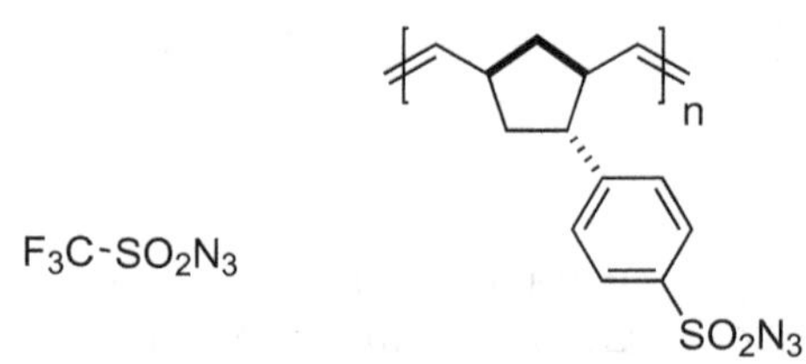

Figure 2.1 New diazo group transfer reagents.

cartridge. This method is particularly valuable when highly unstable α-azocarbonyl compounds need to be prepared (Figure 2.1).

Direct diazo transfer to the methylene group that is only activated by a single carbonyl group usually requires strong base such as LDA to deprotonate. An indirect method is to convert the ester or ketone into the corresponding α-formylated carbonyl compounds **17**, followed by a deformylating diazo transfer. α-Benzoylation and α-trifluoroacetylation have also been utilized for this transformation. In 2005, Taber and co-workers modified their α-benzoylation-based diazo transfer process by using a $TiCl_4$-mediated benzoylation of an ester. The diazo transfer was carried out directly to the crude benzoylated ester (Eq. (2.6)).[27]

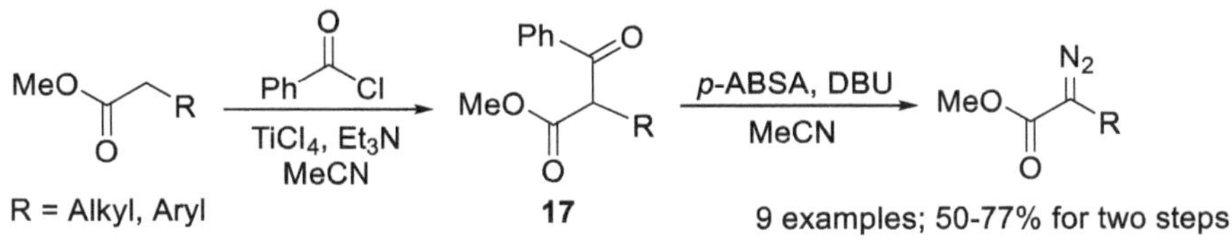

(2.6)

Recently, the upgrading of diazo transfer process has also been pursued. Significant improvement has been achieved for the diazo-transfer with TsN_3 as the reagent using ionic liquid as the solvent,[28] or using molecular sieves as the catalyst.[29] Gram-scale preparation of pure 1,3-dicarbonyl compounds with tosyl azide was reported.[30] Furthermore, a range of bicyclic and monocyclic α-diazo-β-oxosulfoxides has been synthesized.[31]

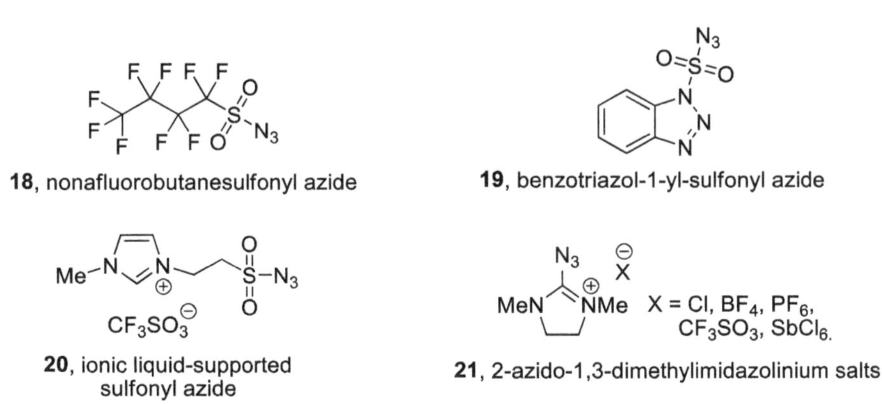

Figure 2.2 Functionalized sulfonyl azides as the diazo transfer reagents.

Along with TsN_3 and *p*-ABSA some newly developed diazo-transfer reagents were explored. Functionalized sulfonyl azides, such as nonafluorobutanesulfonyl azide **18**,[32] benzotriazol-1-yl-sulfonyl azide **19**,[33] ionic liquid-supported sulfonyl azide **20**[34] and 2-azido-1,3-dimethylimidazolinium salts **21**,[35–38] have been prepared and applied to generate α-diazo compounds under mild conditions (Figure 2.2). Moreover, an eco-friendly route to *tetra*-substituted diazodihydrofuranones was developed by using 2,4,6-tris(isopropyl)benzenesulfonyl azide (TIBSA).[39]

Bräse and Schroen synthesized various polymer-bound diazonium salts, such as **23**. These diazonium salts were used for the direct conversion of α-amino esters **22** into the corresponding α-diazoacetic esters (Eq. (2.7)).[40]

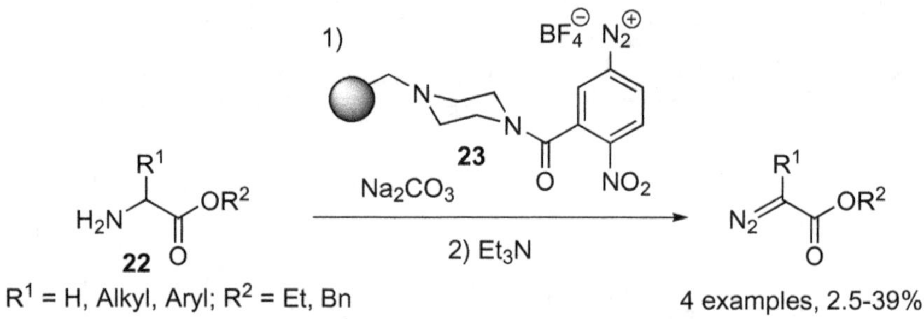

(2.7)

A deacylative oxidation strategy for the introduction of an alkoxy or amino group into the α position relative to a carbonyl group was developed by Brodsky and Du Bois. The trifluoroacetyl-substituted carbonyl compounds were used as the precursors for the generation of diazo compounds. Diazotization with *o*-nitrobenzenesulfonylazide (*o*-NBSA) by using Cs_2CO_3 as base generated the corresponding α-diazocarbonyl compounds, which were catalyzed by rhodium(II) octanoate $[Rh_2(oct)_4]$ *in situ* in the presence of an alcohol or amine (Eq. (2.8)).[41]

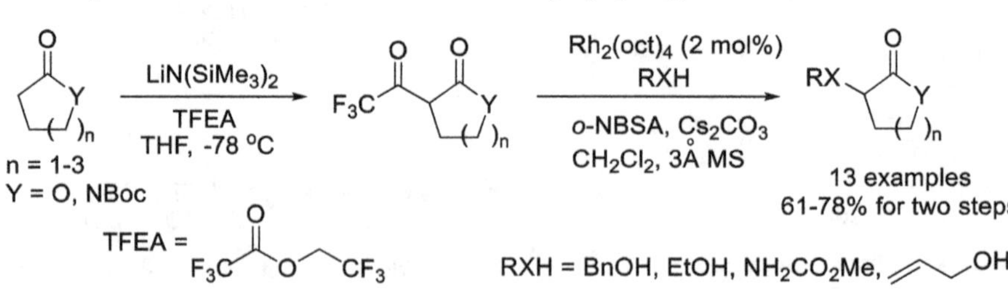

(2.8)

Recently, Kitamura and co-workers reported an interesting synthetic route for *ortho*-diazonaphthoquinones from the corresponding naphthol derivatives (Eq. (2.9)).[42] This diazo transfer reaction, which selectively introduced the diazo group at the *ortho* position of the hydroxy group, was implemented by using readily prepared 2-azido-1,3-dimethylimidazolinium chloride (ADMC) **24** as the diazo transfer reagent.

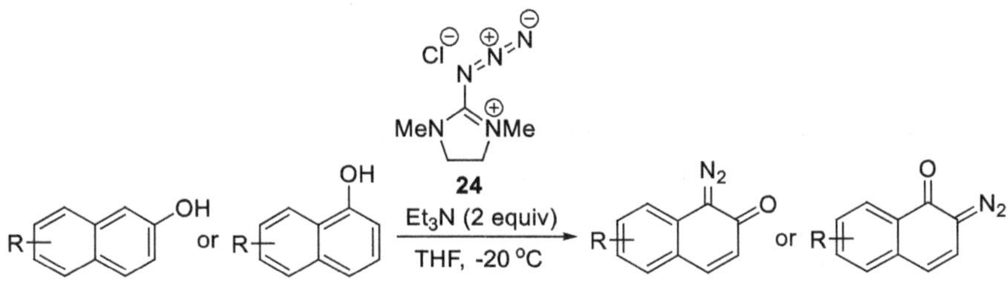

(2.9)

2.5 Diazotization of primary aliphatic amines

In 1943, Jones and co-workers reported the preparation of 2,2,2-trifluorodiazoethane **26** by the treatment of 2,2,2-trifluoroethylamine **25** with sodium nitrite (Eq. (2.10)).[43] This water-phase synthetic route is a useful method toward the facile preparation of 2,2,2-trifluorodiazoethane from readily available 2,2,2-trifluoroethylamine (pathway ***b***, Scheme 2.1).

25 **26** (2.10)

2,2,2-Trifluorodiazoethane **26** as useful synthon containing CF_3 group has attracted attention in recent years. For example, Carreira and co-workers reported the diastereoselective preparation of trifluoromethyl-substituted cyclopropanes,[44] as well as the synthesis of trifluoromethyl-substituted cyclopropenes.[45] Wang and co-workers developed the 2,2,2-trifluoroethylation and *gem*-difluorovinylation of organoboronic acids (Scheme 2.13).[46] From commercially available trifluoroethylamine hydrochloride and sodium nitrite, the solution of CF_3CHN_2 is generated *in situ* in aqueous media under mild conditions.

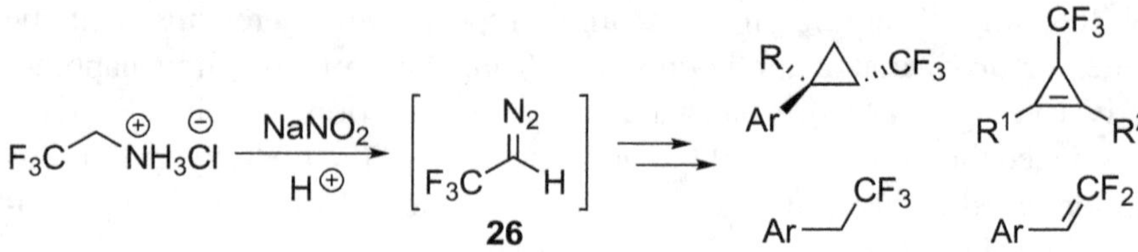

Scheme 2.13 Application of *in situ* generated CF_3CHN_2.

Mykhailiuk reported the *in situ* preparation of difluoromethyl diazomethane **27** (CF_2HCHN_2)[47] and diazoacetonitrile **28** (N_2CHCN)[48] from the corresponding primary amines with diazotization reagents. These active diazo species were further subjected to the reaction with alkynes or electron-deficient alkene to form [3+2] cycloadducts (Scheme 2.14).

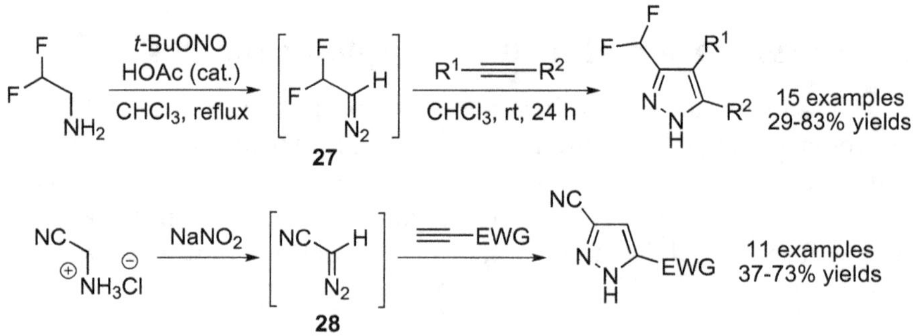

Scheme 2.14 Generation of CF_2HCHN_2 and N_2CHCN.

2.6 Derivatization of diazo compounds through nucleophilic reactions

Nucleophilic addition of diazomethane derivatives to electrophiles represents another important route to prepare functionalized diazo compounds (pathway ***g***, Scheme 2.1). Wang and co-workers reported the base-promoted aldol-type[49,50] and Mannich-type[49,51,52] condensation between diazocarbonyl compounds with aldehydes or imines, affording β-hydroxy- or β-amino-α-diazo carbonyl compounds with high efficiency (Scheme 2.15). The reactions were improved by using quarternary ammonium hydroxide[53] or phase-transfer catalysis.[54]

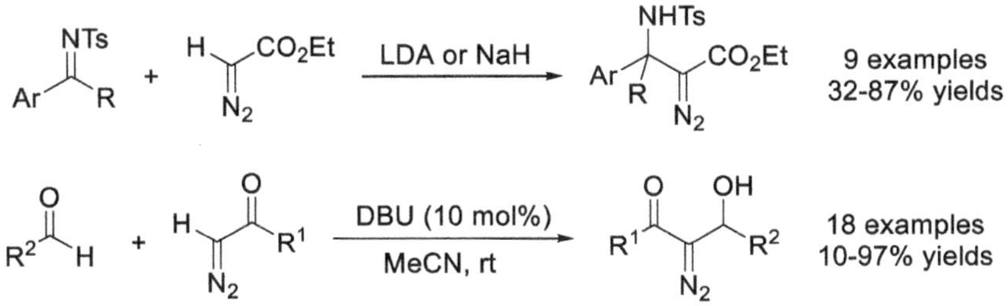

Scheme 2.15 Aldol-type and Mannich-type condensation.

A one-pot preparation of these products is reported by Steel group. The reaction proceeds *via* DBU-promoted aldol-type reaction and by *in situ* oxidation by IBX (Eq. (2.11)).[55] This simple method enabled direct preparation of diazoacetoacetate derivatives **29** from aldehydes.

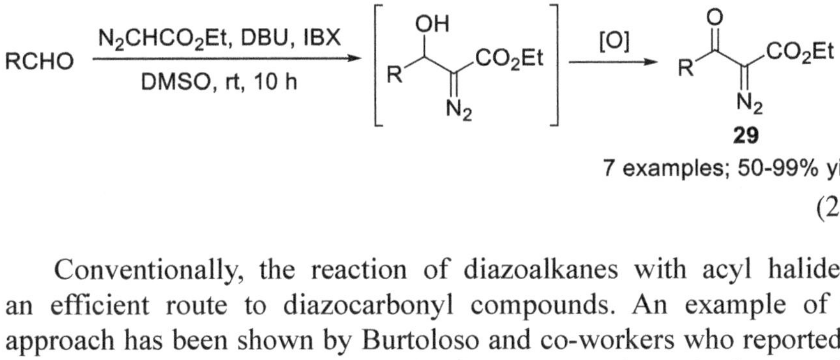

29

7 examples; 50-99% yields

(2.11)

Conventionally, the reaction of diazoalkanes with acyl halides is an efficient route to diazocarbonyl compounds. An example of this approach has been shown by Burtoloso and co-workers who reported the preparation of α,β-unsaturated diazoketones **30** through diazomethane homologation of phosphonate with the leaving group and sequential Horner–Wadsworth–Emmons (HWE) reaction with aldehydes.[56] A series of aryl or alkyl-substituted diazoketones were prepared in moderate to good yields by the two-step operation (Scheme 2.16).

Furthermore, an efficient and safe continuous-flow procedure toward the synthesis of aliphatic and aromatic diazoketones from acyl chloride with $TMSCHN_2$ was reported.[57] The effectiveness of the method was demonstrated by its application in the synthesis of quinoxalines in a multistep sequence without isolation of the potentially explosive diazoketone.

Recently, Rastogi and co-workers reported the synthesis of α-diazo-β-keto esters, phosphonates and sulfones *via* acylation of the corresponding diazomethyl anions with *N*-acylbenzotriazoles (Eq. (2.12)).[58] The reaction between dimethyl diazomethylphosphonate (DAMP or Seyferth–Gilbert

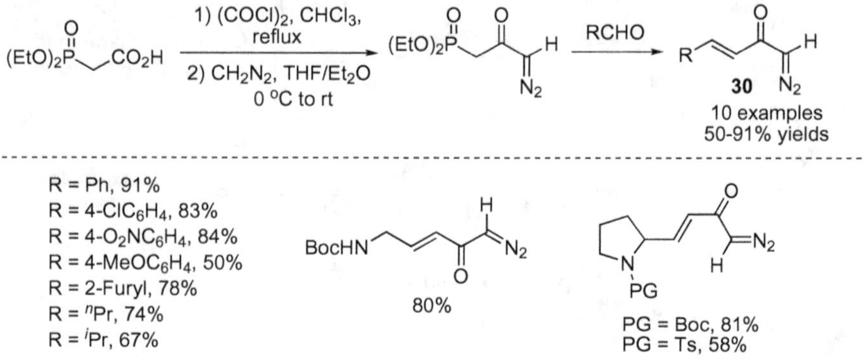

Scheme 2.16 Preparation of α,β-unsaturated diazoketones from aldehydes and HWE reagents.

reagent) with *N-o*-amino-acylbenzotriazoles **31** is a useful transphosphorylation procedure to access diazoacetyl phenylphosphoramidates.

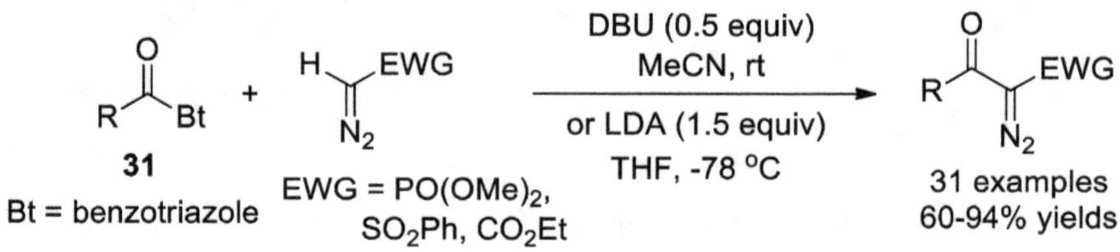

(2.12)

Hansen and co-workers reported a synthesis of α-halogenated analogues of EDA.[59] In the presence of base, various α-halodiazoacetates **32** are obtained with quantitative yields by using *N*-halosuccinimides as the electrophilic halogenation reagents (Eq. (2.13)). These halogenated diazo compounds were used in Rh(II)-catalyzed cyclopropanation of olefins, giving rise to halocyclopropanes in good to excellent yields.

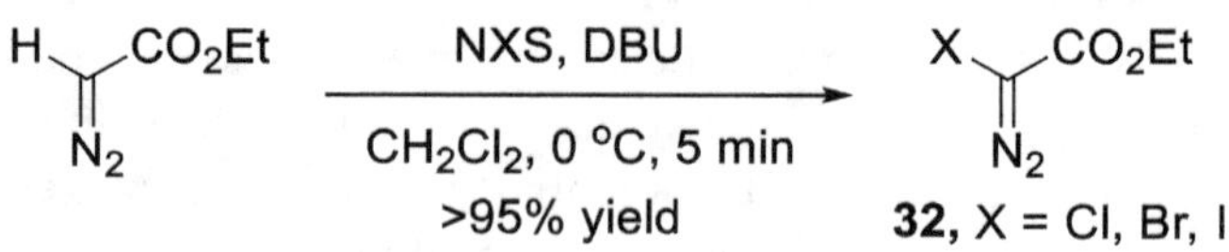

(2.13)

Additionally, the asymmetric catalytic aldol-type reaction of diazo compounds with aldehydes or imines is attractive as an entry into chiral diazo compounds. In 2003, Wang and co-workers reported the first direct aldol-type condensation of aldehydes with EDA in the presence of Zr(IV)-BINOL catalytic system (Eq. (2.14)).[60] The non-racemic β-hydroxy-α-diazo carbonyl compounds were obtained with moderate enantioselectivities (53–87% *ee*).

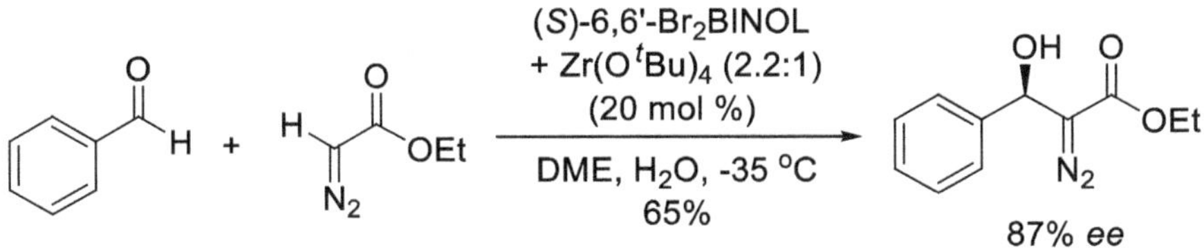

(2.14)

In 2004, the same group further reported the nucleophilic addition of the anion derived from diazocarbonyl compounds bearing a chiral auxiliary to imine derivatives with high level of diastereoselectivity (Eq. (2.15)).[61] Furthermore, the chiral oxazolidinone auxiliary could be easily removed under mild conditions. The diazo products were further converted into *anti*- and *syn*-α-hydroxy-β-amino acid derivatives.

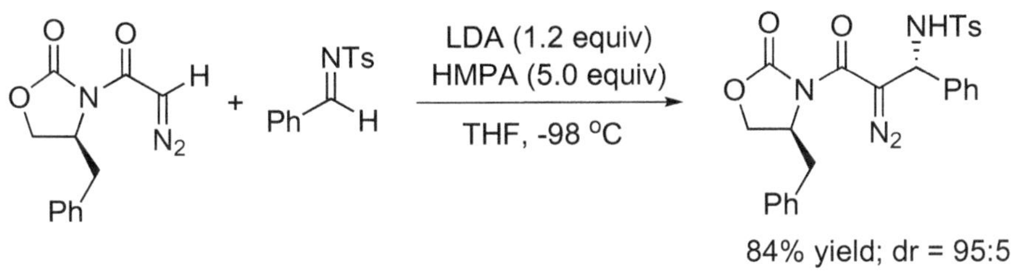

(2.15)

The stereoselective aldol-type and Mannich-type reactions with acyl-diazomethane were later further explored, which is discussed in detail in Chapter 11.

2.7 Synthesis of diazo compounds through Pd-catalyzed coupling reaction

In 2007, Wang and co-workers reported a Pd-catalyzed cross-coupling of iodides with EDA (Eq. (2.16)).[62] Remarkably, the diazo moiety remained intact in the reaction. The reaction afforded the coupling products in moderate to good yields. The reaction provides a new way to introduce diazo functionality into organic compounds. In addition, Pd-catalyzed three-component carbonylative cross-coupling reaction of CO, EDA and iodides was also implemented, affording β-keto-α-diazocarbonyl compounds in moderate yields. The coupling reaction was further optimized by using Ag_2CO_3 as the additive.[63] The usefulness of this reaction has been demonstrated by its application in the synthesis of functionalized diazocarboxylate.[64]

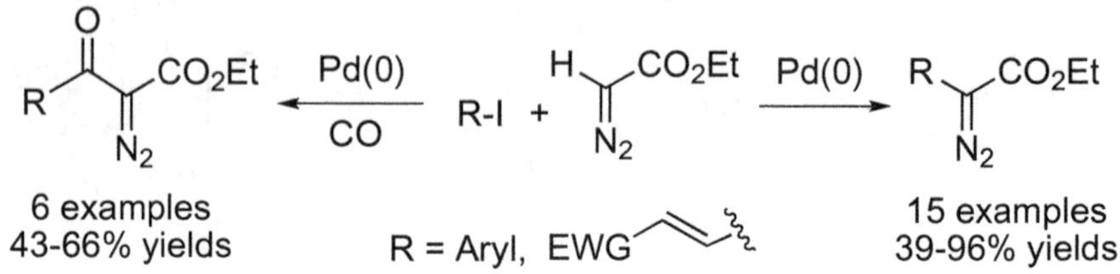

(2.16)

In 2011, Frantz and co-workers reported the concise synthesis of 3,4,5-trisubstituted pyrazoles through tandem cross-coupling/intramolecular electrocyclization procedure.[65] The key steps involve the *in situ* preparation of vinyl diazoesters from the electron-deficient enol triflates, followed by direct electrocyclization of **33** without isolation of the diazo intermediates (Eq. (2.17)).

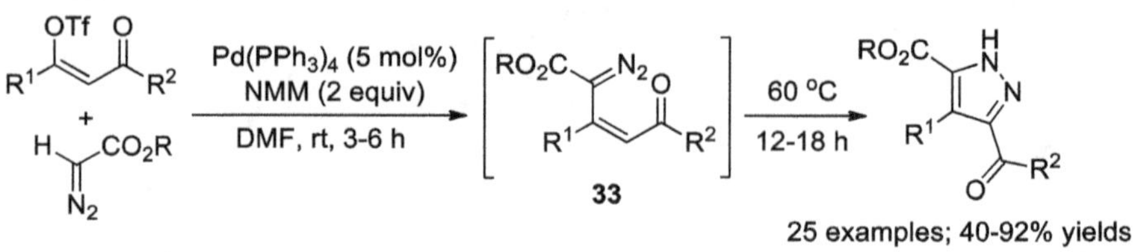

(2.17)

In 2012, Reissig and co-workers reported the Pd-catalyzed diazo coupling reaction of methyl diazoacetate with pyrid-4-yl nonaflates **34** as substrates (Eq. (2.18)).[66] This method has wide substrate scope, enabling facile synthesis of pyrid-4-yl-substituted methyl diazoacetates **35** with good yields.

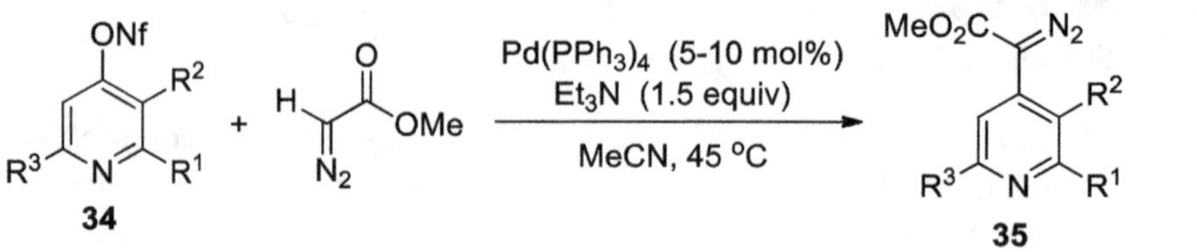

(2.18)

The scope of this type of coupling reaction was further expanded to the preparation of diazophosphonate by Beletskaya and co-workers.[67] The Pd-catalyzed cross-coupling of aryl halides and diethyl diazomethyl phosphonate gave α-aryl-diazophosphonates **36** in moderate to good yields (Eq. (2.19)).

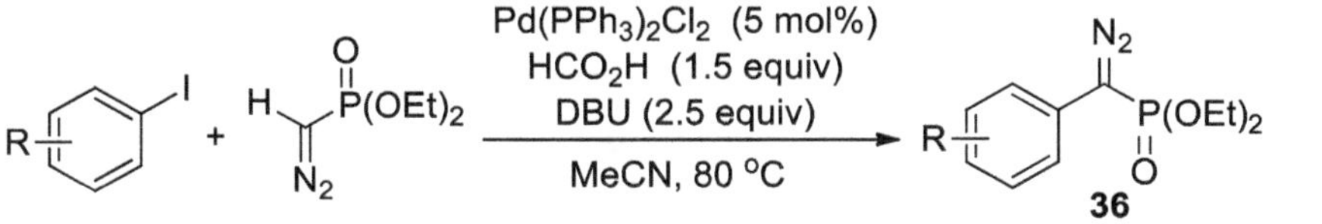

36
8 examples; 25-95% yields

(2.19)

In 2014, Wang and co-workers reported the synthesis of aryldiazoacetates through Pd-catalyzed deacylative cross-coupling from aryliodides with acyldiazoacetates (Eq. (2.20)).[68] This method combined the catalytic deacylative strategy with diazo coupling reaction. Aryldiazoacetates are traditionally prepared by direct diazo transfer to arylacetates. However, some arylacetates are not easily available. Compared to the arylacetates, aryliodides bearing various substitutes are more easily commercially available or can be readily prepared. Therefore, this deacylative cross-coupling provides an alternative access to this type of important donor–acceptor carbene precursor.

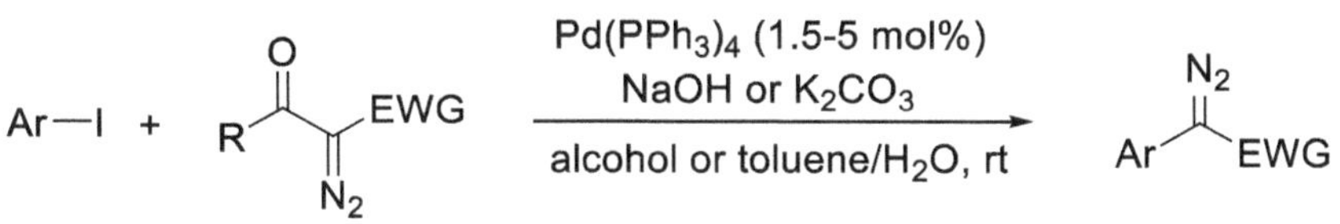

EWG = CO_2Me, $PO(OMe)_2$, COR'
R = Me, CF_3

52 examples
37-94% yields

(2.20)

In conclusion, the generation of versatile functionalized diazo compounds from readily available starting materials under mild reaction conditions still remains an interesting research topic to date. Although some well-established methods have been widely applied in organic synthesis, drawbacks still exist in one way or another. Further exploration of new preparation routes and the development of safe and efficient diazotizing reagents will be continued.

References

1. Curtius, T. *Ber. Dtsch. Chem. Ges.* **1883**, *16*, 2230–2231.
2. von Pechmann, H. *Ber. Dtsch. Chem. Ges.* **1894**, *27*, 1888–1891.
3. Maas, G. *Angew. Chem. Int. Ed.* **2009**, *48*, 8186–8195.
4. Myers, E. L.; Raines, R. T. *Angew. Chem. Int. Ed.* **2009**, *48*, 2359–2363.
5. Chou, H.-H.; Raines, R. T. *J. Am. Chem. Soc.* **2013**, *135*, 14936–14939.
6. Fulton, J. R.; Aggarwal, V. K.; de Vicente, J. *Eur. J. Org. Chem.* **2005**, 1479–1492.
7. Cuevas-Yañez, E.; Serrano, J. M.; Huerta, G.; Muchowski, J. M.; Cruz-Almanza, R. *Tetrahedron* **2004**, *60*, 9391–9396.
8. Thomas, P. S.; Bowling, N. P.; Burrmann, N. J.; McMahon, R. J. *J. Org. Chem.* **2010**, *75*, 6372–6381.
9. Bowling, N. P.; Burrmann, N. J.; Halter, R. J.; Hodges, J. A.; McMahon, R. J. *J. Org. Chem.* **2010**, *75*, 6382–6390.
10. Davis, P. J.; Harris, L.; Karim, A.; Thompson, A. L.; Gilpin, M.; Moloney, M. G.; Pound, M. J.; Thompson, C. *Tetrahedron Lett.* **2011**, *52*, 1553–1556.
11. May, J. A.; Stoltz, B. M. *J. Am. Chem. Soc.* **2002**, *124*, 12426–12427.
12. Furrow, M. E.; Myers, A. G. *J. Am. Chem. Soc.* **2004**, *126*, 12222–12223.
13. Toma, T.; Shimokawa, J.; Fukuyama, T. *Org. Lett.* **2007**, *9*, 3195–3197.
14. Perusquía-Hernández, C.; Lara-Issasi, G. R.; Frontana-Uribe, B. A.; Cuevas-Yañez, E. *Tetrahedron Lett.* **2013**, *54*, 3302–3305.
15. Nicolle, S. M.; Moody, C. J. *Chem. Eur. J.* **2014**, *20*, 4420–4425.
16. Javed, M. I.; Brewer, M. *Org. Lett.* **2007**, *9*, 1789–1792.
17. Nicolau, K. C.; Baran, P. S.; Zhong, Y. L.; Choi, H. S.; Fong, K. C.; He, Y.; Yoon, W. H. *Org. Lett.* **1999**, *1*, 883–886.
18. Aller, E.; Molina, P.; Lorenzo, Á. *Synlett* **2000**, 526–528.
19. Bio, M. M.; Javadi, G.; Song, Z. J. *Synthesis* **2005**, 19–21.
20. Cuevas-Yañez, E.; García, M. A.; Mora, M. A. D. L.; Muchowski, J. M.; Cruz-Almanza, R. *Tetrahedron Lett.* **2003**, *44*, 4815–4817.
21. Pace, V.; Verniest, G.; Sinisterra, J.-V.; Alcántara, A. R.; De Kimpe, N. *J. Org. Chem.* **2010**, *75*, 5760–5763.
22. For a review on Wolff rearrangement, see: Kirmse, W. *Eur. J. Org. Chem.* **2002**, 2193–2256.
23. Siciliano, C.; De Marco, R.; Guidi, L. E.; Spinella, M.; Liguori, A. *J. Org. Chem.* **2012**, *77*, 10575–10582.
24. Charette, A. B.; Wurz, R. P.; Ollevier, T. *J. Org. Chem.* **2000**, *65*, 9252–9254.
25. Wurz, R. P.; Lin, W.; Charette, A. B. *Tetrahedron Lett.* **2003**, *44*, 8845–8848.
26. Harned, A. M.; Sherrill, W. M.; Flynn, D. L.; Hanson, P. R. *Tetrahedron* **2005**, *61*, 12093–12099.
27. Taber, D. F.; Sheth, R. B.; Joshi, P. V. *J. Org. Chem.* **2005**, *70*, 2851–2854.

28. Ramachary, D. B.; Narayana, V. V.; Ramakumar, R. *Tetrahedron Lett.* **2008**, *49*, 2704–2709.
29. Dutra, L. G.; Saibert, C.; Vicentini, D. S.; Sá, M. M. *J. Mol. Catal. A: Chem.* **2014**, 386, 35–41.
30. Presset, M.; Mailhol, D.; Coquerel, Y.; Rodriguez, J. *Synthesis* **2011**, 2549–2552.
31. Collins, S. G.; O'Sullivan, O. C. M.; Kelleher, P. G.; Maguire, A. R. *Org. Biomol. Chem.* **2013**, *11*, 1706–1725.
32. Chiara, J. L.; Suárez, J. R. *Adv. Synth. Catal.* **2011**, *353*, 575–579.
33. Katritzky, A. R.; El Khatib, M.; Bol'shakov, O.; Khelashvili, L.; Steel, P. J. *J. Org. Chem.* **2010**, *75*, 6532–6539.
34. Muthyala, M. K.; Choudhary, S.; Kumar, A. *J. Org. Chem.* **2012**, *77*, 8787–8791.
35. Kitamura, M.; Tashiro, N.; Okauchi, T. *Synlett* **2009**, 2943–2944.
36. Kitamura, M.; Tashiro, N.; Sakata, R.; Okauchi, T. *Synlett* **2010**, 2503–2505.
37. Kitamura, M.; Tashiro, N.; Miyagawa, S.; Okauchi, T. *Synthesis* **2011**, 1037–1044.
38. Kitamura, M.; Miyagawa, S.; Okauchi, T. *Tetrahedron Lett.* **2011**, *52*, 3158–3161.
39. Galkina, O. S.; Maas, G.; Rodina, L. L.; Nikolaev, V. A. *Synthesis* **2015**, *47*, 1469–1478.
40. Schroen, M.; Bräse, S. *Tetrahedron* **2005**, *61*, 12186–12192.
41. Brodsky, B. H.; Du Bois, J. *Org. Lett.* **2004**, *6*, 2619–2621.
42. Kitamura, M.; Sakata, R.; Tashiro, N.; Ikegami, A.; Okauchi, T. *Bull. Chem. Soc. Jpn.* **2015**, *88*, 824–833.
43. Gilman, H.; Jones, R. G. *J. Am. Chem. Soc.* **1943**, *65*, 1458–1460.
44. Morandi, B.; Carreira, E. M. *Angew. Chem. Int. Ed.* **2010**, *49*, 938–941.
45. Morandi, B.; Carreira, E. M. *Angew. Chem. Int. Ed.* **2010**, *49*, 4294–4296.
46. Wu, G.; Deng, Y.; Wu, C.; Wang, X.; Zhang, Y.; Wang, J. *Eur. J. Org. Chem.* **2014**, 4477–4481.
47. Mykhailiuk, P. K. *Angew. Chem. Int. Ed.* **2015**, *54*, 6558–6561.
48. Mykhailiuk, P. K. *Eur. J. Org. Chem.* **2015**, 7235–7239.
49. Jiang, N.; Wang, J. *Tetrahedron Lett.* **2002**, *43*, 1285–1287.
50. Deng, G.; Xu, B.; Wang, J. *Tetrahedron* **2005**, *61*, 10811–10817.
51. Jiang, N.; Qu, Z.; Wang, J. *Org. Lett.* **2001**, *3*, 2989–2992.
52. Jiang, N.; Ma, Z.; Qu, Z.; Xing, X.; Xie, L.; Wang, J. *J. Org. Chem.* **2003**, *68*, 893–900.
53. Varala, R.; Enugala, R.; Nuvula, S.; Adapa, S. R. *Tetrahedron Lett.* **2006**, *47*, 877–880.
54. Hasegawa, K.; Arai, S.; Nishida, A. *Tetrahedron* **2006**, *62*, 1390–1401.
55. Erhunmwunse, M. O.; Steel, P. G. *J. Org. Chem.* **2008**, *73*, 8675–8677.
56. Pinho, V. D.; Burtoloso, A. C. B. *J. Org. Chem.* **2011**, *76*, 289–292.

57. Martin, L. J.; Marzinzik, A. L.; Ley, S. V.; Baxendale, I. R. *Org. Lett.* **2011**, *13*, 320–323.
58. Pramanik, M. M. D.; Rastogi, N. *Org. Biomol. Chem.* **2016**, *14*, 1239–1243.
59. Bonge, H. T.; Pintea, B.; Hansen, T. *Org. Biomol. Chem.* **2008**, *6*, 3670–3672.
60. Yao, W.; Wang, J. *Org. Lett.* **2003**, *5*, 1527–1530.
61. Zhao, Y.; Ma, Z.; Zhang, X.; Zou, Y.; Jin, X.; Wang, J. *Angew. Chem. Int. Ed.* **2004**, *43*, 5977–5980.
62. Peng, C.; Cheng, J.; Wang, J. *J. Am. Chem. Soc.* **2007**, *129*, 8708–8709.
63. Ye, F.; Qu, S.; Zhou, L.; Peng, C.; Wang, C.; Cheng, J.; Hossain, M. L.; Liu, Y.; Zhang, Y.; Wang, Z.-X.; Wang, J. *J. Am. Chem. Soc.* **2015**, *137*, 4435–4444.
64. Krainz, T.; Chow, S.; Korica, N.; Bernhardt, P. V.; Boyle, G. M.; Parsons, P. G.; Davies, H. M. L.; Williams, C. M. *Eur. J. Org. Chem.* **2016**, 41–44.
65. Babinski, D. J.; Aguilar, H. R.; Still, R.; Frantz, D. E. *J. Org. Chem.* **2011**, *76*, 5915–5923.
66. Eidamshaus, C.; Hommes, P.; Reissig, H.-U. *Synlett* **2012**, *23*, 1670–1674.
67. Kosobokov, M. D.; Titanyuk, I. D.; Beletskaya, I. P. *Tetrahedron Lett.* **2014**, *55*, 6791–6794.
68. Ye, F.; Wang, C.; Zhang, Y.; Wang, J. *Angew. Chem. Int. Ed.* **2014**, *53*, 11625–11628.

Chapter 3

Development of New Catalysts for Metal Carbene Transformations

Catalysts play key roles in the transition-metal-catalyzed reactions of α-diazo compounds. Over the years, various catalysts have been developed for effective dinitrogen extrusion and the subsequent carbene transfer reactions, in particular for the asymmetric catalysis. The new catalysts can be devised through the designing of ligands and the use of various different transition metals. Although Cu(I) and Rh(II) complexes are the two most efficient types of catalysts in carbene transfer reactions, other transition metals are also actively explored in the reaction of diazo compounds. For example, metalloporphyrins, *N*-heterocyclic carbene complexes and polymer- supported catalysts have demonstrated high efficiency and unique applications in cyclopropanations, insertions and other types of carbene reactions. In this chapter, a brief introduction to the recent developments of new catalysts will be presented first, and then their applications according to the types of reactions will follow.

Rh(II) complexes have proven to be excellent catalysts for diazo compound-based carbene transfer reactions. A series of achiral Rh(II) catalysts are commercially available, as shown in Figure 3.1. Recently, a series of chiral Rh(II) carboxylates have been well established, and have become easily accessible from chemical suppliers (Figure 3.2). These commercially available chiral Rh(II) catalysts have demonstrated high efficiency and stereoselectivity in catalytic asymmetric carbene transfer reactions.

Corey and co-workers reported a type of unsymmetrical Rh(II) catalysts $[Rh_2(RCO_2)_n(L^*{}_{4-n})]$ (Figure 3.3). Some of these complexes, such

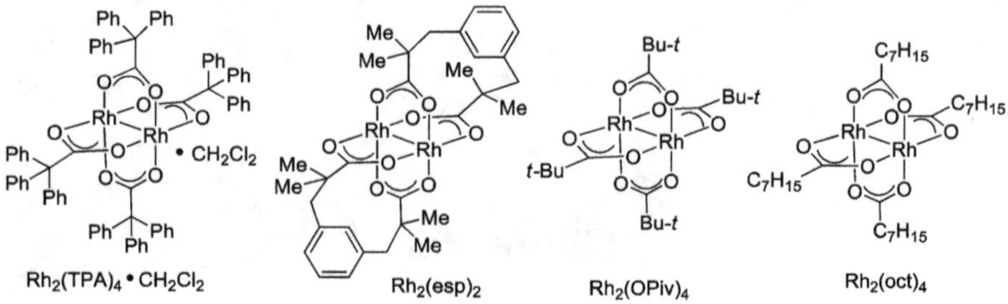

Figure 3.1 Structure of some commercially available Rh(II) complexes.

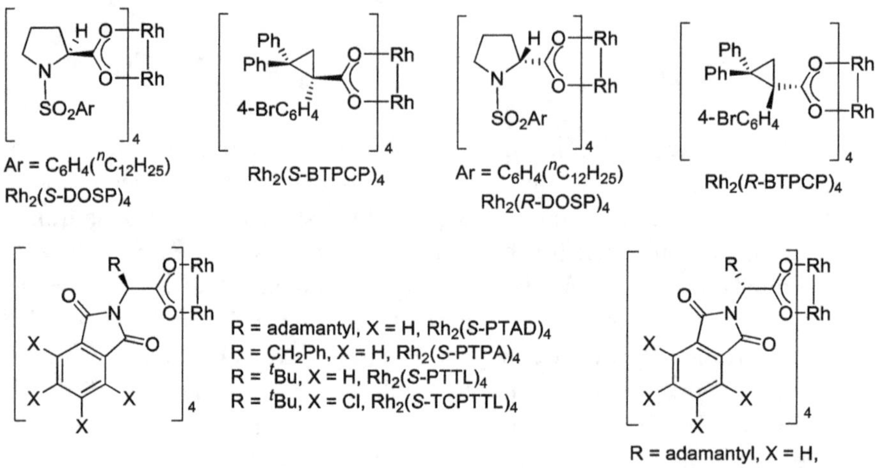

Figure 3.2 Commercially available chiral Rh(II) catalysts.

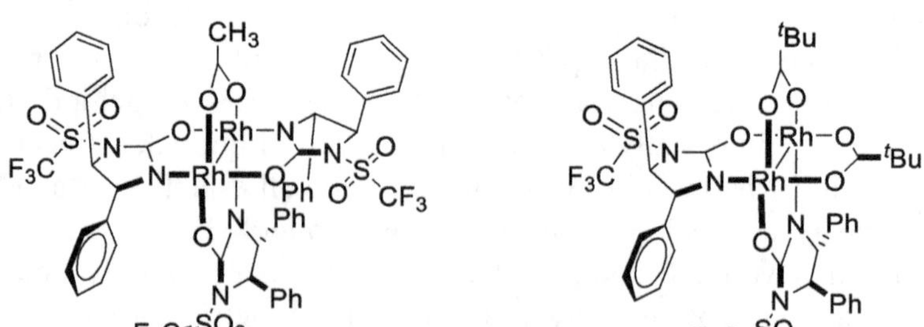

1 2

Figure 3.3 Unsymmetrical Rh(II) catalysts $[Rh_2(RCO_2)_n(L^*_{4-n})]$.

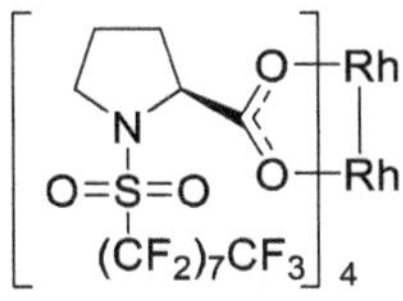

3

Figure 3.4 Fluorous chiral Rh(II) catalyst.

as $Rh_2(OAc)(DPTI)_3$ **1** and $Rh_2(t\text{-}BuCO_2)_2(DPTI)_2$ **2**, are effective for cyclopropanation and cyclopropenation reactions.[1,2]

Although Rh(II) complexes are highly effective for carbene transfer reactions, the high price of rhodium may impede their wide applications in organic synthesis. Thus, efforts have been made in the past few years to recycle the Rh(II) catalysts. For example, Biffis and co-workers have developed a fluorous chiral Rh(II) catalyst **3** (Figure 3.4), which is efficient for cyclopropanation reactions. The catalyst can be easily separated from the reaction mixture and is recyclable. Furthermore, this catalyst enables the reaction to be carried out in fluorous solvent.[3]

In asymmetric carbene transfer reactions, Doyle's Rh(II) carboxamidate complexes have been proven to be particularly effective. Recently, the immobilized chiral Rh(II) carboxamidate complexes have been prepared by attaching the ester moiety of the ligand to a polymer backbone (Scheme 3.1). It has been observed that the immobilization of Rh(II) catalysts with mixed ligands has a significant impact on the reactivity and stereoselectivity of the generated Rh(II) carbene species in both C–H bond insertion and cyclopropanation.[4]

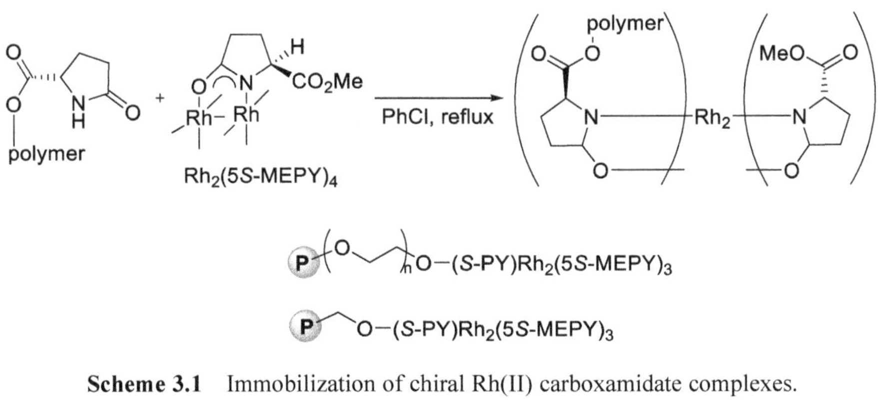

Scheme 3.1 Immobilization of chiral Rh(II) carboxamidate complexes.

Forbes and co-workers reported an ionic liquid metal-conjugate **4** (Eq. (3.1)).[5] The ligand exchange of acetate with an imidazolium group tethered to a carboxylic acid afforded the Rh(II) carboxylate that covalently bonded to an organic salt. This modified Rh(II) catalyst can be used in ionic liquids as an effective catalyst for the cyclopropanation of styrene using ethyl diazoacetate.

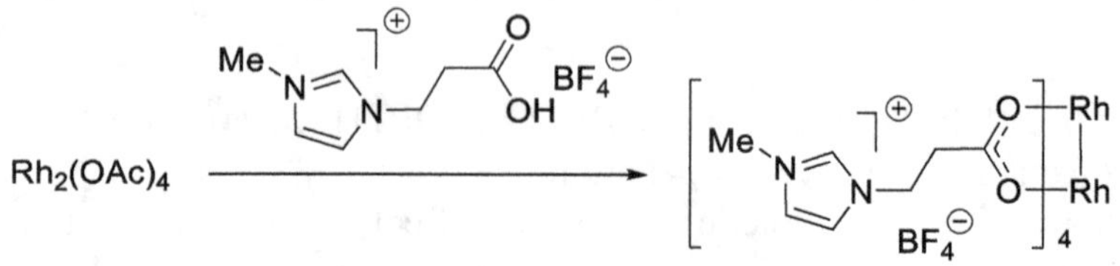

4

(3.1)

Another strategy to immobilize Rh(II) catalysts is to coordinate the solid support to the empty axial sites of the Rh(II) complex. A dirhodium complex has two axial bonding positions. While one axial empty orbital interacts with the diazo carbon to bring dinitrogen extrusion and generation of Rh(II) carbene, the other axial bonding site may interact with the solvent or a ligand. Pyridine is known to strongly bond with rhodium. Taking these advantages, Davies and co-workers have developed a strategy to immobilize the Rh(II) catalyst by using highly cross-linked microporous polystyrene resins with a pyridine attachment. The immobilized catalysts display a similar reactivity and stereoselectivity to their homogeneous counterparts and can be effectively recycled with only a limited loss in stereoselectivity (Scheme 3.2).[6,7]

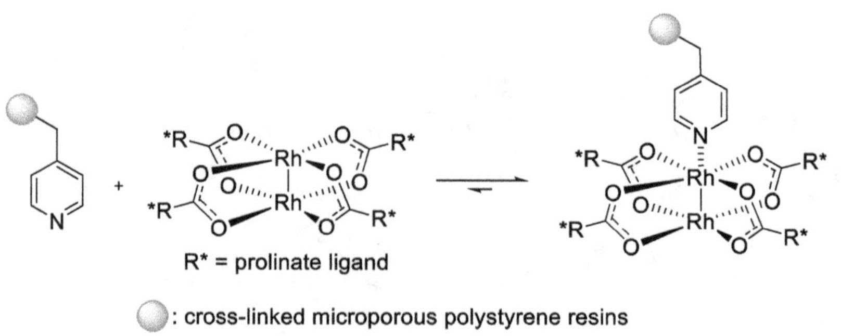

Scheme 3.2 Immobilization of the Rh(II) catalyst with cross-linked microporous polystyrene resins with a pyridine attachment.

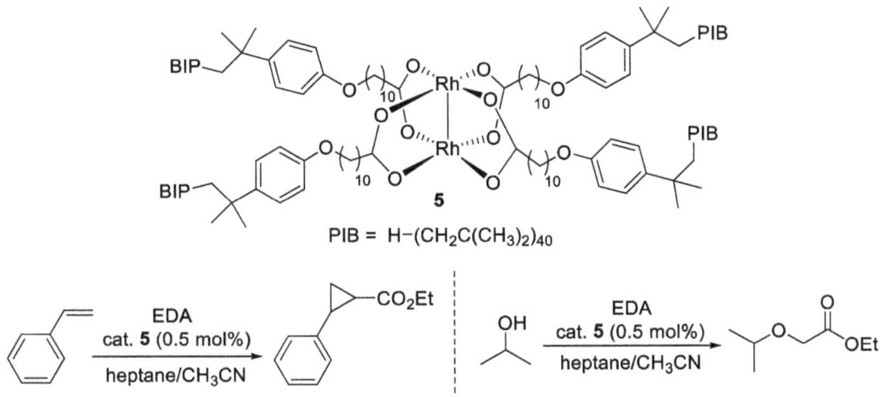

Scheme 3.3 Cyclopropanation and O–H insertion under biphasic conditions.

Recently, a polymer-supported Rh(II) catalyst was synthesized, in which the polymer is soluble to facilitate homogeneous catalysis efficiency.[8] This strategy not only allowed the recycling of the expensive Rh(II) catalyst but also suppressed the dimerization of by-products under the condition that EDA was not slowly added with the syringe pump. The poly-isobutylene-bound Rh(II) catalyst **5** has been applied to both cyclopropanation and O–H insertion (Scheme 3.3). The authors found that adding hydrocarbon polymer (PIB_{2300}) as a co-solvent could further suppress the dimerization of EDA.

On the other hand, the development of novel ligands for Rh(II) catalysts have been explored over the years.[9–19] In 2008, Doyle and co-workers disclosed Rh(II) catalyst Rh_2(1,6-BPGlyc)$_2$(OAc)$_2$, in which 1,6-bis-(*N*-benzyl)-diphenylglycoluril **6** (1,6-BPGlyc) was used as the ligand.[9] The catalytic reactivity between this catalyst and other Rh(II) carboxamidates has been compared for the cyclopropanation of diazo esters with styrene (Scheme 3.4).

Fox and co-workers reported the synthesis of mixed-ligand paddle-wheel complex Rh_2(*S*-PTTL)$_3$TPA **7**, which was applied to cyclopropanation, cyclopropenation and sp^2 C–H bond functionalization in good yields and enantioselectivity (Scheme 3.5).[11] Besides, Rh(II) catalysts with chelating tropolonato ligands[12] and Rh(II) metallo-peptide complex[13] have also been explored, and they have been applied to catalyze C–H insertion, Si–H insertion and cyclopropanations.

In addition to Rh(II) catalyst, Rh(I) complexes bearing chelating imine-functionalized *N*-heterocyclic carbene ligands have also been synthesized

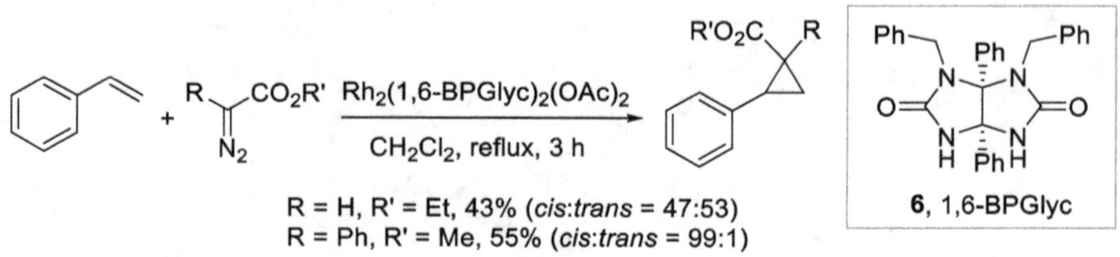

Scheme 3.4 Cyclopropanation of styrene with diazoesters using Rh(II) catalyst **6**.

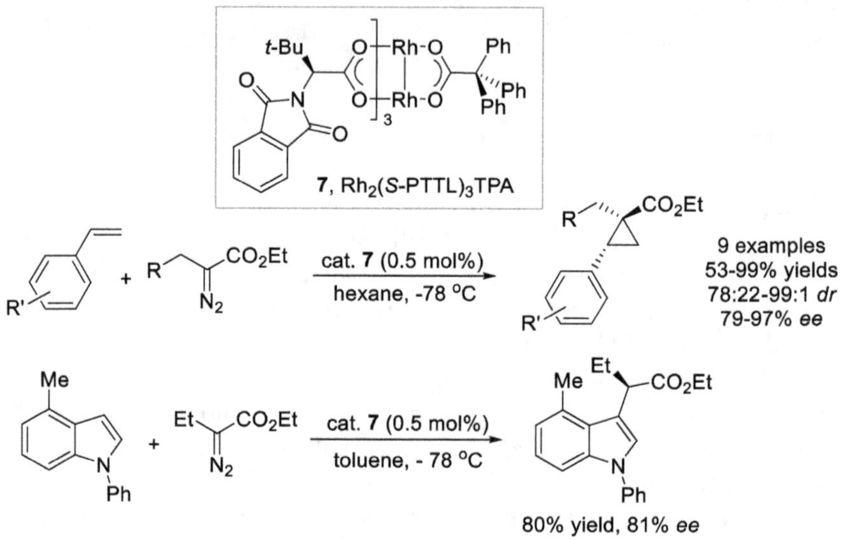

Scheme 3.5 Catalytic cyclopropanation and indole C–H bond functionalization.

and characterized. These series of catalysts have been tested in cyclopropanation between EDA and styrene in order to inspect the steric effect of the ligand.[15]

Apart from rhodium catalysts, copper complexes represent another type of widely applied catalysts for catalyzing diazo compound transformations. Copper complexes with a variety of ligands have been explored in carbene transfer reactions.[20–27] Pérez and co-workers investigated the catalytic behavior of *N*-heterocyclic carbene ligand/Cu(I) catalysts in the diazo decomposition of ethyl diazoacetate. It was found that [IPrCuCl] (IPr = 1,3-bis(diisopropylphenyl)imidazol-2-ylidene) was an effective catalyst for the transfer of carbene from ethyl diazoacetate through cyclopropanation, and O–H and N–H insertion. The formation of

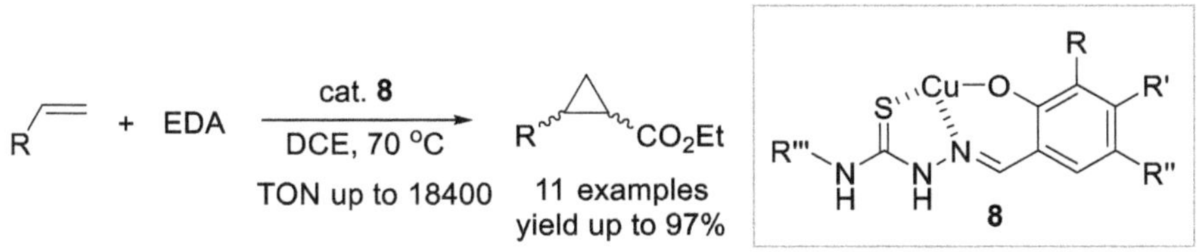

Scheme 3.6 Cyclopropanation catalyzed by thiosemicarbazone/copper complexes.

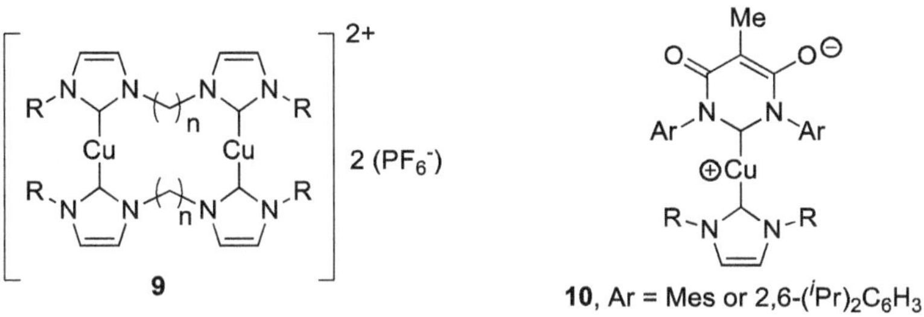

Figure 3.5 Copper complexes with *N*-heterocyclic carbene ligand.

homo-coupling by-products, the common drawback for transition-metal-catalyzed reactions of diazo compounds, could be largely avoided with this catalyst.[20]

In 2012, a series of thiosemicarbazone copper complexes **8** were prepared and characterized by Caselli group.[21] These copper catalysts were examined in the reaction between alkenes with EDA, which yielded cyclopropanes in excellent yields (Scheme 3.6).

Copper complexes coordinating with *N*-heterocyclic carbene ligand have also been synthesized and examined in carbene transfer reactions (Figure 3.5). Poly-nuclear Cu(I) complexes **9** with chelating bis-NHC ligands have been synthesized and evaluated as catalysts in carbene cyclopropanations.[22] Moreover, anionic malonate-derived zwitterionic copper complexes **10** with NHC ligand were prepared and they were found to be efficient catalysts for cyclopropanation.[23]

In addition to traditional Cu(I) and Rh(II) catalysts, ruthenium complexes are known to catalyze carbene transfer reactions of diazo compounds.[24–40] For example, Ru(II) porphyrin complex [Ru^{II}(TDCPP)(CO)] **11** could efficiently catalyze three-component coupling reaction of an α-diazoester with a series of *N*-benzylidene imines and alkenes, affording functionalized pyrrolidines with excellent diastereoselectivity (Scheme 3.7).[24]

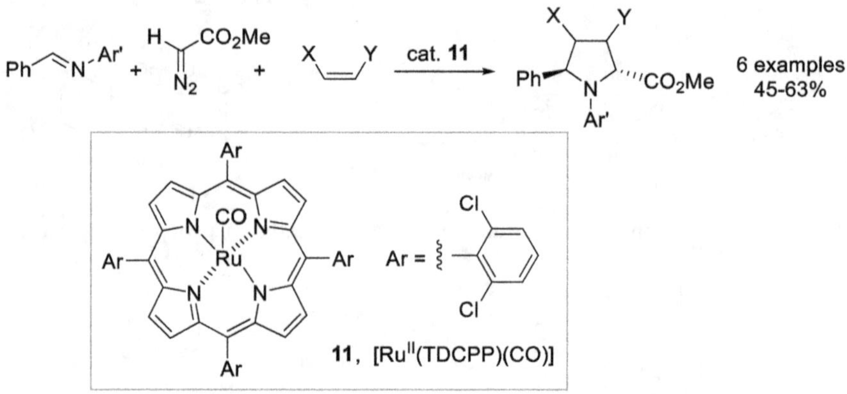

Scheme 3.7 Rh(por)-catalyzed synthesis of functionalized pyrrolidines.

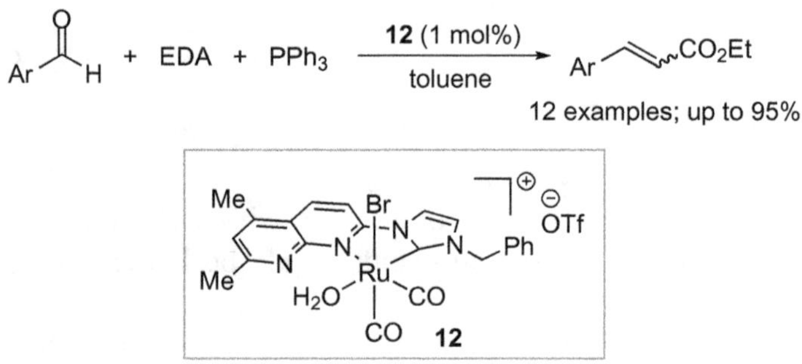

Scheme 3.8 Carbene transfer reaction catalyzed by Ru(II)–NHC complex.

By using *N*-heterocyclic carbene ligand, the mono-nuclear Ru(II)–NHC complex **12** has been prepared from diruthenium(I) precursor with 1,8-naphthyridine functionalized NHC compound.[33] This complex has shown to be an excellent catalyst for diazo compounds dimerization, N–H and O–H bond insertions and cyclopropanation (Scheme 3.8).

In 2008, Che and co-workers reported the preparation of ruthenium nanoparticles supported on non-cross-linked soluble polystyrene.[34] The polymer-supported ruthenium complex demonstrates effective catalytic reactivity in intra- and intermolecular carbene insertion into C–H and N–H bond, cyclopropanation and [2,3]-sigmatropic rearrangement *via* ammonium ylide formation. Subsequently, they synthesized and characterized the ruthenium–salen complex.[35] Those complexes show remarkable catalytic properties toward asymmetric intramolecular cyclopropanation, affording the corresponding products in high yields and good enantioselectivity.

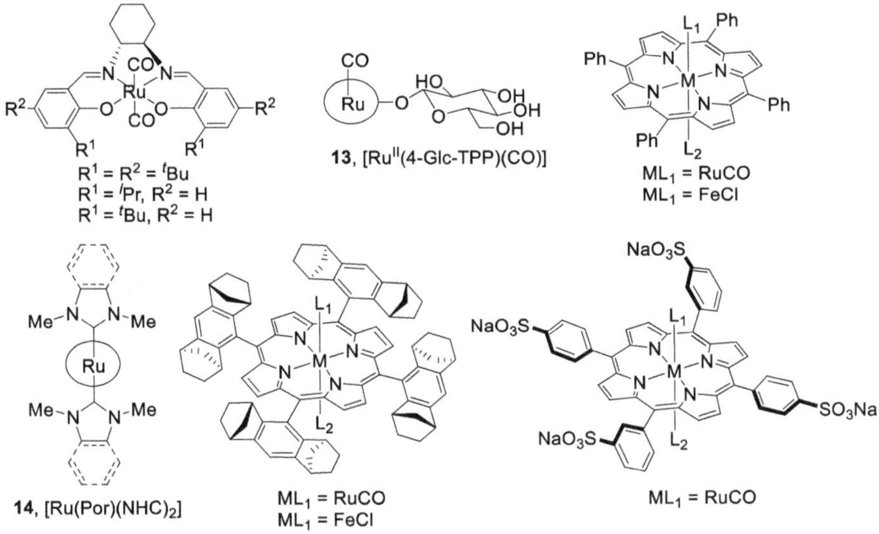

Figure 3.6 Ruthenium–salen and Ru/Fe porphyrin complexes for carbene transfer reactions.

A series of metalloporphyrin complexes, as shown in Figure 3.6, have served as efficient catalysts in carbene transfer reactions. Che and co-workers have explored water-soluble ruthenium glycosylated porphyrin catalyst **13**, which exhibits high activity for various carbene transfer reactions, including inter- and intramolecular cyclopropanation, intermolecular N–H insertion and intramolecular ylide formation/[2,3]-σ rearrangement. The catalyst has also been applied to site-selective modification of peptides and proteins.[36] Besides, they have synthesized bis(NHC)ruthenium(II)-porphyrin complexes **14**, which show high catalytic activity toward cyclopropanation, C–H, N–H, S–H, O–H insertions, and aziridinations.[37] Besides, Ru and iron metalloporphyrins are also reported in catalyzing carbene transformations with 6-diazo-5-oxo-*L*-norleucine as the starting material.[38] Dimerization, cyclopropanation, N–H and S–H insertion reactions are described respectively.

Che and co-workers have also designed chiral iridium porphyrin complex **15** with D_4-symmetricporphyrin ligand.[41] This complex shows excellent reactivity and stereoselectivity toward C–H and Si–H bond insertions, affording the intermolecular insertion products with excellent yields and high enantioselectivities (Scheme 3.9).

Gold complexes as soft Lewis acids have attracted considerable attention.[42] Gold complexes are normally considered not effective in catalyzing

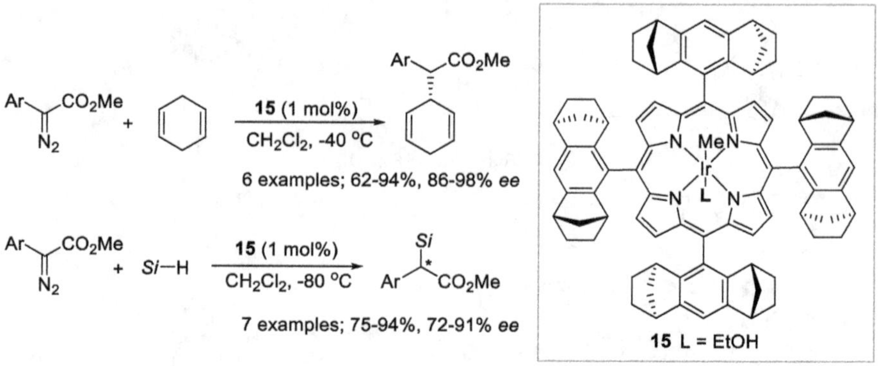

Scheme 3.9 C–H and Si–H bond insertions catalyzed by Ir porphyrin complex.

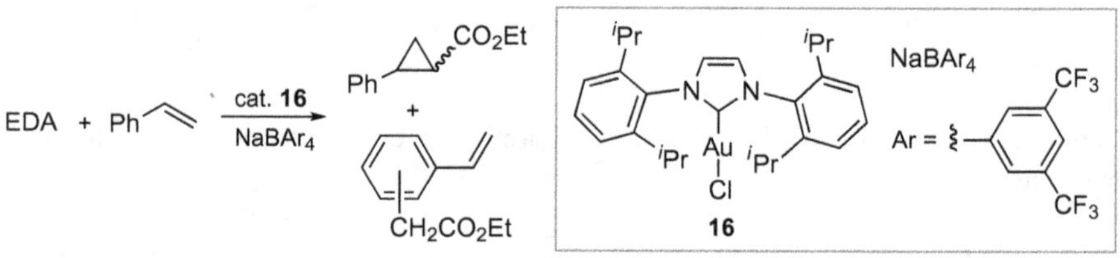

Scheme 3.10 Au(I)-NHC-catalyzed cyclopropanation of olefins.

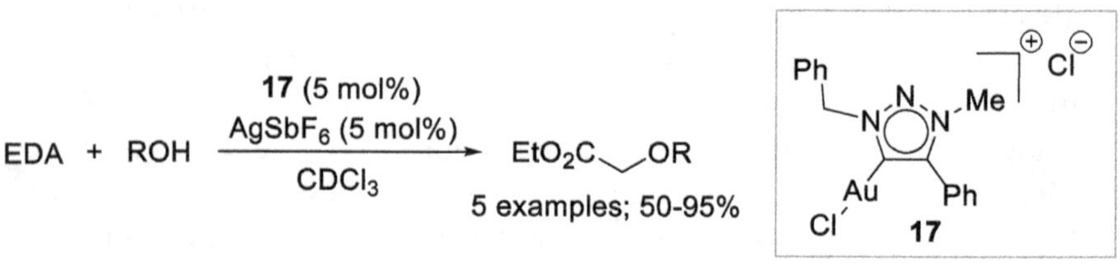

Scheme 3.11 O–H and N–H bond insertions catalyzed by gold(I) complexes with $AgSbF_6$.

the dinitrogen extrusion of diazo compounds. However, a report by Nolan and co-workers demonstrated that Au(I) complex **16** with an *N*-heterocyclic carbene ligand could catalyze the diazo decomposition of ethyl diazoacetate. The cyclopropanation of olefins as well as insertion into N–H and O–H bonds were achieved in this reaction (Scheme 3.10).[43]

Besides the triazapentadienyl gold catalyst, gold(I) "click" carbene (1,2,3-triazolylidene) complex **17** was synthesized and was examined in intermolecular carbene O–H and N–H bond insertions.[44] These transformations were carried out with air and moisture stable gold complexes in the presence of $AgSbF_6$ (Scheme 3.11).

Cobalt complexes have been extensively exploited in cyclopropanation with diazo compounds by Zhang and co-workers, and others.[45–49] Tetra(4-pyridyl)metalloporphyrins (metal = Zn or Co) were reported to be selectively encapsulated in self-assembled cubic M_8L_6 cages.[49] Typically, the cobalt–porphyrin complex was active in radical-type cyclopropanation between styrene with ethyl diazoacetate. This report has combined the co-catalyzed cyclopropanation with supramolecular cage chemistry.

Iron-[50–53] and chromium-based complexes[54] have also been applied in the cyclopropanation and catalytic N–H insertion reaction of α-diazocarbonyl compounds. Aviv and Gross reported the synthesis of non-protected amino acid esters from ammonia and diazoacetates with an iron porphyrin catalyst **18** (Eq. (3.2)).[53]

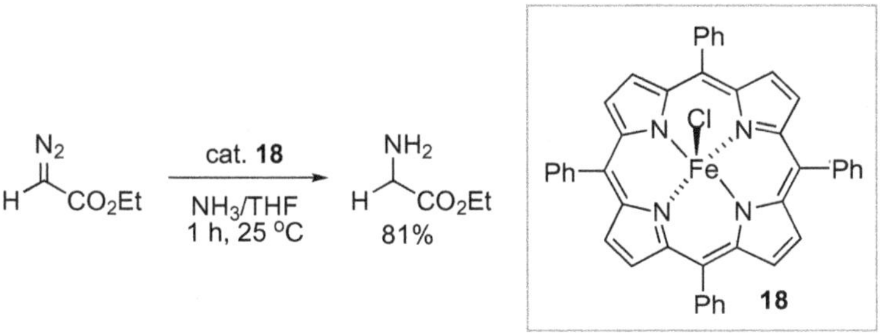

(3.2)

References

1. Lou, Y.; Horikawa, M.; Kloster, R. A.; Hawryluk, N. A.; Corey, E. J. *J. Am. Chem. Soc.* **2004**, *126*, 8916–8918.
2. Lou, Y.; Remarchuk, T. P.; Corey, E. J. *J. Am. Chem. Soc.* **2005**, *127*, 14223–14230.
3. Biffis, A.; Braga, M.; Cadamuro, S.; Tubaro, C.; Basato, M. *Org. Lett.* **2005**, *7*, 1841–1844.
4. Doyle, M. P.; Yan, M.; Gau, H.-M.; Blossey, E. C. *Org. Lett.* **2003**, *5*, 561–563.
5. Forbes, D. C.; Patrawala, S. A.; Tran, K. L. T. *Organometallics* **2006**, *25*, 2693–2695.
6. Nagashima, T.; Davies, H. M. L. *Org. Lett.* **2002**, *4*, 1989–1992.
7. Davies, H. M. L.; Walji, A. M.; Nagashima, T. *J. Am. Chem. Soc.* **2004**, *126*, 4271–4280.
8. Davies, H. M. L.; Walji, A. M. *Org. Lett.* **2005**, *7*, 2941–2944.

9. Nichols, J. M.; Liu, Y.; Zavalij, P.; Isaacs, L.; Doyle, M. P. *Inorg. Chim. Acta* **2008**, *361*, 3309–3314.
10. Xu, X.; Doyle, M. P. *Inorg. Chem.* **2011**, *50*, 7610–7617.
11. Boruta, D. T.; Dmitrenko, O.; Yap, G. P. A.; Fox, J. M. *Chem. Sci.* **2012**, *3*, 1589–1593.
12. Thrun, F.; Butschke, B.; Graening, T. *Chem. Eur. J.* **2012**, *18*, 4854–4858.
13. Sambasivan, R.; Ball, Z. T. *Org. Biomol. Chem.* **2012**, *10*, 8203–8206.
14. Liang, Y.; Harrell, M. L.; Bergbreiter, D. E. *Angew. Chem. Int. Ed.* **2014**, *53*, 8084–8087.
15. Rosenberg, M. L.; Langseth, E.; Krivokapic, A.; Gupta, N. S.; Tilset, M. *New J. Chem.* **2011**, *35*, 2306–2313.

For reviews of dirhodium(II) catalysts, see Refs. 16–18.

16. Hansen, J.; Davies, H. M. L. *Coord. Chem. Rev.* **2008**, *252*, 545–555.
17. Candeias, N. R.; Afonso, C. A. M.; Gois, P. M. P. *Org. Biomol. Chem.* **2012**, *10*, 3357–3378.
18. Trindade, A. F.; Coelho, J. A. S.; Afonso, C. A. M.; Veiros, L. F.; Gois, P. M. P. *ACS Catal.* **2012**, *2*, 370–383.
19. For investigating influence of the nature of ligand on Rh(II) carbene species, see: Lloret, J.; Carbó, J. J.; Bo, C.; Lledós, A.; Pérez-Prieto, J. *Organometallics* **2008**, *27*, 2873–2876.
20. Fructos, M. R.; Belderrain, T. R.; Nicasio, M. C.; Nolan, S. P.; Kaur, H.; Díaz-Requejo, M. M.; Pérez, P. J. *J. Am. Chem. Soc.* **2004**, *126*, 10846–10847.
21. Youssef, N. S.; El-Seidy, A. M. A.; Schiavoni, M.; Castano, B.; Ragaini, F.; Gallo, E.; Caselli, A. *J. Organomet. Chem.* **2012**, *714*, 94–103.
22. Tubaro, C.; Biffis, A.; Gava, R.; Scattolin, E.; Volpe, A.; Basato, M.; Díaz-Requejo, M. M.; Perez, P. J. *Eur. J. Org. Chem.* **2012**, 1367–1372.
23. César, V.; Barthes, C.; Farré, Y. C.; Cuisiat, S. V.; Vacher, B. Y.; Brousses, R.; Lugan, N.; Lavigne, G. *Dalton Trans.* **2013**, *42*, 7373–7385.

For the selected examples of Ru complex-catalyzed reaction of α-diazo carbonyl compounds, see Refs. 24–38.

24. Li, G.-Y.; Chen, J.; Yu, W.-Y.; Hong, W.; Che, C.-M. *Org. Lett.* **2003**, *5*, 2153–2156.
25. Zhou, C.-Y.; Yu, W.-Y.; Chan, P. W. H.; Che, C.-M. *J. Org. Chem.* **2004**, *69*, 7072–7082.
26. Maux, P. L.; Abrunhosa, I.; Berchel, M.; Simonneaux, G.; Gulea, M.; Masson, S. *Tetrahedron: Asymm.* **2004**, *15*, 2569–2573.
27. Ferrand, Y.; Maux, P. L.; Simonneaux, G. *Org. Lett.* **2004**, *6*, 3211–3214.

28. Xu, H.-W.; Li, G.-Y.; Wong, M.-K.; Che, C.-M. *Org. Lett.* **2005**, *7*, 5349–5352.
29. Cornejo, A.; Fraile, J. M.; García, J. I.; Gil, M. J.; Martínez-Merino, V.; Mayoral, J. A.; Salvatella, L. *Organometallics* **2005**, *24*, 3448–3457.
30. Bonaccorsi, C.; Mezzetti, A. *Organometallics* **2005**, *24*, 4953–4960.
31. Wang, M.-Z.; Xu, H.-W.; Liu, Y.; Wong, M.-K.; Che, C.-M. *Adv. Synth. Catal.* **2006**, *348*, 2391–2396.
32. Bonaccorsi, C.; Santoro, F.; Gischig, S.; Mezzetti, A. *Organometallics* **2006**, *25*, 2002–2010.
33. Sinha, A.; Daw, P.; Rahaman, S. M. W.; Saha, B.; Bera, J. K. *J. Organomet. Chem.* **2011**, *696*, 1248–1257.
34. Choi, M. K.-W.; Yu, W.-Y.; So, M.-H.; Zhou, C.-Y.; Deng, Q.-H.; Che, C.-M. *Chem. Asian J.* **2008**, *3*, 1256–1265.
35. Xu, Z.-J.; Fang, R.; Zhao, C.; Huang, J.-S.; Li, G.-Y.; Zhu, N. Che, C.-M. *J. Am. Chem. Soc.* **2009**, *131*, 4405–4417.
36. Ho, C.-M.; Zhang, J.-L.; Zhou, C.-Y.; Chan, O.-Y.; Yan, J. J.; Zhang, F.-Y.; Huang, J.-S.; Che, C.-M. *J. Am. Chem. Soc.* **2010**, *132*, 1886–1894.
37. Chan, K.-H.; Guan, X.; Lo, V. K.-Y.; Che, C.-M. *Angew. Chem. Int. Ed.* **2014**, *53*, 2982–2987.
38. Le Maux, P.; Nicolas, I.; Chevance, S.; Simonneaux, G. *Tetrahedron* **2010**, *66*, 4462–4468.
39. For review of Ru-porphyrin-catalyzed carbenoid transfer reactions, see: Zhou, C.-Y.; Huang, J.-S.; Che, C.-M. *Synlett.* **2010**, 2681–2700.
40. For a review on Ru-catalyzed carbenoids cyclopropanation with diazo compounds, see: Maas, G. *Chem. Soc. Rev.* **2004**, *33*, 183–190.
41. Wang, J.-C.; Xu, Z.-J.; Guo, Z.; Deng, Q.-H.; Zhou, C.-Y.; Wan, X.-L.; Che, C.-M. *Chem. Commun.* **2012**, *48*, 4299–4301.
42. For a review for gold catalysts in organic chemistry, see: Hashmi, A. S. K. *Chem. Rev.* **2007**, *107*, 3180–3211.
43. Fructos, M. R.; Belderrain, T. R.; Frémont, P.; Scott, N. M.; Nolan, S. P.; Díaz-Requejo, M. M.; Pérez, P. J. *Angew. Chem. Int. Ed.* **2005**, *44*, 5284–5288.
44. Kilpin, K. J.; Paul, U. S. D.; Lee, A.-L.; Crowley, J. D. *Chem. Commun.* **2011**, *47*, 328–330.
45. Ikeno, T.; Iwakura, I.; Yabushita, S.; Yamada, T. *Org. Lett.* **2002**, *4*, 517–520.
46. Huang, L.; Chen, Y.; Gao, G.-Y.; Zhang, X. P. *J. Org. Chem.* **2003**, *68*, 8179–8184.
47. Chen, Y.; Zhang, X. P. *J. Org. Chem.* **2004**, *69*, 2431–2435.
48. Chen, Y.; Fields, K. B.; Zhang, X. P. *J. Am. Chem. Soc.* **2004**, *126*, 14718–14719.

49. Otte, M.; Kuijpers, P. F.; Troeppner, O.; Ivanović-Burmazović, I.; Reek, J. N. H.; de Bruin, B. *Chem. Eur. J.* **2013**, *19*, 10170–10178.
50. Li, Y.; Huang, J.-S.; Zhou, Z.-Y.; Che, C.-M.; You, X.-Z. *J. Am. Chem. Soc.* **2002**, *124*, 13185–13193.
51. Du, G.; Andrioletti, B.; Rose, E.; Woo, L. K. *Organometallics* **2002**, *21*, 4490–4495.
52. Edulji, S. K.; Nguyen, S. T. *Organometallics* **2003**, *22*, 3374–3381.
53. Aviv, I.; Gross, Z. *Chem. Commun.* **2006**, 4477–4479.
54. For the selected example of Cr complex-catalyzed reaction of α-diazo carbonyl compounds, see: Hahn, N. D.; Nieger, M.; Dötz, K. H. *Eur. J. Org. Chem.* **2004**, 1049–1056.

Chapter 4

Cyclopropanation and Related Reactions

Along with carbene C–H bond insertions, cyclopropanations with unsaturated bonds through metal carbene represent the characteristic transformations of diazo compounds with transition metal catalysts (Scheme 4.1).[1–6] From the synthetic point of view, three-membered ring products are very important due to their wide applications as synthetic intermediates and their frequent occurrence as structural subunits in biologically active natural/unnatural products.[7–12] Transition-metal-catalyzed asymmetric cyclopropanations with alkenes and alkynes have found applications not only in the synthesis of natural products containing three-membered ring moieties but can also be incorporated with the subsequent tandem transformations for the construction of functional molecules. Even though cyclopropanations with metal carbene have been well established, new advances, such as the application of new chiral catalysts, cascade transformations and substrate scope expansions, have been frequently reported in the literature. This chapter will summarize these advances.

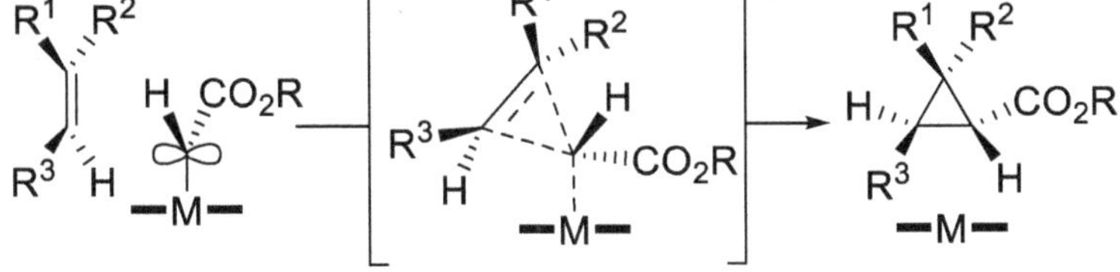

Scheme 4.1 Cyclopropanation with metal carbene.

4.1 Advances in the development of new catalytic systems

Cu(I) and Rh(II) complexes represent the two types of typical catalysts for carbene cyclopropanations. In the past few years, the progress of designing and utilizing new catalysts has experienced considerable growth. Efficient cyclopropanations with diazo compounds have been achieved by a variety of catalytic systems, including homogenous or heterogeneous catalysts, transition-metal complexes and main group metal complexes.

Some new advances have been seen in the catalysts based on copper. For example, homogenous copper catalysts as well as heterogeneous analogues, including dicopper-substituted silicotungstate,[13] copper and gold metal-organic frameworks (MOFs)[14] and copper-plated iron nanoparticles (Cu@FeNPs),[15] demonstrated high reactivity in catalyzing cyclopropanations with olefins. The polyoxometalates containing copper and cobalt have also been utilized in catalytic cyclopropanation of olefins with EDA.[16] Pérez and co-workers synthesized and characterized [(PhBP$_3$)Cu(PPh$_3$)] [PhBP$_3$ = PhB(CH$_2$PPh$_2$)$_3$] complex as the surrogate of Tp^{x}CuL (Tp^x = Hydrotrispyrazolylborate), which demonstrated catalytic activity in carbene transformations, including cyclopropanations and X–H insertions.[17] Moreover, computational studies of Cu-carbene species have been carried out to explore the mechanism of Cu(I)- and Cu(II)-catalyzed cyclopropanation processes.[18–20]

In addition to copper catalysts, new catalysts based on Rh(II) complexes have been explored. Tilset and co-workers reported the synthesis and characterization of novel Rh(I) complex **1** bearing chelating NHC iminocarbene ligand. Remarkably high *cis*-diastereoselective cyclopropanations between EDA and styrenes were achieved with this Rh(I)-NHC compound.[21,22] Davies and co-workers investigated the catalytic activity of heterobimetallic Bi-Rh paddlewheel carboxylates for metal carbene transformations.[23] Interestingly, a series of Pd(0)-NHC complexes **2** could also efficiently catalyze olefin cyclopropanation (Figure 4.1).[26]

While rhodium complexes are highly efficient in carbene transfer reactions, the high price of rhodium obviously impedes their wide applications in organic synthesis. Therefore, immobilization and recovery of this precious metal have attracted considerable attention. Recently, heterogeneous Rh(II) catalysts based on environmentally benign and biocompatible cellulose nano-crystals (CNC) were obtained by ligand exchange of carboxyl groups on the CNC surface, as confirmed by solid-state NMR spectroscopy.[27] This immobilized complex demonstrated high

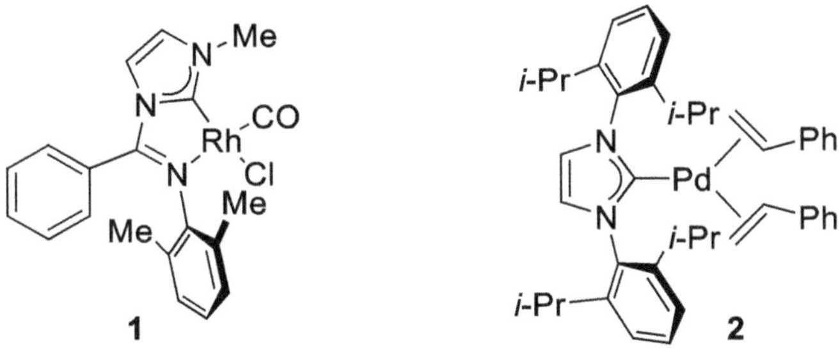

Figure 4.1 Structure of Rh(I)- and Pd(0)-NHC complex for catalytic cyclopropanations.

stability against leaching and allowed the recovery and reuse of catalysts during cyclopropanation between EDA and styrenes.

Coleman and Mack reported a heterogeneous catalytic system by using metallic silver-foil and a stainless-steel vial and ball.[28] This inexpensive silver-foil catalyst under mechanochemical and solvent-free conditions enabled efficient cyclopropanation between a variety of alkenes with diazoacetates with excellent diastereoselectivities (up to 98:2 dr) and in high yields (up to 96%). Moreover, this new catalyst could be easily recyclable without any appreciable loss of activity or selectivity (Eq. (4.1)).

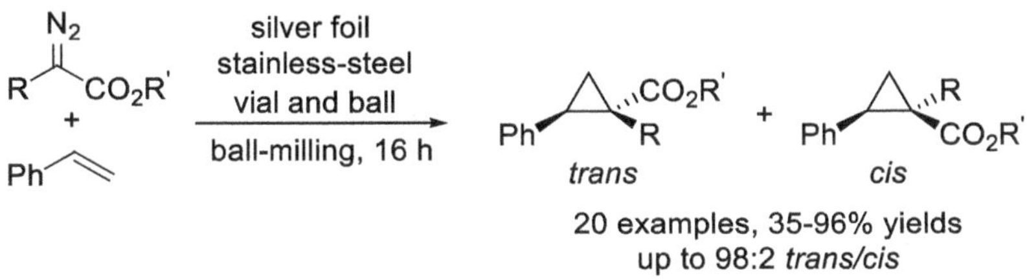

(4.1)

Zhang and de Bruin explored cobalt-porphyrin catalysts for cyclopropanations with diazo compounds.[29–33] In Co(II)-porphyrin-catalyzed olefin cyclopropanation, the reaction is proposed to proceed *via a* stepwise radical mechanism, which involves the species having resonance structure between terminal carbene intermediate and carbon-centered radical (Scheme 4.2).[29] The reaction pathway has been investigated by EPR experiments and DFT calculations. Moreover, Co-carbaporphyrin complexes were reported to catalyze cyclopropanation of styrene to afford selectively the *trans* isomers.[32] This radical-type cyclopropanation strategy can also be carried out in aqueous media with cobalt-porphyrin encapsulated in a "molecular flask".[33] It is worth mentioning that Gronert and

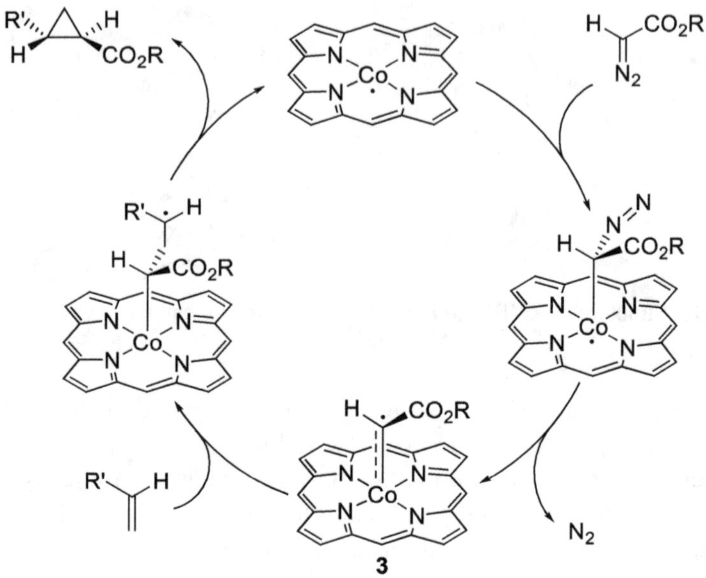

Scheme 4.2 Proposed catalytic cycle for Co(II)(por)-catalyzed radical-type cyclopropanation of olefins.

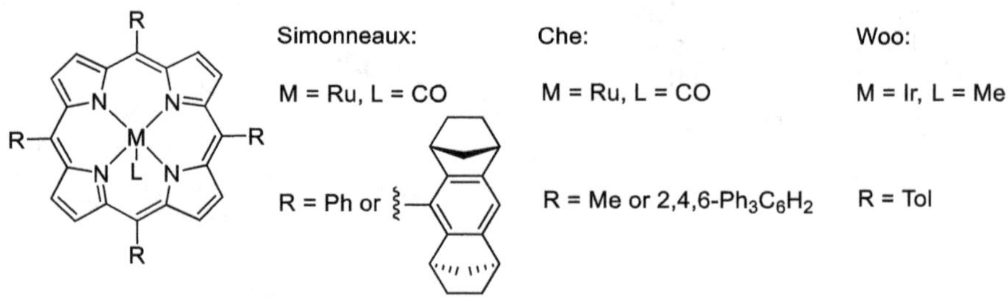

Figure 4.2 The structures of catalytic active metalloporphyrins in cyclopropanation.

co-workers have studied the mechanism of Co(III)-salen-catalyzed carbene reactions in the gas phase.[34]

Furthermore, Ru(II)-[35,36] and Ir(III)-porphyrin[37] complexes have been investigated for cyclopropanations (Figure 4.2). Che and co-workers have determined and elucidated the crystal structures of Ru-carbene intermediates, which are stabilized by sterically encumbered bis-pocket porphyrin ligands.[36] Cyclopropanation with a variety of olefins were attained with high turnover numbers with exceptionally active and robust Ir(III)-porphyrincatalysts.[37]

In addition to transition-metal catalysts, hypervalent tin(IV)-porphyrin complexes have also been used as catalysts for cyclopropanation of styrenes.[38] The Sn(por)-catalyzed transformation induced high specific *trans*-isomers with excellent yields under mild conditions.

4.2 Intermolecular cyclopropanations

Intermolecular cyclopropanations with metal carbene have attracted considerable attention due to the importance of cyclopropanes in organic synthesis. A wide range of diazo compounds and olefins have been employed for the catalytic cyclopropanations with good yields and diastereoselectivities.

As pervasive catalysts for metal carbene transfer reactions, Rh(II)-catalyzed cyclopropanations have demonstrated excellent scope with various diazo compounds, including dimethyl diazomalonate **4**,[39] α-alkyl-α-diazoesters **5**,[40] diethyl [nitro(diazo)methyl]phosphonate **6**,[41] halodiazoacetates **7**,[42,43,45] halodiazophosphonates **8**,[44,45] aryl diazomethanes **9**,[46] diazolactams **10**,[47] cyclic α-diazocarbonyl compounds,[48] enaldiazo esters **11**,[49] quinone diazides **12**,[50] α-aroyl-α-diazoacetates **13**[51] and diazo-bis-2,2,2-trifluoroethylmalonate **14**.[52] The substrate scope of olefins were also expanded to chiral enamides **15**,[53] allenes **16**,[54] glycals with 3-diazo-2-oxindole **17**,[55] 2-aryl-*2H*-chromenes **18**,[56] *N*-vinylacetamide or vinyl acetate with aryldiazoacetates **19**,[57] 1-fluoro-1-(phenylsulfonyl)ethylene **20**[58] and bioactive alkene-containing natural products **21**.[59] In addition, Rh(II)-catalyzed cyclopropanation of *N*-vinyl thymine **22** with diazomalonate has been applied to the preparation of nucleic acid base-substituted diester cyclopropanes, which could be facilely converted to (carbo) nucleoside analogues by sequential [3+2] cycloaddition (Figure 4.3).[60]

Apart from the Rh(II) catalysts, Cu(I) complexes also show high efficiency in carbene transformations of diazoesters with diverse olefins. Although the Cu(I)-catalyzed cyclopropanation is a class of transformations that has been studied extensively previously, new developments still appear in the literature at times. For example, this cyclopropanation strategy has been used to access morpholine-based scaffolds from serine- and threonine-derived oxazines,[61] [n,5]-spiroketals from exocyclic enol ethers[62] and Cu(I)-catalyzed diastereoselective synthesis of spiro-cyclopropanooxindoles from cyclic diazoamides.[63]

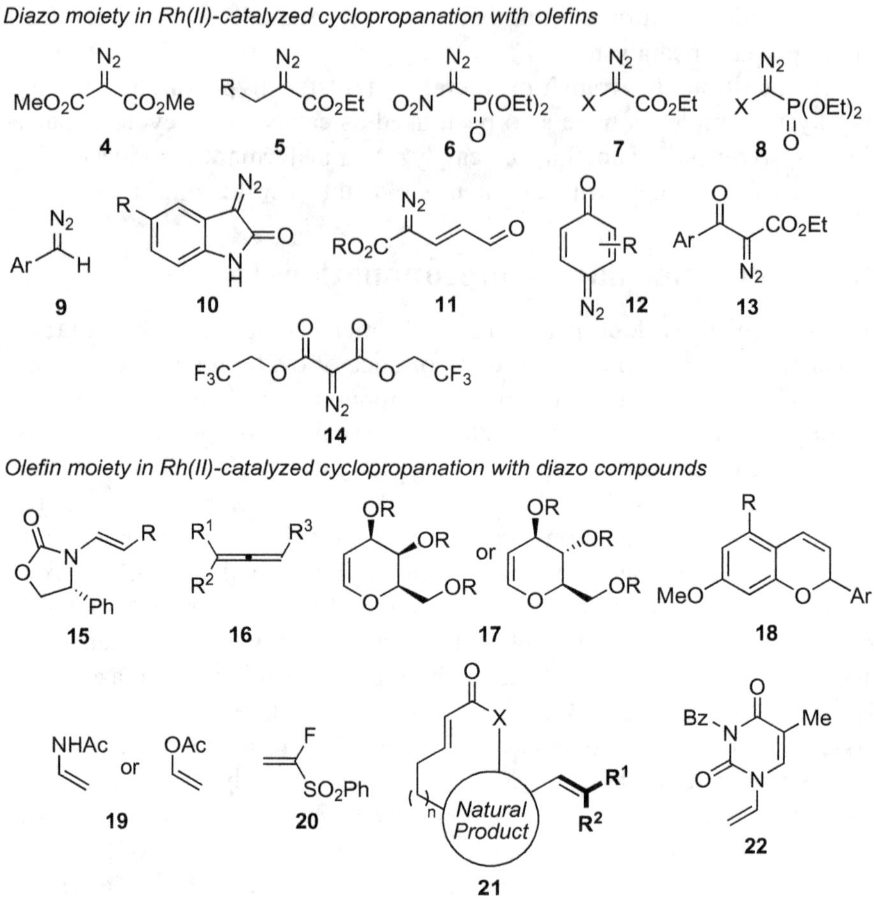

Figure 4.3 The selected diazo compounds and olefin derivatives in Rh(II)-catalyzed cyclopropanations.

Cyclopropanation of 1,3-dienamides with α-aryl diazoesters has been achieved with good regio- and diastereoselectivty.[64] Besides, Cu(I)-catalyzed [2+1] cycloaddition of styrene has been performed with organylthio(silyl)carbenes, which are derived from Si/S-substituted diazo compounds (Scheme 4.3).[65]

In addition to the traditional Rh(II)- and Cu(I)-catalyzed reactions, gold complex was also shown to catalyze carbene cyclopropanations with diazo compounds, as reported by Echavarren and Pérez.[66–69] Moreover, tandem enyne-cross metathesis-cyclopropanation process was completed

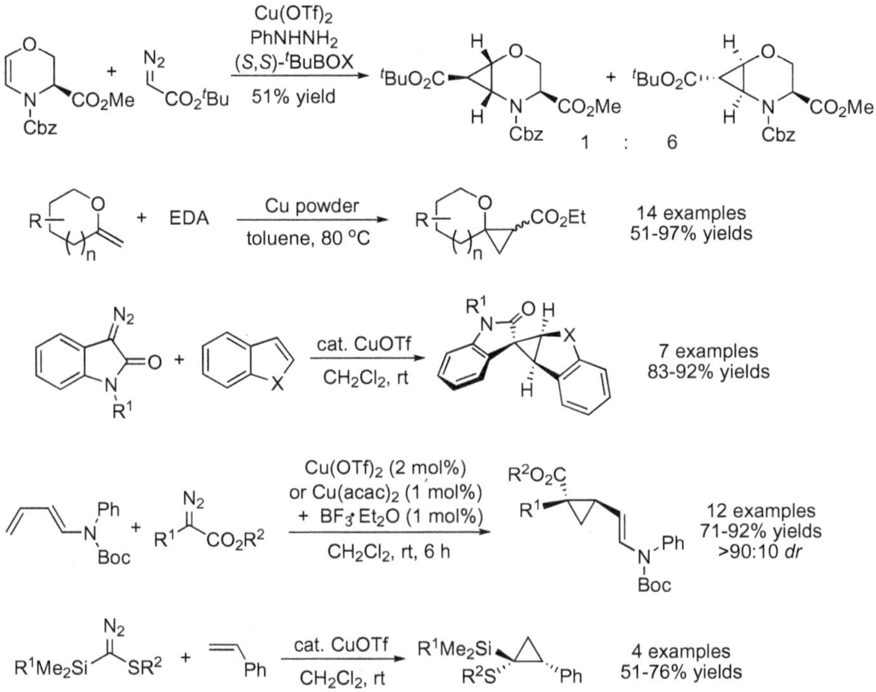

Scheme 4.3 Selected examples of Cu-catalyzed cyclopropanations of diazo compounds.

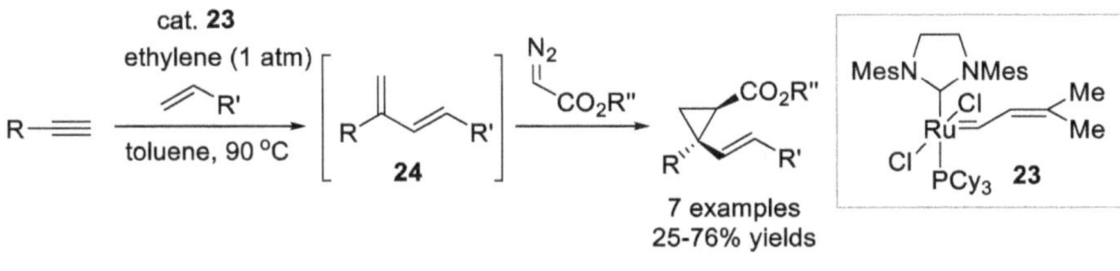

Scheme 4.4 Tandem enyne-cross metathesis-cyclopropanation catalyzed by Ru-NHC complex.

by Grubbs' second generation Ru(II) complexes.[70] Functionalized vinyl cyclopropanes were obtained with good regioselectivity through 1,3-diene intermediate **24** (Scheme 4.4). In most cases, Ru(II)-catalyzed carbene cyclopropanation of diazo compounds will be completed by metathesis and dimerization (diazo homo-coupling).[71,72]

Wang and co-workers reported a Pd(II)-catalyzed cyclopropanation of electron-deficient olefins with α-aryl-α-diazoesters (Eq. (4.2)).[73] Notably, Pd(OAc)$_2$-catalyzed cyclopropanation with diazomethane has been applied to construct three-membered ring skeletons with a series of olefins, including alkenyl boronic ester,[74] alkenyl silanes[75] and oxabenzonorbornadiene.[76] Besides, poly-nuclear Pd(I) carbonyl carboxylate complex also showed catalytic activity for cyclopropanation with EDA.[77]

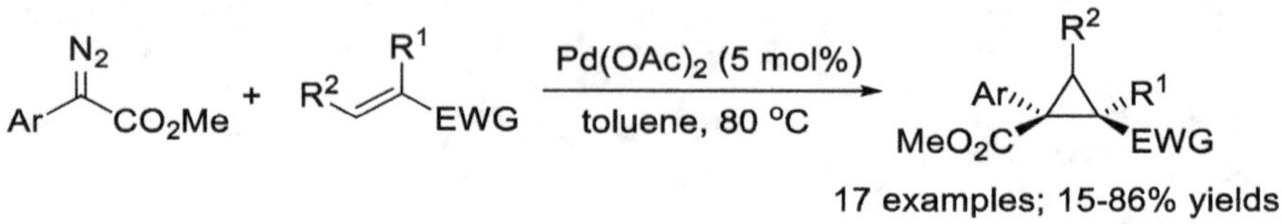

(4.2)

Iron-catalyzed cyclopropanation between olefin derivatives and the *in situ* generated diazo species was explored by Carreira and co-workers (Scheme 4.5). They reported the Fe(por)-catalyzed tandem diazotization/cyclopropanation reaction by employing glycine ethyl ester hydrochloride in aqueous media.[78] The reaction was tolerant with various functional groups, and the *trans*-cyclopropyl esters were obtained in high yields and with good diastereoselectivities. Subsequently, they developed the tandem diazomethane generation/Fe(por)-catalyzed cyclopropanation under mild conditions.[79] This biphasic reaction medium provided a convenient and safe protocol for the use of diazomethane, which is explosive and toxic.

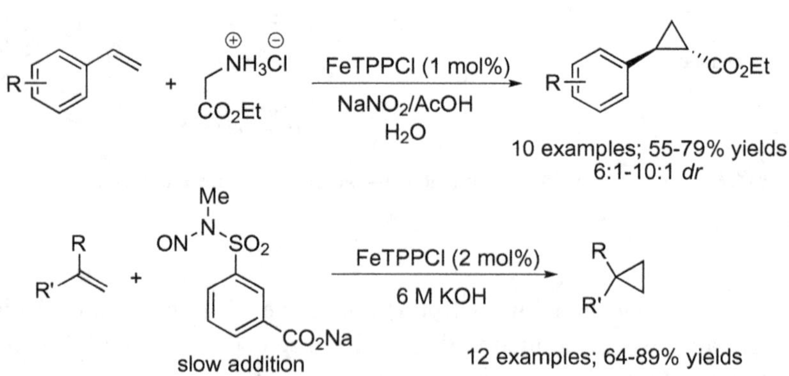

Scheme 4.5 FeTPPCl-catalyzed cyclopropanation with the *in situ* generated diazo intermediates in aqueous media.

Recently, iron(III) phthalocyanine complex has been explored as a highly active and biocompatible catalyst for olefin cyclopropanation in the presence of a living microorganism.[80] This strategy demonstrated the integration of metallocarbene with the metabolism of living organisms, providing a new approach for the construction of non-natural molecules.

The traditional Simmons–Smith reaction is a powerful method for the construction of cyclopropane rings. Recently, Charette and co-workers developed an improved Zn-catalyzed Simmons–Smith reaction, which converts allylic ethers and styrenes into phenyl cyclopropanes with high diastereoselectivities.[81] This transformation is initiated by the generation of aryl-substituted zinc carbenoids from phenyldiazomethane, followed by sequential cyclopropanation to give various tri-substituted cyclopropanes (Eq. (4.3)).

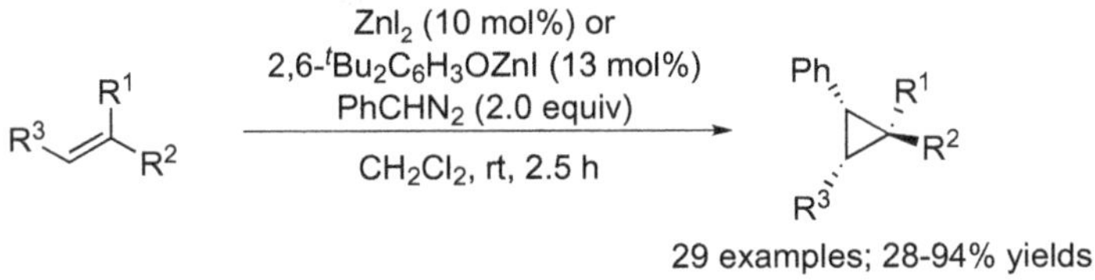

(4.3)

4.3 Metal-free intermolecular cyclopropanations

In addition to the catalytic cyclopropanations with metal carbenes, catalyst-free cyclopropanes of olefins can also be developed through thermolysis or photolysis of diazo compounds or their precursors. For example, by using spiro-oxadiazoline **25** as the carbene precursor, the cyclopropanation of alkenes bearing electron-withdrawing groups occurs readily.[82] The *trans* cyclopropanation product was generated exclusively from stabilized nucleophilic carbene species (Scheme 4.6).

Davies and co-workers reported thermally induced cyclopropanation or [2+2] cycloaddition of donor/acceptor diazo compounds with electron-neutral styrene derivatives (Scheme 4.7).[83] The *cis* cyclopropanes were obtained in high yields and with good diastereoselectivity from aryldiazoacetates bearing electron-donating group on aromatic rings. In contrast, cyclobutanones **27** were obtained in high yields and with good diastereo-control from aryldiazoketones **26**. The reaction pathway is proposed to involve generation of ketene intermediate through Wolff rearrangement and subsequent [2+2] cycloaddition.

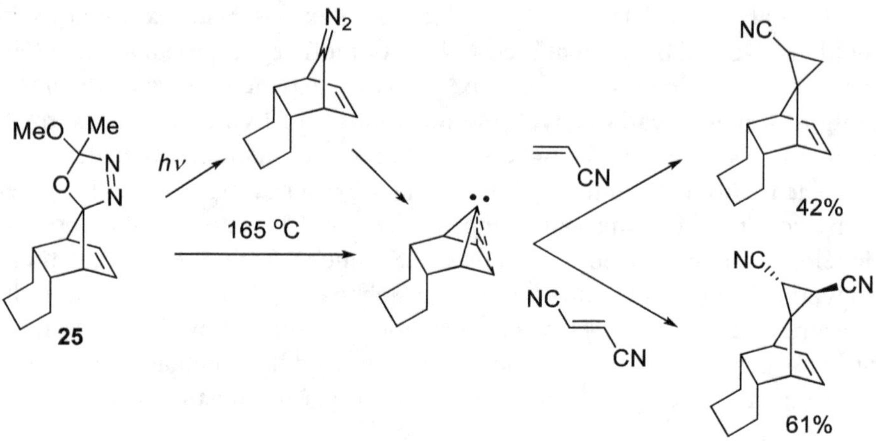

Scheme 4.6 Thermolysis and photolysis of oxadiazoline and cyclopropanation with olefins.

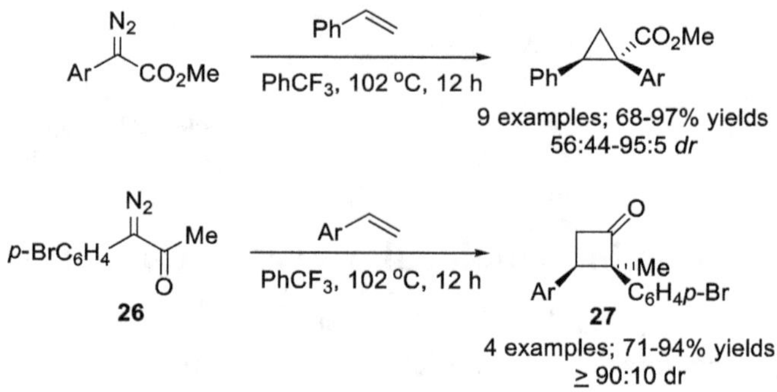

Scheme 4.7 Thermally induced cycloadditions of donor/acceptor carbenes with styrenes.

Recently, trifluoromethyl-substituted diazomethane has attracted attention because it can be used as a versatile reagent for introducing trifluoromethyl group. Ma and co-workers reported the cyclopropanation of tri-substituted olefinic azlactones **28** with CF_3CHN_2.[84] Under reflux conditions, the reaction afforded trifluoromethyl-substituted cyclopropanes bearing azlactone rings with good yields and excellent diastereoselectivities. From the key intermediates **29**, *α*-amino acids **30**, which bear CF_3-substituted cyclopropane backbone, and the relative dipeptides **31** could be obtained upon sequential transformations (Scheme 4.8).

Intermolecular catalyst-free cyclopropanation of alkenes with *in situ* generated diazo intermediate from *N*-tosylhydrazones was explored

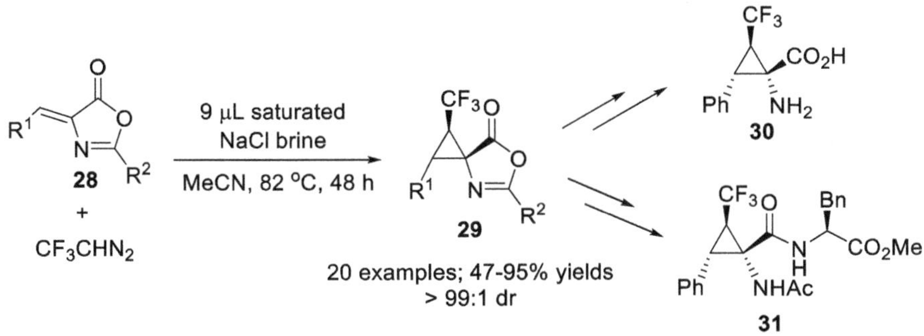

Scheme 4.8 Cyclopropanation of CF_3CHN_2 with alkenes under thermolytic conditions.

by Barluenga and co-workers (Eq. (4.4)).[85] The *N*-tosylhydrazones, which are readily derived from ketones or aldehydes, react efficiently with a wide range of alkenes (styrenes, dienes, enynes and electron-deficient olefins).

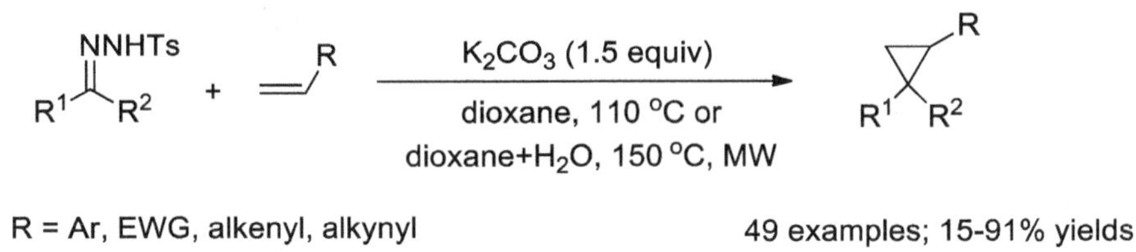

(4.4)

Moreover, catalyst-free cyclopropanation by using *N*-tosylhydrazones or the corresponding salts has been applied to the modification of nanotubes.[86]

4.4 Asymmetric intermolecular cyclopropanations

Although asymmetric catalytic cyclopropanation *via* metal carbene transfer has a long history, it still attracts attention and there have been new developments in the past few years. With the emerging of new chiral metal catalysts, enantioselective synthesis of cyclopropanes has experienced significant advances. A wide range of transition-metal complexes and organic catalysts have been designed, characterized and then applied into asymmetric cyclopropanations. In this section, these transformations are summarized according to the type of catalysts as well as the starting materials.

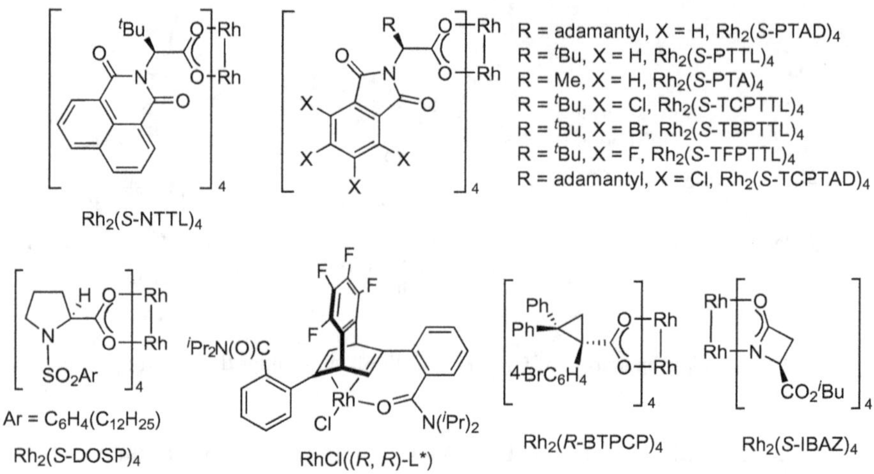

Figure 4.4 Some selected chiral Rh(II) carboxylates.

The traditional Rh(II) catalysts have been proven highly effective in enantioselective cyclopropanations. Some selected chiral dirhodium carboxylates, which have been proven effective for enantioselective cyclopropanations, are summarized in Figure 4.4.

The diazo substrates that have been utilized in asymmetric cyclopropanations with chiral Rh(II) catalysts are summarized in Figure 4.5. These include α-amido-α-diazoacetates **32**,[87] 2-diazo-2-phenylacetonitrile **33**,[88] α-cyano/nitrodiazoacetamide **34**,[89,93] α-alkyl-α-diazoesters **35**,[90,99] α-nitro diazoacetophenones **36**,[91,97] α-pyrrolidine amido-α-diazoesters **37**,[92] α-aryl-α-diazoketones **38**,[109] donor/acceptor carbene precursor of α-aryl-α-diazoesters **39**,[94,103] stable dimethyl diazomalonate substrate **40**,[95] *tert*-butyl α-diazopropionate **41**,[96] acceptor/acceptor diazo compounds containing α-PMP-ketone group **42**,[97] donor/acceptor carbenoids of styryldiazoacetates **43**,[98] diazooxindole **44**,[100] acceptor/acceptor diazo compounds of α-cyano diazophosphonate and α-cyano diazoesters **45**,[101] vinyldiazoacetates,[102] trimethylsilylethyl and trichloroethyl aryldiazoacetates.[104]

On the other hand, the scope of olefins has been expanded beyond styrenes and simple alkenes. The olefins which have been applied in enantioselective cyclopropanations include tri-substituted alkenes containing allylic C–H bond,[110] mono-substituted allenes,[105] unsaturated cyclic lactams,[106] dihydronaphthalenes,[111] substituted azido alkenes[107] and alkenyl fluorides.[108] Consequently, a series of chiral cyclopropanes

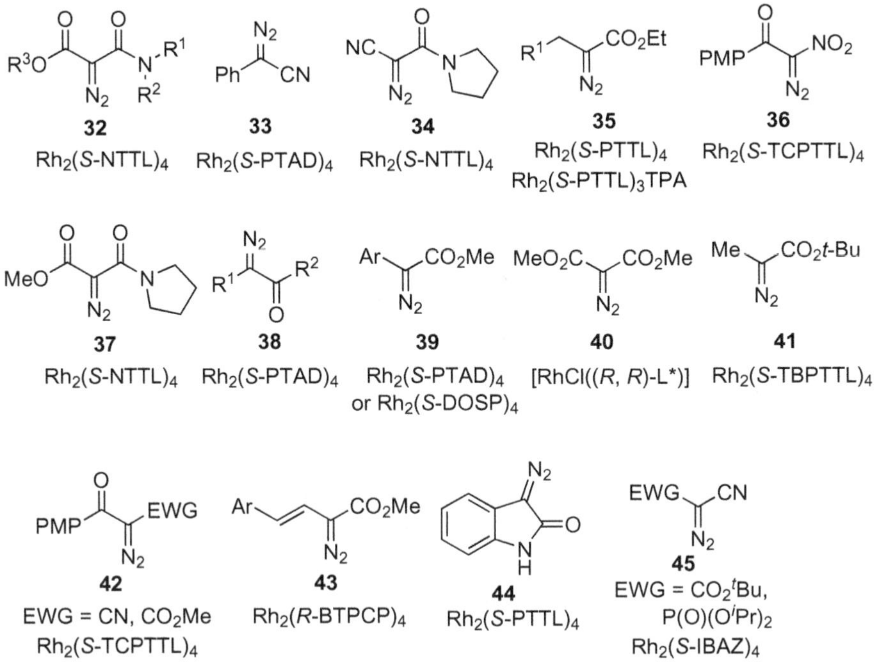

Figure 4.5 Diazo substrates and chiral Rh(II) catalysts in Rh(II)-catalyzed asymmetric cyclopropanations with olefins.

bearing various functionalized groups can be obtained in a straightforward manner with chiral Rh(II) complexes.

As shown in Chapter 5, the prolinate chiral Rh(II) catalysts developed by McKervey and Davies have been very successful for asymmetric C–H insertions with donor/acceptor-substituted diazo compounds. The catalytic systems are also effective for cyclopropanations and cyclopropenations. For example, Davies and co-workers have demonstrated that Rh_2(*S*-DOSP)$_4$ is an effective chiral catalyst for the enantioselective cyclopropenation of alkynes by methyl aryldiazoacetates.[112]

Furthermore, Davies and co-workers have designed a bridged Rh(II) tetraprolinate Rh_2(*S*-biTISP)$_2$ **46**, which has proven to be an excellent catalyst in asymmetric cyclopropanations. It was used for α-diazophosphonates, affording a high level of diastereocontrol and a moderate-to-good level of enantiocontrol (Scheme 4.9).[113] This catalyst also afforded a high turnover number and turnover frequency.[114] Moreover, the catalyst was immobilized

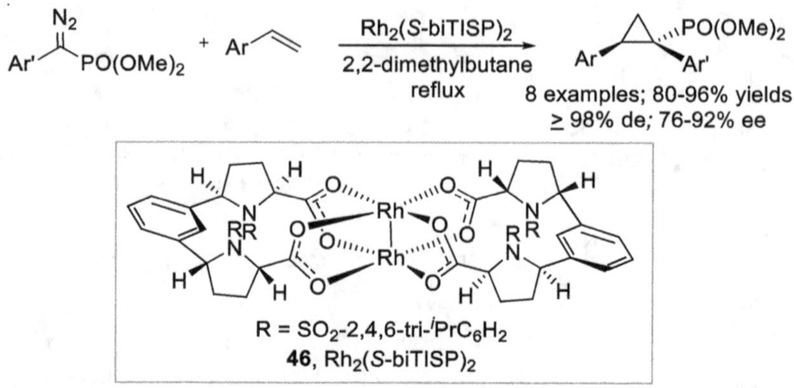

Scheme 4.9 $Rh_2(S\text{-}biTISP)_2$-catalyzed asymmetric cyclopropanation.

by highly cross-linked macroporous polystyrene resins with a pyridine attachment.[115–117]

Doyle's Rh(II) carboxamidate complexes are also effective chiral catalysts in intramolecular cyclopropanation reactions. Recently, the diastereoselectivity in Rh(II) carboxamidate-catalyzed intramolecular cyclopropanation was evaluated with diazoacetates bearing achiral linkers to the remote double bond.[118,119]

Cyclopropanation of substituted double bonds by metal carbenes usually favors the formation of *trans*-substituted cyclopropanes. Obtaining high selectivity for the *cis* isomer remains a challenge. To tackle this problem, Doyle and co-workers developed a new azetidine-ligated Rh(II) catalyst **47** that possesses an *L*-menthyl ester attachment. This chiral Rh(II) catalyst provides significant diastereocontrol and high enantiocontrol for the formation of *cis*-cyclopropane products from the reactions of substituted styrenes with diazo esters. The usefulness of this reaction has been demonstrated by the synthesis of a cyclopropane-configured phenylethylthiazoylthiourea (PETT) analogue **48** (Scheme 4.10).[120]

Charette and co-workers have focused on the transition-metal-catalyzed reaction of diazo compounds bearing an adjacent nitro or cyano group. A new azetidine-based Rh(II) catalyst $Rh_2[4S\text{-}(4')\text{-}FBNAZ]_4$ was prepared and was found to be effective in the intramolecular cyclopropanation of substituted allylic cyanodiazoacetates with up to 91% *ee* (Scheme 4.11).[121]

While Rh(II) catalysts are highly efficient for carbene transfer reactions, the high price of these catalysts impose limitations for their wide

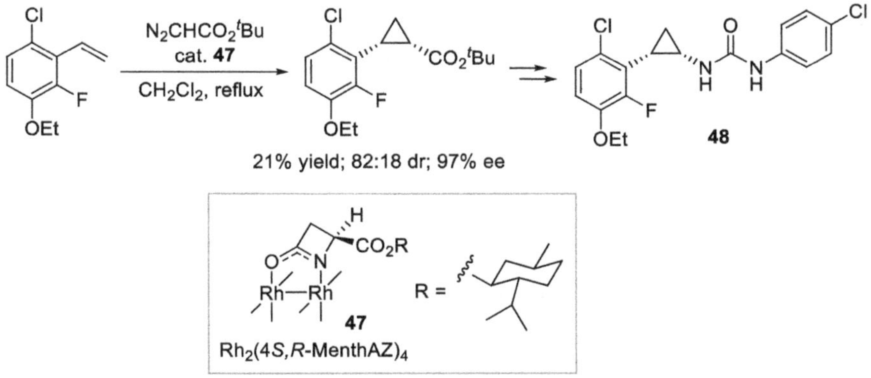

Scheme 4.10 Rh(II)-catalyzed *cis*-selective asymmetric cyclopropanations.

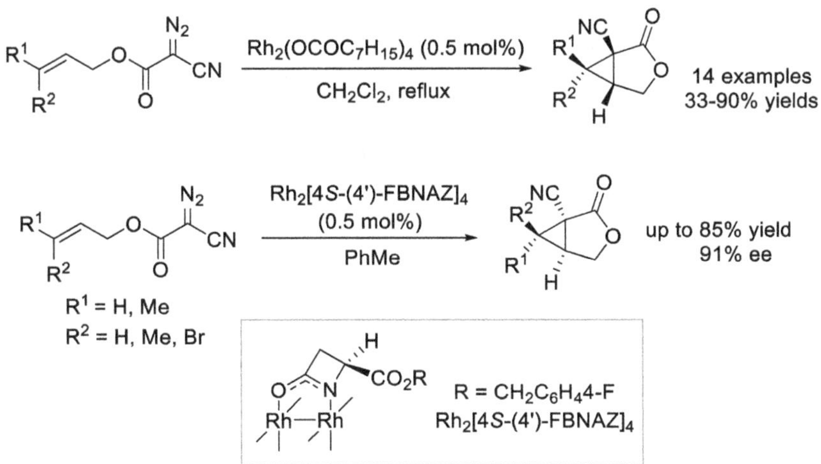

Scheme 4.11 Intramolecular asymmetric cyclopropanation of allyl cyanodiazoacetates.

applications. To overcome such problems, immobilized homogeneous as well as heterogeneous Rh(II) complexes have attracted much attention.[122–124] Chiral Rh(II) carboxylates, which contain *ortho*-metalated *tris*-styryl phosphines moiety, were efficiently immobilized to polymer linker through radical copolymerization.[122] Asymmetric cyclopropanations were carried out to evaluate these heterogeneous catalysts. The corresponding cyclopropanes were obtained in improved yields, however, the diastereo- and enantioselectivities were low.

Davies and co-workers designed and synthesized silica-immobilized chiral Rh(II) catalysts, which demonstrated comparable results to those obtained with their homogeneous counterpart in a series of carbene transformations, including cyclopropanation, cyclopropenation, ylide formation/[2,3]-σ rearrangement and combined C–H functionalization/Cope rearrangement.[123] Besides, Ball and co-workers demonstrated that the Rh(II) complexes coordinating to tripodal peptide ligand could be synthesized and serve as efficient catalysts for asymmetric cyclopropanation with higher efficiency than the original soluble bis-peptide $Rh_2(peptide)_2$ catalysts.[124]

Chiral Cu(I) complexes are also efficient catalysts for asymmetric cyclopropanation in carbene transfer reactions. In particular, C_2 symmetric bisoxazoline copper complexes have proven to be very effective chiral catalysts for cyclopropanations. Modification of these chiral ligands has been one of the main areas of interest in recent years. Heterogeneous bis-oxazolines have been tested with the employment of various immobilizations and recycling methods.[125] For example, immobilization of the chiral ligands by a fluorous-phase approach has been examined for bisoxazoline ligands. The fluorous bisoxazoline ligands were tested in the Cu(I)-catalyzed cyclopropanation of styrene with various α-diazoacetates (Figure 4.6).[126] The catalyst can be easily separated from the products by simple precipitation using hexane, and can be recycled without loss of diastereo- and enantioselectivity.

On the other hand, the structure of bis-oxazolines may be tuned through modification of their steric and electronic features. For example, 3,3′-bithiophene backbone has been studied and tested in asymmetric cyclopropanation.[127] This investigation suggests that steric factors and geometrical features of the catalyst may be more significant than the electronic properties of the chiral ligands.

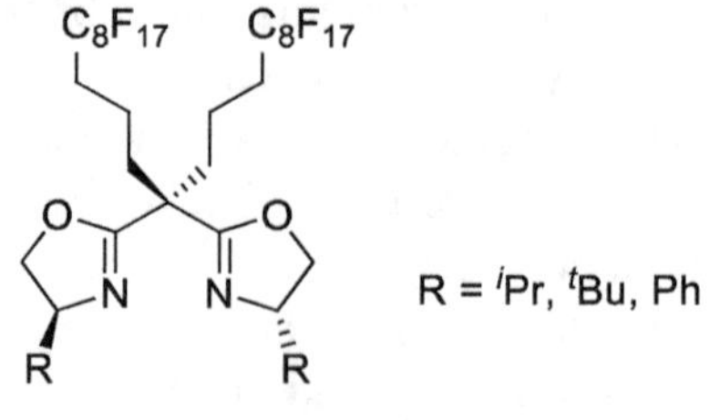

Figure 4.6 Fluorous bisoxazoline ligands for Cu-catalyzed asymmetric cyclopropanation.

Yamamoto and co-workers prepared bisoxazoline ligands **49** with an aryl group at the 4-position and a *gem*-dimethyl group at the 5-position on the oxazoline ring (Eq. (4.5)).[128] Enhancement of the *trans*-selectivity (*trans*/*cis* 87/13) and the enantioselectivity (96% *ee* for the *trans* product) was demonstrated for the asymmetric cyclopropanation of 2,5-dimethyl-2,4-hexadiene by *tert*-butyl diazoacetate with CuOTf/chiral ligand.

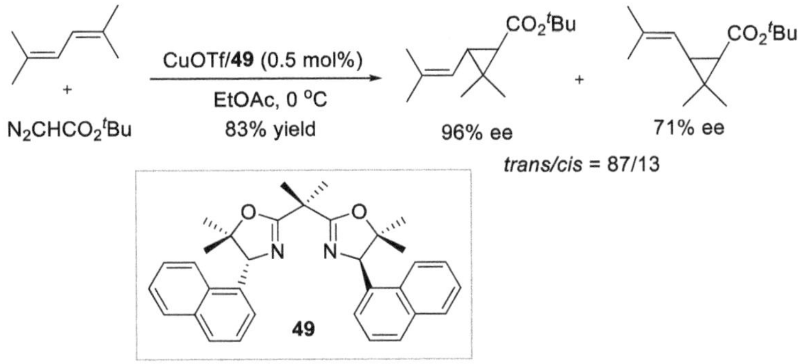

(4.5)

Ikeda and co-workers introduced a biaryl backbone into the bisoxazoline ligands.[129] Heteroatoms have also been introduced into the backbone. In one example, Pfaltz and co-workers developed anionic, boron-bridged analogues of the chiral bisoxazoline ligands **50** and **51**. The interesting feature of this type of bisoxazoline ligands is that the negative charge is located in the backbone. The Cu(I)-catalyzed asymmetric cyclopropanation with these ligands afforded the products with high enantioselectivities (Eq. (4.6)).[130]

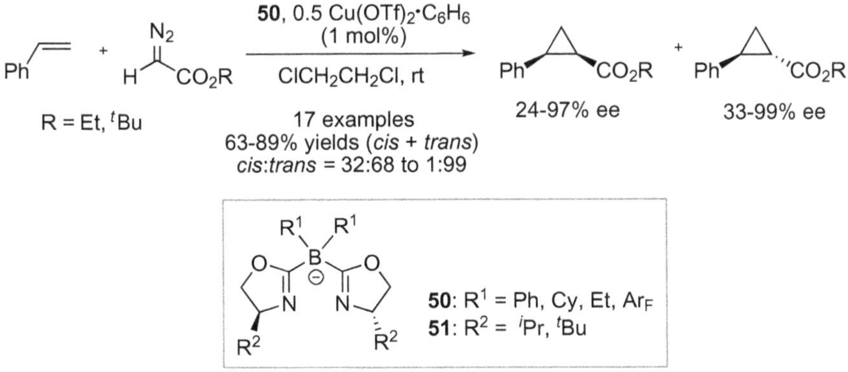

(4.6)

Some other C_2 symmetric ligands have also been explored.[131,132] Wilson and co-workers have reported a new chiral bipyridyl ligand **52** (Eq. (4.7)).[131] This ligand was evaluated in Cu(I)-catalyzed asymmetric cyclopropanation of a series of alkenes and diazoesters. High diastereoselectivity and enantioselectivity were observed (>95:5 *dr* and up to 99% *ee*).

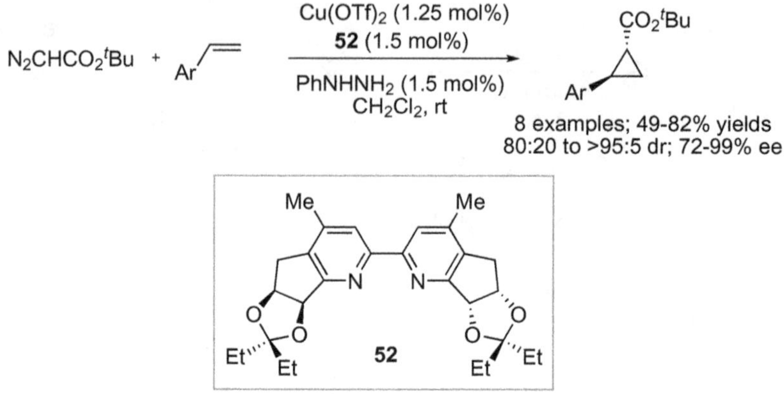

(4.7)

Tang and co-workers described Cu-catalyzed enantioselective cyclopropanations by using bis(oxazoline) ligand **53**.[133] Tri-substituted cyclopropanes were obtained with high level of diastereo- and enantioselectivity from 1,2-disubstituted alkenes with diazoesters (Scheme 4.12). Particularly, high *trans*/*cis* ratio was achieved for *trans* alkenes.

In addition to the traditional Rh(II) and Cu(I) complexes, Ru(II) complexes have also gained attention in asymmetric cyclopropanation, as demonstrated by Iwasa and others. As shown in Figure 4.7, catalytic asymmetric intermolecular cyclopropanations have been achieved with homogeneous chiral Ru(II) complexes, including chiral Ru(II) porphyrins **54**,[134] Ru(II)-bis(oxazolinyl)phenyl **55**,[135] Ru(II)-Pheox complex **56**[136–140] and Ru(salen) complexes **57**.[141] A wide range of substrates from diazoketones, diazoesters, succinimidyl diazoacetate and methyl (diazoacetoxy)acetate to diethyl diazomethylphosphonate were subjected to the Ru(II)-catalyzed stereoselective cyclopropanations with styrenes, alkenes, α,β-unsaturated carbonyl compounds, vinyl carbamates and allenes.

Moreover, heterogeneous chiral Ru(II) catalysts, such as reusable macroporous polymer-supported Ru(II)-pheox,[147] recyclable mesoporous silica-supported Ru(II)-(NHC)NN-pincer catalysts[148] and Ru(II)/salen-derived chiral MOFs,[149] have shown excellent reactivity and

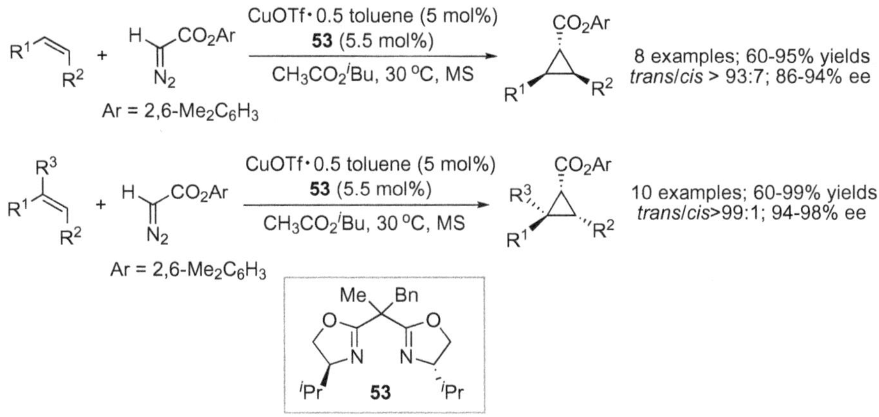

Scheme 4.12 Catalytic asymmetric cyclopropanations with Cu(I)/**53**.

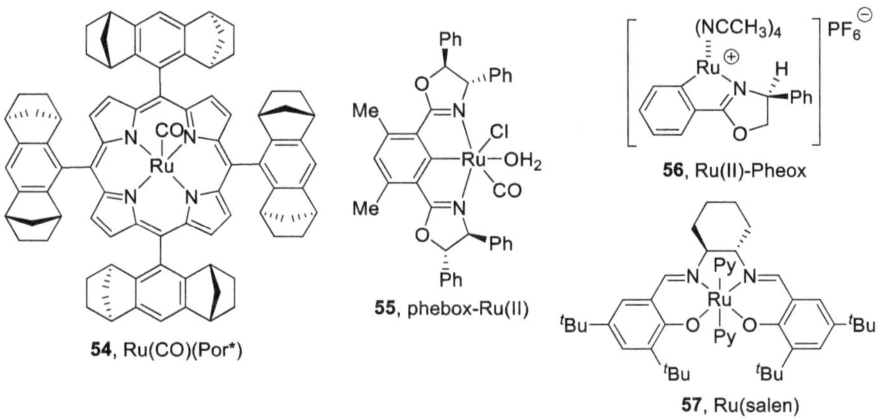

Figure 4.7 Chiral Ru(II) complexes for asymmetric cyclopropanations.

enantioselectivity in the intra- and intermolecular asymmetric carbene cyclopropanations.

Co(II)-catalyzed asymmetric cyclopropanation with diazo substrates were explored with various types of chiral Co(II) complexes, including Co(II)-porphyrins and Co(II)-salens. Zhang and co-workers demonstrated high enantioselectivities in Co(II)-porphyrin-catalyzed cyclopropanations with diazosulfones,[150] acceptor/acceptor carbene precursor of α-nitrodiazoacetates,[151] succinimidyl diazoacetate,[152] acceptor/acceptor-type diazo reagent of α-cyanodiazoacetates[153] and α-ketodiazoacetates.[154] The remarkable catalytic effectiveness was attributed to both the rigidification of

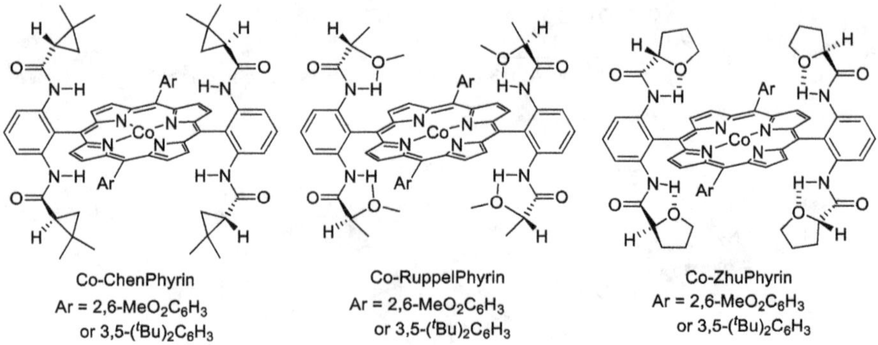

Figure 4.8 Structures of D_2-symmetric chiral Co(II) porphyrins.

porphyrin ligand and polarity of chiral environment through intramolecular hydrogen bonding as well as planar cyclic structure (Figure 4.8).

White and co-workers have reported the preparation of Co(II)-salen catalyst **58** based on *cis*-2,5-diaminobicyclo[2.2.2]octane as chiral scaffold. Asymmetric cyclopropanation from 1,1-disubstituted ethylenes and diazoesters with potassium thioacetate as the promoter was performed with a good yield and excellent diastereo- and enantioselectivities.[159] This strategy was applied to facile and straightforward enantioselective synthesis of (+)-synosutine (Scheme 4.13).

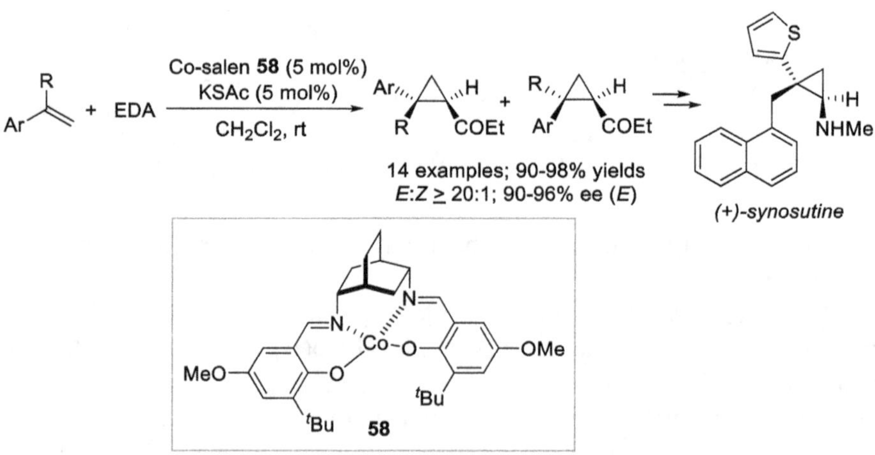

Scheme 4.13 Co(II)-salen-catalyzed asymmetric cyclopropanation of 1,1-disubstituted ethylenes with EDA.

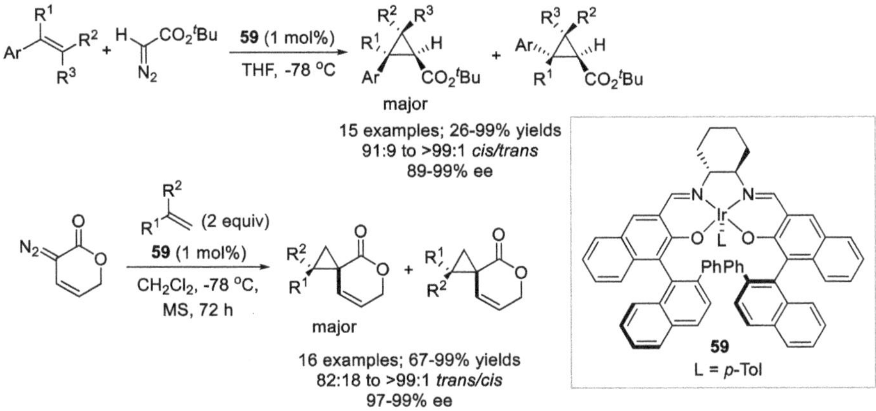

Scheme 4.14 Ir(III)-salen-catalyzed asymmetric cyclopropanation with olefins.

Chiral complexes with other transition-metals, such as those of iron[160] and chromium,[161] have also been investigated in asymmetric cyclopropanations. On the other hand, in addition to the C_2 symmetric ligands, C_1 symmetric ligands,[162,163] C_3 symmetric ligands[164,165] and multidentate *N*-donor ligands[166] have been developed for catalytic asymmetric cyclopropanations.

Katsuki and co-workers explored aryliridium(III)salen **59**-catalyzed asymmetric cyclopropanation with diazoesters with high *cis*-selectivity and excellent enantioselectivity.[167] The substrates scope was wide, including mono-, di- and tri-substituted olefins as well as non-conjugated terminal olefins. They applied this transformation using vinyl diazolactone as the substrates to access α-spirocyclopropyl lactones.[168] Functionalized 5-oxaspiro[2.5]oct-7-en-4-ones were obtained in highly enantio- and *trans*-selective manner (Scheme 4.14).

Metalloporphyrin compounds have demonstrated wide applications in catalytic cyclopropanations. Apart from the chiral Ru(II)-, Co(II)-porphyrins, Simonneaux successfully synthesized water-soluble chiral Fe(II)-porphyrin complexes, which were evaluated in catalytic asymmetric cyclopropanation of styrene with EDA.[169] The same group also explored the synthesis of optically active cyclopropyl ketones by using iron porphyrin in homogeneous catalytic asymmetric cyclopropanation of diazoketones with styrenes.[170] Furthermore, Boitrel and Gallo group reported a robust C_2-symmetrical chiral iron porphyrin

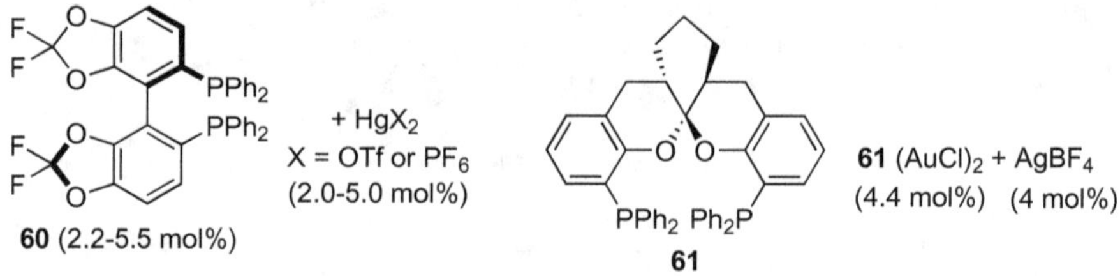

Figure 4.9 Hg(II)- and Au(I)-catalyst for cyclopropanation with diazooxindoles.

catalyst. With this catalyst, the cyclopropanation between methylstyrene and EDA was performed with excellent yields, and good *trans*-diastereo- and enantioselectivity.[171] More recently, an engineered myoglobin-based catalyst demonstrated robust and efficient cyclopropanations of aryl-substituted alkenes with good diastereo- and enantioselectivity (98–99.9% *ee*).[172]

Stereoselective cyclopropanation of diazooxindoles with styrene derivatives was investigated by Zhou and co-workers (Figure 4.9).[173–175] They reported the first examples of Hg(II)-catalyzed asymmetric cyclopropanation with chiral difluorphos ligand **60**.[173] In order to avoid using toxic mercury reagent, the C_2-symmetrical spiroketal bisphosphine (**61**)/Au(I) catalytic system was explored.[174] These transformations have provided effective and powerful methods for the highly diastereo- and enantioselective synthesis of spirocyclopropyl oxindoles.

In addition to the chiral transition-metal complexes, enantioselective synthesis of cyclopropanes promoted by chiral organo-catalyst was also explored (Scheme 4.15). In 2009, Charette and co-workers succeeded in asymmetric synthesis of 1,2,3-substituted cyclopropanes from allylic alcohol. The reaction proceeded through zinc carbene generated *in situ* from aryldiazomethane.[176] The high diastereo- and enantioselectivity was induced by chiral dioxaborolane **62**. Furthermore, Ryu and Hwang developed the oxazaborolidinium ion **63**-catalyzed asymmetric cyclopropanation.[177] Functionalized cyclopropanes containing two chiral quaternary stereocenters were obtained from α-aryl/alkyl-α-diazoacetates with substituted acroleins in high yields and with excellent diastereo- and enantioselectivities. This reaction was assumed to occur through Lewis acid-promoted Michael-initiated ring closure with the extrusion of N_2.

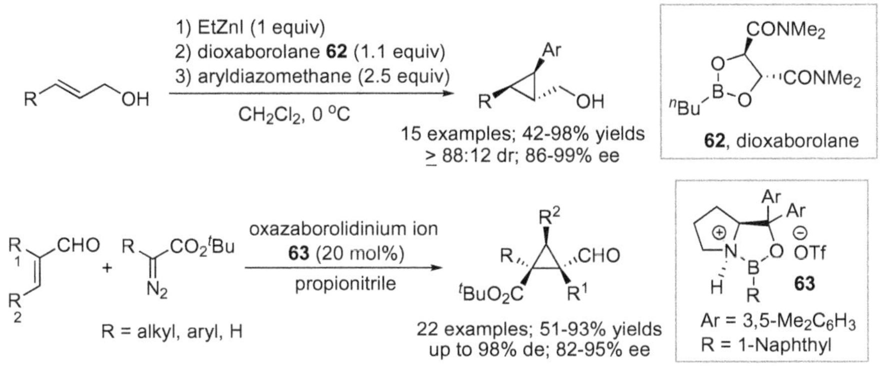

Scheme 4.15 Asymmetric cyclopropanations promoted or catalyzed by organic molecules.

4.5 Cyclopropanation with CF_3CHN_2

Trifluoromethyldiazomethane (CF_3CHN_2) has attracted attention as a reagent for introducing CF_3 group. CF_3CHN_2 can be usually generated *in situ* from trifluoroethyl amine hydrochloride with sodium nitrite in aqueous media. The use of this diazo compound for cyclopropanation is a straightforward way to access CF_3-substituted cyclopropanations.

Cu(I)- or Rh(II)-catalyzed cyclopropanation of electron-rich olefins with CF_3CHN_2 was explored by Komarov and co-workers.[178,179] This protocol was applied to synthesize CF_3-substituted proline analogues which served as ^{19}F NMR labels for peptides,[178] as well as 2-(trifluoromethyl)cyclopropyl boronic compounds.[180,181] For example, the skeleton of 6-trifluoromethyl-3-azabicyclo[3.1.0]hexanes was constructed by [3+2] cycloaddition of CF_3CHN_2 with *N*-benzyl maleimide, followed by extrusion of N_2 under thermolytic conditions (Scheme 4.16).[182]

Carreira and co-workers developed a series of transition-metal-catalyzed cyclopropanations of CF_3CHN_2 (Scheme 4.17). Thus, iron(III)-porphyrin-catalyzed diastereoselective carbene cyclopropanation was performed with styrene derivatives in aqueous media.[183] This strategy afforded single *trans*-CF_3-substituted cyclopropanes under operationally safer conditions with the employment of *in situ* generated CF_3CHN_2. Then, substituted conjugated 1,3-dienes and 1,3-enynes were subjected to this convenient procedure with the generation of vinyl- and alkynyl cyclopropanes chemo- and diastereoselectively.[184]

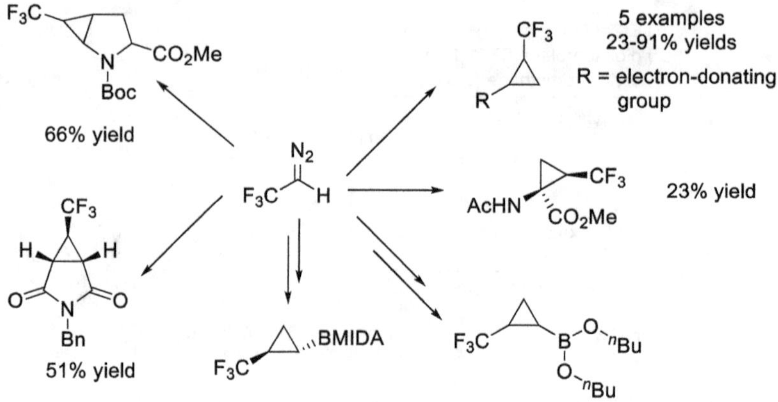

Scheme 4.16 Cyclopropanations of CF_3CHN_2 with various olefins.

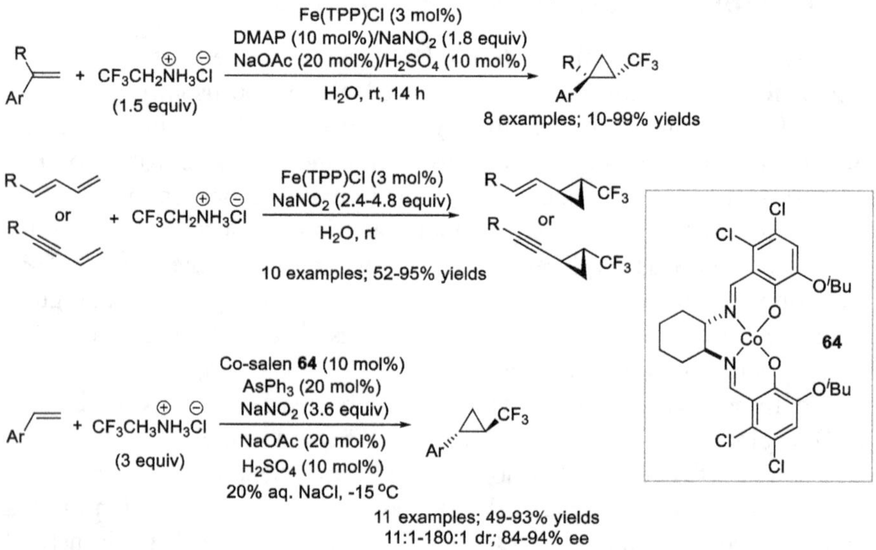

Scheme 4.17 Fe(por)- and Co-salen-catalyzed olefin cyclopropanation with CF_3CHN_2.

Furthermore, the same group developed an asymmetric version of the reaction catalyzed by chiral Co(II)-salen complexes **64**.[185] This reaction provided a practical synthesis of asymmetric disubstituted cyclopropanes containing a CF_3 group, having a preference for *trans*-isomers scaffold formation.

4.6 Intramolecular cyclopropanation

Catalytic intramolecular cyclopropanations of substrates containing diazo group and olefin moiety have been explored for the construction of a fused ring system, including the corresponding asymmetric synthesis and sequential transformations.[187–191] An example is the enantioselective synthesis of bicyclo[6.1.0]nonane-9-carboxylic acids **65** *via* Rh(II)-catalyzed intramolecular cyclopropanation followed by Me_2AlOTf-promoted intramolecular Friedel–Crafts alkylation of the aromatic nucleophile with the *g*-lactone moiety in bicyclic products (Scheme 4.18).[187]

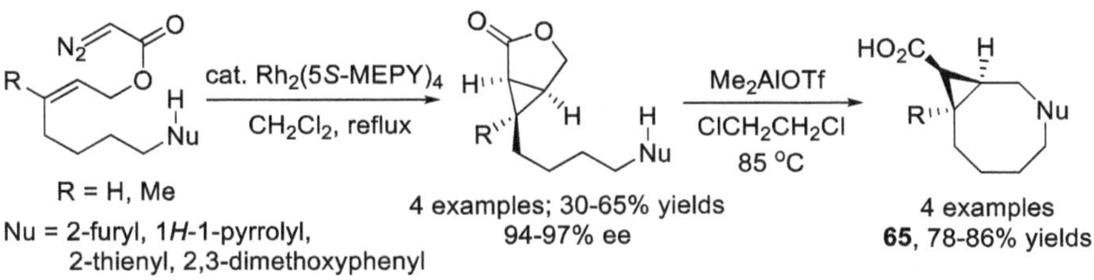

Scheme 4.18 Rh(II)-Catalyzed intramolecular cyclopropanation followed by Me_2AlOTf-promoted intramolecular Friedel–Crafts alkylation.

While intramolecular cyclopropanation is a powerful tool to construct bicyclic systems, chemoselectivity is a commonly encountered problem. In most cases, intramolecular C–H insertion is the major competing pathway, although other competing pathways such as sigmatropic rearrangement are also possible. The factors that influence the selectivity of these reactions are complex. Doyle and co-workers developed a strategy to build macrocycles by using intramolecular cyclopropanation.[190] They also systematically studied various factors that may influence the chemoselectivity of the reaction. As shown in Scheme 4.19, both the transition metals and the ligands can exert a significant influence on the ratio of cyclopropanation/[2,3] sigmatropic rearrangement.[191]

Doyle and co-workers further reported the divergent outcomes of catalyst-dependent transformations from vinyldiazoacetates with *α,β*-unsaturated aldehydes. The chemoselectivity of the reaction is highly dependent on the Lewis acidity and coordinating ability of these catalysts.[192] Bicyclization products **66** were obtained in the presence of cooperative Rh(II) and Cu(I) catalysts. The reaction was proposed to proceed *via* Cu(II)-catalyzed Mukaiyama-aldol reaction, followed by

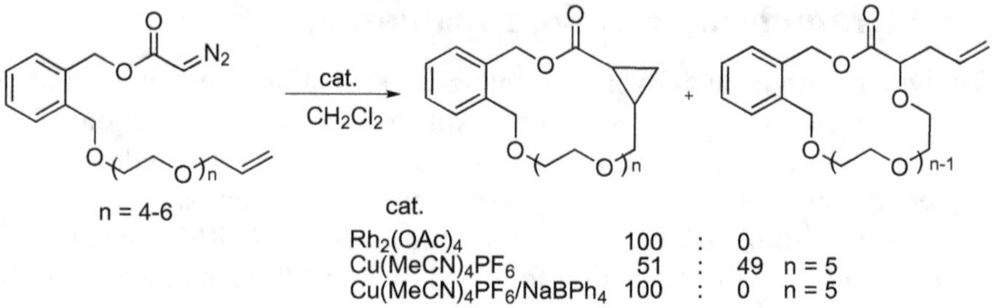

Scheme 4.19 The effect of transition metal and ligand on the ratio of cyclopropanation *versus* [2,3] sigmatropic rearrangement.

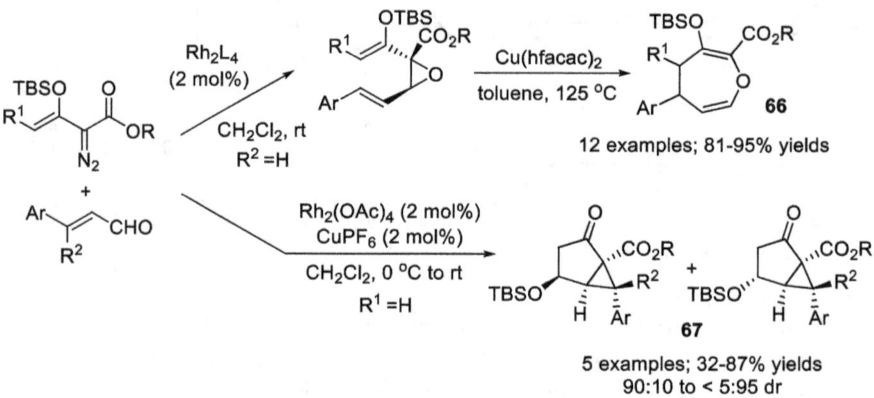

Scheme 4.20 Divergent outcomes from Cu(II)- and Rh(II)-catalyzed reactions with cinnamaldehydes.

subsequent Rh(II)-catalyzed intramolecular cyclopropanation. On the other hand, two-step one-pot formal [4+3] cycloaddition afforded substituted dihydrooxepines **67** through epoxidation with tandem Cope rearrangement (Scheme 4.20).

Moreover, the intramolecular cyclopropanation was extended to a series of substrates, including allene-containing α-diazoesters,[193] N–O tethered diazoamide **68**,[194] *cis*-1,2-dihydrocatechol derivatives incorporating C3-tethered diazoketone **69**,[195] diazoesters containing alkene moiety **70**[196] and allylic bromodiazoacetates **71**.[197] Functionalized fused-ring systems, such as bicyclic methylenecyclopropanes, cyclopropyl-fused pyrrolidines, spiro[5.5]undecanes, polycyclic lactones and cyclopropyl bromo lactones, could be constructed, respectively (Scheme 4.21).

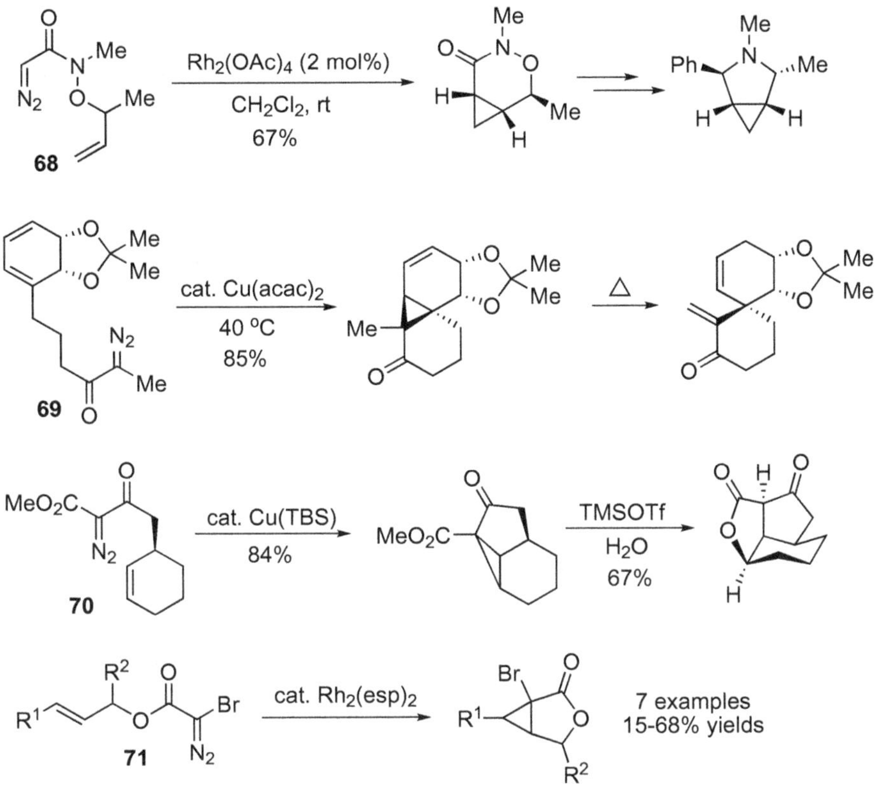

Scheme 4.21 Synthesis of fused-ring derivatives through intramolecular cyclopropanation.

In addition to Rh(II) and Cu(I) catalysts, other transition metals are also explored in catalytic intramolecular cyclopropanations. Pd(0)-catalyzed intramolecular cyclopropanation from allylic *N*-tosylhydrazones has been reported, which provided a facile and efficient method for constructing bicyclo[3.1.0] hexane scaffold.[198] Che and co-workers reported intramolecular cyclopropanation of *N*-alkyl indoles/pyrroles with alkyl carbenes using Co(II)-fluorinated porphyrin complex as the catalyst (Scheme 4.22).[199] The employment of bulky *N*-2,4,6-triisopropylbenzenesulfonyl hydrazones allowed efficient construction of *N*-heterocycles under mild conditions.

Furthermore, the intramolecular cyclopropanation strategy can be incorporated into subsequent ring-open tandem transformation. Wee and co-workers investigated the Rh(II)-catalyzed reaction of 2-indolyl diazoamide to form cyclopropyl intermediates, which rearranged to

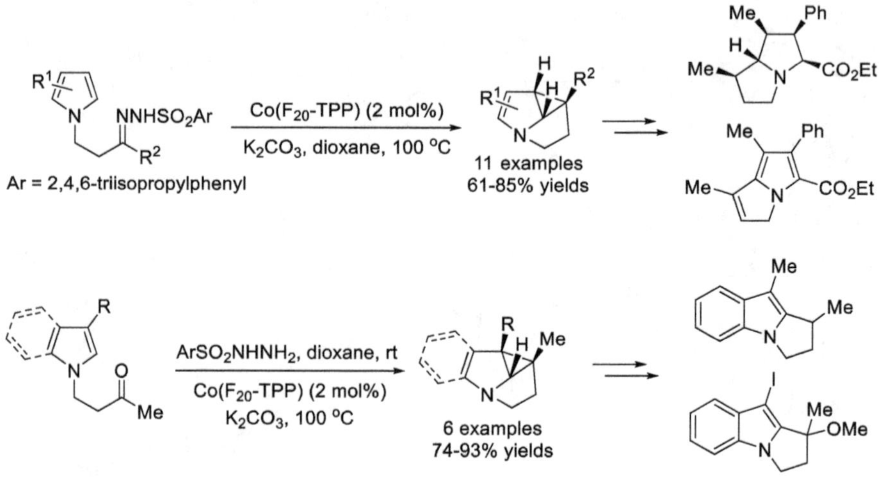

Scheme 4.22 Co(Por)-catalyzed intramolecular cyclopropanation of *N*-alkyl indoles/pyrroles.

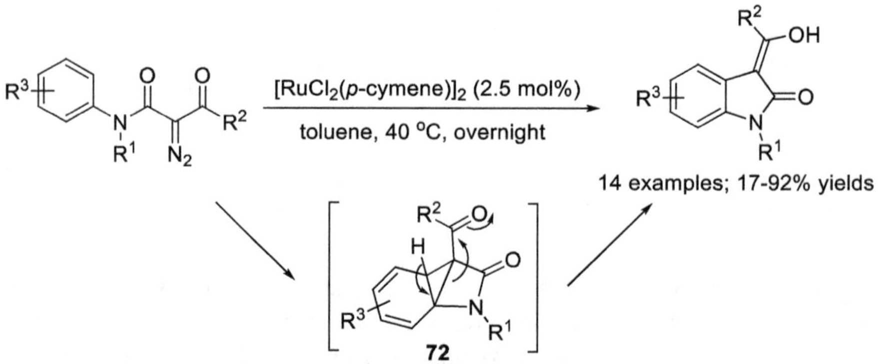

Scheme 4.23 Ru(II)-catalyzed intramolecular cyclization of diazo-β-ketoanilides.

β-carboline ring system.[200,201] Moreover, Yu and co-workers reported Ru(II)-catalyzed intramolecular cyclization of diazo-β-ketoanilide to form **72** with subsequent ring-opening (Scheme 4.23).[202]

Furthermore, α-vinyl-α-diazoester has been employed in tandem cyclopentannulation of α,β-unsaturated enones with high efficiency and diastereoselectivity.[203] This reaction was initiated by sequential

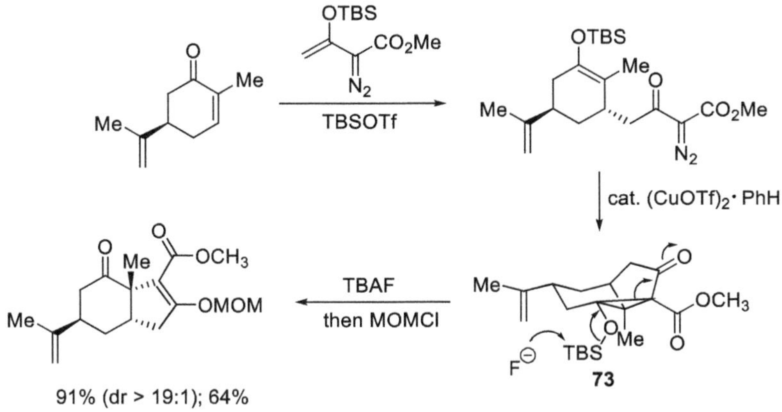

Scheme 4.24 Cyclopentannulation of conjugated enones with α-vinyl-α-diazoester.

Mukaiyama–Michael addition and intramolecular cyclopropanation to afford **73**, followed by fluoride-induced ring-open enolization to afford the cyclization products (Scheme 4.24).

As expected, the asymmetric intramolecular cyclopropanation are a powerful approach to access optically pure polycycle skeletons containing cyclopropane moieties. Chiral transition-metal complexes, such as Co(II)-porphyrin complexes,[204,205] Ru(II)-phenyl oxazoline complexes[140,206,207] and DNA-based Cu(I)-dppz complex in aqueous media,[208] have been implemented in the enantioselective synthesis of fused-ring system for a wide scope of diazo substrates. Notably, catalytic enantioselective cyclopropanation of electron-deficient alkenes was successfully implemented with Ru(II)-Pheox as the catalyst, affording cyclopropane-fused γ-lactones with high yields and excellent enantioselectivity.[207] Besides, chiral bisoxazoline ligands have been proven efficient for Cu(I)-catalyzed intramolecular cyclopropanation with α-diazo-β-keto phenyl-5-hexenyl sulfone,[209] as well as Fe(II)-catalyzed efficient protocol approaching chiral [3.1.0]bicycloalkanes with high yields and enantioselectivities (Figure 4.10).[210]

C_2 Symmetric bisoxazoline ligands are mostly used in Cu(I)-catalyzed intermolecular cyclopropanation. A recent example reported by Nakada and co-workers on the asymmetric cyclopropanation of α-diazo-β-keto sulfones demonstrated the successful application of C_2 symmetric bisoxazoline ligands (**74**) in intramolecular cyclopropanation (Eq. (4.8)).[211–214]

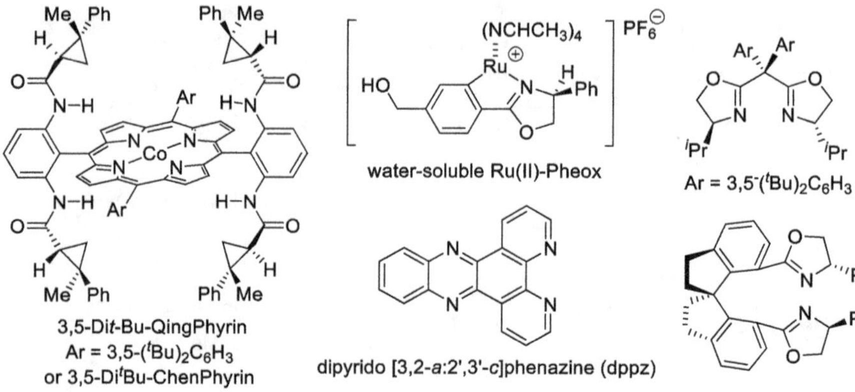

Figure 4.10 Transition-metal complexes and chiral ligands used for asymmetric intramolecular cyclopropanations.

This example is the first highly enantioselective cyclopropanation for α-diazo-β-keto sulfones.

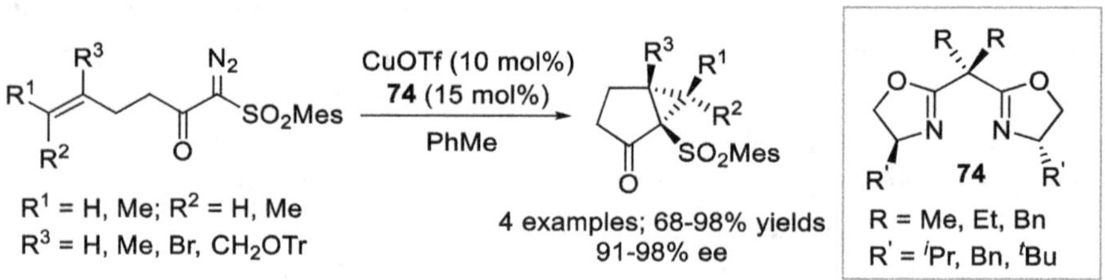

(4.8)

As for Rh(II)-catalyzed asymmetric intramolecular cyclopropanations, Fox and co-workers reported the synthesis of bicyclobutane **75** from (*E*)-2-diazo-5-arylpent-4-enolates, which were subjected to Cu(I)-catalyzed homoconjugate addition/enolate trapping to afford the substituted cyclobutanes with high enantioselectivity.[215] Additionally, Davies and co-workers reported a protocol for preparing 2-arylbicyclo[1.1.0]butane carboxylates **76** through intramolecular cyclopropanation with extremely low catalyst loading.[216] Interestingly, with $Rh_2(TPA)_4$ as the catalyst, the reaction afforded cyclohexene derivatives **77** as the products, which were formed through [4+2] cycloaddition dimerization of diene intermediates generated from 1,2-styryl migration (Scheme 4.25).

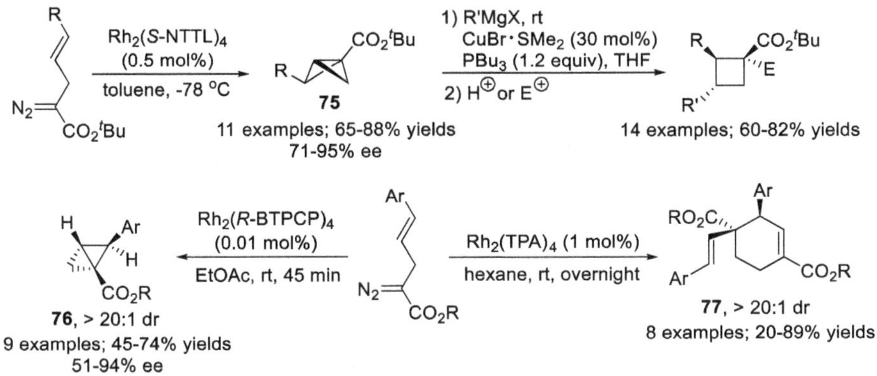

Scheme 4.25 Chiral Rh(II)-catalyzed asymmetric intramolecular cyclopropanations.

4.7 Tandem transformation consisting of cyclopropanation and the subsequent reactions

An interesting development seen in the past few years is the combination of transition-metal-catalyzed cyclopropanations with other types of reactions in single or cascade operations. Owing to the high strain of the three-membered ring system, there are various possible pathways for the ring-opening of cyclopropanation products. On the other hand, many functional groups can be tolerated in the transition-metal-catalyzed reactions of α-diazo compounds. Consequently, it is possible to design the diazo substrates bearing the necessary functional groups for further transformations following cyclopropanations.

4.7.1 *Buchner reaction and the synthesis of cycloheptatrienes*

Cyclopropanation of carbene generated from diazo substrates with aromatic rings under thermal or photochemical conditions followed by ring-opening is a unique method to prepare cycloheptatriene derivatives.[3–5] This dearomatization process, known as Buchner reaction, has been known for a long time. Recently, the transition-metal-catalyzed Buchner reaction has been explored for the synthesis of functional molecules with seven-membered ring moiety.

Maguire and co-workers explored enantioselective Cu(I)-catalyzed Buchner reaction (Scheme 4.26).[217–219] The investigation opens up the

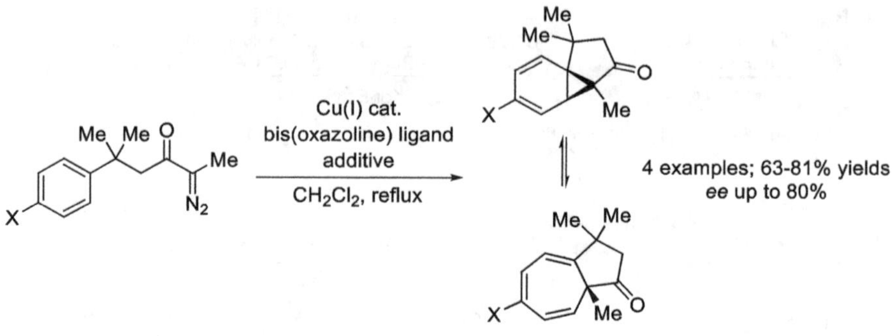

78

Scheme 4.26 Cu(I)-catalyzed asymmetric Buchner reaction.

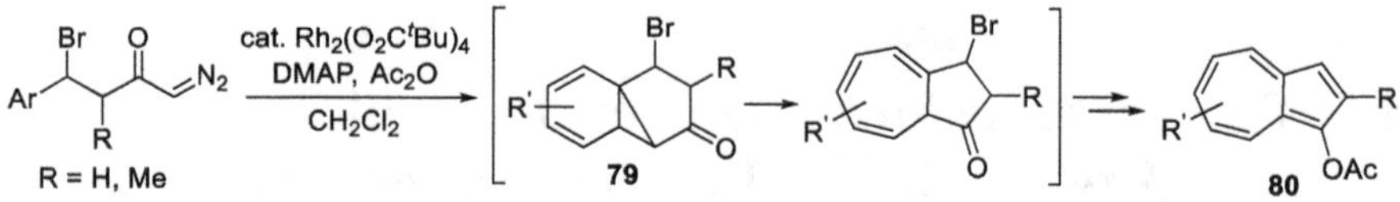

79 80

Scheme 4.27 Rh(II)-catalyzed Buchner reaction for the synthesis of substituted azulenes.

possibility for the synthesis of enantioenriched azulenones compounds **78** from diazoketone substrates.

In another example, Rh(II)-catalyzed reaction of diazo compounds affords the intramolecular cyclopropanation products **79**, which is followed by a 6-electron electrocyclic ring-opening to give the substituted azulenes **80** as the final products (Scheme 4.27).[220]

Cyclopropanation also occurs in the reactions of metal carbenes with furan and pyrrole. In the case of furan, the initial cyclopropanation product **81** is followed by a consecutive ring-opening process, leading to the ring-opening of the furan moiety.[221–223] This reaction has been recently utilized in the construction of the central seven-membered ring of guanacastepene (Scheme 4.28).[223]

Unlike furan, the cyclopropanation products from pyrrole undergo a rearrangement that leads to the formal C–H insertion products. The formal C–H insertion gives a mixture of regio-isomers (Eq. (4.9)).[224] Other aromatic compounds explored in cyclopropanation include activated quinolines, isoquinolines[225] and benzopyrylium triflates.[226] In these

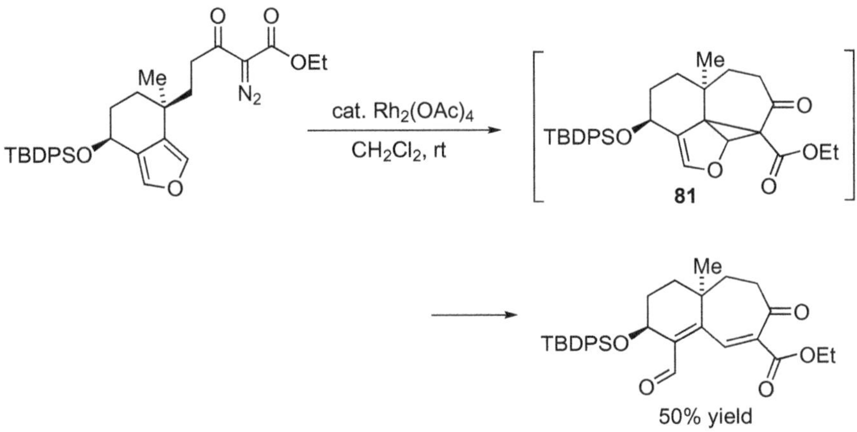

Scheme 4.28 Rh(II)-catalyzed intramolecular reaction of furan with carbene.

cases, cyclopropanation is followed by ring enlargement, similar to the Buchner reaction.

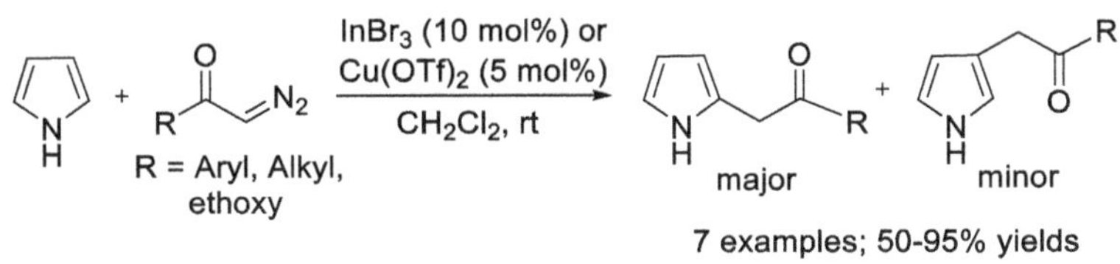

(4.9)

Chemo- and stereoselective Rh(II)-catalyzed Buchner reaction has attracted attentions recently. For example, Buchner reaction was carried out with diazocarbonyl compounds bearing flexible chiral 2,4-pentanediol tether.[227] Moreover, Maguire and co-workers reported the synthesis of daucane sesquiterpene derivatives with Rh(II)-catalyzed Buchner reaction as the key step for the construction of bicyclo[5.3.0]decane skeleton.[228] The same group also investigated the catalyst and substituent effects of Rh(II)-catalyzed aromatic addition from α-diazoketones, including steric, electronic and conformational influence.[229] Apart from intramolecular reaction, Rh(II)-catalyzed intermolecular Buchner-type reaction of aryl halides with EDA was reported to afford halogen-substituted cycloheptatrienes in regioselective manner.[230]

In 2013, Reisman and co-workers revealed the unique reactivity of α-diazo-β-ketonitriles in Rh(II)-catalyzed intramolecular cyclopropanations

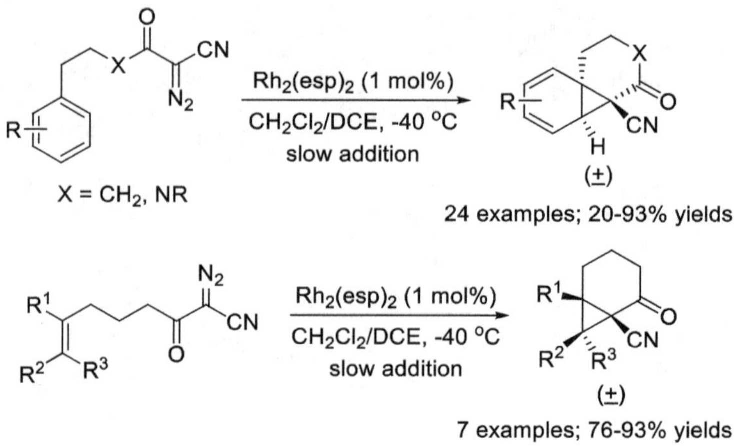

Scheme 4.29 Rh(II)-catalyzed arene and alkene cyclopropanations of α-diazo-β-ketonitriles.

with arenes and tri-/tetra-substituted alkenes (Scheme 4.29).[231] Interestingly, other similar acceptor/acceptor diazo species favored C–H insertion or dimerization process, rather than cyclopropanations. This transformation provides a concise and efficient route to access highly functionalized cyclopropanes with cyano group, especially for those with three contiguous all-carbon quaternary centers.

Doyle and co-workers examined the catalytic Buchner reaction with cascade transformation of various diazo compounds.[232–234] For example, the synthesis of tricyclic heterocycles from vinyldiazoacetate with a wide scope of nitrones was achieved with high yields and regioselectivity.[232] This multistep cascade reaction was proposed to proceed through [3+2] cycloaddition/Buchner reaction/[1,7]-oxygen migration with N–O bond cleavage simultaneously (Scheme 4.30).

Moreover, Doyle and co-workers reported an efficient construction of natural product-like scaffolds from diazoacetoacetate enones, which involved nucleophilic additions followed by Rh(II)-catalyzed carbene transformations.[233] Tetralone, dihydrocarbazole and dihydroindole derivatives were constructed through Rh(II)-catalyzed Buchner reaction to generate cycloheptatriene intermediate **82**, followed by Lewis acid-induced rearrangement (Scheme 4.31). Recently, they extensively reported chiral Rh(II)-catalyzed stereoselective desymmetrization of δ,δ-diaryl α-diazo-β-ketoesters, which afforded aryltetrahydroazulene or 4-aryltetralones with high enantioselectivities (up to 97% ee).[234]

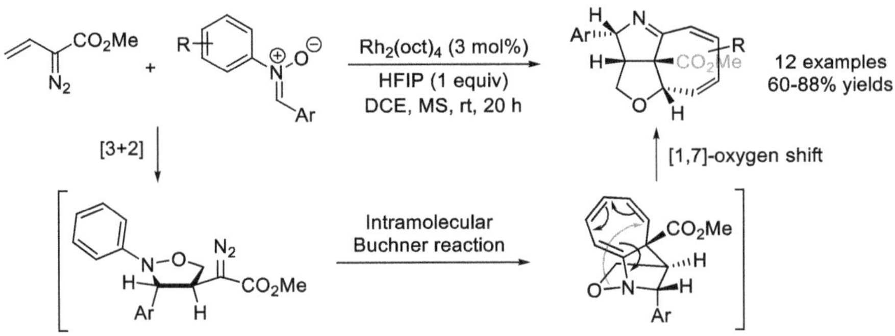

Scheme 4.30 Rh(II)-catalyzed synthesis of multi-functionalized tricyclic heterocycles from vinyldiazoacetate and nitrones.

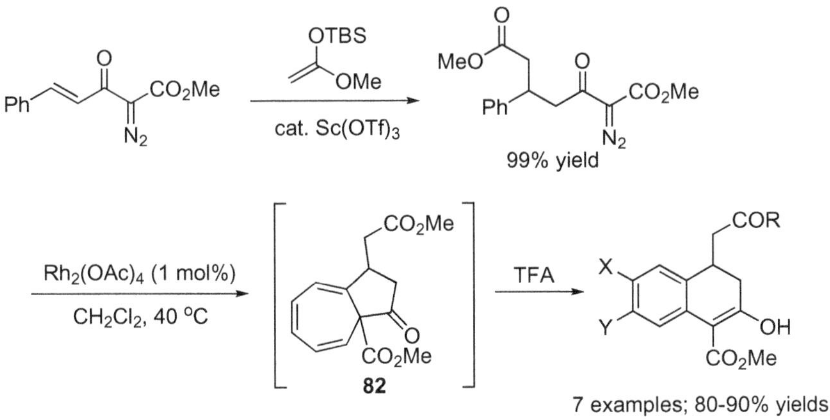

Scheme 4.31 Synthesis of tetralone from diazoacetoacetate enones.

In addition to Rh(II) and Cu(I) complexes, Pérez and co-workers evaluated Au(I)-NHC ligand-catalyzed intermolecular carbene transformation of diazoesters with aromatic compounds in the presence of $NaBAr_4$ as halide scavenger.[235–237] Both naphthalene and benzene derivatives were explored in this carbene transfer reaction, resulting in aromatic C–H bond functionalization and competing cyclopropanation process (Scheme 4.32). In particular, higher chemoselectivity was observed by using Au(III) catalyst, in which C–H bond functionalization was favored while carbene addition was suppressed.[236] Furthermore, a theoretical study toward trinuclear Ag(I) complex catalyzed Buchner reactions of EDA with several substituted arenes was also reported by Mindiola and co-workers.[237]

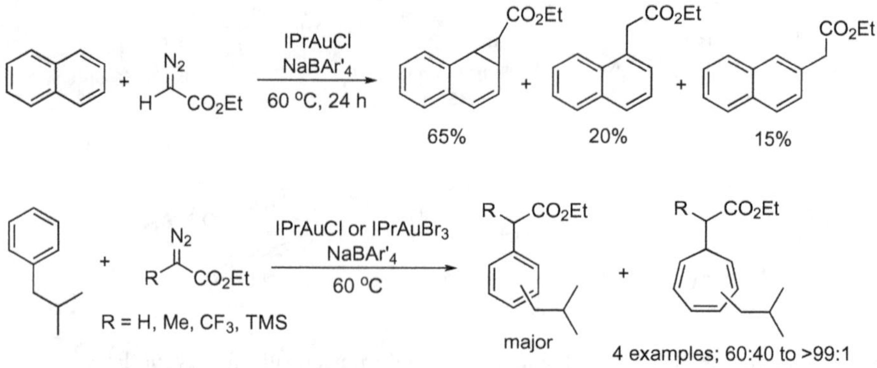

Scheme 4.32 Au(I)-IPr complex-catalyzed carbene transformation of diazoesters.

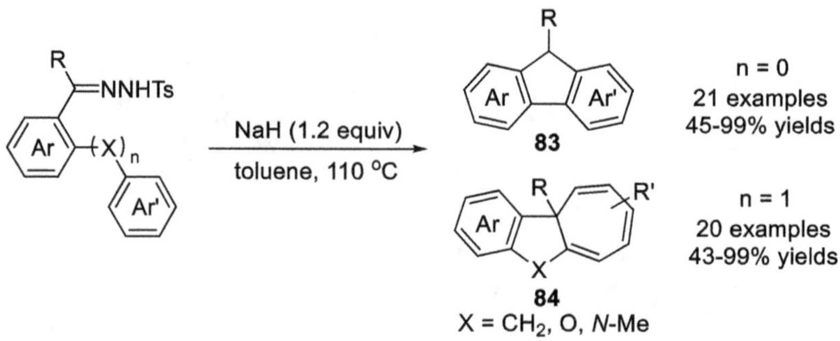

Scheme 4.33 Transition-metal-free synthesis of fluorenes and [6,5,7]benzo-fused compounds.

Wang and co-workers reported Buchner reaction with *N*-tosylhydrazones bearing trifluoromethyl group as the substrates.[238] This intermolecular transition-metal-free Buchner reaction afforded trifluoromethylated cycloheptatrienes in moderate yields.

Recently, the same group developed a substrate-dependent transition-metal-free intramolecular carbene aromatic substitution (formal C–H insertion) or Buchner reaction from *N*-tosylhydrazones.[239] Depending on the substrates, either fluorene derivatives **83** or [6,5,7]benzo-fused aromatic compounds **84** could be obtained with high efficiency under simple reaction conditions. In most cases, the corresponding products could be synthesized with up to 90% yields (Scheme 4.33).

4.7.2 *C–H bond functionalization*

Formal carbene C–H functionalization with diazo compounds can be achieved through the generation of cyclopropane intermediates followed by cascade ring-opening sequence. For example, Yu and co-workers reported the directing-group-free C2-selective carbene functionalization of indoles in the presence of Ru(II) catalyst.[240] With α-aryldiazoesters as the substrates, highly regioselective 2-alkylated indoles were obtained in good to excellent yields. The author proposed that this reaction occurred *via* the formation of cyclopropylindoline intermediate **85**, followed by ring-opening at C3-position, to afford the final product (Scheme 4.34).

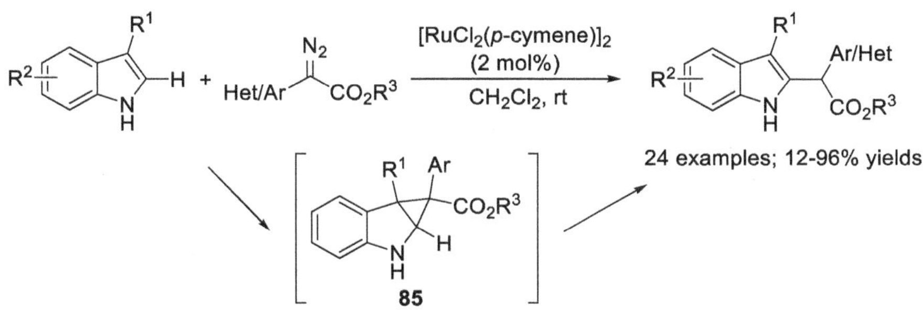

Scheme 4.34 Ru-catalyzed C2-selective functionalization of indoles.

Tang and co-workers reported a series of Fe(por)-catalyzed formal olefinic C–H bond insertions of allylic phosphoranes.[241,242] The three-component synthetic method provides efficient access to tri-substituted 1,3-butadienes **86** as well as vinyl cyclopropanes **87**. Instead of direct iron-carbene insertion into olefinic C–H bond, the reaction was proposed to be initiated by cyclopropanation of methyl diazoacetate (MDA) to form **88** with sequential ring-opening process (Scheme 4.35).

4.7.3 *Cyclopropanation with furans and related reactions*

Catalytic cyclopropanation of furans with diazo species demonstrates unique features. The Cu(I)- and Rh(II)-catalyzed reaction of substituted furans with a series of diazo compounds, including diazomethane sulfonate,[243] diazomethane phosphonate[243] and cyclic diazoamide,[244] have shown applications for synthesizing 1,4-dicarbonyl-1,3-butadienes (Scheme 4.36). The catalytic cyclopropanation/ring-opening strategy

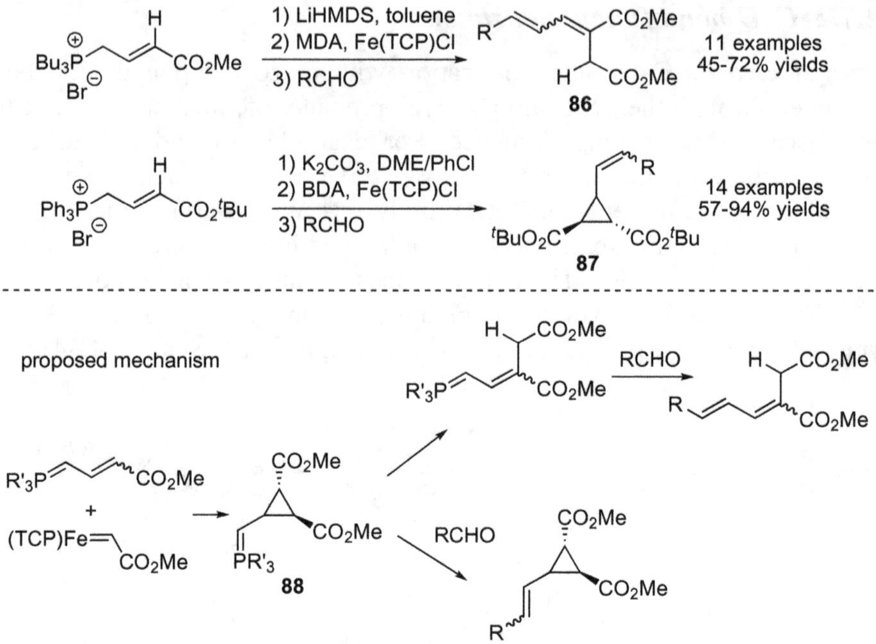

Scheme 4.35 One-pot synthesis of vinylcyclopropanes and 1,3-butadienes *via* formal carbene insertion of allylic ylides.

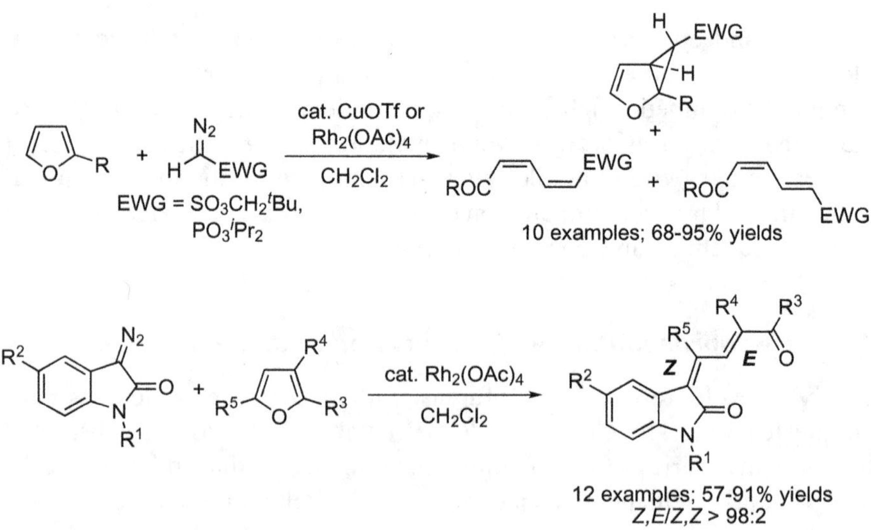

Scheme 4.36 Cyclopropanation/ring-opening reaction with furans.

represents a complementary method for the synthesis of functionalized 1,3-butadiene compounds with good regioselectivities and stereoselective control.

Werz and co-workers reported a domino reaction of diazo compounds with furans toward the synthesis of a wide range of functional heterocycles, including nonacyclic octaacetal **90**, bispyrroles **91**, terpyrroles **92** and pyrrolo[3,2-*e*]indoles **93**.[245–247] The key steps of these transformations are the ring enlargement of donor–acceptor-substituted cyclopropanes **89**, which is generated from catalytic cyclopropanation of furans with diazoesters (Scheme 4.37). These cascade transformations demonstrated potential value in synthesizing fused- or multi-heterocyclic compounds from commercially available starting materials.

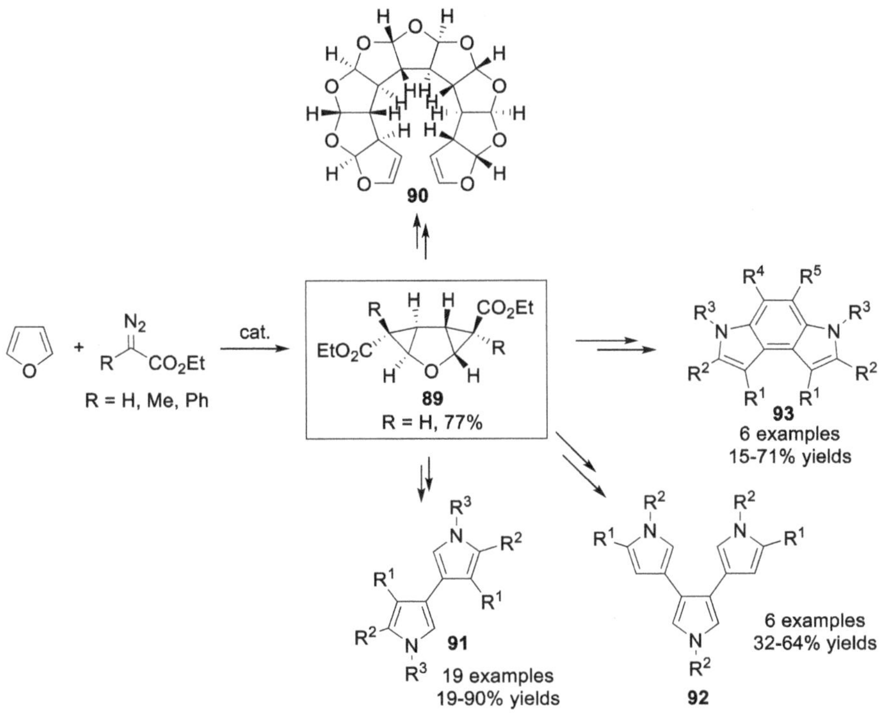

Scheme 4.37 Domino reaction of donor–acceptor-substituted cyclopropanes and synthetic applications for preparing diverse heterocycles.

4.7.4 *Other cyclopropane-ring-open transformations*

Barluenga and co-workers demonstrated that Cu(I) complex could discriminate between simple diazo compounds **94** and vinyl diazoesters **95** toward metal-carbene formation (Scheme 4.38).[248] Based on this feature, a cyclobutene synthesis was developed through catalytic cyclopropanation to give intermediate **96**, followed by ring-enlargement. Furthermore, the Cu(I) carbene species **98** generated from propargylic esters substrate **97** was also integrated into this strategy, providing a new approach for furyl-substituted cyclobutene derivatives **99**.

Highly substituted pyrrole **102** could be synthesized from α-diazo oxime ether **100** with 3-aminoalkenoate **101**.[249] The reaction was proposed to involve the reaction of copper carbene intermediate **103** with **101** to generate intermediate **104**. Subsequently, two plausible

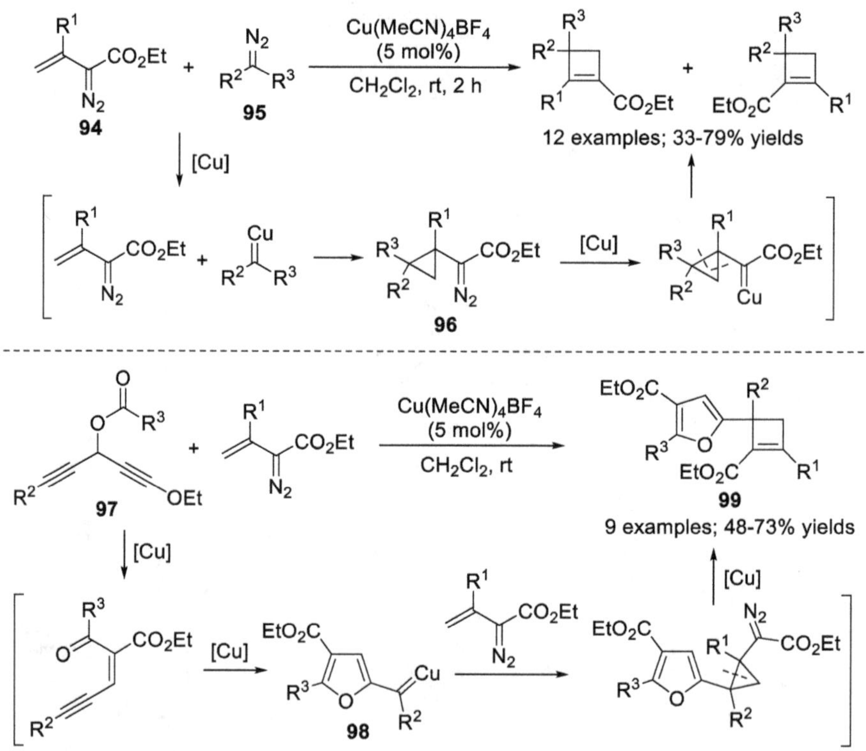

Scheme 4.38 Cu(I)-catalyzed carbene cyclopropanation/ring-enlargement sequence.

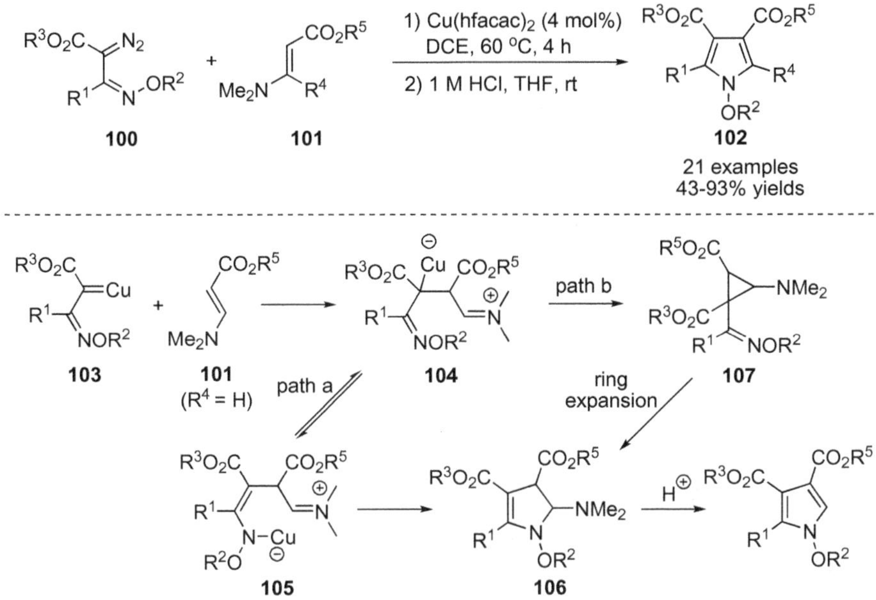

Scheme 4.39 Cu(I)-catalyzed [3+2] cycloaddition of α-diazo oxime ethers with 3-aminoalkenoates to access *N*-alkoxy pyrroles.

pathways have been hypothesized: one involves the generation of *N*-metalated intermediate **105**, followed by intramolecular nucleophilic addition to iminium species to afford **106**; the other pathway involves cyclopropanation to generate **107**, followed by ring-expansion sequence (Scheme 4.39). Moreover, catalytic cyclopropanation of diazoesters with dikene was applied to generate cyclopropane spiro-β-lactones in good diastereoselectivity, which further underwent Cu(I)-catalyzed rearrangement to produce pyranone derivatives.[250]

Ru(II)-catalyzed carbene cascade transformation of cyclic diazo dicarbonyl compounds with olefins was reported to give multisubstituted dihydrofurans.[251] With dihydrofurans as the substrates, the reaction was proposed to proceed *via* cyclopropanation/rearrangement process. As for other electron-rich or poor olefins, the reaction pathway involving cyclopropanation/ring-opening/cyclization or 1,3-dipolar cycloaddition was proposed, respectively.

Charette and co-workers explored the α-nitro-α-diazoesters or α-cyano-α-diazoesters as the carbene precursors in Rh(II)-catalyzed

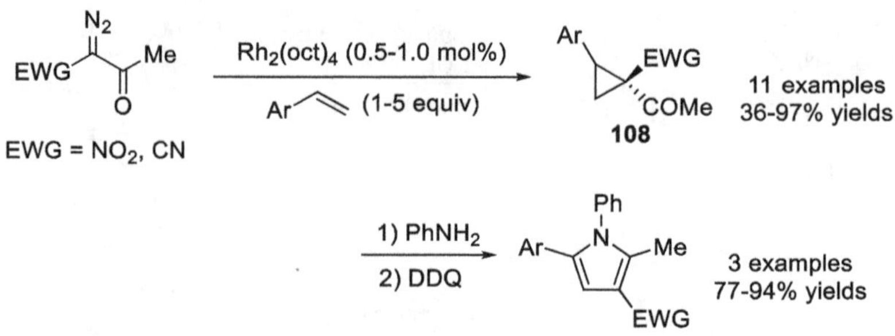

Scheme 4.40 Ring-opening of the cyclopropanes and sequential oxidation.

cyclopropanations. Through the combination of the cyclopropanation with reduction, a series of cyclopropane α-amino acids and amines could be synthesized.[252,253] With subsequent ring-opening of the cyclopropanation products **108** by amine, followed by DDQ oxidation, a series of polysubstituted pyrroles could be obtained (Scheme 4.40).[254]

Cyclopropanes bearing both electron-donating and electron-withdrawing groups are prone to undergo ring-opening reactions.[255] Thus, to access this type of cyclopropanes through cyclopropanation is useful. As an example, Rh(II)-catalyzed reaction of α-diazoketone with vinyl ether afforded the cyclopropanation product **109**, which bears an electron-withdrawing carbonyl group and an electron-donating alkoxy group. The cyclopropane product **109** was found to undergo ring-opening in the presence of $Rh_2(OAc)_4$, affording the 1,4-diketones **110**. The $Rh_2(OAc)_4$-catalyzed ring-opening presumably facilitates the ring opening through the coordination of Rh(II) catalyst with the carbonyl oxygen (Eq. (4.10)).[256]

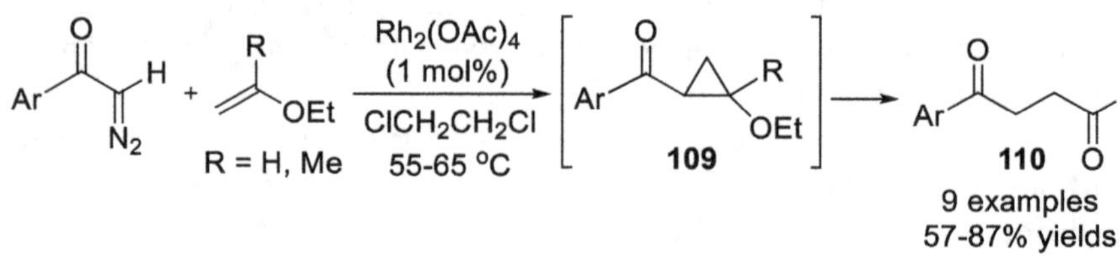

(4.10)

The sequential catalytic process involving cyclopropanation can be developed in combination with other well-established transition-metal-catalyzed reactions, such as Pd-catalyzed cross-coupling and Ru-catalyzed olefin metathesis.[257–260] The substrates used in the Pd-catalyzed cross-coupling reactions are organohalides and triflates,

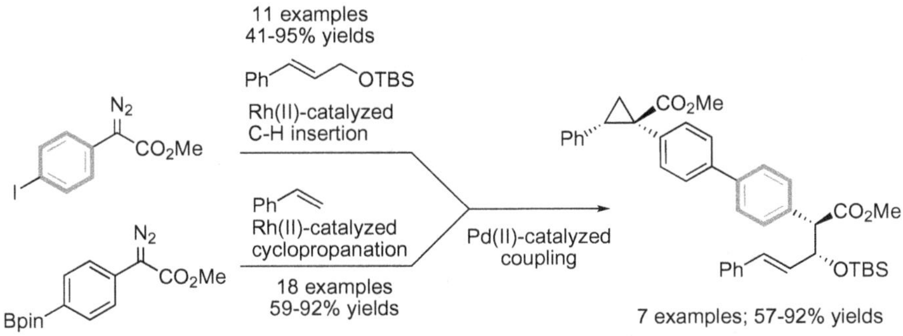

Scheme 4.41 Rh(II)-catalyzed cyclopropanation followed by Pd-catalyzed coupling reaction.

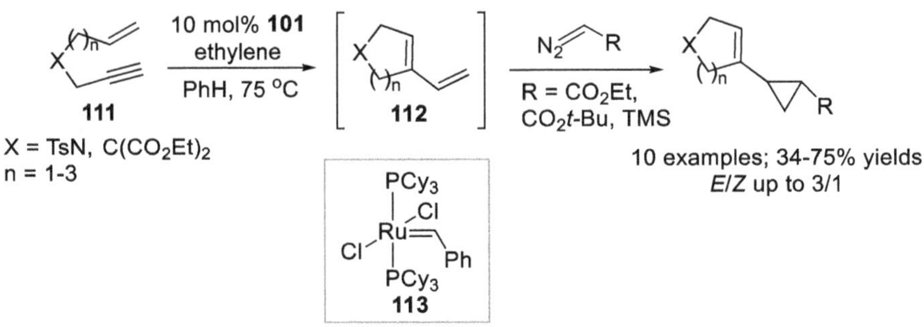

Scheme 4.42 Ru(II)-catalyzed metathesis with sequential cyclopropanations.

organoborons and organostannanes. Although Rh(II) complexes are highly reactive in the dediazoniation of diazo compounds, they do not react with these functional groups. A demonstration of the usefulness of this strategy is the combination of the Rh(II)-catalyzed cyclopropanation and C–H insertion with a subsequent Pd(II)-catalyzed Suzuki–Miyaura coupling (Scheme 4.41).[258]

Cyclopropanation can also be incorporated into tandem catalytic process involving olefin generation. A recent example is the tandem enyne metathesis/cyclopropanation catalyzed by Grubbs' ruthenium catalyst. With single Grubbs' ruthenium catalyst **113**, diazo compounds could undergo cyclopropanation in regioselective manner with 1,3-diene intermediates **112**, generated *in situ* from the enynes **111** (Scheme 4.42).[259]

Panek and co-workers have demonstrated sequential processes including Rh(II)-catalyzed cyclopropanation of vinyl diazoester **114** with

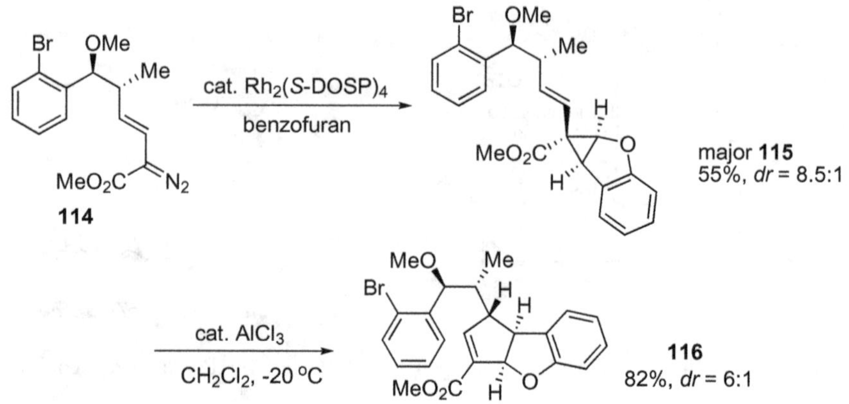

Scheme 4.43 Sequential reactions of Rh(II)-catalyzed cyclopropanation with ring-open rearrangement.

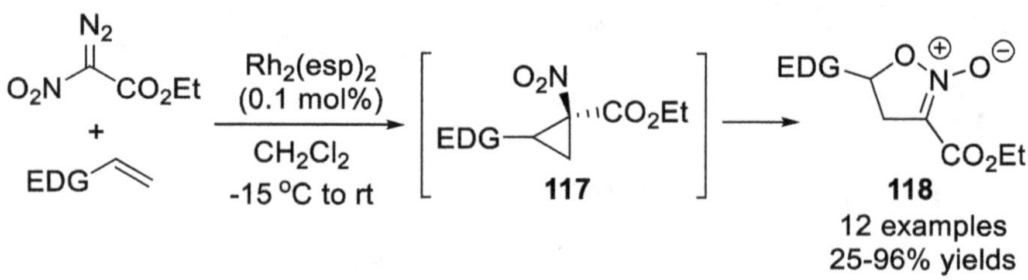

Scheme 4.44 Synthesis of cyclic nitronates by cyclopropanation and subsequent ring-enlargement process.

benzofuran.[260] Further elaboration of the product **115** derived from cyclopropane ring-opening rearrangement allowed for the construction of polycyclic scaffold of **116**, which finally led to a diversity-oriented synthesis of a series of alkaloid natural products. The key steps in these transformations are the chiral Rh(II)-catalyzed asymmetric cyclopropanation followed by Lewis acid-induced rearrangement process (Scheme 4.43).

Werz and co-workers applied the catalytic cyclopropanation/ring-enlargement protocol to the synthesis of cyclic nitronates from α-diazo-α-nitro ethyl acetate and the alkenes bearing electron-donating group.[261] Owing to the preference in forming 1,3-zwitterionic intermediates from donor–acceptor cyclopropane **117**, the nitrocyclopropane product readily undergoes intramolecular ring-enlargement to generate cyclic nitronate **118** as the final product (Scheme 4.44).

4.8 Cyclopropenation of alkynes

In parallel with cyclopropanation, catalytic carbene cyclopropenations of alkynes with diazo compounds have been proven valuable as synthetic methods for preparing cyclopropenes, which contain highly strained structures and show unique reactivities.[262] As shown in Scheme 4.45, versatile diazo compounds including EDA,[263,264] *in situ* generated CF_3CHN_2[265] and cyclic α-diazocarbonyl compounds[48] were subjected to Rh(II)-catalyzed cyclopropenation for the synthesis of 3-(cycloprop-2-en-1-oyl)oxazolidinones **119**, CF_3-substituted cyclopropenes **120**, as well as diazirine- (**121**) and cyclopropene-containing photo-cross-linkers (**122**). Moreover, a facile synthetic protocol for 3-aryl-3-methoxycarbonyl cyclopropenes with non-substituted double bond was developed through Rh(II)-catalyzed cyclopropenation of trimethylsilylacetylene with aryldiazoacetates.[266]

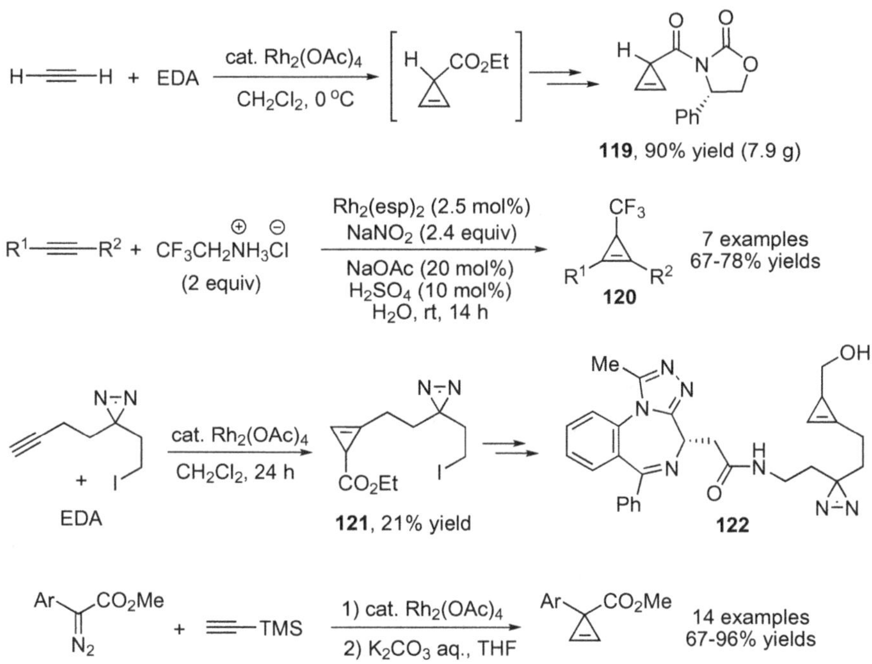

Scheme 4.45 Rh(II)-catalyzed intermolecular cyclopropenation of diazo compounds with alkynes.

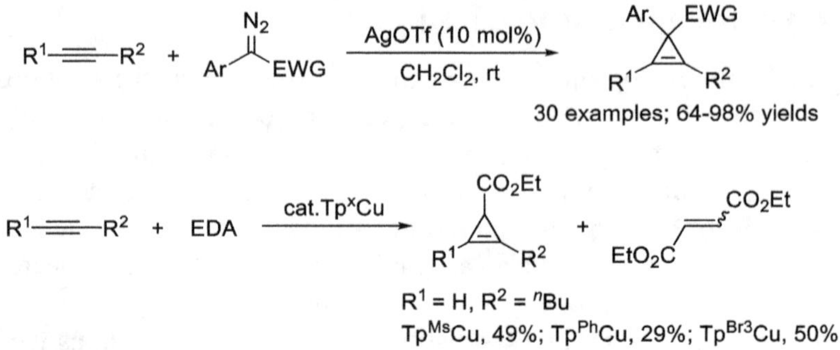

Scheme 4.46 Ag(I)- and Cu(I)-catalyzed intermolecular cyclopropenations.

In addition to the classic Rh(II) catalysts, Davies and co-workers reported the Ag(I)-catalyzed cyclopropenation from α-aryl-α-diazoesters with internal alkynes,[267] which allowed for the preparation of highly functionalized cyclopropenes. Moreover, Pérez and co-workers reported the synthesis and characterization of hydrotris(3-mesitylpyrazolyl)borato-copper(I) alkyne complexes (Tp^{Ms}Cu-alkyne), which were able to catalyze cyclopropenation of alkynes with EDA (Scheme 4.46).[268]

Enantioselective synthesis of substituted cyclopropenes has been explored in a series of asymmetric catalytic systems. Davies and co-workers reported the chiral Rh(II) carboxylates-catalyzed cyclopropenations of terminal alkynes with donor/acceptor carbenes. Chiral vinylcyclopropenes **123** with quaternary carbons were obtained from styryldiazoacetates with good yields and excellent enantioselectivities with $Rh_2(S\text{-DOSP})_4$.[269] Moreover, optically active cyclopropenes containing germinal acceptor groups were also readily accessible from aryl acetylenes with siloxyvinyl-diazoacetates **124** catalyzed by $Rh_2(S\text{-PTAD})_4$.[270] In addition to utilizing donor/acceptor diazo substrates, Hashimoto and co-workers preformed asymmetric cyclopropenation of α-alkyl-α-diazoesters with terminal alkynes catalyzed by $Rh_2(S\text{-TBPTTL})_4$, which induced exceptionally high level of enantioselectivity and good yields (Scheme 4.47).[271]

The most commonly applied chiral Rh(II) catalysts in the reaction of α-diazocarbonyl compounds have been the Rh_2-bridged dimers having four identical chiral ligands. In contrast, Corey and co-workers have designed and synthesized a series of mixed Rh(II) catalysts $[Rh_2(RCO_2)_n(L^*_{4-n})]$.[272,273] Especially, $Rh_2(OAc)(DPTI)_3$ and $Rh_2(O_2C\text{-}^tBu)_2(DPTI)_2$ are very effective in asymmetric cyclopropenation

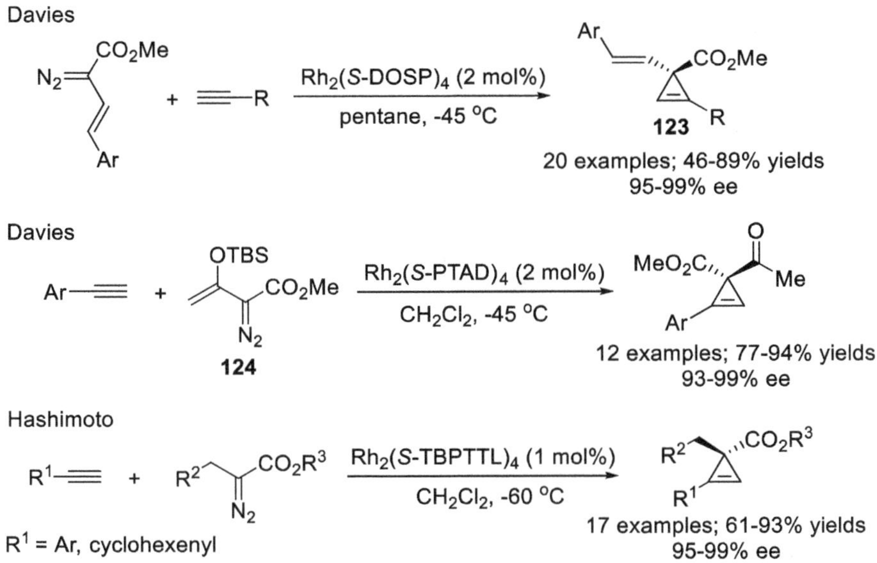

Scheme 4.47 Chiral Rh(II) carboxylate catalyzed cyclopropenations of terminal alkynes with variety of diazo compounds.

(Eq. (4.11)). The reaction mechanism of the Rh(II)-catalyzed cyclopropenation was discussed based on the experimental data obtained with the mixed Rh(II) complexes.[272–274]

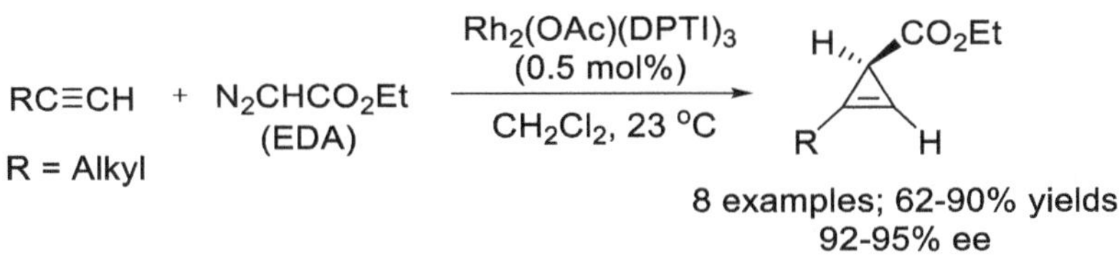

(4.11)

In addition to Rh(II) carboxylates, other transition-metal complexes have also been proven efficient catalysts for asymmetric cyclopropenations. Katsuki and co-workers explored Ir(salen) **125** for cyclopropenations with a wide scope of diazo substrates, including both acceptor/acceptor- and acceptor/donor-substituted diazocarbonyl compounds (Scheme 4.48).[275]

Zhang and co-workers explored their Co(II)-porphyrin complexes for enantioselective cyclopropenations of terminal aromatic alkynes with

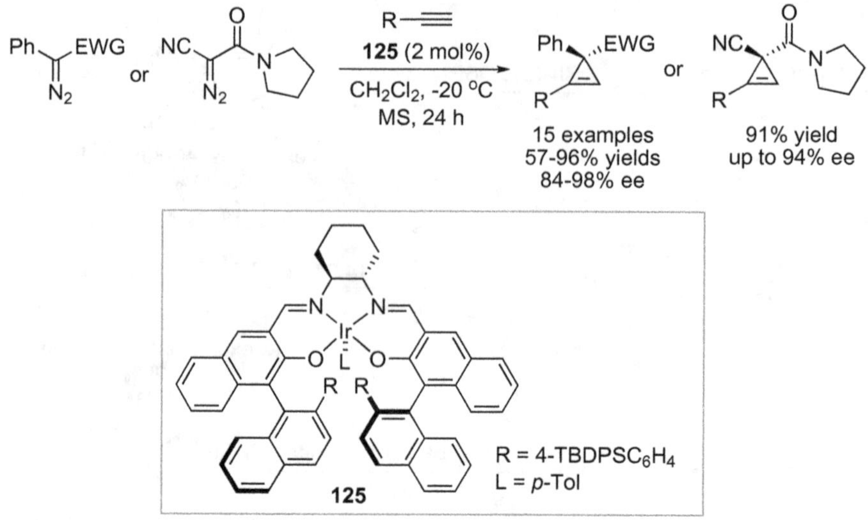

Scheme 4.48 Ir(III)-salen-catalyzed asymmetric cyclopropenations of terminal alkynes.

acceptor/acceptor diazo species, including α-cyanodiazoacetamides and α-cyanodiazoacetates.[276] Moreover, Davies and co-workers studied chiral di-gold cationic complex **126** as efficient catalysts for asymmetric cyclopropenations of internal alkynes with aryldiazoacetates (Scheme 4.49).[277]

The cyclopropenes are highly strained molecules and thus easily undergo ring-opening and subsequent transformations. This can be useful in organic synthesis, in particular for the development of tandem transformations in combination with carbene cyclopropenation. For example, Rh(II)-catalyzed cyclopropenations were reported to be carried out with subsequent ring-opening reactions, which resulted in the efficient synthesis of annulation products. One-pot synthesis of highly substituted 2-amido-furans was reported by Hsung and co-workers through catalytic cyclopropenation of ynamides to form **127**, followed by generation of zwitterions *via* ring-opening process (Scheme 4.50).[278] Besides, a variety of furo[2,3-*b*]pyran-6-one derivatives were synthesized by Rh(II)-catalyzed one-pot reactions, which were assumed to proceed *via* carbene cyclopropenation/ketene formation/[2+2]cycloaddition/ring-expansion.[279]

For this type of tandem transformations, it is usually unnecessary to isolate the cyclopropene intermediates. For another example, Lewis acid or transition-metal induced cycloisomerizations of cyclopropenes **128**/**129** were performed to allow the facile synthesis of functionalized

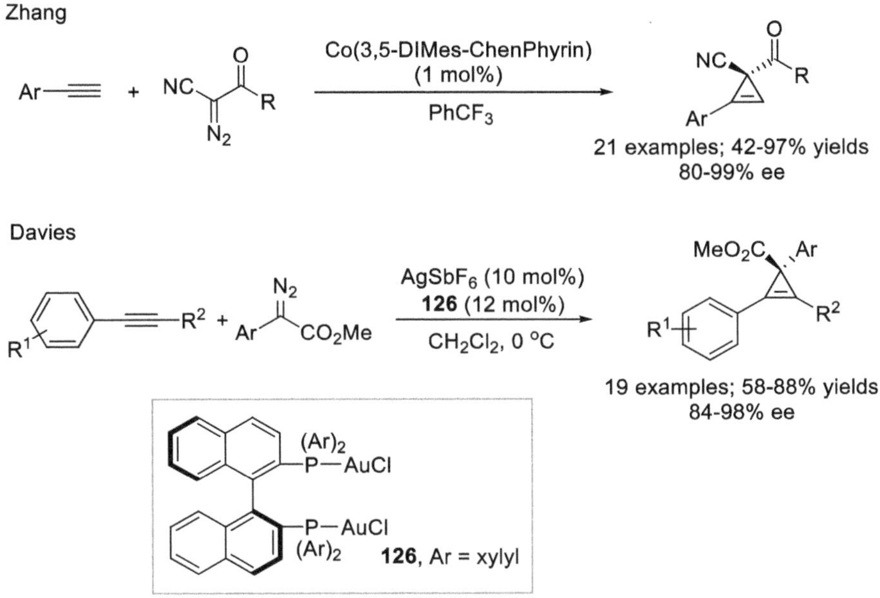

Scheme 4.49 Co(II)(por)- and Au(I)-catalyzed asymmetric cyclopropenations.

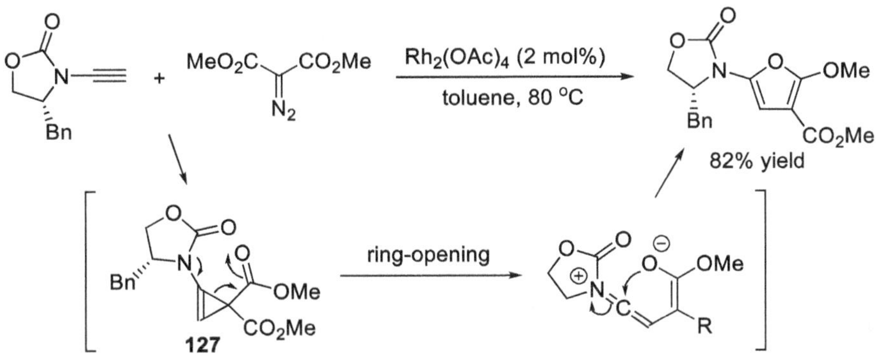

Scheme 4.50 Rh(II)-Catalyzed cyclopropenation with one-pot ring-open reactions to access 2-amido-furans.

benzo-fused heteroaromatics and heterobiaryls,[280] as well as catalyst-dependent regioselective synthesis of 2-substituted or 3-substituted *4H*-furo[3,2-*c*]chromenes (Scheme 4.51).[281]

In conclusion, as typical carbene transformations of *α*-diazocarbonyl compounds, cyclopropanation with olefins and alkynes has been one of the

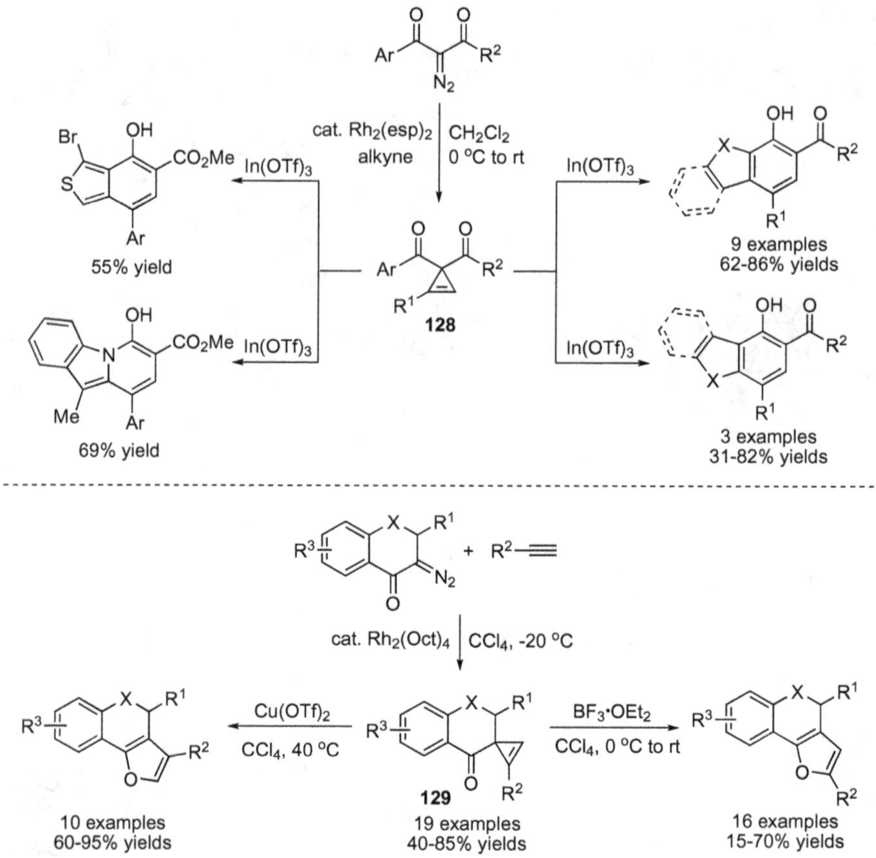

Scheme 4.51 Lewis acid-induced or Cu(I)-catalyzed cycloisomerizations of cyclopropenes.

most important methods for the construction of three-membered ring scaffolds. In recent years, much attention has been paid to the asymmetric synthesis of cyclopropane derivatives, which have been achieved through developing novel chiral catalytic systems. Moreover, cascade cyclopropanation with ring-opening strategy has demonstrated practical usefulness of the cyclopropanation in the rapid assembly of complex molecular structures.

References

For reviews on cyclopropanation of diazo compounds, see Refs. 1–6.

1. Honma, M.; Takeda, H.; Takano, M.; Nakada, M. *Synlett* **2009**, 1695–1712.
2. Doyle, M. P. *Angew. Chem. Int. Ed.* **2009**, *48*, 850–852.

3. Foley, D. A.; Maguire, A. R. *Tetrahedron* **2010**, *66*, 1131–1175.
4. McNamara, O. A.; Maguire, A. R. *Tetrahedron* **2011**, *67*, 9–40.
5. Reisman, S. E.; Nani, R. R.; Levin, S. *Synlett* **2011**, 2437–2442.
6. Chen, D. Y.-K.; Pouwer, R. H.; Richard, J.-A. *Chem. Soc. Rev.* **2012**, *41*, 4631–4642.

For recent reviews on three-membered ring compounds, see Refs. 7–9.

7. Lebel, H.; Marcoux, J.-F.; Molinaro, C.; Charette, A. B. *Chem. Rev.* **2003**, *103*, 977–1050.
8. Rubin, M.; Rubina, M.; Gevorgyan, V. *Synthesis* **2006**, 1221–1245.
9. Rubin, M.; Rubina, M.; Gevorgyan, V. *Chem. Rev.* **2007**, *107*, 3117–3179.

For selected recent examples for cyclopropanations in total synthesis, see: Refs. 10–12.

10. Taber, D. F.; Xu, M.; Hartnett, J. C. *J. Am. Chem. Soc.* **2002**, *124*, 13121–13126.
11. Yang, J.; Song, H.; Xiao, X.; Wang, J.; Qin, Y. *Org. Lett.* **2006**, *8*, 2187–2190.
12. Song, H.; Yang, J.; Chen, W.; Qin, Y. *Org. Lett.* **2006**, *8*, 6011–6014.
13. Kamata, K.; Kimura, T.; Mizuno, N. *Chem. Lett.* **2010**, *39*, 702–703.
14. Corma, A.; Iglesias, M.; Llabrés I. Xamena, F. X.; Sánchez, F. *Chem. Eur. J.* **2010**, *16*, 9789–9795.
15. Ishikawa, S.; Hudson, R.; Masnadi, M.; Bateman, M.; Castonguay, A.; Braidy, N.; Moores, A.; Li, C.-J. *Tetrahedron* **2014**, *70*, 6162–6168.
16. Boldini, I.; Guillemot, G.; Caselli, A.; Proust, A.; Gallo, E. *Adv. Synth. Catal.* **2010**, *352*, 2365–2370.
17. Fuentes, M. Á.; Álvarez, E.; Caballero, A.; Pérez, P. J. *Organometallics* **2012**, *31*, 959–965.
18. Özen, C.; Tüzün, N. Ş. *Organometallics* **2008**, *27*, 4600–4610.
19. Shishkov, I. V.; Rominger, F.; Hofmann, P. *Organometallics* **2009**, *28*, 1049–1059.
20. García, J. I.; Jiménez-Osés, G.; Mayoral, J. A. *Chem. Eur. J.* **2011**, *17*, 529–539.
21. Rosenberg, M. L.; Krivokapic, A.; Tilset, M. *Org. Lett.* **2009**, *11*, 547–550.
22. Rosenberg, M. L.; Krapp, A.; Tilset, M. *Organometallics* **2011**, *30*, 6562–6571.
23. Hansen, J.; Li, B.; Dikarev, E.; Autschbach, J.; Davies, H. M. L. *J. Org. Chem.* **2009**, *74*, 6564–6571.
24. For computational study of Rh(II) carbenoid transformations, see: Bonge, H. T.; Hansen, T. *Tetrahedron Lett.* **2010**, *51*, 5298–5301.
25. Bonge, H. T.; Hansen, T. *Tetrahedron Lett.* **2010**, *51*, 5378–5381.

26. Martín, C.; Molina, F.; Alvarez, E.; Belderrain, T. R. *Chem. Eur. J.* **2011**, *17*, 14885–14895.
27. Liu, J.; Plog, A.; Groszewicz, P.; Zhao, L.; Xu, Y.; Breitzke, H.; Stark, A.; Hoffmann, R.; Gutmann, T.; Zhang, K.; Buntkowsky, G. *Chem. Eur. J.* **2015**, *21*, 12414–12420.
28. Chen, L.; Bovee, M. O.; Lemma, B. E.; Keithley, K. S. M.; Pilson, S. L.; Coleman, M. G.; Mack, J. *Angew. Chem. Int. Ed.* **2015**, *54*, 11084–11087.
29. Dzik, W. I.; Xu, X.; Zhang, X. P.; Reek, J. N. H.; de Bruin, B. *J. Am. Chem. Soc.* **2010**, *132*, 10891–10902.
30. Belof, J. L.; Cioce, C. R.; Xu, X.; Zhang, X. P.; Space, B.; Woodcock, H. L. *Organometallics* **2011**, *30*, 2739–2746.
31. Lu, H.; Dzik, W. I.; Xu, X.; Wojtas, L.; de Bruin, B.; Zhang, X. P. *J. Am. Chem. Soc.* **2011**, *133*, 8518–8521.
32. Fields, K. B.; Engle, J. T.; Sripothongnak, S.; Kim, C.; Zhang, X. P.; Ziegler, C. J. *Chem. Commun.* **2011**, *47*, 749–751.
33. Otte, M.; Kuijpers, P. F.; Troeppner, O.; Ivanović-Burmazović, I.; Reek, J. N. H.; de Bruin, B. *Chem. Eur. J.* **2014**, *20*, 4880–4884.
34. Koehn, S. K.; Gronert, S.; Aldajaei, J. T. *Org. Lett.* **2010**, *12*, 676–679.
35. Le Maux, P.; Roisnel, T.; Nicolas, I.; Simonneaux, G. *Organometallics* **2008**, *27*, 3037–3042.
36. Deng, Q.-H.; Chen, J.; Huang, J.-S.; Chui, S. S.-Y.; Zhu, N.; Li, G.-Y.; Che, C.-M. *Chem. Eur. J.* **2009**, *15*, 10707–10712.
37. Anding, B. J.; Ellern, A.; Woo, L. K. *Organometallics* **2012**, *31*, 3628–3635.
38. Gharaati, S.; Moghadam, M.; Tangestaninejad, S.; Mirkhani, V.; Mohammadpoor-Baltork, I.; Barati, B.; Sadegh, F. *J. Organomet. Chem.* **2013**, *741–742*, 78–82.
39. González-Bobes, F.; Fenster, M. D. B.; Kiau, S.; Kolla, L.; Kolotuchin, S.; Soumeillant, M. *Adv. Synth. Catal.* **2008**, *350*, 813–816.
40. Panne, P.; DeAngelis, A.; Fox, J. M. *Org. Lett.* **2008**, *10*, 2987–2989.
41. Chemagin, A. V.; Yashin, N. V.; Grishin, Y. K.; Kuznetsova, T. S.; Zefirov, N. S. *Synthesis* **2010**, 259–266.
42. Bonge, H. T.; Pintea, B.; Bonge-Hansen, T. *Org. Biomol. Chem.* **2008**, *6*, 3670–3672.
43. Bonge, H. T.; Hansen, T. *J. Org. Chem.* **2010**, *75*, 2309–2320.
44. Schnaars, C.; Hansen, T. *Org. Lett.* **2012**, *14*, 2794–2797.
45. Schnaars, C.; Hennum, M.; Hansen, T. *J. Org. Chem.* **2013**, *78*, 7488–7497.
46. Verdecchia, M.; Tubaro, C.; Biffis, A. *Tetrahedron Lett.* **2011**, *52*, 1136–1139.
47. Kumari, G.; Nutan; Modi, M.; Gupta, S. K.; Singh, R. K. *Eur. J. Med. Chem.* **2011**, *46*, 1181–1188.
48. DeAngelis, A.; Dmitrenko, O.; Fox, J. M. *J. Am. Chem. Soc.* **2012**, *134*, 11035–11043.

49. Rathore, K. S.; Katukojvala, S. *Tetrahedron Lett.* **2014**, *55*, 6370–6372.
50. Dao, H. T.; Baran, P. S. *Angew. Chem. Int. Ed.* **2014**, *53*, 14382–14386.
51. Reyes, Y.; Mead, K. T. *Synthesis* **2015**, *47*, 3020–3026.
52. Armstrong, E. L.; Kerr, M. A. *Org. Chem. Front.* **2015**, *2*, 1045–1047.
53. Lu, T.; Song, Z.; Hsung, R. P. *Org. Lett.* **2008**, *10*, 541–544.
54. Gregg, T. M.; Algera, R. F.; Frost, J. R.; Hassan, F.; Stewart, R. J. *Tetrahedron Lett.* **2010**, *51*, 6429–6432.
55. Reddy, B. V. S.; Rajasekaran, T.; Karthik, G.; Rao, T. P. *Tetrahedron Lett.* **2012**, *53*, 3416–3420.
56. Stokes, S.; Mustain, R.; Pickle, L.; Mead, K. T. *Tetrahedron Lett.* **2012**, *53*, 3890–3893.
57. Hughes, R. A.; Tølløfsrud, M.; Bryant, N.; Kaboli, M.; Hennum, M.; Bonge-Hansen, T. *Tetrahedron Lett.* **2013**, *54*, 318–320.
58. Shibue, T.; Fukuda, Y. *J. Org. Chem.* **2014**, *79*, 7226–7231.
59. Robles, O.; Serna-Saldívar, S. O.; Gutiérrez-Uribe, J. A.; Romo, D. *Org. Lett.* **2012**, *14*, 1394–1397.
60. Racine, S.; de Nanteuil, F.; Serrano, E.; Waser, J. *Angew. Chem. Int. Ed.* **2014**, *53*, 8484–8487.
61. Sladojevich, F.; Trabocchi, A.; Guarna, A. *Org. Biomol. Chem.* **2008**, *6*, 3328–3336.
62. Brand, C.; Rauch, G.; Zanoni, M.; Dittrich, B.; Werz, D. B. *J. Org. Chem.* **2009**, *74*, 8779–8786.
63. Muthusamy, S.; Azhagan, D.; Gnanaprakasam, B.; Suresh, E. *Tetrahedron Lett.* **2010**, *51*, 5662–5665.
64. Tayama, E.; Horikawa, K.; Iwamoto, H.; Hasegawa, E. *Tetrahedron Lett.* **2014**, *55*, 3041–3044.
65. Wagner, T.; Lange, J.; Grote, D.; Sander, W.; Schaumann, E.; Adiwidjaja, G.; Adam, A.; Kopf, J. *Eur. J. Org. Chem.* **2009**, 5198–5207.
66. Prieto, A.; Fructos, M. R.; Díaz-Requejo, M. M.; Pérez, P. J.; Pérez-Galán, P.; Delpont, N.; Echavarren, A. M. *Tetrahedron* **2009**, *65*, 1790–1793.
67. Solorio-Alvarado, C. R.; Echavarren, A. M. *J. Am. Chem. Soc.* **2010**, *132*, 11881–11883.
68. Pérez, P. J.; Díaz-Requejo, M. M.; Rivilla, I. *Beilstein J. Org. Chem.* **2011**, *7*, 653–657.
69. For review of Au-catalyzed cyclopropanations, see: Qian, D.; Zhang, J. *Chem. Soc. Rev.* **2015**, *44*, 677–698.
70. Murelli, R. P.; Catalán, S.; Gannon, M. P.; Snapper, M. L. *Tetrahedron Lett.* **2008**, *49*, 5714–5717.
71. Basato, M.; Tubaro, C.; Biffis, A.; Bonato, M.; Buscemi, G.; Lighezzolo, F.; Lunardi, P.; Vianini, C.; Benetollo, F.; Del Zotto, A. *Chem. Eur. J.* **2009**, *15*, 1516–1526.
72. Dragutan, I.; Dragutan, V.; Verpoort, F. *Appl. Organometal. Chem.* **2014**, *28*, 211–215.

73. Chen, S.; Ma, J.; Wang, J. *Tetrahedron Lett.* **2008**, *49*, 6781–6783.
74. Pietruszka, J.; Solduga, G. *Eur. J. Org. Chem.* **2009**, 5998–6008.
75. Berthon-Gelloz, G.; Marchant, M.; Straub, B. F.; Marko, I. E. *Chem. Eur. J.* **2009**, *15*, 2923–2931.
76. McKee, M.; Haner, J.; Carlson, E.; Tam, W. *Synthesis* **2014**, *46*, 1518–1524.
77. Shishilov, O. N.; Stromnova, T. A.; Cámpora, J.; Palma, P.; Cartes, M. Á.; Martínez-Prieto, L. M. *Dalton Trans.* **2009**, 6626–6633.
78. Morandi, B.; Dolva, A.; Carreira, E. M. *Org. Lett.* **2012**, *14*, 2162–2163.
79. Morandi, B.; Carreira, E. M. *Science* **2012**, *335*, 1471–1474.
80. Wallace, S.; Balskus, E. P. *Angew. Chem. Int. Ed.* **2015**, *54*, 7106–7109.
81. Lévesque, É.; Goudreau, S. R.; Charette, A. B. *Org. Lett.* **2014**, *16*, 1490–1493.
82. Mieusset, J.-L.; Abraham, M.; Brinker, U. H. *J. Am. Chem. Soc.* **2008**, *130*, 14634–14639.
83. Ovalles, S. R.; Hansen, J. H.; Davies, H. M. L. *Org. Lett.* **2011**, *13*, 4284–4287.
84. Zhu, C.-L.; Yang, L.-J.; Li, S.; Zheng, Y.; Ma, J.-A. *Org. Lett.* **2015**, *17*, 3442–3445.
85. Barluenga, J.; Quiñones, N.; Tomás-Gamasa, M.; Cabal, M.-P. *Eur. J. Org. Chem.* **2012**, 2312–2317.
86. Syrgiannis, Z.; Hadad, C.; Prato, M. *Eur. J. Org. Chem.* **2014**, 4225–4229.
87. Marcoux, D.; Charette, A. B. *Angew. Chem. Int. Ed.* **2008**, *47*, 10155–10158.
88. Denton, J. R.; Cheng, K.; Davies, H. M. L. *Chem. Commun.* **2008**, 1238–1240.
89. Marcoux, D.; Azzi, S.; Charette, A. B. *J. Am. Chem. Soc.* **2009**, *131*, 6970–6972.
90. DeAngelis, A.; Dmitrenko, O.; Yap, G. P. A.; Fox, J. M. *J. Am. Chem. Soc.* **2009**, *131*, 7230–7231.
91. Lindsay, V. N. G.; Lin, W.; Charette, A. B. *J. Am. Chem. Soc.* **2009**, *131*, 16383–16385.
92. Marcoux, D.; Goudreau, S. R.; Charette, A. B. *J. Org. Chem.* **2009**, *74*, 8939–8955.
93. Marcoux, D.; Lindsay, V. N. G.; Charette, A. B. *Chem. Commun.* **2010**, *46*, 910–912.
94. Pelphrey, P.; Hansen, J.; Davies, H. M. L. *Chem. Sci.* **2010**, *1*, 254–257.
95. Nishimura, T.; Maeda, Y.; Hayashi, T. *Angew. Chem. Int. Ed.* **2010**, *49*, 7324–7327.
96. Goto, T.; Takeda, K.; Anada, M.; Ando, K.; Hashimoto, S. *Tetrahedron Lett.* **2011**, *52*, 4200–4203.
97. Lindsay, V. N. G.; Nicolas, C.; Charette, A. B. *J. Am. Chem. Soc.* **2011**, *133*, 8972–8981.

98. Qin, C.; Boyarskikh, V.; Hansen, J. H.; Hardcastle, K. I.; Musaev, D. G.; Davies, H. M. L. *J. Am. Chem. Soc.* **2011**, *133*, 19198–19204.
99. Boruta, D. T.; Dmitrenko, O.; Yap, G. P. A.; Fox, J. M. *Chem. Sci.* **2012**, *3*, 1589–1593.
100. Awata, A.; Arai, T. *Synlett* **2013**, *24*, 29–32.
101. Lindsay, V. N. G.; Fiset, D.; Gritsch, P. J.; Azzi, S.; Charette, A. B. *J. Am. Chem. Soc.* **2013**, *135*, 1463–1470.
102. Wang, H.; Guptill, D. M.; Varela-Alvarez, A.; Musaev, D. G.; Davies, H. M. L. *Chem. Sci.* **2013**, *4*, 2844–2850.
103. Chepiga, K. M.; Qin, C.; Alford, J. S.; Chennamadhavuni, S.; Gregg, T. M.; Olson, J. P.; Davies, H. M. L. *Tetrahedron* **2013**, *69*, 5765–5771.
104. Negretti, S.; Cohen, C. M.; Chang, J. J.; Guptill, D. M.; Davies, H. M. L. *Tetrahedron* **2015**, *71*, 7415–7420.
105. Gregg, T. M.; Farrugia, M. K.; Frost, J. R. *Org. Lett.* **2009**, *11*, 4434–4436.
106. del Villar, I. S.; Gradillas, A.; Pérez-Castells, J. *Eur. J. Org. Chem.* **2010**, 5850–5862.
107. Gu, P.; Su, Y.; Wu, X.-P.; Sun, J.; Liu, W.; Xue, P.; Li, R. *Org. Lett.* **2012**, *14*, 2246–2249.
108. Su, Y.; Bai, M.; Qiao, J.-B.; Li, X.-J.; Li, R.; Tu, Y.-Q.; Gu, P. *Tetrahedron Lett.* **2015**, *56*, 1805–1807.
109. Denton, J. R.; Davies, H. M. L. *Org. Lett.* **2009**, *11*, 787–790.
110. Ventura, D. L.; Li, Z.; Coleman, M. G.; Davies, H. M. L. *Tetrahedron* **2009**, *65*, 3052–3061.
111. Nadeau, E.; Ventura, D. L.; Brekan, J. A.; Davies, H. M. L. *J. Org. Chem.* **2010**, *75*, 1927–1939.
112. Davies, H. M. L.; Lee, G. H. *Org. Lett.* **2004**, *6*, 1233–1236.
113. Davies, H. M. L.; Lee, G. H. *Org. Lett.* **2004**, *6*, 2117–2120.
114. Davies, H. M. L.; Venkataramani, C. *Org. Lett.* **2003**, *5*, 1403–1406.
115. Nagashima, T.; Davies, H. M. L. *Org. Lett.* **2002**, *4*, 1989–1992.
116. Davies, H. M. L.; Walji, A. M.; Nagashima, T. *J. Am. Chem. Soc.* **2004**, *126*, 4271–4280.
117. Davies, H. M. L.; Walji, A. M. *Org. Lett.* **2005**, *7*, 2941–2944.
118. Weathers Jr., T. M.; Doyle, M. P.; Carducci, M. D. *Adv. Synth. Catal.* **2006**, *348*, 449–455.
119. Doyle, M. P.; Jr., T. M. W.; Wang, Y. *Adv. Synth. Catal.* **2006**, *348*, 2403–2409.
120. Hu, W.; Timmons, D. J.; Doyle, M. P. *Org. Lett.* **2002**, *4*, 901–904.
121. Lin, W.; Charette, A. B. *Adv. Synth. Catal.* **2005**, *347*, 1547–1552.
122. Lloret, J.; Stern, M.; Estevan, F.; Sanaú, M.; Úbeda, M. A. *Organometallics* **2008**, *27*, 850–856.
123. Chepiga, K. M.; Feng, Y.; Brunelli, N. A.; Jones, C. W.; Davies, H. M. L. *Org. Lett.* **2013**, *15*, 6136–6139.

124. Sambasivan, R.; Zheng, W.; Burya, S. J.; Popp, B. V.; Turro, C.; Clementi, C.; Ball, Z. T. *Chem. Sci.* **2014**, *5*, 1401–1407.
125. Rechavi, D.; Lemaire, M. *Chem. Rev.* **2002**, *102*, 3467–3494.
126. Bayardon, J.; Holczknecht, O.; Pozzi, G.; Sinou, D. *Tetrahedron: Asymm.* **2006**, *17*, 1568–1572.
127. Benaglia, M.; Benincori, T.; Mussini, P.; Pilati, T.; Rizzo, S.; Sannicolò, F. *J. Org. Chem.* **2005**, *70*, 7488–7495.
128. Itagaki, M.; Masumoto, K.; Yamamoto, Y. *J. Org. Chem.* **2005**, *70*, 3292–3295.
129. Zhang, W.; Xie, F.; Matsuo, S.; Imahori, Y.; Kida, T.; Nakatsuji, Y.; Ikeda, I. *Tetrahedron: Asymm.* **2006**, *17*, 767–777.
130. Mazet, C.; Köhler, V.; Pfaltz, A. *Angew. Chem. Int. Ed.* **2005**, *44*, 4888–4891.
131. Lyle, M. P. A.; Wilson, P. D. *Org. Lett.* **2004**, *6*, 855–857.
132. Llewellyn, D. B.; Arndtsen, B. A. *Organometallics* **2004**, *23*, 2838–2840.
133. Li, J.; Liao, S.-H.; Xiong, H.; Zhou, Y.-Y.; Sun, X.-L.; Zhang, Y.; Zhou, X.-G.; Tang, Y. *Angew. Chem. Int. Ed.* **2012**, *51*, 8838–8841.
134. Nicolas, I.; Le Maux, P.; Simonneaux, G. *Tetrahedron Lett.* **2008**, *49*, 2111–2113.
135. Ito, J.-I.; Ujiie, S.; Nishiyama, H. *Chem. Eur. J.* **2010**, *16*, 4986–4990.
136. Chanthamath, S.; Phomkeona, K.; Shibatomi, K.; Iwasa, S. *Chem. Commun.* **2012**, *48*, 7750–7752.
137. Chanthamath, S.; Takaki, S.; Shibatomi, K.; Iwasa, S. *Angew. Chem. Int. Ed.* **2013**, *52*, 5818–5821.
138. Chanthamath, S.; Nguyen, D. T.; Shibatomi, K.; Iwasa, S. *Org. Lett.* **2013**, *15*, 772–775.
139. Chanthamath, S.; Ozaki, S.; Shibatomi, K.; Iwasa, S. *Org. Lett.* **2014**, *16*, 3012–3015.
140. Chanthamath, S.; Chua, H. W.; Kimura, S.; Shibatomi, K.; Iwasa, S. *Org. Lett.* **2014**, *16*, 3408–3411.
141. Shi, T.; Luo, Y.; Wang, X.-L.; Lu, S.; Zhao, Y.-L.; Zhang, J. *Organometallics* **2014**, *33*, 3673–3682.

For other Ru-catalyzed examples, see Refs. 142–146.
142. Maux, P. L.; Abrunhosa, I.; Berchel, M.; Simonneaux, G.; Gulea, M.; Masson, S. *Tetrahedron: Asymm.* **2004**, *15*, 2569–2573.
143. Bonaccorsi, C.; Mezzetti, A. *Organometallics* **2005**, *24*, 4953–4960.
144. Li, G.-Y.; Zhang, J.; Chan, P. W. H.; Xu, Z.-J.; Zhu, N.; Che, C.-M. *Organometallics* **2006**, *25*, 1676–1688.
145. Bonaccorsi, C.; Santoro, F.; Gischig, S.; Mezzetti, A. *Organometallics* **2006**, *25*, 2002–2010.

146. Ferrand, Y.; Maux, P. L.; Simonneaux, G. *Org. Lett.* **2004**, *6*, 3211–3214.
147. Abu-Elfotoh, A.-M.; Phomkeona, K.; Shibatomi, K.; Iwasa, S. *Angew. Chem. Int. Ed.* **2010**, *49*, 8439–8443.
148. del Pozo, C.; Corma, A.; Iglesias, M.; Sánchez, F. *Green Chem.* **2011**, *13*, 2471–2481.
149. Falkowski, J. M.; Wang, C.; Liu, S.; Lin, W. *Angew. Chem. Int. Ed.* **2011**, *50*, 8674–8678.
150. Zhu, S.; Ruppel, J. V.; Lu, H.; Wojtas, L.; Zhang, X. P. *J. Am. Chem. Soc.* **2008**, *130*, 5042–5043.
151. Zhu, S.; Perman, J. A.; Zhang, X. P. *Angew. Chem. Int. Ed.* **2008**, *47*, 8460–8463.
152. Ruppel, J. V.; Gauthier, T. J.; Snyder, N. L.; Perman, J. A.; Zhang, X. P. *Org. Lett.* **2009**, *11*, 2273–2276.
153. Zhu, S.; Xu, X.; Perman, J. A.; Zhang, X. P. *J. Am. Chem. Soc.* **2010**, *132*, 12796–12799.
154. Xu, X.; Zhu, S.; Cui, X.; Wojtas, L.; Zhang, X. P. *Angew. Chem. Int. Ed.* **2013**, *52*, 11857–11861.
155. For review for investigating ligand effect, see: Zhu, S.; Cui, X.; Zhang, X. P. *Eur. J. Inorg. Chem.* **2012**, 430–434.

For other Co-catalyzed examples, see Refs. 156–158.

156. Huang, L.; Chen, Y.; Gao, G.-Y.; Zhang, X. P. *J. Org. Chem.* **2003**, *68*, 8179–8184.
157. Chen, Y.; Zhang, X. P. *J. Org. Chem.* **2004**, *69*, 2431–2435.
158. Chen, Y.; Fields, K. B.; Zhang, X. P. *J. Am. Chem. Soc.* **2004**, *126*, 14718–14719.
159. White, J. D.; Shaw, S. *Org. Lett.* **2014**, *16*, 3880–3883.
160. Edulji, S. K.; Nguyen, S. T. *Organometallics* **2003**, *22*, 3374–3381.
161. Hahn, N. D.; Nieger, M.; Dötz, K. H. *Eur. J. Org. Chem.* **2004**, 1049–1056.
162. Suenobu, K.; Itagaki, M.; Nakamura, E. *J. Am. Chem. Soc.* **2004**, *126*, 7271–7280.
163. Tepfenhart, D.; Moisan, L.; Dalko, P. I.; Cossy, J. *Tetrahedron Lett.* **2004**, *45*, 1781–1783.
164. Beaufort, L.; Demonceau, A.; Noels, A. F. *Tetrahedron* **2005**, *61*, 9025–9030.
165. Urbano, J.; Izarra, R.; Gómez-Ariza, J. L.; Trofimenko, S.; Díaz-Requejo, M.; Pérez, P. J. *Chem. Commun.* **2006**, 1000–1002.
166. Sun, W.; Xia, C.-G.; Wang, H.-W. *New J. Chem.* **2002**, *26*, 755–758.
167. Suematsu, H.; Kanchiku, S.; Uchida, T.; Katsuki, T. *J. Am. Chem. Soc.* **2008**, *130*, 10327–10337.

168. Ichinose, M.; Suematsu, H.; Katsuki, T. *Angew. Chem. Int. Ed.* **2009**, *48*, 3121–3123.
169. Nicolas, I.; Le Maux, P.; Simonneaux, G. *Tetrahedron Lett.* **2008**, *49*, 5793–5795.
170. Nicolas, I.; Roisnel, T.; Le Maux, P.; Simonneaux, G. *Tetrahedron Lett.* **2009**, *50*, 5149–5151.
171. Intrieri, D.; Le Gac, S.; Caselli, A.; Rose, E.; Boitrel, B.; Gallo, E. *Chem. Commun.* **2014**, *50*, 1811–1813.
172. Bordeaux, M.; Tyagi, V.; Fasan, R. *Angew. Chem. Int. Ed.* **2015**, *54*, 1744–1748.
173. Cao, Z.-Y.; Zhou, F.; Yu, Y.-H.; Zhou, J. *Org. Lett.* **2013**, *15*, 42–45.
174. Cao, Z.-Y.; Wang, X.; Tan, C.; Zhao, X.-L.; Zhou, J.; Ding, K. *J. Am. Chem. Soc.* **2013**, *135*, 8197–8200.
175. For review of catalytic asymmetric synthesis of spirocyclopropyl oxindoles, see: Cao, Z.-Y.; Zhou, J. *Org. Chem. Front.* **2015**, *2*, 849–858.
176. Goudreau, S. R.; Charette, A. B. *J. Am. Chem. Soc.* **2009**, *131*, 15633–15635.
177. Gao, L.; Hwang, G.-S.; Ryu, D. H. *J. Am. Chem. Soc.* **2011**, *133*, 20708–20711.
178. Mykhailiuk, P. K.; Afonin, S.; Palamarchuk, G. V.; Shishkin, O. V.; Ulrich, A. S.; Komarov, I. V. *Angew. Chem. Int. Ed.* **2008**, *47*, 5765–5767.
179. Mykhailiuk, P. K.; Afonin, S.; Ulrich, A. S.; Komarov, I. V. *Synthesis* **2008**, 1757–1760.
180. Duncton, M. A. J.; Ayala, L.; Kaub, C.; Janagani, S.; Edwards, W. T.; Orike, N.; Ramamoorthy, K.; Kincaid, J.; Kelly, M. G. *Tetrahedron Lett.* **2010**, *51*, 1009–1011.
181. Duncton, M. A. J.; Singh, R. *Org. Lett.* **2013**, *15*, 4284–4287.
182. Artamonov, O. S.; Slobodyanyuk, E. Y.; Shishkin, O. V.; Komarov, I. V.; Mykhailiuk, P. K. *Synthesis* **2013**, *45*, 225–230.
183. Morandi, B.; Carreira, E. M. *Angew. Chem. Int. Ed.* **2010**, *49*, 938–941.
184. Morandi, B.; Cheang, J.; Carreira, E. M. *Org. Lett.* **2011**, *13*, 3080–3081.
185. Morandi, B.; Mariampillai, B.; Carreira, E. M. *Angew. Chem. Int. Ed.* **2011**, *50*, 1101–1104.
186. For review of construction of stereogenic carbon centers featuring CF_3 group, see: Nie, J.; Guo, H.-C.; Cahard, D.; Ma, J.-A. *Chem. Rev.* **2011**, *111*, 455–529.
187. Fillion, E.; Beingessner, R. L. *J. Org. Chem.* **2003**, *68*, 9485–9488.
188. Srikrishna, A.; Dinesh, C. *Tetrahedron: Asymm.* **2005**, *16*, 2203–2207.
189. Renslo, A. R.; Gao, H.; Jaishankar, P.; Venkatachalam, R.; Gordeev, M. F. *Org. Lett.* **2005**, *7*, 2627–2630.
190. For an account, see: Doyle, M. P.; Hu, W. *Synlett* **2001**, 1364–1370.
191. Thomas M. Weathers, J.; Wang, Y.; Doyle, M. P. *J. Org. Chem.* **2006**, *71*, 8183–8189.

192. Xu, X.; Hu, W.-H.; Zavalij, P. Y.; Doyle, M. P. *Angew. Chem. Int. Ed.* **2011**, *50*, 11152–11155.
193. Phelps, A. M.; Dolan, N. S.; Connell, N. T.; Schomaker, J. M. *Tetrahedron* **2013**, *69*, 5614–5621.
194. Kalia, D.; Merey, G.; Guo, M.; Sintim, H. O. *J. Org. Chem.* **2013**, *78*, 6131–6142.
195. Reekie, T. A.; Banwell, M. G.; Willis, A. C. *J. Org. Chem.* **2013**, *78*, 7100–7111.
196. Newhouse, T. R.; Kaib, P. S. J.; Gross, A. W.; Corey, E. J. *Org. Lett.* **2013**, *15*, 1591–1593.
197. Bolsønes, M.; Bonge-Hansen, H. T.; Bonge-Hansen, T. *Synlett* **2014**, *25*, 221–224.
198. Yang, Y.; Li, J.; Du, B.; Yuan, C.; Liu, B.; Qin, S. *Chem. Commun.* **2015**, *51*, 6179–6182.
199. Reddy, A. R.; Hao, F.; Wu, K.; Zhou, C.-Y.; Che, C.-M. *Angew. Chem. Int. Ed.* **2016**, *55*, 1810–1815.
200. Zhang, B.; Wee, A. G. H. *Chem. Commun.* **2008**, 4837–4839.
201. Zhang, B.; Wee, A. G. H. *Org. Biomol. Chem.* **2012**, *10*, 4597–4608.
202. Chan, W.-W.; Kwong, T.-L.; Yu, W.-Y. *Org. Biomol. Chem.* **2012**, *10*, 3749–3755.
203. Del Bel, M.; Rovira, A.; Guerrero, C. A. *J. Am. Chem. Soc.* **2013**, *135*, 12188–12191.
204. Xu, X.; Lu, H.; Ruppel, J. V.; Cui, X.; de Mesa, S. L.; Wojtas, L.; Zhang, X. P. *J. Am. Chem. Soc.* **2011**, *133*, 15292–15295.
205. Ruppel, J. V.; Cui, X.; Xu, X.; Zhang, X. P. *Org. Chem. Front.* **2014**, *1*, 515–520.
206. Abu-Elfotoh, A.-M.; Nguyen, D. P. T.; Chanthamath, S.; Phomkeona, K.; Shibatomi, K.; Iwasa, S. *Adv. Synth. Catal.* **2012**, *354*, 3435–3439.
207. Nakagawa, Y.; Chanthamath, S.; Shibatomi, K.; Iwasa, S. *Org. Lett.* **2015**, *17*, 2792–2795.
208. Oelerich, J.; Roelfes, G. *Chem. Sci.* **2013**, *4*, 2013–2017.
209. Sawada, T.; Nakada, M. *Tetrahedron: Asymm.* **2012**, *23*, 350–356.
210. Shen, J.-J.; Zhu, S.-F.; Cai, Y.; Xu, H.; Xie, X.-L.; Zhou, Q.-L. *Angew. Chem. Int. Ed.* **2014**, *53*, 13188–13191.
211. Honma, M.; Sawada, T.; Fujisawa, Y.; Utsugi, M.; Watanabe, H.; Umino, A.; Matsumura, T.; Hagihara, T.; Takano, M.; Nakada, M. *J. Am. Chem. Soc.* **2003**, *125*, 2860–2861.
212. Honma, M.; Nakada, M. *Tetrahedron Lett.* **2003**, *44*, 9007–9011.
213. Sawada, T.; Nakada, M. *Adv. Synth. Catal.* **2005**, *347*, 1527–1532.
214. Takeda, H.; Nakada, M. *Tetrahedron: Asymm.* **2006**, *17*, 2896–2906.
215. Panish, R.; Chintala, S. R.; Boruta, D. T.; Fang, Y.; Taylor, M. T.; Fox, J. M. *J. Am. Chem. Soc.* **2013**, *135*, 9283–9286.

216. Qin, C.; Davies, H. M. L. *Org. Lett.* **2013**, *15*, 310–313.
217. O'Neill, S.; O'Keeffe, S.; Harrington, F.; Maguire, A. R. *Synlett* **2009**, 2312–2314.
218. Slattery, C. N.; Clarke, L.-A.; O'Neill, S.; Ring, A.; Ford, A.; Maguire, A. R. *Synlett* **2012**, *23*, 765–767.
219. Slattery, C. N.; O'Keeffe, S.; Maguire, A. R. *Tetrahedron: Asymm.* **2013**, *24*, 1265–1275.
220. Crombie, A. L.; John L. Kane, J.; Shea, K. M.; Danheiser, R. L. *J. Org. Chem.* **2004**, *69*, 8652–8667.
221. Caballero, A.; Díaz-Requejo, M. M.; Trofimenko, S.; Belderrain, T. R.; Pérez, P. J. *J. Org. Chem.* **2005**, *70*, 6101–6104.
222. Hahn, N. D.; Nieger, M.; Dötz, K. H. *J. Organomet. Chem.* **2004**, *689*, 2662–2673.
223. Hughes, C. C.; Kennedy-Smith, J. J.; Trauner, D. *Org. Lett.* **2003**, *5*, 4113–4115.
224. Yadav, J. S.; Reddy, B. V. S.; Satheesh, G. *Tetrahedron Lett.* **2003**, *44*, 8331–8334.
225. Yadav, J. S.; Reddy, B. V. S.; Gupta, M. K.; Prabhakar, A.; Jagadeesh, B. *Chem. Commun.* **2004**, 2124–2125.
226. Rotzoll, S.; Appel, B.; Langer, P. *Tetrahedron Lett.* **2005**, *46*, 4057–4059.
227. Im, C. Y.; Sugimura, T. *Tetrahedron* **2012**, *68*, 3744–3749.
228. Foley, D. A.; O'Leary, P.; Buckley, N. R.; Lawrence, S. E.; Maguire, A. R. *Tetrahedron* **2013**, *69*, 1778–1794.
229. McNamara, O. A.; Buckley, N. R.; O'Leary, P.; Harrington, F.; Kelly, N.; O'Keeffe, S.; Stack, A.; O'Neill, S.; Lawrence, S. E.; Slattery, C. N.; Maguire, A. R. *Tetrahedron* **2014**, *70*, 6870–6878.
230. Wyatt, E. E.; Galloway, W. R. J. D.; Spring, D. R. *Synlett* **2011**, 1449–1453.
231. Nani, R. R.; Reisman, S. E. *J. Am. Chem. Soc.* **2013**, *135*, 7304–7311.
232. Wang, X.; Abrahams, Q. M.; Zavalij, P. Y.; Doyle, M. P. *Angew. Chem. Int. Ed.* **2012**, *51*, 5907–5910.
233. Shanahan, C. S.; Truong, P.; Mason, S. M.; Leszczynski, J. S.; Doyle, M. P. *Org. Lett.* **2013**, *15*, 3642–3645.
234. Liu, Y.; Deng, Y.; Zavalij, P. Y.; Liu, R.; Doyle, M. P. *Chem. Commun.* **2015**, *51*, 565–568.
235. Pérez, P. J.; Díaz-Requejo, M. M.; Rivilla, I. *Beilstein J. Org. Chem.* **2011**, *7*, 653–657.
236. Rivilla, I.; Gómez-Emeterio, B. P.; Fructos, M. R.; Díaz-Requejo, M. M.; Pérez, P. J. *Organometallics* **2011**, *30*, 2855–2860.
237. Komine, N.; Flores, J. A.; Pal, K.; Caulton, K. G.; Mindiola, D. J. *Organometallics* **2013**, *32*, 3185–3191.

238. Zhang, Z.; Feng, J.; Xu, Y.; Zhang, S.; Ye, Y.; Li, T.; Wang, X.; Chen, J.; Zhang, Y.; Wang, J. *Synlett* **2015**, *26*, 59–62.
239. Liu, Z.; Tan, H.; Wang, L.; Fu, T.; Xia, Y.; Zhang, Y.; Wang, J. *Angew. Chem. Int. Ed.* **2015**, *54*, 3056–3060.
240. Chan, W.-W.; Yeung, S.-H.; Zhou, Z.; Chan, A. S. C.; Yu, W.-Y. *Org. Lett.* **2010**, *12*,604–607.
241. Wang, S. R.; Zhu, C.-Y.; Sun, X.-L.; Tang, Y. *J. Am. Chem. Soc.* **2009**, *131*, 4192–4193.
242. Wang, P.; Ling, L.; Liao, S.-H.; Zhu, J.-B.; Wang, S. R.; Li, Y.-X.; Tang, Y. *Chem. Eur. J.* **2013**, *19*, 6766–6773.
243. Marinozzi, M.; Fulco, M. C.; Amori, L.; Fiumi, M.; Pellicciari, R. *Tetrahedron* **2009**, *65*, 7092–7098.
244. Muthusamy, S.; Azhagan, D. *Tetrahedron Lett.* **2011**, *52*, 6732–6735.
245. Schneider, T. F.; Kaschel, J.; Dittrich, B.; Werz, D. B. *Org. Lett.* **2009**, *11*, 2317–2320.
246. Kaschel, J.; Schneider, T. F.; Kratzert, D.; Stalke, D.; Werz, D. B. *Angew. Chem. Int. Ed.* **2012**, *51*, 11153–11156.
247. Kaschel, J.; Schneider, T. F.; Kratzert, D.; Stalke, D.; Werz, D. B. *Org. Biomol. Chem.* **2013**, *11*, 3494–3509.
248. Barluenga, J.; Riesgo, L.; López, L. A.; Rubio, E.; Tomás, M. *Angew. Chem. Int. Ed.* **2009**, *48*, 7569–7572.
249. Lourdusamy, E.; Yao, L.; Park, C.-M. *Angew. Chem. Int. Ed.* **2010**, *49*, 7963–7967.
250. Geraghty, N. W. A.; McArdle, P.; Mullen, L. M. A. *Tetrahedron* **2011**, *67*, 3546–3552.
251. Xia, L.; Lee, Y. R. *Adv. Synth. Catal.* **2013**, *355*, 2361–2374.
252. Wurz, R. P.; Charette, A. B. *Org. Lett.* **2002**, *4*, 4531–4533.
253. Wurz, R. P.; Charette, A. B. *J. Org. Chem.* **2004**, *69*, 1262–1269.
254. Wurz, R. P.; Charette, A. B. *Org. Lett.* **2005**, *7*, 2313–2316.
255. For a review, see: Reissig, H.-U.; Zimmer, R. *Chem. Rev.* **2003**, *103*, 1151–1196.
256. For a selected example, see: Muthusamy, S.; Srinivasan, P. *Tetrahedron Lett.* **2006**, *47*, 6297–6300.
257. Deng, L.; Giessert, A. J.; Gerlitz, O. O.; Dai, X.; Diver, S. T.; Davies, H. M. L. *J. Am. Chem. Soc.* **2005**, *127*, 1342–1343.
258. Ni, A.; France, J. E.; Davies, H. M. L. *J. Org. Chem.* **2006**, *71*, 5594–5598.
259. Kim, B. G.; Snapper, M. L. *J. Am. Chem. Soc.* **2006**, *128*, 52–53.
260. Wu, J.; Becerril, J.; Lian, Y.; Davies, H. M. L.; Porco, J. A., Jr.; Panek, J. S. *Angew. Chem. Int. Ed.* **2011**, *50*, 5938–5942.
261. Schmidt, C. D.; Kaschel, J.; Schneider, T. F.; Kratzert, D.; Stalke, D.; Werz, D. B. *Org. Lett.* **2013**, *15*, 6098–6101.

262. For recent review, see: Zhu, Z.-B.; Wei, Y.; Shi, M. *Chem. Soc. Rev.* **2011**, *40*, 5534–5563.
263. Yan, N.; Liu, X.; Pallerla, M. K.; Fox, J. M. *J. Org. Chem.* **2008**, *73*, 4283–4286.
264. Li, Z.; Wang, D.; Li, L.; Pan, S.; Na, Z.; Tan, C. Y. J.; Yao, S. Q. *J. Am. Chem. Soc.* **2014**, *136*, 9990–9998.
265. Morandi, B.; Carreira, E. M. *Angew. Chem. Int. Ed.* **2010**, *49*, 4294–4296.
266. Edwards, A.; Rubin, M. *Tetrahedron* **2015**, *71*, 3237–3246.
267. Briones, J. F.; Davies, H. M. L. *Org. Lett.* **2011**, *13*, 3984–3987.
268. Martín, C.; Sierra, M.; Alvarez, E.; Belderrain, T. R.; Pérez, P. J. *Dalton Trans.* **2012**, *41*, 5319–5325.
269. Briones, J. F.; Hansen, J.; Hardcastle, K. I.; Autschbach, J.; Davies, H. M. L. *J. Am. Chem. Soc.* **2010**, *132*, 17211–17215.
270. Briones, J. F.; Davies, H. M. L. *Tetrahedron* **2011**, *67*, 4313–4317.
271. Goto, T.; Takeda, K.; Shimada, N.; Nambu, H.; Anada, M.; Shiro, M.; Ando, K.; Hashimoto, S. *Angew. Chem. Int. Ed.* **2011**, *50*, 6803–6808.
272. Lou, Y.; Horikawa, M.; Kloster, R. A.; Hawryluk, N. A.; Corey, E. J. *J. Am. Chem. Soc.* **2004**, *126*, 8916–8918.
273. Lou, Y.; Remarchuk, T. P.; Corey, E. J. *J. Am. Chem. Soc.* **2005**, *127*, 14223–14230.
274. Nowlan, D. T.; Singleton, D. A. *J. Am. Chem. Soc.* **2005**, *127*, 6190–6191.
275. Uehara, M.; Suematsu, H.; Yasutomi, Y.; Katsuki, T. *J. Am. Chem. Soc.* **2011**, *133*, 170–171.
276. Cui, X.; Xu, X.; Lu, H.; Zhu, S.; Wojtas, L.; Zhang, X. P. *J. Am. Chem. Soc.* **2011**, *133*, 3304–3307.
277. Briones, J. F.; Davies, H. M. L. *J. Am. Chem. Soc.* **2012**, *134*, 11916–11919.
278. Li, H.; Hsung, R. P. *Org. Lett.* **2009**, *11*, 4462–4465.
279. Somai Magar, K. B.; Lee, Y. K.; Kim, S. H. *Tetrahedron* **2013**, *69*, 9294–9302.
280. Phun, L. H.; Aponte-Guzman, J.; France, S. *Angew. Chem. Int. Ed.* **2012**, *51*, 3198–3202.
281. Gong, J.; Zhao, Z.; Zhang, F.; Wu, S.; Yan, G.; Quan, Y.; Ma, B. *Org. Lett.* **2014**, *16*, 5524–5527.

Chapter 5

Insertion Reactions

Along with cyclopropanations, X–H bond (X = C, O, S, N, Si, etc.) insertions represent another type of characteristic transformations for metal carbenes. Chemo-, regio- and stereoselective insertions with various carbene precursors, in particular the diazo compounds, have been well explored and used in wide applications in organic synthesis. The past achievements have been summarized in several excellent review articles.[1–23] Recent endeavors in the field have mainly focused on the expansion of the scope of the reaction and the asymmetric catalytic reactions for polar X–H bonds, such as O–H insertions and N–H insertions. In this chapter, we will summarize the most recent advances in insertion reactions according to the type of the targeting bonds. The applications of insertion reactions in the synthesis of natural products and functional molecules are discussed in Chapter 15.

5.1 C–H insertion

Carbene insertion into C–H bonds has been well studied over the years, including their applications in organic synthesis. In particular, the inert C–H bond insertion by metal carbenes provides a unique approach for C–H bond functionalization (Scheme 5.1).[1]

Regardless of the numerous studies in the literature published on carbene C–H bond insertions, some challenging problems still remain. Moreover, C–H bond insertion has become frequently applied as a key step in natural product synthesis. In this section, the intramolecular C–H bond insertion reactions are organized according to the ring size produced

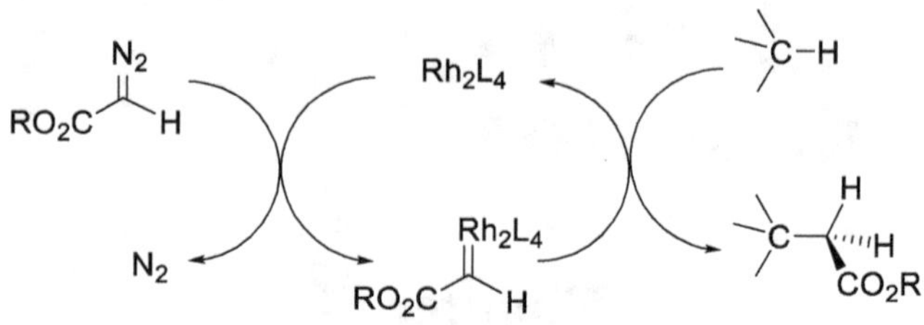

Scheme 5.1 Rh(II)-catalyzed carbene insertion to C–H bonds.

by cyclization. This type of reaction presents a general approach for the synthesis of carbocyclic and heterocyclic structures in regio- and stereo-controlled manner. As for intermolecular carbene C–H insertions, the major challenging issue is the control of chemo- and regioselectivity. The new advances for intermolecular C–H insertion are summarized in the last part of this section.

5.1.1 *Four-membered ring formation*

Although intramolecular 1,5-C–H insertion is overwhelmingly predominant, 1,4-C–H insertions are also observed, in particular, when the target C–H bond is activated by a neighboring heteroatom, in most cases oxygen or nitrogen.[24–28]

Rh(II)-catalyzed 1,4-C–H insertion was commonly utilized to prepare β-lactam derivatives from diazoacetamides. In addition to the classic Rh(II) catalysts, some new Rh(II) complexes and Rh(III) catalysts were also utilized for 1,4-C–H insertion (Figure 5.1). Afonso and Gois reported the studies on NHC-Rh(II) complexes (**1**)-catalyzed C–H insertions[29,30] and asymmetric Rh(II)-catalyzed C–H insertions with natural α-amino acids as the chiral ligands (**2**).[31] Rh(III) porphyrin complexes (**3**) were also utilized by Che and co-workers in catalytic asymmetric carbene C–H insertion to afford *cis*-β-lactams and *trans*-γ-lactams in excellent yields and enantioselectivities.[32] Besides, Afonso and Gois demonstrated the photolytic induced C–H insertion of α-diazoacetamides in water, which afforded the corresponding β- and γ-lactams in the absence of metal catalysts.[33] Recently, asymmetric *cis*-β-lactams preparation was realized with high enantioselectivity (up to 99% *ee*) by Rh_2(*S*-PTTL)$_4$-catalyzed intramolecular C–H insertion of enoldiazoacetamides.[34]

Davies and co-workers developed the Rh(II)-catalyzed stereoselective synthesis of allylsilanes **5** through the reaction with 2-(trialkylsilyl)ethyl

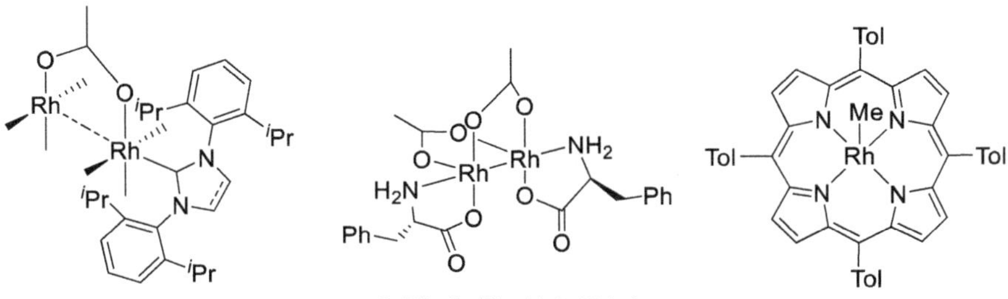

1, $Rh_2(OAc)_4(NHC)$ 2, $Rh_2(L\text{-}PheAla)_2(OAc)_2$ 3, [Rh(TTP)Me]

Figure 5.1 Rh(II) and Rh(III) catalysts used for 1,4-C–H bond insertion.

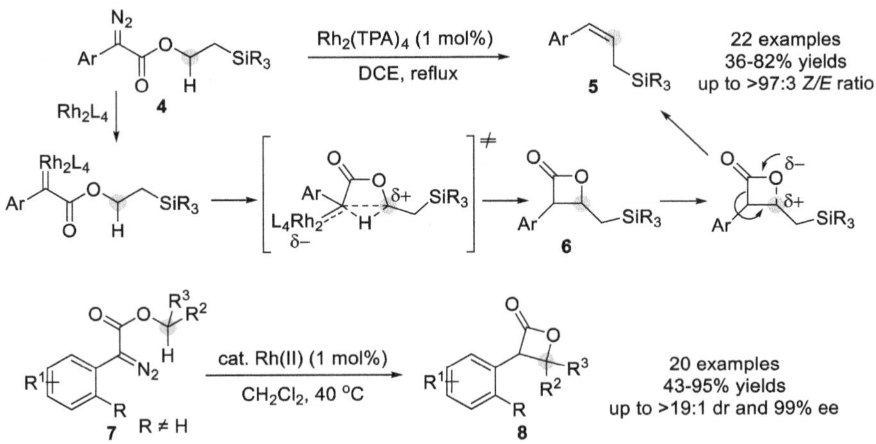

Scheme 5.2 Rh(II)-catalyzed intramolecular C–H insertion of aryldiazoacetate.

aryl- or vinyldiazoacetates.[35] The reaction is rationalized by a mechanism involving silicon-directed regioselective intramolecular C–H insertion to form β-lactones **6**, followed by silicon-facilitated olefination by means of extrusion of carbon dioxide. When the trialkylsilyl group was absent, the method could be applied to synthesize β-lactones **8** in high yields and with high levels of diastereo- and enantioselectivities (Scheme 5.2).[36] The *ortho*-substituent of aryldiazoacetate moiety enhances the efficiency of intramolecular C–H insertion while minimizing the corresponding intermolecular reactions. Similar strategy has also been reported with analogous α-(2-bromophenyl)-α-diazocarboxylates as the substrates.[37]

Padwa and co-workers investigated a series of transformations of Rh(II)-carbene from 2-alkynyl 2-diazo amido-substituted esters.[38] The ratio of C–H insertion product and Buchner-type 3-oxocyclohepta[*c*] pyrrole was found to be dependent on both the solvent and the catalyst ligand, as well as the substituent on the nitrogen atom.

5.1.2 *Five-membered ring formation*

In most cases, intramolecular 1,5-C–H insertion of diazo compounds is predominant and thus the reaction is widely used as a unique method to construct five-membered carbocycles or γ-lactam derivatives. While new developments in the field mainly include asymmetric catalysis, the chemo- and regioselectivity are also extensively studied. For example, in order to achieve remote C–H functionalization, an N–O moiety has been applied as tether in the diazo substrate **9** as shown in Scheme 5.3. Both 1,5- and 1,7-C–H bond insertion products **10** and **11** were observed when one of the substituents on 1,5-C–H bond carbon was methyl (R = Me). Switching the substituent to electron-withdrawing ester group (R = CO_2Me) led to the exclusive formation of 1,7-C–H insertion product (Scheme 5.3).[39] This observation clearly reflects the inference of both steric and electronic effects on the regioselectivity of C–H insertion.

The regio- and stereoselective 1,5-C–H bond insertion has been extensively employed as an efficient method for the construction of cyclopentanones,[15,40–43] dihydrofuranones,[44,45] γ-lactones,[46] γ-lactams,[47–50,27] tetrahydrofurans[51,52] and tetrahydrothiophenes,[53] as summarized in Scheme 5.4.

Rh(II)-catalyzed intramolecular C–H insertions were applied to the synthesis of a series of functional molecules.[54–58] For examples, Lecourt and co-workers developed Rh(II)-catalyzed carbene insertion of anomeric C–H bond of carbohydrates toward the synthesis of substituted ketopyranosides.[59–62] Besides, one-pot C–H insertion/olefination cascade reaction was explored for accessing α-alkylidene-γ-butyrolactone **12** with

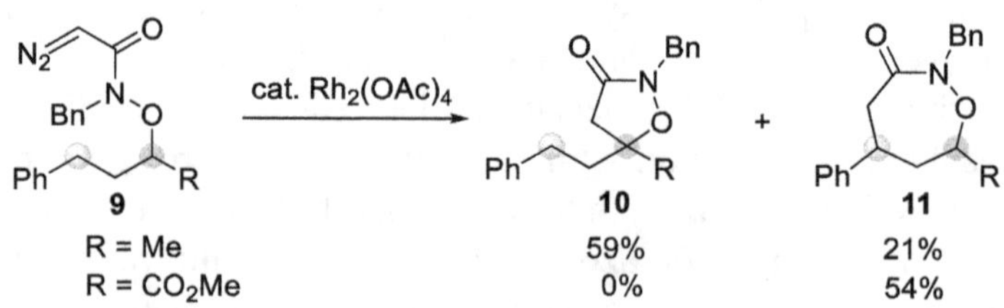

Scheme 5.3 Regioselectivity in C–H bond insertion: 1,5- versus 1,7-C–H insertions.

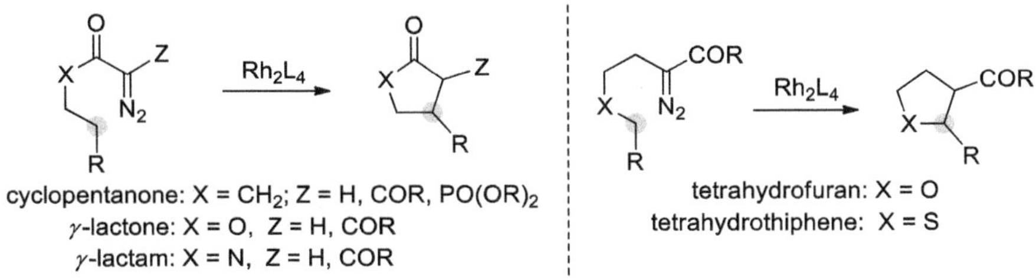

Scheme 5.4 Rh(II)-catalyzed intramolecular C–H insertion to access divergent five-membered rings.

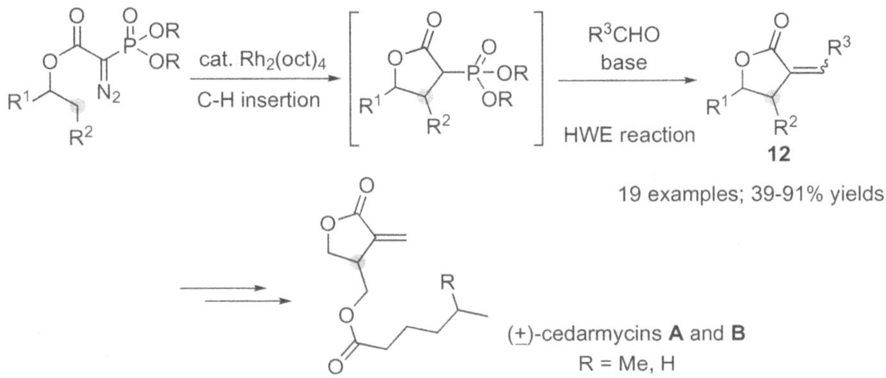

Scheme 5.5 Synthesis of α-alkylidene-γ-butyrolactone *via* C–H insertion/olefination.

high efficiency. The method was utilized in total synthesis of cedarmycins **A** and **B** (Scheme 5.5).[63,64]

Although Rh(II) complexes have been proved as the most effective catalysts for carbene C–H insertion, other metal complexes can also be effective, such as copper or ruthenium complexes. In two examples, Yu and Che reported the intramolecular C–H insertion *via* ruthenium porphyrin or $RuCl_2(p\text{-cymene})_2$ catalysis.[65,66]

In addition, intramolecular carbene 1,5-insertion into C–H bond was catalyzed by several highly efficient metal complexes, including gold complex,[67] iridium porphyrins $TPP[Ir(CO)_3]_2$,[68] (*p*-cymene) Ru(II) carboxylate complex for carbene insertion into primary C–H bond[69] and Ru-porphyrin for diastereoselective C–H insertion with *N*-tosylhydrazones.[70]

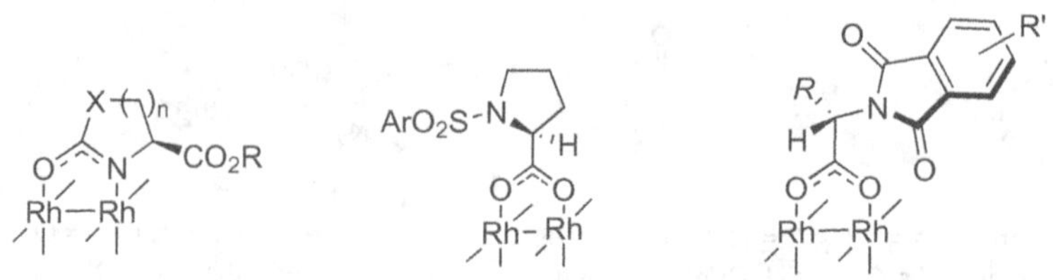

Figure 5.2 The main types of chiral Rh(II) catalysts.

On the other hand, asymmetric catalytic carbene C–H bond insertion has attracted significant attention over the years. The three major types of chiral Rh(II) catalysts developed in the 1990s,[71] namely chiral Rh(II) carboxamidates introduced by Doyle,[72–279] proline-based chiral Rh(II) carboxylates developed by the groups of Davies and McKervey[23,80–92] and phthalimide derivatives of amino acid-based chiral Rh(II) carboxylates developed by Hashimoto,[93–98] have found wide applications in asymmetric C–H insertions (Figure 5.2). Although further elaboration of these catalysts to reach higher level of reactivity and selectivity continues to be an important part of research activities, more attention has been directed to the application of these catalysts in organic synthesis. Topics such as expanding the substrate scope, application to total synthesis and development of new synthetic methodologies in connection with other reactions (tandem processes) have been the major focus in recent years.

Doyle's chiral Rh(II) carboxamidate complexes are particularly successful for the asymmetric intramolecular C–H insertion of acceptor-substituted carbenes. These catalysts have been extensively used in the asymmetric synthesis of lactones[75,78] and lactams.[73,74,76,77] An example is the total synthesis of (+)-imperanene *via* the asymmetric synthesis of γ-butyrolactone **13** by intramolecular C–H insertion catalyzed by $Rh_2(4S\text{-}MPPIM)_4$ (Scheme 5.6).[72]

The chiral Rh(II) carboxamidate complexes are also found to influence the chemoselectivity and regioselectivity of the C–H insertions.[75] An interesting observation made by Doyle and co-workers is that the $Rh_2(4S\text{-}MEOX)_4$-catalyzed asymmetric C–H insertion of diazoamide **14** contains a bis(trimethylsilyl)methyl protective group (Scheme 5.7).[74] This is an unusual example in which the chiral catalysts of opposite configuration induce different outcomes with the same achiral substrate.

The performance of chiral Rh(II) catalysts may be significantly modified by tuning electronic properties of the chiral ligands. Hashimoto and co-workers

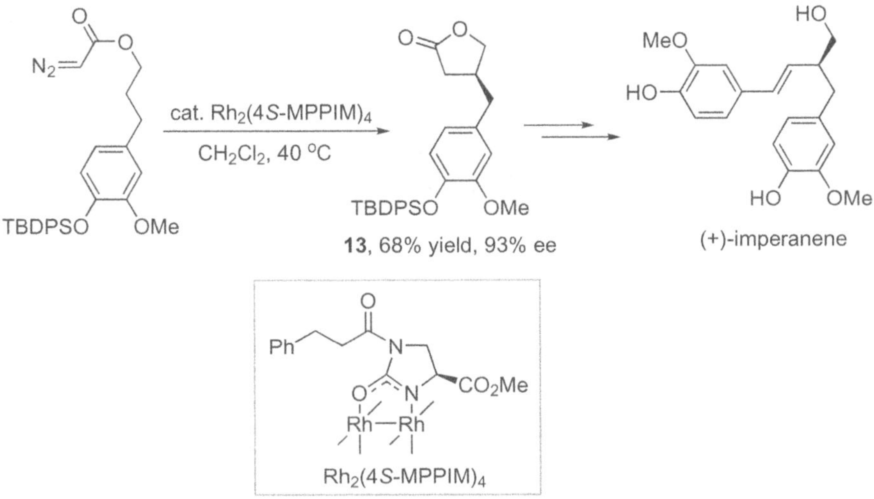

Scheme 5.6 Rh(II)-catalyzed asymmetric intramolecular C–H insertion in the synthesis of (+)-imperanene.

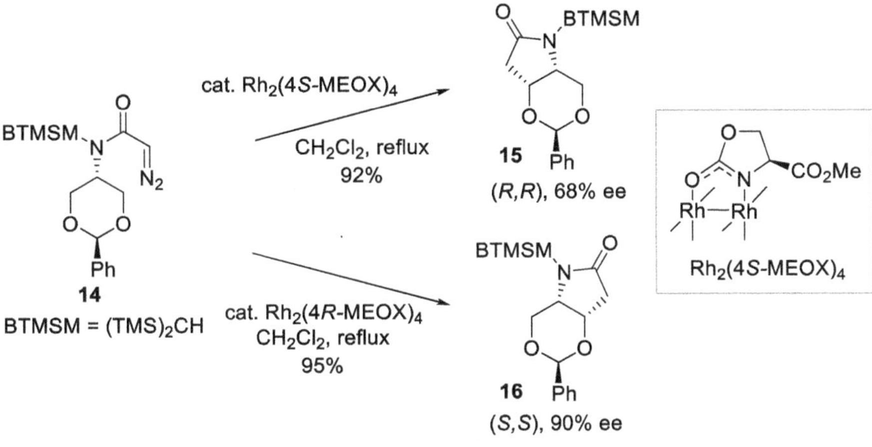

Scheme 5.7 Rh(II)-catalyzed asymmetric C–H insertion of diazoamide.

have introduced fluorine into their phthalimide derivatives of amino acid-based chiral rhodium(II) carboxylates. Both the reactivity and enantioselectivity are dramatically enhanced with the fluorinated Rh(II) catalyst, Rh_2[(*S*)-TFPTTL]$_4$ **17** [dirhodium(II) tetrakis[*N*-tetrafluorophthaloyl-(*S*)-*tert*-leucinate]. This catalyst has achieved a very high turnover number (up to 98000; when R = Me) with only 0.001 mol% loading, while without compromising the yield or enantioselectivity of the process (Eq. (5.1)).[94]

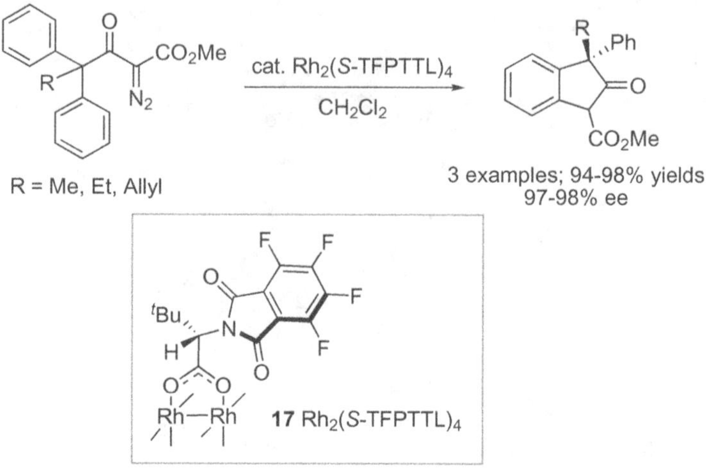

(5.1)

Davies and co-workers developed an adamantyl glycine-derived dirhodium tetracarboxylate, Rh_2(*S*-PTAD)$_4$ **18**. This Rh(II) catalyst was found to be very effective for carbene reactions and a high asymmetric induction was obtained in intermolecular cyclopropanation (99% *ee*), intermolecular C–H insertion (92% *ee*) and intramolecular C–H insertion (95% *ee*) (Eq. (5.2)). Interestingly, the chiral ligand, adamantyl glycine, itself was synthesized from Rh_2(*S*-DOSP)$_4$-catalyzed C–H insertion reaction of the vinyldiazoacetate with adamantane.[97]

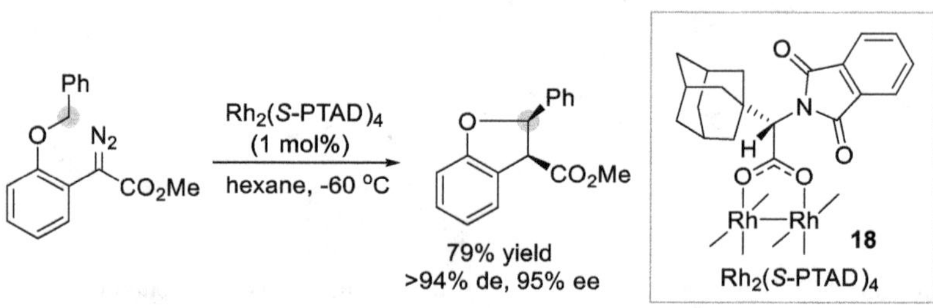

(5.2)

Immobilization of Rh(II) catalysts are important in view of the fact that rhodium is an expensive metal. In this context, Hashimoto and co-workers developed a heterogeneous catalytic system for asymmetric carbene insertion to C–H bond with polymer-supported chiral Rh(II) catalyst.[98] This immobilized Rh(II) complex was synthesized by copolymerization from dirhodium(II)-complex-containing monomer with

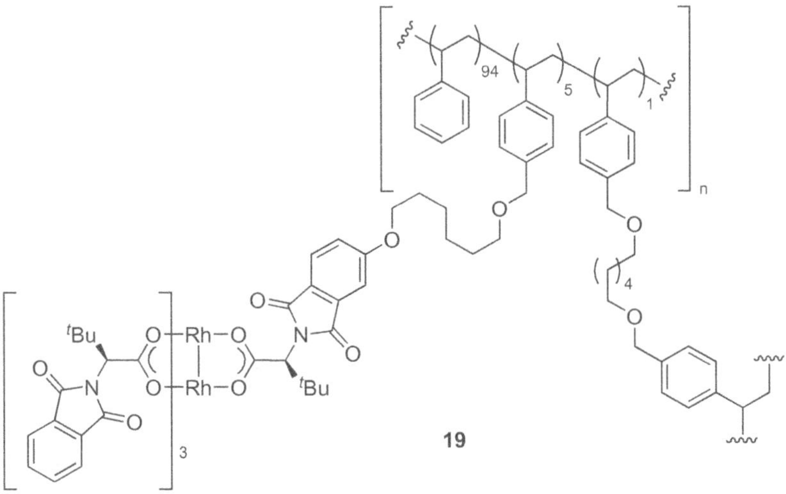

19

Figure 5.3 Polymer-supported chiral Rh(II) complex.

styrene and 1,6-bis(4-vinylbenzyloxy)hexane **19** (Figure 5.3). The catalyst showed high catalytic efficiency with high enantioselectivity.

As a synthetic application, enantioselective Rh(II)-catalyzed C–H insertion of donor–donor carbenes enabled the synthesis of functionalized dihydrobenzofurans **20** with high efficiency and stereoselectivities, which has potential synthetic value in synthesizing oligoresveratrol natural products.[99] The tandem transformation proceeded through oxidation of the hydrazone intermediate, followed by asymmetric intramolecular C–H insertion, leading to *E*-δ-viniferin **21** (Scheme 5.8).

While Rh(II) catalysts are predominant in carbene 1,5-C–H insertions, other transition-metal-catalysts have also been explored. Maguire and coworkers reported Cu(I)-catalyzed enantioselective C–H insertion reactions of α-diazosulfonyl compounds and other substrates.[100–107] In these cases, high enantioselectivity was induced by chiral bis(oxazoline)ligands **22** or **23** in the presence of $NaBAr_F$ as the additive (Scheme 5.9).[100,102] For expanding substrates scope, α-diazo-β-ketoesters and α-diazo-β-ketophosphonates were also subjected to the asymmetric intramolecular C–H insertions.[105] The same group also demonstrated enantioselective C–H insertions of 2-sulfonyl-2-diazoacetamides to access *trans*-γ-lactams by utilizing Cu(II)-bisoxazoline-$NaBAr_F$ catalytic system.[107]

Formal catalytic asymmetric $C(sp^2)$–H insertion from *N*-aryl diazoamides was catalyzed by Ti(IV)-BINOL as the means to achieve the

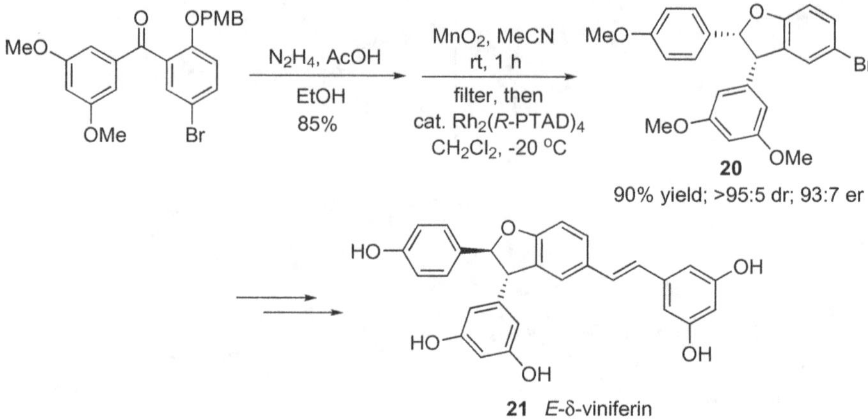

Scheme 5.8 Rh(II)-catalyzed asymmetric C–H insertion of donor–donor carbene.

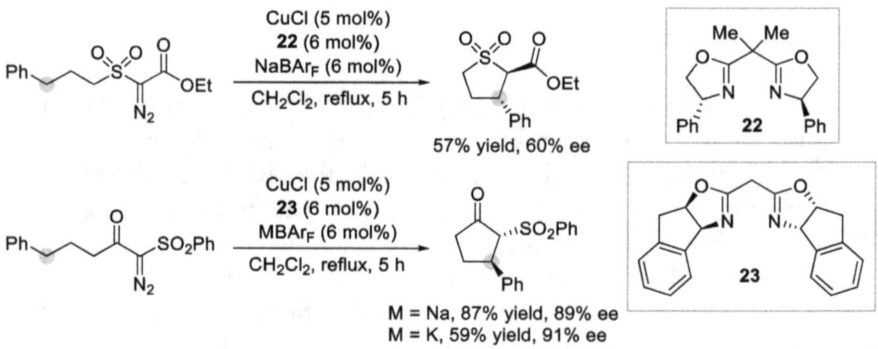

Scheme 5.9 Cu(I)-catalyzed asymmetric C–H insertion of diazosulfonyl compounds.

preparation of chiral oxindoles **24** (Eq. (5.3)).[108] In contrast to the conventional metal carbene C–H insertions, this cyclization reaction was assumed to proceed through Ti-BINOL-assisted asymmetric protonation of diazo carbon, followed by electrophilic aromatic substitution to afford oxindoles as the final products.

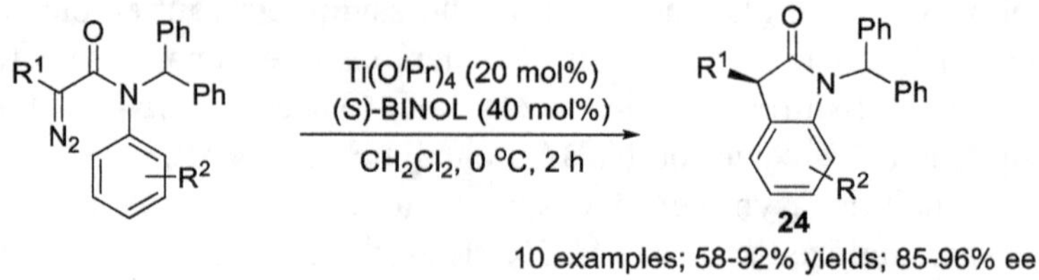

(5.3)

Zhang and co-workers demonstrated Co(II)-catalyzed asymmetric intramolecular C–H insertion of acceptor/acceptor-substituted diazo substrates.[109] The Co(II)-based metalloradical catalytic system bearing D_2-symmetric chiral amidoporphyrin served as an effective catalyst for the asymmetric intramolecular 1,5-C–H alkylation process. The five-membered sulfolanes were produced in high yields as well as with excellent diastereo- and enantioselectivities.

Another interesting development of the intramolecular carbene 1,5-C–H insertion is the application of "green" solvents. The Rh(II)-catalyzed carbene C–H insertion usually uses less coordinative solvents, such as dichloromethane or 1,2-dichloroethane. Afonso and co-workers demonstrated that Rh(II)-catalyzed intramolecular C–H insertion on α-diazoacetamides could be carried out effectively in ionic liquids or in water.[110–112] The use of water as the solvent in Rh(II) carbene C–H insertion allows a simple work up of the reaction and recycling of the catalyst. The fact that the highly feasible O–H insertion into water cannot compete with C–H insertion in this case implies that the reaction might actually occur "on water", rather than "in water".

5.1.3 *Six-membered ring formation*

While carbene 1,6-C–H insertion is less favorable as compared with its 1,5 counterpart, 1,6-C–H insertion has been reported for some structurally rigid systems.[17,117] Maguire and co-workers investigated Cu(I)-catalyzed asymmetric C–H insertion from diazosulfonyl compounds **25**, affording thiopyrans **26** with excellent enantioselectivities (Eq. (5.4)).[100,102,103]

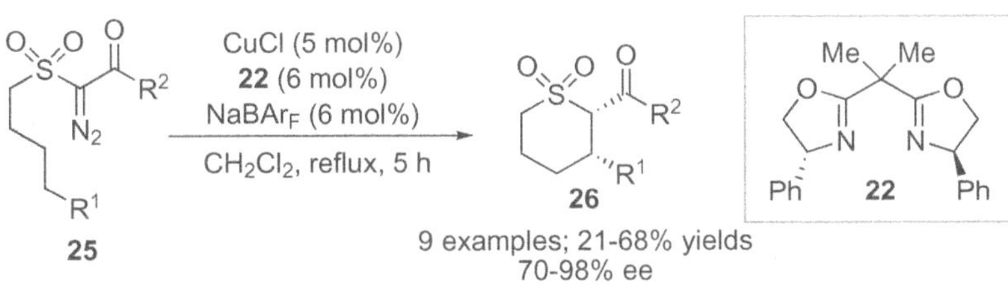

(5.4)

Rh(II)-catalyzed 1,6-C–H insertion was reported in some cases. Zhang and co-workers reported the reaction of α-diazocarbonyl glucosides **27** to afford 1,6-C–H insertion product **28** in the presence of Rh(II) catalyst, while intermolecular dimerization occurred when using Cu(I) catalyst (Scheme 5.10).[113] This 1,6 insertion selectivity is obviously due to the electronic effects. Another example of carbene 1,6-C–H insertion is demonstrated by Taber and co-workers (Scheme 5.10).[114] The preferred

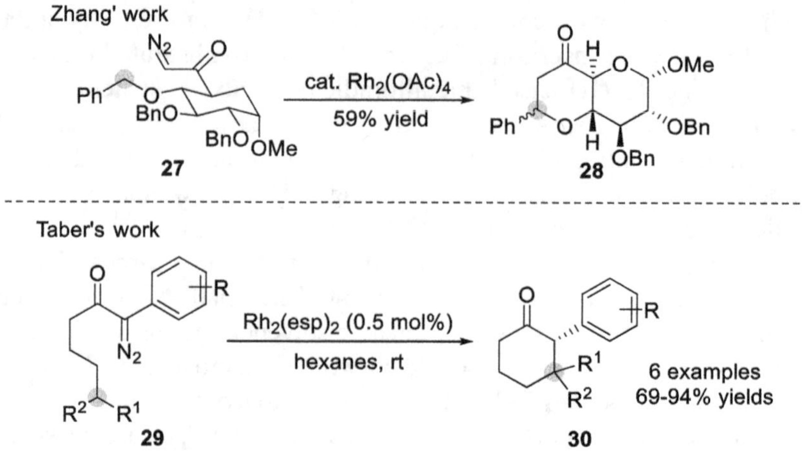

Scheme 5.10 Rh(II)-catalyzed carbene 1,6-C–H insertion.

formation of the carbene 1,6-C–H insertion product **30** from diazo substrate **29** is a result of both electronic and steric effects. Moreover, Rh(II)-catalyzed carbene 1,6-C–H insertions have been applied in the synthesis of chromane and flavanone skeletons.[115]

Recently, the asymmetric intramolecular 1,6-C–H insertion of α-diazoesters was achieved with $Rh_2(S\text{-PTTL})_4$. Instead of the competitive ylide formation, highly efficient C–H insertion afforded 2-alkenyltetrahydropyran-3-carboxylates in both chemo- and stereoselective manner with up to 95% *ee*.[116]

5.1.4 *Three-membered ring formation*

Carbene insertion into 1,3 C–H bond is very rare. Wang and co-workers reported an example of 1,3 C–H insertion in the Rh(II)-catalyzed reactions of β-tosyl-α-diazo carbonyl compounds **31**, affording cyclopropane derivatives **32** (Eq. (5.5)).[118] This study underscores the dramatic effect of neighboring groups on Rh(II) carbene reactions.

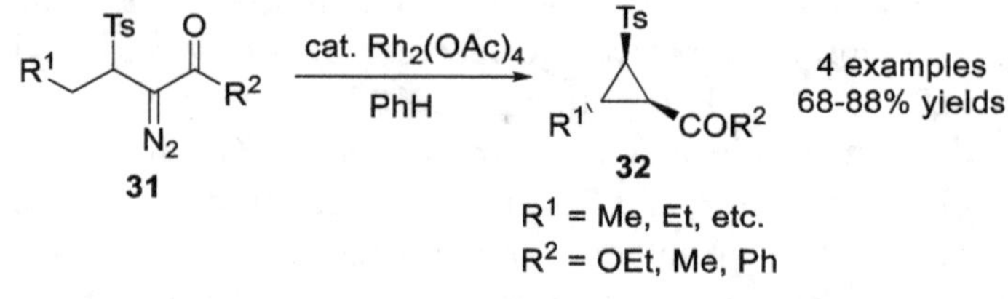

(5.5)

5.1.5 *Intermolecular C–H insertion*

The intermolecular reaction had been considered to be of limited synthetic utility because of the poor chemoselectivity. The situation has changed to some extent with the development of the new catalytic systems and the donor/acceptor-substituted carbene precusors.[71]

Regioselective intermolecular carbene insertions into C–H bond were performed by employing a variety of transition-metal catalysts, including Fe, Cu, Ag and Rh complexes. Iron(III)porphyrin and iron(II)-bis(imino) pyridine demonstrated catalytic reactivity in C–H insertions.[119,120] Particularly, highly selective intermolecular insertion into primary C–H bond could be catalyzed by Rh-porphyrin complexes **33** and **34** (Figure 5.4).[121] Besides, intermolecular C–H insertion catalyzed by tetra-tolylporphyrinato methyliridium(III) complex was carried out in order to gain insights into the reaction mechanism.[122]

To avoid carbene dimerization, one strategy is to increase the bulk of the catalyst ligand. In this context, high yields in intermolecular C–H insertion using bulky copper catalysts, such as **35** and **36**, have been reported by Pérez and co-workers (Scheme 5.11).[123–125] In the reaction of ethyl diazoacetate and tetrahydrofuran, the yield of the C–H activation product could be improved to 98% with Cu(I) catalyst **35**.[123] The C–H insertion of alkanes catalyzed by this catalyst also demonstrated a high chemoselectivity.[124] The main advantage of these catalysts appears is their

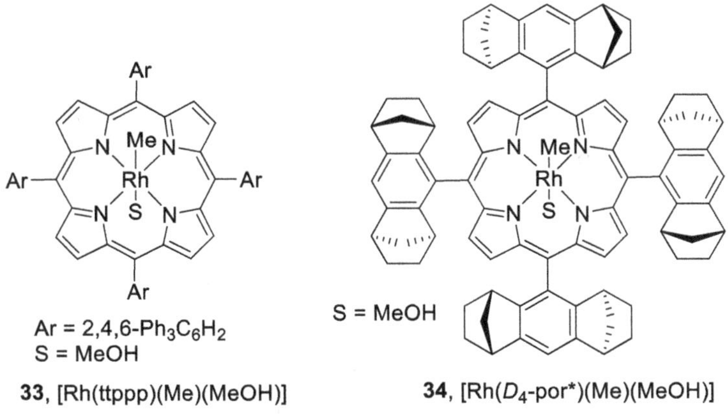

33, [Rh(ttppp)(Me)(MeOH)] **34**, [Rh(D_4-por*)(Me)(MeOH)]

Figure 5.4 Rh-porphyrin complexes for catalytic intermolecular C–H insertion.

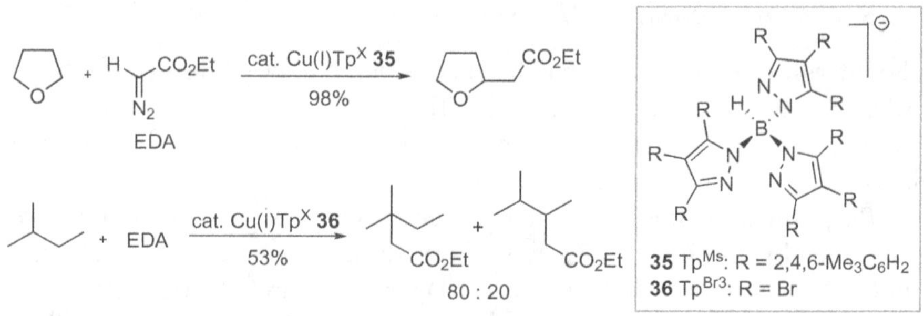

Scheme 5.11 Cu(I)Tp^x catalyst for the intermolecular C–H insertion with EDA.

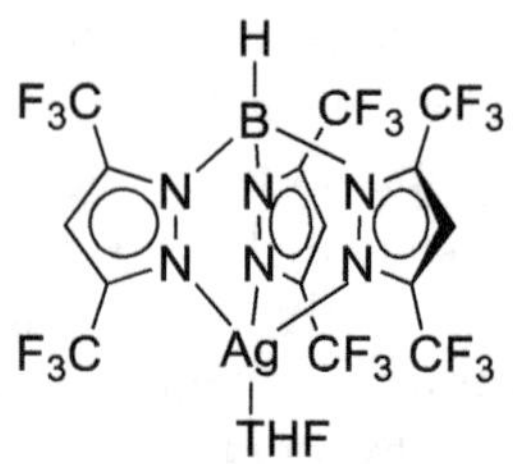

37, {HB[3,5-$(CF_3)_2$Pz]$_3$}Ag(THF)

Figure 5.5 Structure of silver tris(pyrazolyl)borate complex.

extremely bulky nature, yet the catalytic activity still remains. In addition, Pérez and co-workers discovered silver(I) scorpionate[126–128] and (NHC) CuCl (NHC = *N*-heterocyclic carbene)[129] catalyst systems for the intermolecular C–H insertion of ethyl diazoacetate.

With fluorinated borate ligands, Ag(I)-catalyzed intermolecular carbene insertion into inert alkanes has been proved to be efficient and selective.[130–135] In 2009, Dias and Lovely reported the silver tris(pyrazolyl) borate complex **37** catalyzed C–H insertion of donor–acceptor carbene with alkanes and ethers.[130] This transformation proceeded in good yields and exhibited improved selectivity as compared with the reaction with simple EDA (Figure 5.5).

Furthermore, the Pérez group explored the methane C–H bond insertion with Ag(I)-Tp^x catalyst **38** with supercritical carbon dioxide as the solvent. With perfluorinated indazolylborates as the ligands, the intermolecular C–H insertion of methane as well as other alkanes was successfully implemented (Eq. (5.6)).[131] Furthermore, investigations were focused on solvent and ligand effects to optimize this transformation for a wide range of alkanes.[132,133]

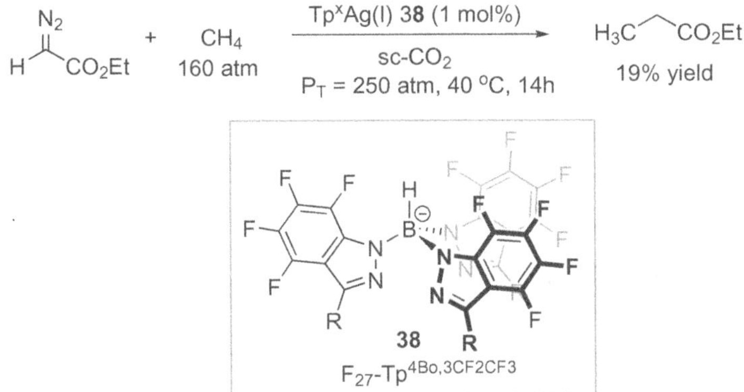

(5.6)

With donor–acceptor carbenes, the chemo- and regioselectivity of intermolecular C–H insertion have been significantly improved. In particular, asymmetric catalysis has been achieved. In this regard, Davies and co-workers have extensively exploited the asymmetric catalytic reactions with aryldiazoacetates and vinyldiazoacetates.[71,80–92] The Rh(II) carbenes derived from aryldiazoacetates and vinyldiazoacetates are generally highly selective in intermolecular C–H insertion reactions. Due to electronic effects, C–H insertion prefers to occur at the α-position of oxygen[81,84] or nitrogen,[80,85,88,89,91] which in general affords high chemoselectivity. This property can be valuable in organic synthesis, because the insertion reactions can be considered as the surrogates of classic aldol, Michael and Mannich reactions. A typical example is the enantioselective synthesis of β-amino esters **40** with $Rh_2(S\text{-DOSP})_4$-catalyzed C–H insertion reactions of the bis-silylmethylamine **39** with various aryldiazoacetates (Eq. (5.7)).[91]

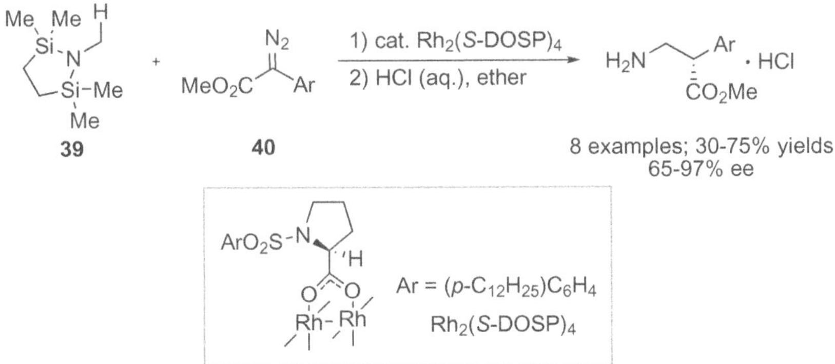

(5.7)

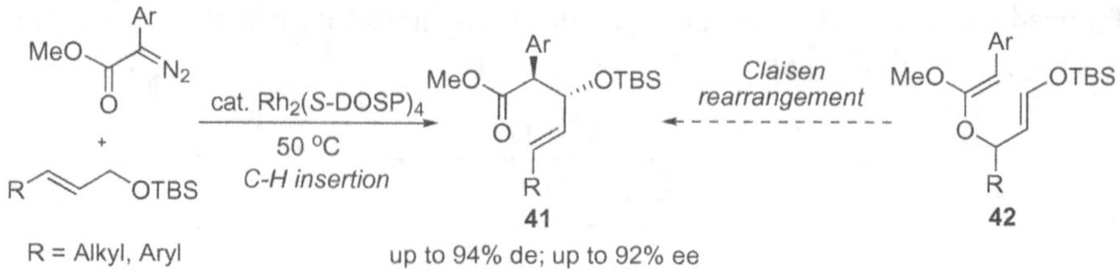

Scheme 5.12 Rh(II)-Catalyzed asymmetric allylic C–H insertion.

C–H insertion is also favored for allylic and benzylic sites, and the carbene derived from aryldiazoacetate is far more prone toward C–H insertion than cyclopropanation. Asymmetric allylic C–H insertion of alkenes *via* metal-catalyzed diazoacetate decomposition offers a unique approach to γ,δ-unsaturated esters **41**, which contain two stereocenters. This type of compound is classically synthesized through a Claisen rearrangement of **42** (Scheme 5.12).[90] The corresponding asymmetric benzylic C–H insertion has also been developed by Davies and co-workers.[83]

Vinyldiazolactone as a vinylcarbene precursor has also been explored recently. The Rh(II) carbene generated from this type of diazo compound is more favored for intermolecular C–H insertion than cyclopropanation (ratio 9:1). A moderately high level of enantioselectivity has been achieved with chiral Rh(II) carboxamidate catalyst **43** (Eq. (5.8)).[87]

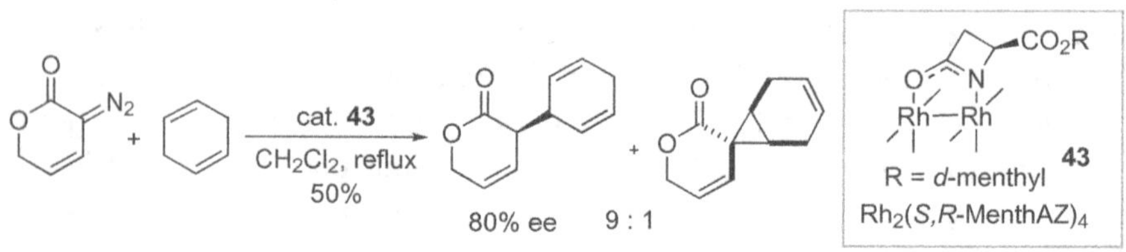

(5.8)

Katsuki and co-workers demonstrated Ir(III)-salen-catalyzed asymmetric C–H bond functionalization in high enantioselectivities and diastereoselectivities.[136] The insertion into the C–H bond of THF or 1,4-cyclohexadiene with various α-substituted-α-diazoesters proceeded with high enantioselectivity (Scheme 5.13). Moreover, the Ir(III)-bis(oxazolinyl) phenyl complexes were also shown to be effective for asymmetric intermolecular C–H insertions from donor/acceptor carbene precursors.[137]

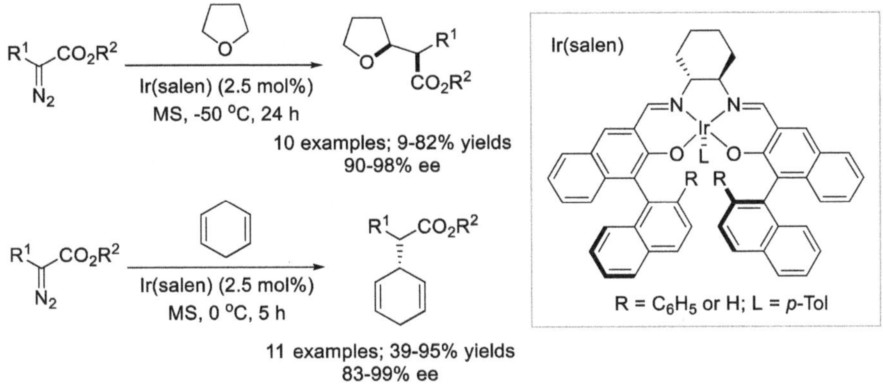

Scheme 5.13 Ir-Salen-catalyzed asymmetric intermolecular C–H insertions.

Rh(II) carbenes derived from aryldiazoacetates and vinyldiazoacetates have been extensively studied by Davies and co-workers.[71,138–145] The reaction with donor/acceptor-substituted carbene precursors are far less susceptible to dimer formation, presumably due to the relatively low nucleophilicity of the diazo substrates. As an example, the Rh(II)-catalyzed asymmetric insertion into the C–H of 1,4-cyclohexadiene with α-aryl-α-diazoketones showed high enantioselectivities (71–89%) (Eq. (5.9)).[139]

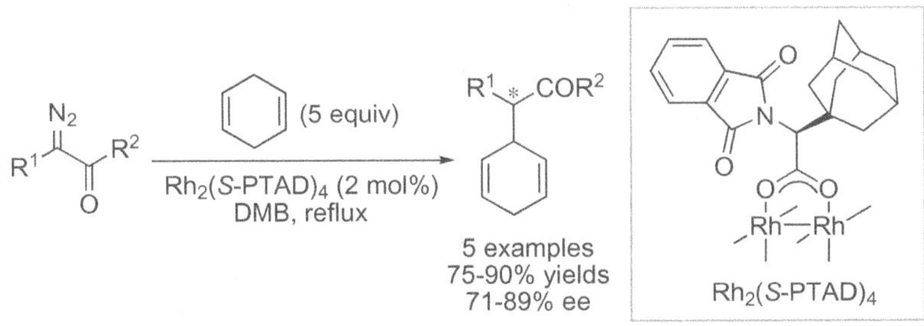

(5.9)

The reaction between aryldiazoacetates with tri-substituted alkenes catalyzed by Rh(II) complex was optimized to control C–H insertion versus cyclopropanation.[141] The chemo-, diastereo- and enantioselectivity were profoundly influenced by the structure of the substrates and the rhodium catalysts. Moreover, C–H insertion of aryldiazoacetates with TBS-protected benzyl alcohols was achieved in good yields and excellent stereocontrol (Scheme 5.14).[142] The products can be applied in the

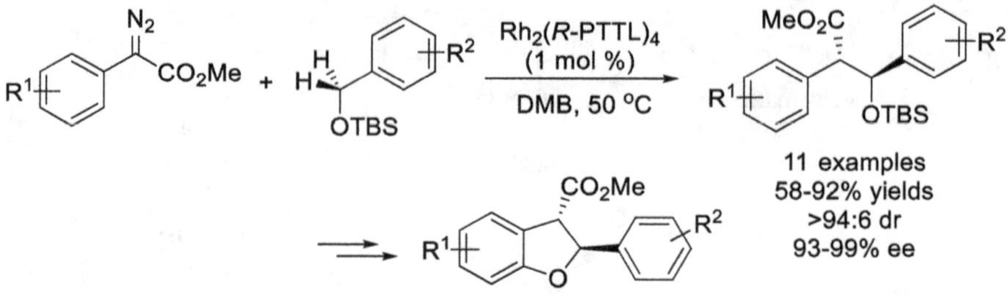

Scheme 5.14 Rh(II)-catalyzed C–H insertion and sequential enantioselective synthesis of 2,3-dihydrobenzofurans.

enantioselective synthesis of highly functionalized 2,3-dihydrobenzofurans **44** through sequential Pd-catalyzed C–H activation.

In addition to Rh(II)-catalyzed tertiary C–H insertion,[140] the same group recently reported the asymmetric C–H functionalization of activated primary C–H bonds catalyzed by sterically demanding Rh(II) carboxylates [Rh_2(*R*-BPCP)$_4$] **45**.[143] High preference for primary benzylic C–H bonds is presumably attributed to steric and electronic effects of bulky Rh(II) catalysts. Furthermore, they have recently reported the high level of regio- and enantioselective intermolecular C–H insertion of methyl ethers with 2,2,2-trichloroethyl aryldiazoacetates as reagents (Scheme 5.15).[144] With the employment of IR vibrations, Davies and Sigman quantitatively described and predicted site-selectivity of multivariate catalytic benzylic C–H functionalization in the presence of Rh_2(*S*-DOSP)$_4$ and Rh_2(*R*-BPCP)$_4$.[145]

The selective C–H bond insertion products could be utilized in the subsequent transformations. One of such examples is the Rh(II)-catalyzed C–H insertion reaction of ethyl phenyldiazoacetate into dihydroaromatic compound (Eq. (5.10)). The reaction was highly regioselective, affording polyfunctionalized aromatic compound upon oxidation with DDQ.[147]

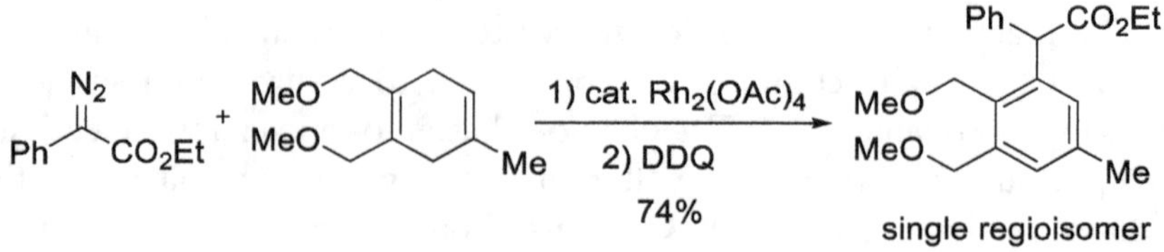

(5.10)

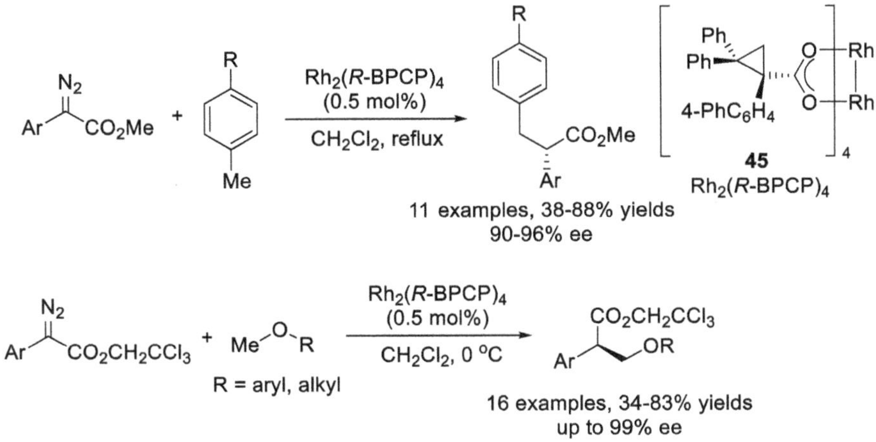

Scheme 5.15 Rh(II)-catalyzed site-selective carbene insertion of primary C–H bonds.

Davies and co-workers have established the cascade C–H functionalization of donor–acceptor carbenoids/Cope Rearrangement strategy. The Rh_2(*S*-DOSP)$_4$-mediated reaction of vinyl-substituted diazo compound with allylic substrate resulted in a reaction sequence of C–H insertion/Cope rearrangement.[86,87] For an example, vinyl-substituted diazo compound **46** was used in asymmetric intermolecular C–H insertion. The vinylcarbene generated from diazo substrate underwent allylic C–H bond insertion of cyclohexenes, which was followed by a Cope rearrangement to yield rearranged product **47**. The C–H insertion product **48** was also detected (Scheme 5.16).[86] This process is highly efficient and stereoselective, it has been successfully utilized in the total synthesis of the natural products, (−)-colombiasin A and (−)-elisapterosin B.[23]

Furthermore, this cascade process using vinyl diazoesters and the substrates containing allylic C–H bonds has been considered as a synthetic equivalent to a series of synthetically useful reactions.[148] For example, the strategy has been utilized for the synthesis of a series of enantiomerically enriched 1-naphthyl and 4-indolyl esters, which could be transformed to arylalkylamines as selective monoamine reuptake inhibitors.[149] The chemoselectivity of cyclopropanation versus vinylic C–H bond insertion was studied in the reaction of vinyldiazoacetates and dihydronaphthalenes.[150] The examples of C–H insertion/Cope rearrangement

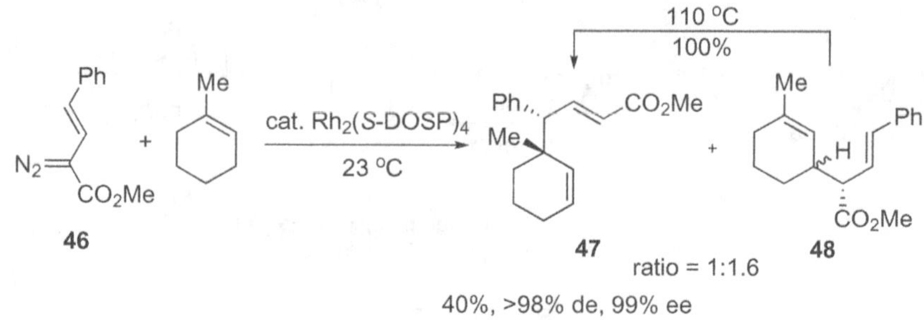

Scheme 5.16 Rh_2(*S*-DOSP)$_4$-mediated reactions of vinyl-substituted carbene with allylic substrates.

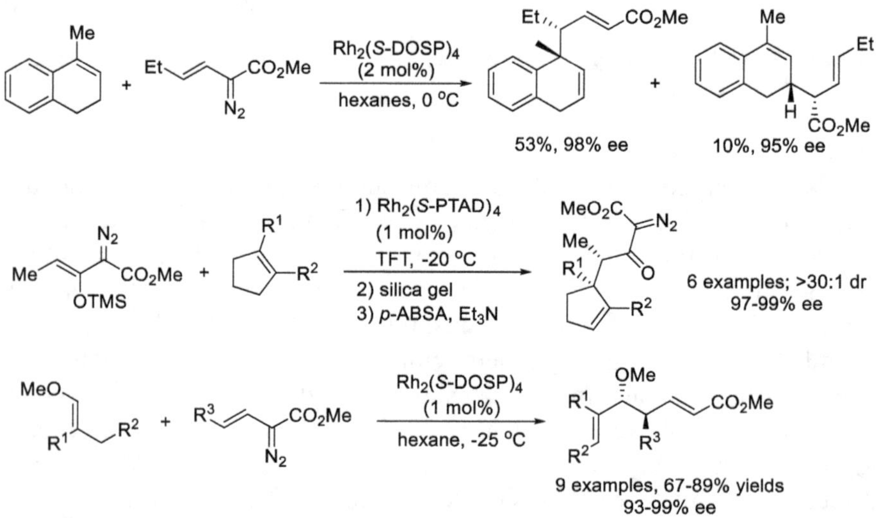

Scheme 5.17 Application of C–H insertion/Cope rearrangement in the synthesis of dihydronaphthalenes and cyclopentenes.

are shown in Scheme 5.17.[151–153] The reaction with vinyl ethers can be considered as the surrogate for vinylogous Mukaiyama-aldol reaction.[153]

5.2 Heteroatom–hydrogen insertions

5.2.1 *B–H insertion*

The electron-deficient B–H bond is not favored for insertions with metal carbenes. One way to circumvent this problem is by the adding of Lewis

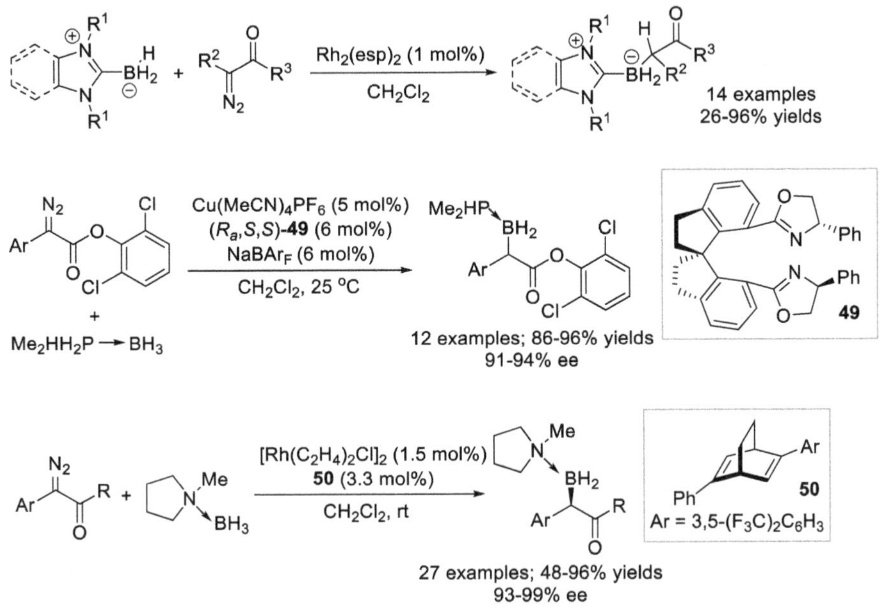

Scheme 5.18 Catalytic B–H bond insertions with diazo compounds.

base that can effectively coordinate with the electron-deficient boron center, thus altering the electronic property of the B–H bond. With this strategy, the B–H bond is activated and carbene insertion becomes possible. Recently, based on such a strategy the groups of Curran,[154] Zhou[155] and Xu[156] reported breakthroughs in the field, through the catalysis with Rh(II) or Cu(I) catalysts, respectively (Scheme 5.18). Curran and co-workers achieved the carbene B–H insertions of NHC-borane adducts by Rh(II) catalysts. On the other hand, Zhou and co-workers developed highly efficient enantioselective C–B bond formation with amine- or phosphine-borane adducts, catalyzed by Cu(I) with chiral spiro-bisoxazoline ligand **49**.[155,157] More recently, Rh(I)-catalyzed asymmetric B–H insertion with amine-borane adducts was reported by Xu and co-workers, using C_1-symmetric chiral diene **50** as the chiral ligand.[156] Both protocols enabled the highly efficient C–B bond formation with excellent enantioselectivities.

5.2.2 *N–H insertions*

N–H insertion of amines with diazo compounds has been extensively explored as a practical method for the construction of C–N bonds. In particular, intramolecular N–H insertion can be served as a straightforward method for synthesizing nitrogen containing heterocycles. Five-membered ring is usually favored in intramolecular N–H insertion, providing pyrrolidine derivatives.[158–161] However, there are also many examples of metal carbene insertions into 1,4-, 1,6- and 1,7-N–H bonds, which afford four-,[162] six-[163] and seven-[164] membered nitrogen-containing rings, respectively. It should be emphasized that the intermolecular carbene N–H bond insertion is also highly selective and thus synthetically useful, due to the high reactivity of the nitrogen as Lewis base toward electron-deficient metal carbene carbon center.

While the field has been well explored over the years, new developments have been continually documented in the literature recently. One of the new developments is the expansion of the scope of diazo substrates. For example, Taran and co-workers expanded the N–H insertion using tetraethyl diphosphonodiazomethane **51** with Rh(II) catalyst.[165] This reaction enabled facile synthesis of aminomethylene bisphosphonates **52** (Scheme 5.19). Moody and co-workers investigated the carbene reactivity of 4-diazo-*4H*-imidazoles in a range of insertion reactions and other transformations.[166] The carbene precursor diazoimidazole carboxylates could undergo N–H, O–H and C–H bond insertions.

This approach has been utilized for the synthesis of amino acid-derived oxazole building blocks of the natural products diazonamide A[167,168] and amythiamicin D.[169] Janda and co-workers further developed a solid-phase variant of these reactions.[170,171] The synthesis of indoles in a similar strategy has also been reported.[172–176] Furthermore, this reaction sequence can be served as an effective method to synthesize

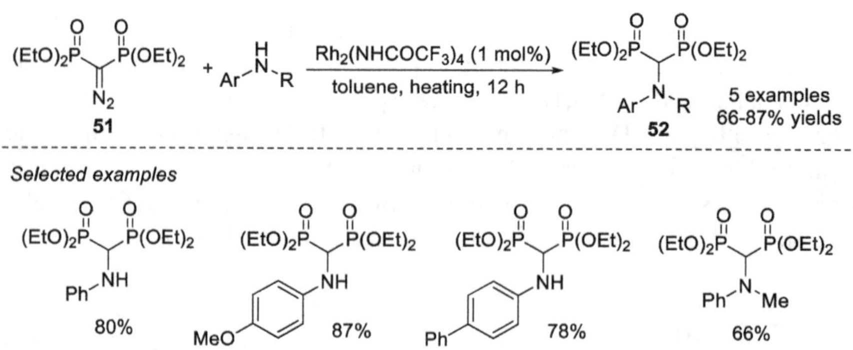

Scheme 5.19 N–H insertion of aromatic amines with diazodiphosphonates.

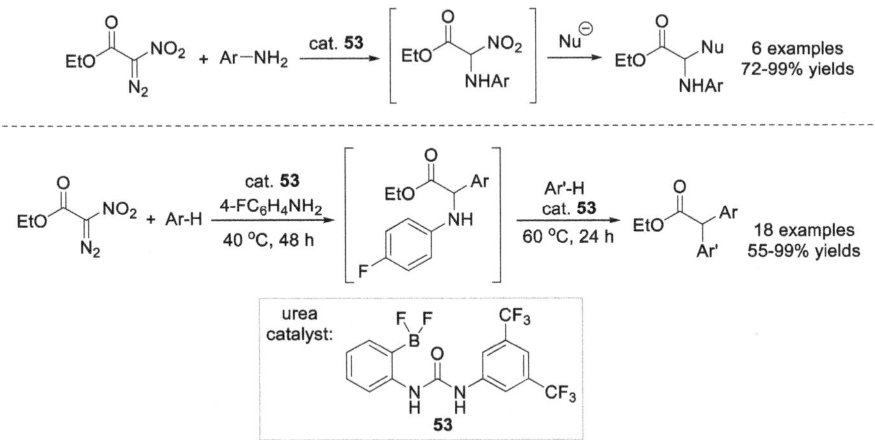

Scheme 5.20 Urea-catalyzed N–H insertion and sequential coupling of nitrodiazoesters.

oxazolones and other heterocycles when α-diazo-β-ketoesters are used as substrates.[177–179]

On the other hand, Mattson and co-workers developed an organocatalytic strategy toward N–H insertion from α-nitrodiazoesters with anilines by using urea **53** (Scheme 5.20).[180,181] In contrast to the transition-metal-catalyzed reactions, this transformation represents a new concept of hydrogen bond donor-activation of the α-nitrodiazoester. Moreover, they have further developed the one-pot double arylation of nitrodiazoesters through a transient N–H insertion intermediate.[180] Interestingly, the urea catalyst is essential for both the N–H insertion process and the second α-arylation cycle for the activation of α-aminoester intermediate. In 2013, Ren and co-workers also performed the catalyst-free N–H insertion/Mannich-type cascade synthesis of α-amino-α-arylesters from α-nitrodiazoesters in protic solvent.[182]

In addition to the expansion of substrates scope with α-diazocarbonyl components, new catalytic system has also been applied to N–H insertions. Glycine esters are obtained through N–H insertion catalyzed by nano-crystalline Cu(II) oxide, which is a heterogeneous catalyst.[183] Ruthenium complexes also show high reactivity in catalyzing N–H insertion of diazo compounds. For example, in addition to the well-explored Ru(II)-porphyrin catalysts,[184] N–H insertion in the reaction of α-diazoacetamides with anilines or secondary amines were catalyzed by Ru(II)-*dm*-Pheox **54** with good to excellent yields.[185] Iron(III)-corroles **55**,

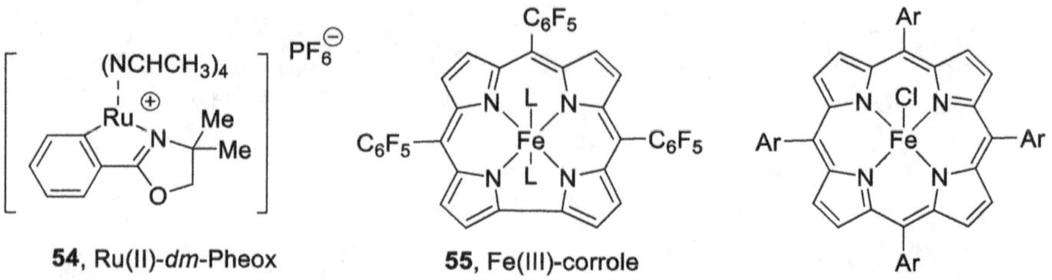

54, Ru(II)-*dm*-Pheox **55**, Fe(III)-corrole Ar = C_6H_5 or C_6F_5 **56**, Fe(III)-porphyrin

Figure 5.6 Structures of Ru(II)-*dm*-Pheox, Fe(III)-corrole and Fe(III)-porphyrin.

porphyrins **56**, cytochrome P450-BM3 and myoglobin have demonstrated catalytic activity, respectively, for the insertion of carbene into N–H bonds (Figure 5.6).[186–188] The mechanism of Ir(III)-porphyrin-catalyzed N–H insertion from various aryl, aliphatic, primary and secondary amines with α-diazoacetates was investigated, which indicated that N–H insertion proceeded through stepwise transformation with metal-ammonium ylide intermediate, similar to the traditional Cu(I)- or Rh(II)-catalyzed reactions.[189]

Sivasankar and co-workers designed and synthesized a phosphine-coordinated air-stable Cu(I) catalyst **57**, which enabled chemoselective carbene N–H insertion under mild condition over O–H insertion for the reaction of α-aryl diazoesters with various aminophenols (Eq. (5.11)).[190] The ligand in this complex played a crucial role for selective carbene insertion only into N–H bonds.

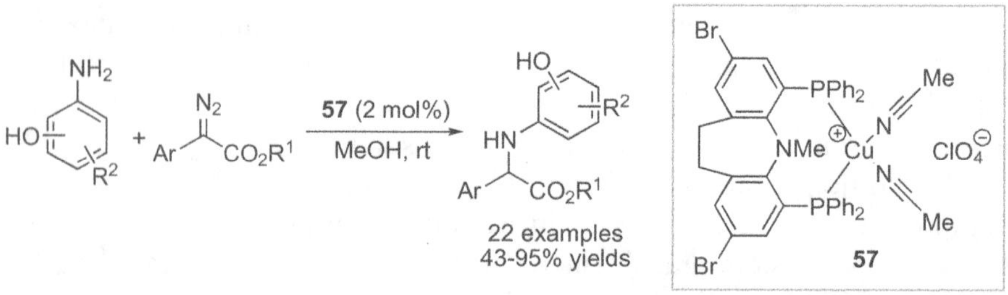

(5.11)

Recently, an aqueous hemin-catalyzed, cyclodextrin-assisted N–H insertion of α-diazoesters with anilines was reported.[191] This aqueous catalytic system showed good compatibility with various arylamines, and afforded the desired insertion products in high yields.

In addition to transition-metal catalysts, reusable acidic ionic liquid of 1-methylimidazolium tetrafluoroborate was recently investigated to catalyze N–H bond insertion of amines with EDA.[192] The ionic liquid can be recovered without noticeable decrease in catalytic activity. Moreover, catalyst-free N–H bond formation with primary and secondary amines was also carried out with aryldiazoacetates under thermal decomposition conditions.[193] DFT calculation of the reaction mechanism suggests that the reaction proceeds through the generation of free donor/acceptor carbene and subsequent ammonium ylide formation.

Intramolecular N–H bond insertion catalyzed by $[RuCl_2(p\text{-cymene})]_2$ was performed as well as the intermolecular N–H insertions.[194] Aryl- or aliphatic-substituted 3-hydroxyprolines were prepared with high diastereoselectivities by one-pot synthesis under mild conditions (Eq. (5.12)).

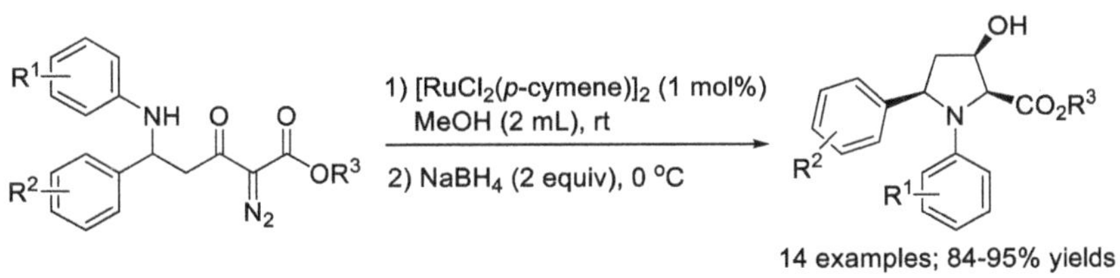

(5.12)

Gillingham and co-workers have applied N–H insertion of diazo compounds in chemical biology research.[195–197] They established the structure-specific catalytic alkylation of DNA and RNA with metal carbene, in which the amino group of purine nucleobases underwent Rh-catalyzed N–H insertion with aryldiazoesters.[195] Subsequently, the same group developed an auto-tandem catalytic process including the Cu-catalyzed azide–alkyne cycloaddition and N–H insertion in aqueous media.[196] The substrate scope includes small molecule anilines as well as the nucleobases in DNA and RNA (Scheme 5.21). Moreover, the modification of nucleobases in cyclic purine dinucleotides was achieved by N–H insertion, which can be applied to introduce fluorophores and photo-crosslinking groups.[197]

One of the most significant progresses in carbene N–H insertion in recent years has been the realization of highly enantioselective asymmetric catalysis, developed by Zhou and co-workers. In 2010, the Feng group reported the Cu(I)-catalyzed asymmetric N–H insertion of primary and secondary arylamines with α-substituted-α-diazoesters with BINOL ligands **58**.[198] However, the enantioselective control of secondary anilines is poor. Recently, they improved the enantioselectivity of N–H insertion

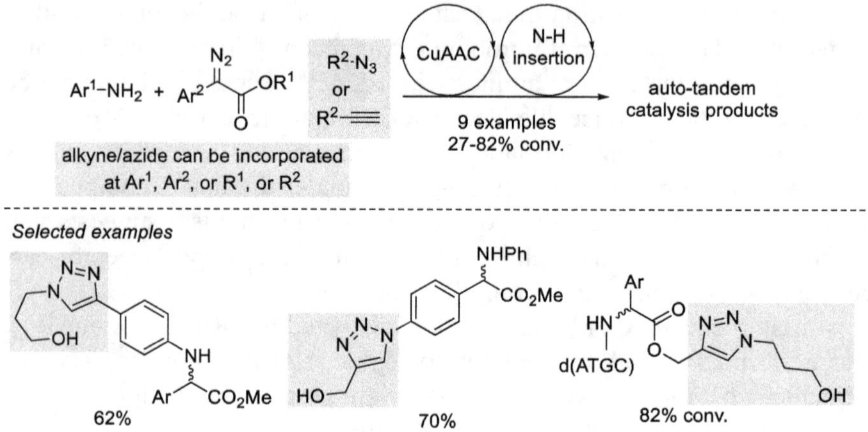

Scheme 5.21 Auto-tandem catalytic CuAAC/N–H insertion.

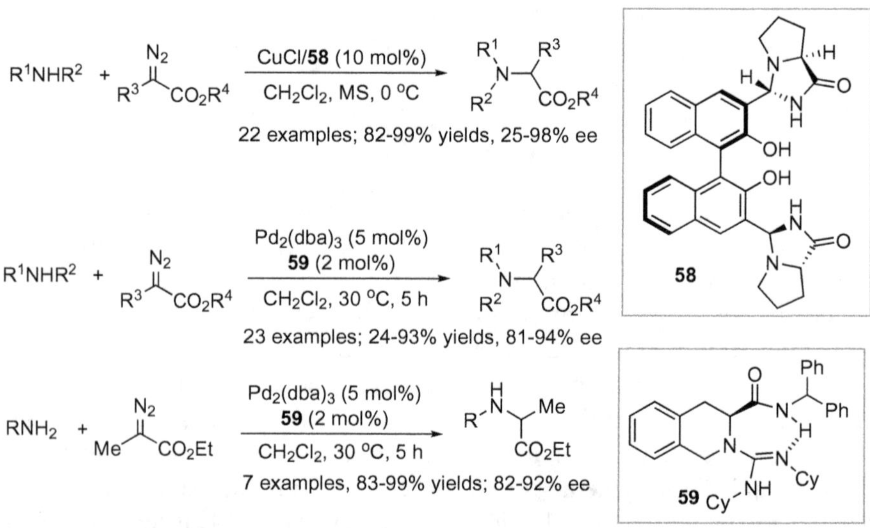

Scheme 5.22 Asymmetric N–H insertion catalyzed by Cu-BINOL and Pd-guanidine ligand.

for both secondary and primary anilines with catalysis of palladium and chiral guanidine ligand **59** (Scheme 5.22).[199]

On the other hand, Zhou and co-workers made a breakthrough in Cu(I)-catalyzed asymmetric N–H insertion by using their chiral spiro-bisoxazoline **60** as the ligand. For ethyl 2-diazopropionate, generally high level of enantioselectivity could be obtained, except for ethyl phenyldiazoacetate (Eq. (5.13)).[200] Furthermore, asymmetric N–H insertion of α-alkyl-α-diazoacetates with anilines was achieved with excellent enantioselectivity, catalyzed by binuclear chiral spiro copper(I) complex.[201] These transformations show the effectiveness of chiral spiro ligands in catalytic asymmetric carbene N–H insertions.

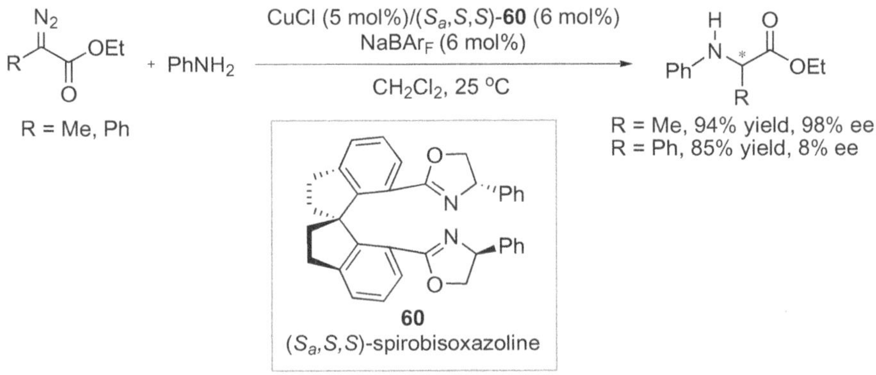

(5.13)

Furthermore, a mechanistically different approach by combining organic catalyst and metal catalyst has been recently proved to be highly efficient for asymmetric N–H insertions by the Zhou group. Thus, a series of chiral spiro ligands **61**, **62** were reported to promote enantioselective N–H insertion of Rh(II) carbenes generated from diazo compounds (Scheme 5.23).[202,203] They reported the asymmetric N–H insertion of α-aryldiazoesters[202] or α-aryldiazoketones[203] with $BocNH_2$, catalyzed cooperatively by achiral Rh(II) catalyst and phosphoric acids bearing chiral spiro backbone, in good yields and excellent enantioselectivities.

Rh(II)-catalyzed asymmetric N–H insertion reactions of α-diazoesters and carbamates were developed with cooperative catalytic system of chiral sulfonamide urea and achiral sulfonic acid.[204] Mechanistic insights toward the cooperative dual-catalytic strategy involving dirhodium carbene species were investigated through DFT studies.[205]

Recently, Zhou and Zhu established an efficient synthetic method for chiral α-alkenyl α-amino acid derivatives by means of N–H insertion of vinyldiazoacetates and *tert*-butyl carbamates cooperatively catalyzed by achiral Rh(II) complex and chiral spiro phosphoric acids (Eq. (5.14)).[206]

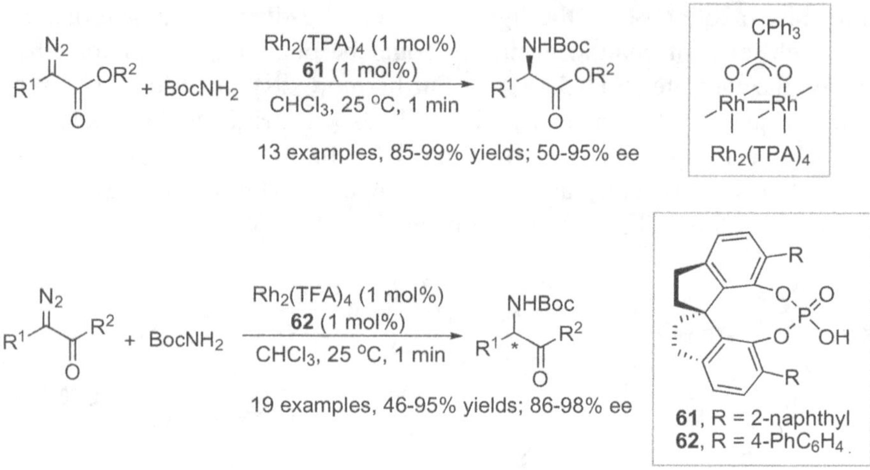

Scheme 5.23 Asymmetric N–H insertion catalyzed by achiral Rh(II) catalyst/chiral phosphoric acids.

This protocol afforded carbene insertion products with high yields (61–99%) and good enantioselectivities (83–98% *ee*).

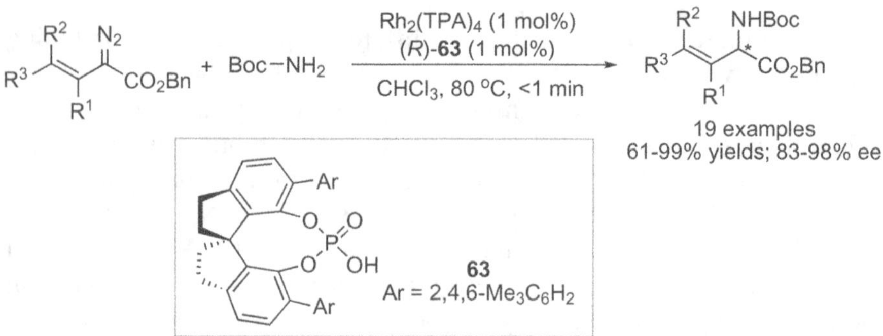

(5.14)

The metal carbene insertion reactions have been successfully incorporated into the synthesis of various compounds of synthetic value, such as aromatics[207–209] and heteroaromatics,[210–217] 1,2,4-triazines,[218] *N*-methylindoles,[219] aminoimidazoles[220] and *N*-heterocycles.[221] For example, Moody and co-workers have developed the synthesis of oxazoles **64** and thiazoles **65** *via* Rh-catalyzed N–H insertions and cyclodehydration (Eq. (5.15)).[222–225] This strategy has been applied to the formal total synthesis of telomestatin, in which the key step was the Rh(II)-catalyzed reaction of diazocarbonyl compounds to generate six oxazole rings.[224] The same group developed a synthetic method for trifluoromethyl-substituted heterocycles through Cu/Rh-catalyzed carbene N–H insertions.[225]

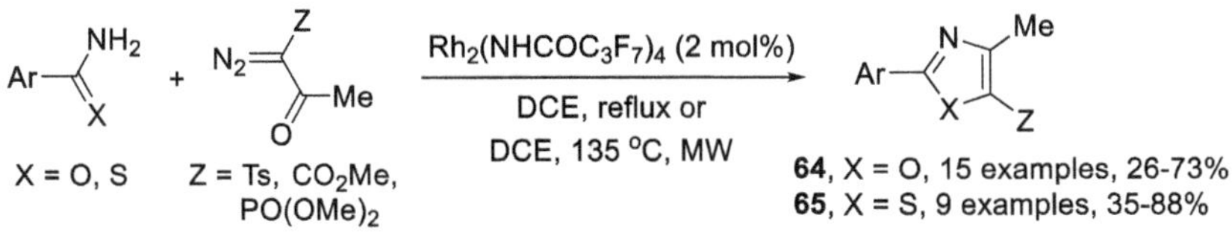

(5.15)

Finally, the metal carbene insertions have also been explored with *N*-tosylhydrazones as the carbene precursors, in which unstable diazo species are generated *in situ*. Alami and co-workers reported Cu-catalyzed reductive coupling of *N*-tosylhydrazones with aliphatic amines[226] and anilines (Scheme 5.24).[227] This protocol can be considered as a unique strategy for C(sp^3)–N bond formation. Moreover, a one-pot synthesis of primary amine was reported by Yadav and co-workers based on Cu-catalyzed N–H insertion phthalimide followed by hydrazinolysis

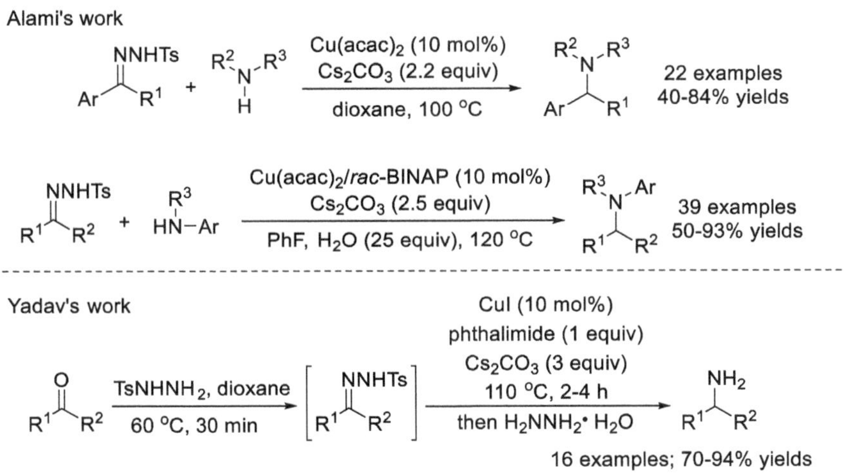

Scheme 5.24 Cu-catalyzed N–H insertion with *N*-tosylhydrazones.

process.[228] This reductive method provides an alternative approach for deoxygenative hydroamination of carbonyl compounds.

5.2.3 *O–H insertion*

Catalytic O–H insertion provides a straightforward method for the C–O bond formation, which can be served as a useful route to acyclic and cyclic ethers.[229] Transition-metal-catalyzed O–H insertion is generally considered as a stepwise process *via* an initial oxonium ylide formation followed by a rapid 1,2-hydrogen shift.

Typically, carbene O–H insertions are catalyzed by Rh(II) catalysts; however, carbene O–H insertions can be catalyzed by other metal complexes. For example, diruthenium(II)tetrakis(acetate)[230] and Tp^{x}Cu[231] have been reported as effective catalysts for O–H insertion. Tp^{x}Cu has also been found to be an effective catalyst for N–H insertion.[232]

The O–H insertions can also be promoted by Lewis acids or protonic acids.[233–235] This makes O–H insertions potentially useful in chemical biology. Recently, Raines and co-workers reported the metal-free O–H insertion of carboxylic acids with diazo compound derived from phenylglycinamide.[236] With the employment of optimized diazo scaffold, protein esterification was efficiently realized in an aqueous environment.

Catalytic O–H insertion has been applied to the synthesis of molecules of biochemical interest.[237–239] Maguire and co-workers reported Rh(II)-catalyzed O–H insertion by employing α-diazophosphonate **66** for the synthesis of nucleoside derivatives **67**, including protected

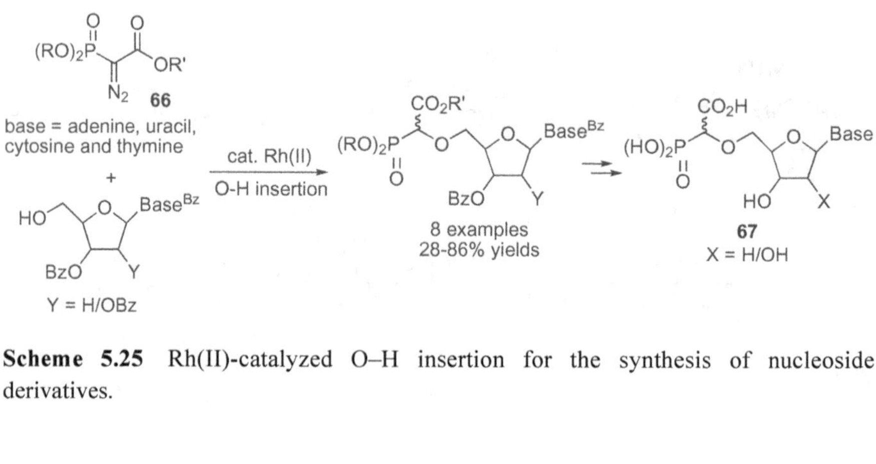

Scheme 5.25 Rh(II)-catalyzed O–H insertion for the synthesis of nucleoside derivatives.

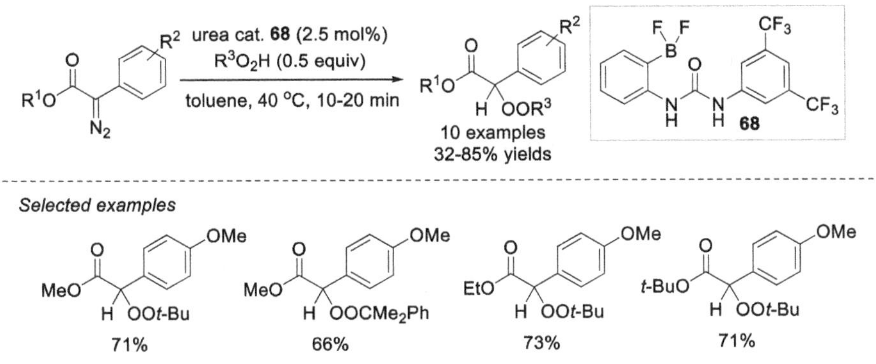

Scheme 5.26 Urea-catalyzed O–H insertion with hydroperoxides.

thymidine, uridine, cytosine, adenosine and guanosine derivatives (Scheme 5.25).[237,238] Upon deprotection, a series of phosphono nucleosides bearing a carboxylic acid moiety adjacent to the phosphonate could be obtained. Moreover, the solid phase materials capable of enriching phosphorylated peptides were prepared by O–H insertion from polymer-supported diazo substrates.[239]

Carbene insertions into O–H bond of water, alcohol and carboxylic acids are common, moreover, the reactions with other types of O–H bonds have also been explored. Mattson and co-workers reported the O–H insertion of α-aryl-α-diazoesters with hydroperoxides in the presence of urea catalyst **68**.[240] This organocatalytic strategy permitted the efficient synthesis of α-peroxyesters in high yields (Scheme 5.26). Moreover, difluoroboronate urea-induced O–H insertion of aryldiazoacetates to carboxylic acids was performed with high yields.[241]

Controlling of stereoselectivity in carbene O–H insertion is an intriguing problem. High diastereoselectivity could be achieved with achiral auxiliary approach, as demonstrated by Sugimura and co-workers.[242,243] In contrast to the corresponding C–H and Si–H insertions, metal carbene O–H insertion and N–H insertion are believed to follow a ylide generation/1,2-proton shift mechanism. Asymmetric catalytic O–H insertion and N–H insertion have been essentially unsuccessful until very recently. A breakthrough in asymmetric catalytic O–H insertion was made by Fu and co-workers.[244] With $Cu(OTf)_2$/(+)-bisazaferrocene (**69**) as the chiral catalyst and 2-trimethylsilylethanol as the O–H bond donor, a high level of enantioselectivity was obtained with a series of methyl aryldiazoacetates (Eq. (5.16)).

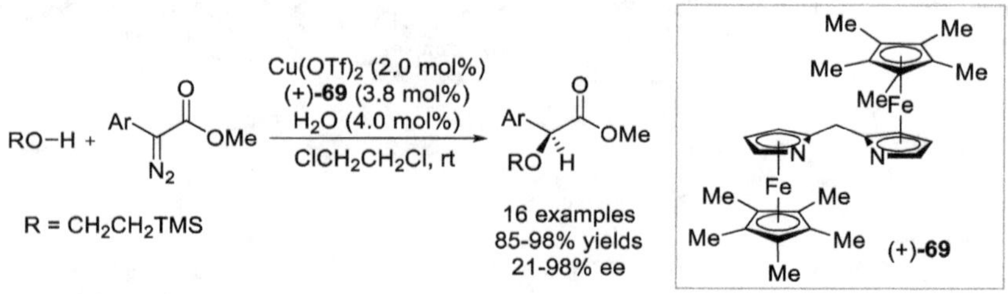

(5.16)

Subsequently, highly efficient asymmetric catalytic inter- or intramolecular O–H carbene insertions over a wide range of substrates were demonstrated by Zhou and co-workers. With spiro-bisoxazoline ligands **70/71**, a series of α-diazoesters, in particular α-aryl-α-diazoesters, were subjected to transition-metal-catalyzed asymmetric intermolecular O–H insertions of water,[245,246] alcohol[246,247] and phenols.[248] Remarkably, in addition to Cu(I) catalysts, Pd(II) and Fe(II) catalysts also show high reactivity and selectivity for O–H insertions (Scheme 5.27).

The intramolecular version of the catalytic asymmetric O–H insertions has been achieved by Zhou and co-workers. The reaction led to the formation of chiral cyclic ethers in high enantioselective manner (Scheme 5.28).[249,250] Asymmetric carbene O–H insertion has also been explored with other catalytic systems.[251–253]

The O–H insertion products are highly useful in organic synthesis, in particular the insertion products may be subjected to subsequent transformations.[254–259] For example, Bull and co-workers reported stepwise Rh(II)-catalyzed O–H insertion/cyclization strategy toward the synthesis of di-, tri- and tetra-substituted oxetanes **74** (Eq. (5.17)).[257]

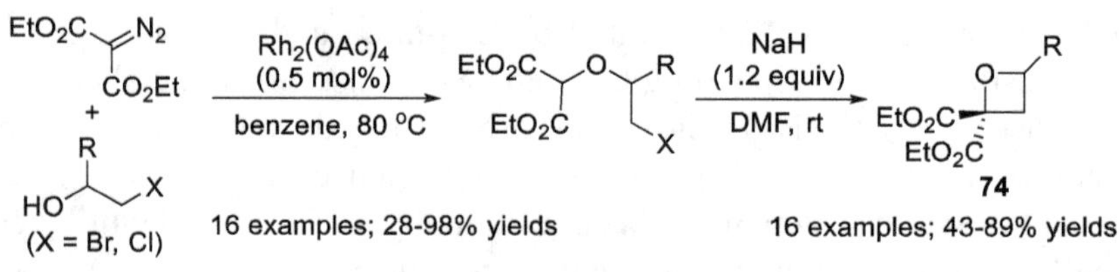

(5.17)

Similarly, *N*-tosylhydrazones have been used as the precursors of diazo compounds in carbene O–H insertions. In 2010, metal-free C–O bond formation of *N*-tosylhydrazones with alcohols or phenols was

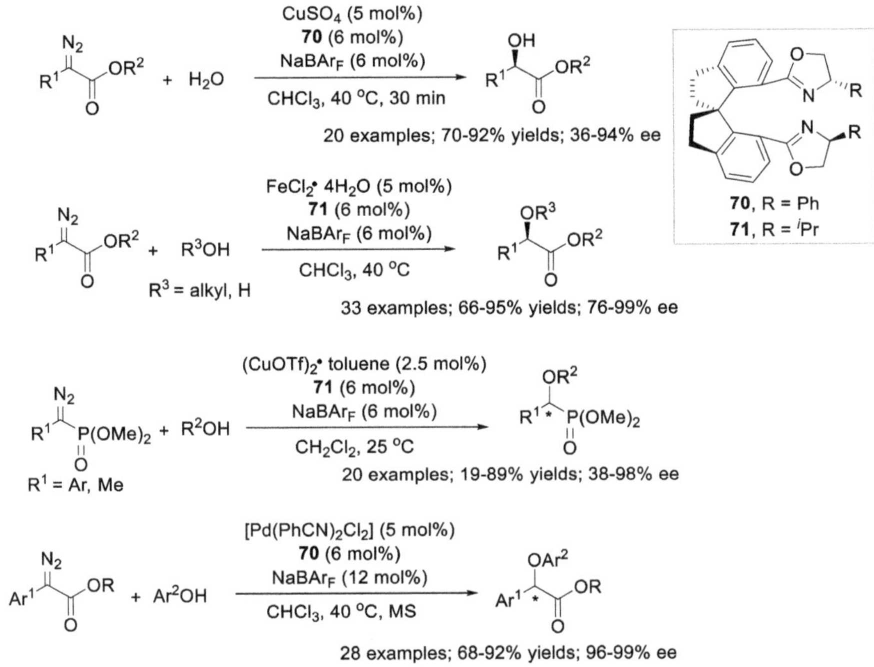

Scheme 5.27 Catalytic asymmetric intermolecular O–H insertion with chiral spiro-bisoxazoline ligands.

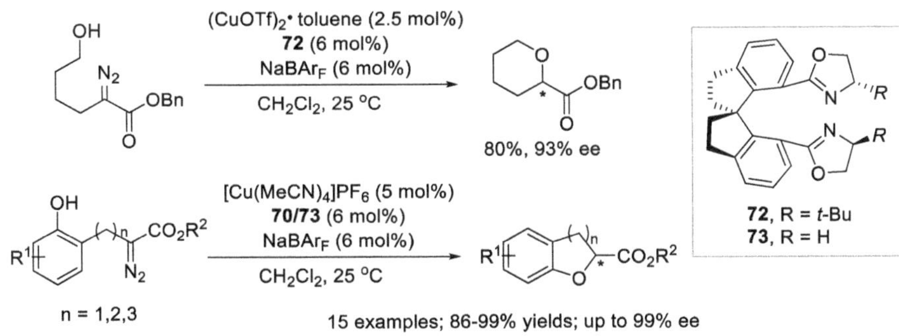

Scheme 5.28 Catalytic asymmetric intramolecular O–H insertions.

reported by Barluenga and Valdés (Scheme 5.29).[260] The reaction can also be performed in one-pot and stepwise manner, starting from ketones or aldehydes.

Moreover, the same group reported the reductive esterification from *N*-tosylhydrazones with carboxylic acids through O–H insertion of *in situ*

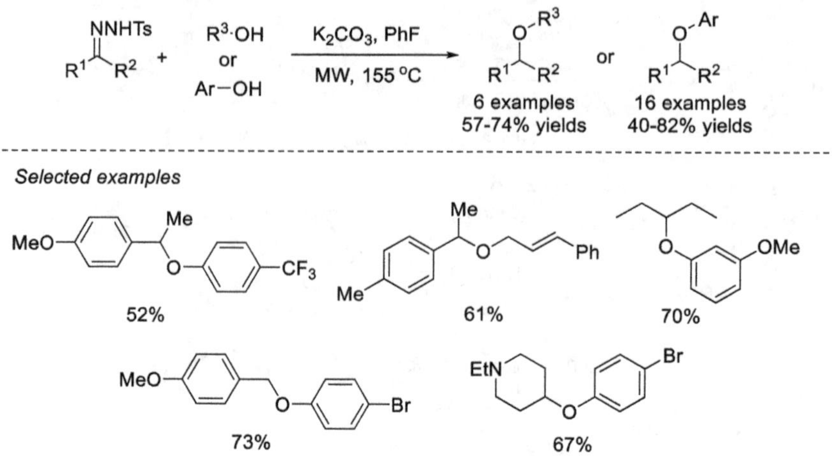

Scheme 5.29 Metal-free reductive coupling between *N*-tosylhydrazones and alcohols or phenols.

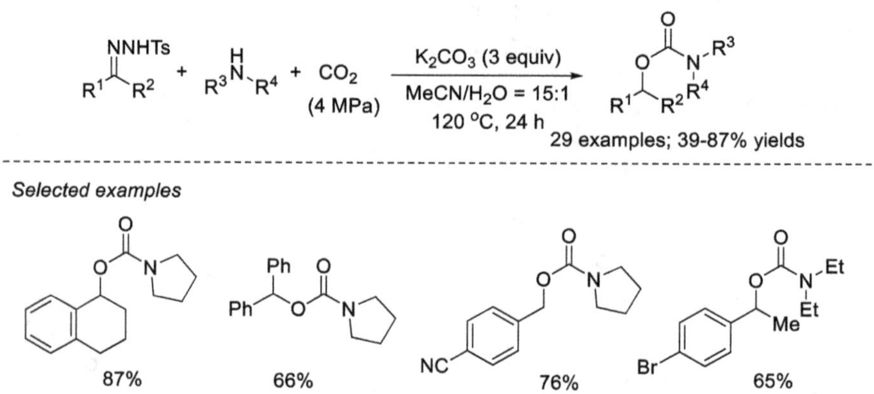

Scheme 5.30 Synthesis of carbamates from *N*-tosylhydrazones, CO_2 and amines.

generated diazo compounds.[261,262] The transition-metal-free reaction between *N*-tosylhydrazones and water was also reported, which was a unique approach for the synthesis of benzyl alcohols.[263]

Jiang and co-workers reported the formal O–H insertion of *N*-tosylhydrazones with CO_2 and amines.[264] The base-promoted three-component reaction is a straightforward synthetic approach toward a variety of alkyl carbamates (Scheme 5.30). The reaction pathway involves the initial thermal generation of diazo species in the presence of base. Then the protonated diazo substrate undergoes nucleophilic attack by carbamate anion, which is generated *in situ* from carbon dioxide with amines.

Finally, the mechanistic studies on carbene O–H insertion have made some progress in the past few years. DFT calculations for Cu(I)- or Rh(II)-catalyzed O–H insertions have been investigated by the groups of Yu and Fang.[265,266] Besides, with Cu(I)-catalyzed O–H bond carbene insertion with water, the Cu(I)-carbene intermediate and Cu-diazo adducts have been detected and characterized by using α-aryl-α-diazoacetate as the carbene precursor.[267]

5.2.4 *Formal X(halogen)–H insertion*

The reaction between α-diazocarbonyl compounds with hydrogen halide affords formal X–H insertion products. This convenient strategy paves the way to access α-halo carbonyl derivatives, which serve as synthetic precursors for biologically relevant compounds, including potent water-soluble tissue transglutaminase inhibitors,[268] transition-state analog inhibitors against carboxypeptidase A[269] and sequence independent peptidyl-fluoromethyl ketones.[270]

In 2013, Yadav reported efficient and easy access to fluoroalkanes by using diazo compounds generated *in situ* from *N*-tosylhydrazones.[271] A wide range of monofluoroalkanes was prepared by deoxygenative hydrofluorination of carbonyl compounds through *N*-tosylhydrazones (Scheme 5.31).

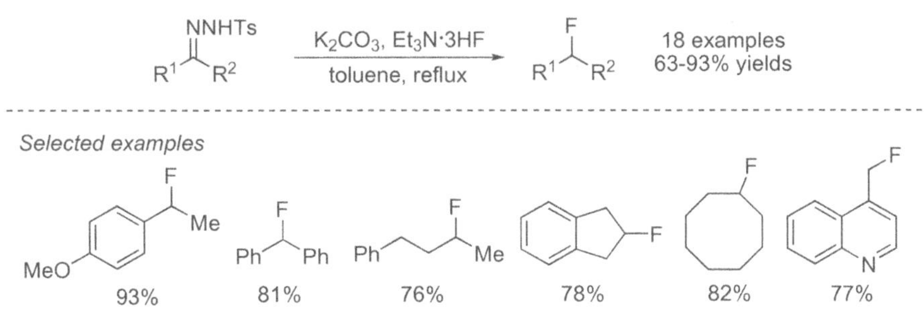

Scheme 5.31 Reductive fluorination of *N*-tosylhydrazones with $Et_3N{\cdot}3HF$.

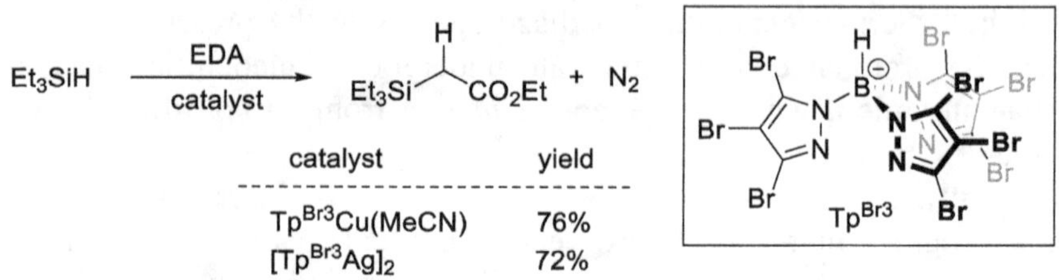

Scheme 5.32 Transition-metal-Tp^{Br3}-catalyzed insertion from EDA to silanes.

5.2.5 *Si–H insertion*

Si–H insertion is similar to C–H insertion in terms of reaction mechanism. Recent studies of catalytic carbene Si–H insertion have been primarily focused on the asymmetric construction of C–Si bond, as well as new catalytic systems and the expanding of substrates' scope. In addition to aforementioned utilization of Ir(III)-porphyrin complex for carbene insertion to C–H and Si–H bond,[272] Cu(I)- or Ag(I)-hydro-tris(3,4,5-tribromopyrazolyl)borate (Tp^{Br3}) complex was reported to catalyze the transfer of carbene units from EDA to Si–H bonds (Scheme 5.32).[273]

Bonge and Hansen explored a series of transformations by employing halodiazoacetates as the substrates.[274,275] They reported Rh(II)-catalyzed C–H and Si–H insertions from ethyl diazobromoacetate.[274] However, all the attempts of separating the Si–H insertion products failed due to their easy decomposition.

Catalytic enantioselective Si–H insertion from diazo compounds has attracted significant attention over the years. Moody and co-workers screened a series of chiral Rh(II) catalysts, which bear the chiral ligands derived from α-hydroxy acids and *N*-arenesulfonyl α-amino acids, for asymmetric Si–H insertion. However, the enantioselectivity remains only low to moderate for all of the chiral Rh(II) catalysts.[276] In 2006, Corey and co-workers achieved a high level of enantioselectivity with diazo-2-cyclohexenone **75** by using a fluorinated proline-based chiral Rh(II) catalyst **77** (Eq. (5.18)).[277]

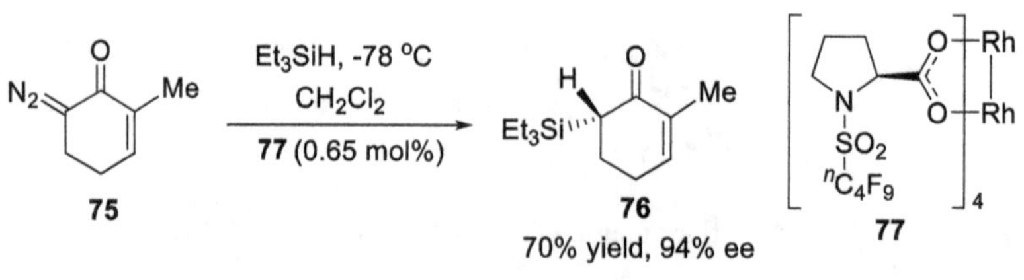

(5.18)

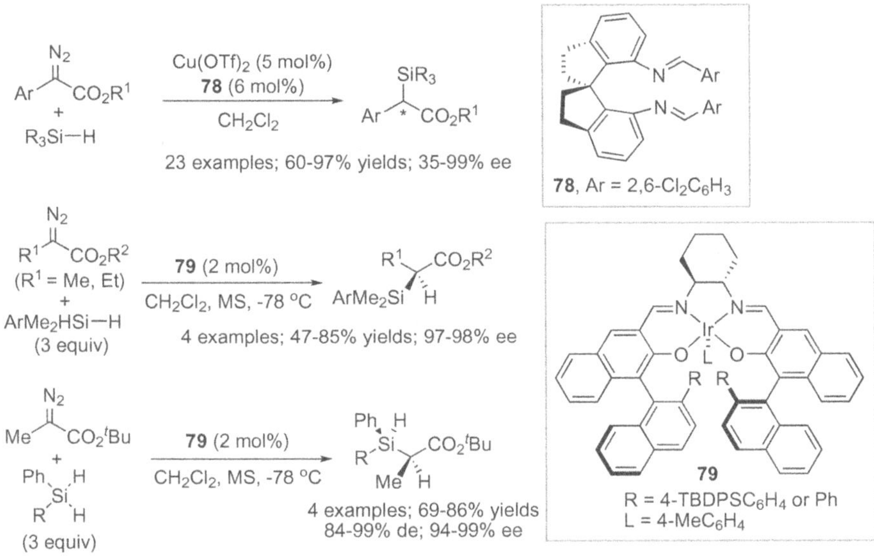

Scheme 5.33 Cu-spirodiimine and Ir-salen catalyzed asymmetric Si–H insertion.

In 2008, Zhou and co-workers reported a highly enantioselective Si–H insertion by using Cu(I) catalyst with chiral spirodiimine ligand **78**.[278] The reaction afforded a series of α-aryl-α-silylesters in high yields and excellent enantioselectivities. Subsequently, Katsuki reported the Ir(III)-salen complex catalyzed Si–H insertion with di- or tri-substituted silanes in high enantioselectivity (Scheme 5.33).[279]

Sambasivan and Ball explored asymmetric carbene insertion into Si–H bond using the Rh(II) catalysis with polypeptides as the chiral ligands, which can be quickly screened through a peptide library.[280] The same group also investigated the kinetic and steric effects of phosphite ligands, which act as axial ligands for Rh(II)-tetracarboxylate complexes in the Si–H insertions.[281] Lacour and co-workers synthesized and characterized Rh(II)-tetrakis(binaphthyl or octahydrobinaphthyl phosphate) complexes, which demonstrated high reactivity both in enantioselective Si–H insertion and cyclopropanation.[282] Finally, Panek and co-workers reported that Cu(I)- or Rh(II)-catalyzed Si–H insertion of vinyl diazoesters proceeded with a high *ee* value, which served as building blocks for vinylogous aldol reactions.[283]

5.2.6 *P–H insertion*

Insertion of carbene into P–H bond has been explored with diazo compounds or the corresponding *N*-tosylhydrazone surrogates. In most cases, copper catalysts are effective to achieve these transformations. Zhu and co-workers reported a catalyst-controlled switchable reaction between phosphonate and α-diazoesters.[284] In the presence of Cu(I) catalyst, carbene P–H bond insertion occurred effectively to afford **80**. However, when DBU was applied as the organo-catalyst, *N*-phosphination products **81** were formed selectively (Scheme 5.34). The latter process involves the base-promoted nucleophilic attack from phosphonate to the terminal nitrogen atom on diazo moiety.

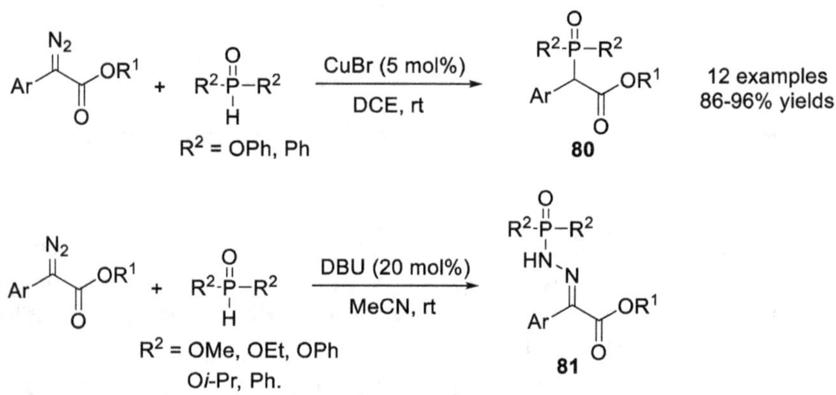

Scheme 5.34 Cu(I)-Catalyzed P–H insertion and DBU-catalyzed *N*-phosphination.

Cu(I)-catalyzed reductive coupling of *N*-tosylhydrazones with *H*-phosphorus oxides was reported by the groups of Zhao, Wu and Liang.[285–287] A wide range of *N*-tosylhydrazones derived from aldehydes or ketones were subjected to this P–H insertion reaction, in which C(sp^3)–P bonds were constructed in high efficiency. The synthetic method for phosphine oxides can also be carried out in one-pot stepwise procedure with ketones or aldehydes as the starting materials.

5.2.7 *S–H insertion*

The S–H insertion of α-diazocarbonyl compounds with thiols is a synthetically useful reaction that can introduce a sulfur-containing substituent adjacent to the carbonyl group of ketones or esters. One of the latest developments in S–H insertion of diazo compounds is the new catalytic system, especially with metalloporphyrin.[184,288] Another interesting development is the application of S–H insertion in chemical biology research. For example, Rh(II)-catalyzed S–H insertion has been applied to modify cysteine, affording alkylated cysteine through insertion of thiol groups with biotinylated diazo compounds.[289]

Intriguingly, myoglobin-based biocatalysts showed high activity (1100–5400 TON) in intermolecular S–H insertion for a wide range of aryl thiols with α-diazoesters.[290] Besides, the use of urea catalyst was reported as the essential acidity amplifier to promote S–H insertion of α-aryldiazoacetates.[241]

Catalytic asymmetric carbene S–H bond insertion has been a challenging topic over the last several years. Significant progress has been made in the past years by Zhou and co-workers.[291,292] They first attempted the asymmetric carbene insertion into S–H bonds by using Cu(I)/chiral spiro bisoxazoline catalysts (**70**). However, α-mercaptoesters were obtained with only moderate enantioselectivity.[291] Subsequently, they improved the reaction by using the collaborative catalytic system with achiral Rh(II) catalysts and chiral spiro phosphoric acid (**82**) (Scheme 5.35).[292] Mechanistic study indicates that the chiral spiro phosphoric acid plays crucial roles in proton transfer as protons shuttle to achieve chiral induction efficiently.

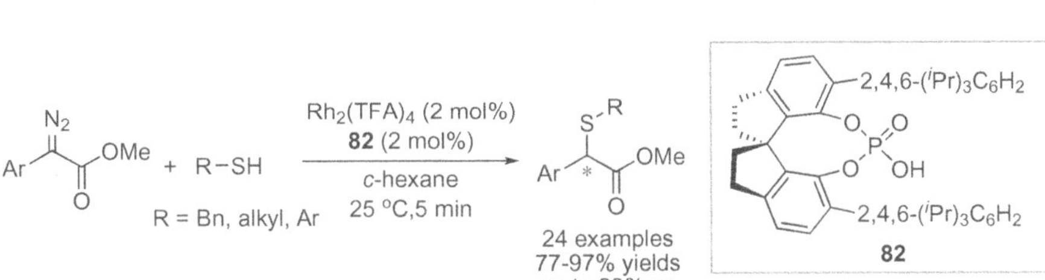

Scheme 5.35 Catalytic asymmetric carbene S–H insertions.

Similar to the case of carbene O–H insertions, the corresponding S–H insertion can also proceed with *N*-tosylhydrazones and thiols under transition-metal-free conditions. A wide range of sulfides could be easily obtained with this method (Eq. (5.19)).[293]

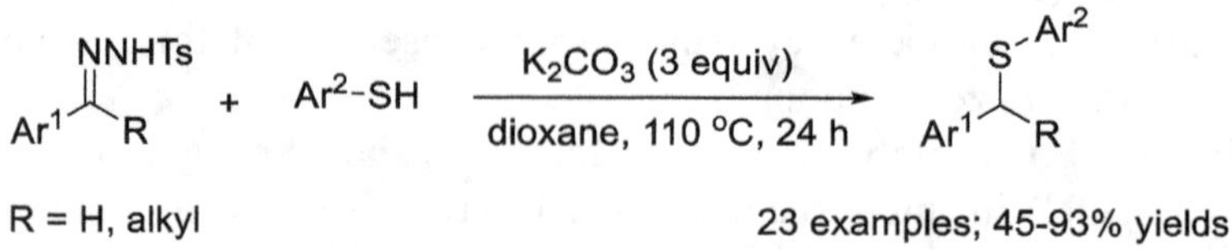

(5.19)

Besides, the transition-metal-free one-pot synthesis of alkyl dithiocarbamates was realized by the reaction of *N*-tosylhydrazones, CS_2 and amines,[294] similar to the previously described reaction with CO_2.[264]

5.3 *α,α*-Disubstitution

In addition to carbene insertion into X–H bonds, it is also possible to form two C–X bonds on the carbonic carbon of diazo compounds, the so-called *α,α*-disubstitution. The *α,α*-disubstitution with diazo compounds provides a straightforward entry to *gem*-dihalogenation of carbonyl substrates. In 2013, Murphy and co-workers reported pyridine-catalyzed dichlorination and BF_3-catalyzed difluorination of *α*-aryl-*α*-diazoesters.[295] The halogenation reagent iodanes can be easily prepared in gram-scale from readily available materials, and this method has the potential as a practical method for synthesizing *α,α*-dihalocarbonyl derivatives. Recently, the same group further demonstrated facile and efficient *α,α*-dichlorination of oxindoles[296] and isatin-3-*p*-tosylhydrazones,[297] as well as cyclic dicarbonyls derivatives (Scheme 5.36).[298]

Moreover, Gouverneur and co-workers explored a series of H–X insertions and *α,α*-disubstitution of 1-(diazo-2,2,2-trifluoroethyl)arenes.[299] A variety of perfluoroalkyl-substituted arenes were prepared from these donor/acceptor diazo compounds through formal insertion of H-F, F/Br, F/F, H-CF_3 and H-SCF_3 sources. The method was applied for the preparation of [^{18}F] labeling compounds.

In 2015, Cu(II)-catalyzed *gem*-bromochlorination of *α*-diazocarbonyl compounds was reported by using tetrabutylammonium dichlorobromide under mild reaction conditions.[300] Furthermore, a new approach toward

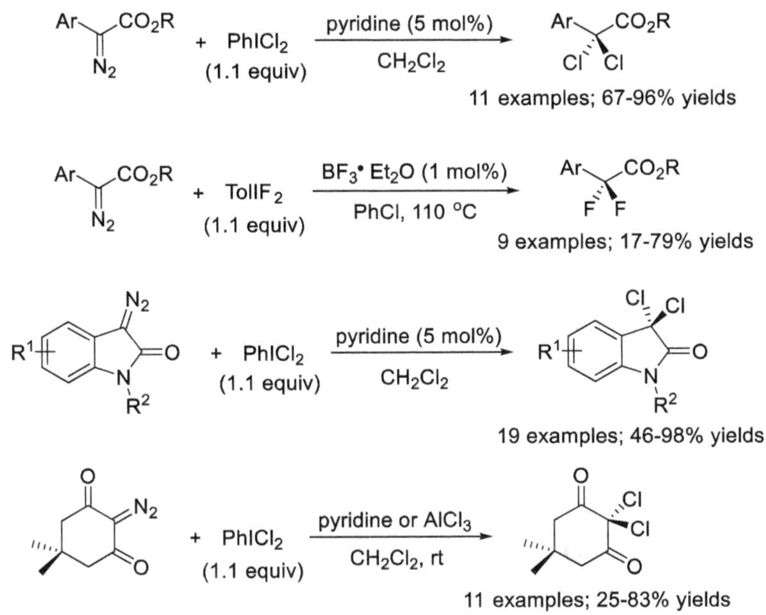

Scheme 5.36 α,α-Dihalogenation of α-diazoesters and diazo oxindoles.

o-fluorophenols was realized by the treatment of α-diazocyclohexenones **83** with electrophiles and fluorides (Eq. (5.20)).[301] Reaction pathway involves initial α,α-disubstitution to form **84**, followed by elimination and tautomerization to afford the fluorinated aromatic products **85**. This method has also been used for preparing [^{18}F] labeling compounds.

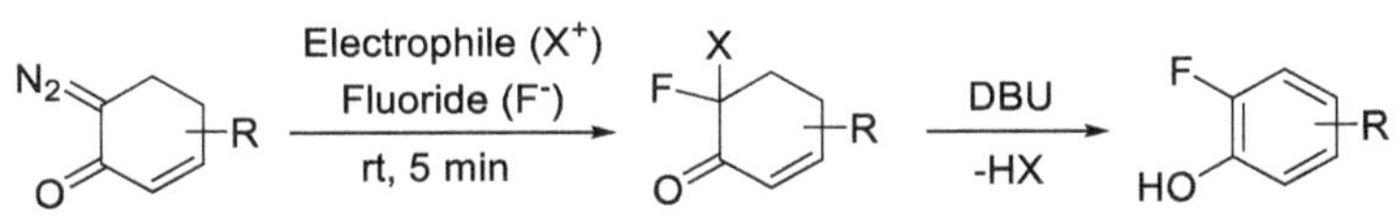

83 **84** **85**

3 examples; 60-82% yields

(5.20)

In addition to α,α-dihalogenation, the procedure can be extended to the substitution of different groups. A recent example is the transition-metal-free formal N–F insertion of α-aryl-α-diazo esters with *N*-fluorobenzenesulfonimide (NFSI), affording *gem*-aminofluorination products **86** (Eq. (5.21)).[302] Mechanistic studies indicate that the reaction is initiated by the nucleophilic attack of diazo carbon on electrophilic fluorine atom, followed by the substitution with nitrogen nucleophile.

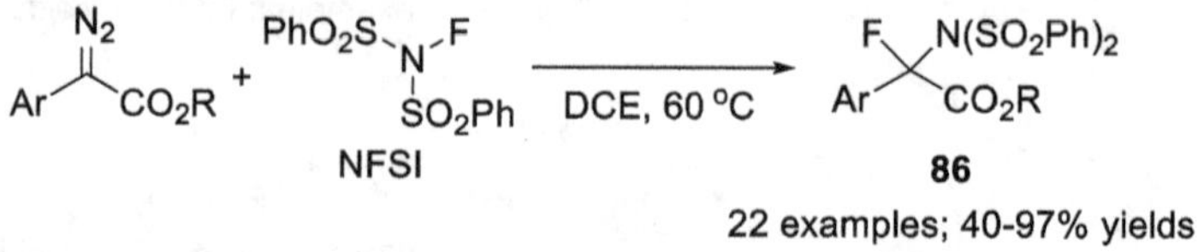

(5.21)

It is also worth mentioning that α,α-disubstitution of diazo compounds can also be extended to two C–O bond formations. For example, Taylor and co-workers reported an intramolecular insertion of carbenes into the peroxide O–O bond in the presence of Rh(II) catalyst.[303]

5.4 Formal C(sp^2)–H insertions

In most cases, transition-metal-catalyzed formal C–H insertion of aromatic compounds with diazo compounds took place through metal–carbene mechanism. In the initial step, catalytic decomposition of diazo substrates generates metal carbene species **87**, which is readily attacked by aromatic π-electron to form a zwitterionic intermediate **88**.[304] Upon deprotonation and protonation in either intra- or intermolecular manner, the corresponding re-aromatized product is generated (Scheme 5.37).

Intramolecular formal C(sp^2)–H carbene insertion shows wide applications in synthesizing a variety of functional molecules. Moreover, Rh(II) complexes have been useful and efficient catalysts for this type of transformation. Various isoquinolinone derivatives could be obtained with high regio- and diastereoselectivity through formal aromatic C–H

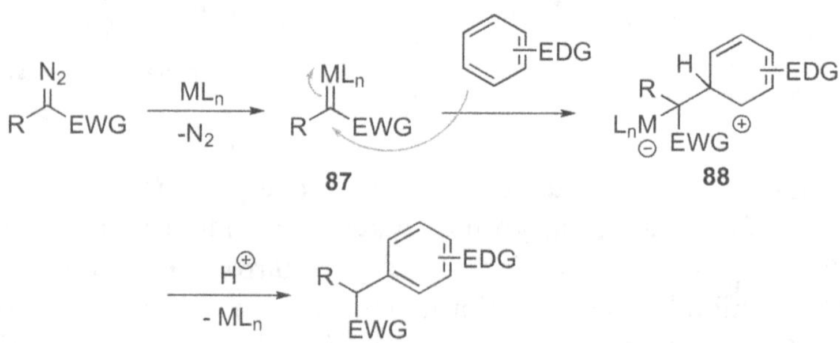

Scheme 5.37 General pathway of transition-metal-catalyzed formal C(sp^2)–H insertion.

insertion of Rh(II) carbenoid.[305] Reaction mechanism was proposed to involve electrophilic aromatic substitution.

Hu and co-workers reported a tunable aromatic C–H insertion of α-diazo-β-ketoester bearing aniline moiety.[306] Experimental results showed that the directing group on nitrogen played a crucial role on the chemoselectivity: sulfonamide substituents afforded dihydroquinolines **89** in high yields in the presence of Rh(II) catalyst; while alkyl substituents underwent Ag(I)-catalyzed Wolff rearrangement to afford amides **90** (Scheme 5.38).

Besides, Rh(II)-catalyzed chemoselective C–H insertion of α-diazo-β-ketoesters was applied to synthesize tetralins **91**[307] and 4-carbonyl chromans derivatives **92**[308] (Scheme 5.39).

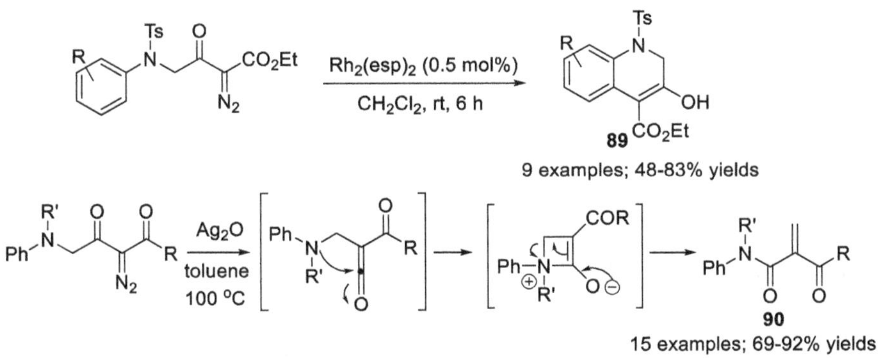

Scheme 5.38 Rh(II)-catalyzed aromatic C–H insertion and Ag(I)-catalyzed Wolff rearrangement.

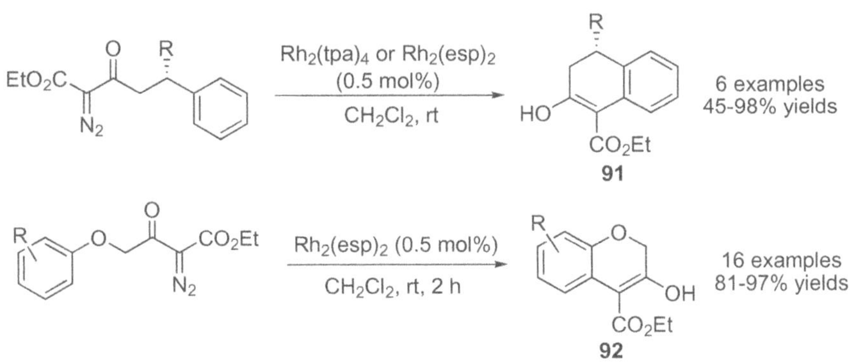

Scheme 5.39 Rh(II)-catalyzed aromatic C–H insertion of α-diazo-β-ketoesters.

Chang and co-workers reported the Rh(II)- or Cu(II)-catalyzed aromatic carbene C–H insertion of α-biaryl-α-diazoacetates **93**, providing a convenient and regioselective route to substituted fluorene carboxylates **94** (Eq. (5.22)).[309] Under thermal conditions, the selectivity of two regio-isomeric products was not satisfactory when using substituted biaryl diazoacetates as the substrates. In contrast, the reaction under catalytic conditions afforded excellent regioselectivity as well as good yields, presumably owing to steric hinders. Both the KIE and substituent effect experiment indicated that the reaction proceeded *via* intramolecular electrophilic aromatic substitution mechanism.

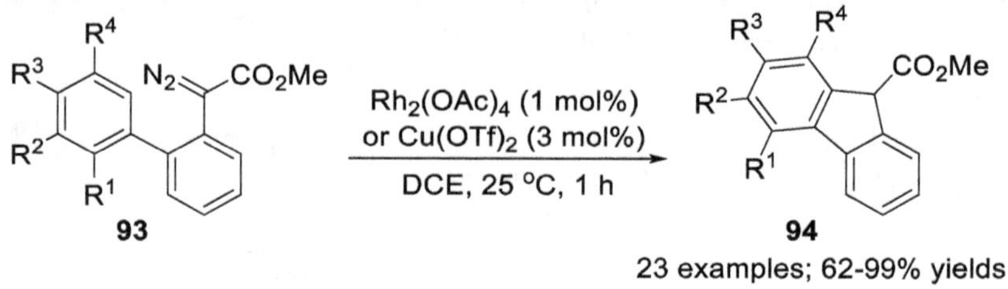

(5.22)

Apart from Rh(II) and Cu(II) complexes, Ag(I) salt has also been proven to be an efficient catalyst for promoting aromatic C–H functionalization/cyclization of *N*-aryl-α-diazoamides **95** (Eq. (5.23)).[310] The transformation provides a novel protocol for the preparation of various 3-alkylideneoxindoles with high yields.

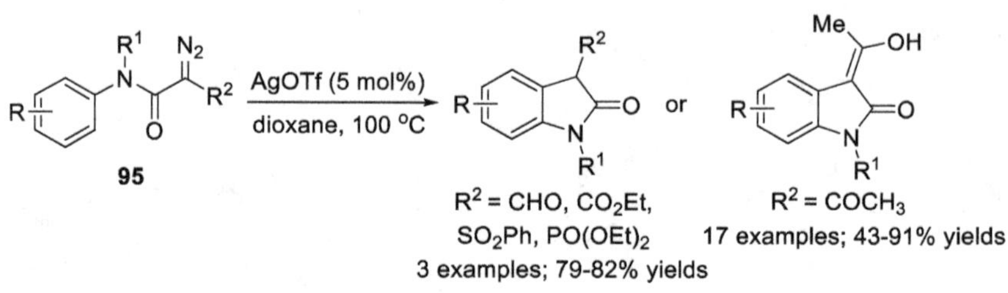

(5.23)

Besides, Pd(II)-catalyzed reaction of ethyl 2-diazo-4-(4-indolyl)-3-oxobutanoate **96** through formal carbene C–H insertion was realized to afford tetrahydroindole β-ketoester scaffold **97**, followed by sequential rearrangement to give thermodynamically more stable naphthalene derivative **98** (Scheme 5.40).[311] This reaction represents a rare case of Pd(II) carbene intramolecular C(sp^2)–H functionalization.

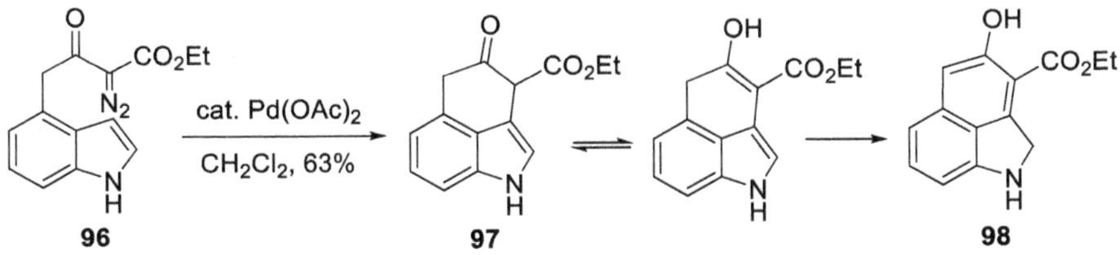

Scheme 5.40 Pd(II)-catalyzed intramolecular C–H functionalization.

Transition-metal-catalyzed intermolecular carbene insertion to C(sp^2)–H bond has also attracted attention. Typically, metal carbene C–H functionalization of indoles and pyrroles with diazo compounds provides direct alkylation strategy for constructing highly functionalized heterocycles. In these transformations, Rh(II) complexes demonstrate excellent reactivity for promoting selective indole or pyrrole C–H insertions with various diazo substrates, including macrocyclic diazocarbonyl compounds,[312] 3-diazopiperidin-2-one[313] and cyclic diazoamides or bis-diazoamides.[314]

Enantioselective and regioselective alkylation of indoles is a challenging topic. A catalytic asymmetric C–H functionalization of indoles with α-alkyl-α-diazoesters was developed by Fox group with Rh_2 (S-NTTL)$_4$.[315] C3-alkylated indole derivatives were obtained with high yields and excellent enantioselectivity (up to 99% *ee*) (Scheme 5.41). DFT calculations suggest that the reaction mechanism involves the generation of Rh(II)-ylide intermediate with oxocarbenium character. Another case of enantioselective C–H functionalization with α-diazopropionates was reported by Hashimoto and co-workers with chiral Rh(II) carboxylate catalysts.[316] They allowed the asymmetric synthesis of the (+)-α-methyl-3-indolylacetic acid fragment of acremoauxin A. Notably, asymmetric C–H functionalization of indoles has also been investigated by utilizing $Rh_2(OAc)_4$ with chiral phosphoric acid co-catalytic system *via* enantioselective protonation.[317]

In addition to Rh(II) complexes as the traditional catalysts in diazo transformations, the copper complexes have also been widely utilized in metal-catalyzed indole carbene C–H insertions with a wide range of diazo substrates, such as aryldiazoacetates,[318] dimethyl diazomalonate[319] and methyl 3,3,3-trifluoro-2-diazopropionate (Scheme 5.42).[320] Moreover, inorganic copper oxide on silica (CuO/SiO_2) has been developed as recoverable heterogeneous catalyst for direct alkylation of indole and pyrrole derivatives with aryldiazoacetate.[321]

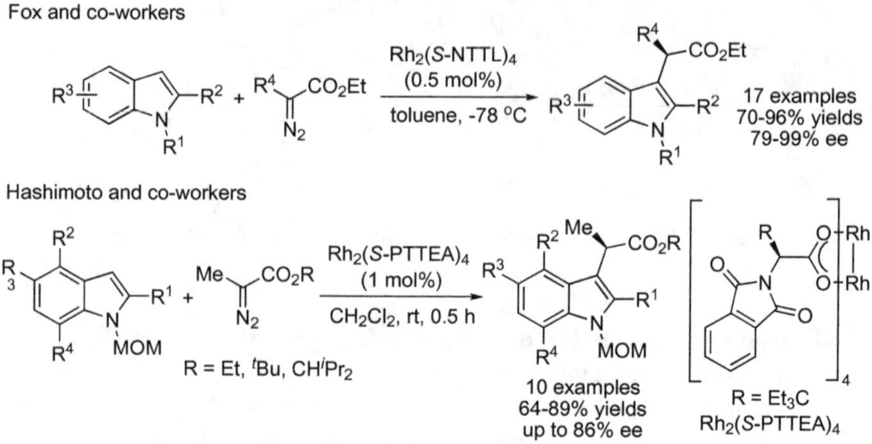

Scheme 5.41 Rh(II)-catalyzed enantioselective C–H functionalization of indoles.

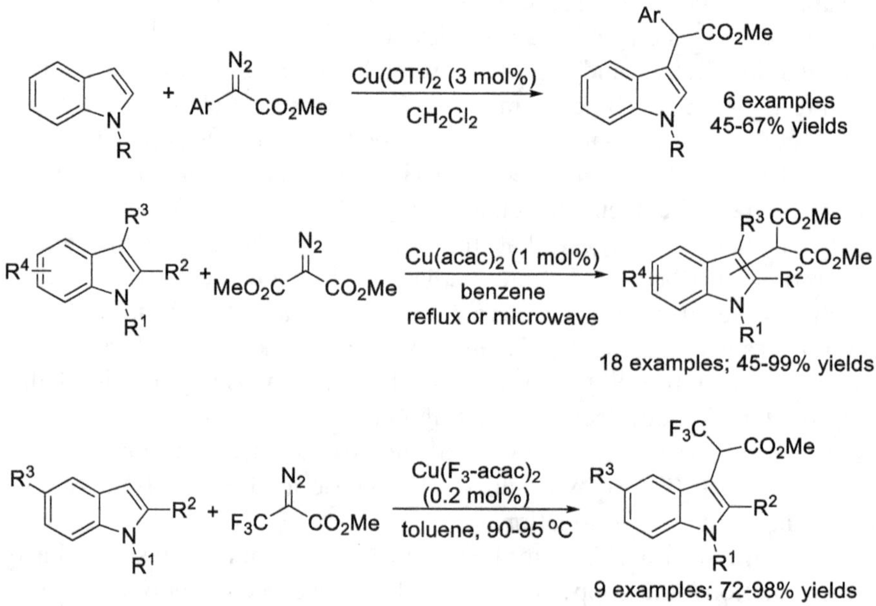

Scheme 5.42 Cu-catalyzed C–H functionalization of indoles with diazo compounds.

In addition to Rh(II) and Cu(I) complexes, easily available iron catalysts were explored by Zhou and co-workers to achieve C–H functionalization of indoles with α-aryl-α-diazoesters (Scheme 5.43).[322] Furthermore, the asymmetric version of this reaction was realized by

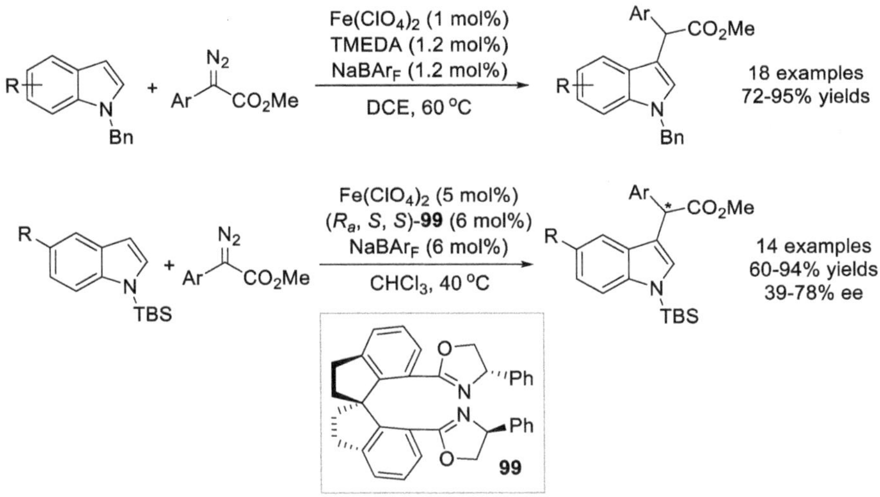

Scheme 5.43 Fe(II)-catalyzed asymmetric C–H functionalization of indoles.

using chiral spiro bisoxazoline ligands **99**, providing an efficient and mild method for the synthesis of chiral α-aryl-α-indolyl acetates with good enantioselectivity (up to 78% *ee*).

Recently, by using axially chiral 2,2'-bipyridine ligand **100**, Zhou and co-workers further reported the Pd(II)-catalyzed indole asymmetric C–H functionalization with α-aryl-α-diazoesters, which afforded indol-3-acetates with high level of enantiocontrol (up to 98% *ee*).[323] This catalytic system was also applicable for highly effective O–H insertion of phenol derivatives with up to 99% *ee* (Scheme 5.44).

Apart from indole and pyrrole derivatives, other electron-rich aromatic compounds were also subjected to formal carbene C–H insertion reactions. The Rh(II)-catalyzed C–H insertion reaction of 3-trifluoromethyl-4-diazopyrazolinones with anisoles or xylenes was reported, which afforded CF_3-substituted pyrazolols.[324] Furthermore, intermolecular aromatic substitution of *N,N*-disubstituted anilines with α-aryl-α-diazoesters was developed by Tayama and co-workers with Cu(II) triflate as the catalyst. This mild protocol allowed the regioselective C–H functionalization with high efficiency (up to 86% yield).[325] The same group expanded this strategy to alkoxybenzenes by using the Cu(II)-acid co-catalytic system (Scheme 5.45).[326]

Recently, the arylation of 5-diazobarbituric acids with arenes was realized with commercially available Rh(II) complex, which allowed the

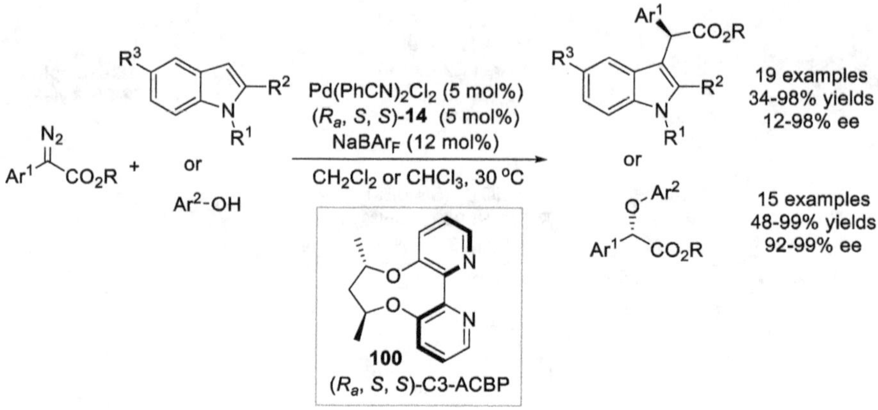

Scheme 5.44 Pd(II)-catalyzed asymmetric C–H functionalization and O–H insertion of indoles with diazoesters.

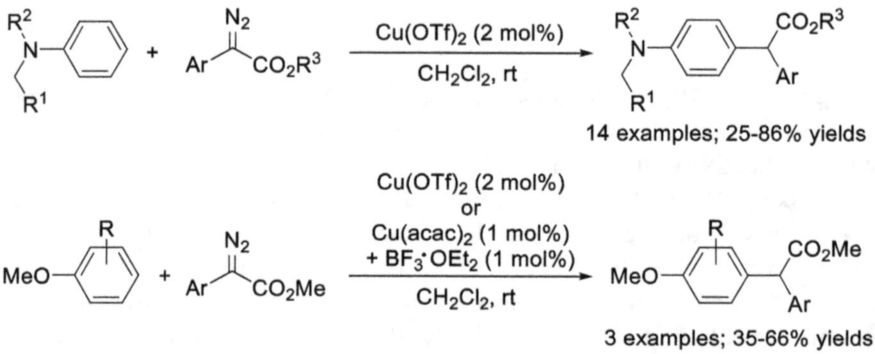

Scheme 5.45 Cu(II)-catalyzed formal C(sp^2)-H insertion of *N*,*N*-dialkylanilines and alkoxybenzenes.

synthesis of 5-aryl barbituric acids in a highly regioselective manner.[327] This mild and straightforward protocol demonstrated the application in the concise synthesis of potent matrix metalloproteinase inhibitors.

Zhou and co-workers realized the asymmetric arylation of diazo compounds with aniline derivatives cooperatively catalyzed by achiral dirhodium(II) carboxylate with chiral spiro phosphoric acid **101** (Scheme 5.46).[328] The high level of stereocontrol has been rationalized due to the chiral phosphoric acid-promoted proton shift of metal-free zwitterionic intermediate, which is produced through nucleophilic attack of anilines to Rh(II) carbene.

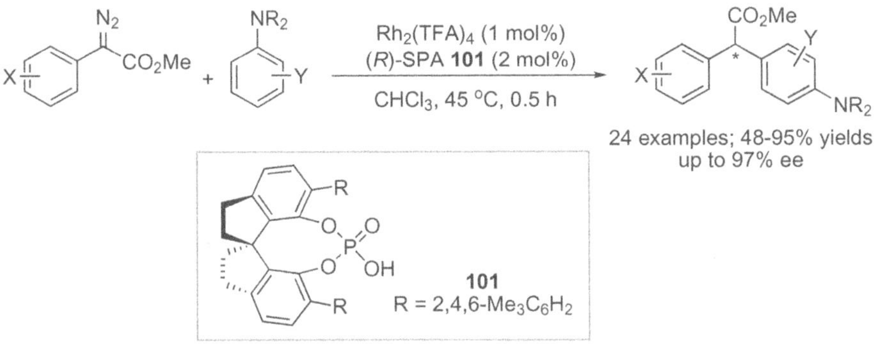

Scheme 5.46 Rh(II)-catalyzed asymmetric C(sp^2)–H functionalization of anilines with diazoesters.

Hu and co-workers reported a metal-free Brønsted acid-catalyzed C–H functionalization of electron-rich aromatic compounds with 3-diazooxindoles, affording 3-aryloxindoles as single regio-isomers (Eq. (5.24)).[329] Reaction pathway was proposed to proceed *via* acid-catalyzed protonation of diazo component into diazonium ions, followed by Friedel–Crafts alkylation of arenes.

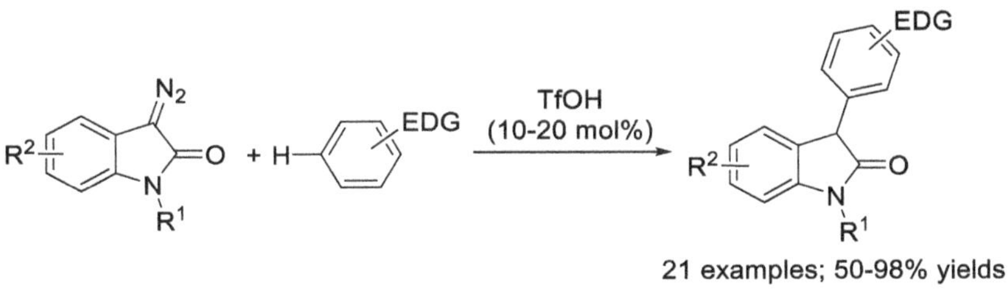

(5.24)

Recently, Au(I) complexes have been developed as efficient catalysts for selective direct C–H functionalization of arenes with diazo compounds. In this context, Pérez and co-workers reported the exclusive aromatic C–H bond insertion of isobutylbenzene with diazo compounds in the presence of gold catalyst with NHC ligand.[330] In 2014, Zhang and co-workers reported a formal C–H insertion of unprotected phenols with α-aryl-α-diazoacetates and diazooxindoles, furnishing functionalized diarylacetates with excellent yields and high chemo- and site-selectivity.[331] Moreover, this transformation could also be applied to phenols and anilines (Scheme 5.47). Notably, in this reaction formal C(sp^2)–H

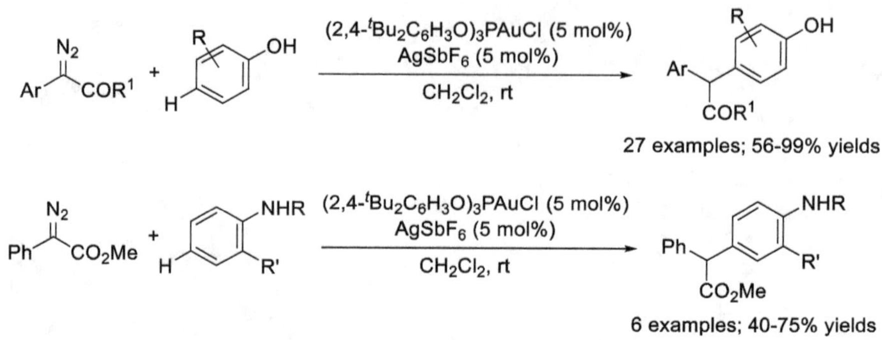

Scheme 5.47 Au(I)-catalyzed C–H functionalization of phenols and anilines with α-aryl-α-diazoacetates.

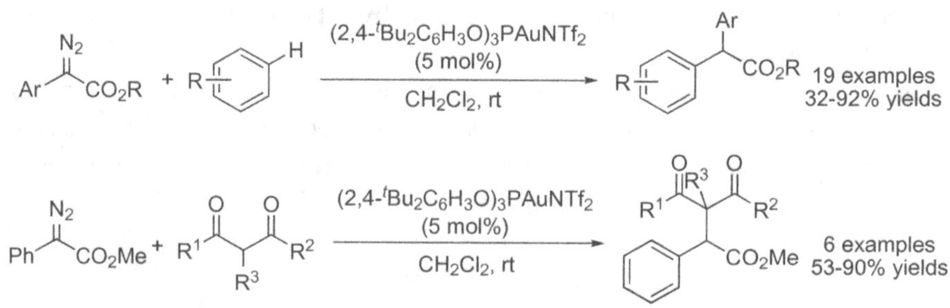

Scheme 5.48 Au(I)-catalyzed chemoselective carbophilic addition of arenes and 1,3-diketones with α-diazoesters.

functionalization is favored over competitive X–H insertion as well as Buchner ring expansion.

Almost simultaneously, Shi and co-workers demonstrated a series of chemoselective carbophilic additions to α-aryl-α-diazoesters with inert arenes and 1,3-diketones catalyzing by gold(I) catalyst with an electron-deficient phosphite as the necessary ligand (Scheme 5.48).[332] Both experimental and theoretical studies indicated that the reaction pathway involved Friedel–Crafts-type electrophilic aromatic substitution as the key step.

Zhang and co-workers carried out theoretical and experimental studies toward the mechanism of Au–carbene insertion to C–H or O–H bonds.[333] The result suggests that the ligand $(ArO)_3P$ significantly affected the chemoselectivity of C–H bond functionalization. Moreover, water was also found to play a crucial role as a proton shuttle in hydrogen transfer process from enol intermediate to the final products.

In summary, inter- and intramolecular insertions into X–H bond with metal carbenes provide important approaches for the construction of C–X bonds with high efficiency. Versatile functional molecules can be accessed from diazo substrates and their surrogates under transition-metal catalyzed conditions. Among the transition-metal catalysts that are effective for X–H insertions, Rh(II) complexes play the most important roles, particularly for C–H bond insertions. Thus, considerable efforts have been devoted to the design and synthesis of chiral Rh(II) complexes for accomplishing asymmetric C–H insertions with high enantioselectivities. Besides, metal carbene insertions to C–H bonds in combination with cascade process such as Cope rearrangement show potential applications in organic synthesis.

References

1. Davies, H. M. L.; Manning, J. R. *Nature* **2008**, *451*, 417–424.
2. Doyle, M. P.; Duffy, R.; Ratnikov, M.; Zhou, L. *Chem. Rev.* **2010**, *110*, 704–724.
3. Slattery, C. N.; Ford, A.; Maguire, A. R. *Tetrahedron* **2010**, *66*, 6681–6705.
4. Davies, H. M. L.; Morton, D. *Chem. Soc. Rev.* **2011**, *40*, 1857–1869.
5. Lu, H.; Zhang, X. P. *Chem. Soc. Rev.* **2011**, *40*, 1899–1909.
6. Che, C.-M.; Lo, V. K.-Y.; Zhou, C.-Y.; Huang, J.-S. *Chem. Soc. Rev.* **2011**, *40*, 1950–1975.
7. Doyle, M. P.; Ratnikov, M.; Liu, Y. *Org. Biomol. Chem.* **2011**, *9*, 4007–4016.
8. Zhu, S.-F.; Zhou, Q.-L. *Acc. Chem. Res.* **2012**, *45*, 1365–1377.
9. Boultadakis-Arapinis, M.; Lescot, C.; Micouin, L.; Lecourt, T. *Synlett* **2013**, *24*, 2477–2491.
10. Gillingham, D.; Fei, N. *Chem. Soc. Rev.* **2013**, *42*, 4918–4931.
11. Wang, J.-C.; Zhang, Y.; Xu, Z.-J.; Lo, V. K.-Y.; Che, C.-M. *ACS Catal.* **2013**, *3*, 1144–1148.
12. Egger, J.; Carreira, E. M. *Nat. Prod. Rep.* **2014**, *31*, 449–455.
13. Zheng, C.; You, S.-L. *RSC Adv.* **2014**, *4*, 6173–6214.
14. Lombard, F. J.; Coster, M. J. *Org. Biomol. Chem.* **2015**, *13*, 6419–6431.
15. Srikrishna, A.; Kumar, P. R. *Tetrahedron Lett.* **2002**, *43*, 1109–1111.
16. Spiegel, D. A.; Njardarson, J. T.; Wood, J. L. *Tetrahedron* **2002**, *58*, 6545–6554.
17. Hinman, A.; Du Bois, J. *J. Am. Chem. Soc.* **2003**, *125*, 11510–11511.

18. Kurosawa, W.; Kan, T.; Fukuyama, T. *J. Am. Chem. Soc.* **2003**, *125*, 8112–8113.
19. Kurosawa, W.; Kobayashi, H.; Kan, T.; Fukuyama, T. *Tetrahedron* **2004**, *60*, 9615–9628.
20. Chavan, S. P.; Thakkar, M.; Kharul, R. K.; Pathak, A. B.; Bhosekar, G. V.; Bhadbhade, M. M. *Tetrahedron* **2005**, *61*, 3873–3879.
21. Taber, D. F.; Frankowski, K. J. *J. Org. Chem.* **2005**, *70*, 6417–6421.
22. Tambar, U. K.; Ebner, D. C.; Stoltz, B. M. *J. Am. Chem. Soc.* **2006**, *128*, 11752–11753.
23. Davies, H. M. L.; Dai, X.; Long, M. S. *J. Am. Chem. Soc.* **2006**, *128*, 2485–2490.
24. Lall, M. S.; Ramtohul, Y. K.; James, M. N. G.; Vederas, J. C. *J. Org. Chem.* **2002**, *67*, 1536–1547.
25. Basak, A.; Mandal, S. *Tetrahedron Lett.* **2002**, *43*, 4241–4243.
26. Wee, A. G. H.; Duncan, S. C. *Tetrahedron Lett.* **2002**, *43*, 6173–6176.
27. Gois, P. M. P.; Afonso, C. A. M. *Eur. J. Org. Chem.* **2003**, 3798–3810.
28. Wee, A. G. H.; Duncan, S. C. *J. Org. Chem.* **2005**, *70*, 8372–8380.
29. Gomes, L. F. R.; Trindade, A. F.; Candeias, N. R.; Gois, P. M. P.; Afonso, C. A. M. *Tetrahedron Lett.* **2008**, *49*, 7372–7375.
30. Gomes, L. F. R.; Trindade, A. F.; Candeias, N. R.; Veiros, L. F.; Gois, P. M. P.; Afonso, C. A. M. *Synthesis* **2009**, 3519–3526.
31. Candeias, N. R.; Carias, C.; Gomes, L. F. R.; André, V.; Duarte, M. T.; Gois, P. M. P.; Afonso, C. A. M. *Adv. Synth. Catal.* **2012**, *354*, 2921–2927.
32. Lo, V. K.-Y.; Thu, H.-Y.; Chan, Y.-M.; Lam, T.-L.; Yu, W.-Y.; Che, C.-M. *Synlett* **2012**, *23*, 2753–2757.
33. Candeias, N. R.; Gois, P. M. P.; Veiros, L. F.; Afonso, C. A. M. *J. Org. Chem.* **2008**, *73*, 5926–5932.
34. Xu, X.; Deng, Y.; Yim, D. N.; Zavalij, P. Y.; Doyle, M. P. *Chem. Sci.* **2015**, *6*, 2196–2201.
35. Guptill, D. M.; Cohen, C. M.; Davies, H. M. L. *Org. Lett.* **2013**, *15*, 6120–6123.
36. Fu, L.; Wang, H.; Davies, H. M. L. *Org. Lett.* **2014**, *16*, 3036–3039.
37. Wamser, M.; Bach, T. *Synlett* **2014**, *25*, 1081–1084.
38. Padwa, A.; Zou, Y. *J. Org. Chem.* **2015**, *80*, 1802–1808.
39. Wang, J.; Stefane, B.; Jaber, D.; Smith, J. A. I.; Vickery, C.; Diop, M.; Sintim, H. O. *Angew. Chem. Int. Ed.* **2010**, *49*, 3964–3968.
40. Andrei, M.; Römming, C.; Undheim, K. *Tetrahedron: Asymm.* **2004**, *15*, 2711–2717.
41. Srikrishna, A.; Beeraiah, B.; Satyanarayana, G. *Tetrahedron: Asymm.* **2006**, *17*, 1544–1548.
42. Deng, G.; Tian, X.; Wang, J. *Tetrahedron Lett.* **2003**, *44*, 587–589.

43. Pattenden, G.; Blake, A. J.; Constandinos, L. *Tetrahedron Lett.* **2005**, *46*, 1913–1915.
44. Clark, J. S.; Dossetter, A. G.; Wong, Y.-S.; Townsend, R. J.; Whittingham, W. G.; Russell, C. A. *J. Org. Chem.* **2004**, *69*, 3886–3898.
45. Wardrop, D. J.; Forslund, R. E.; Landrie, C. L.; Velter, A. I.; Wink, D.; Surve, B. *Tetrahedron: Asymm.* **2003**, *14*, 929–940.
46. Berndt, D. F.; Norris, P. *Tetrahedron Lett.* **2002**, *43*, 3961–3962.
47. Yoon, C. H.; Nagle, A.; Chen, C.; Gandhi, D.; Jung, K. W. *Org. Lett.* **2003**, *5*, 2259–2262.
48. Chen, Z.; Chen, Z.; Jiang, Y.; Hua, W. *Tetrahedron* **2005**, *61*, 1579–1586.
49. Flanigan, D. L.; Yoon, C. H.; Jung, K. W. *Tetrahedron Lett.* **2005**, *46*, 143–146.
50. Grohmann, M.; Buck, S.; Schäffler, L.; Maas, G. *Adv. Synth. Catal.* **2006**, *348*, 2203–2211.
51. Swain, N. A.; Brown, R. C. D.; Bruton, G. *J. Org. Chem.* **2004**, *69*, 122–129.
52. García-Muñoz, S.; Álvarez-Corral, M.; Jiménez-González, L.; López-Sánchez, C.; Rosales, A.; Muñoz-Dorado, M.; Rodríguez-García, I. *Tetrahedron* **2006**, *62*, 12182–12190.
53. Skerry, P. S.; Swain, N. A.; Harrowven, D. C.; Smyth, D.; Bruton, G.; Brown, R. C. D. *Chem. Commun.* **2004**, 1772–1773.
54. Gauthier, D.; Dodd, R. H.; Dauban, P. *Tetrahedron* **2009**, *65*, 8542–8555.
55. Wang, J.; Boyarskikh, V.; Rainier, J. D. *Org. Lett.* **2011**, *13*, 700–702.
56. Deasy, R. E.; O'Riordan, N.; Maguire, A. R. *Catalysts* **2014**, *4*, 186–195.
57. Yang, Z.; Xu, J. *Chem. Commun.* **2014**, *50*, 3616–3618.
58. Budny, M.; Nowak, M.; Wojtczak, A.; Wolan, A. *Eur. J. Org. Chem.* **2014**, 6361–6365.
59. Boultadakis-Arapinis, M.; Lemoine, P.; Turcaud, S.; Micouin, L.; Lecourt, T. *J. Am. Chem. Soc.* **2010**, *132*, 15477–15479.
60. Boultadakis-Arapinis, M.; Lescot, C.; Micouin, L.; Lecourt, T. *Synthesis* **2012**, *44*, 3731–3734.
61. Boultadakis-Arapinis, M.; Prost, E.; Gandon, V.; Lemoine, P.; Turcaud, S.; Micouin, L.; Lecourt, T. *Chem. Eur. J.* **2013**, *19*, 6052–6066.
62. Boultadakis-Arapinis, M.; Gandon, V.; Prost, E.; Micouin, L.; Lecourt, T. *Adv. Synth. Catal.* **2014**, *356*, 2493–2505.
63. Lloyd, M. G.; Taylor, R. J. K.; Unsworth, W. P. *Org. Lett.* **2014**, *16*, 2772–2775.
64. Lloyd, M. G.; D'Acunto, M.; Taylor, R. J. K.; Unsworth, W. P. *Tetrahedron* **2015**, *71*, 7107–7123.
65. Cheung, W.-H.; Zheng, S.-L.; Yu, W.-Y.; Zhou, G.-C.; Che, C.-M. *Org. Lett.* **2003**, *5*, 2535–2538.
66. Choi, M. K.-W.; Yu, W.-Y.; Che, C.-M. *Org. Lett.* **2005**, *7*, 1081–1084.

67. Bhunia, S.; Ghorpade, S.; Huple, D. B.; Liu, R.-S. *Angew. Chem. Int. Ed.* **2012**, *51*, 2939–2942.
68. López-Sánchez, C.; Álvarez-Corral, M.; Muñoz-Dorado, M.; Rodríguez-García, I. *Synlett* **2012**, *23*, 2469–2472.
69. Lo, V. K.-Y.; Guo, Z.; Choi, M. K.-W.; Yu, W.-Y.; Huang, J.-S.; Che, C.-M. *J. Am. Chem. Soc.* **2012**, *134*, 7588–7591.
70. Reddy, A. R.; Zhou, C.-Y.; Guo, Z.; Wei, J.; Che, C.-M. *Angew. Chem. Int. Ed.* **2014**, *53*, 14175–14180.
71. Davies, H. M. L.; Beckwith, R. E. J. *Chem. Rev.* **2003**, *103*, 2861–2904.
72. Doyle, M. P.; Hu, W.; Valenzuela, M. V. *J. Org. Chem.* **2002**, *67*, 2954–2959.
73. Doyle, M. P.; Yan, M.; Phillips, I. M.; Timmons, D. J. *Adv. Synth. Catal.* **2002**, *344*, 91–95.
74. Doyle, M. P.; Hu, W.; Wee, A. G. H.; Wang, Z.; Duncan, S. C. *Org. Lett.* **2003**, *5*, 407–410.
75. Doyle, M. P.; Wang, Y.; Ghorbani, P.; Bappert, E. *Org. Lett.* **2005**, *7*, 5035–5038.
76. Liu, W.-J.; Chen, Z.-L.; Chen, Z.-Y.; Hu, W.-H. *Tetrahedron: Asymm.* **2005**, *16*, 1693–1698.
77. Wee, A. G. H.; Duncan, S. C.; Fan, G.-J. *Tetrahedron: Asymm.* **2006**, *17*, 297–307.
78. Fan, G.-J.; Wang, Z.; Wee, A. G. H. *Chem. Commun.* **2006**, 3732–3734.
79. Bykowski, D.; Wu, K.-H.; Doyle, M. P. *J. Am. Chem. Soc.* **2006**, *128*, 16038–16039.
80. Davies, H. M. L.; Venkataramani, C. *Angew. Chem. Int. Ed.* **2002**, *41*, 2197–2199.
81. Davies, H. M. L.; Yang, J. *Adv. Synth. Catal.* **2003**, *345*, 1133–1138.
82. Davies, H. M. L.; Walji, A. M. *Org. Lett.* **2003**, *5*, 479–482.
83. Davies, H. M. L.; Jin, Q. *Tetrahedron: Asymm.* **2003**, *14*, 941–949.
84. Davies, H. M. L.; Beckwith, R. E. J.; Antoulinakis, E. G.; Jin, Q. *J. Org. Chem.* **2003**, *68*, 6126–6132.
85. Davies, H. M. L.; Venkataramani, C.; Hansen, T.; Hopper, D. W. *J. Am. Chem. Soc.* **2003**, *125*, 6462–6468.
86. Davies, H. M. L.; Beckwith, R. E. J. *J. Org. Chem.* **2004**, *69*, 9241–9247.
87. Davies, H. M. L.; Jin, Q. *J. Am. Chem. Soc.* **2004**, *126*, 10862–10863.
88. Davies, H. M. L.; Hopper, D. W.; Hansen, T.; Liu, Q.; Childers, S. R. *Bioorg. Med. Chem. Lett.* **2004**, *14*, 1799–1802.
89. Davies, H. M. L.; Jin, Q. *Org. Lett.* **2004**, *6*, 1769–1772.
90. Davies, H. M. L.; Yang, J.; Nikolai, J. *J. Organomet. Chem.* **2005**, *690*, 6111–6124.
91. Davies, H. M. L.; Ni, A. *Chem. Commun.* **2006**, 3110–3112.

92. Davies, H. M. L.; Yang, J.; Manning, J. R. *Tetrahedron: Asymm.* **2006**, *17*, 665–673.
93. Saito, H.; Oishi, H.; Kitagaki, S.; Nakamura, S.; Anada, M.; Hashimoto, S. *Org. Lett.* **2002**, *4*, 3887–3890.
94. Tsutsui, H.; Yamaguchi, Y.; Kitagaki, S.; Nakamura, S.; Anada, M.; Hashimoto, S. *Tetrahedron: Asymm.* **2003**, *14*, 817–821.
95. Chiu, P.; Zhang, X.; Ko, R. Y. Y. *Tetrahedron Lett.* **2004**, *45*, 1531–1534.
96. Minami, K.; Saito, H.; Tsutsui, H.; Nambu, H.; Anada, M.; Hashimoto, S. *Adv. Synth. Catal.* **2005**, *347*, 1483–1487.
97. Reddy, R. P.; Lee, G. H.; Davies, H. M. L. *Org. Lett.* **2006**, *8*, 3437–3440.
98. Takeda, K.; Oohara, T.; Anada, M.; Nambu, H.; Hashimoto, S. *Angew. Chem. Int. Ed.* **2010**, *49*, 6979–6983.
99. Soldi, C.; Lamb, K. N.; Squitieri, R. A.; González-López, M.; Di Maso, M. J.; Shaw, J. T. *J. Am. Chem. Soc.* **2014**, *136*, 15142–15145.
100. Flynn, C. J.; Elcoate, C. J.; Lawrence, S. E.; Maguire, A. R. *J. Am. Chem. Soc.* **2010**, *132*, 1184–1185.
101. Slattery, C. N.; Maguire, A. R. *Org. Biomol. Chem.* **2011**, *9*, 667–669.
102. Slattery, C. N.; Clarke, L.-A.; O'Neill, S.; Ring, A.; Ford, A.; Maguire, A. R. *Synlett* **2012**, *23*, 765–767.
103. Slattery, C. N.; Clarke, L.-A.; Ford, A.; Maguire, A. R. *Tetrahedron* **2013**, *69*, 1297–1301.
104. Slattery, C. N.; O'Keeffe, S.; Maguire, A. R. *Tetrahedron: Asymm.* **2013**, *24*, 1265–1275.
105. Slattery, C. N.; Maguire, A. R. *Tetrahedron Lett.* **2013**, *54*, 2799–2801.
106. Slattery, C. N.; Ford, A.; Eccles, K. S.; Lawrence, S. E.; Maguire, A. R. *Synlett* **2014**, *25*, 591–595.
107. Clarke, L.-A.; Ring, A.; Ford, A.; Sinha, A. S.; Lawrence, S. E.; Maguire, A. R. *Org. Biomol. Chem.* **2014**, *12*, 7612–7628.
108. Hashimoto, T.; Yamamoto, K.; Maruoka, K. *Chem. Commun.* **2014**, *50*, 3220–3223.
109. Cui, X.; Xu, X.; Jin, L.-M.; Wojtas, L.; Zhang, X. P. *Chem. Sci.* **2015**, *6*, 1219–1224.
110. Gois, P. M. P.; Afonso, C. A. M. *Tetrahedron Lett.* **2003**, *44*, 6571–6573.
111. Candeias, N. R.; Gois, P. M. P.; Afonso, C. A. M. *Chem. Commun.* **2005**, 391–393.
112. Candeias, N. R.; Gois, P. M. P.; Afonso, C. A. M. *J. Org. Chem.* **2006**, *71*, 5489–5497.
113. Guan, Z.; Zhang, L.-H.; Sinaÿ, P.; Zhang, Y. *J. Org. Chem.* **2012**, *77*, 8888–8895.
114. Taber, D. F.; Paquette, C. M.; Gu, P.; Tian, W. *J. Org. Chem.* **2013**, *78*, 9772–9780.

115. Rosales, A.; Rodríguez-García, I.; López-Sánchez, C.; Álvarez-Corral, M.; Muñoz-Dorado, M. *Tetrahedron* **2011**, *67*, 3071–3075.
116. Ito, M.; Kondo, Y.; Nambu, H.; Anada, M.; Takeda, K.; Hashimoto, S. *Tetrahedron Lett.* **2015**, *56*, 1397–1400.
117. John, J. P.; Novikov, A. V. *Org. Lett.* **2007**, *9*, 61–63.
118. Shi, W.; Zhang, B.; Zhang, J.; Liu, B.; Zhang, S.; Wang, J. *Org. Lett.* **2005**, *7*, 3103–3106.
119. Mbuvi, H. M.; Woo, L. K. *Organometallics* **2008**, *27*,637–645.
120. Varela-Álvarez, A.; Musaev, D. G. *Chem. Sci.* **2013**, *4*, 3758–3764.
121. Thu, H.-Y.; Tong, G. S.-M.; Huang, J.-S.; Chan, S. L.-F.; Deng, Q.-H.; Che, C.-M. *Angew. Chem. Int. Ed.* **2008**, *47*, 9747–9751.
122. Anding, B. J.; Brgoch, J.; Miller, G. J.; Woo, L. K. *Organometallics* **2012**, *31*, 5586–5590.
123. Díaz-Requejo, M. M.; Belderraín, T. R.; Nicasio, M. C.; Trofimenko, S.; Pérez, P. J. *J. Am. Chem. Soc.* **2002**, *124*, 896–897.
124. Caballero, A.; Díaz-Requejo, M. M.; Belderraín, T. R.; Nicasio, M. C.; Trofimenko, S.; Pérez, P. J. *J. Am. Chem. Soc.* **2003**, *125*, 1446–1447.
125. Caballero, A.; Díaz-Requejo, M. M.; Belderraín, T. R.; Nicasio, M. C.; Trofimenko, S.; Pérez, P. J. *Organometallics* **2003**, *22*, 4145–4150.
126. Dias, H. V. R.; Browning, R. G.; Richey, S. A.; Lovely, C. J. *Organometallics* **2004**, *23*, 1200–1202.
127. Urbano, J.; Belderraín, T. R.; Nicasio, M. C.; Trofimenko, S.; Díaz-Requejo, M. M.; Pérez, P. J. *Organometallics* **2005**, *24*, 1528–1532.
128. Braga, A. A. C.; Maseras, F.; Urbano, J.; Caballero, A.; Díaz-Requejo, M. M.; Pérez, P. J. *Organometallics* **2006**, *25*, 5292–5300.
129. Fructos, M. R.; De Frémont, P.; Nolan, S. P.; Díaz-Requejo, M. M.; Pérez, P. J. *Organometallics* **2006**, *25*, 2237–2241.
130. Lovely, C. J.; Flores, J. A.; Meng, X.; Dias, H. V. R. *Synlett* **2009**, 129–132.
131. Caballero, A.; Despagnet-Ayoub, E.; Díaz-Requejo, M. M.; Díaz-Rodríguez, A.; González-Núñez, M. E.; Mello, R.; Muñoz, B. K.; Ojo, W.-S.; Asensio, G.; Etienne, M.; Pérez, P. J. *Science* **2011**, *332*, 835–838.
132. Fuentes, M. Á.; Muñoz, B. K.; Jacob, K.; Vendier, L.; Caballero, A.; Etienne, M.; Pérez, P. J. *Chem. Eur. J.* **2013**, *19*, 1327–1334.
133. Fuentes, M. Á.; Olmos, A.; Muñoz, B. K.; Jacob, K.; González-Núñez, M. E.; Mello, R.; Asensio, G.; Caballero, A.; Etienne, M.; Pérez, P. J. *Chem. Eur. J.* **2014**, *20*, 11013–11018.
134. For developing and characterization of perfluorinated scorpionate ligand, see: Despagnet-Ayoub, E.; Jacob, K.; Vendier, L.; Etienne, M.; Álvarez, E.; Caballero, A.; Díaz-Requejo, M. M.; Pérez, P. J. *Organometallics* **2008**, *27*, 4779–4787.

135. For review of methane in organometallic chemistry, see: Cavaliere, V. N.; Mindiola, D. J. *Chem. Sci.* **2012**, *3*, 3356–3365.
136. Suematsu, H.; Katsuki, T. *J. Am. Chem. Soc.* **2009**, *131*, 14218–14219.
137. Owens, C. P.; Varela-Álvarez, A.; Boyarskikh, V.; Musaev, D. G.; Davies, H. M. L.; Blakey, S. B. *Chem. Sci.* **2013**, *4*, 2590–2596.
138. Hansen, J.; Autschbach, J.; Davies, H. M. L. *J. Org. Chem.* **2009**, *74*, 6555–6563.
139. Denton, J. R.; Davies, H. M. L. *Org. Lett.* **2009**, *11*, 787–790.
140. Nadeau, E.; Li, Z.; Morton, D.; Davies, H. M. L. *Synlett* **2009**, 151–154.
141. Ventura, D. L.; Li, Z.; Coleman, M. G.; Davies, H. M. L. *Tetrahedron* **2009**, *65*, 3052–3061.
142. Wang, H.; Li, G.; Engle, K. M.; Yu, J.-Q.; Davies, H. M. L. *J. Am. Chem. Soc.* **2013**, *135*, 6774–6777.
143. Qin, C.; Davies, H. M. L. *J. Am. Chem. Soc.* **2014**, *136*, 9792–9796.
144. Guptill, D. M.; Davies, H. M. L. *J. Am. Chem. Soc.* **2014**, *136*, 17718–17721.
145. Bess, E. N.; Guptill, D. M.; Davies, H. M. L.; Sigman, M. S. *Chem. Sci.* **2015**, *6*, 3057–3062.
146. For review of application of donor/acceptor-carbenoids, see: Davies, H. M. L.; Denton, J. R. *Chem. Soc. Rev.* **2009**, *38*, 3061–3071.
147. Hilt, G.; Galbiati, F. *Org. Lett.* **2006**, *8*, 2195–2198.
148. Davies, H. M. L.; Lian, Y. *Acc. Chem. Res.* **2012**, *45*, 923–935.
149. Manning, J. R.; Sexton, T.; Childers, S. R.; Davies, H. M. L. *Bioorg. Med. Chem. Lett.* **2009**, *19*, 58–61.
150. Nadeau, E.; Ventura, D. L.; Brekan, J. A.; Davies, H. M. L. *J. Org. Chem.* **2010**, *75*, 1927–1939.
151. Lian, Y.; Hardcastle, K. I.; Davies, H. M. L. *Angew. Chem. Int. Ed.* **2011**, *50*, 9370–9373.
152. Hansen, J. H.; Gregg, T. M; Ovalles, S. R.; Lian, Y.; Autschbach, J.; Davies, H. M. L. *J. Am. Chem. Soc.* **2011**, *133*, 5076–5085.
153. Lian, Y.; Davies, H. M. L. *J. Am. Chem. Soc.* **2011**, *133*, 11940–11943.
154. Li, X.; Curran, D. P. *J. Am. Chem. Soc.* **2013**, *135*, 12076–12081.
155. Cheng, Q.-Q.; Zhu, S.-F.; Zhang, Y.-Z.; Xie, X.-L.; Zhou, Q.-L. *J. Am. Chem. Soc.* **2013**, *135*, 14094–14097.
156. Chen, D.; Zhang, X.; Qi, W.-Y.; Xu, B.; Xu, M.-H. *J. Am. Chem. Soc.* **2015**, *137*, 5268–5271.
157. For recent review of chiral spiro ligands, see: Xie, J.; Zhou, Q.-L. *Acta Chim. Sinica* **2014**, *72*, 778–797.
158. Davis, F. A.; Fang, T.; Goswami, R. *Org. Lett.* **2002**, *4*, 1599–1602.
159. Davis, F. A.; Yang, B.; Deng, J. *J. Org. Chem.* **2003**, *68*, 5147–5152.
160. Davis, F. A.; Deng, J. *Tetrahedron* **2004**, *60*, 5111–5115.
161. Davis, F. A.; Wu, Y.; Xu, H.; Zhang, J. *Org. Lett.* **2004**, *6*, 4523–4525.

162. Burtoloso, A. C. B.; Correia, C. R. D. *Tetrahedron Lett.* **2004**, *45*, 3355–3358.
163. Jung, J.-C.; Avery, M. A. *Tetrahedron Lett.* **2006**, *47*, 7969–7972.
164. Williams, R. M.; Lee, B. H.; Miller, M. M.; Anderson, O. P. *J. Am. Chem. Soc.* **1989**, *111*, 1073–1081.
165. Lecerclé, D.; Gabillet, S.; Gomis, J.-M.; Taran, F. *Tetrahedron Lett.* **2008**, *49*, 2083–2087.
166. Smith, M. R.; Blake, A. J.; Hayes, C. J.; Stevens, M. F. G.; Moody, C. J. *J. Org. Chem.* **2009**, *74*, 9372–9380.
167. Davies, J. R.; Kane, P. D.; Moody, C. J. *J. Org. Chem.* **2005**, *70*, 7305–7316.
168. Palmer, F. N.; Lach, F.; Poriel, C.; Pepper, A. G.; Bagley, M. C.; Slawin, A. M. Z.; Moody, C. J. *Org. Biomol. Chem.* **2005**, *3*, 3805–3811.
169. Hughes, R. A.; Thompson, S. P.; Alcaraz, L.; Moody, C. J. *J. Am. Chem. Soc.* **2005**, *127*, 15644–15651.
170. Clapham, B.; Spanka, C.; Janda, K. D. *Org. Lett.* **2001**, *3*, 2173–2176.
171. Clapham, B.; Lee, S.-H.; Koch, G.; Zimmermann, J.; Janda, K. D. *Tetrahedron Lett.* **2002**, *43*, 5407–5410.
172. Bashford, K. E.; Cooper, A. L.; Kane, P. D.; Moody, C. J.; Muthusamy, S.; Swann, E. *J. Chem. Soc., Perkin Trans. 1* **2002**, 1672–1687.
173. Lee, S.-H.; Clapham, B.; Koch, G.; Zimmermann, J.; Janda, K. D. *J. Comb. Chem.* **2003**, *5*, 188–196.
174. Yamazaki, K.; Kondo, Y. *Chem. Commun.* **2002**, 210–211.
175. Yamazaki, K.; Nakamura, Y.; Kondo, Y. *J. Org. Chem.* **2003**, *68*, 6011–6019.
176. Nakamura, Y.; Ukita, T. *Org. Lett.* **2002**, *4*, 2317–2320.
177. Lee, S.-H.; Clapham, B.; Koch, G.; Zimmermann, J.; Janda, K. D. *Org. Lett.* **2003**, *5*, 511–514.
178. Matsushita, H.; Lee, S.-H.; Yoshida, K.; Clapham, B.; Koch, G.; Zimmermann, J.; Janda, K. D. *Org. Lett.* **2004**, *6*, 4627–4629.
179. Yamashita, M.; Lee, S.-H.; Koch, G.; Zimmermann, J.; Clapham, B.; Janda, K. D. *Tetrahedron Lett.* **2005**, *46*, 5495–5498.
180. Auvil, T. J.; So, S. S.; Mattson, A. E. *Angew. Chem. Int. Ed.* **2013**, *52*, 11317–11320.
181. Nickerson, D. M.; Angeles, V. V.; Auvil, T. J.; So, S. S.; Mattson, A. E. *Chem. Commun.* **2013**, *49*, 4289–4291.
182. Yang, T.; Zhuang, H.; Lin, X.; Xiang, J.-N.; Elliott, J. D.; Liu, L.; Ren, F. *Tetrahedron Lett.* **2013**, *54*, 4159–4163.
183. Kantam, M. L.; Laha, S.; Yadav, J.; Jha, S. *Tetrahedron Lett.* **2009**, *50*, 4467–4469.
184. Chan, K.-H.; Guan, X.; Lo, V. K.-Y.; Che, C.-M. *Angew. Chem. Int. Ed.* **2014**, *53*, 2982–2987.

185. Chanthamath, S.; Thongjareun, S.; Shibatomi, K.; Iwasa, S. *Tetrahedron Lett.* **2012**, *53*, 4862–4865.
186. Aviv, I.; Gross, Z. *Chem. Eur. J.* **2008**, *14*, 3995–4005.
187. Wang, Z. J.; Peck, N. E.; Renata, H.; Arnold, F. H. *Chem. Sci.* **2014**, *5*, 598–601.
188. Sreenilayam, G.; Fasan, R. *Chem. Commun.* **2015**, *51*, 1532–1534.
189. Anding, B. J.; Woo, L. K. *Organometallics* **2013**, *32*, 2599–2607.
190. Ramakrishna, K.; Murali, M.; Sivasankar, C. *Org. Lett.* **2015**, *17*, 3814–3817.
191. Xu, X.; Li, C.; Tao, Z.; Pan, Y. *Adv. Synth. Catal.* **2015**, *357*, 3341–3345.
192. Akbari, J.; Ebrahimi, A.; Heydari, A. *Tetrahedron Lett.* **2014**, *55*, 5417–5419.
193. Hansen, S. R.; Spangler, J. E.; Hansen, J. H.; Davies, H. M. L. *Org. Lett.* **2012**, *14*, 4626–4629.
194. Deng, Q.-H.; Xu, H.-W.; Yuen, A. W.-H.; Xu, Z.-J.; Che, C.-M. *Org. Lett.* **2008**, *10*, 1529–1532.
195. Tishinov, K.; Schmidt, K.; Häussinger, D.; Gillingham, D. G. *Angew. Chem. Int. Ed.* **2012**, *51*, 12000–12004.
196. Tishinov, K.; Fei, N.; Gillingham, D. *Chem. Sci.* **2013**, *4*, 4401–4406.
197. Fei, N.; Häussinger, D.; Blümli, S.; Laventie, B.-J.; Bizzini, L. D.; Zimmermann, K.; Jenal, U.; Gillingham, D. *Chem. Commun.* **2014**, *50*, 8499–8502.
198. Hou, Z.; Wang, J.; He, P.; Wang, J.; Qin, B.; Liu, X.; Lin, L.; Feng, X. *Angew. Chem. Int. Ed.* **2010**, *49*, 4763–4766.
199. Zhu, Y.; Liu, X.; Dong, S.; Zhou, Y.; Li, W.; Lin, L.; Feng, X. *Angew. Chem. Int. Ed.* **2014**, *53*, 1636–1640.
200. Liu, B.; Zhu, S.-F.; Zhang, W.; Chen, C.; Zhou, Q.-L. *J. Am. Chem. Soc.* **2007**, *129*, 5834–5835.
201. Zhu, S.-F.; Xu, B.; Wang, G.-P.; Zhou, Q.-L. *J. Am. Chem. Soc.* **2012**, *134*, 436–442.
202. Xu, B.; Zhu, S.-F.; Xie, X.-L.; Shen, J.-J.; Zhou, Q.-L. *Angew. Chem. Int. Ed.* **2011**, *50*, 11483–11486.
203. Xu, B.; Zhu, S.-F.; Zuo, X.-D.; Zhang, Z.-C.; Zhou, Q.-L. *Angew. Chem. Int. Ed.* **2014**, *53*, 3913–3916.
204. Ni, Y.; Guo, X.; Hu, W.; Liu, S. *Chin. J. Org. Chem.* **2014**, *34*, 107–111.
205. Kisan, H. K.; Sunoj, R. B. *J. Org. Chem.* **2015**, *80*, 2192–2197.
206. Guo, J.-X.; Zhou, T.; Xu, B.; Zhu, S.-F.; Zhou, Q.-L. *Chem. Sci.* **2016**, *7*, 1104–1108.
207. Liu, G.; Li, J.; Qiu, L.; Liu, L.; Xu, G.; Ma, B.; Sun, J. *Org. Biomol. Chem.* **2013**, *11*, 5998–6002.
208. Ding, D.; Lv, X.; Li, J.; Qiu, L.; Xu, G.; Sun, J. *Org. Biomol. Chem.* **2014**, *12*, 4084–4088.

209. Ding, D.; Lv, X.; Li, J.; Xu, G.; Ma, B.; Sun, J. *Chem. Asian J.* **2014**, *9*, 1539–1542.
210. Ding, D.; Liu, G.; Xu, G.; Li, J.; Wang, G.; Sun, J. *Org. Biomol. Chem.* **2014**, *12*, 2533–2537.
211. Pinho, V. D.; Burtoloso, A. C. B. *J. Org. Chem.* **2011**, *76*, 289–292.
212. Rosset, I. G.; Burtoloso, A. C. B. *J. Org. Chem.* **2013**, *78*, 9464–9470.
213. Yadav, J. S.; Reddy, B. V. S.; Rao, Y. G.; Narsaiah, A. V. *Chem. Lett.* **2008**, *37*, 348–349.
214. Peng, C.; Cheng, J.; Wang, J. *Adv. Synth. Catal.* **2008**, *350*, 2359–2364.
215. Zhu, Y.; Wang, S.; Wen, S.; Lu, P.; Wang, Y. *Tetrahedron Lett.* **2010**, *51*, 4763–4766.
216. Hong, D.; Zhu, Y.; Li, Y.; Lin, X.; Lu, P.; Wang, Y. *Org. Lett.* **2011**, *13*, 4668–4671.
217. Shi, B.; Lewis, W.; Campbell, I. B.; Moody, C. J. *Org. Lett.* **2009**, *11*, 3686–3688.
218. Honey, M. A.; Blake, A. J.; Campbell, I. B.; Judkins, B. D.; Moody, C. J. *Tetrahedron* **2009**, *65*, 8995–9001.
219. Su, Z.; Peng, L.; Melander, C. *Tetrahedron Lett.* **2012**, *53*, 1204–1206.
220. Pavlyuk, O.; Teller, H.; McMills, M. C. *Tetrahedron Lett.* **2009**, *50*, 2716–2718.
221. Taber, D. F.; Berry, J. F.; Martin, T. J. *J. Org. Chem.* **2008**, *73*, 9334–9339.
222. Shi, B.; Blake, A. J.; Campbell, I. B.; Judkins, B. D.; Moody, C. J. *Chem. Commun.* **2009**, 3291–3293.
223. Shi, B.; Blake, A. J.; Lewis, W.; Campbell, I. B.; Judkins, B. D.; Moody, C. J. *J. Org. Chem.* **2010**, *75*, 152–161.
224. Linder, J.; Garner, T. P.; Williams, H. E. L.; Searle, M. S.; Moody, C. J. *J. Am. Chem. Soc.* **2011**, *133*, 1044–1051.
225. Honey, M. A.; Pasceri, R.; Lewis, W.; Moody, C. J. *J. Org. Chem.* **2012**, *77*, 1396–1405.
226. Hamze, A.; Tréguier, B.; Brion, J.-D.; Alami, M. *Org. Biomol. Chem.* **2011**, *9*, 6200–6204.
227. Aziz, J.; Brion, J.-D.; Hamze, A.; Alami, M. *Adv. Synth. Catal.* **2013**, *355*, 2417–2429.
228. Yadav, A. K.; Yadav, L. D. S. *RSC Adv.* **2014**, *4*, 34764–34767.
229. Liao, M.; Dong, S.; Deng, G.; Wang, J. *Tetrahedron Lett.* **2006**, *47*, 4537–4540.
230. Cenini, S.; Cravotto, G.; Giovenzana, G. B.; Palmisano, G.; Penoni, A.; Tollari, S. *Tetrahedron Lett.* **2002**, *43*, 3637–3640.
231. Morilla, M. E.; Molina, M. J.; Díaz-Requejo, M. M.; Belderraín, T. R.; Nicasio, M. C.; Trofimenko, S.; Pérez, P. J. *Organometallics* **2003**, *22*, 2914–2918.

232. Morilla, M. E.; Díaz-Requejo, M. M.; Belderraín, T. R.; Nicasio, M. C.; Trofimenko, S.; Pérez, P. J. *Chem. Commun.* **2002**, 2998–2999.
233. Petursson, S. *Tetrahedron: Asymm.* **2009**, *20*, 887–891.
234. Krishna, P. R.; Prapurna, Y. L.; Alivelu, M. *Tetrahedron Lett.* **2011**, *52*, 3460–3462.
235. Dumitrescu, L.; Azzouzi-Zriba, K.; Bonnet-Delpon, D.; Crousse, B. *Org. Lett.* **2011**, *13*, 692–695.
236. Mix, K. A.; Raines, R. T. *Org. Lett.* **2015**, *17*, 2358–2361.
237. Debarge, S.; Balzarini, J.; Maguire, A. R. *J. Org. Chem.* **2011**, *76*, 105–126.
238. Hladezuk, I.; Chastagner, V.; Collins, S. G.; Plunkett, S. J.; Ford, A.; Debarge, S.; Maguire, A. R. *Tetrahedron* **2012**, *68*, 1894–1909.
239. Cass, S. M. F.; Reid, G. E.; Tepe, J. J. *Org. Biomol. Chem.* **2009**, *7*, 3291–3299.
240. Fisher, T. J.; Mattson, A. E. *Org. Lett.* **2014**, *16*, 5316–5319.
241. Couch, E. D.; Auvil, T. J.; Mattson, A. E. *Chem. Eur. J.* **2014**, *20*, 8283–8287.
242. Sugimura, T.; Nagai, T. *Chem. Lett.* **2008**, *37*, 286–287.
243. Im, C. Y.; Okuyama, T.; Sugimura, T. *Eur. J. Org. Chem.* **2008**, 285–294.
244. Maier, T. C.; Fu, G. C. *J. Am. Chem. Soc.* **2006**, *128*, 4594–4595.
245. Zhu, S.-F.; Chen, C.; Cai, Y.; Zhou, Q.-L. *Angew. Chem. Int. Ed.* **2008**, *47*, 932–934.
246. Zhu, S.-F.; Cai, Y.; Mao, H.-X.; Xie, J.-H.; Zhou, Q.-L. *Nat. Chem.* **2010**, *2*, 546–551.
247. Zhu, S.-F.; Chen, W.-Q.; Zhang, Q.-Q.; Mao, H.-X.; Zhou, Q.-L. *Synlett* **2011**, 919–922.
248. Xie, X.-L.; Zhu, S.-F.; Guo, J.-X.; Cai, Y.; Zhou, Q.-L. *Angew. Chem. Int. Ed.* **2014**, *53*, 2978–2981.
249. Zhu, S.-F.; Song, X.-G.; Li, Y.; Cai, Y.; Zhou, Q.-L. *J. Am. Chem. Soc.* **2010**, *132*, 16374–16376.
250. Song, X.-G.; Zhu, S.-F.; Xie, X.-L.; Zhou, Q.-L. *Angew. Chem. Int. Ed.* **2013**, *52*, 2555–2558.
251. Saito, H.; Iwai, R.; Uchiyama, T.; Miyake, M.; Miyairi, S. *Chem. Pharm. Bull.* **2010**, *58*, 872–874.
252. Osako, T.; Panichakul, D.; Uozumi, Y. *Org. Lett.* **2012**, *14*, 194–197.
253. Kitagaki, S.; Sugisaka, K.; Mukai, C. *Org. Biomol. Chem.* **2015**, *13*, 4833–4836.
254. Sosa, J. R.; Tudjarian, A. A.; Minehan, T. G. *Org. Lett.* **2008**, *10*, 5091–5094.
255. Austeri, M.; Rix, D.; Zeghida, W.; Lacour, J. *Org. Lett.* **2011**, *13*, 1394–1397.
256. Cai, Y.; Lu, Y.; Yu, C.; Lyu, H.; Miao, Z. *Org. Biomol. Chem.* **2013**, *11*, 5491–5499.

257. Davis, O. A.; Bull, J. A. *Angew. Chem. Int. Ed.* **2014**, *53*, 14230–14234.
258. Zhou, L.; Liu, Y.; Zhang, Y.; Wang, J. *Beilstein J. Org. Chem.* **2011**, *7*, 631–637.
259. Urabe, F; Miyamoto, S; Takahashi, K.; Ishihara, J.; Hatakeyama, S. *Org. Lett.* **2014**, *16*, 1004–1007.
260. Barluenga, J.; Tomás-Gamasa, M.; Aznar, F.; Valdés, C. *Angew. Chem. Int. Ed.* **2010**, *49*, 4993–4996.
261. García-Muñoz, A.-H.; Tomás-Gamasa, M.; Pérez-Aguilar, M. C.; Cuevas-Yañez, E.; Valdés, C. *Eur. J. Org. Chem.* **2012**, 3925–3928.
262. Zhou, A.; Wu, L.; Li, D.; Chen, Q.; Zhang, X.; Xia, W. *Chin. J. Chem.* **2012**, *30*, 1862–1866.
263. García-Muñoz, Á.; Ortega-Arizmendi, A. I.; García-Carrillo, M. A.; Díaz, E.; Gonzalez-Rivas, N.; Cuevas-Yañez, E. *Synthesis* **2012**, *44*, 2237–2242.
264. Xiong, W.; Qi, C.; He, H.; Ouyang, L.; Zhang, M.; Jiang, H. *Angew. Chem. Int. Ed.* **2015**, *54*, 3084–3087.
265. Liang, Y.; Zhou, H.; Yu, Z.-X. *J. Am. Chem. Soc.* **2009**, *131*, 17783–17785.
266. Xie, Z.-Z.; Liao, W.-J.; Cao, J.; Guo, L.-P.; Verpoort, F.; Fang, W. *Organometallics* **2014**, *33*, 2448–2456.
267. Pereira, A.; Champouret, Y.; Martín, C.; Álvarez, E.; Etienne, M.; Belderraín, T. R.; Pérez, P. J. *Chem. Eur. J.* **2015**, *21*, 9769–9775.
268. Griffin, M.; Mongeot, A.; Colligham, R.; Saint, R. E.; Jones, R. A.; Coutts, I. G. C.; Rathbone, D. L. *Bioorg. Med. Chem. Lett.* **2008**, *18*, 5559–5562.
269. Jin, J.-Y.; Wang, S.-F.; Xuan, W.; Sheng, J.-W.; Wang, S.-H.; Tian, G.-R. *Chin. J. Chem.* **2008**, *26*, 153–157.
270. Roiban, G.-D.; Matache, M; Hădade, N. D.; Funeriu, D. P. *Org. Biomol. Chem.* **2012**, *10*, 4516–4523.
271. Yadav, A. K.; Srivastava, V. P.; Yadav, L. D. S. *Chem. Commun.* **2013**, *49*, 2154–2156.
272. Wang, J.-C.; Xu, Z.-J.; Guo, Z.; Deng, Q.-H.; Zhou, C.-Y.; Wan, X.-L.; Che, C.-M. *Chem. Commun.* **2012**, *48*, 4299–4301.
273. Iglesias, M. J.; Nicasio, M. C.; Caballero, A.; Pérez, P. J. *Dalton Trans.* **2013**, *42*, 1191–1195.
274. Bonge, H. T.; Hansen, T. *Synthesis* **2009**, 91–96.
275. Bonge, H. T.; Hansen, T. *Eur. J. Org. Chem.* **2010**, 4355–4359.
276. Buck, R. T.; Coe, D. M.; Drysdale, M. J.; Ferris, L.; Haigh, D.; Moody, C. J.; Pearson, N. D.; Sanghera, J. B. *Tetrahedron: Asymm.* **2003**, *14*, 791–816.
277. Ge, M.; Corey, E. J. *Tetrahedron Lett.* **2006**, *47*, 2319–2321.
278. Zhang, Y.-Z.; Zhu, S.-F.; Wang, L.-X.; Zhou, Q.-L. *Angew. Chem. Int. Ed.* **2008**, *47*, 8496–8498.

279. Yasutomi, Y.; Suematsu, H.; Katsuki, T. *J. Am. Chem. Soc.* **2010**, *132*, 4510–4511.
280. Sambasivan, R.; Ball, Z. T. *J. Am. Chem. Soc.* **2010**, *132*, 9289–9291.
281. Zaykov, A. N.; Ball, Z. T. *Tetrahedron* **2011**, *67*, 4397–4401.
282. Hrdina, R.; Guénée, L.; Moraleda, D.; Lacour, J. *Organometallics* **2013**, *32*, 473–479.
283. Wu, J.; Chen, Y.; Panek, J. S. *Org. Lett.* **2010**, *12*, 2112–2115.
284. Jiang, H.; Jin, H.; Abdukader, A.; Lin, A.; Cheng, Y.; Zhu, C. *Org. Biomol. Chem.* **2013**, *11*, 3612–3615.
285. Miao, W.; Gao, Y.; Li, X.; Gao, Y.; Tang, G.; Zhao, Y. *Adv. Synth. Catal.* **2012**, *354*, 2659–2664.
286. Wu, L.; Zhang, X.; Chen, Q.-Q.; Zhou, A.-K. *Org. Biomol. Chem.* **2012**, *10*, 7859–7862.
287. Chen, Z.-S.; Zhou, Z.-Z.; Hua, H.-L.; Duan, X.-H.; Luo, J.-Y.; Wang, J.; Zhou, P.-X.; Liang, Y.-M. *Tetrahedron* **2013**, *69*, 1065–1068.
288. Le Maux, P.; Nicolas, I.; Chevance, S.; Simonneaux, G. *Tetrahedron* **2010**, *66*, 4462–4468.
289. Kundu, R.; Ball, Z. T. *Chem. Commun.* **2013**, *49*, 4166–4168.
290. Tyagi, V.; Bonn, R. B.; Fasan, R. *Chem. Sci.* **2015**, *6*, 2488–2494.
291. Zhang, Y.-Z.; Zhu, S.-F.; Cai, Y.; Mao, H.-X.; Zhou, Q.-L. *Chem. Commun.* **2009**, 5362–5364.
292. Xu, B.; Zhu, S.-F.; Zhang, Z.-C.; Yu, Z.-X.; Ma, Y.; Zhou, Q.-L. *Chem. Sci.* **2014**, *5*, 1442–1448.
293. Ding, Q.; Cao, B.; Yuan, J.; Liu, X.; Peng, Y. *Org. Biomol. Chem.* **2011**, *9*, 748–751.
294. Sha, Q.; Wei, Y.-Y. *Org. Biomol. Chem.* **2013**, *11*, 5615–5620.
295. Tao, J.; Tran, R.; Murphy, G. K. *J. Am. Chem. Soc.* **2013**, *135*, 16312–16315.
296. Murphy, G. K.; Abbas, F. Z.; Poulton, A. V. *Adv. Synth. Catal.* **2014**, *356*, 2919–2923.
297. Hepples, C.; Murphy, G. K. *Tetrahedron Lett.* **2015**, 56, 4971–4974.
298. Coffey, K. E.; Murphy, G. K. *Synlett* **2015**, *26*, 1003–1007.
299. Emer, E.; Twilton, J.; Tredwell, M.; Calderwood, S.; Collier, T. L.; Liégault, B.; Taillefer, M.; Gouverneur, V. *Org. Lett.* **2014**, *16*, 6004–6007.
300. Shariff, N.; Mathi, S.; RameshKumar, C.; Emmanuvel, L. *Tetrahedron Lett.* **2015**, *56*, 934–936.
301. Yasui, N.; Mayne, C. G.; Katzenellenbogen, J. A. *Org. Lett.* **2015**, *17*, 5540–5543.
302. Chen, G.; Song, J.; Yu, Y.; Luo, X.; Li, C.; Huang, X. *Chem. Sci.* **2016**, *7*, 1786–1790.
303. Zvarec, O.; Avery, T. D.; Taylor, D. K. *J. Org. Chem.* **2010**, *75*, 450–454.
304. For review of metal carbenoid mediated zwitterionic intermediate trapping process, see: Xing, D.; Hu, W. *Tetrahedron Lett.* **2014**, *55*, 777–783.

305. Park, C. P.; Nagle, A.; Yoon, C. H.; Chen, C.; Jung, K. W. *J. Org. Chem.* **2009**, *74*, 6231–6236.
306. Ma, B.; Chen, F.-L.; Xu, X.-Y.; Zhang, Y.-N.; Hu, L.-H. *Adv. Synth. Catal.* **2014**, *356*, 416–420.
307. Rodriguez-Cárdenas, E.; Sabala, R.; Romero-Ortega, M.; Ortiz, A.; Olivo, H. F. *Org. Lett.* **2012**, *14*, 238–240.
308. Zhang, X.; Lei, M.; Zhang, Y.-N.; Hu, L.-H. *Tetrahedron* **2014**, *70*, 3400–3406.
309. Kim, J.; Ohk, Y.; Park, S. H.; Jung, Y.; Chang, S. *Chem. Asian J.* **2011**, *6*, 2040–2047.
310. Wang, H.-L.; Li, Z.; Wang, G.-W.; Yang, S.-D. *Chem. Commun.* **2011**, *47*, 11336–11338.
311. Rosenberg, M. L.; Aasheim, J. H. F.; Trebbin, M.; Uggerud, E.; Hansen, T. *Tetrahedron Lett.* **2009**, *50*, 6506–6508.
312. Muthusamy, S.; Gnanaprakasam, B. *Tetrahedron Lett.* **2008**, *49*, 475–480.
313. Muthusamy, S.; Srinivasan, P. *Tetrahedron* **2009**, *65*, 1567–1573.
314. Muthusamy, S.; Karikalan, T.; Gunanathan, C.; Suresh, E. *Tetrahedron* **2012**, *68*, 1595–1605.
315. DeAngelis, A.; Shurtleff, V. W.; Dmitrenko, O.; Fox, J. M. *J. Am. Chem. Soc.* **2011**, *133*, 1650–1653.
316. Goto, T.; Natori, Y.; Takeda, K.; Nambu, H.; Hashimoto, S. *Tetrahedron: Asymm.* **2011**, *22*, 907–915.
317. Qiu, H.; Zhang, D.; Liu, S.; Qiu, L.; Zhou, J.; Qian, Y.; Zhai, C.; Hu, W. *Acta Chim. Sinica* **2012**, *70*, 2484–2488.
318. Zhang, X.-J.; Liu, S.-P.; Yan, M. *Chin. J. Chem.* **2008**, *26*, 716–720.
319. Johansen, M. B.; Kerr, M. A. *Org. Lett.* **2010**, *12*, 4956–4959.
320. Tsyshchuk, I. E.; Vorobyeva, D. V.; Peregudov, A. S.; Osipov, S. N. *Eur. J. Org. Chem.* **2014**, 2480–2486.
321. Fraile, J. M.; Le Jeune, K.; Mayoral, J. A.; Ravasio, N.; Zaccheria, F. *Org. Biomol. Chem.* **2013**, *11*, 4327–4332.
322. Cai, Y.; Zhu, S.-F.; Wang, G.-P.; Zhou, Q.-L. *Adv. Synth. Catal.* **2011**, *353*, 2939–2944.
323. Gao, X.; Wu, B.; Huang, W.-X.; Chen, M.-W.; Zhou, Y.-G. *Angew. Chem. Int. Ed.* **2015**, *54*, 11956–11960.
324. Fan, H.; Zhang, Z.; Li, X.; Zhao, J.; Gao, J.; Zhu, S. *Tetrahedron* **2013**, *69*, 1978–1986.
325. Tayama, E.; Yanaki, T.; Iwamoto, H.; Hasegawa, E. *Eur. J. Org. Chem.* **2010**, 6719–6721.
326. Tayama, E.; Ishikawa, M.; Iwamoto, H.; Hasegawa, E. *Tetrahedron Lett.* **2012**, *53*, 5159–5161.
327. Best, D.; Burns, D. J.; Lam, H. W. *Angew. Chem. Int. Ed.* **2015**, *54*, 7410–7413.

328. Xu, B.; Li, M.-L.; Zuo, X.-D.; Zhu, S.-F.; Zhou, Q.-L. *J. Am. Chem. Soc.* **2015**, *137*, 8700–8703.
329. Zhai, C.; Xing, D.; Jing, C.; Zhou, J.; Wang, C.; Wang, D.; Hu, W. *Org. Lett.* **2014**, *16*, 2934–2937.
330. Rivilla, I.; Gómez-Emeterio, B. P.; Fructos, M. R.; Díaz-Requejo, M. M.; Pérez, P. J. *Organometallics* **2011**, *30*, 2855–2860.
331. Yu, Z.; Ma, B.; Chen, M.; Wu, H.-H.; Liu, L.; Zhang, J. *J. Am. Chem. Soc.* **2014**, *136*, 6904–6907.
332. Xi, Y.; Su, Y.; Yu, Z.; Dong, B.; McClain, E. J.; Lan, Y.; Shi, X. *Angew. Chem. Int. Ed.* **2014**, *53*, 9817–9821.
333. Liu, Y.; Yu, Z.; Zhang, J. Z.; Liu, L.; Xia, F.; Zhang, J. *Chem. Sci.* **2016**, *7*, 1988–1995.

Chapter 6

Ylide Formation and Related Reactions

Owing to the electrophilic characters of the metal carbene derived from α-diazocarbonyl compounds, transition-metal-associated or free ylide intermediate can be generated through nucleophilic attack of different heteroatoms. The ylide species thus generated shows versatile activities and has found many applications in organic synthesis.

Mechanistically, the interaction of the electron-deficient carbenic carbon of the metal carbene intermediate with a pair of non-bonding electrons contributed by a Lewis base (B:) generates a metal complex-associated ylide **1** or a free ylide **2**. The ylide intermediate thus generated is usually highly reactive and undergoes subsequent reactions to give stable products (Scheme 6.1).

The common Lewis bases that are utilized to generate ylides include ethers, sulfides, amines, carbonyl compounds and imines. The typical reactions of the ylides include: (1) [2,3]-sigmatropic rearrangement of allylic, propargylic and allenic ylides; (2) 1,2-shift (Stevens rearrangement); (3) 1,3-dipolar cycloaddition of the ylides generated from carbonyl compounds or imines with dipolarophiles, usually C=C or C≡C bonds; (4) nucleophilic addition/elimination, leading to the formation of epoxides or cyclopropanes.

The ylide formation from a metal carbene and the subsequent reactions can occur in either inter- or intramolecular manner. With these cascade transformations, it is possible to rapidly assemble organic compounds with considerable complexity from relatively simple starting

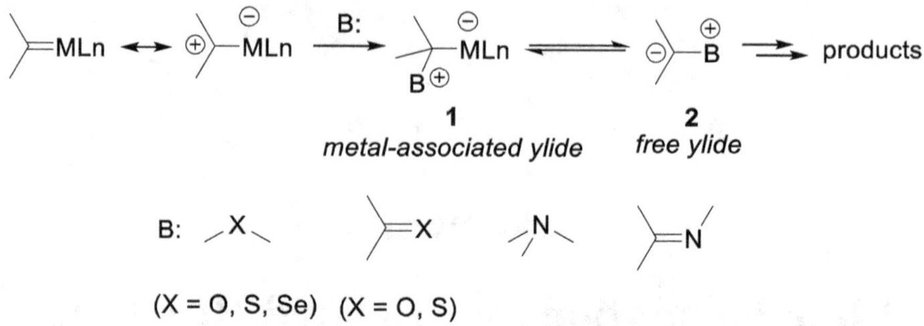

Scheme 6.1 General procedure of ylide-type reaction of diazo compounds.

materials. Besides, some of these reactions show excellent chemo-, regio- and stereoselectivity. Recent advances in asymmetric catalysis in this field add further merit to these transformations. Since this area has been summarized by several reviews,[1–13] in this chapter we will focus on the most recent advances in the past several years.

6.1 [2,3]-Sigmatropic rearrangement

Sulfide groups are typically easy to combine with metal carbenes to generate the corresponding sulfonium ylides. The subsequent reactions of the sulfonium ylides mainly fall into two classes: [2,3]-sigmatropic rearrangements and 1,2-shifts. In particular, [2,3]-σ rearrangement of ylides represents a unique bond reorganization method for synthesizing functional molecules. Thus, the [2,3]-σ rearrangement of sulfonium ylides derived from the transition-metal-catalyzed reaction of diazo compounds and allyl sulfides, known as Doyle–Kirmse reaction, has attracted considerable attention in recent years.

Crich and co-workers have used this rearrangement as a tool for amino-acid and peptide modification (Eq. (6.1)).[14] Notably, even though there are many potential reactive sites toward Rh(II) carbene in both the allyl sulfide and the diazo substrate, such as the N–H bonds and carbonyl groups, the sulfur ylide [2,3]-sigmatropic rearrangement product is still formed with reasonable yield.

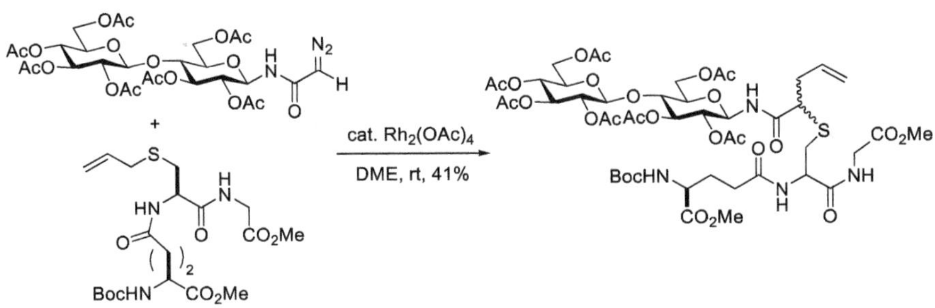

(6.1)

The Doyle–Kirmse reaction is usually performed in organic solvent under an inert atmosphere. However, Wang and co-workers demonstrated that [2,3]-sigmatropic rearrangement of sulfur ylides derived from Rh(II) carbene and sulfides could be efficiently carried out in water. No O–H insertion product could be identified.[15] This might be an example of an "on water" reaction.

The [2,3]-sigmatropic rearrangement of sulfonium ylides generated from metal carbene and propargyl sulfide affords allenyl sulfides, which are reactive substrates toward transition-metal catalysis, thus making cascade transformation possible. For example, Wang and co-workers have shown that the allenyl sulfides **3**, obtained from the [2,3]-sigmatropic rearrangement of sulfonium ylides generated from Rh(II) carbene and propargyl sulfide, can be further catalyzed by a Ru(II) catalyst to afford the furan derivatives through a 1,4 migration of the sulfonyl group and generation of Ru carbene intermediates (Scheme 6.2).[16]

Asymmetric catalysis in sulfonium ylide [2,3]-sigmatropic rearrangement reaction has progressed slowly in the past few years. High stereoselectivity has been achieved using a double asymmetric induction approach with diazo substrates bearing a chiral camphorsultam auxiliary, as shown in Scheme 6.3. With either chiral diimine ligand **4** or achiral ligand **5**, comparable enantioselectivities were obtained. Notably, the sense of the asymmetric induction is dictated by the chiral camphorsultam auxiliary, rather than by the chiral catalyst.[17]

In addition to the traditional Rh(II) and Cu(I) catalysts, other transition-metal-catalytic systems, including Ag(I)[18] and Fe(0) complex,[19] have been utilized in sulfur ylide [2,3]-σ rearrangements from substrates

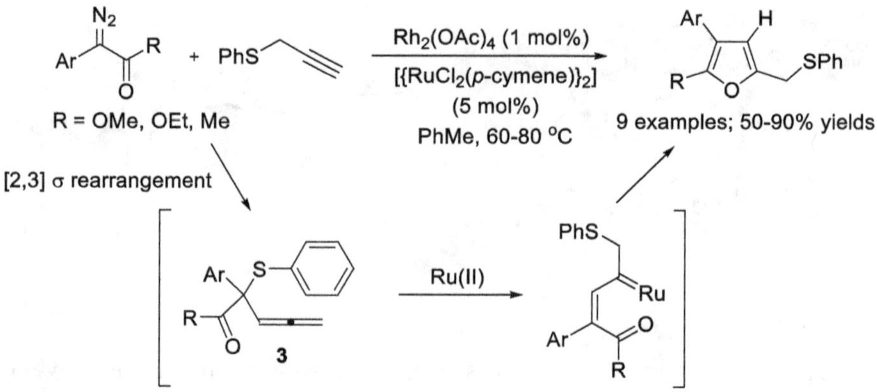

Scheme 6.2 Rh(II)-catalyzed sulfur ylide rearrangement and Ru(II)-catalyzed sequential cyclization.

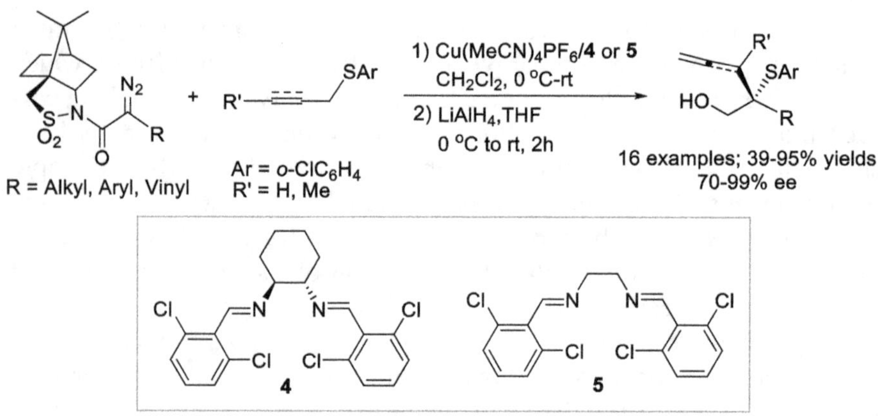

Scheme 6.3 Cu(I)-catalyzed sulfonium ylide [2,3]-sigmatropic rearrangement with asymmetric induction.

of α-aryl-α-diazoesters and EDA (Eqs. (6.2) and (6.3)). Besides, the *N*-tosylhydrazones as the precursors for non-stabilized diazo substrates have also been utilized in sulfur ylide [2,3]-σ rearrangements (Eq. (6.4)).[20]

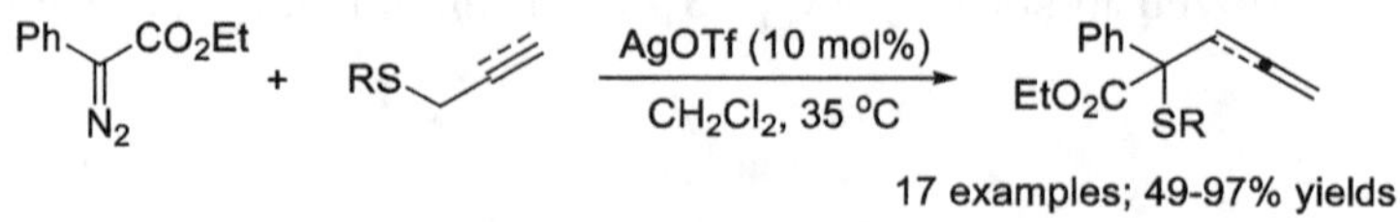 (6.2)

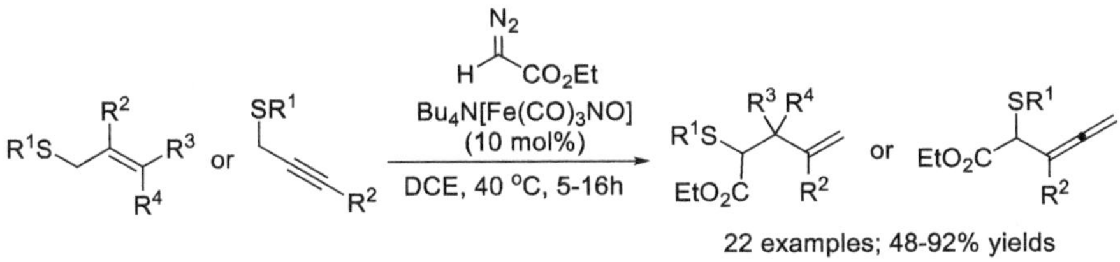

(6.3)

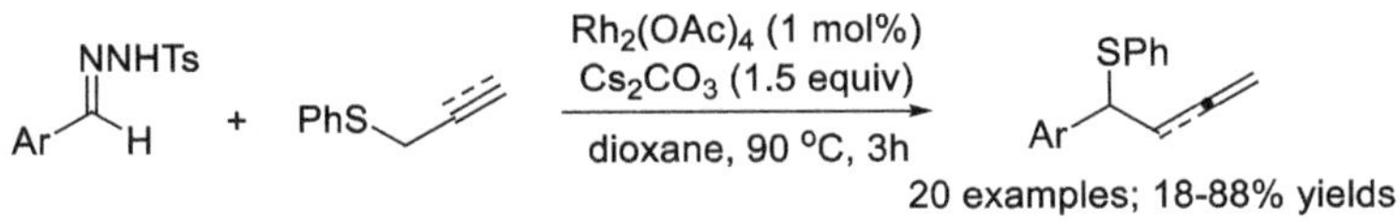

(6.4)

A special case of [2,3]-σ rearrangement involving an aromatic system is called Sommelet–Hauser rearrangement. Wang and co-workers reported the Rh(II)-catalyzed Sommelet–Hauser rearrangement reactions of diazoesters with sulfenamides or sulfides (Scheme 6.4). From the substrates of aryldiazoacetates with ethyl benzylthioacetate, arylacetates derivatives were effectively induced through sulfonium ylide [2,3]-σ rearrangements with rearomatization sequence.[21] In another example, a facile and efficient approach to 3-arylthio-1,3-disubstituted-oxindoles was developed, which was assumed to proceed *via* thia-Sommelet–Hauser rearrangement.[22]

In addition to sulfonium ylides, catalytic ammonium ylide formation from diazo compounds followed by [2,3]-σ rearrangement sequence was also reported.[23-26] With the utilization of copper catalysts, [2,3]-σ rearrangement of allylic ammonium ylides[23] or tetrahydropyridinium ylides[24] was performed with diazoesters. Moreover, insertion and [2,3]-σ rearrangement of diazo compounds catalyzed by Fe(III) corroles and porphyrins were proved to proceed through generation of ylide as the key intermediate.[25]

Halonium ylides can be generated through the interaction of halides with metal carbene. When allyl halides or propargyl halides are employed in the reaction, similar [2,3]-σ rearrangements may occur. One such example was reported by Dias and co-workers. Ag(I) tris(pyrazolyl)borate (**6**)-catalyzed reaction of diazoacetate in the presence of primary allylic or propargylic halides affords the [2,3]-sigmatropic rearrangement products in good yields (Scheme 6.5).[26]

The oxygen heteroatom in ethers is a weak-to-moderate Lewis base. Nevertheless, a highly reactive metal carbene complex, such as Rh(II) carbene or Cu(I) carbene complex, can interact with the oxygen to

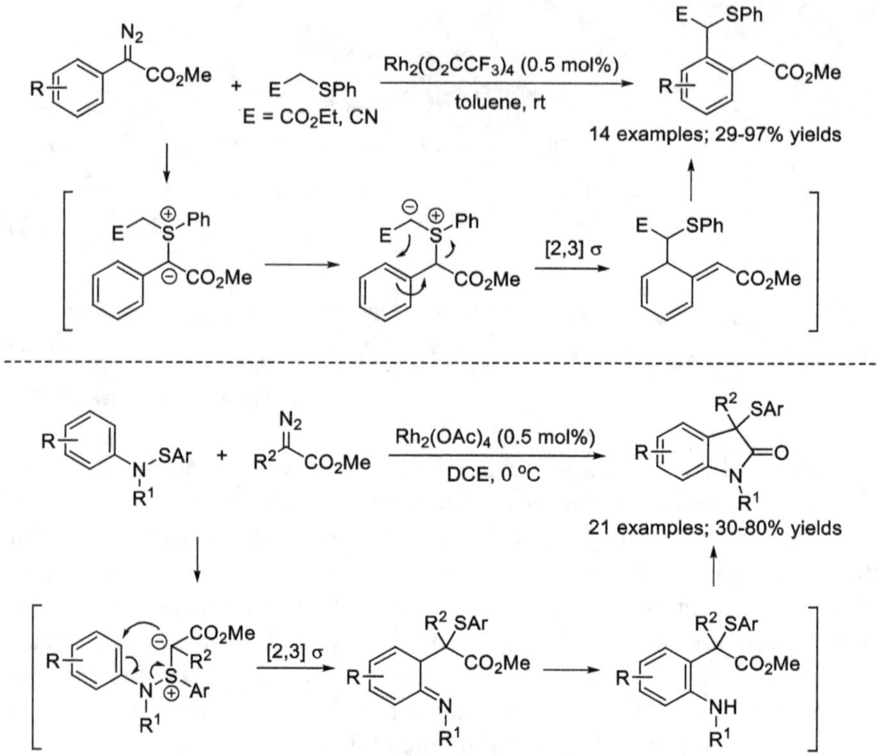

Scheme 6.4 Rh(II)-catalyzed Sommelet–Hauser rearrangement of sulfur ylides.

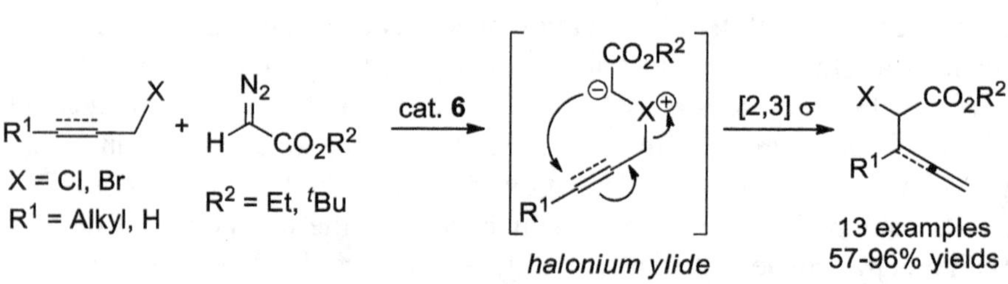

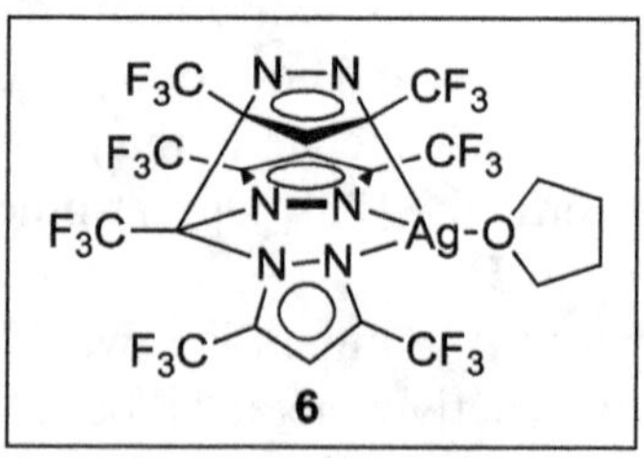

6

Scheme 6.5 [2,3]-Sigmatropic rearrangement involving halonium ylides.

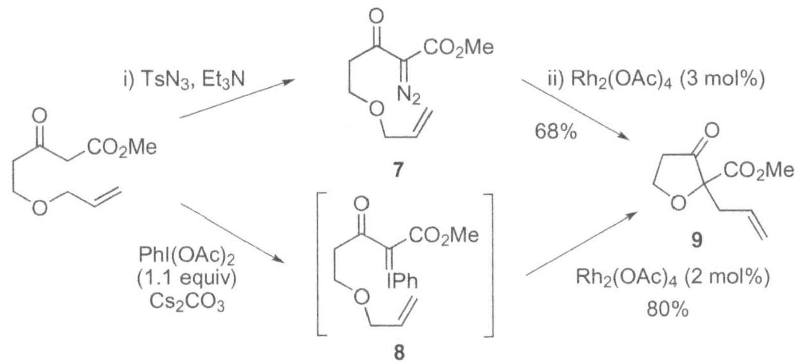

Scheme 6.6 [2,3]-Sigmatropic rearrangements by using iodonium ylide or diazoketones to form oxonium ylide.

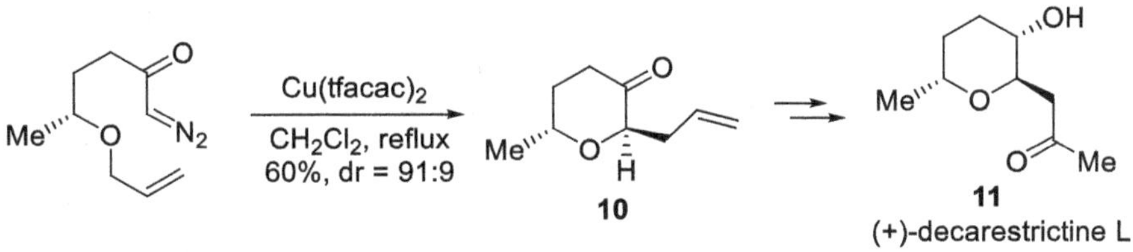

Scheme 6.7 Synthesis of (+)-decarestrictine L involving Cu-catalyzed ylide-type rearrangement.

generate an oxygen ylide. Similar to those of the sulfonium ylides, the major reaction pathways for oxonium ylides generated from metal carbene complexes are also [2,3]-sigmatropic rearrangements and 1,2-shifts. A recent report in [2,3]-sigmatropic rearrangements of oxonium ylides is the utilization of iodonium ylide **8** as diazoketone surrogate for the generation of onium ylide intermediate and the subsequent rearrangement. The one-step procedure results in higher yields compared to the corresponding two-step route employing diazoketone intermediate **7**, affording the same product **9** (Scheme 6.6).[27]

Stereoselective [2,3]-sigmatropic rearrangement of oxonium ylides has found applications in organic synthesis. A recent example is the synthesis of the fungal metabolite (+)-decarestrictine L **11**. Tandem oxonium ylide formation and the subsequent rearrangement are devised to construct the tetrahydropyranylcore in **10**, the key intermediate leading to **11** (Scheme 6.7).[28] With $Cu(tfacac)_2$ as the catalyst, the reaction of diazoketone afforded the tetrahydropyrane with good diastereoselectivity.

Asymmetric catalytic oxonium ylide formation/[2,3]-sigmatropic rearrangement was explored by Hashimoto and co-workers with their chiral Rh(II) complexes. 2,8-Dioxabicyclo [3.2.1]octane scaffold of zaragozic acids could be obtained with 72% yield and 93% *ee* (Eq. (6.5)).[29]

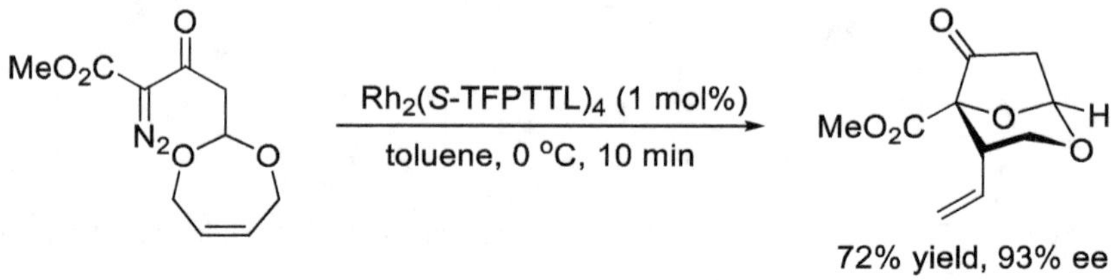

(6.5)

A recent significant progress in oxonium ylide chemistry is the generation of ylide from allylic or propargylic alcohols. Instead of conventional O–H insertions, oxonium ylide [2,3]-sigmatropic rearrangement occurs effectively. Davies and co-workers extensively explored asymmetric Rh-catalyzed intermolecular oxonium ylide [2,3]-sigmatropic rearrangement between donor/acceptor diazo compounds with racemic alcohols (Eq. (6.6)),[30] chiral allylic alcohols (Eq. (6.7))[31] and tertiary propargylic alcohols (Eq. (6.8)).[32] Highly substituted homoallylic alcohols containing vicinal stereogenic centers as well as α-hydroxy allenes were produced with high diastereo- and enantioselectivity.

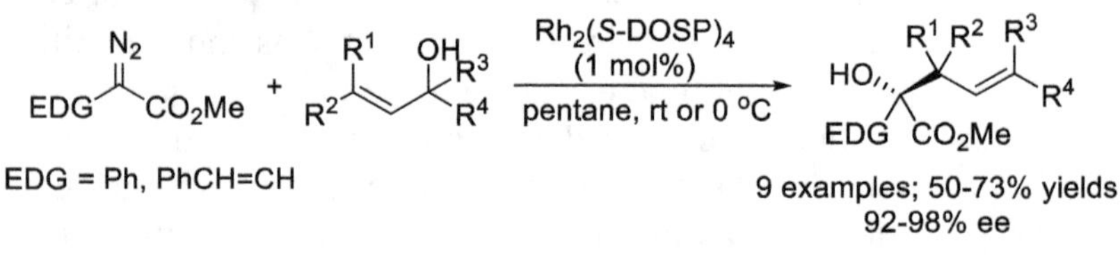

(6.6)

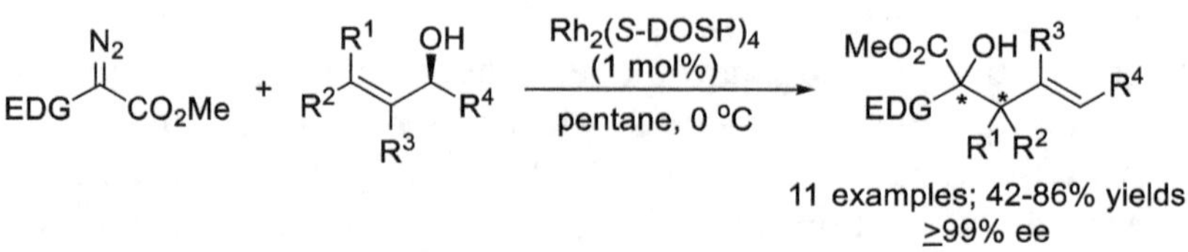

(6.7)

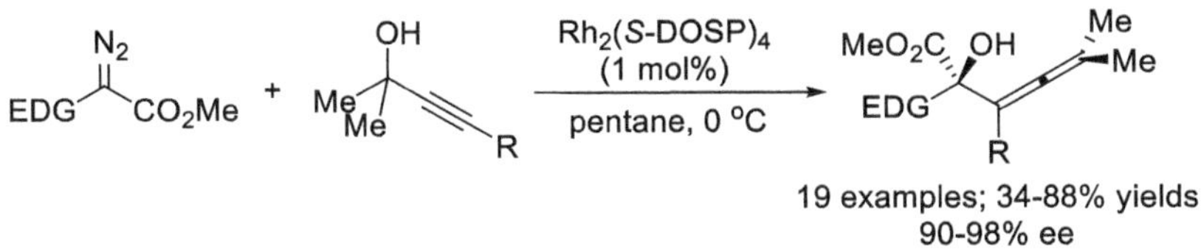

(6.8)

The same group also accomplished a straightforward stereoselective synthesis of functionalized cyclopentanes from vinyldiazoacetates and allyl alcohols.[33,34] This carbene-initiated cascade reaction proceeds through Rh(II)-catalyzed oxonium ylide formation/[2,3]-sigmatropic rearrangement to form **13**, oxy-Cope rearrangement, enol-keto tautomerization to form **14** with eventually carbonyl-ene reaction. With this domino sequence annulation, highly functionalized cyclopentanes **12** with four stereogenic centers were generated effectively in high yield with excellent diastereo- and enantioselectivity. Furthermore, they achieved the stereoselective synthesis of polyfunctionalized cyclohexanes **15** containing four stereogenic centers through Rh(II)-catalyzed domino strategy (Scheme 6.8).[35]

6.2 [1,2]-Migration

[1,2]-Migration of sulfonium or ammonium ylide, which is known as Stevens rearrangement, represents another important type of ylide reaction. With sulfonium ylide [1,2]-migration process, Cu(I)- and Rh(II)-catalyzed synthesis of thia-heterocyclic compounds were developed with high efficiency under mild conditions.[36–40] Such rearrangements may occur *via* a homolysis-recombination mechanism with the generation of radical species. The [1,2]-migration represents a useful methodology for the formal carbene insertion into C–S bonds. This reaction is particularly useful for the synthesis of cyclic thioethers. A recent example is a macrocyclic ring expansion by a double Stevens rearrangement (Eq. (6.9)).[37]

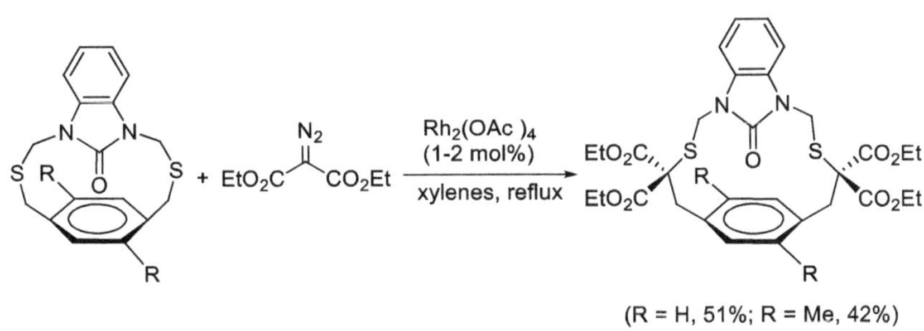

(6.9)

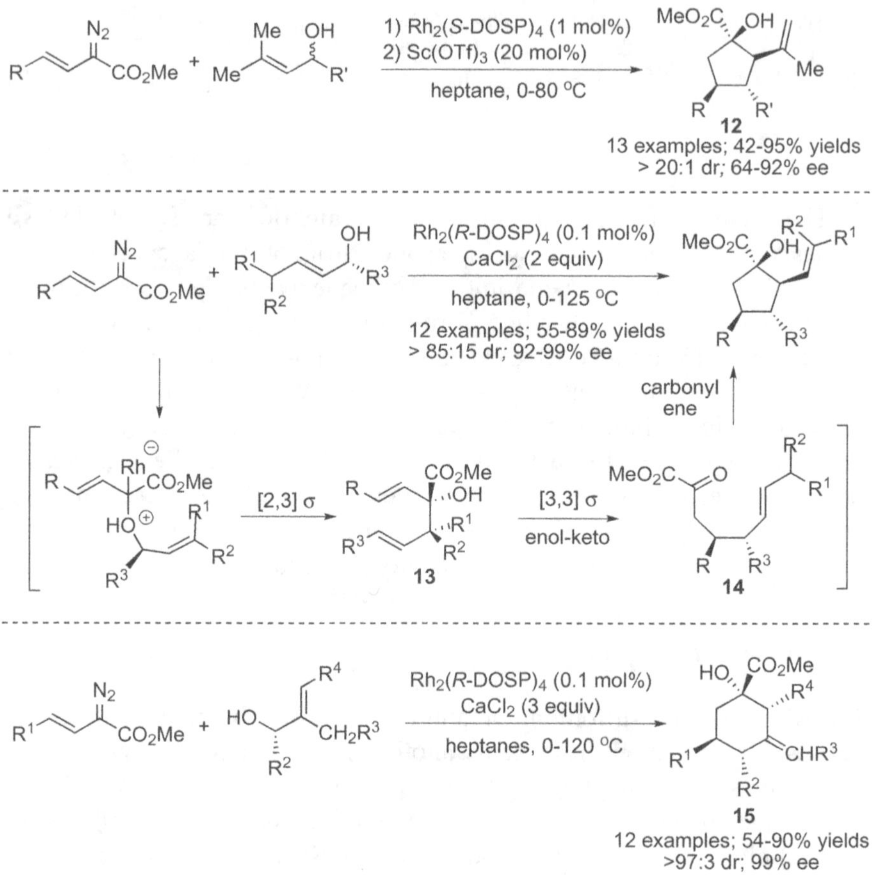

Scheme 6.8 Rh(II)-catalyzed domino sequence for asymmetric synthesis of cyclopentanes.

Tang and co-workers reported the catalytic asymmetric synthesis of optically active 1,4-oxathianes from 1,3-oxathiolanes with diazomalonates (Eq. (6.10)).[40] In this case, Cu(II)-bisoxazoline (**16**) complex has emerged as an effective catalyst to afford the desired products in high yields with good enantioselectivity (up to 90% *ee*).

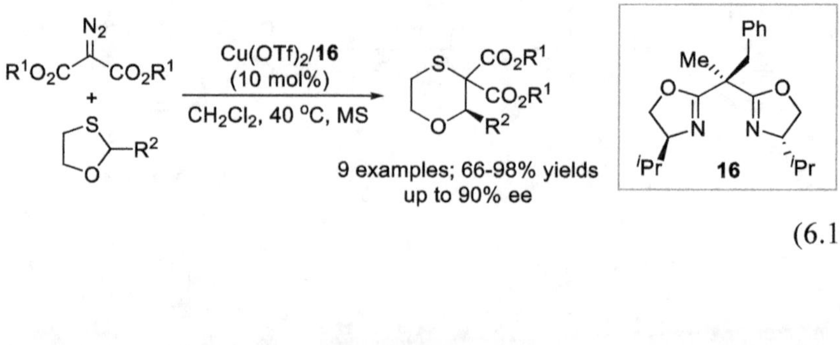

(6.10)

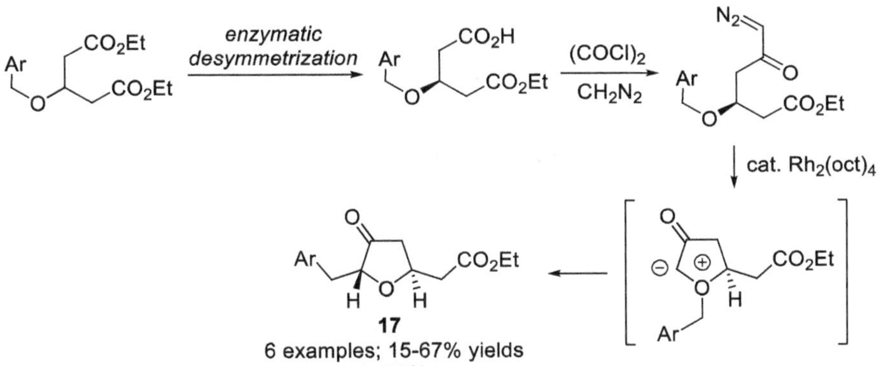

17
6 examples; 15-67% yields
up to 99% ee

Scheme 6.9 Oxonium ylide [1,2]-migration of enzymatically desymmetrized glutarates.

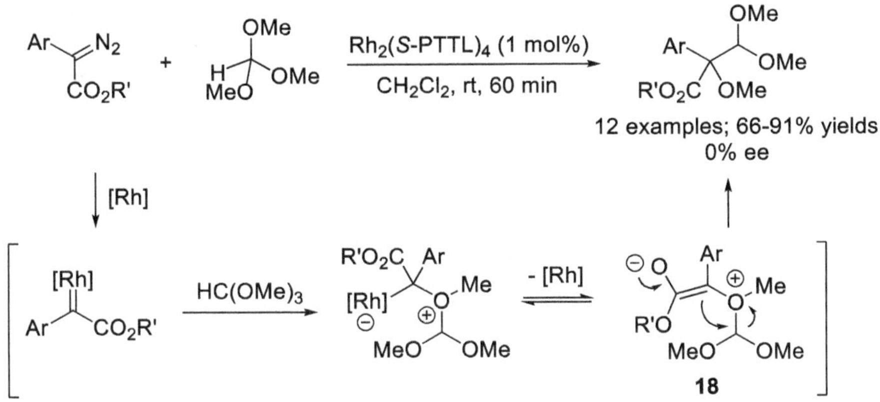

Scheme 6.10 Rh(II)-catalyzed alkoxylation/acetalization of diazo compounds.

Rh(II)-catalyzed oxonium ylide [1,2]-migration has been extensively explored for the synthesis of oxygen-containing cyclic compounds.[41–43] For example, enzymatic desymmetrization/Rh(II)-catalyzed oxonium ylide rearrangement from symmetric glutarates provided a concise and highly enantioselective synthetic method for functionalized tetrahydrofurans **17** (Scheme 6.9).[43]

Beller and co-workers reported a Rh(II)-catalyzed alkoxylation/acetalization of donor/acceptor substituted diazo compounds, affording densely functionalized α-alkoxy-β-oxo-esters with good yields (Scheme 6.10).[44] This protocol was benefited from the employment of readily available trimethyl orthoformate serving as C1-building block, enabling the construction of functionalized quaternary carbon center in a one-step manner. Mechanistically, the lack of enantiomeric excess as well as the deuterium labeling experiments indicated that this reaction might proceed *via* [1,2]-Stevens rearrangement process of the enolate anion **18**.

Apart from transition-metal-catalyzed reactions, diverse rearrangement products were obtained through oxonium ylide intermediate, which could be generated from diaryldiazomethane with *ortho*-acetal moiety under photochemical excitation.[45]

In the cases of ammonium ylide transformation, Cu(I)-catalyzed ring expansion of azetidines with diazo compounds has allowed for the synthesis of substituted pyrrolidines *via* Stevens rearrangement.[46] Doyle and co-workers achieved efficient synthesis of multi-functionalized 3-hydroxypyrroles by the combination of vinyldiazoacetate with nitrones in one-pot process. The transformation proceeded through catalytic Mannich addition, ylide formation/[1,2]-migration to form **19**, followed by acid-promoted aromatization sequence (Scheme 6.11).[47] This example demonstrates the use of Lewis acid in combination with Rh(II) carboxylate as a powerful catalytic system in carbene transformations of versatile diazo compounds.

Hamada and co-workers developed a general approach toward divergent nitrogen-bridged bicyclic frameworks through Rh(II)-catalyzed selective [1,2]-acyl migration (Scheme 6.12).[48] This protocol enabled the synthesis of a series of azabicyclo[X.Y.Z]alkane derivatives with good to excellent yields. DFT calculation indicates that this formal carbene insertion to C–N bond proceeds through Rh(II)-associated *N*-ylide generation

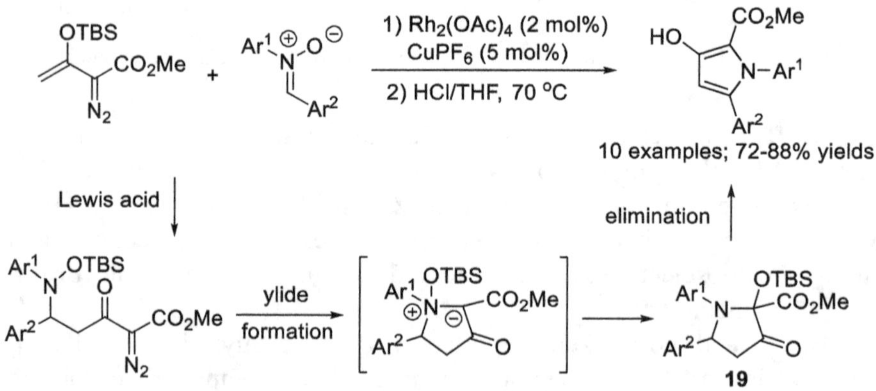

Scheme 6.11 One-pot synthesis of pyrroles from vinyldiazoacetate with nitrones.

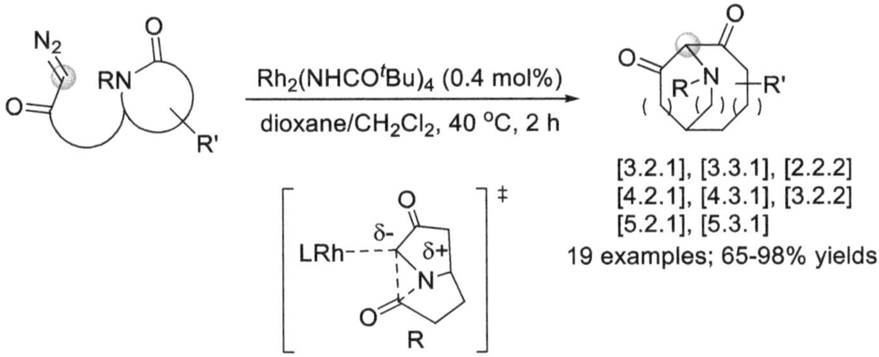

Scheme 6.12 Rh(II)-catalyzed formal carbene insertion into amide C–N bond.

and sequential acyl group [1,2]-shift through a concerted addition/elimination process.

Lacour and co-workers evaluated the Rh(II)- and Cu(I)-catalyzed carbene reaction of methylene-bridged Tröger base with diazo compounds, resulting in the highly enantio-enriched synthesis of configurationally stable ethylene-bridged Tröger derivatives.[49,50] This ring-enlargement process was rationalized with the generation of ammonium ylide **20**, followed by Stevens rearrangement to afford the ethylene-bridged Tröger base **21** with good diastereo- and enantioselectivity (Scheme 6.13). Furthermore, they continued to present the asymmetric ring expansion of binaphthyl azepines and ferrocenophanes with diazomalonates *via* catalytic [1,2]-Stevens rearrangement.[51]

The ammonium ylides generated from metal carbenes and amines undergo 1,2-shift or [2,3]-sigmatropic rearrangement similar to that of the corresponding oxonium and sulfonium ylides.[52,53] One of the recent examples was reported by Vanecko and West, who explored a ring-expansion reaction of spiroazetidinium ylide **22** in the synthesis of pyrrolizidine alkaloids **23** and **24** (Scheme 6.14).[53]

Synthesis for *2H*-1,3-oxazines through Rh(II)-carbene-induced ring expansion of isoxazoles has been reported by Davies and Khlebnikov.[54–56] Mechanistic hypothesis indicated the formation of isoxazolium *N*-ylide **25** through the attack of the isoxazole nitrogen onto the Rh(II)-carbene. One plausible pathway from **25** is the direct [1,2]-migration to afford oxazines **26**. Alternative pathway is the ring-opening process to generate 1-oxa-5-azahexa-1,3,5-trienes **27**, followed by 6π-electrocyclization to give the

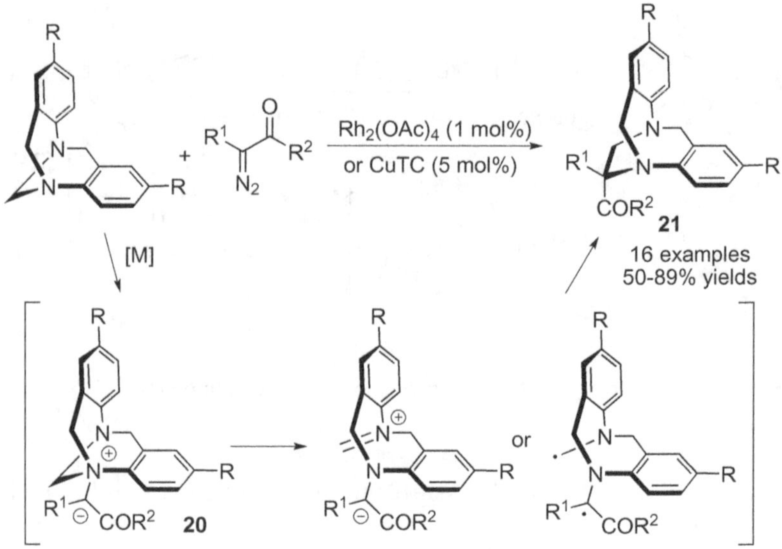

Scheme 6.13 Synthesis of ethylene-bridged Tröger base from diazo compounds.

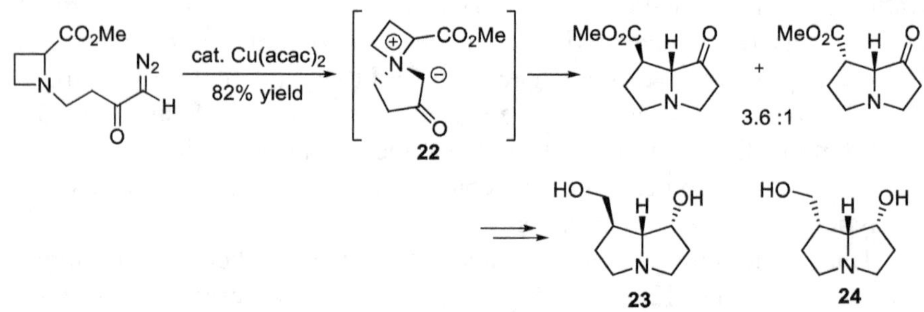

Scheme 6.14 Ring-expansion involving [1,2]-migration of spiroazetidinium ylide.

final product (Scheme 6.15). This formal N–O insertion strategy with vinyldiazo substrates was applied to the one-pot synthesis of functionalized pyridines, through a multistep reaction sequence involving ring expansion, rearrangement cascade, tautomerization and oxidation.[55]

Rainier and co-workers reported an unusual sulfonium ylide [3,3]-sigmatropic rearrangement from 2-thioindoles **28** with substituted vinyl diazoacetates.[57,58] This reaction was assumed to proceed through

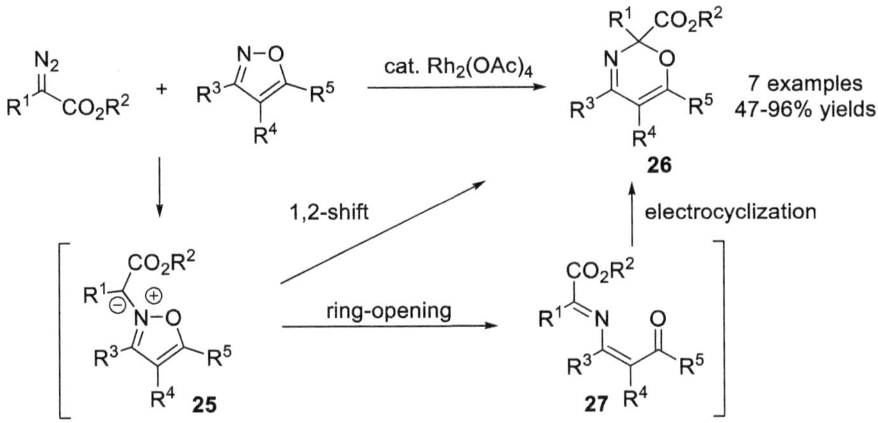

Scheme 6.15 Synthesis of *2H*-1,3-oxazines *via* ring expansion of isoxazoles with diazoesters.

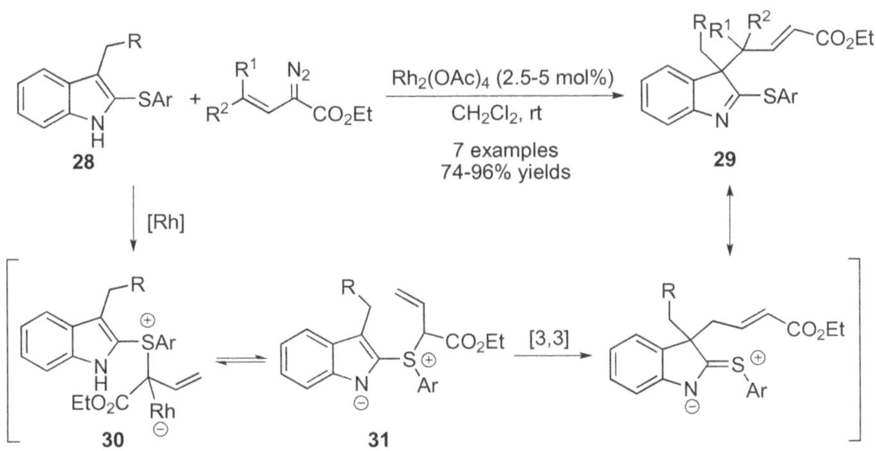

Scheme 6.16 Rh(II)-catalyzed indoline synthesis from 2-thioindole with vinyldiazo compounds.

Rh(II)-catalyzed sulfonium ylide **30** formation/proton transfer/[3,3]-sigmatropic rearrangement of **31** to deliver thioindoline **29** with high diastereoselectivity (Scheme 6.16). This cascade transformation provided a unique synthesis of quaternary substituted indolines.

6.3 Ylide cycloaddition

The electrons of oxygen lone-pair in a carbonyl group can interact with the electron-deficient carbonic carbon of a metal carbene complex to generate carbonyl ylide. Unlike the oxonium ylide, a positive charge in such a carbonyl ylide is mainly localized at the carbonyl carbon. Consequently, carbonyl ylide **32** behaves like 1,3-dipolar species **33** (Scheme 6.17). Carbonyl ylides possess versatile reactivities, among which the 1,3-dipolar cycloaddition is the most common and important reaction. The free carbonyl ylide and the corresponding metal-associated ylide should be in equilibrium.

The carbonyl ylides generated from metal carbenes can add not only to alkenes and alkynes but also to C=O and C=N bonds. In addition, the ylide formation and the subsequent 1,3-dipolar cycloaddition can occur in either inter- or intramolecular manner. Carbene-mediated ylide formation with tandem cycloaddition, especially carbonyl ylide derived from diazo component, spans broad and great applications in synthesizing natural products. Both intramolecular and intermolecular dipolar cycloaddition of *in situ* generated ylide intermediate can be utilized as attractive strategies for the construction of oxygen-containing heterocycles.

Cyclic carbonyl ylides can be generated through intramolecular reaction of metal carbene complex with pendant carbonyl group. Then the ylide can be trapped by a dipolarophile to give an oxygen-bridged bicyclic product. This tandem ylide formation/1,3-dipolar cycloaddition sequence has been extensively explored in organic synthesis by Padwa,[59–67] Hashimoto,[68–70] Hodgson[71–73] and other groups.[74–79]

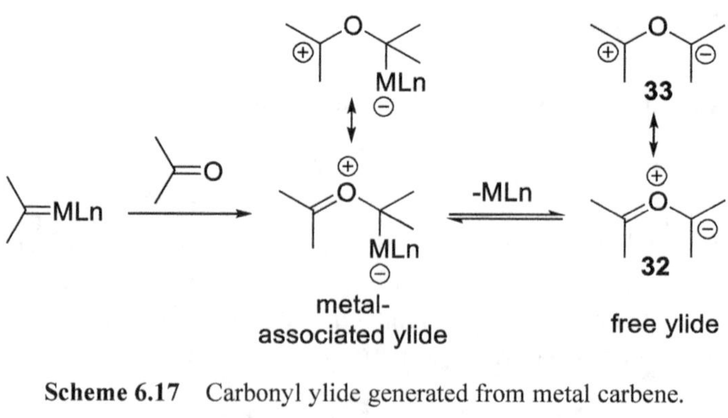

Scheme 6.17 Carbonyl ylide generated from metal carbene.

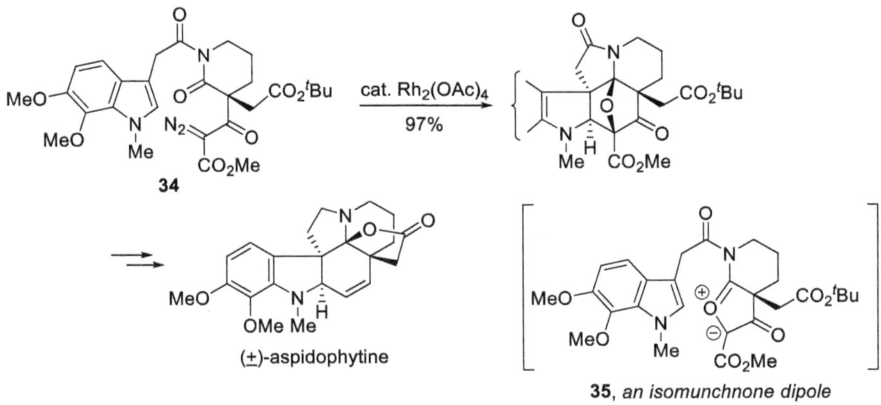

Scheme 6.18 Concise total synthesis of (±)-aspidophytine *via* Rh(II)-catalyzed ylide cycloaddition.

Padwa and co-workers developed an isomünchnone dipole **35**, a push–pull carbonyl ylide, as an important intermediate in organic synthesis. The isomünchnone dipole can be trapped efficiently by various dipolarophiles to give the nitrogen-containing polycyclic compounds. The specific report by Padwa and co-workers describes the concise total synthesis of (±)-aspidophytine by a Rh(II)-catalyzed carbene cyclization/cycloaddition cascade of a suitably substituted α-diazoimide **34** as the key step (Scheme 6.18).[64]

Hashimoto and co-workers have focused on the application of ylide formation/1,3-dipolar cycloaddition cascade in the total synthesis of zaragozic acids, a family of polyketide natural products. They have achieved the total synthesis of zaragozic acids A and C (Scheme 6.19).[69] The key step in their synthesis was the generation of a carbonyl ylide (a 1,3-dipole) from the $Rh_2(OAc)_4$-catalyzed reaction of diazo ester **36**. The carbonyl ylide further reacted with an electron-deficient triple bond through **37** to give cycloadduct **38**, which contains the core structure of zaragozic acid.

Versatile dipolarophiles, including allene carboxylates **39**,[80] methylwglyoxylate **40**[81] and 1,2-benzoquinones **41**[82] have been utilized in Rh(II)-catalyzed carbonyl ylide [3+2] cycloaddition (Scheme 6.20). Recently, the generation of α,β-unsaturated carbonyl ylides derived from enones and their sequential Rh(II)-catalyzed intra- and intermolecular cycloaddition were reported, which provided the facile construction to functionalized oxapolycyclic rings.[83]

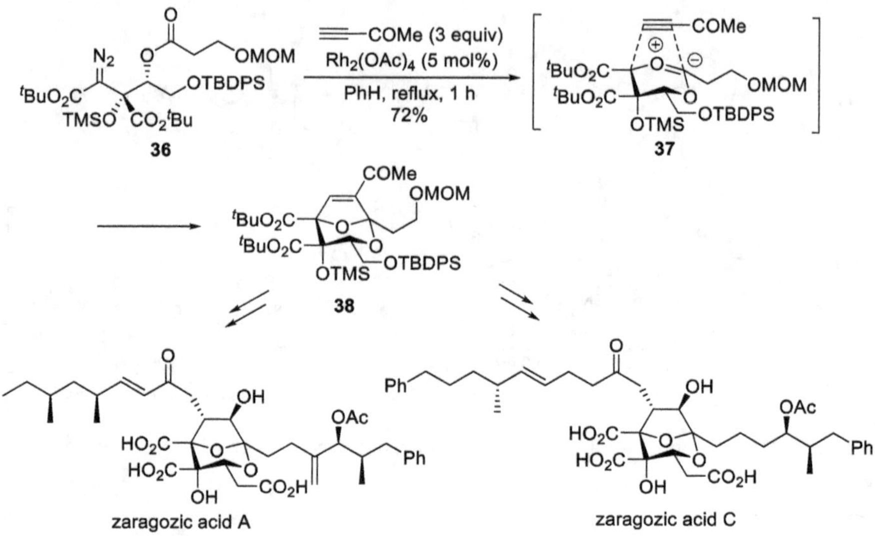

Scheme 6.19 Rh(II)-catalyzed carbonyl ylide cycloaddition for the total synthesis of zaragozic acids A and C.

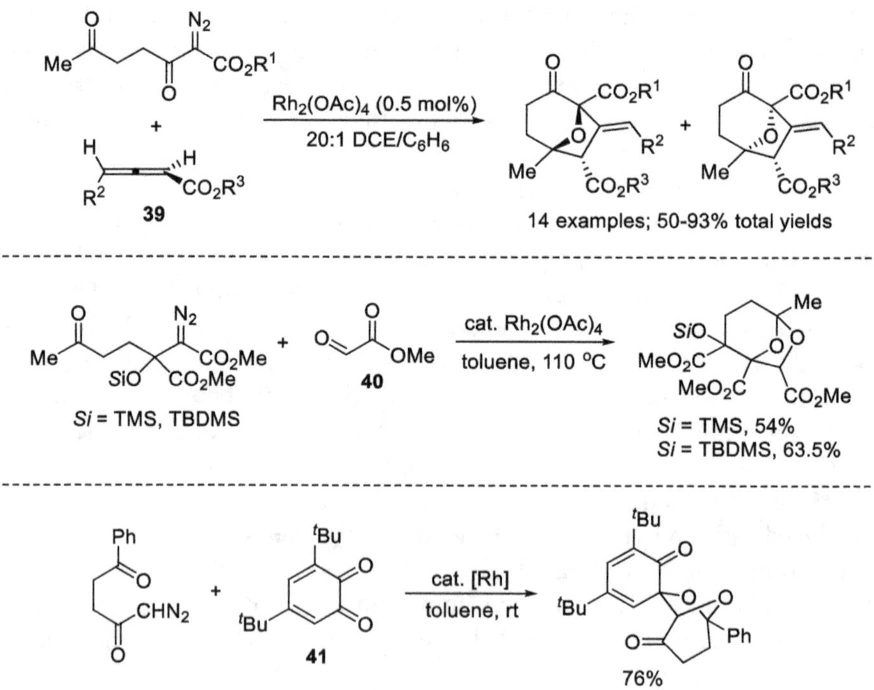

Scheme 6.20 Rh(II)-catalyzed carbonyl ylide cycloaddition with versatile dipolarophiles.

Recently, Wood and co-workers reported a Rh(II)-catalyzed chemoselective generation of carbonyl ylide derived from electronically differentiated diazo malonates.[84] The carbonyl ylide generated from the electron-rich ester was trapped by a series of alkynes to afford substituted oxabicyclic derivatives in a chemo-, regio- and diastereoselective manner.

In addition to alkenes and alkynes, aldehydes, ketones and imines can also act as dipolarophiles to react with carbonyl ylide dipoles. Various electron-deficient multiple bonds, such as N=N, N=O, C≡C bonds and singlet oxygen, are found to react with carbonyl ylides. These cycloaddition reactions provide efficient routes to various heterocycles.[85-89]

Asymmetric catalysis of the cascade carbonyl ylide formation-cycloaddition with chiral Rh(II) catalysts has been extensively investigated by Hashimoto, Hodgson and other groups. These investigations have used chiral metal complexes, mostly chiral Rh(II) complexes, to generate chiral metal carbenes from diazocarbonyl compounds. An alternative and interesting approach, developed by Suga and co-workers, is to achieve the asymmetric induction by using an achiral Rh(II) catalyst to generate the carbonyl ylide and a chiral Lewis acid to activate the C=O bond. High levels of enantioselectivity can be obtained by this approach with the chiral Lewis acids, Sc(III)-Pybox-*i*-Pr and Yb(III)-Pybox-Ph.[90,91]

Other developments in stereoselective 1,3-dipole cycloaddition include a chiral auxiliary approach by using dipolarophiles bearing chiral sulfinyl groups, as reported by Ruano and co-workers.[92] A moderately high level of diastereoselectivity was observed.

In 2007, Son and Fu reported the first examples of diastereo- and enantioselective copper-catalyzed asymmetric [4+1] cycloadditions of enones with diazo compounds through the use of a planar-chiral bipyridine ligand (-)-bpy **42**. The carbonyl ylide intermediate **43** was considered as a reactive intermediate in this reaction (Scheme 6.21).[93]

On the other hand, Padwa and co-workers demonstrated the chemoselectivity of bis(diazo)-substituted molecules within Rh(II)-catalyzed carbene reactions.[94] Donor/acceptor diazo moiety of **44** preferentially undergoes cyclopropanation/insertion process to form **45** with the retention of di-acceptor diazo component. Furthermore, dipolar cycloaddition reaction was performed subsequently with acceptor/acceptor diazo group (Scheme 6.22). This strategy provided an efficient method for preparing polycyclic structures. Very recently, the same group also realized the synthesis of azapolycycle through Rh(II)-catalyzed one-pot cyclopropanation/carbonyl ylide cycloaddition of bis(diazo)piperidin-2-one derivatives.[95]

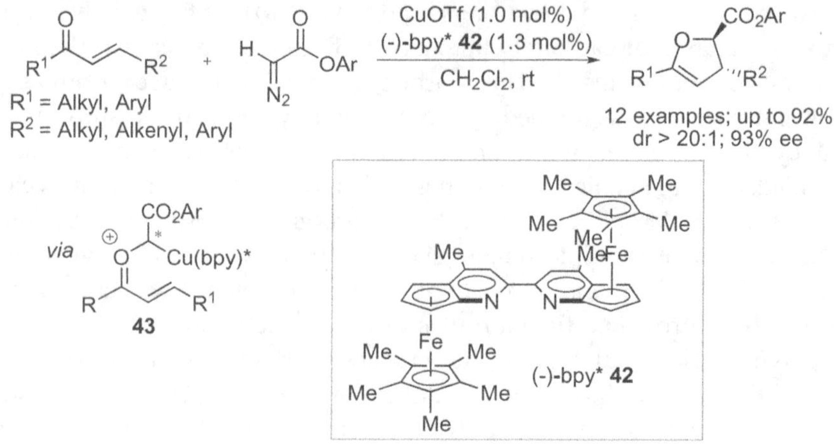

Scheme 6.21 Cu(I)-catalyzed asymmetric formal [4+1] cycloaddition.

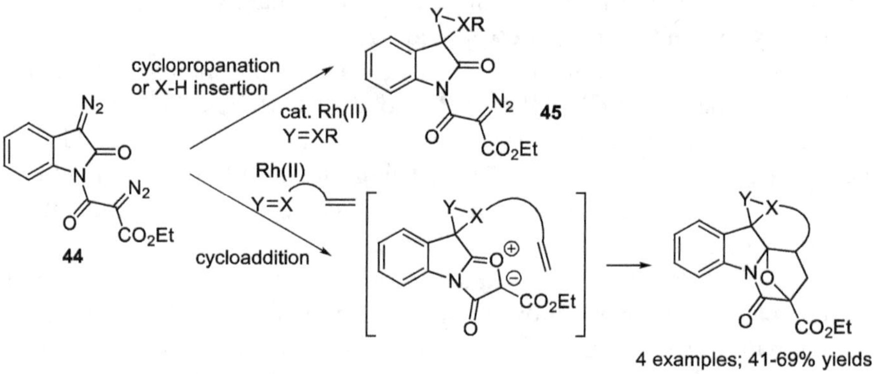

Scheme 6.22 Substituted bis(diazo) molecules in Rh(II)-catalyzed carbene reactions.

As expected, the asymmetric catalytic version of the carbonyl ylide cycloaddition attracted significant attention. Hashimoto and co-workers developed a series of asymmetric carbonyl ylide cycloadditions of 2-diazo-3,6-diketoester with alkynes/styrenes,[96] indoles[97] and arylallenes[98] by using Rh_2(*S*-TCPTTL)$_4$ as efficient chiral catalyst. A wide range of cycloadducts could be obtained in good yields with excellent diastereo- and enantioselectivities. Moreover, asymmetric carbonyl ylide inverse electron-demand cycloaddition of vinyl ethers was achieved with high level of enantioselectivity by chiral Lewis acid.[99]

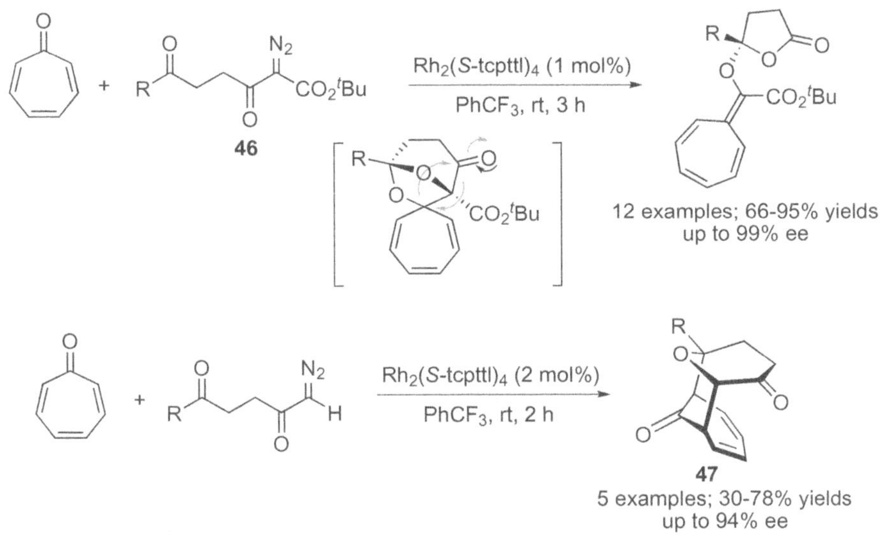

Scheme 6.23 Rh(II)-catalyzed enantioselective synthesis of troponoids.

Recently, an enantioselective synthetic method of troponoids was realized by asymmetric Rh(II)-catalyzed 1,3-dipolar cycloaddition between the carbonyl moiety of tropone and diazo diketoester **46**.[100] Besides, facile [6+3] cycloaddition of α-diazoketone-derived carbonyl ylides with tropone was also demonstrated to furnish bridged tricycles **47** with excellent stereoselectivity (Scheme 6.23).

Tang and co-workers reported an efficient protocol to access optically active multi-functionalized dihydrofurans from α-benzylidene-β-ketoester with diazoester.[101] By using side-armed BOX ligand **48**, this Cu(I)-catalyzed asymmetric carbonyl ylide [4+1] cycloaddition was performed in good to excellent yields and with high stereoselective control (Scheme 6.24).

An alternative approach for achieving catalytic asymmetric 1,3-dipolar cycloaddition is to use a dual catalyst system.[90,91] Such a strategy has been demonstrated by Suga and co-workers in the reaction of *N*-methylindoles with cyclic carbonyl ylides derived from diazodiketone or diazoketoester in the presence of achiral Rh(II) complexes and chiral lanthanoid metal catalysts (Eq. (6.11)).[102] For the six-membered cyclic carbonyl ylides, this strategy afforded polycyclic compounds with good yields and high enantioselectivities. For the five-membered cyclic carbonyl ylide, the cycloaddition reaction resulted in relatively good

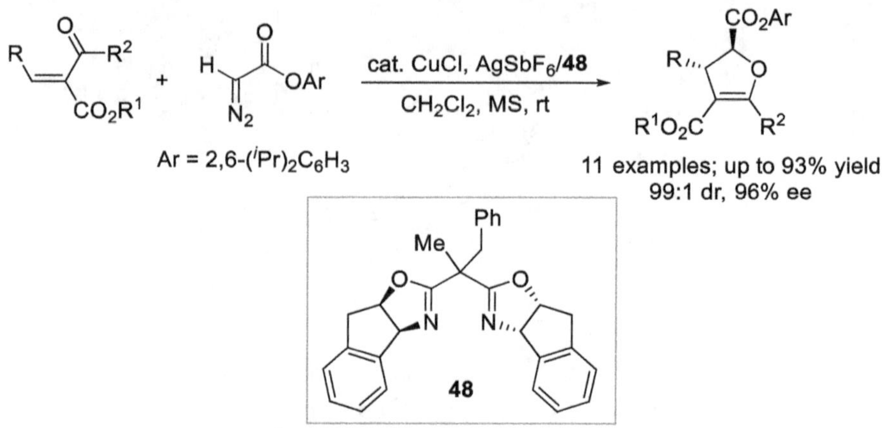

Scheme 6.24 Cu(I)-catalyzed asymmetric [4+1] carbonyl ylide cycloaddition.

enantioselectivity and *endo*-selectivity. As evidenced by X-ray crystal analysis, it was proposed that the coordination between chiral Pybox-Ph_2-Lu complex and carbonyl ylide played a crucial role in asymmetric induction.

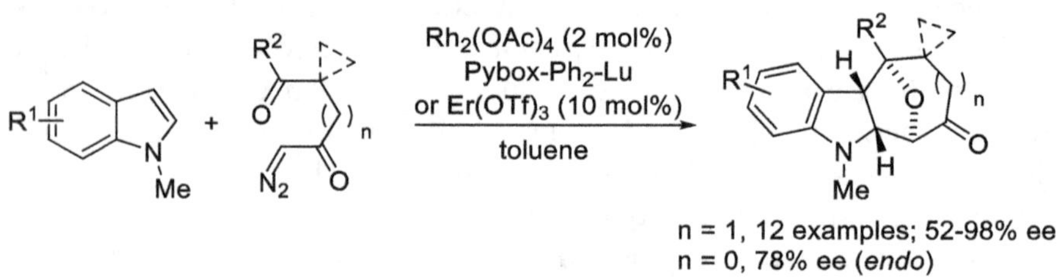

(6.11)

Recently, a Pd-catalyzed formal [4+1] cycloaddition of diazoacetates and aryl propargyl alcohols was reported, which afforded 2,5-dihydrofuran derivatives **49** as the dominant product, instead of the possible 2,3-σ rearrangement or cyclopropenation products (Scheme 6.25).[103] This auto-tandem catalytic process was assumed to proceed *via* Pd(II)-catalyzed intermolecular oxonium ylide **50** formation and subsequent intramolecular trapping of the ylide with activated alkynes.

Rh(II)-catalyzed intramolecular carbonyl ylide [3+2] cycloaddition was performed with unsaturated α-diazo-β,ε-diketo sulfones,[106] or using vinylsulfonates as the dipolarophiles.[107] These transformations provide asymmetric synthetic approach for oxapolycycle and polycyclic sultones.

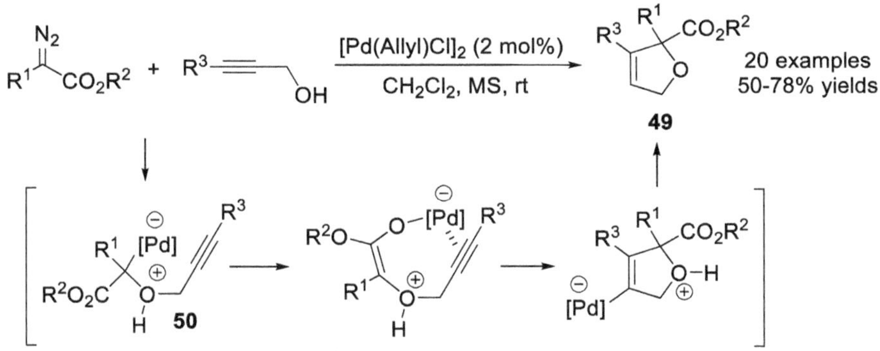

Scheme 6.25 Pd(II)-catalyzed formal [4+1] cycloaddition of diazoacetates and aryl propargyl alcohols.

Besides, both intra- and intermolecular Rh(II)-catalyzed synthesis of polycyclic spirolactones were achieved with γ-alkylidenebutenolides as highly competent dipolarophiles.[108]

Catalytic multicomponent cycloaddition provides an efficient strategy toward the synthesis of highly functionalized oxa-heterocyclic compounds. Fox[109] and Muthusamy's group[110] reported the three-component carbonyl ylide cycloadditions by utilizing two molecules of aryl aldehyde with diazoesters or cyclic diazoamide. For the two aromatic aldehyde substrates, one serves as the carbonyl ylide source and the other serves as the dipolarophile. Furthermore, the scope of this transformation has been extended to propenoate/propiolate,[111] acetylene dicarboxylate,[111,112] 3-(2-alkenoyl)-2-oxazolidinones,[113] β-nitrostyrenes[114] and aldimines[115] (Scheme 6.26). Recently, three-component Rh(II)-catalyzed [3+2] cycloaddition of 3-diazooxindoles, aryl aldehydes with 3-arylideneoxindoles as dipolarophiles has been reported to afford a series of functionalized dispirooxindoles with good yields and excellent stereocontrol.[116]

Similar to the reaction with carbonyl compounds, a metal carbene can react with imines to generate the corresponding azomethine ylides. The major reaction pathways of azomethine ylides are ring closure and 1,3-dipolar addition, affording aziridines and pyrrolidines, respectively. Transition-metal-complex-catalyzed three-component reactions, which include an imine, a diazo compound and a dipolarophile, have been extensively studied in recent years.[117–120]

In addition to [3+2] cycloaddition of carbonyl ylide, intramolecular1,3-dipolar cycloaddition of azomethine ylides was realized with moderate

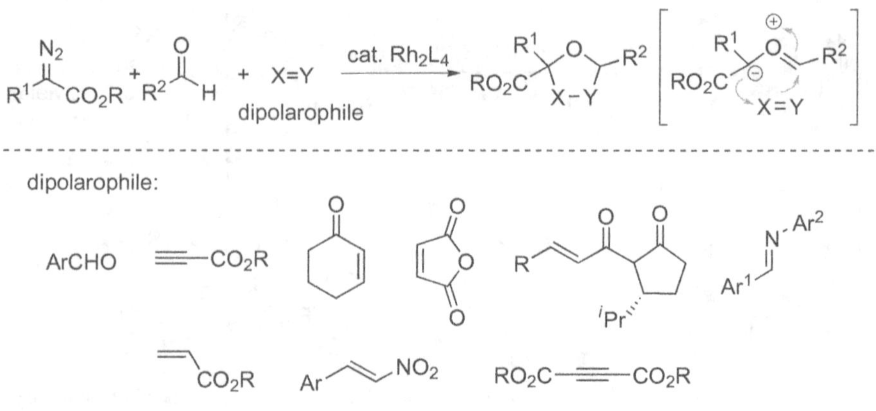

Scheme 6.26 Versatile dipolarophiles in three-component carbonyl ylide cycloaddition.

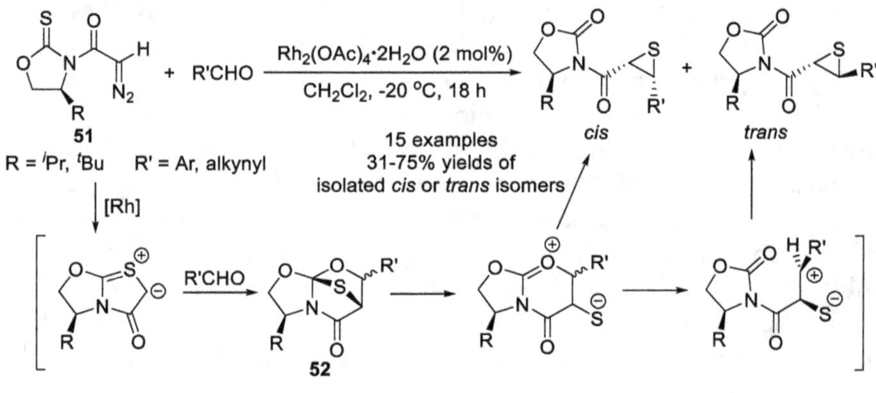

Scheme 6.27 Asymmetric synthesis of thiiranes through thiocarbonyl ylide formation.

yields with Cu(II) catalyst.[121] Langhals and co-workers observed the competition between dithiolane generation through thiocarbonyl ylide cycloaddition (Schönberg reaction) with expected formation of thiirane rings.[122] Furthermore, Rh(II)-catalyzed reaction of *N*-(diazoacetyl)oxazolidin-2-thiones **51** with aldehydes was performed to synthesize thiiranes instead of oxiranes.[123] Theoretical calculation indicated that this intriguing synthetic approach proceeded through thiocarbonyl ylide [3+2] cycloaddition to generate tricyclic adduct **52**, followed by ring-opening sequence to afford thiiranes as the final products (Scheme 6.27).

Ma and co-workers reported the Cu(II)-catalyzed three-component reaction of CF_3CHN_2, nitriles and aldehydes.[124] The reaction leads to the facile and efficient synthesis of trifluoromethylated oxazolines **53** with high yields and excellent diastereoselectivities under mild conditions (Eq. (6.12)). Moreover, the corresponding cycloadducts could be readily converted into CF_3-substituted vicinal amino alcohols. The reaction proceeded through generation of ammonium ylide intermediates, followed by stepwise [3+2] cycloaddition of aldehyde to furnish trifluoromethylated oxazolines.

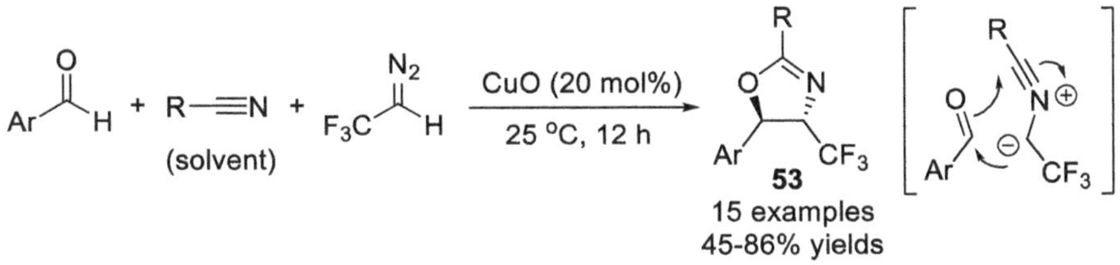

(6.12)

On the other hand, Che and co-workers have recently studied the stereoselective ruthenium porphyrin-catalyzed three-component coupling reaction of α-diazo esters with a series of *N*-benzylidene imines and alkenes, alkynes or azodicarboxylates to form the corresponding functionalized pyrrolidines, pyrrolines or 1,2,4-triazolidines.[117,118,125]

6.4 Trapping of the ylides

6.4.1 *Trapping of oxonium ylides generated from alcohols*

One of the important developments of carbene related ylides is the trapping of oxonium ylides generated from carbene and alcohols. In many cases, the 1,2-hydrogen shift is completely suppressed by the trapping of the ylide with electrophiles. Hu and co-workers have extensively explored the catalytic ylide formation derived from diazo compounds with anilines/alcohols, followed by tandem nucleophilic addition to unsaturated bonds.[8,126–129] The Rh(II)-catalyzed three-component reaction of aryl diazoacetate, alcohol and aldehyde/imine, provided evidence of oxonium ylide formation **54** and its trapping with electrophiles (Scheme 6.28).[128,129] In this case, the proposed oxonium ylide intermediate can be trapped by aldehyde or imine, effectively competing with the 1,2-H shift for generating formal O–H insertion products. The three-component reaction afforded additive products in high yields, and excellent diastereoselectivity was achieved.[129]

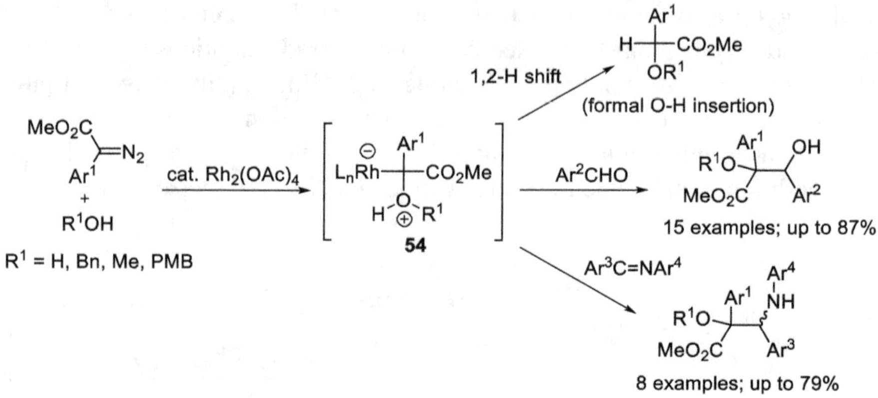

Scheme 6.28 Rh(II)-catalyzed oxonium ylide trapping versus formal O–H insertion.

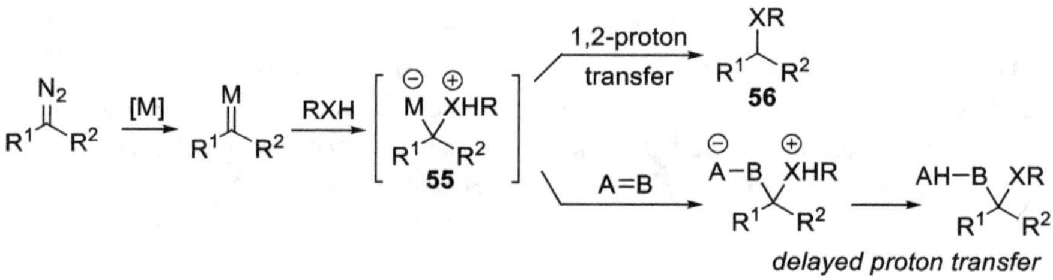

Scheme 6.29 Ylide trapping process *versus* direct proton transfer.

Experimental evidence indicated the existence of the transient ylide intermediate generated from diazoesters with alcohol.[130] Thus, these multicomponent reactions were assumed to proceed *via* trapping of active ammonium or oxonium ylide intermediate **55** with a wide range of electrophiles. In this case, a "delayed proton transfer" process occurred after the ylide trapping process rather than direct 1,2-H transfer to give formal insertion products **56** (Scheme 6.29).

Three-component reactions of diazo compounds, alcohols and imines, as well as the corresponding asymmetric version of the reaction, were explored by using Rh(II) catalysts cooperated with chiral Lewis acids (Scheme 6.30).[143–153] A wide range of diazoesters and diazoketones substrates were subjected to oxonium ylide transformations with alcohols, Ti(IV) isopropoxide,[145] water[146,150,151] and carboxylic acids,[152] followed by trapping with divergent imines, including *in situ* generated

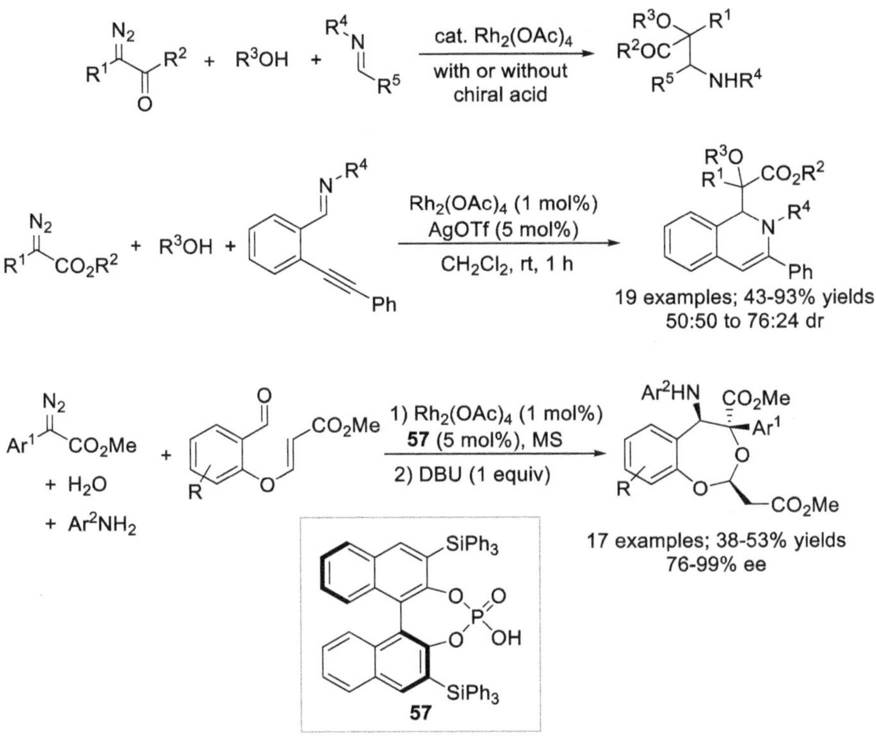

Scheme 6.30 Rh(II)-catalyzed three-component reaction of diazo compounds, alcohols and imines.

imines[143,145,150,151] and 2-alkynylarylaldimines.[148] With the cooperative catalytic strategy of Rh(II) with chiral phosphoric acids, these multicomponent ylide trapping reactions were implemented in good to excellent diastereo- and enantioselectivity. Chiral Brønsted acid **57** played important roles both in activating carbonyl/imine group and also in accelerating proton-transfer process, serving as proton shuttle within different intermediates. Furthermore, the tandem protocol involving cyclization or oxo-Michael addition was reported respectively to synthesize 1,2-dihydroisoquinolines[148] or chiral cyclic acetals.[151]

Hu and co-workers reported the enantioselective three-component reaction of alcohols, α-diazoesters, aldimines/aldehydes with one-pot subsequent cyclization sequence, providing a divergent and novel synthetic protocol for chiral heterocycles (Scheme 6.31).[153] The employment of Rh(III)/chiral ligand (**58**)-Zr(IV) co-catalytic system achieved high

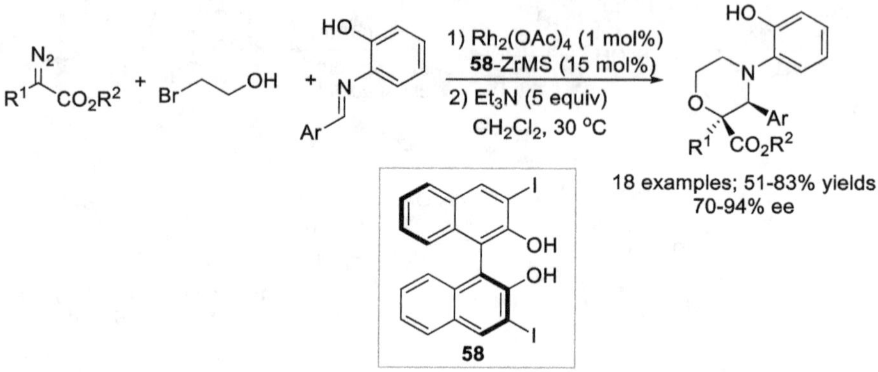

Scheme 6.31 Three-component reaction for the synthesis of chiral morpholines.

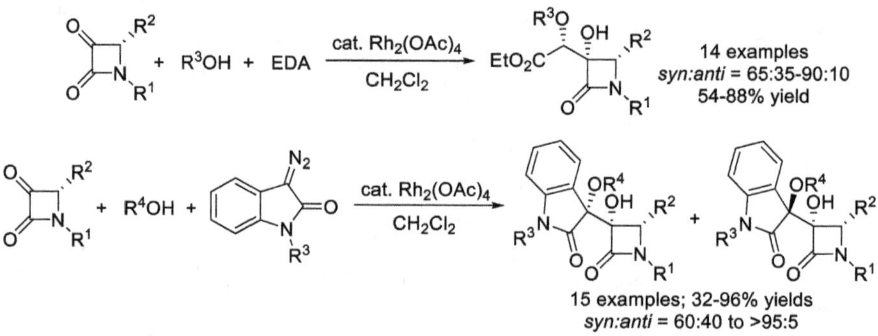

Scheme 6.32 Rh(II)-catalyzed aldol-type three-component reaction involving ylide formation.

efficiency and excellent diastereo- and enantioselectivity. This sequencing strategy combined the oxonium ylide trapping process with the subsequent cyclization in one-pot fashion under asymmetric catalytic conditions.

Moreover, transition-metal-catalyzed reactions of oxonium ylide generation/trapping by carbonyl compounds were also developed. Alcaide and co-workers reported the synthesis of 3-hydroxy-β-lactams through Rh(II)-catalyzed three-component reaction of alcohols, azetidine-2,3-diones with EDA[154] or 3-diazooxindoles.[155] This aldol-type method provided an efficient preparation for substituted β-lactam derivatives with high diastereomeric excess (Scheme 6.32).

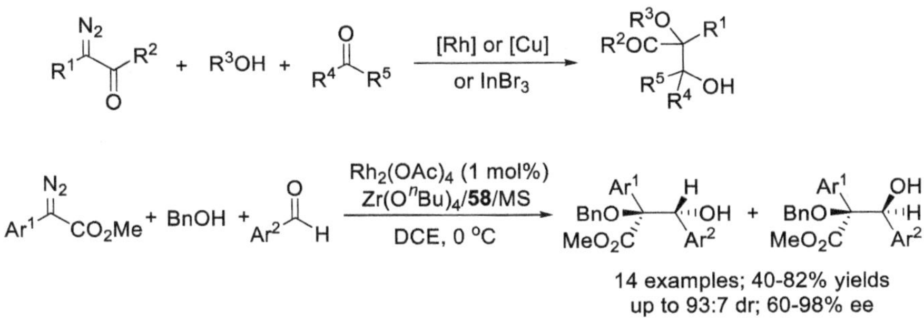

Scheme 6.33 Three-component reaction of diazo compounds, alcohols with carbonyl derivatives.

Simultaneously, Hu and others groups developed multicomponent oxonium ylide trapping strategy with alcohol[156–161] or water,[162] diazoesters or 3-diazooxindoles,[161,162] and a series of carbonyl compounds, including aldehydes,[156,157,160,162] alkynals,[158] β,γ-unsaturated α-ketoesters[159] and isatins/ninhydrin/acenaphthenequinone[161] (Scheme 6.33). In these cases, in addition to the Rh(II) carboxylates, Cu(I) complexes[157] and $InBr_3$[160] demonstrated high reactivity in the three-component reaction of aldol-type addition. They also designed and utilized cooperative catalytic strategy in Rh(II)/Zr(IV)-catalyzed enantioselective alcoholic oxonium trapping process with aryl aldehydes.[156] Furthermore, 3-substituted 1,4-dioxan-2-one scaffold was effectively achieved through Cu(II)-catalyzed trapping of intramolecularly generated protic oxonium ylide intermediate.[163]

Recently, Moody and co-workers disclosed Rh(II)- and Cu(I)-catalyzed carbene reaction of diazocarbonyl compounds with β-hydroxyketones involving oxonium ylide intermediates. The reaction afforded highly substituted tetrahydrofurans with excellent diastereoselectivities (Scheme 6.34).[164] The reaction is proposed to proceed through the generation of metal-associated ylide **59** or free oxonium ylide **60**, which was followed by intramolecular aldol reaction to afford the tetrahydrofuran product.

Muthusamy and co-workers reported a multicomponent reaction of the diazoketone with an alcohol and aldehyde. This reaction involves a tricyclooxonium ylide intermediate (Scheme 6.35).[165] The rearrangement of oxonium ylide **61** results in the generation of dipole cyclic intermediates **62**. Subsequent trapping of the positive charge by the alcohol and the negative charge by the aldehyde give the final adduct.

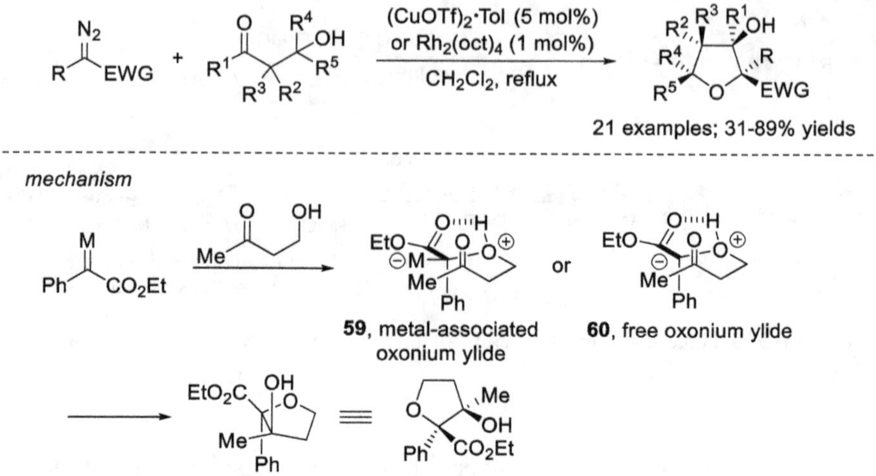

Scheme 6.34 Diastereoselective synthesis of highly substituted tetrahydrofurans from β-hydroxyketones with diazocarbonyl compounds.

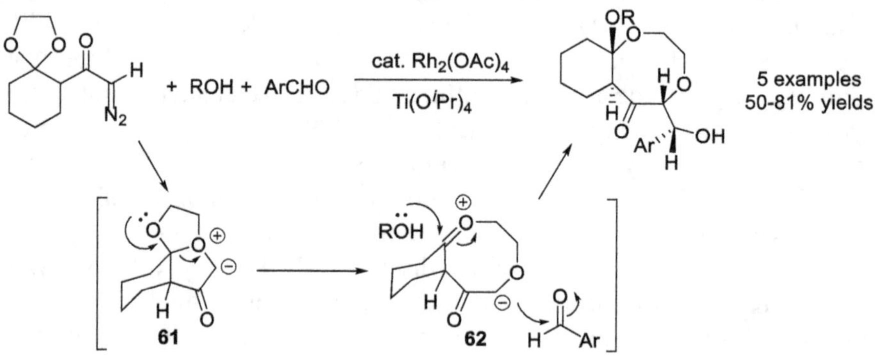

Scheme 6.35 Three-component reaction of diazoketone, alcohol and aldehyde involving ylide rearrangement and sequential addition cascade.

In 2015, Hu and co-workers developed a divergent strategy for the synthesis of multisubstituted tetrahydrofurans and pyrrolidines through Rh(II)-catalyzed oxonium ylide or ammonium ylide generation/intramolecular aldol-type trapping process.[166] From readily available substrate of β-hydroxyketones or β-aminoketones with diazoesters, this method afforded functionalized heterocyclic compounds with good yields and good to excellent diastereoselectivities (Scheme 6.36).

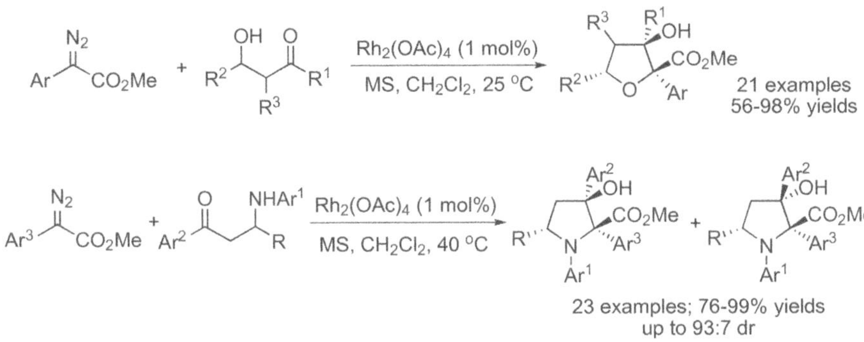

Scheme 6.36 Synthesis of multisubstituted tetrahydrofurans and pyrrolidines.

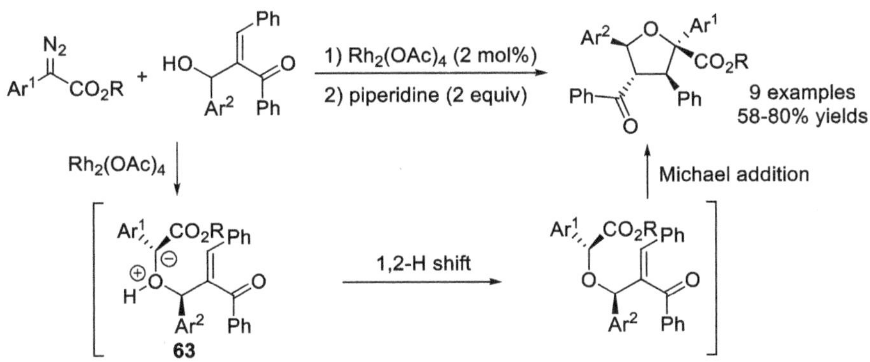

Scheme 6.37 Rh(II)-catalyzed one-pot cascade reactions of aryldiazoacetates with allyl alcohols.

Furthermore, Hu and co-workers extensively explored the oxonium ylide generation from diazo component with alcohols, which was then followed by trapping with α,β-unsaturated compounds.[167–172] Firstly, they achieved the diastereoselective synthesis of highly functionalized tetrahydrofurans from α-aryl-α-diazoesters with allyl alcohols through one-pot ylide **63** formation/1,2-shift/Michael cascade (Scheme 6.37).[167] A mechanism involving formal O–H insertion and subsequent base-promoted Michael reaction is proposed by the authors.

By utilizing Rh(II) or Cu(II) catalysts, the oxonium ylide could be trapped by intermolecular 1,4-addition with versatile Michael acceptors, including α,β-unsaturated 2-acyl imidazoles,[168] chalcones,[169] benzylidene

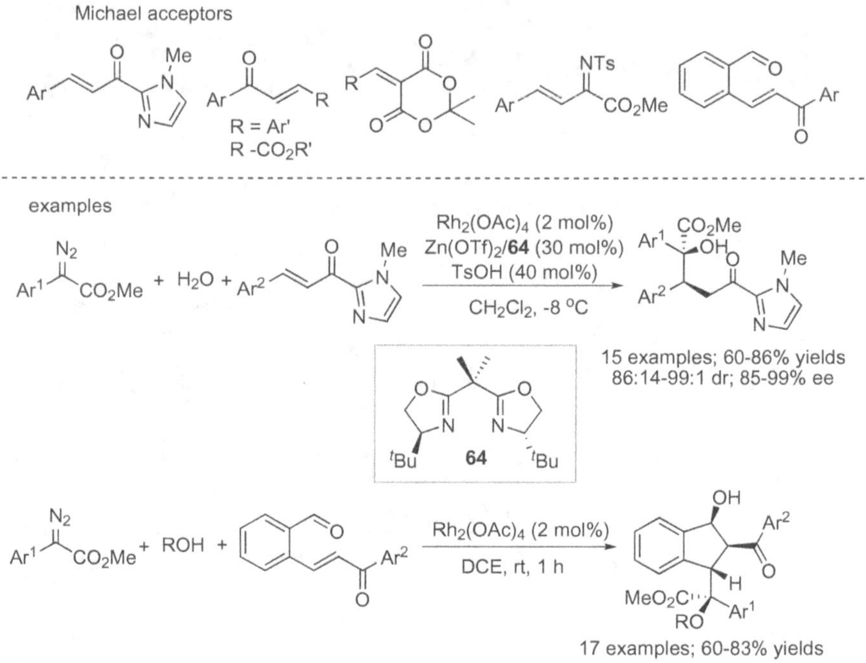

Scheme 6.38 Three-component reaction of diazo compounds, alcohols with Michael acceptors.

meldrum's acids/4-oxo-enoates,[170] 2-formyl chalcones[171] and *β,γ*-unsaturated *N*-sulfonylimines[172] (Scheme 6.38). Notably, enantioselective synthesis of *γ*-hydroxyketones was accomplished by using $Rh_2(OAc)_4$/$Zn(OTf)_2$ cooperative catalytic system in the presence of chiral ligand.[168] Furthermore, they also achieved diastereoselective synthesis of functionalized 1-indanols through Rh(II)-catalyzed three-component cascade.[171]

6.4.2 *Ammonium ylide trapping*

Similar to the trapping of oxonium ylides, the corresponding trapping of ammonium ylides provides efficient access to nitrogen-containing organic compounds. Hu and co-workers have established three-component Mannich-type reaction of diazoesters or ketones, with amines and functionalized imines.[173–181] In most cases, the enantioselective ammonium ylide trapping process was successfully achieved by using Rh(II) or

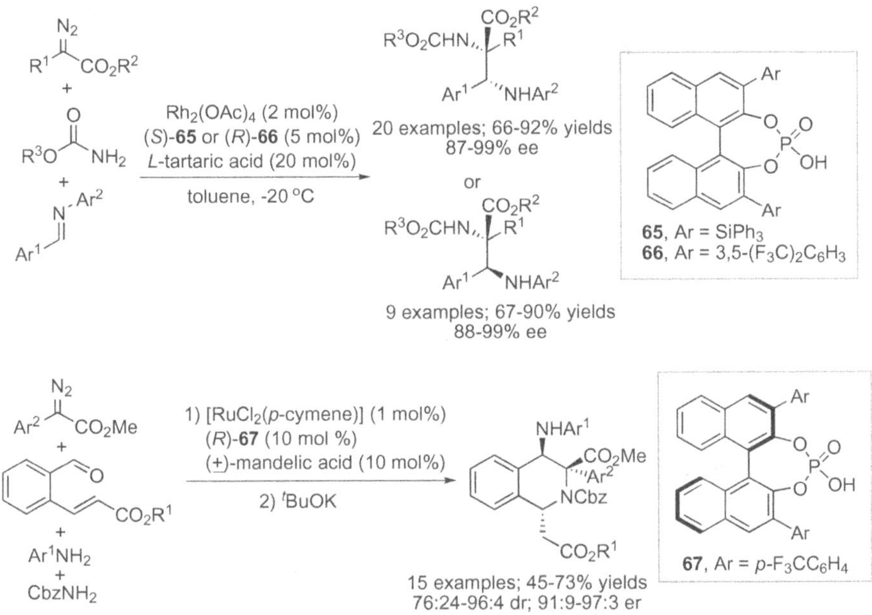

Scheme 6.39 Examples of three-component Mannich-type reaction of diazo compounds, amines and imines.

Ru(II) complexes co-catalyzed with chiral Brønsted acids. The scope of amino derivatives include carbamate,[173,178] arylamines[174,176,177] and phosphoramidate,[175] while the scope of C=N bond containing substrates included aryl aldehyde derived imines,[173,174] imino esters,[175,176] *N*-sulfonyl aldimines[177] and *in situ* generated imine.[178] Particularly, the diastereoselective switchable enantioselective ammonium ylide trapping reaction opened a superior avenue for synthesizing α-substituted α,β-diaminoesters in a controlled manner.[173] Moreover, the intriguing Ru(II)/chiral Brønsted acid cooperative catalytic system demonstrated high enantioselectivity toward ammonium ylide trapping by iminium/aza-Michael cascade.[178] Efficient preparation of optically pure 1,3,4-tetrasubstituted tetrahydroquinolines can be achieved through this four-component asymmetric synergistic synthetic method (Scheme 6.39).

Moreover, this nucleophilic addition reaction has been extended to azo compounds. Thus, the ammonium ylide intermediate **68** added to diethyl azodicarboxylate (DEAD) to give the desired adducts in moderate to good yields (Eq. (6.13)).[181]

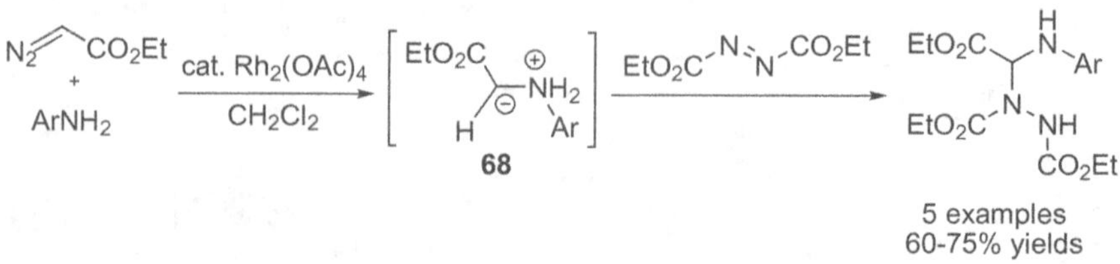

(6.13)

Aldol-type trapping of ammonium ylide with divergent carbonyl compounds were extensively explored by Hu and others.[182–189] Hu and co-workers successfully demonstrated a series of Rh(II)-catalyzed three-component reactions of diazo compounds, anilines with ketones/isatins[182] as well as formaldehyde[162,183] as the substrates. Moreover, they also established the intramolecular ammonium ylides trapping process with aryl ketones, which produced substituted 3-hydroxy-2,2,3-tri-substituted indolines **69** with good yields and high diastereoselectivity (Eq. (6.14)).[184] Notably, when choosing 2′-aminochalcones as the trapping reagents, the preference of aldol-type addition rather than Michael-type reaction was observed.

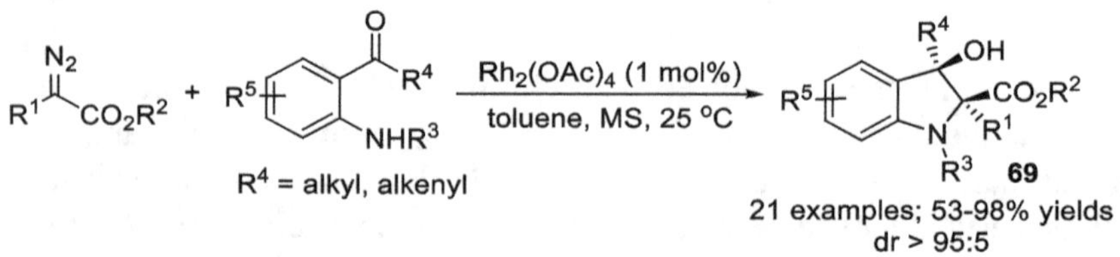

(6.14)

Furthermore, Hu group achieved three-component aldol-type reactions of EDA, aliphatic amines with β,γ-unsaturated α-keto esters by using iron porphyrin complex.[185] With the Fe(TPP) catalyst, this transformation overcomes the catalyst poisoning issue observed in the corresponding Rh(II)-catalyzed reactions. They also reported the diversity-oriented ammonium ylide trapping process of diazoesters, anilines with 4-oxo-enoates.[186] Through controlling the addition sequence of the three components, this pathway-switchable synthetic method provided either pyrrolidines derivatives through aza-Michael/ammonium ylide generation/intramolecular aldol addition, or linear α-amino esters through ammonium ylide generation/Michael addition (Scheme 6.40).

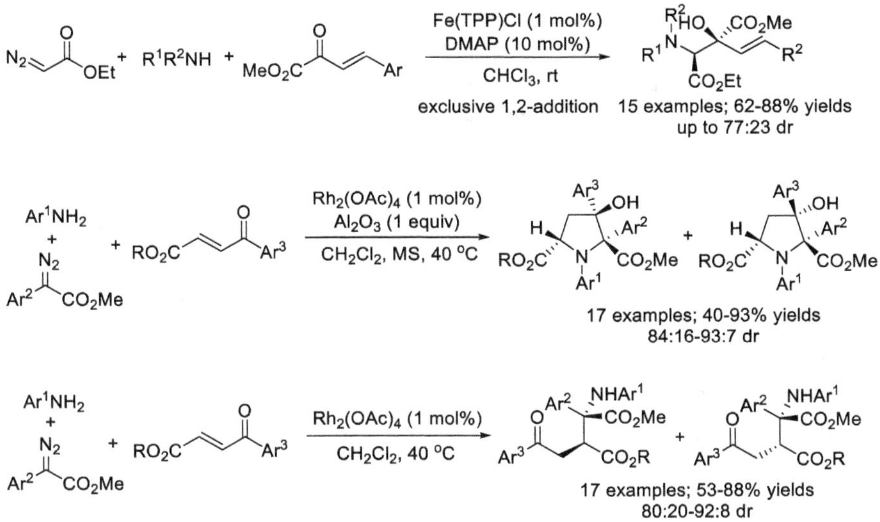

Scheme 6.40 1,2-addition *versus* 1,4-addition in the catalytic three-component reaction.

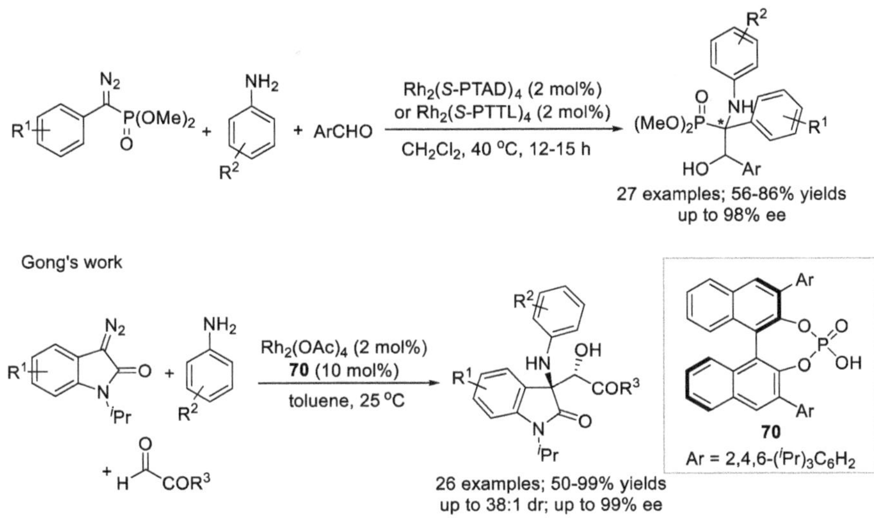

Scheme 6.41 Asymmetric catalytic three-component reaction through ammonium ylide.

As for the asymmetric catalysis in ammonium ylide trapping, the groups of Che and Gong utilized different strategies to tackle this problem (Scheme 6.41). Che and co-workers realized chiral Rh(II)-catalyzed enantioselective three-component coupling of diazophosphonates, anilines and electron-deficient aromatic aldehydes.[187] This asymmetric catalysis protocol explored the ylide trapping process with α-diazophosphonates as the substrates, resulting in the formation of α-amino-β-hydroxyphosphonates with moderate to high enantioselectivities. Alternatively, Gong and co-workers applied the Rh(II)/chiral Brønsted acid (**70**) in the three-component aldol-type reaction of 3-diazo oxindoles, anilines with glyoxylates.[188]

Recently, Hu and co-workers realized the aldol-type Rh(II)/In(III) co-catalyzed diastereoselective ammonium ylide trapping process with aldehydes.[189] This three-component method of diazo compounds, carbamates and aldehydes allowed the synthesis of β-hydroxyl-α-amino acid derivatives in. a highly stereoselective manner.

In addition, they realized Rh(II)-catalyzed one-pot three-component tandem Michael-type reactions of α-diazo compounds, anilines with β,γ-unsaturated α-keto esters (Scheme 6.42).[190–192] This multistep

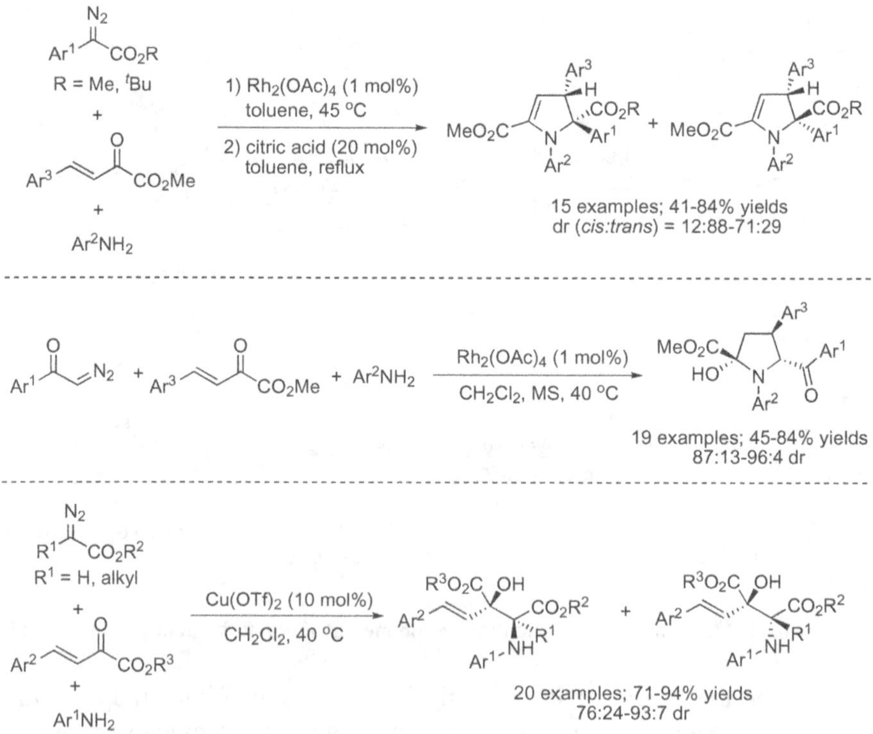

Scheme 6.42 Rh(II)- and Cu(II)-catalyzed three-component reaction of α-diazo compounds, anilines with β,γ-unsaturated α-keto esters.

strategy afforded polysubstituted pyrrolidines in good diastereoselectivity through ammonium ylide generation/1,4-addition/cyclization sequence, or 2,3-dihydropyrrole derivatives in moderate to good yields *via* subsequent dehydration. Notably, when α-unsubstituted or α-alkyl-substituted diazoacetates were employed as the substrates, the densely functionalized β-hydroxy-α-amino acids were obtained in high yields through Cu(I)-catalyzed ylide formation, followed by regioselective 1,2-addition.[191]

Through Rh(II)/Brønsted acid cooperative catalytic system, Hu and co-workers extended this ammonium ylide intramolecular trapping process with enones, leading to the formation of 2,2,3-tri-substituted indolines in a diastereoselective manner.[193] Furthermore, they applied the co-catalytic strategy in asymmetric synthesis. Chiral Rh(I)-diene (**71**) system has emerged as an efficient cooperative catalyst in enantioselective three-component reactions of aryldiazoacetates, anilines with β-nitroacrylates, affording polyfunctional amino succinic acids with high diastereo- and enantioselectivity (Scheme 6.43).[194]

Recently, a domino annulation toward poly-functionalized *N*-aryl pyrrolidines was achieved by the reaction of α,β-unsaturated δ-amino esters with Rh(II)-carbene.[195] This reaction proceeds *via* ammonium ylide generation, followed by intramolecular Michael addition to conjugate enolate.

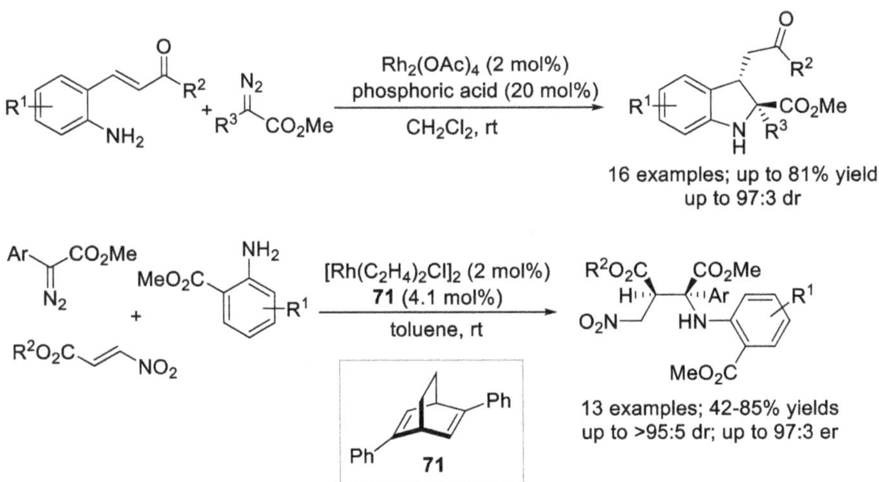

Scheme 6.43 Rh(II)/Brønsted acid and chiral Rh(I)-diene catalysis in Michael addition of ammonium ylides.

6.5 Ylides in stepwise cyclization reactions

The ylides generated from metal carbene have been explored in cyclization reactions, in which the transformations are normally initiated by the process where the ylides function as nucleophiles. For example, Kitamura and co-workers developed Rh(II)-catalyzed synthesis of 1,2-naphthalenediol derivatives from 1,2-diazonaphthoquinones. 1,2-naphthalenediol diacetates were obtained through Rh(II)-catalyzed diazo decomposition with acetic anhydride, which involved the formation of ylide with subsequent acetoxy migration.[196] Interestingly, Pd(II) carboxylate was found to be an efficient catalyst for macrocyclization of 1,2-diazonaphthoquinones with cyclic ethers.[197] The transformation is proposed to proceed through oxonium ylide **72** or aromatic palladium naphtholate, followed by intramolecular cyclization (Scheme 6.44).

Lacour and co-workers have studied the synthesis of various macrocycles with Rh(II)-catalyzed oxonium ylide reactions.[198–201] As shown in Scheme 6.45, densely functionalized polyether macrocycles were prepared through one-pot four-component reactions of diazoesters with cyclic ethers.[198,199] Mechanism hypothesis includes the synchronous dimerization of *in situ* generated oxonium ylide intermediate **73**.

Subsequently, they reported the one-pot synthesis of 15-membered unsaturated macrocycle through Rh(II)-catalyzed macrocyclization of oxetane with α-diazo-biscarbonyl compounds.[200] Besides, instead of the macrocycle formation, polyoxygenated medium-sized rings were synthesized through Rh(II)-catalyzed one-pot reaction of α-diazo-β-ketoesters

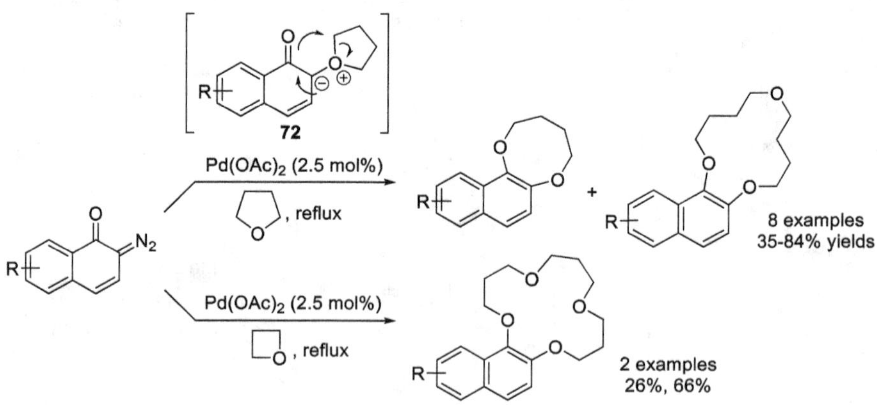

Scheme 6.44 Synthesis of 1,2-naphthalenediol derivatives from 1,2-diazonaphthoquinones.

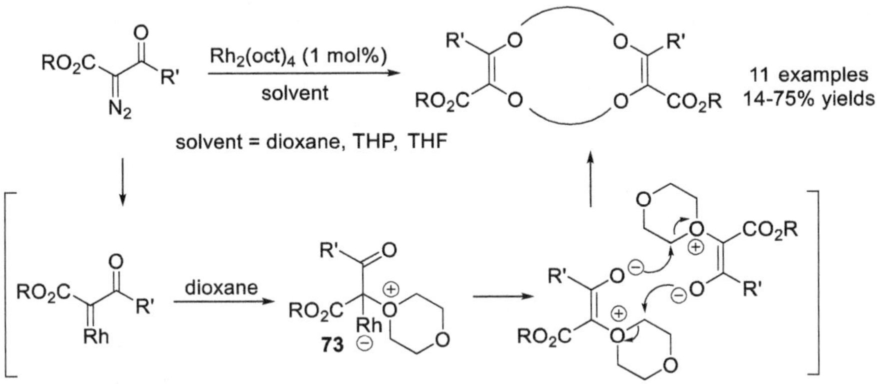

Scheme 6.45 Rh(II)-catalyzed one-pot four-component synthesis of polyether macrocycles.

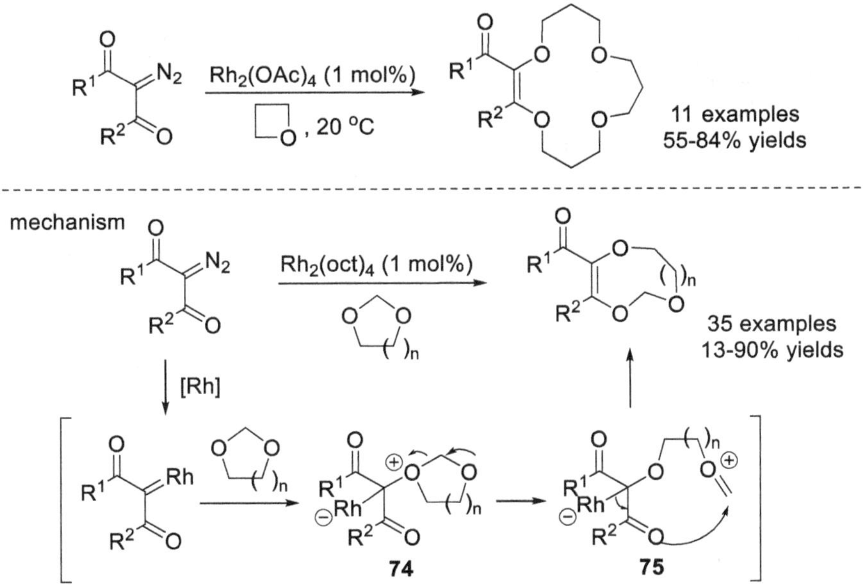

Scheme 6.46 Rh(II)-catalyzed oxonium ylide reaction from α-diazo-biscarbonyl compounds.

and diketones with cyclic acetals.[201] This reaction provided a facile approach for 8- or 9-membered rings in good yields (up to 90%) under mild reaction conditions (Scheme 6.46). The reaction is proposed to proceed through the generation of oxonium ylide **74**, followed by nucleophilic attack of lone-pair electron of unbound oxygen atom to lead to the formation of unsaturated oxocarbenium ions **75**, and finally intramolecular cyclization sequence to give the products.

With the combination of Ru(II) catalyst with 1,10-phenanthroline ligand, Lacour and co-workers demonstrated formal carbene insertion of α-diazo-β-ketoesters into epoxides.[202] 1,4-Dioxene products were obtained as single regio- and stereoisomers in high yields (Eq. (6.15)). Notably, the complete retention of original stereo carbon center was observed in this ring-opening synthesis.

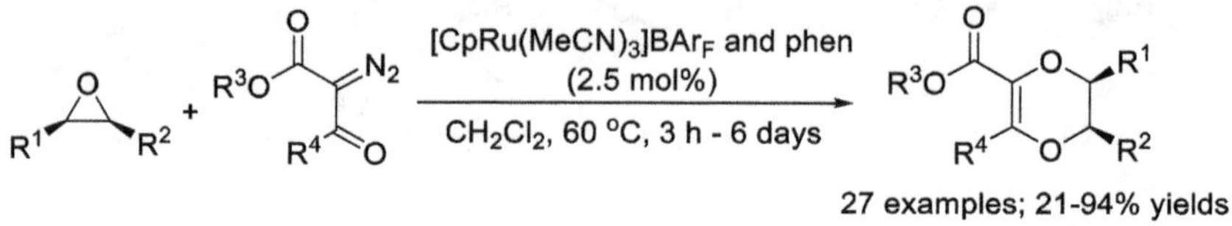

(6.15)

Moreover, Lacour and co-workers reported a carbonyl ylide generation/sequential cyclization process from α-diazo-β-ketoesters with cyclic ketones, lactones and carbonates (Scheme 6.47).[203] In the presence of Ru(II) catalyst and 1,10-phenanthroline, various spiro ketals, orthoesters and orthocarbonates were obtained effectively under mild conditions. Mechanistic rationale involves the generation of Ru(II)-carbene and carbonyl ylide intermediate, followed by intramolecular ring-closure to deliver the spiro products.

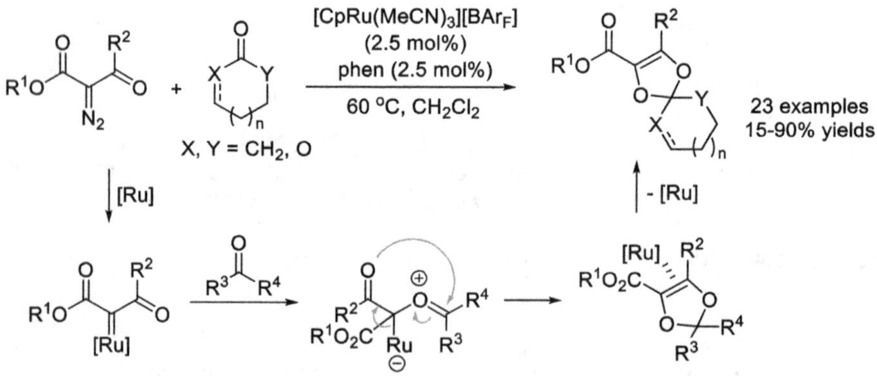

Scheme 6.47 Ru(II)-catalyzed reaction of α-diazo-β-ketoesters with cyclic carbonyls to form spiro bicyclic compounds.

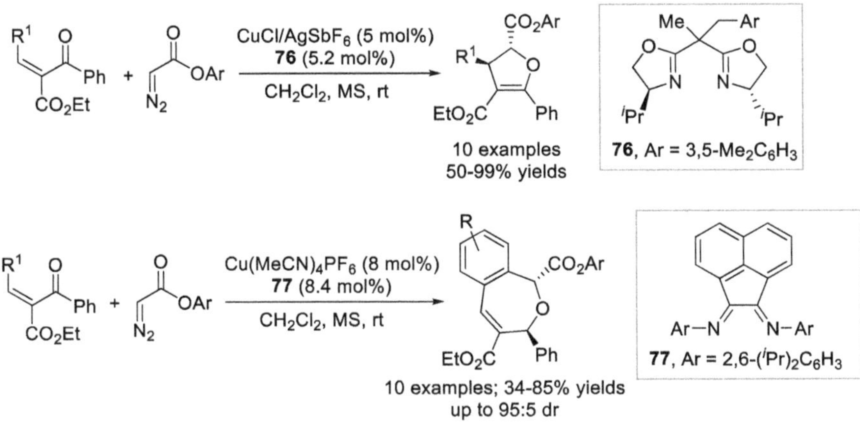

Scheme 6.48 Cu(I)-catalyzed synthesis of dihydrofurans or dihydrobenzoxepines.

For the stereoselective cyclization, Tang and co-workers reported the Cu(I)-catalyzed carbonyl ylide cyclization of α-benzylidene-β-dicarbonyl compounds with diazoacetate.[205] This reaction gave dihydrofurans or dihydrobenzoxepines with excellent diastereoselectivities (Scheme 6.48). Computational studies indicated that reaction pathway involved carbonyl ylide formation, followed by 1,5-electrocyclization; or 1,3-Cu migration, followed by 1,7-electrocyclization/1,5-H shift sequence.

Recently, a new synthesis of 2,5-dihydrofurans was disclosed, which involved Au(I)-catalyzed formal [4+1] cycloaddition of α-diazoesters and propargyl alcohols (Scheme 6.49).[206] This transformation was initiated by the nucleophilic attack of the propargyl alcohol on the carbene center to form oxonium ylide species **78**. Preliminary mechanistic investigation indicated that the reaction most likely occurred through a 5-*endo*-dig cyclization of α-hydroxy allene intermediate **79**.

Sun and co-workers reported a catalyst-dependent tandem annulation procedure between diazo carbonyl compounds and amino alkynes, leading to the stereodivergent synthesis of a series of five-membered *N*-heterocycles, including 2,3-dihydropyrroles, and 2-methylene and 3-methylene pyrrolidines (Scheme 6.50).[207] Mechanistically, the Cu(I)-catalyzed reaction proceeds *via* the generation of allenoate, followed by Cu(I)-catalyzed intramolecular hydroamination, while the Rh(II)-catalyzed reaction may be initiated by formal carbene N–H insertion and subsequent Zn-mediated Conia-ene cyclization.

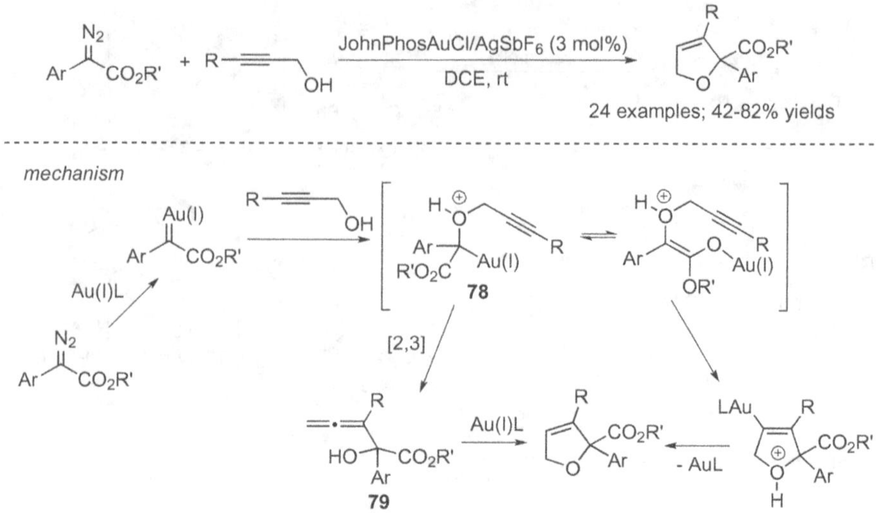

Scheme 6.49 Au(I)-catalyzed formal [4+1] cycloaddition of α-diazoesters and propargyl alcohols.

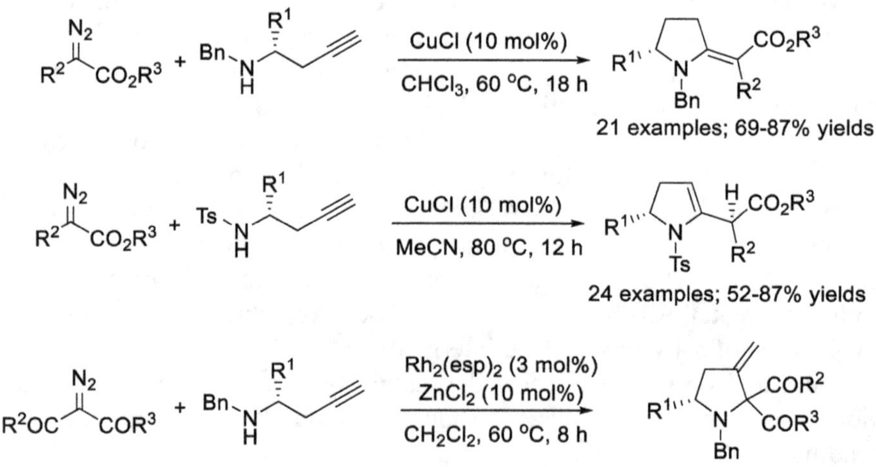

Scheme 6.50 Catalyst-dependent tandem annulation of diazo compounds with amino alkynes.

6.6 Azirinium ylide transformation

2*H*-Azirine as a unique type of nitrogen-based Lewis base has also been explored in the generation of ylide from metal carbene. Novikov and co-workers explored a series of transformations through azirinium ylide intermediate.[208] In 2009, the Rh(II)-catalyzed synthesis of 2*H*-1,4-oxazines was reported from 2*H*-azirines as the substrates with ethyl 2-acyl-2-diazoacetates (Scheme 6.51).[209,210] Mechanistic rationale involves the azirinium ylides **80** formation, followed by ring-opening process leading to azadienes **81** with subsequent 1,6-electrocyclization. On the other hand, the synthesis of 4-halo-2-azabuta-1,3-dienes could also be achieved through Rh(II)-catalyzed diazo ester-azirine coupling.[211]

Subsequently, the same group expanded this Rh(II)-catalyzed reaction with 2*H*-azirine-2-carbaldehydes and dimethyl diazomalonate, which afforded 2*H*-1,3-oxazines exclusively (Scheme 6.52).[212] Besides, this

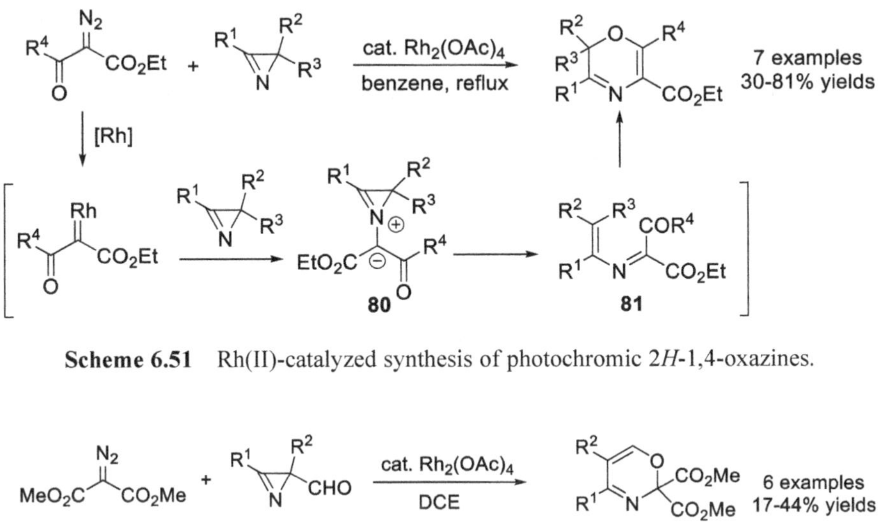

Scheme 6.51 Rh(II)-catalyzed synthesis of photochromic 2*H*-1,4-oxazines.

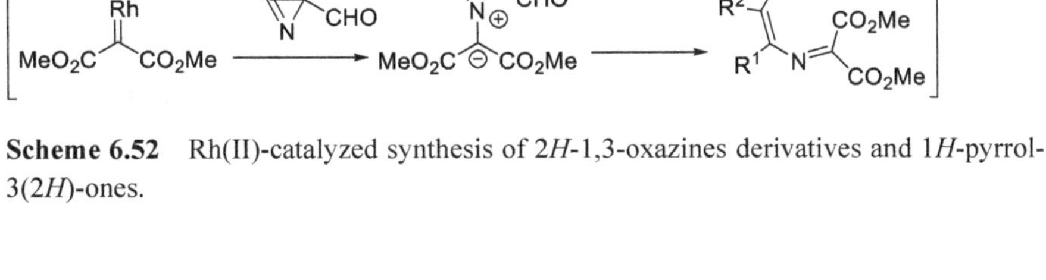

Scheme 6.52 Rh(II)-catalyzed synthesis of 2*H*-1,3-oxazines derivatives and 1*H*-pyrrol-3(2*H*)-ones.

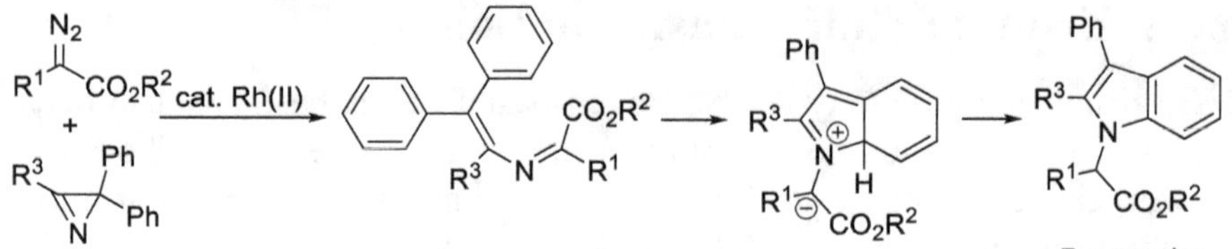

82

7 examples
61-99% yields
(from azadiene)

Scheme 6.53 Rh(II)-catalyzed azirinium ylide transformation from 2,2-diphenyl-2*H*-azirines and α-diazocarbonyl compounds.

Rh(II)-catalyzed strategy was also applied to the synthesis of polyfunctionalized 2*H*-1,3-oxazines and 1*H*-pyrrol-3(2*H*)-ones from cyano- or CF_3-substituted diazocarbonyl compounds.

Novikov and co-workers also demonstrated that the 2-azabuta-1,3-dienes, which were readily accessed by Rh(II)-catalyzed azirinium ylide transformation from 2,2-diphenyl-2*H*-azirines and α-diazocarbonyl compounds, could undergo thermal cyclization to afford indoles in good yields.[214] In the presence of electron-withdrawing group of C^1-substituents, this reaction first proceeded through pseudopericyclic 1,5-electrocyclization to generate 7a*H*-indolium ylide intermediate **82**, followed by 1,8-prototropic shift (Scheme 6.53).

With vinyl-substituted diazo substrates, Park and co-workers reported an efficient synthesis of pyridines through Rh(II)-catalyzed carbene mediated ring opening of 2*H*-azirines.[216] Reaction pathway was proposed to involve the formation of azirinium ylide **83** derived from vinyl diazo compounds, followed by ring-opening rearrangement to generate 3-azatriene **84**, 1,6-electrocyclization to afford 3,4-dihydropyridine **85**. Finally, the oxidation gave the pyridine derivatives (Scheme 6.54).

Furthermore, Park and co-workers developed a Cu(II)-catalyzed synthesis of unsymmetrically substituted pyrazines from α-diazo oxime ethers with 2*H*-azirines.[217] This transformation was initiated by azirinium ylide **86** formation from α-diazo carbene, followed by isomerization to 1,4-diazahexatriene **87**. Then under high temperature, electrocyclization occurred to generate 1,4-dihydropyrazine **88**, with spontaneous elimination of MeOH to provide pyrazine derivatives (Scheme 6.55).

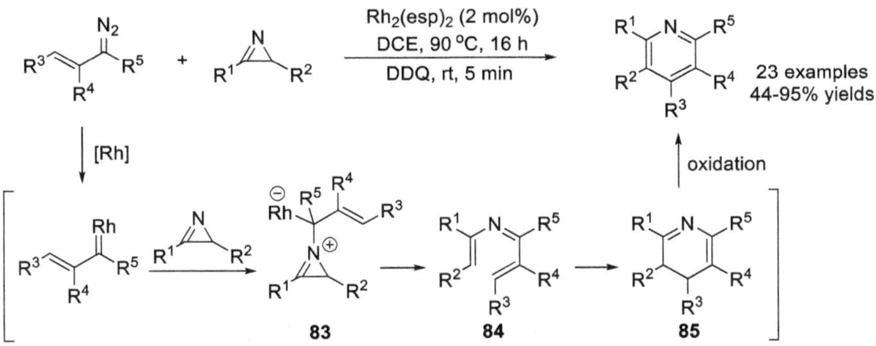

Scheme 6.54 Rh(II)-catalyzed synthesis of pyridines through azirinium ylide intermediate.

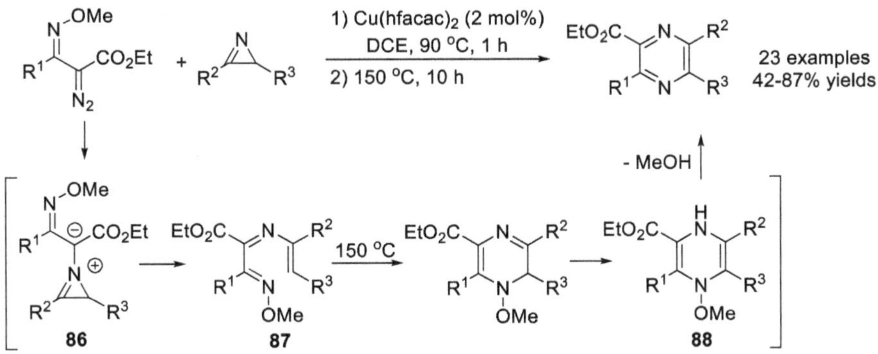

Scheme 6.55 Cu(II)-catalyzed synthesis of pyrazines.

6.7 Other ylide-related transformations

In addition to the reactions discussed above, there are some other ylide-related transformations that have been documented in the literature. For example, Hu and co-workers reported a C–S bond formation from diazoesters with *tert*-butyl sulfoxides in the presence of Cu(I) catalyst.[218] The S-centered chirality is transferred to C-centered chirality without racemization. Thus, the reaction represents a facile synthetic method for chiral sulfoxides. The reaction was proposed to proceed through intramolecular α',β-elimination of *in situ* generated sulfoxonium ylide intermediate **89** (Scheme 6.56).

By using trifluoromethyl thioether with dimethyl diazomalonate, Shen and Lu's group developed the Rh(II)-catalyzed synthetic approach for trifluoromethyl-substituted sulfonium ylide.[219] This ylide intermediate

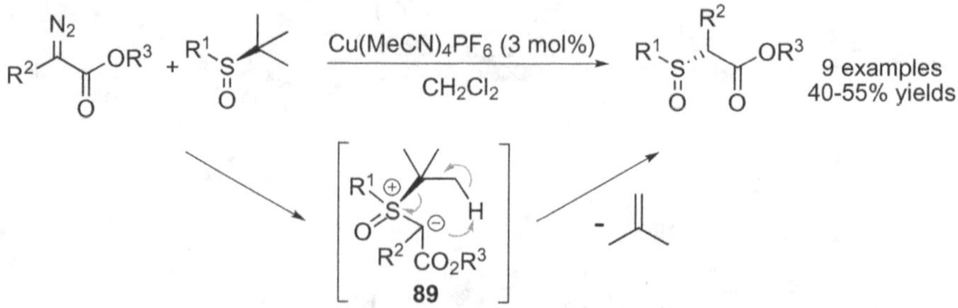

Scheme 6.56 Cu(II)-catalyzed synthesis of chiral sulfoxides through α',β-elimination.

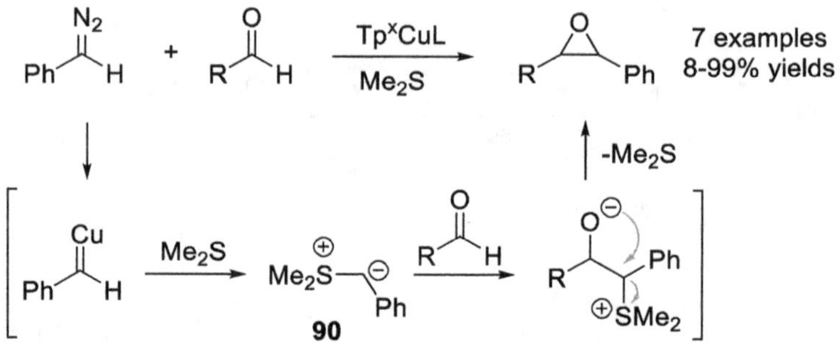

Scheme 6.57 Cooperative Cu(I) and sulfide-catalyzed epoxidation of aldehyde.

can act as an efficient electrophilic trifluoromethylation reagent, allowing direct trifluoromethylation of β-ketoesters and aryl iodides.

Nucleophilic addition of sulfonium ylides to a C=O or C=N bond is a route to produce epoxides or aziridines, with the release of sulfides. Aggarwal and co-workers developed a catalytic cycle to achieve asymmetric epoxidation with a sulfonium ylide as a reactive intermediate.[220]

Pérez revealed the efficient cooperative Cu- and sulfur-catalytic system for epoxidation of aldehydes by using prepared or *in situ* generated phenyldiazomethane reagent.[221] This protocol provided the corresponding epoxides in high yields with wide scope of the substituted aldehydes. The proposed reaction mechanism involved the formation of sulfur ylide **90** derived from dimethyl sulfide with Cu(I)-carbene species, followed by epoxidation process similar to Corey–Chaykovsky reaction (Scheme 6.57).

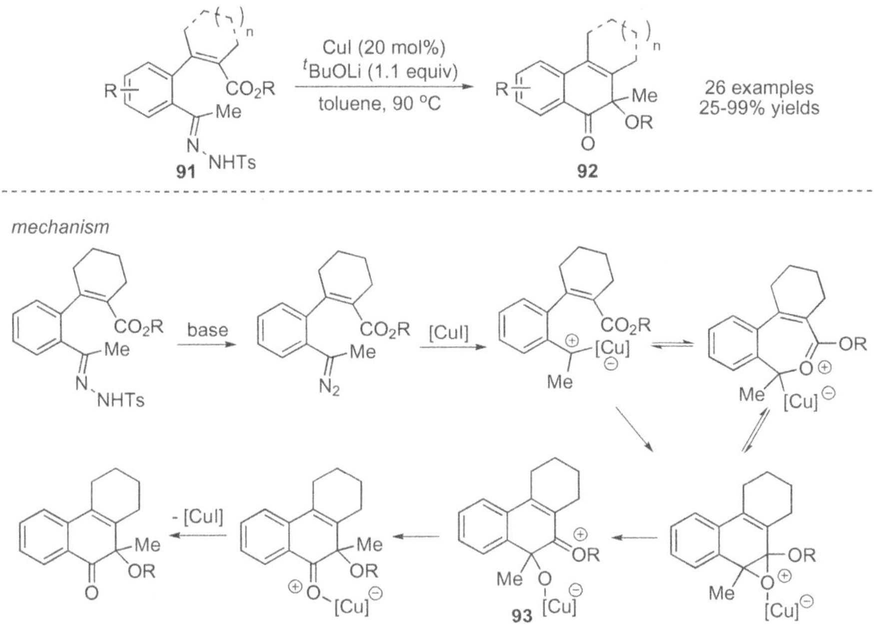

Scheme 6.58 Cu(I)-catalyzed 2*H*-naphthalenone formation of α-alkoxy cycloalkenones from *N*-tosylhydrazones.

Recently, Driver and co-workers reported that the efficient preparation of a broad range of functionalized α-alkoxy 2*H*-naphthalenones **92** from *N*-tosylhydrazones **91** could be triggered by inexpensive copper catalysts in the presence of base (Scheme 6.58).[222] This reaction was proposed to proceed *via* the generation of carbonyl ylide **91** and the sequential epoxidation. Subsequently, ring-opening process of the epoxide generated the oxocarbenium ion **93**, which triggered a [1,2]-methyl shift. This transformation has revealed a new reaction pattern for metal aryl carbenes toward pendant carboxylates, which transforms readily accessible aryl *N*-tosylhydrazones into functionalized α-alkoxy 2*H*-naphthalenone derivatives.

The 1,3-dipolar intermediates may follow pathways other than cycloaddition to double bonds. These pathways include rearrangement, ring closure to give epoxides and nucleophilic addition.[11] An interesting report is the addition of nucleophiles to the carbonyl ylides. As shown in

Eq. (6.16), the $Rh_2(OAc)_4$-catalyzed reaction of diazoketone generates the carbonyl ylide intermediate **94**, which is attacked by nucleophiles, such as alcohols or amines, to afford the bicyclic products.[223] The reaction gave only one diastereomeric product.

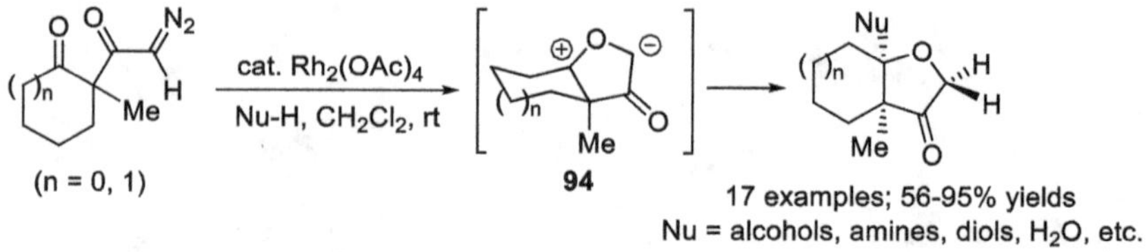

(6.16)

Furthermore, asymmetric reduction of carbonyl ylide intermediates was realized by using Rh(II)/chiral Brønsted acid binary catalytic system, which provided multisubstituted isochromene derivatives in high enantioselectivities.[224] This consecutive transformation proceeded through Rh(II)-catalyzed diazo decomposition along with carbonyl ylide **95** formation/tautomerization to oxidopyrylium **96**/isobenzopyrylium **97** generation/asymmetric reduction with Hantzsch ester (Scheme 6.59).

Finally, halonium ylides can be generated from the reaction of metal carbene with halides and subsequently a similar reaction may occur.[225–227] Recently, a synthesis of functionalized halonaphthalenyl ethers and esters was attained with Rh(II)-catalyzed reaction of 1-diazonaphthalen-2(1*H*)-ones with benzyl halides or acyl halides (Scheme 6.60).[228] The key step in this process was the nucleophilic attack of halogen atom on the

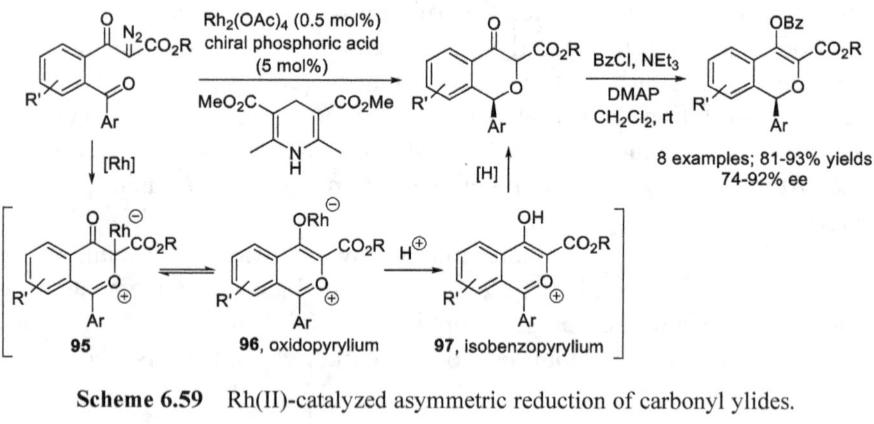

Scheme 6.59 Rh(II)-catalyzed asymmetric reduction of carbonyl ylides.

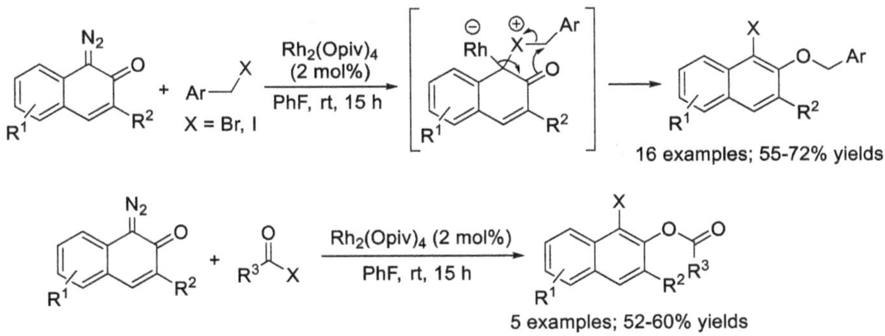

Scheme 6.60 Rh(II)-catalyzed synthesis of halonaphthalenyl ethers and esters.

Rh(II)-carbene species to generate halonium ylide, followed by intramolecular attack of oxygen on the benzylic carbon atom.

In summary, the ylide formation derived from diazo compounds and subsequent reactions represent a type of versatile synthetic approach toward the construction of various organic molecules. Among them, ylide-type [2,3]-sigmatropic rearrangement, Stevens rearrangement and intra- or intermolecular 1,3-dipolar cycloaddition reactions have been applied for the rapid assembly of polycyclic scaffolds. This type of transformation has served as an attractive disconnection strategy for the synthetic design of natural products. Besides, the multicomponent trapping reactions of the various ylides have high synthetic efficiency and atom economy, thus having potential applications in synthetic organic chemistry and drug-discovery research.

References

1. Wang, J. In *Comprehensive Organometallic Chemistry III*; Mingos, D. M. P.; Crabtree, R. H. ed.; *Applications II: Transition Metal Compounds in Organic Synthesis 2*, V. 11, Elsevier, Oxford, 2007, p. 151.
2. Vanecko, J. A.; Wan, H.; West, F. G. *Tetrahedron* **2006**, *62*, 1043–1062.
3. Padwa, A. *J. Org. Chem.* **2009**, *74*, 6421–6441.
4. Padwa, A. *Chem. Soc. Rev.* **2009**, *38*, 3072–3081.
5. Sweeney, J. B. *Chem. Soc. Rev.* **2009**, *38*, 1027–1038.
6. Zhang, Y.; Wang, J. *Coord. Chem. Rev.* **2010**, *254*, 941–953.
7. Padwa, A. *Tetrahedron* **2011**, *67*, 8057–8072.
8. Guo, X.; Hu, W. *Acc. Chem. Res.* **2013**, *46*, 2427–2440.

9. Murphy, G. K.; Stewart, C.; West, F. G. *Tetrahedron* **2013**, *69*, 2667–2686.
10. Padwa, A.; Cheng, B.; Zou, Y. *Aust. J. Chem.* **2014**, *67*, 343–353.
11. Jones, A. C.; May, J. A.; Sarpong, R.; Stoltz, B. M. *Angew. Chem. Int. Ed.* **2014**, *53*, 2556–2591.
12. West, T. H.; Spoehrle, S. S. M.; Kasten, K.; Taylor, J. E.; Smith, A. D. *ACS Catal.* **2015**, *5*, 7446–7479.
13. Sheng, Z.; Zhang, Z.; Chu, C.; Zhang, Y.; Wang, J. *Tetrahedron* **2017**, *73*, 4011–4022.
14. Crich, D.; Zou, Y.; Brebion, F. *J. Org. Chem.* **2006**, *71*, 9172–9177.
15. Liao, M.; Wang, J. *Green Chem.* **2007**, *9*, 184–188.
16. Peng, L.; Zhang, X.; Ma, M.; Wang, J. *Angew. Chem. Int. Ed.* **2007**, *46*, 1905–1908.
17. Ma, M.; Peng, L.; Li, C.; Zhang, X.; Wang, J. *J. Am. Chem. Soc.* **2005**, *127*, 15016–15017.
18. Davies, P. W.; Albrecht, S. J.-C.; Assanelli, G. *Org. Biomol. Chem.* **2009**, *7*, 1276–1279.
19. Holzwarth, M. S.; Alt, I.; Plietker, B. *Angew. Chem. Int. Ed.* **2012**, *51*, 5351–5354.
20. Li, Y.; Huang, Z.; Wu, X.; Xu, P.-F.; Jin, J.; Zhang, Y.; Wang, J. *Tetrahedron* **2012**, *68*, 5234–5240.
21. Liao, M.; Peng, L.; Wang, J. *Org. Lett.* **2008**, *10*, 693–696.
22. Li, Y.; Shi, Y.; Huang, Z.; Wu, X.; Xu, P.-F.; Wang, J.; Zhang, Y. *Org. Lett.* **2011**, *13*, 1210–1213.
23. Honda, K.; Shibuya, H.; Yasui, H.; Hoshino, Y.; Inoue, S. *Bull. Chem. Soc. Jpn.* **2008**, *81*, 142–147.
24. Sançon, J.; Sweeney, J. B. *Synlett* **2010**, 664–666.
25. Aviv, I.; Gross, Z. *Chem. Eur. J.* **2008**, *14*, 3995–4005.
26. Krishnamoorthy, P.; Browning, R. G.; Singh, S.; Sivappa, R.; Lovely, C. J.; Dias, H. V. R. *Chem. Commun.* **2007**, 731–733.
27. Murphy, G. K.; West, F. G. *Org. Lett.* **2006**, *8*, 4359–4361.
28. Clark, J. S.; Fessard, T. C.; Whitlock, G. A. *Tetrahedron* **2006**, *62*, 73–78.
29. Shimada, N.; Nakamura, S.; Anada, M.; Shiro, M.; Hashimoto, S. *Chem. Lett.* **2009**, *38*, 488–489.
30. Li, Z.; Davies, H. M. L. *J. Am. Chem. Soc.* **2010**, *132*, 396–401.
31. Li, Z.; Parr, B. T.; Davies, H. M. L. *J. Am. Chem. Soc.* **2012**, *134*, 10942–10946.
32. Li, Z.; Boyarskikh, V.; Hansen, J. H.; Autschbach, J.; Musaev, D. G.; Davies, H. M. L. *J. Am. Chem. Soc.* **2012**, *134*, 15497–15504.
33. Parr, B. T.; Li, Z.; Davies, H. M. L. *Chem. Sci.* **2011**, *2*, 2378–2382.
34. Parr, B. T.; Davies, H. M. L. *Nat. Commun.* **2014**, *5*, article number: 4455.
35. Parr, B. T.; Davies, H. M. L. *Org. Lett.* **2015**, *17*, 794–797.
36. Ioannou, M.; Porter, M. J.; Saez, F. *Tetrahedron* **2005**, *61*, 43–50.

37. Ellis-Holder, K. K.; Peppers, B. P.; Kovalevsky, A. Y.; Diver, S. T. *Org. Lett.* **2006**, *8*, 2511–2514.
38. Stepakov, A. V.; Molchanov, A. P.; Magull, J.; Vidović, D.; Starova, G. L.; Kopfc, J.; Kostikov, R. R. *Tetrahedron* **2006**, *62*, 3610–3618.
39. Zhu, S.; Xing, C.; Zhu, S. *Tetrahedron* **2006**, *62*, 829–832.
40. Qu, J.-P.; Xu, Z.-H.; Zhou, J.; Cao, C.-L.; Sun, X.-L.; Dai, L.-X.; Tang, Y. *Adv. Synth. Catal.* **2009**, *351*, 308–312.
41. Bhujbal, N. N.; Kumar, K. S. A.; Dhavale, D. D. *Synthesis* **2009**, 2423–2429.
42. Jaber, D. M.; Burgin, R. N.; Hepler, M.; Zavalij, P.; Doyle, M. P. *Chem. Commun.* **2011**, *47*, 7623–7625.
43. Skrobo, B.; Deska, J. *Org. Lett.* **2013**, *15*, 5998–6001.
44. Pospech, J.; Lennox, A. J. J.; Beller, M. *Chem. Commun.* **2015**, *51*, 14505–14508.
45. Garcia-Garibay, M. A.; Dang, H. *Org. Biomol. Chem.* **2009**, *7*, 1106–1114.
46. Bott, T. M.; Vanecko, J. A.; West, F. G. *J. Org. Chem.* **2009**, *74*, 2832–2836.
47. Xu, X.; Ratnikov, M. O.; Zavalij, P. Y.; Doyle, M. P. *Org. Lett.* **2011**, *13*, 6122–6125.
48. Harada, S.; Kono, M.; Nozaki, T.; Menjo, Y.; Nemoto, T.; Hamada, Y. *J. Org. Chem.* **2015**, *80*, 10317–10333.
49. Sharma, A.; Guénée, L.; Naubron, J.-V.; Lacour, J. *Angew. Chem. Int. Ed.* **2011**, *50*, 3677–3680.
50. Sharma, A.; Besnard, C.; Guénée, L.; Lacour, J. *Org. Biomol. Chem.* **2012**, *10*, 966–969.
51. Harthong, S.; Bach, R.; Besnard, C.; Guénée, L.; Lacour, J. *Synthesis* **2013**, *45*, 2070–2078.
52. Roberts, E.; Sançon, J. P.; Sweeney, J. B. *Org. Lett.* **2005**, *7*, 2075–2078.
53. Vanecko, J. A.; West, F. G. *Org. Lett.* **2005**, *7*, 2949–2952.
54. Manning, J. R.; Davies, H. M. L. *Tetrahedron* **2008**, *64*, 6901–6908.
55. Manning, J. R.; Davies, H. M. L. *J. Am. Chem. Soc.* **2008**, *130*, 8602–8603.
56. Khlebnikov, A. F.; Novikov, M. S.; Gorbunova, Y. G.; Galenko, E. E.; Mikhailov, K. I.; Pakalnis, V. V.; Avdontceva, M. S. *Beilstein J. Org. Chem.* **2014**, *10*, 1896–1905.
57. Boyarskikh, V.; Nyong, A.; Rainier, J. D. *Angew. Chem. Int. Ed.* **2008**, *47*, 5374–5377.
58. Wang, J.; Boyarskikh, V.; Rainier, J. D. *Org. Lett.* **2011**, *13*, 700–702.
59. Padwa, A.; Weingarten, M. D. *Chem. Rev.* **1996**, *96*, 223–269.
60. Padwa, A. *J. Organomet. Chem.* **2001**, *617-618*, 3–16.
61. Padwa, A. *Helv. Chim. Acta* **2005**, *88*, 1357–1374.
62. Padwa, A. *J. Organomet. Chem.* **2005**, *690*, 5533–5540.
63. Hong, X.; Mejía-Oneto, J. M.; Padwa, A. *Tetrahedron Lett.* **2006**, *47*, 8387–8390.

64. Mejía-Oneto, J. M.; Padwa, A. *Org. Lett.* **2006**, *8*, 3275–3278.
65. Hong, X.; Mejía-Oneto, J. M.; France, S.; Padwa, A. *Tetrahedron Lett.* **2006**, *47*, 2409–2412.
66. Hong, X.; France, S.; Mejía-Oneto, J. M.; Padwa, A. *Org. Lett.* **2006**, *8*, 5141–5144.
67. Hong, X.; France, S.; Padwa, A. *Tetrahedron* **2007**, *63*, 5962–5976.
68. Nakamura, S.; Sugano, Y.; Kikuchi, F.; Hashimoto, S. *Angew. Chem. Int. Ed.* **2006**, *45*, 6532–6535.
69. Hirata, Y.; Nakamura, S.; Watanabe, N.; Kataoka, O.; Kurosaki, T.; Anada, M.; Kitagaki, S.; Shiro, M.; Hashimoto, S. *Chem. Eur. J.* **2006**, *12*, 8898–8925.
70. Snider, B. B.; Wu, X.; Nakamura, S.; Hashimoto, S. *Org. Lett.* **2007**, *9*, 873–874.
71. Hodgson, D. M.; Pierard, F. Y. T. M.; Stupple, P. A. *Chem. Soc. Rev.* **2001**, *30*, 50–61.
72. Hodgson, D. M.; Angrish, D.; Labande, A. H. *Chem. Commun.* **2006**, 627–628.
73. Hodgson, D. M.; Angrish, D. *Adv. Synth. Catal.* **2006**, *348*, 2509–2514.
74. Oguri, H.; Schreiber, S. L. *Org. Lett.* **2005**, *7*, 47–50.
75. Looper, R. E.; Pizzirani, D.; Schreiber, S. L. *Org. Lett.* **2006**, *8*, 2063–2066.
76. Muthusamy, S.; Krishnamurthi, J.; Nethaji, M. *Chem. Commun.* **2005**, 3862–3864.
77. Anderson, R. J.; Raolji, G. B.; Kanazawa, A.; Greene, A. E. *Org. Lett.* **2005**, *7*, 2989–2991.
78. Muthusamy, S.; Gnanaprakasam, B. *Tetrahedron* **2007**, *63*, 3355–3362.
79. Molchanov, A. P.; Diev, V. V.; Magull, J.; Vidović, D.; Kozhushkov, S. I.; Meijere, A. d.; Kostikov, R. R. *Eur. J. Org. Chem.* **2005**, 593–599.
80. Rout, L.; Harned, A. M. *Chem. Eur. J.* **2009**, *15*, 12926–12928.
81. Hodgson, D. M.; Villalonga-Barber, C.; Goodman, J. M.; Pellegrinet, S. C. *Org. Biomol. Chem.* **2010**, *8*, 3975–3984.
82. Domingo, L. R.; Aurell, M. J.; Pérez, P.; Sáez, J. A. *RSC Adv.* **2012**, *2*, 1334–1342.
83. Yu, Y.; Cornelissen, L.; Wong, W.-T.; Chiu, P. *Synlett* **2015**, *26*, 1553–1556.
84. Nakhla, M. C.; Lee, C.-W.; Wood, J. L. *Org. Lett.* **2015**, *17*, 5760–5763.
85. Muthusamy, S.; Krishnamurthi, J.; Babu, S. A.; Suresh, E. *J. Org. Chem.* **2007**, *72*, 1252–1262.
86. Torssell, S.; Somfai, P. *Adv. Synth. Catal.* **2006**, *348*, 2421–2430.
87. Suga, H.; Ebiura, Y.; Fukushima, K.; Kakehi, A.; Baba, T. *J. Org. Chem.* **2005**, *70*, 10782–10791.
88. Hamaguchi, M.; Tomida, N.; Iyama, Y.; Oshima, T. *J. Org. Chem.* **2006**, *71*, 5162–5170.

89. Hamaguchi, M.; Tomida, N.; Iyama, Y. *J. Org. Chem.* **2007**, *72*, 1326–1334.
90. Suga, H.; Inoue, K.; Inoue, S.; Kakehi, A.; Shiro, M. *J. Org. Chem.* **2005**, *70*, 47–56.
91. Suga, H.; Suzuki, T.; Inoue, K.; Kakehi, A. *Tetrahedron* **2006**, *62*, 9218–9225.
92. Ruano, J. L. G.; Fraile, A.; Martín, M. R.; Neúñz, A. *J. Org. Chem.* **2006**, *71*, 6536–6541.
93. Son, S.; Fu, G. C. *J. Am. Chem. Soc.* **2007**, *129*, 1046–1047.
94. Bonderoff, S. A.; Padwa, A. *Org. Lett.* **2013**, *15*, 4114–4117.
95. Bonderoff, S. A.; Padwa, A. *Tetrahedron Lett.* **2015**, *56*, 3127–3129.
96. Shimada, N.; Anada, M.; Nakamura, S.; Nambu, H.; Tsutsui, H.; Hashimoto, S. *Org. Lett.* **2008**, *10*, 3603–3606.
97. Shimada, N.;Oohara, T.; Krishnamurthi, J.; Nambu, H.; Hashimoto, S. *Org. Lett.* **2011**, *13*, 6284–6287.
98. Krishnamurthi, J.; Nambu, H.;Takeda, K.; Anada, M.; Yamano, A.; Hashimoto, S.*Org. Biomol. Chem.* **2013**, *11*, 5374–5382.
99. Suga, H.; Higuchi, S.; Ohtsuka, M.; Ishimoto, D.; Arikawa, T.; Hashimoto, Y.; Misawa, S.; Tsuchida, T.; Kakehi, A.; Baba, T. *Tetrahedron* **2010**, *66*, 3070–3089.
100. Murarka, S.; Jia, Z.-J.; Merten, C.; Daniliuc, C.-G.; Antonchick, A. P.; Waldmann, H. *Angew. Chem. Int. Ed.* **2015**, *54*, 7653–7656.
101. Zhou, J.-L.; Wang, L.-J.; Xu, H.; Sun, X.-L.; Tang, Y. *ACS Catal.* **2013**, *3*, 685–688.
102. Suga, H.; Sekikawa, Y.; Misawa, S.; Kinugawa, D.; Oda, R.; Itoh, K.; Toda, Y.; Kiyono, R. *J. Org. Chem.* **2015**, *80*, 6687–6696.
103. Shi, T.; Guo, X.; Teng, S.; Hu, W. *Chem. Commun.* **2015**, *51*, 15204–15207.
104. For review of asymmetric 1,3-dipolar cycloadditions, see: Hashimoto, T.; Maruoka, K. *Chem. Rev.* **2015**, *115*, 5366–5412.
105. For review of formal [4+1] annulation reactions involving diazo components, see: Chen, J.-R.; Hu, X.-Q.; Lu, L.-Q.; Xiao, W.-J. *Chem. Rev.* **2015**, *115*, 5301–5365.
106. Hodgson, D. M.; Glen, R.; Redgrave, A. J. *Tetrahedron: Asymm.* **2009**, *20*, 754–757.
107. Shi, B.; Merten, S.; Wong, D. K. Y.; Chu, J. C. K.; Liu, L. L.; Lam, S. K; Jäger, A.; Wong, W.-T.; Chiu, P.; Metz, P. *Adv. Synth. Catal.* **2009**, *351*, 3128–3132.
108. Rodier, F.; Rajzmann, M.; Parrain, J.-L.; Chouraqui, G.; Commeiras, L. *Chem. Eur. J.* **2013**, *19*, 2467–2477.
109. DeAngelis, A.; Panne, P.; Yap, G. P. A.; Fox, J. M. *J. Org. Chem.* **2008**, *73*, 1435–1439.
110. Muthusamy, S.; Ramkumar, R.; Mishra, A. K. *Tetrahedron Lett.* **2011**, *52*, 148–150.

111. DeAngelis, A.; Taylor, M. T.; Fox, J. M. *J. Am. Chem. Soc.* **2009**, *131*, 1101–1105.
112. Zhu, S.; Chen, L.; Wang, C.; Liang, R.; Wang, X.; Ren, Y.; Jiang, H. *Tetrahedron* **2011**, *67*, 5507–5515.
113. Hashimoto, Y.; Itoh, K.; Kakehi, A.; Shiro, M.; Suga, H. *J. Org. Chem.* **2013**, *78*, 6182–6195.
114. Qiu, L.; Guo, X.; Zhou, J.; Liu, S.; Yang, L.; Wu, X.; Hu, W. *RSC Adv.* **2013**, *3*, 20065–20070.
115. Xu, X.; Guo, X.; Han, X.; Yang, L.; Hu, W. *Org. Chem. Front.* **2014**, *1*, 181–185.
116. Reddy, B. V. S.; Karthik, G.; Rajasekaran, T.; Sridhar, B. *Eur. J. Org. Chem.* **2015**, 2038–2041.
117. Xu, H.-W.; Li, G.-Y.; Wong, M.-K.; Che, C.-M. *Org. Lett.* **2005**, *7*, 5349–5352.
118. Wang, M.-Z.; Xu, H.-W.; Liu, Y.; Wong, M.-K.; Che, C.-M. *Adv. Synth. Catal.* **2006**, *348*, 2391–2396.
119. Galliford, C. V.; Martenson, J. S.; Stern, C.; Scheidt, K. A. *Chem. Commun.* **2007**, 631–633.
120. Galliford, C. V.; Scheidt, K. A. *J. Org. Chem.* **2007**, *72*, 1811–1813.
121. Kadina, A. P.; Khlebnikov, A. F.; Novikov, M. S.; Pérez, P. J.; Yufit, D. S. *Org. Biomol. Chem.* **2012**, *10*, 5582–5591.
122. Langhals, H.; Pust, T. *Eur. J. Org. Chem.* **2010**, 3140–3145.
123. Cano, I.; Gómez-Bengoa, E.; Landa, A.; Maestro, M.; Mielgo, A.; Olaizola, I.; Oiarbide, M.; Palomo, C. *Angew. Chem. Int. Ed.* **2012**, *51*, 10856–10860.
124. Cai, A.-J.; Zheng, Y.; Ma, J.-A. *Chem. Commun.* **2015**, *51*, 8946–8949.
125. Li, G.-Y.; Chen, J.; Yu, W.-Y.; Hong, W.; Che, C.-M. *Org. Lett.* **2003**, *5*, 2153–2156.
126. Lv, F.; Liu, S.; Hu, W. *Asian J. Org. Chem.* **2013**, *2*, 824–836.
127. Tang, M.; Xing, D.; Cai, M.; Hu, W. *Chin. J. Org. Chem.* **2014**, *34*, 1268–1276.
128. Lu, C.-D.; Liu, H.; Chen, Z.-Y.; Hu, W.-H.; Mi, A.-Q. *Org. Lett.* **2005**, *7*, 83–86.
129. Huang, H.; Guo, X.; Hu, W. *Angew. Chem. Int. Ed.* **2007**, *46*, 1337–1339.
130. Xue, J.; Luk, H. L.; Platz, M. S. *J. Am. Chem. Soc.* **2011**, *133*, 1763–1765.
131. Wang, J.; Kubicki, J.; Gustafson, T. L.; Platz, M. S. *J. Am. Chem. Soc.* **2008**, *130*, 2304–2313.
132. Wang, J.; Burdzinski, G.; Kubicki, J.; Gustafson, T. L.; Platz, M. S. *J. Am. Chem. Soc.* **2008**, *130*, 5418–5419.
133. Wang, J.; Burdzinski, G.; Kubicki, J.; Platz, M. S. *J. Am. Chem. Soc.* **2008**, *130*, 11195–11209.
134. Zhang, Y.; Kubicki, J.; Platz, M. S. *J. Am. Chem. Soc.* **2009**, *131*, 13602–13603.

135. Burdzinski, G.; Zhang, Y.; Selvaraj, P.; Sliwa, M.; Platz, M. S. *J. Am. Chem. Soc.* **2010**, *132*, 2126–2127.
136. Rudolf, P.; Buback, J.; Aulbach, J.; Nuernberger, P.; Brixner, T. *J. Am. Chem. Soc.* **2010**, *132*, 15213–15222.
137. Dye, D. F.; Köpke, T.; Ramabhadran, R. O.; Raghavachari, K.; Zaleski, J. M. *J. Am. Chem. Soc.* **2011**, *133*, 13110–13120.
138. Pharr, C. R.; Kopff, L. A.; Bennett, B.; Reid, S. A.; McMahon, R. J. *J. Am. Chem. Soc.* **2012**, *134*, 6443–6454.
139. Wentrup, C.; Bibas, H.; Kuhn, A.; Mitschke, U.; McMills, M. C. *J. Org. Chem.* **2013**, *78*, 10705–10717.
140. Costa, P.; Trosien, I.; Fernandez-Oliva, M.; Sanchez-Garcia, E.; Sander, W. *Angew. Chem. Int. Ed.* **2015**, *54*, 2656–2660.
141. For photo-induced ultrafast Wolff rearrangement, see: Cui, G.; Thiel, W. *Angew. Chem. Int. Ed.* **2013**, *52*, 433–436.
142. For review of perspective on physical organic chemistry, see: Platz, M. S. *J. Org. Chem.* **2014**, *79*, 2341–2353.
143. Xu, X.; Zhou, J.; Yang, L.; Hu, W. *Chem. Commun.* **2008**, 6564–6566.
144. Hu, W.; Xu, X.; Zhou, J.; Liu, W.-J.; Huang, H.; Hu, J.; Yang, L.; Gong, L.-Z. *J. Am. Chem. Soc.* **2008**, *130*, 7782–7783.
145. Zhang, X.; Zhang, N.; Guo, X.; Yang, L.; Hu, W. *Tetrahedron* **2009**, *65*, 8277–8282.
146. Guo, Z.; Shi, T.; Jiang, J.; Yang, L.; Hu, W. *Org. Biomol. Chem.* **2009**, *7*, 5028–5033.
147. Qian, Y.; Xu, X.; Jiang, L.; Prajapati, D.; Hu, W. *J. Org. Chem.* **2010**, *75*, 7483–7486.
148. Guo, Z.; Cai, M.; Jiang, J.; Yang, L.; Hu, W. *Org. Lett.* **2010**, *12*, 652–655.
149. Xu, X.; Qian, Y.; Yang, L.; Hu, W. *Chem. Commun.* **2011**, *47*, 797–799.
150. Qian, Y.; Jing, C.; Liu, S.; Hu, W. *Chem. Commun.* **2013**, *49*, 2700–2702.
151. Qiu, L.; Guo, X.; Ma, C.; Qiu, H.; Liu, S.; Yang, L.; Hu, W. *Chem. Commun.* **2014**, *50*, 2196–2198.
152. Zhai, C.; Xing, D.; Qian, Y.; Ji, J.; Ma, C.; Hu, W. *Synlett* **2014**, *25*, 1745–1750.
153. Tang, M.; Xing, D.; Huang, H.; Hu, W. *Chem. Commun.* **2015**, *51*, 10612–10615.
154. Alcaide, B.; Almendros, P.; Aragoncillo, C.; Callejo, R.; Ruiz, M. P.; Torres, M. R. *J. Org. Chem.* **2009**, *74*, 8421–8424.
155. Alcaide, B.; Almendros, P.; Aragoncillo, C.; Callejo, R.; Ruiz, M. P.; Torres, M. R. *Eur. J. Org. Chem.* **2012**, 2359–2366.
156. Zhang, X.; Huang, H.; Guo, X.; Guan, X.; Yang, L.; Hu, W. *Angew. Chem. Int. Ed.* **2008**, *47*, 6647–6649.
157. Yue, Y.; Guo, X.; Chen, Z.; Yang, L.; Hu, W. *Tetrahedron Lett.* **2008**, *49*, 6862–6865.

158. Han, X.; Jiang, L.; Tang, M.; Hu, W. *Org. Biomol. Chem.* **2011**, *9*, 3839–3843.
159. Ji, J.; Zhang, X.; Zhu, Y.; Qian, Y.; Zhou, J.; Yang, L.; Hu, W. *J. Org. Chem.* **2011**, *76*, 5821–5824.
160. Ji, J.; Zhang, X.; Jiang, L.; Hu, W. *Tetrahedron Lett.* **2012**, *53*, 182–185.
161. Lakshmi, N. V.; Sivakumar, P. M.; Muralidharan, D.; Doble, M.; Perumal, P. T. *RSC Adv.* **2013**, *3*, 496–507.
162. Wang, C.; Xing, D.; Wang, D.; Wu, X.; Hu, W. *J. Org. Chem.* **2014**, *79*, 3908–3916.
163. Guo, X.; Liu, W.; Hu, W. *Chem. Asian J.* **2014**, *9*, 117–120.
164. Nicolle, S. M.; Lewis, W.; Hayes, C. J.; Moody, C. J. *Angew. Chem. Int. Ed.* **2015**, *54*, 8485–8489.
165. Muthusamy, S.; Krishnamurthia, J.; Suresh, E. *Chem. Commun.* **2007**, 861–863.
166. Jing, C.; Xing, D.; Gao, L.; Li, J.; Hu, W. *Chem. Eur. J.* **2015**, *21*, 19202–19207.
167. Xu, X.; Han, X.; Yang, L.; Hu, W. *Chem. Eur. J.* **2009**, *15*, 12604–12607.
168. Guan, X.-Y.; Yang, L.-P.; Hu, W. *Angew. Chem. Int. Ed.* **2010**, *49*, 2190–2192.
169. Zhu, Y.; Zhai, C.; Yang, L.; Hu, W. *Chem. Commun.* **2010**, *46*, 2865–2867.
170. Han, X.; Gan, M.; Qiu, H.; Ji, J.; Zhang, X.; Jiang, L.; Hu, W. *Synlett* **2011**, 1717–1722.
171. Jiang, J.; Guan, X.; Liu, S.; Ren, B.; Ma, X.; Guo, X.; Lv, F.; Wu, X.; Hu, W. *Angew. Chem. Int. Ed.* **2013**, *52*, 1539–1542.
172. Qiu, L.; Gao, L.; Tang, J.; Wang, D.; Guo, X.; Liu, S.; Yang, L.; Li, J.; Hu, W. *J. Org. Chem.* **2014**, *79*, 4142–4147.
173. Jiang, J.; Xu, H.-D.; Xi, J.-B.; Ren, B.-Y.; Lv, F.-P.; Guo, X.; Jiang, L.-Q.; Zhang, Z.-Y.; Hu, W.-H. *J. Am. Chem. Soc.* **2011**, *133*, 8428–8431.
174. Jiang, L.; Zhang, D.; Wang, Z.; Hu, W. *Synthesis* **2013**, *45*, 452–458.
175. Jiang, J.; Ma, X.; Liu, S.; Qian, Y.; Lv, F.; Qiu, L.; Wu, X.; Hu, W. *Chem. Commun.* **2013**, *49*, 4238–4240.
176. Jing, C.; Xing, D.; Qian, Y.; Hu, W. *Synthesis* **2014**, *46*, 1348–1354.
177. Qiu, L.; Wang, D.; Lv, F.; Guo, X.; Hu, W.; Yang, L.; Liu, S. *Tetrahedron* **2014**, *70*, 1471–1477.
178. Jiang, J.; Ma, X.; Ji, C.; Guo, Z.; Shi, T.; Liu, S.; Hu, W. *Chem. Eur. J.* **2014**, *20*, 1505–1509.
179. Wang, Y.; Zhu, Y.; Chen, Z.; Mi, A.; Hu, W.; Doyle, M. P. *Org. Lett.* **2003**, *5*, 3923–3926.
180. Wang, Y.; Chen, Z.; Mi, A.; Hu, W. *Chem. Commun.* **2004**, 2486–2487.
181. Huang, H.; Wang, Y.; Chen, Z.; Hu, W. *Adv. Synth. Catal.* **2005**, *347*, 531–534.
182. Guo, X.; Yue, Y.; Hu, G.; Zhou, J.; Zhao, Y.; Yang, L.; Hu, W. *Synlett* **2009**, 2109–2114.

183. Wang, C.; Liu, S.; Xing, D.; Wang, X.; Wu, X.; Hu, W. *Tetrahedron* **2013**, *69*, 11203–11208.
184. Jing, C.; Xing, D.; Hu, W. *Chem. Commun.* **2014**, *50*, 951–953.
185. Ma, C.; Xing, D.; Zhai, C.; Che, J.; Liu, S.; Wang, J.; Hu, W. *Org. Lett.* **2013**, *15*, 6140–6143.
186. Jing, C.; Xing, D.; Qian, Y.; Shi, T.; Zhao, Y.; Hu, W. *Angew. Chem. Int. Ed.* **2013**, *52*, 9289–9292.
187. Zhou, C.-Y.; Wang, J.-C.; Wei, J.; Xu, Z.-J.; Guo, Z.; Low, K.-H.; Che, C-.M. *Angew. Chem. Int. Ed.* **2012**, *51*, 11376–11380.
188. Ren, L.; Lian, X.-L.; Gong, L.-Z. *Chem. Eur. J.* **2013**, *19*, 3315–3318.
189. Xi, J.-B.; Ma, M.-L.; Hu, W. *Tetrahedron* **2015**, *72*, 579–583.
190. Zhu, Y.; Zhai, C.; Yue, Y.; Yang, L.; Hu, W. *Chem. Commun.* **2009**, 1362–1364.
191. Zhu, Y.; Zhai, C.; Yang, L.; Hu, W. *Eur. J. Org. Chem.* **2011**, 1113–1124.
192. Zhang, X.; Ji, J.; Zhu, Y.; Jing, C.; Li, M.; Hu, W. *Org. Biomol. Chem.* **2012**, *10*, 2133–2138.
193. Jiang, L.; Xu, R.; Kang, Z.; Feng, Y.; Sun, F.; Hu, W. *J. Org. Chem.* **2014**, *79*, 8440–8446.
194. Ma, X.; Jiang, J.; Lv, S.; Yao, W.; Yang, Y.; Liu, S.; Xia, F.; Hu, W. *Angew. Chem. Int. Ed.* **2014**, *53*, 13136–13139.
195. Medvedev, J. J.; Galkina, O. S.; Klinkova, A. A.; Giera, D. S.; Hennig, L.; Schneider, C.; Nikolaev, V. A. *Org. Biomol. Chem.* **2015**, *13*, 2640–2651.
196. Kitamura, M.; Kisanuki, M.; Okauchi, T. *Eur. J. Org. Chem.* **2012**, 905–907.
197. Kitamura, M.; Kisanuki, M.; Kanemura, K.; Okauchi, T. *Org. Lett.* **2014**, *16*, 1554–1557.
198. Zeghida, W.; Besnard, C.; Lacour, J. *Angew. Chem. Int. Ed.* **2010**, *49*, 7253–7256.
199. Vishe, M.; Hrdina, R.; Guénée, L.; Besnard, C.; Lacour, J. *Adv. Synth. Catal.* **2013**, *355*, 3161–3169.
200. Rix, D.; Ballesteros-Garrido, R.; Zeghida, W.; Besnard, C.; Lacour, J. *Angew. Chem. Int. Ed.* **2011**, *50*, 7308–7311.
201. Ballesteros-Garrido, R.; Rix, D.; Besnard, C.; Lacour, J. *Chem. Eur. J.* **2012**, *18*, 6626–6631.
202. Achard, T.; Tortoreto, C.; Poblador-Bahamonde, A. I.; Guénée, L.; Bürgi, T.; Lacour, J. *Angew. Chem. Int. Ed.* **2014**, *53*, 6140–6144.
203. Tortoreto, C.; Achard, T.; Egger, L.; Guénée, L.; Lacour, J. *Org. Lett.* **2016**, *18*, 240–243.
204. For research highlight, see: Dias, F. R. F. *Synlett* **2013**, *24*, 397–398.
205. Zhou, J.-L.; Liang, Y.; Deng, C.; Zhou, H.; Wang, Z.; Sun, X.-L.; Zheng, J.-C.; Yu, Z.-X.; Tang, Y. *Angew. Chem. Int. Ed.* **2011**, *50*, 7874–7878.

206. Wang, J.; Yao, X.; Wang, T.; Han, J.; Zhang, J.; Zhang, X.; Wang, P.; Zhang, Z. *Org. Lett.* **2015**, *17*, 5124–5127.
207. Liu, K.; Zhu, C.; Min, J.; Peng, S.; Xu, G.; Sun, J. *Angew. Chem. Int. Ed.* **2015**, *54*, 12962–12967.
208. Khlebnikov, A. F.; Novikov, M. S. *Russ. J. Gen. Chem.* **2010**, *80*, 1652–1666.
209. Khlebnikov, V. A.; Novikov, M. S.; Khlebnikov, A. F.; Rostovskii, N. V. *Tetrahedron Lett.* **2009**, *50*, 6509–6511.
210. Rostovskii, N. V.; Novikov, M. S.; Khlebnikov, A. F.; Khlebnikov, V. A.; Korneev, S. M. *Tetrahedron* **2013**, *69*, 4292–4301.
211. Novikov, M. S.; Smetanin, I. A.; Khlebnikov, A. F.; Rostovskii, N. V.; Yufit, D. S. *Tetrahedron Lett.* **2012**, *53*, 5777–5780.
212. Zavyalov, K. V.; Novikov, M. S.; Khlebnikov, A. F.; Yufit, D. S. *Tetrahedron* **2013**, *69*, 4546–4551.
213. Zavyalov, K. V.; Novikov, M. S.; Khlebnikov, A. F.; Pakalnis, V. V. *Tetrahedron* **2014**, *70*, 3377–3384.
214. Novikov, M. S.; Khlebnikov, A. F.; Rostovskii, N. V.; Tcyrulnikov, S.; Suhanova, A. A.; Zavyalov, K. V.; Yufit, D. S. *J. Org. Chem.* **2015**, *80*, 18–29.
215. Smetanin, I. A.; Novikov, M. S.; Rostovskii, N. V.; Khlebnikov, A. F.; Starova, G. L.; Yufit, D. S. *Tetrahedron* **2015**, *71*, 4616–4628.
216. Loy, N. S. Y.; Singh, A.; Xu, X.; Park, C.-M. *Angew. Chem. Int. Ed.* **2013**, *52*, 2212–2216.
217. Loy, N. S. Y.; Kim, S.; Park, C.-M. *Org. Lett.* **2015**, *17*, 395–397.
218. Guan, X.-Y.; Liu, Z.-Q.; Huang, H.-X.; Yang, L.-P.; Jian, L.; Hu, W.-H. *Synlett* **2009**, 2183–2187.
219. Liu, Y.; Shao, X.; Zhang, P.; Lu, L.; Shen, Q. *Org. Lett.* **2015**, *17*, 2752–2755.
220. Aggarwal, V. K.; Winn, C. L. *Acc. Chem. Res.* **2004**, *37*, 611–620.
221. Pereira, A.; Martín, C.; Maya, C.; Belderrain, T. R.; Pérez, P. J. *Adv. Synth. Catal.* **2013**, *355*, 2942–2951.
222. Su, N.; Theorell, J. A.; Wink, D. J.; Driver, T. G. *Angew. Chem. Int. Ed.* **2015**, *54*, 12942–12946.
223. Muthusamy, S.; Gnanaprakasam, B.; Suresh, E. *Org. Lett.* **2005**, *7*, 4577–4580.
224. Terada, M.; Toda, Y. *Angew. Chem. Int. Ed.* **2012**, *51*, 2093–2097.
225. Doyle, M. P.; Forbes, D. C.; Vasbinder, M. M.; Peterson, C. S. *J. Am. Chem. Soc.* **1998**, *120*, 7653–7654.
226. Xu, B.; Tambar, U. K. *J. Am. Chem. Soc.* **2016**, *138*, 12073–12076.
227. Xu, B.; Tambar, U. K. *Angew. Chem. Int. Ed.* **2017**, *56*, 9868–9871.
228. Baral, E. R.; Lee, Y. R.; Kim, S. H.; Wee, Y.-J. *Synthesis* **2016**, *48*, 579–587.

Chapter 7

Wolff Rearrangement and Related 1,2-Shift Reactions

Under photochemical, thermal or catalytic conditions, α-diazocarbonyl compounds may undergo Wolff rearrangement with the generation of active ketene intermediate **1**, followed by subsequent transformations, which include [2+2], [4+2] cycloadditions or nucleophilic additions (Scheme 7.1). The Wolff rearrangement has been widely used in organic synthesis for the homologation of carboxylic acids (Arndt–Eistert reaction), and for ring contraction leading to strained cyclic systems. The Wolff rearrangement has been reviewed in detail recently.[1–4] Herein we will introduce some selected advances in the last few years. Moreover, the related 1,2-shift reactions of carbenes will also be summarized at the end of this chapter.

7.1 [2+2] Cycloaddition

The [2+2] cycloadditions of ketene intermediates generated *in situ* from Wolff rearrangement with unsaturated compounds, such as alkenes, alkynes and imines, represent a type of classic method for the synthesis of four-membered ring systems. In particular, the cycloaddition between ketene with imines (Staudinger reaction) has been a powerful method for the synthesis of poly-functionalized β-lactams. Recently, Xu and co-workers expanded the method by employing α-diazocarbonyl substrates substituted with functional groups.[5–9] For example, by using ethoxycarbonyl(phenylthio)ketene **2**, which is generated by thia-Wolff

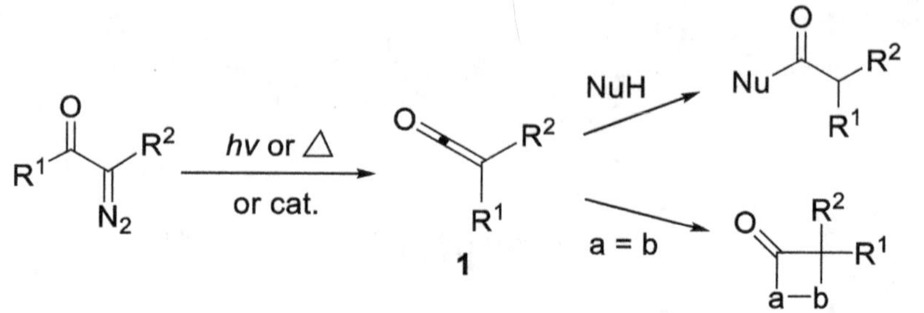

Scheme 7.1 Wolff rearrangement and sequential transformations.

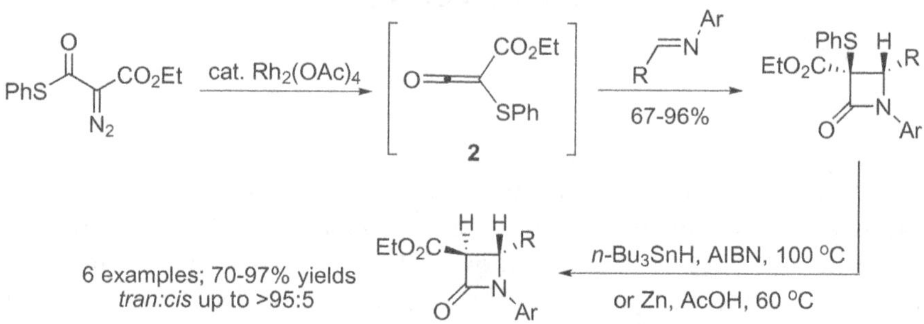

Scheme 7.2 Thia-Wolff rearrangement catalyzed by $Rh_2(OAc)_4$.

rearrangement in $Rh_2(OAc)_4$-catalyzed reaction of 3-phenylthio-2-diazo-3-oxopropionate, the synthesis of 3-alkoxycarbonyl β-lactam derivatives was developed (Scheme 7.2).[5] Notably, in this thia-Wolff rearrangement the 1,2-migrating group is the phenylthio group.

Moreover, a series of α-amino acid-derived β-lactams,[10,11] 3-alkoxy/aryloxy-β-lactams[12] and spiro(oxindolyl)-β-lactams[13] were efficiently prepared by catalytic or photo-irradiated [2+2] cycloaddition from α-diazocarbonyl substrates with imines. The substituent effects on the diastereoselective control in Staudinger reaction have also been evaluated.[14,15]

Danheiser and co-workers have extensively explored benzannulation method based on Wolff rearrangement. From α'-silyl-α'-diazo-α,β-unsaturated ketones **3**, (trialkylsilyl)vinylketenes **4** was generated through photochemical Wolff rearrangement. The (trialkylsilyl)vinylketenes **4** are found stable at 25°C and amenable to purification using conventional silica gel chromatography. The reaction of ketene **4** with ynolates afforded the benzannulation products **7**. The reaction is proceeded from the initial *C*-acylation of the ynolates by the vinylketene intermediates, followed by [4+2] 6π electrocyclization (Scheme 7.3).[16]

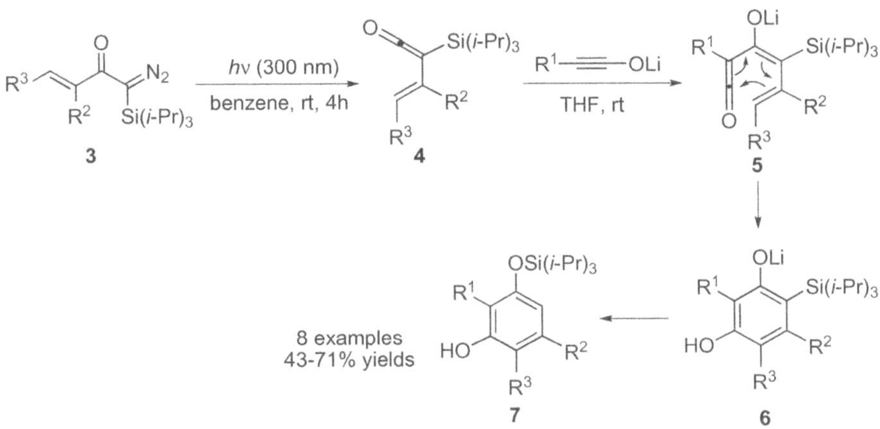

Scheme 7.3 Benzannulation through the reaction of (trialkylsilyl)vinylketene with ynolate.

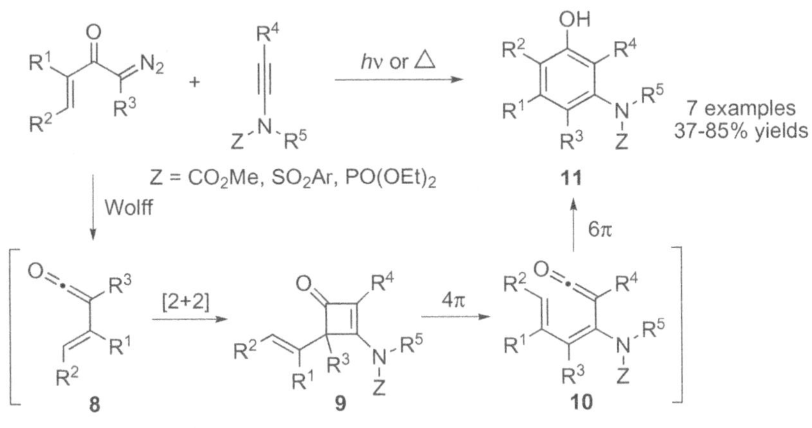

Scheme 7.4 Benzannulation based on the reaction of diazoketone with ynamides.

Furthermore, the same group reported the benzannulation cascade from diazoketones with substituted ynamides under batch or continuous-flow conditions.[17] The reaction proceeds through the following sequence: (1) Wolff rearrangement to generate ketene **8**; (2) [2+2] cycloaddition to generate **9**; (3) electrocyclic ring-opening of **9** to generate intermediate **10**; (4) finally, 6π electrocyclic ring-closure to give product **11** (Scheme 7.4). The strategy provided a straightforward synthetic approach to densely substituted polycyclic aromatic rings.

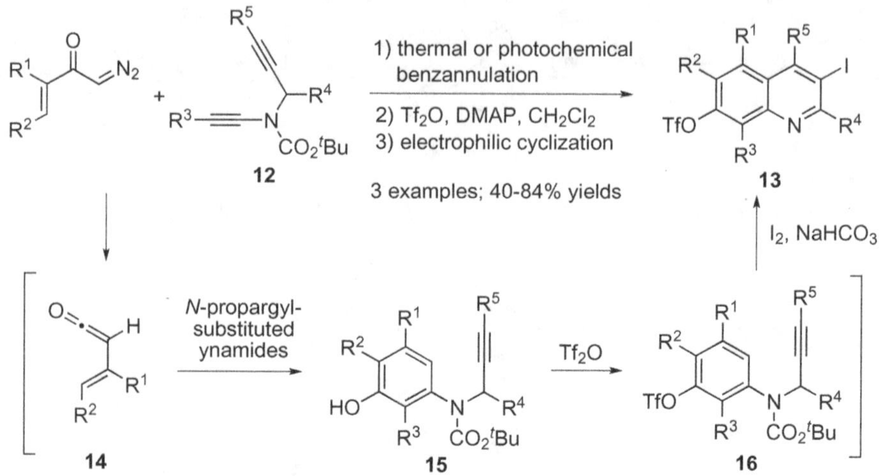

Scheme 7.5 Synthesis of highly substituted quinolines *via* Wolff rearrangement, benzannulation and iodocyclization sequence.

In 2015, Danheiser and co-workers reported the synthesis of highly substituted quinolines through benzannulation/iodocyclization process (Scheme 7.5).[18] Initially, this tandem reaction proceeds *via* cascade pericyclic benzannulation of Wolff-rearranged ketene **14** with *N*-propargyl-substituted ynamide **12**, affording highly substituted aniline derivative **15**. Subsequently, the triflate derivative **16** undergoes iodocyclization to afford poly-substituted quinolines **13**.

Recently, a synthesis of β-lactam-triflones was reported by catalyst-free reaction of 2-diazo-1-aryl-2-(trifluoromethylsulfonyl)ethanones with imines (Eq. (7.1)).[19] This Wolff rearrangement/Staudinger [2+2] cycloaddition cascade afforded substituted β-lactam-triflones **17** with high regio- and diastereoselectivity. The author indicated that both electronic effect and steric hindrance of R group in imine moiety, as well as the electronic effect of the aryl group, played crucial roles in the eventual [2+2] ring-closure process.

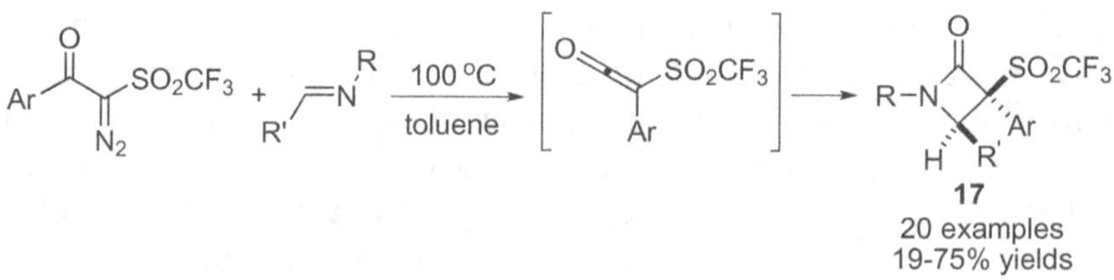

(7.1)

7.2 Nucleophilic addition to ketenes

Known as Arndt–Eistert homologation, the traditional homologation of carboxylic acids involves Wolff rearrangement of α-diazoketone to generate ketene intermediate, followed by nucleophilic addition of water or alcohol. A recent example was shown in Scheme 7.6 for the synthesis of 3-substituted homotaurines **18** from naturally occurring amino acids.[20]

As a well-established synthetic methodology, the Arndt–Eistert homologation has continued to find applications in organic synthesis in the past few years.[21–26] A remarkable example of a Wolff rearrangement-based ring contraction in synthesis was reported by Mascitti and Corey in the total synthesis of pentacycloanammoxic acid **20**, a highly unusual fatty acid with a rigid ladder structure. To construct the unusual fused four-membered ring system **19**, ring contraction by the Wolff rearrangement was repeated two times (Scheme 7.7).[24,25]

Nucleophilic addition of ketene intermediates generated by Wolff rearrangement with alcohols or amines provided the method for the synthesis of esters or amides. For example, light-induced Wolff rearrangement of *N*-diazoacetylpyrrole[27] or β-alkynyl-α-diazo-β-ketoesters[28] was carried out in alcoholic solvents. Thermal Wolff rearrangement of cyclic or acyclic α-diazocarbonyl compounds with subsequent nucleophilic attack by nitrogen nucleophiles afforded *N*-acyl isatin[29] or cyclic β-enaminoamides.[30] Furthermore, commercially available LED lamps were explored as simple and inexpensive alternatives for carrying out classic Wolff rearrangement reactions.[31]

Banert and co-workers reported a 1,3-dipolar cycloaddition of nitrous oxide with strained cyclic alkynes **21** for the generation of diazoketones.[32] The *in situ* generated diazo species **22** underwent various transformations, including Wolff rearrangement (Scheme 7.8). Ring-contracted ester and

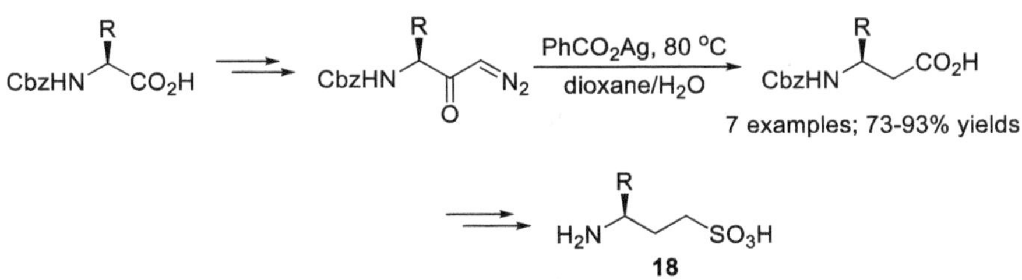

Scheme 7.6 Synthesis of 3-substituted homotaurines from naturally occurring amino acids.

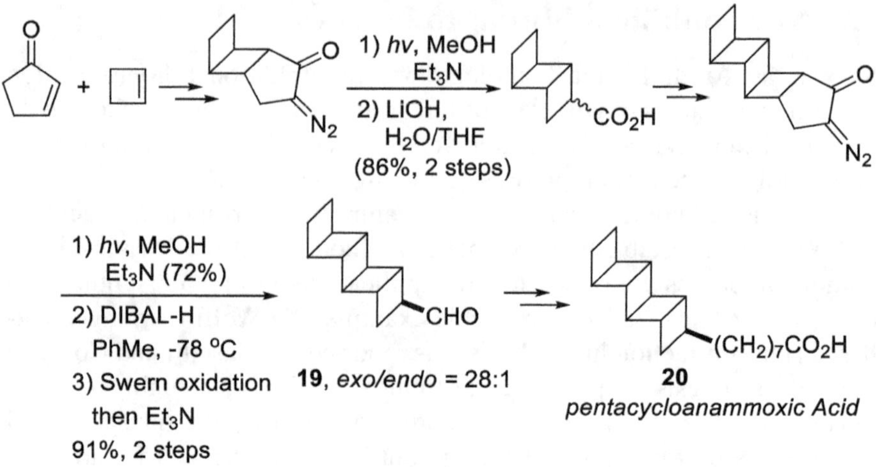

Scheme 7.7 Wolff rearrangement-based ring contraction in the total synthesis of pentacycloanammoxic acid.

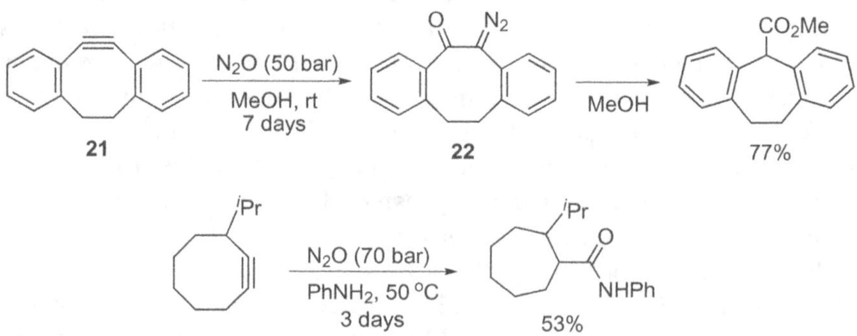

Scheme 7.8 Diazoketone generated by 1,3-dipolar cycloaddition of nitrous oxide with strained cyclic alkynes and the corresponding transformations.

amide could be obtained through the tandem 1,3-dipolar cycloaddition/ Wolff/nucleophilic addition strategy.

Photoinduced ring-contraction of diazo dihydrofuranones was demonstrated to afford the substituted four-membered ring derivatives, including oxetane carboxylic acid derivatives and benzophenone through secondary photochemical decomposition,[33,34] as well as α-carboxy-β-lactones.[35] Besides, photolysis of 2-diazo-3-oxochlorins generated pyrrole ring-contracted azeteoporphyrinoid derivatives through nucleophilic attack of the ketene generated from Wolff rearrangement.[36]

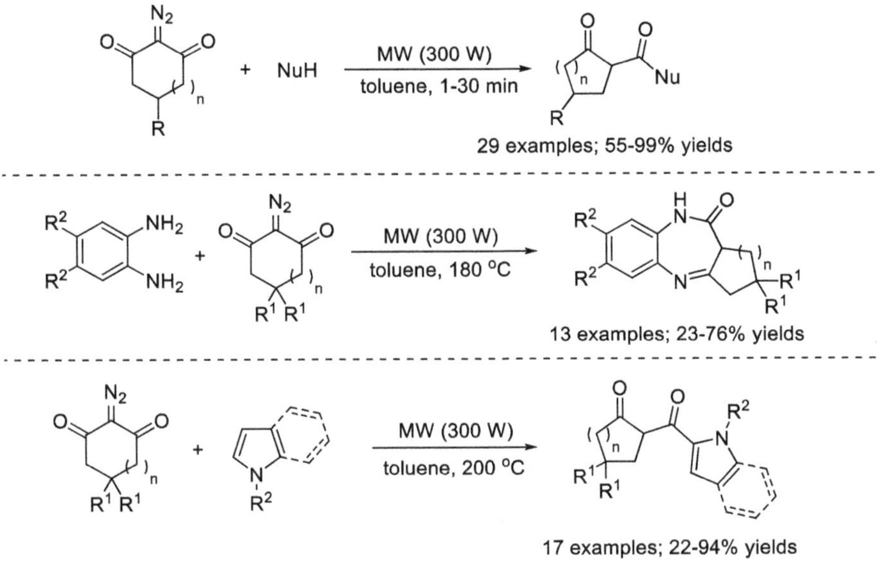

Scheme 7.9 Microwave-assisted Wolff rearrangement of cyclic 2-diazo-1,3-diketones.

Rodriguez and co-workers explored a range of microwave-assisted Wolff rearrangements of medium-sized cyclic 2-diazo-1,3-diketones.[37–40] This microwave activation strategy was found effective for the synthesis of α-carbonylated cycloalkanones,[37] α-spirolactones/lactams through Wolff/ketene trapping/cross metathesis/Michael cascade,[38] 1,3-dihydro-2*H*-1,5-benzodiazepin-2-ones through domino benzannulation cascade[39] and heterocycles containing 1,3-diketone moiety through Friedel–Crafts α-ketoacylation (Scheme 7.9).[40]

Trapping of ketene generated by thermal Wolff rearrangement with carbon nucleophiles was also explored. When the carbon nucleophiles are α-isocyanoacetamides **23**[41] or enol ethers **25**,[42] the reaction was followed by tandem transformations to afford 5-aminooxazoles **24** or γ,δ-unsaturated β-ketoesters **26** (Scheme 7.10).

With imine as the nucleophiles, Rodriguez and co-workers established microwave-irradiated cascade synthesis of spiro compounds from cyclic 2-diazo-1,3-diketones (Scheme 7.11).[43,44] α-Spiro-δ-lactam derivatives **27** were obtained by three-component reaction of diazo substrate, amine and α,β-unsaturated ketone, which proceeded through imination/Wolff/formal [4+2] cycloaddition.[43] Furthermore, normal electron-demand Povarov-type cycloadducts of spiro hydropyrid-4-ones **28** were obtained through Wolff/Friedel–Crafts/Mannich cascade process.[44]

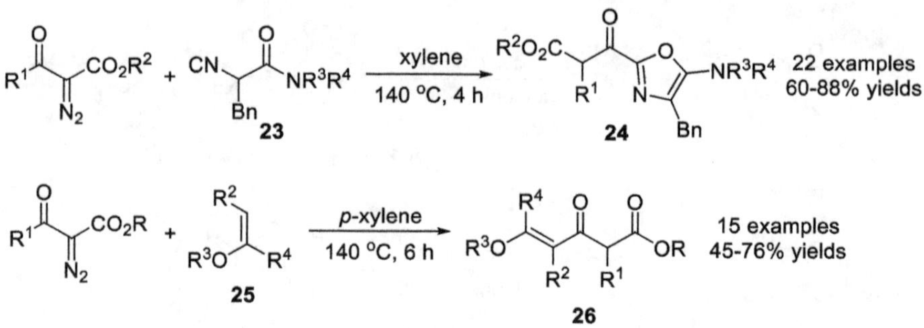

Scheme 7.10 Synthesis of 5-aminooxazoles and γ,δ-unsaturated β-ketoesters through thermal Wolff rearrangement.

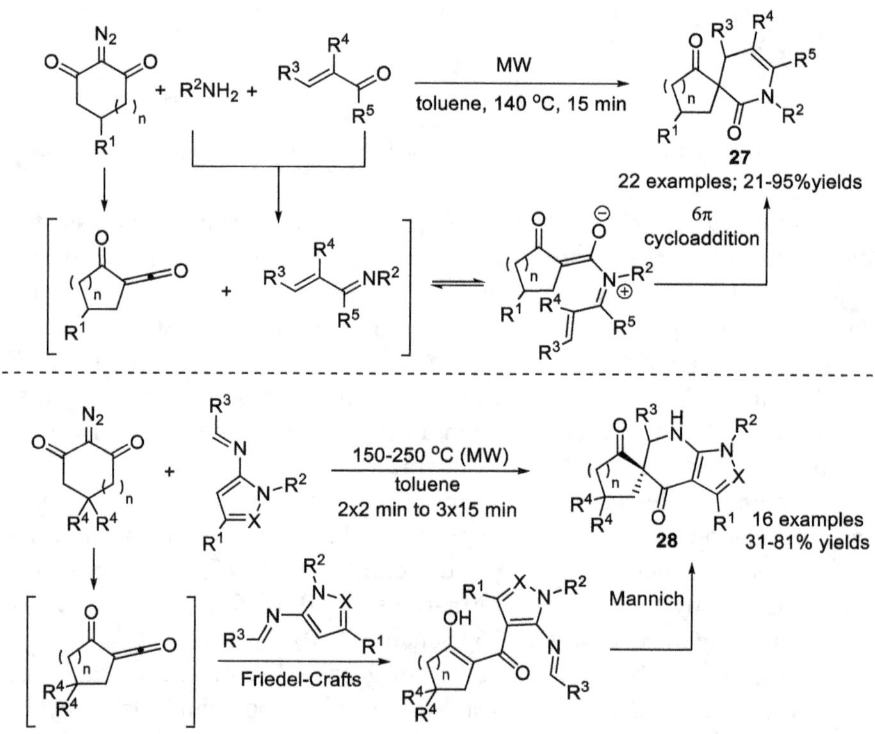

Scheme 7.11 Microwave-irradiated cascade reaction for the synthesis of spiro compounds from cyclic 2-diazo-1,3-diketones as substrates.

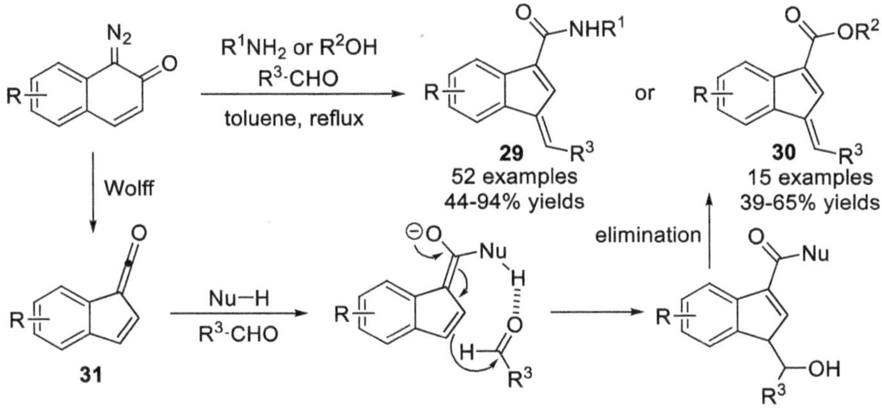

Scheme 7.12 Thermal cascade reaction for the synthesis of 1-diazonaphthalen-2(1*H*)-ones.

Cascade transformations can be devised based on the ketene intermediates generated from Wolff rearrangement. For example, catalytic-free thermolysis of 1-diazonaphthalen-2(1*H*)-ones with amines or alcohols afforded 1*H*-indene-3-carboxamides **29** or 1*H*-indene-3-carboxylates **30** in the presence of various aldehydes.[45] The cascade reaction involved the trapping of ketene **31** with nucleophilic reagents of either amines or alcohols, followed by aldol-type addition/elimination (Scheme 7.12).

In another example, a Cu(II)-catalyzed cascade cyclization of two molecules α-diazo-β-oxoamides with amines was furnished to synthesize poly-substituted pyrrol-3(2*H*)-ones **32**.[46] The reaction pathway was assumed to proceed *via* divergent generation of α-diazoimine and ketene intermediate **33** in the presence of copper catalyst, followed by formal [3+2] cycloaddition sequence to construct pyrrole rings (Scheme 7.13). Notably, intramolecular hydrogen bond of the substrates was essential in this cyclization process.

Basso and co-workers reported the Passerini-type three-component reaction of carboxylic acids, isocyanides with ketene derived *in situ* from diazoketones (Eq. (7.2)).[47] As a result, captodative olefin derivatives, serving as versatile synthons, were synthesized in a stereoselective manner under photo-irradiation condition. Furthermore, this multicomponent synthetic approach for acyloxyacrylamides was extensively achieved both in batch and under continuous-flow conditions.[48]

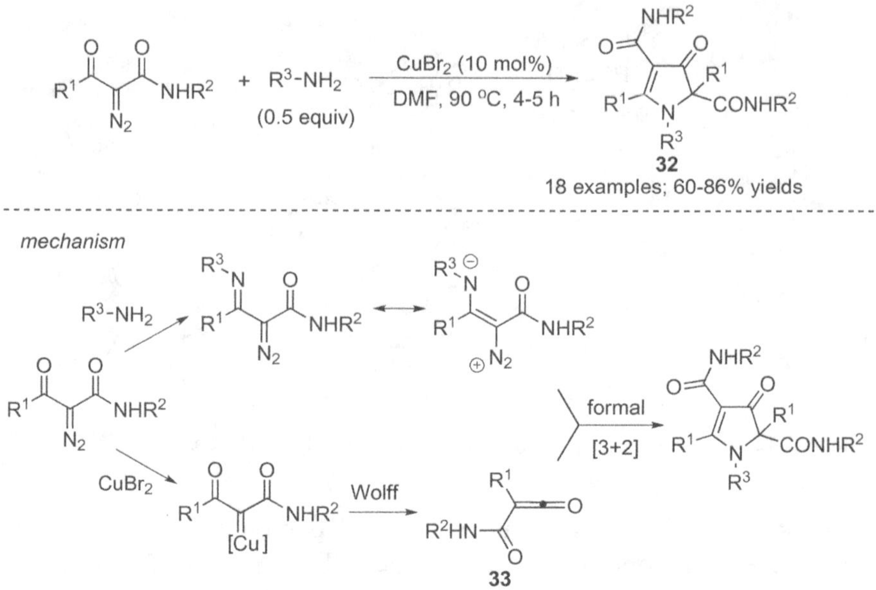

Scheme 7.13 Copper-catalyzed [3+2] cycloaddition of α-diazo-β-oxoamides with amines.

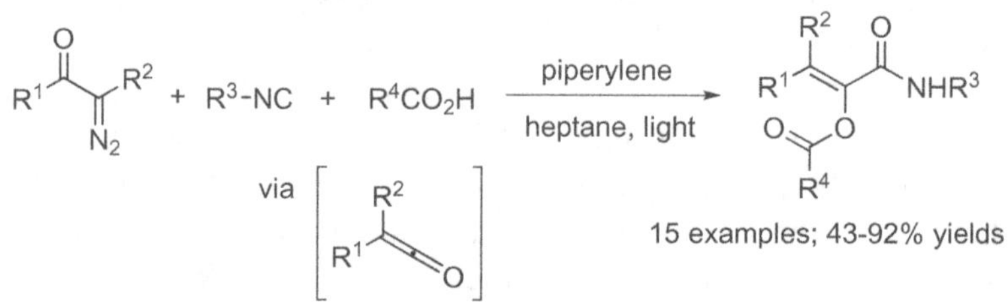

(7.2)

Nikolaev and co-workers reported cascade procedure involving Wolff rearrangement of β-keto α-diazocarbonyl compounds **35** with α,β-unsaturated δ-amino esters **34** under thermolysis condition in the presence of NaH.[49] The reaction afforded 2-oxopiperidines **36**, 5,6-dihydropyridin-2(1*H*)-ones **37**, depending on the structure of the initial reagents. The domino sequence was initiated by ketene formation, base-promoted intramolecular Michael addition and in some cases additional processes of cycloelimination and subsequent intramolecular Claisen condensation (Scheme 7.14).

Apart from the trapping of ketene intermediates by intermolecular nucleophiles, intramolecular trapping was also explored. In the cases of

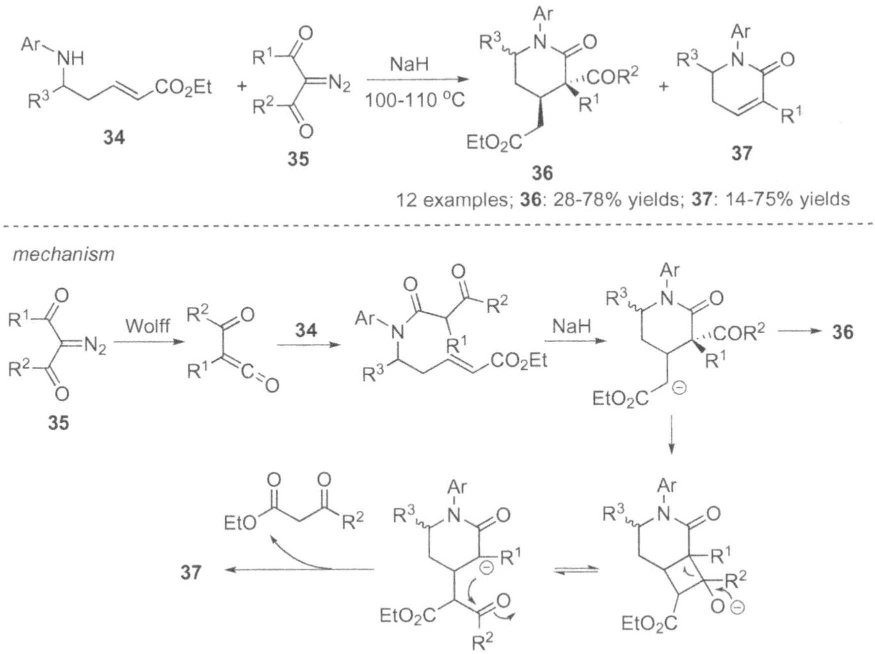

Scheme 7.14 Domino reaction of β-keto α-diazocarbonyl compounds with α,β-unsaturated δ-amino esters.

intramolecular trapping by oxygen nucleophile, the eight-membered ring lactones were constructed by Rh(II)-catalyzed Wolff rearrangement of pyruvate ketal-derived diazoacetoacetates.[50] For trapping with nitrogen nucleophile, the Rh(II)-catalyzed reaction of α-diazo-β-ketoamide allowed the synthesis of β-lactam nucleus **38** (Eq. (7.3)).[51]

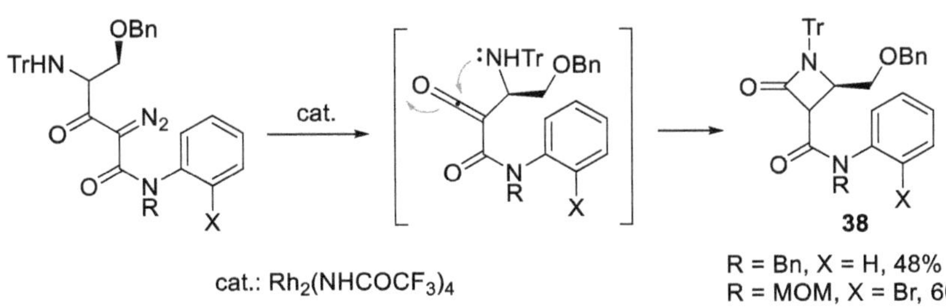

(7.3)

Another unexpected Rh(II)-catalyzed carbene reaction of diazoketone with pyrrolidin-1-yl moiety was reported by Diehl and Brückner (Scheme 7.15).[52] Instead of the formation of pyrrolizidinone through C–H insertion, acrylamide derivatives **39** were generated through Wolff rearrangement/azetidine enolate formation/electrocyclic ring-opening sequence. Moreover, Ag(I)-catalyzed synthesis of 2-carbamoylacrylate motifs was reported with azetidine enolate **40** as the key intermediate. The substituents of nitrogen played essential roles in this carbene transformation, which is in competition with aromatic C–H insertion (Scheme 7.15).[53]

Seki and Georg reported a synthesis of cyclic enaminones **41** through a cascade of Wolff rearrangement followed by an intramolecular nucleophilic attack by vinylogous amides (Scheme 7.16).[54] The reaction

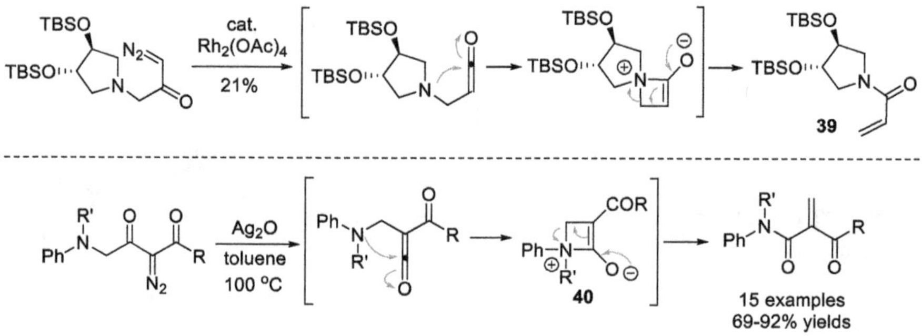

Scheme 7.15 Rh(II)- or Ag(I)-catalyzed Wolff rearrangement with azetidine enolate generation/ring-opening sequence.

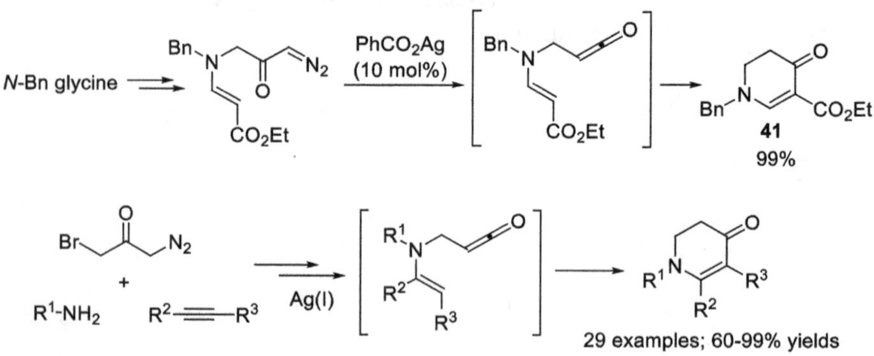

Scheme 7.16 Ag(I)-catalyzed synthesis of cyclic enaminones *via* ketene cyclization.

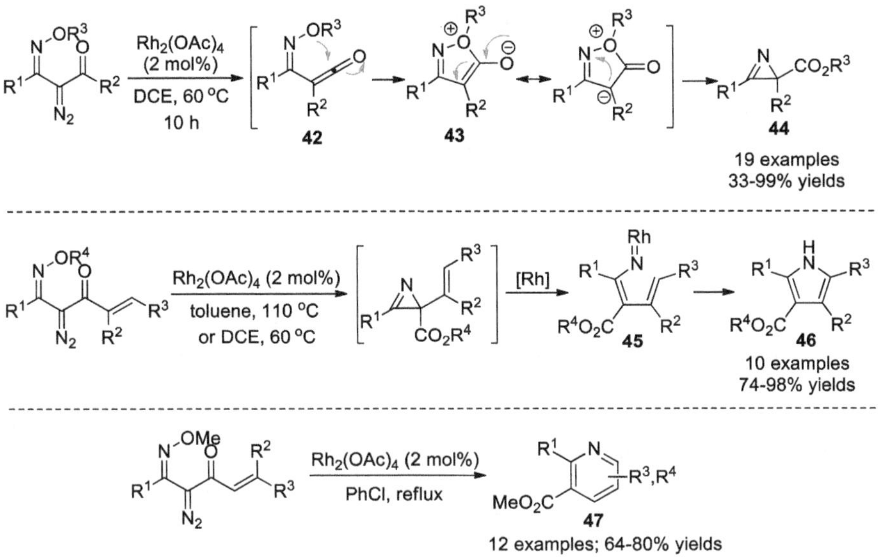

Scheme 7.17 Catalytic cascade reaction of α-diazo oxime ethers for the synthesis of 2*H*-azirines, pyrroles and pyridines.

pathway involves Ag(I)-mediated Wolff rearrangement with sequential nucleophilic 6-*exo*-dig ketene cyclization. Furthermore, the reaction could be carried out in a three-component manner with bromodiazoacetone, primary amine and alkyne.[55]

Park and co-workers explored transition-metal-catalyzed rearrangement cascade reaction of α-diazo oxime ethers for the synthesis of divergent heterocycles (Scheme 7.17). At first, they furnished the efficient synthesis of 2*H*-azirines **44** and polysubstituted pyrroles **46** by Rh(II)-catalyzed tandem cascade involving Wolff rearrangement.[56] The reaction was proposed to proceed *via* the generation of ketene intermediate **42**, followed by nucleophilic attack by oxygen atom of the oxime ether moiety to give **43**. Subsequent sequential rearrangement afforded 2*H*-azirines. They also reported that 2*H*-azirines intermediates underwent Rh(II)-nitrene **45** generation with subsequent ring-opening rearrangement to afford highly functionalized pyrroles. Furthermore, the same group developed the catalyst-controlled synthesis of pyridines **47** *via* hydride or proton transfer with 6π-electrocyclization sequence.[57]

7.3 Other transformations of ketenes

Maas and co-workers reported the generation of aryl silyl ketenes **48** from α-silyl α-diazoketones through triflic acid-induced Wolff rearrangement.[58] In this report, they observed the formation of aryl silyl ketenes and also examined their reactivities. In this unusual reaction, 3-silyl-1-silyloxyallenes **49** were generated through cyclopropanation of ketene with 1,3-Brook-rearranged 1-diazo-2-silyloxyethenes **50**, followed by extrusion of CO gas (Scheme 7.18).

Wang and co-workers reported a one-pot synthesis of pyridine derivatives **53** through the reaction of 2-azido-2,4-dienoates **51** with α-diazocarbonyl compounds **52**, promoted by triphenylphosphine.[59] This thermal cascade transformation proceeded through Staudinger–Meyer reaction of azide moiety and Wolff rearrangement of diazo components, followed by aza-Wittig reaction of the ketene intermediate with **54** to generate intermediate **55**, and finally electrocyclic ring-closure of **54** to afford the final pyridine derivatives **53** (Scheme 7.19).

In addition to the [2+2] and [4+2] cycloaddition reaction, ketenes can also undergo other cycloaddition reactions, such as [2+1], and [4+1] cycloadditions. Danheiser and co-workers have extensively investigated the [4+1] cycloaddition. A recent example is a new [4+1] annulation route to cyclopentenones **58** *via* the reactions of $TMSCHN_2$ with (trialkylsilyl) arylketenes **57**, which are generated photochemically from diazo compounds **56** (Scheme 7.20).[60]

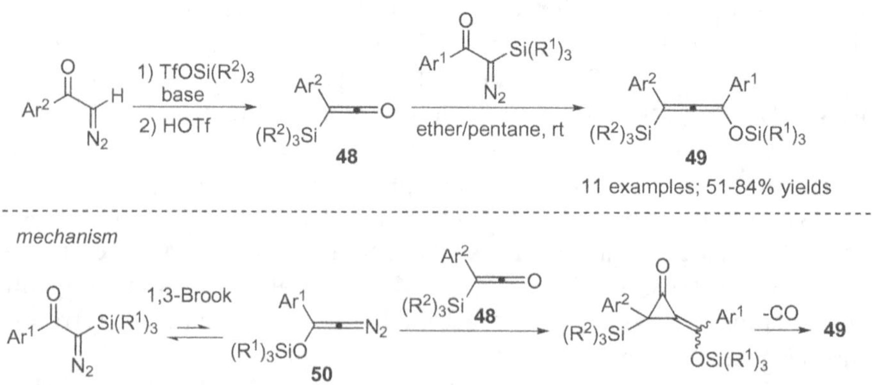

Scheme 7.18 Generation of aryl silyl ketenes and the formation of 3-silyl-1-silyloxyallenes.

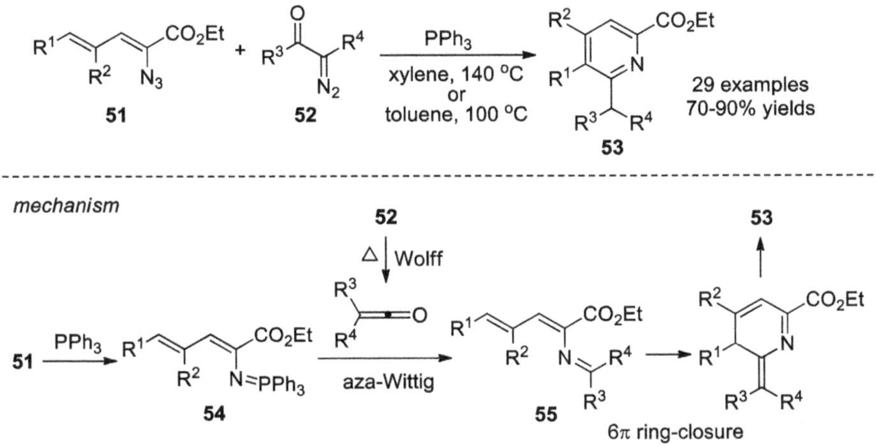

Scheme 7.19 Formation of pyridines from 2-azido-2,4-dienoates with α-diazocarbonyl compounds.

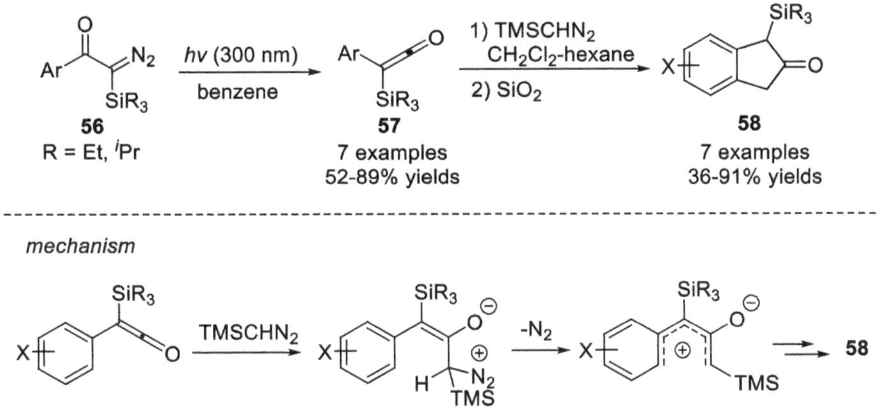

Scheme 7.20 [4+1] Annulation through the reaction of $TMSCHN_2$ with (trialkylsilyl) arylketene.

Stoltz and co-workers developed a method for the construction of a highly functionalized cycloheptadienone **61** and vinylcyclopentenone **63** by tandem Wolff/Cope rearrangement sequence with vinyl cyclopropyl diazoketone **59**. The interesting feature is that under Ag(I)-catalyzed conditions the tandem reaction afforded the cycloheptadienone derivative **61**, while under photochemical conditions it gave the vinylcyclopentenone **63** (Scheme 7.21).[61] In the latter case, the formation of bicycles [3,3,0] is due to a secondary photochemical reaction of the primary product **62**.

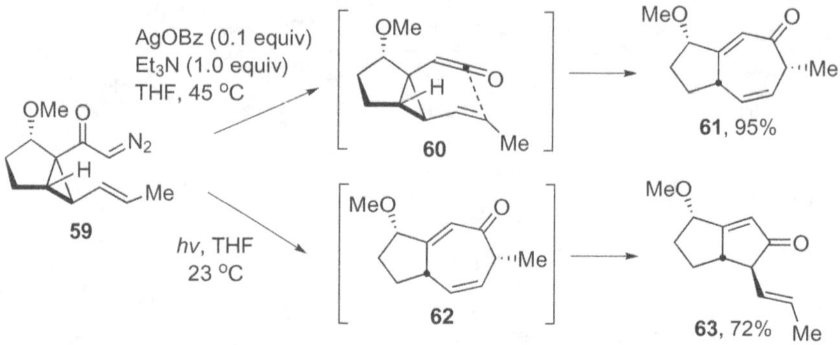

Scheme 7.21 Tandem Wolff/Cope rearrangement sequence with vinyl cyclopropyl diazoketone.

Other recent developments in the Wolff rearrangement include the application of new catalytic systems, such as *in situ*-generated silver nanoclusters,[62,63] mechanistic studies,[64–67] and applications of Wolff rearrangement in the processes of patterning polymers[68,69] and for the synthesis of UV light-sensitive micelles.[70]

7.4 1,2-Shift reactions

Wolff rearrangement belongs to a general reaction of carbene species, namely the 1,2-shift of the adjacent group, a process also commonly encountered in transition metal-catalyzed reactions of α-diazocarbonyl compounds.[71] As shown in Scheme 7.22, the 1,2-migration of R' group on carbonyl moiety leads to the formation of ketene intermediates (Wolff rearrangement). Alternatively, 1,2-shift of R group on the β-carbon atom generates α,β-unsaturated carbonyl compounds or 1,3-dicarbonyl compounds. Although in many cases 1,2-shift is considered as a side reaction

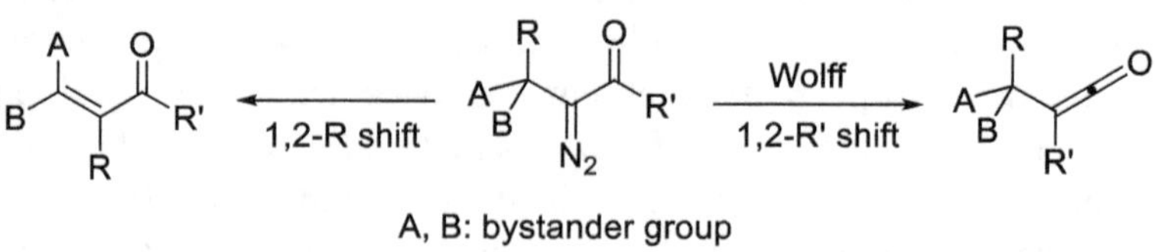

Scheme 7.22 1,2-Shift of α-diazocarbonyl compounds.

that competes with other typical carbene reactions such as X–H insertion and cyclopropanation, it may also find useful applications in organic synthesis.

In 2001, Wang and co-workers reported the Rh(II)-catalyzed reaction of *N*-tosyl diazoketamines **64**, which was prepared by the addition of the ethyl α-diazoacetate anion to *N*-sulfonylimines (Eq. (7.4)).[72] The reaction resulted in aryl migration to give α-aryl-β-enamino esters **65** in good yields and with high stereoselectivity. The migratory aptitude of 1,2-aryl over 1,2-hydride migration was found to be affected by catalysts.

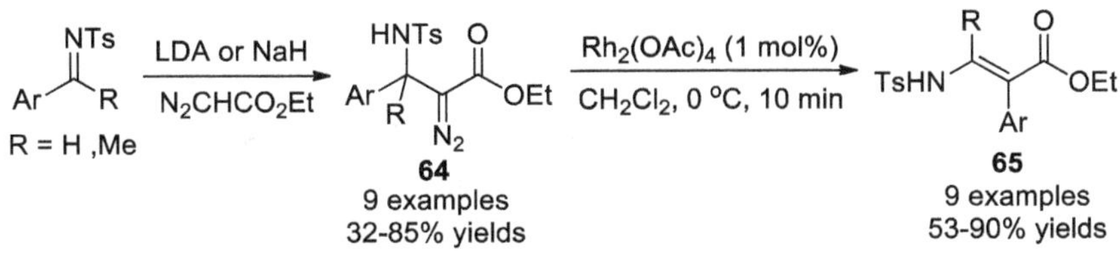

(7.4)

Wang and co-workers subsequently demonstrated that this 1,2-migratory aptitude was significantly affected by the electronic and steric effects of the non-migrating substituents (the bystander groups A and B as shown in Scheme 7.22).[73–75] As shown in Scheme 7.23, 1,2-phenyl migration becomes predominant over 1,2-H migration when the bystander group is changed from a hydroxyl group to a tosylamino group and to a trichloroacetylamino group. This change of migratory aptitude is rationalized by the stabilizing effect of the partial positive charge developed in the transition state by the amino group.[73,74] It is assumed that the 1,2-hydrogen migration largely resembles a hydride

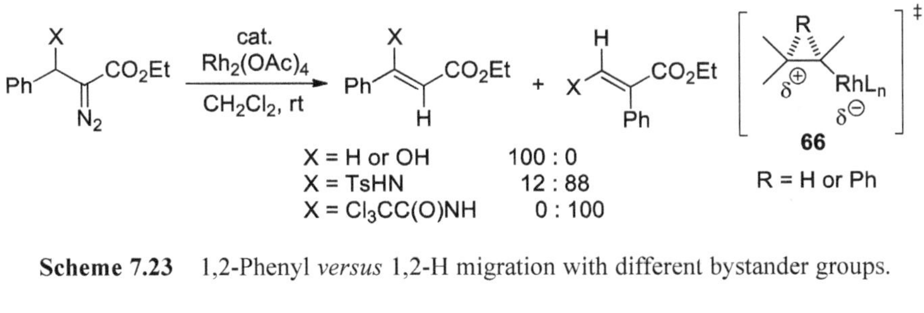

Scheme 7.23 1,2-Phenyl *versus* 1,2-H migration with different bystander groups.

shift with significant charge separation in the transition state **66**, while the 1,2-phenyl migration proceeds with less charge separation in the transition state.

A similar change of migratory aptitude due to the bystander group can also be seen in the case of 1,2-H *versus* 1,2-vinyl or 1,2-acetylenyl migration (Eq. (7.5)).[74]

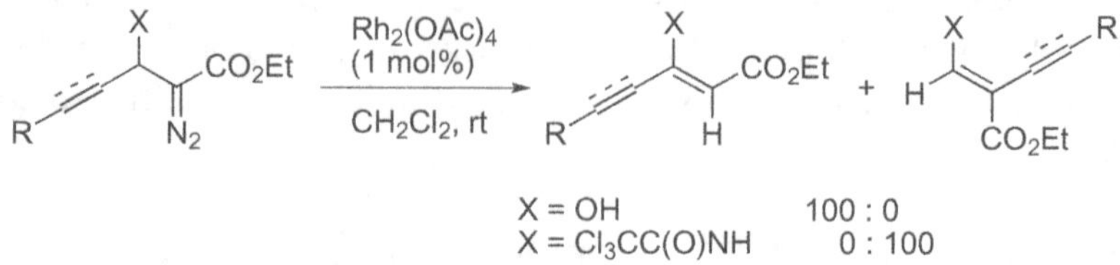

(7.5)

The steric effect also plays a vital role in affecting the migratory aptitude. A sterically bulkier group in the migrating origin obviously favors a large group, such as phenyl, to migrate, because the steric hindrance can thus be relieved. On the other hand, when there is a bulkier group attached to the carbene carbon, a small group (H) will migrate in order to avoid the steric congestion in the product (Eq. (7.6)).[75]

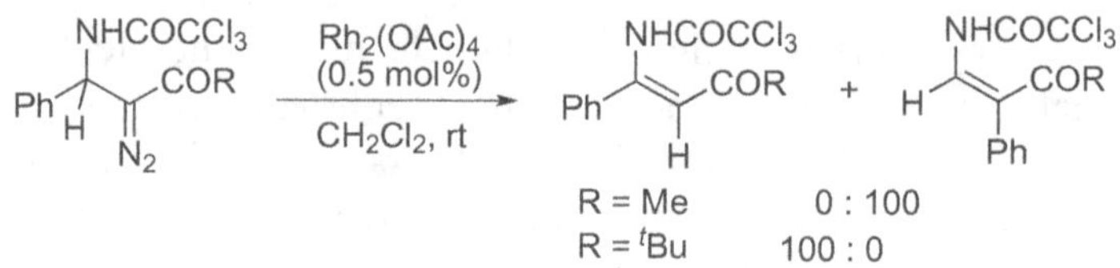

(7.6)

It has also been observed that a thio group has a higher migratory aptitude over hydrogen in the $Rh_2(OAc)_4$-catalyzed reactions of diazo compounds (Eq. (7.7)).[76] This should be attributed to the relatively large size of the sulfur, which allows the interaction of the lone pair of sulfur with the electron-deficient carbene center.

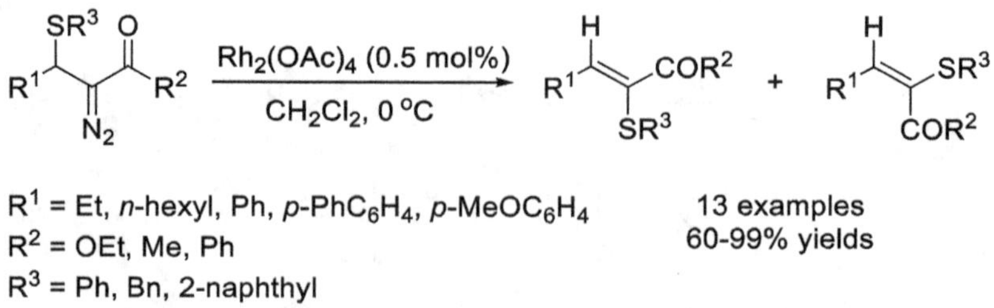

(7.7)

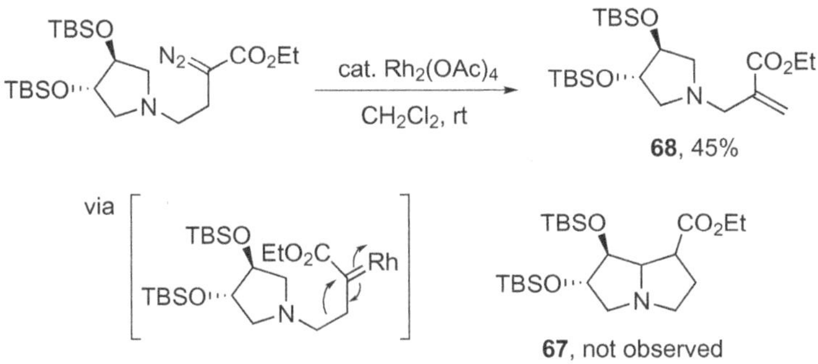

Scheme 7.24 An unusual Rh(II)-catalyzed carbene 1,2-alkyl migration.

The ligands of Rh(II) catalysts have been known to change the reaction pathway of Rh(II) carbenes. Aggarwal and co-workers showed that, by tuning the ligands on Rh(II) from acetate to trifluoroacetate, the normally facile 1,2-H migration process could be completely reversed to favor 1,2-alkyl migration in the reaction with diazo compounds (Eq. (7.8)).[77]

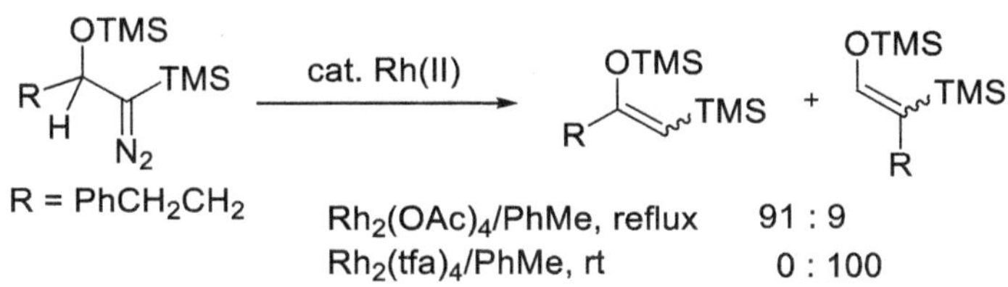

(7.8)

An unusual rearrangement of Rh(II) carbene containing pyrrolidine moiety was observed by Diehl and Brückner.[78] The expected intramolecular C–H bond insertion product **67** could not be observed. Instead, α-diazoester undergoes Rh(II)-catalyzed 1,2-alkyl migration, resulting in the formation of methacrylic ester **68** with moderate yield (Scheme 7.24).

In addition to the Rh(II) catalysts, the Cu(I) complex-catalyzed 1,2-H migration was also studied. Recently, Miao and co-workers reported the regiospecific and stereoselective 1,2-shift of dialkyl α-diazophosphonates by using $[Cu(MeCN)_4PF_6]$ as the catalyst and DMF as the additive.[79]

Ag(I)-catalyzed ring expansion of cyclopropanes **66** was reported by Tang and co-workers.[80] This transformation proceeds through the generation of Ag(I) carbene followed by 1,2-shift of one of the cyclopropane C–C bonds to form ring-expansion products. Functionalized cyclobutenoates **70**

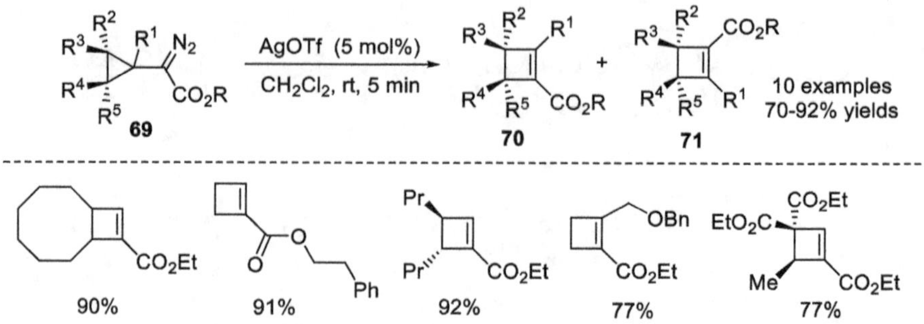

Scheme 7.25 Scope of Ag(I)-catalyzed synthesis of poly-substituted cyclobutenoates.

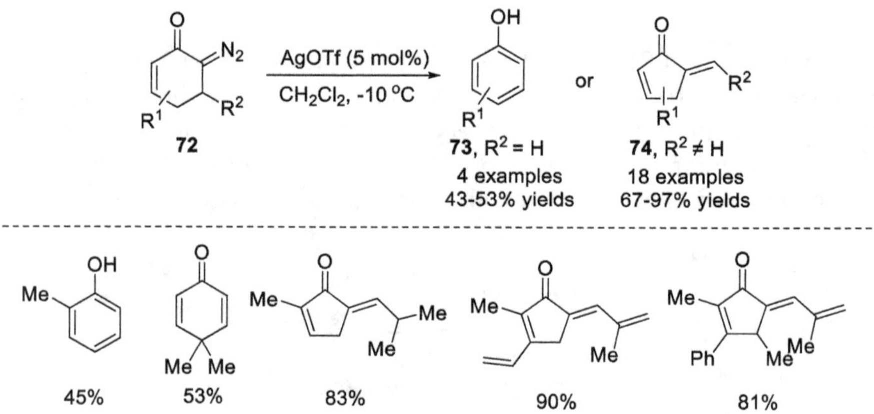

Scheme 7.26 Ag(I)-catalyzed 1,2-H shift or ring-contractive rearrangement of substituted 6-diazo-2-cyclohexenones.

and **71** were obtained from α-cyclopropyl-α-diazoesters in good yields with high chemo-, regio- and stereoselectivity (Scheme 7.25).

Zhai and co-workers reported the Ag(I)-catalyzed substrate-dependent migration process of substituted 6-diazo-2-cyclohexenones **72**.[81] The reaction was presumably initiated by silver-carbene formation followed by [1,2]-H shift to afford phenol derivatives **73**, or [1,2]-endocyclic allyl migration to afford ring-contractive products **74**, respectively (Scheme 7.26). This protocol provides an efficient access to 5-alkylidene-2-cyclopentenones **74** in high yields with broad substrates scope.

In 2013, Doyle and co-workers reported a Lewis acid-catalyzed [3+2] cycloaddition from azomethine imines with enoldiazoacetates to prepare β-methylene-β-silyloxy-β-amido-α-diazoacetates **75**.[82] These diazo

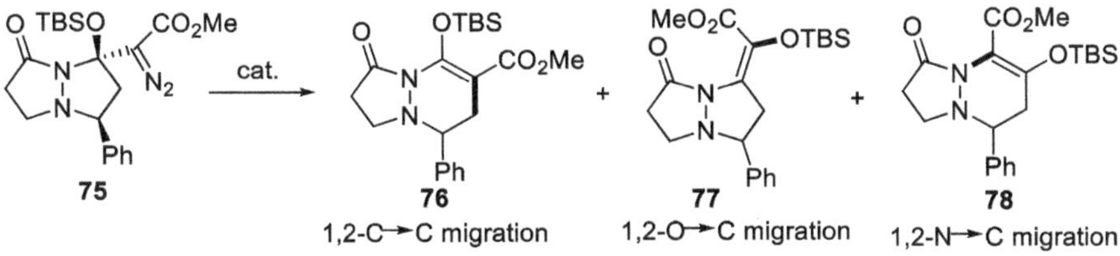

cat.	**76**:**77**:**78**
$Rh_2(OAc)_4$	58:25:17
$Rh_2(tfa)_4$	78:22:0
$Rh_2(cap)_4$	78:9:13
$Cu(OTf)_2$	0:0:100
$CuPF_6$	0:0:100

Scheme 7.27 Catalyst-dependent competitive 1,2-C→C, -O→C and -N→C migration.

intermediates have shown unique catalyst-dependent competitive migration process. Variety of 1,2-C→C (**76**), -O→C (**77**) and -N→C (**78**) migration products were observed, depending on the steric and stereoelectronic factors of the catalytically generated metal carbenes (Scheme 7.27). The copper and silver catalysts have displayed exclusive selectivity for the formation of -N→C (**78**) migration products.

In summary, as a typical reaction of α-diazocarbonyl compounds, Wolff rearrangement has continued to demonstrate its new applications in organic synthesis. In particular, attention has been paid to the synthesis of functional molecules as well as natural products through the ketene intermediates with the combination of [2+2], [4+2] cycloadditions, or with trapping by a wide range of nucleophiles. Further attempts will focus on divergent types of sequential transformations by utilizing *in situ* generated ketenes. In addition, the studies in the past few years have revealed that for 1,2-shift reactions, the migratory aptitude is affected by various factors, including the bystander groups, the catalysts and the nature of the migrating groups.

References

1. For a recent review on Wolff rearrangement, see: Kirmse, W. *Eur. J. Org. Chem.* **2002**, 2193–2256.
2. Tuba, R. *Org. Biomol. Chem.* **2013**, *11*, 5976–5988.
3. For reviews on ketene chemistry and applications to β-lactams preparation, see: Brandi, A.; Cicchi, S.; Cordero, F. M. *Chem. Rev.* **2008**, *108*, 3988–4035.

4. Allen, A. D.; Tidwell, T. T. *Chem. Rev.* **2013**, *113*, 7287–7342.
5. Jiao, L.; Zhang, Q.; Liang, Y.; Zhang, S.; Xu, J. *J. Org. Chem.* **2006**, *71*, 815–818.
6. Liang, Y.; Jiao, L.; Zhang, S.; Xu, J. *J. Org. Chem.* **2005**, *70*, 334–337.
7. Jiao, L.; Liang, Y.; Xu, J. *J. Am. Chem. Soc.* **2006**, *128*, 6060–6069.
8. Wang, Y.; Liang, Y.; Jiao, L.; Du, D.-M.; Xu, J. *J. Org. Chem.* **2006**, *71*, 6983–6990.
9. Li, B.; Wang, Y.; Du, D.-M.; Xu. J. *J. Org. Chem.* **2007**, *72*, 990–997.
10. Taubinger, A. A.; Fenske, D.; Podlech, J. *Tetrahedron* **2008**, *64*, 8659–8667.
11. Taubinger, A. A.; Fenske, D.; Podlech, J. *Synlett* **2008**, 539–542.
12. Qi, H.; Yang, Z.; Xu, J. *Synthesis* **2011**, 723–730.
13. Reddy, B. V. S.; Karthik, G.; Rajasekaran, T.; Antony, A.; Sridhar, B. *Tetrahedron Lett.* **2012**, *53*, 2396–2401.
14. Yang, Z.; Xu, J. *Tetrahedron Lett.* **2012**, *53*, 786–789.
15. Qi, H.; Li, X.; Xu, J. *Org. Biomol. Chem.* **2011**, *9*, 2702–2714.
16. Austin, W. F.; Zhang, Y.; Danheiser, R. L. *Org. Lett.* **2005**, *7*, 3905–3908.
17. Willumstad, T. P.; Haze, O.; Mak, X. Y.; Lam, T. Y.; Wang, Y.-P.; Danheiser, R. L. *J. Org. Chem.* **2013**, *78*, 11450–11469.
18. Willumstad, T. P.; Boudreau, P. D.; Danheiser, R. L. *J. Org. Chem.* **2015**, *80*, 11794–11805.
19. Huang, Z.; Wang, C.; Tokunaga, E.; Sumii, Y.; Shibata, N. *Org. Lett.* **2015**, *17*, 5610–5613.
20. Zheng, Y.; Xu, J. *Tetrahedron* **2014**, *70*, 5197–5206.
21. Mander, L. N.; McLachlan, M. M. *J. Am. Chem. Soc.* **2003**, *125*, 2400–2401.
22. Lee, D. J.; Kim, K. *J. Org. Chem.* **2004**, *69*, 4867–4869.
23. Snyder, S. A.; Corey, E. J. *J. Am. Chem. Soc.* **2006**, *128*, 740–742.
24. Mascitti, V.; Corey, E. J. *J. Am. Chem. Soc.* **2004**, *126*, 15664–15665.
25. Mascitti, V.; Corey, E. J. *J. Am. Chem. Soc.* **2006**, *128*, 3118–3119.
26. Thijs, L.; Zwanenburg, B. *Tetrahedron Lett.* **2004**, *60*, 5237–5252.
27. Islami, M. R.; Allen, A. D.; Vukovic, S.; Tidwell, T. T. *Org. Lett.* **2011**, *13*, 494–497.
28. Zhegalova, N. G.; Popik, V. V. *J. Phys. Org. Chem.* **2011**, *24*, 969–975.
29. Singh, G. S.; Masutlha, L. L. *Proc. Natl. Acad. Sci., India, Sect. A Phys. Sci.* **2012**, *82*, 147–149.
30. Neupane, P.; Li, X.; Jung, J. H.; Lee, Y. R.; Kim, S. H. *Tetrahedron* **2012**, *68*, 2496–2508.
31. Bernardim, B.; Hardman-Baldwin, A. M.; Burtoloso, A. C. B. *RSC Adv.* **2015**, *5*, 13311–13314.
32. Banert, K.; Plefka, O. *Angew. Chem. Int. Ed.* **2011**, *50*, 6171–6174.
33. Malashikhin, S. A.; Linden, A.; Heimgartner, H.; Rodina, L. L.; Nikolaev, V. A. *Helv. Chim. Acta* **2008**, *91*, 1662–1669.
34. Nikolaev, V. A.; Galkina, O. S.; Sieler, J.; Rodina, L. L. *Tetrahedron Lett.* **2010**, *51*, 2713–2716.

35. Beneke, J.; Schobert, R. *Synthesis* **2013**, *45*, 773–776.
36. Köpke, T.; Pink, M.; Zaleski, J. M. *J. Am. Chem. Soc.* **2008**, *130*, 15864–15871.
37. Presset, M.; Coquerel, Y.; Rodriguez, J. *J. Org. Chem.* **2009**, *74*, 415–418.
38. Boddaert, T.; Coquerel, Y.; Rodriguez, J. *Eur. J. Org. Chem.* **2011**, 5061–5070.
39. Castillo, J.-C.; Presset, M.; Abonia, R.; Coquerel, Y.; Rodriguez, J. *Eur. J. Org. Chem.* **2012**, 2338–2345.
40. Mohanan, K.; Presset, M.; Mailhol, D.; Coquerel, Y.; Rodriguez, J. *Chem. Eur. J.* **2012**, *18*, 9217–9220.
41. Wu, J.; Chen, W.; Hu, M.; Zou, H.; Yu, Y. *Org. Lett.* **2010**, *12*, 616–618.
42. Pandit, R. P.; Lee, Y. R. *Org. Biomol. Chem.* **2014**, *12*, 4407–4411.
43. Presset, M.; Coquerel, Y.; Rodriguez, J. *Org. Lett.* **2010**, *12*, 4212–4215.
44. Galvez, J.; Castillo, J.-C.; Quiroga, J.; Rajzmann, M.; Rodriguez, J.; Coquerel, Y. *Org. Lett.* **2014**, *16*, 4126–4129.
45. Magar, K. B. S.; Lee, Y. R. *Org. Lett.* **2013**, *15*, 4288–4291.
46. Wang, Z.; Bi, X.; Liao, P.; Liu, X.; Dong, D. *Chem. Commun.* **2013**, *49*, 1309–1311.
47. Basso, A.; Banfi, L.; Garbarino, S.; Riva, R. *Angew. Chem. Int. Ed.* **2013**, *52*, 2096–2099.
48. Garbarino, S.; Banfi, L.; Riva, R.; Basso, A. *J. Org. Chem.* **2014**, *79*, 3615–3622.
49. Medvedev, J. J.; Meleshina, M. V.; Panikorovskii, T. L.; Schneider, C.; Nikolaev, V. A. *Org. Biomol. Chem.* **2015**, *13*, 9107–9117.
50. Erhunmwunse, M. O.; Steel, P. G. *Tetrahedron Lett.* **2009**, *50*, 3568–3570.
51. Gerstenberger, B. S.; Lin, J.; Mimieux, Y. S.; Brown, L. E.; Oliver, A. G.; Konopelski, J. P. *Org. Lett.* **2008**, *10*, 369–372.
52. Diehl, J.; Brückner, R. *Tetrahedron Lett.* **2014**, *55*, 2629–2632.
53. Ma, B.; Chen, F.-L.; Xu, X.-Y.; Zhang, Y.-N.; Hu, L.-H. *Adv. Synth. Catal.* **2014**, *356*, 416–420.
54. Seki, H.; Georg, G. I. *J. Am. Chem. Soc.* **2010**, *132*, 15512–15513.
55. Seki, H.; Georg, G. I. *Org. Lett.* **2011**, *13*, 2147–2149.
56. Jiang, Y.; Chan, W. C.; Park, C.-M. *J. Am. Chem. Soc.* **2012**, *134*, 4104–4107.
57. Jiang, Y.; Park, C.-M. *Chem. Sci.* **2014**, *5*, 2347–2351.
58. Bucher, S. M.; Brückmann, R.; Maas, G. *Eur. J. Org. Chem.* **2008**, 4426–4433.
59. Chen, Z.-B.; Hong, D.; Wang, Y.-G. *J. Org. Chem.* **2009**, *74*, 903–905.
60. Dalton, A. M.; Zhang, Y.; Davie, C. P.; Danheiser, R. L. *Org. Lett.* **2002**, *4*, 2465–2468.
61. Sarpong, R.; Su, J. T.; Stoltz, B. M. *J. Am. Chem. Soc.* **2003**, *125*, 13624–13625.
62. Sudrik, S. G.; Maddanimath, T.; Chaki, N. K.; Chavan, S. P.; Chavan, S. P.; Sonawane, H. R.; Vijayamohanan, K. P. *Org. Lett.* **2003**, *5*, 2355–2358.

63. Sudrik, S. G.; Sharma, J.; Chavan, V. B.; Chaki, N. K.; Sonawane, H. R.; Vijayamohanan, K. P. *Org. Lett.* **2006**, *8*, 1089–1092.
64. Julian, R. R.; May, J. A.; Stoltz, B. M.; Beauchamp, J. L. *J. Am. Chem. Soc.* **2003**, *125*, 4478–4486.
65. Chiang, Y.; Gaplovsky, M.; Kresge, A. J.; Leung, K. H.; Ley, C.; Mac, M.; Persy, G.; Phillips, D. L.; Popik, V. V.; Rödig, C.; Wirz, J.; Zhu, Y. *J. Am. Chem. Soc.* **2003**, *125*, 12872–12880.
66. Bogdanova, A.; Popik, V. V. *J. Am. Chem. Soc.* **2003**, *125*, 14153–14162.
67. Chang, J. A.; Chiang, Y.; Keeffe, J. R.; Kresge, A. J.; Nikolaev, V. A.; Popik, V. V. *J. Org. Chem.* **2006**, *71*, 4460–4467.
68. Urdabayev, N. K.; Popik, V. V. *J. Am. Chem. Soc.* **2004**, *126*, 4058–4059.
69. Mizokuro, T.; Mochizuki, H.; Tanigaki, N.; Hiraga, T. *Polym. Adv. Technol.* **2006**, *17*, 841–844.
70. For the synthesis of UV light-sensitive micelles, see: Goodwin, A. P.; Mynar, J. L.; Ma, Y.; Fleming, G. R.; Fréchet, J. M. J. *J. Am. Chem. Soc.* **2005**, *127*, 9952–9953.
71. Zhang, Z.; Wang, J. *Tetrahedron* **2008**, *64*, 6577–6605.
72. Jiang, N.; Qu, Z.; Wang, J. *Org. Lett.* **2001**, *3*, 2989–2992.
73. Jiang, N.; Ma, Z.; Qu, Z.; Xing, X.; Xie, L.; Wang, J. *J. Org. Chem.* **2003**, *68*, 893–900.
74. Shi, W.; Xiao, F.; Wang, J. *J. Org. Chem.* **2005**, *70*, 4318–4322.
75. Xiao, F.; Wang, J. *J. Org. Chem.* **2006**, *71*, 5789–5791.
76. Xu, F.; Shi, W.; Wang, J. *J. Org. Chem.* **2005**, *70*, 4191–4194.
77. Vitale, M.; Lecourt, T.; Sheldon, C. G.; Aggarwal, V. K. *J. Am. Chem. Soc.* **2006**, *128*, 2524–2525.
78. Diehl, J.; Brückner, R. *Tetrahedron Lett.* **2014**, *55*, 2629–2632.
79. Ge, H.; Liu, S.; Cai, Y.; Sun, Y.; Miao, Z. *Synthesis* **2016**, *48*, 448–454.
80. Xu, H.; Zhang, W.; Shu, D.; Werness, J. B.; Tang, W. *Angew. Chem. Int. Ed.* **2008**, *47*, 8933–8936.
81. Zhao, L.; Wang, J.; Zheng, H.; Li, Y.; Yang, K.; Cheng, B.; Jin, X.; Yao, X.; Zhai, H. *Org. Lett.* **2014**, *16*, 6378–6381.
82. Xu, X.; Qian, Y.; Zavalij, P. Y.; Doyle, M. P. *J. Am. Chem. Soc.* **2013**, *135*, 1244–1247.

Chapter 8

Olefination with Diazocarbonyl Compounds

Diazo compounds have been explored as a straightforward strategy for C=C bond construction. In this chapter, several pervasive reaction types, including dimerization, carbene metathesis and catalytic ylide-type reaction, will be summarized. The C=C bond formation based on transition-metal-catalyzed cross-coupling reaction of diazo compounds will be introduced in detail in Chapter 12.

8.1 Dimerization

In most cases, carbene dimerization of diazo compounds occurs as an undesirable side reaction in typical metal-catalyzed diazo transformations, such as cyclopropanation or X–H insertion. Nevertheless, this process can be explored as a concise and efficient approach for preparing symmetric alkenes. Many transition-metal complexes, such as nickel, copper, iridium, osmium, tantalum, rhenium and chromium complexes, have been shown to catalyze the generation of olefins from diazo compounds. For recent examples, a wide range of transition-metal complexes, including Os(VIII) alkylidene,[1] bulk gold complexes,[2,3] ruthenium,[4–7] and copper catalysts,[8] were used to catalyze homo-coupling of diazocarbonyl compounds. An auto-tandem catalytic process involving Rh(III)-catalyzed aromatic C–H bond activation with the *in situ* generated electron-deficient alkenes through diazoacetate dimerization catalyzed by Rh(III) has also been reported.[9]

However, when these reactions were used as the synthetic method for olefins, the *cis*/*trans* selectivity usually became a crucial issue. In general, the thermodynamically more stable *trans* isomer is predominant. Recently, Hodgson and co-workers have demonstrated that the highly *cis* selective hetero-coupling of α-diazoacetates can be achieved by Grubbs' second-generation catalyst **1** (Eq. (8.1)).[10,11]

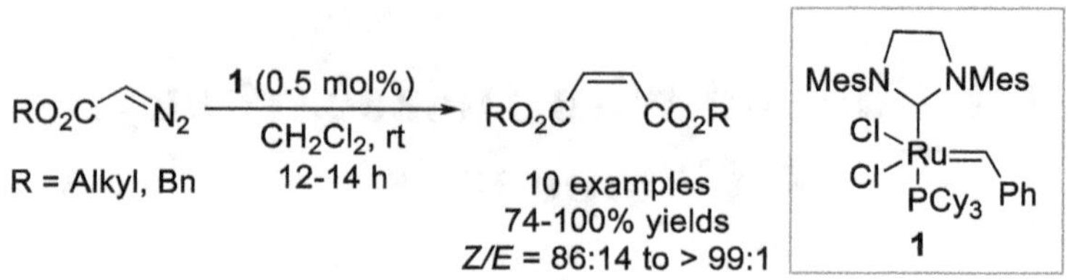

(8.1)

Li and Che have also reported high *cis* selectivity in the coupling reactions of α-diazoacetates with their ruthenium porphyrin catalyst **3**. This catalytic system was especially efficient for the synthesis of macrocyclic compounds bearing *cis* alkene motif, as demonstrated by the cyclization with bis-diazo compound **2** (Eq. (8.2)).[12]

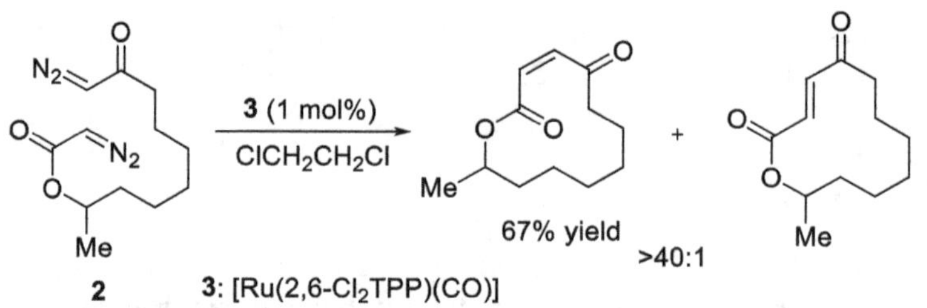

(8.2)

Apparently, the challenging issue in the carbene olefination is the selective cross-coupling of two different diazo compounds. Several coinage metal complexes were investigated by Pérez and co-workers toward the catalytic reactivity in cross-coupling dimerization.[13] Davies and co-workers reported the Rh(II)-catalyzed selective cross-dimerization of α-aryl-α-diazo compounds/vinyldiazoacetates with EDA/diazoketones to provide a variety of tri-substituted alkenes.[14] The selectivity was attributed to the difference of decomposition rate between two diazo species as well as the preference for trapping of the initial Rh(II) carbene. Selective cross-coupling synthesis for tetra-substituted alkenes was achieved with the employment of NHC-Au(I) catalyst.[15] Furthermore, the Au(I)-initiated tandem benzannulation of diverse vinyldiazoesters was developed by Sun and co-workers through cross-dimerization/6π-electrocyclization/oxidation cascade (Scheme 8.1).

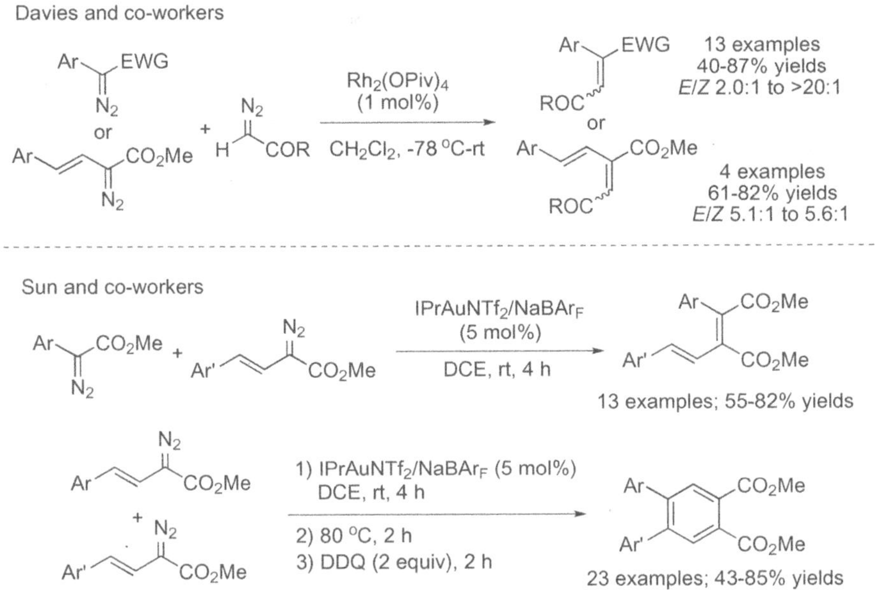

Scheme 8.1 Rh(II)- and Au(I)-catalyzed cross-coupling of diazo compounds.

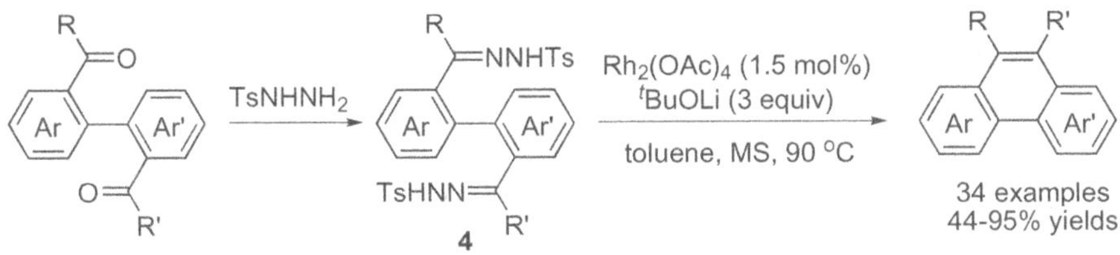

Scheme 8.2 Rh(II)-catalyzed intramolecular carbene cyclization of bis(*N*-tosylhydrazone)s.

Wang and co-workers developed a Rh(II)-catalyzed intramolecular carbene dimerization of bis(*N*-tosylhydrazone)s **4**.[16] This catalytic cyclization process, as well as the one-pot procedure from corresponding carbonyl substrates, provided a general and highly efficient synthetic method for polycyclic aromatic compounds, including chrysene, helicene, picene and pentaphenes (Scheme 8.2). The reaction is complementary to the classic McMurry reaction and Wittig/RCM cascade for the preparation for poly-aromatic ring systems.

Sun and co-workers successfully established an efficient strategy for synthesizing cyclic olefins through Au(I)-catalyzed intramolecular diazo coupling reaction (Eq. (8.3)).[17] This macrocyclization afforded medium (8-membered) to large sized (24-membered) cyclic olefins with moderate to good yields. It is noteworthy that high dilution of concentration was unnecessary for this reaction. As expected, the *Z*/*E* selectivity of the olefins is highly dependent on the ring size.

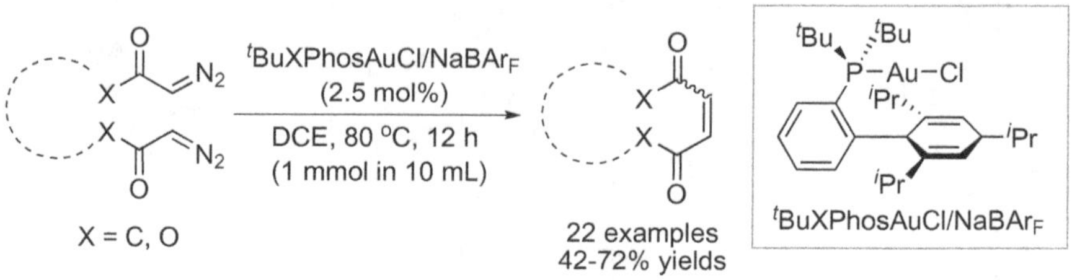

(8.3)

Recently, Sun and co-workers developed selective and controllable reactions of bis-*N*-tosylhydrazones (Scheme 8.3).[18] Under transition-metal-free conditions, the base-promoted intramolecular selective detosylative cyclization occurred, affording nitrogen-containing heterocycles including pyridazines, diazepines and other related heterocyclic compounds with high efficiency. Alternatively, in the presence of Au(I) catalyst and strong base, poly-substituted cyclic olefins were obtained.

Moreover, Sun and co-workers reported a selective intermolecular cross- and homo-coupling reaction of aryl–aryl or aryl–alkyl diazoesters in the presence of inexpensive Cu(II) catalyst and commercially available ligand.[19] The key step of the proposed mechanism involves nucleophilic

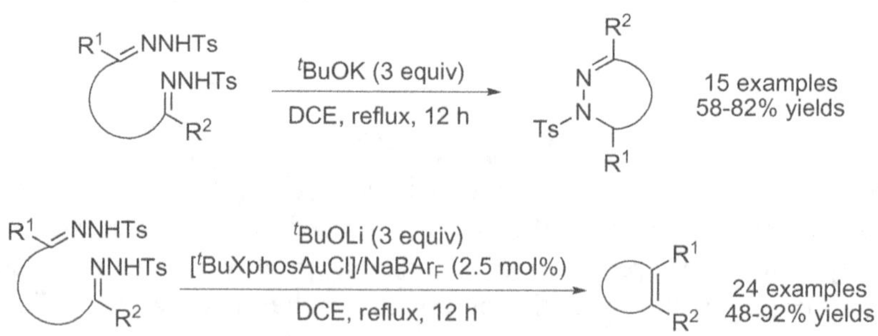

Scheme 8.3 Base-promoted/Au-catalyzed decomposition of bis-*N*-tosylhydrazones.

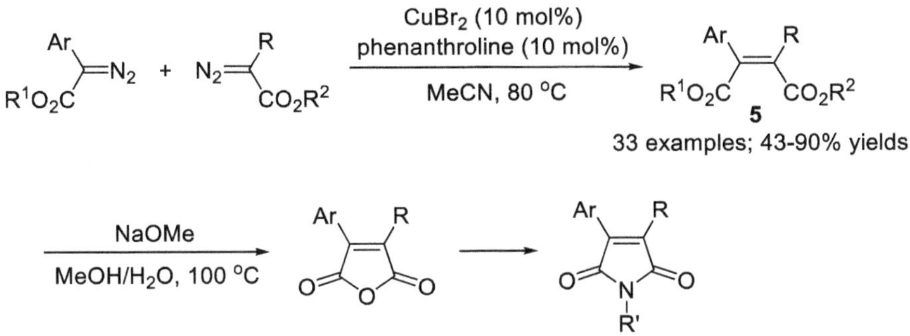

6, R' = H, OH, Bn

Scheme 8.4 Copper-catalyzed coupling reaction of diazo compounds to synthesize tetra-substituted olefins and maleimide derivatives.

attack of the carbon atom of diazo compound on the active Cu(I)-carbene species. This highly efficient and stereoselective strategy enables the concise synthesis of *tetra*-substituted olefins **5** with moderate to good yields and with high level of *Z*-selectivity (Scheme 8.4). The corresponding alkenes also allowed the rapid assembly of maleimide libraries **6**.

In view of the difficulty in controlling the chemo-selectivity of cross-dimerization of two diazo compounds, other carbene precursors were also investigated for the cross-coupling with diazo substrates. In this context, enynone derivatives **7** were found to be efficient carbene precursors in the reaction with diazo substrates catalyzed by $ZnCl_2$.[20] A series of functionalized 2-alkenylfurans derivatives **8** could be obtained with high efficiency through Zn(II)-carbene species **9** (Scheme 8.5).

Wang and co-workers reported the reaction of *N*-tosylhydrazones with chromium Fischer carbene complexes **10**, providing a new method for preparing diarylethanones **11** (Scheme 8.6).[21] This formal carbene dimerization approach proceeds *via* base-mediated deprotonation to generate diazo species **12** *in situ*, followed by nucleophilic attack on the electron deficient carbenic center to generate **13**, and sequential elimination of chromium carbonyl fragment and N_2. Alternatively, direct nucleophilic attack of anion with Fischer carbene also resulted in the formation of olefin intermediates **14** along with the extrusion of N_2, Ts group and $Cr(CO)_5$.

For other carbene precursors, Hu and co-workers developed a transition-metal-free reaction of diazoester with difluorocarbene, namely the

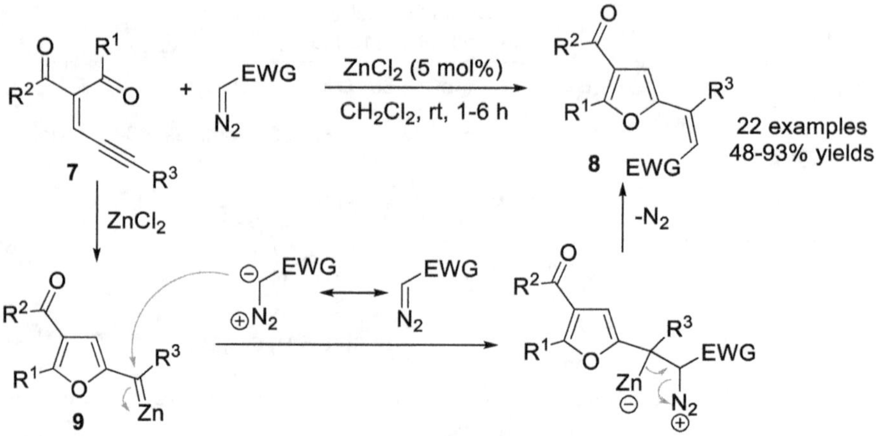

Scheme 8.5 Zn(II)-catalyzed synthesis of 2-alkenylfurans through carbene cross-coupling of enynones with diazo compounds.

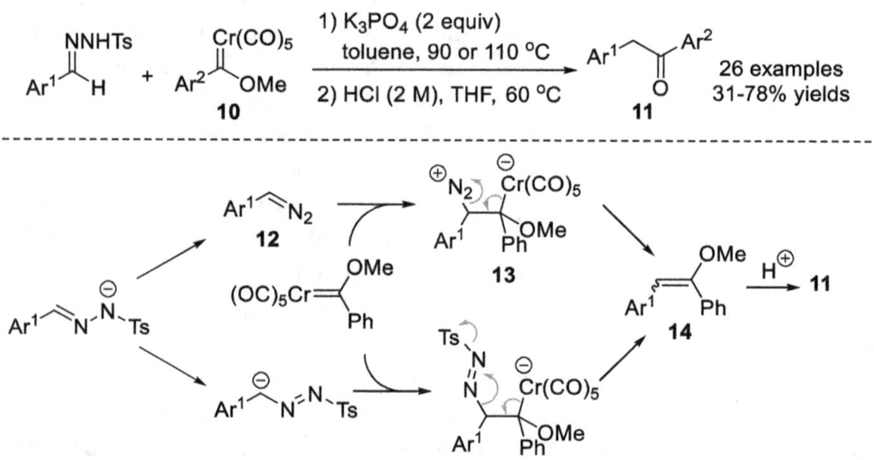

Scheme 8.6 Formal carbene dimerization of *N*-tosylhydrazones with chromium Fischer carbene complexes.

gem-difluoroolefination of diazo compounds by using $TMSCF_3$ or $TMSCF_2Br$ as the difluorocarbene precursors (Scheme 8.7).[22] The reaction allowed the synthesis of a series of 1,1-difluoroalkenes from diaryl diazomethanes as well as diazirines as their surrogates. A mechanism has been proposed, which involves nucleophilic attack of diazo substrate on

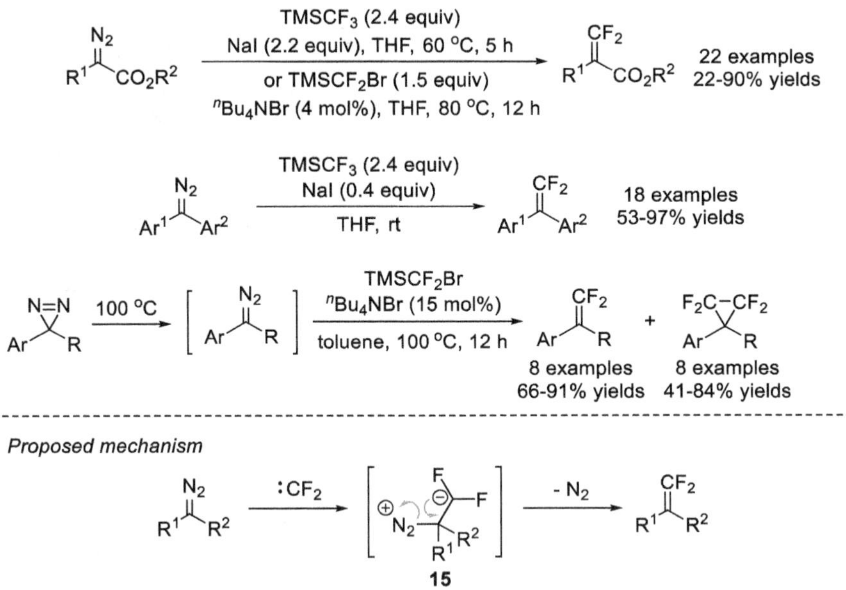

Scheme 8.7 Transition-metal-free *gem*-difluoroolefination of diazo compounds.

difluorocarbene to form intermediate **15**, followed by the release of nitrogen gas. The chemoselectivity observed for the formal carbene dimerization is attributed to the electronic properties and the stability of the difluorocarbene species.

On the other hand, Wang and co-workers reported the transition-metal-free difluoromethylenation of various diazo compounds under mild conditions, affording functionalized 1,1-difluoroolefins with high efficiency (Eq. (8.4)).[23]

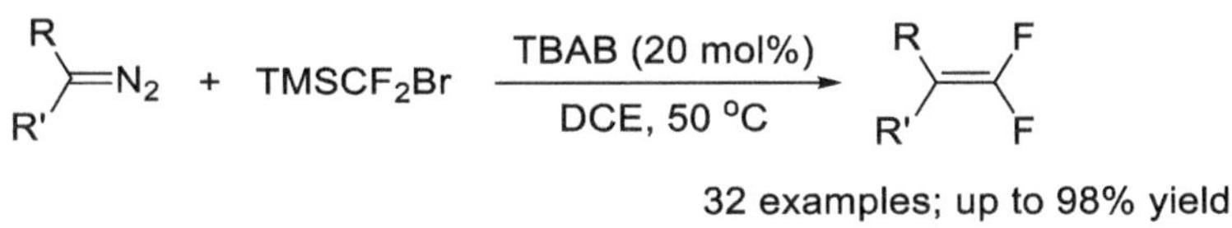

(8.4)

At the same time, with difluoromethylene phosphobetaine ($Ph_3P^+CF_2CO_2^-$), Xiao and co-workers developed the synthesis of steric hindered *gem*-difluoroolefins through cross-coupling reaction of difluorocarbene with diazo compounds.[24] This formal carbene dimerization process between difluorocarbene and α-aryl-α-diazoacetates was carried out in the presence of CuBr.

Transition-metal-catalyzed carbene coupling of diazo compounds with azides provides an efficient C=N bond construction method.[25–29] Intramolecular Rh(II)- or Cu(I)-catalyzed azide-diazo coupling reaction of azide-tethered diazocarbonyl compounds afforded *N*-heterocyclic frameworks directly, including aminoglycoside mimics[25] and functionalized indolines.[26] The first asymmetric intramolecular azide-diazo coupling of 1,3-diazido-2-propanol was catalyzed by Cu(I)-bisoxazoline (**16**) system.[27] This desymmetrization reaction provided cyclic α-imino esters with excellent yields and good enantioselectivity (up to 97% *ee*) (Eq. (**8.5**)).

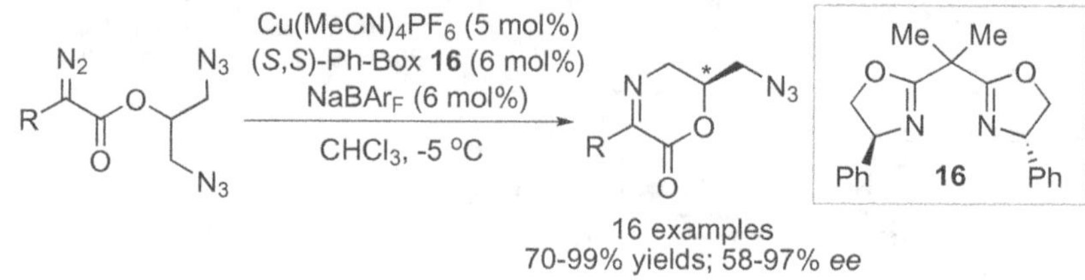

(8.5)

Moreover, intermolecular cross-coupling of α-diazocarbonyl compounds with alkyl or aryl azides was also achieved with the employment of Rh(II) complexes. Interception of alkyl azides with diazo(aryl)acetates resulted in the generation of α-imino esters, which finally afforded α-keto esters with high efficiency upon acidic workup process.[28] Doyle and coworkers also developed a mild and effective Rh(II)-catalyzed synthetic method for α-carbonylimines **17** with high yields (up to 97%), which underwent one-pot or stepwise Wolff rearrangement/Staudinger [2+2] reaction with enonediazoacetate **18** to provide poly-functionalized β-lactams **19** in excellent yields and with good diastereoselectivity (Scheme **8.8**).[29]

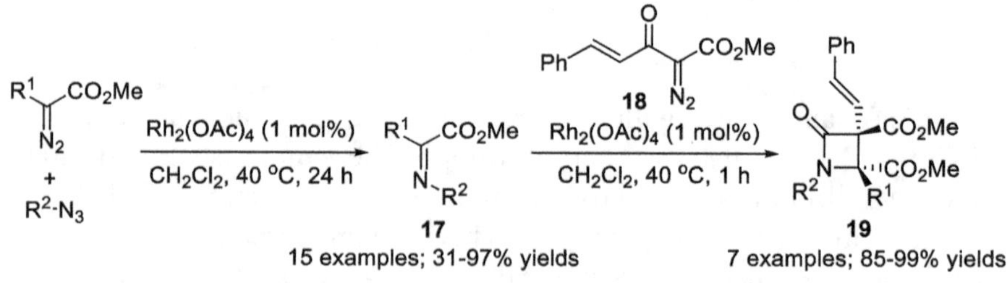

Scheme 8.8 Rh(II)-catalyzed α-carbonylimines synthesis from α-diazoesters and azides.

8.2 Transition-metal-catalyzed metathesis

Ruthenium- and other transition-metal-catalyzed carbene metathesis of diazo compounds provides an effective C=C double bond construction strategy for the synthesis of substituted alkenes, conjugated dienes and polycyclic rings. For a range of ruthenium complex-catalyzed reactions between diazoesters with olefins, the generation of substituted alkenes was achieved as the result of the metathesis process.[30,5,6]

Cp*Ru(cod)Cl has been proven to be an efficient catalyst for the addition of two carbene units of diazo compounds to alkynes, furnishing a general approach to functionalized conjugated dienes with simultaneous elimination of two equivalents of N_2.[31,32] This ene–yne cross-metathesis of diazo compounds with a wide range of alkynes, including terminal and internal, as well as the 1,3-diynes, created two C=C double bonds directly with one simple step. Ru-vinyl-carbene species **20** was considered as the key intermediate in this transformation (Scheme 8.9).

Dixneuf and Dérien demonstrated another functionalized dienes synthesis through Ru(II)-catalyzed formal carbene metathesis of silyl diazo compounds with propargylic esters[33] or propargylic carbonates.[34] Functionalized dienyl esters and carbonates were concisely synthesized *via* this catalytic cross-coupling of two carbene fragments. The proposed mechanism involves the enyne metathesis to generate ruthenacyclobutene **21** and then the ruthenium-vinylcarbene intermediate **22**, followed by 1,2-migration onto carbene carbon to furnish the final product (Scheme 8.10).

Subsequently, they reported the formation of functionalized bicyclic amino esters through Ru(II)-catalyzed tandem carbene

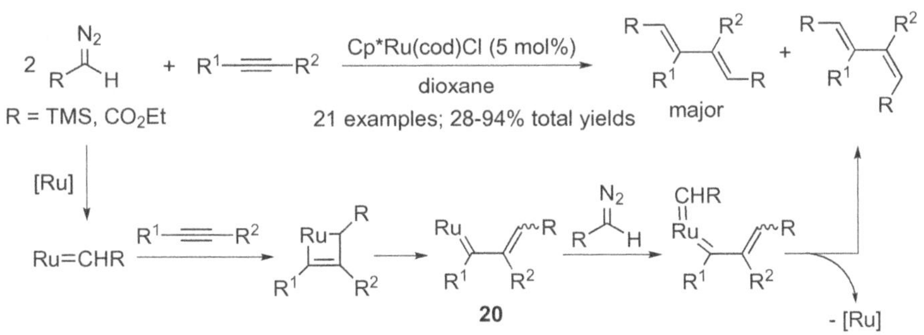

Scheme 8.9 Ru(II)-catalyzed metathesis of diazo compounds with alkynes to synthesize conjugated dienes.

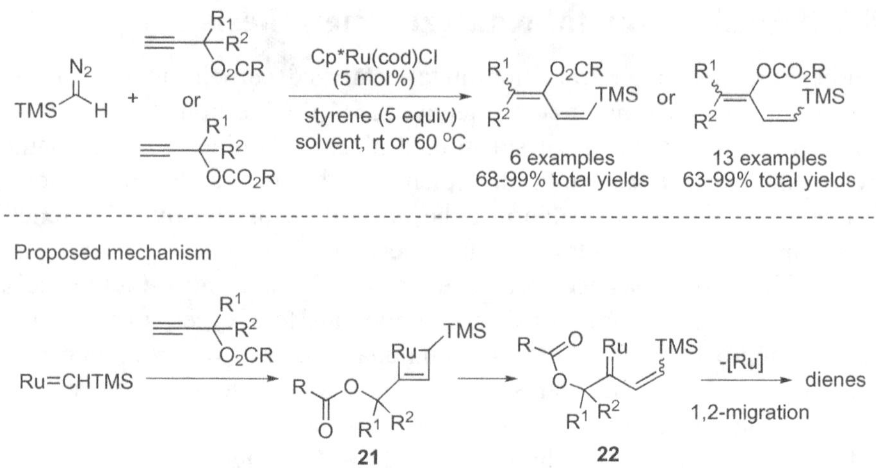

Scheme 8.10 Ru(II)-catalyzed cross-metathesis of propargylic esters/carbonates with diazo compounds.

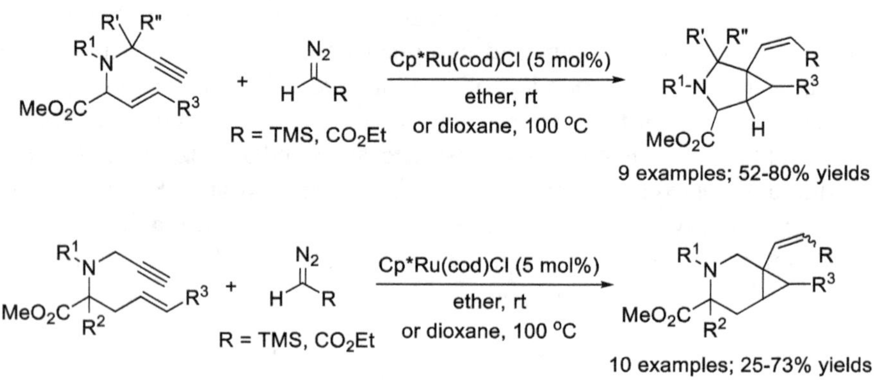

Scheme 8.11 Ru(II)-catalyzed synthesis of bicyclic amino esters through tandem carbene addition/cyclopropanation of enynes.

metathesis/cyclopropanation of 1,6- and 1,7-enynes with diazo compounds (Scheme 8.11).[35,36] The proposed reaction pathway involves Ru(II)-mediated decomposition of diazo species to generate Ru(II)-carbene, followed by sequential enyne-metathesis to generate Ru(II) vinylcarbene. Finally, cyclopropanation of alkene moiety with the Ru(II) carbene affords the final product bearing bicyclic scaffold.

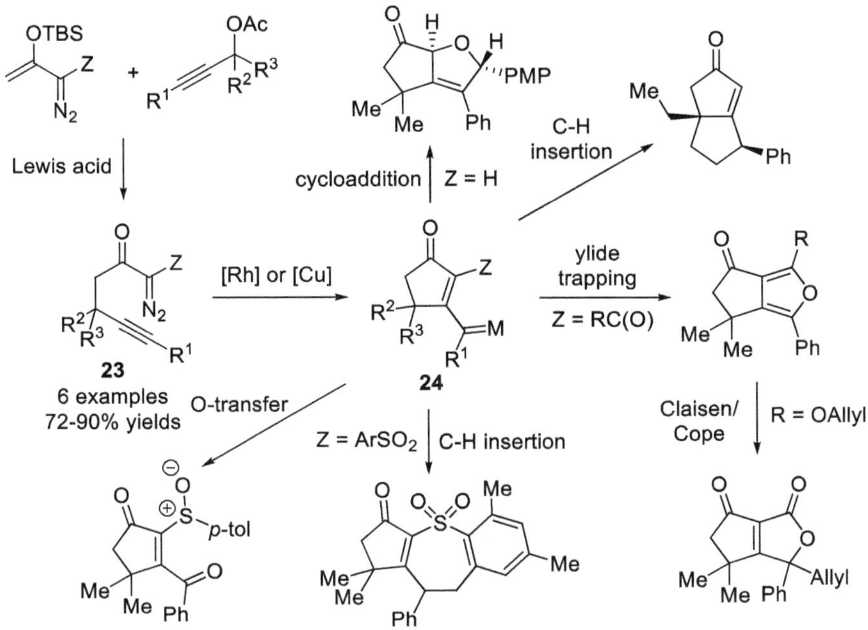

Scheme 8.12 Carbene metathesis cascade transformations of alkynyl-tethered diazoketones

Doyle and co-workers developed a method for accessing alkynyl-tethered diazoketones **23** *via* nucleophilic addition of enol-diazo compounds with propargyl acetates.[37] Further, **23** was subjected to Rh(II)- or Cu(I)-catalyzed diazo-alkyne metathesis to generate metal carbene species **24** through cyclization, followed by sequential metal carbene transformations to afford a wide range of diverse heterocyclic building blocks (Scheme 8.12). Besides, Ru(II)-catalyzed intermolecular diazo-alkyne metathesis/cyclization cascade of diazodicarbonyl compounds with terminal alkynes leads to the formation of functionalized furans.[38]

May and co-workers developed a general approach toward the construction of bridged polycyclic ring skeleton through Rh(II)-catalyzed carbene cascade reaction.[39,40] From alkynyl diazoester **25**, this cascade transformation followed Rh(II)-catalyzed diazo decomposition/carbene-alkyne cross-metathesis to form **26**/transannular C–H insertion to readily afford a variety of complex polycyclic targets **27** (Eq. (8.6)).

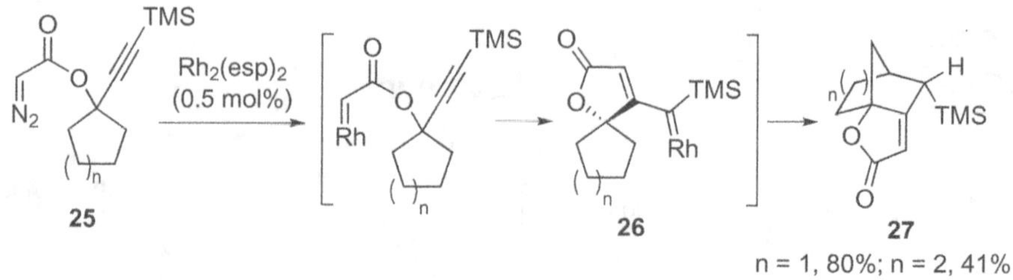

(8.6)

Xu and co-workers reported a Rh(II)-catalyzed intramolecular carbene cascade transformation of alkynyl-tethered styryl diazoesters **28** (Scheme 8.13).[41] This substrate-dependent carbene/alkyne cascade reaction afforded bicyclic cyclopentadiene scaffold **29** through the termination of formal [3+2] cycloaddition/[1,5]-H shift. In contrast, in the case of alkyl alkyne-tethered substrates, β-H elimination was the dominating process to form triene product **30**.

In addition, Saá and co-workers reported the construction of spiro- and fused-bicyclic structures from linear alkynyl compounds with silyl diazomethane by means of Ru(II)-catalyzed metathesis and termination with intramolecular carbene C–H insertion.[42] The same group also developed a synthesis of 2-vinyl dihydropyrans and dihydrooxazines from alkynals and alkynones in the presence of Ru(II) complex. The proposed reaction mechanism involves the generation of Ru(II) vinyl carbene as the key intermediate, which is followed by sequential intramolecular nucleophilic attack of carbonyl group to form Ru(II)-zwitterionic species,

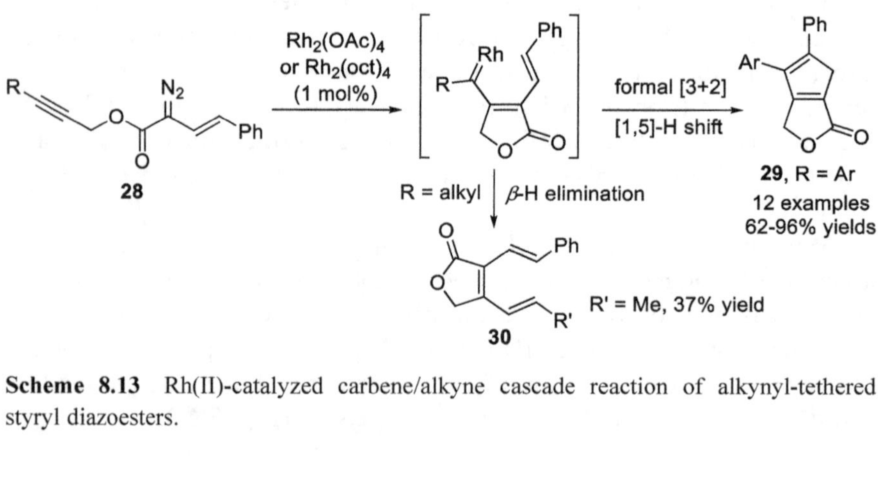

Scheme 8.13 Rh(II)-catalyzed carbene/alkyne cascade reaction of alkynyl-tethered styryl diazoesters.

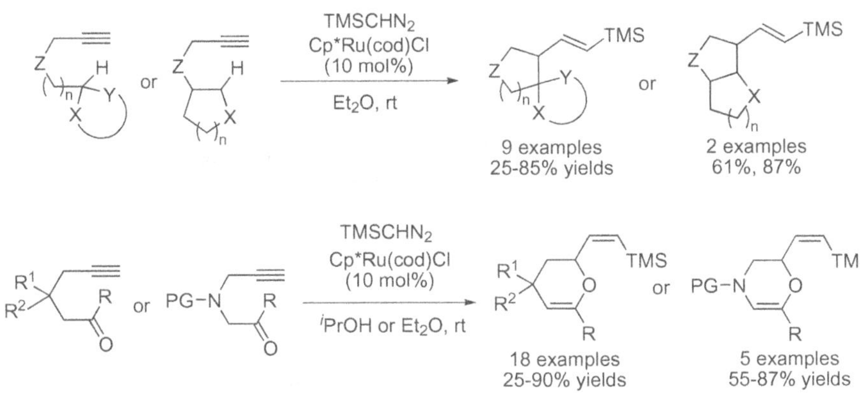

Scheme 8.14 Ru(II)-catalyzed synthesis of spiro- and fused-bicyclic rings, 2-vinyl dihydropyrans and dihydrooxazines.

followed by deprotonation and re-protonation to furnish the cyclization products (Scheme 8.14).[43]

The same group developed an entry to functionalized dihydro-1,3-benzoxazines **32** *via* Ru(II)-catalyzed cyclization of *o*-(alkynyloxy) benzylamines **31**.[44] This intriguing cascade protocol proceeded through the direct nucleophilic addition of amine with *in situ* generated Ru(II) vinyl carbene intermediate **33**, followed by ring-opening rearrangement to form enamine **34** and the final product 1,3-benzoxazines **32** (Scheme 8.15). In addition to its synthetic value, this cyclization reveals an intriguing type of Ru(II)-vinyl carbene transformation through intramolecular amine nucleophilic addition.

8.3 Phosphorus or Arsenic ylide

Catalytic olefination of carbonyl substrates with diazo compounds provides an important complementary method for the synthesis of functionalized alkenes, which avoids the use of phosphorus ylide reagents as these under ordinary Wittig conditions. The reaction is usually carried out in the presence of a phosphine, in most cases Ph_3P. When aldehydes are used as the substrates, the olefins are usually produced with high selectivity for the *E* geometry. Since the first report by Herrmann and co-workers on the reaction with $MeReO_3$ (methylrhenium trioxide, MTO) in 1991,[45] this type of reaction has been extensively explored and efficient catalytic

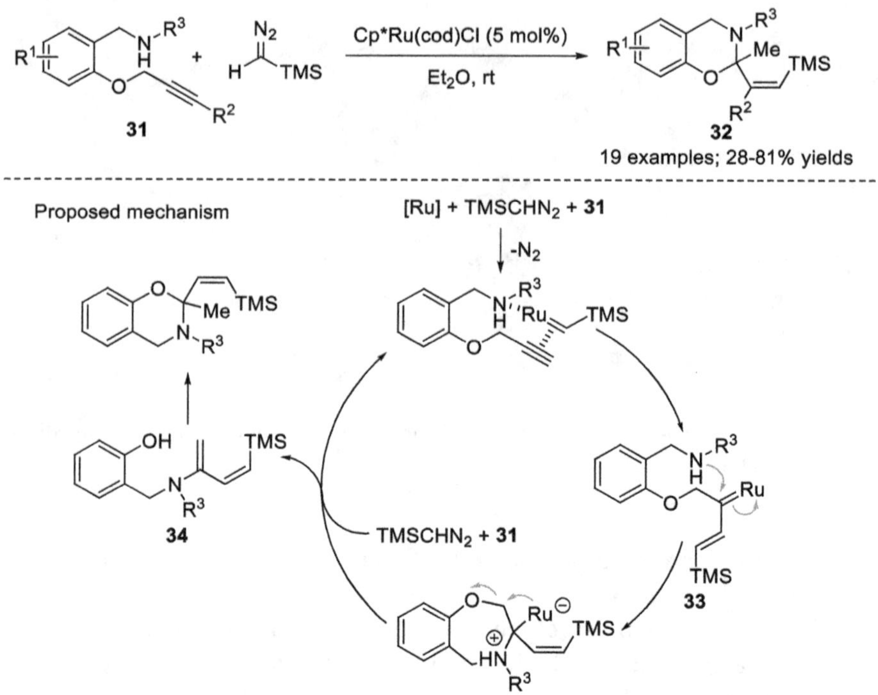

Scheme 8.15 Nucleophilic addition of amines to Ru(II) vinyl carbene to access the synthesis of 1,3-benzoxazines.

olefination reactions with good stereoselectivity have been reported based on Re,[45–52] Ru,[53–56] Rh,[57–60] Fe,[61–68] Cu[69,70] and Co[71] complexes.

Depending on the metal complex, there are two different mechanistic pathways proposed for the olefination of aldehydes with diazo compounds (Scheme 8.16). In the first place, the metal-carbene reacts with the carbonyl compound to form a metal-oxo species **35**, which is then reduced by phosphorus reagent followed by the reaction with EDA to generate the metal carbene **36**. The metal carbene **36** reacts with aldehyde in a manner which is similar to olefin metathesis to afford the olefination product, with regeneration of the metal oxo species. With electron spray mass spectrometry, NMR and other spectroscopic methods, evidence to support both mechanisms has been obtained. It has been suggested that, in the reaction with methylrhenium trioxide (MTO) and similar rhenium oxides as catalysts, the reaction follows the mechanism shown in the scheme, with a metallaoxetane as the intermediate.[45–52]

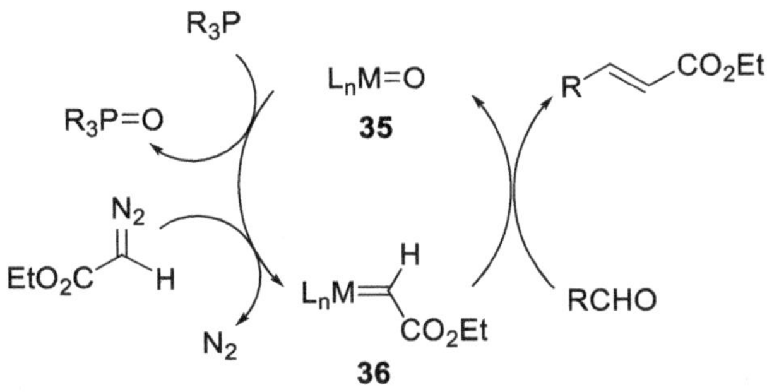

Scheme 8.16 Olefination of diazo compounds involving metal oxo species.

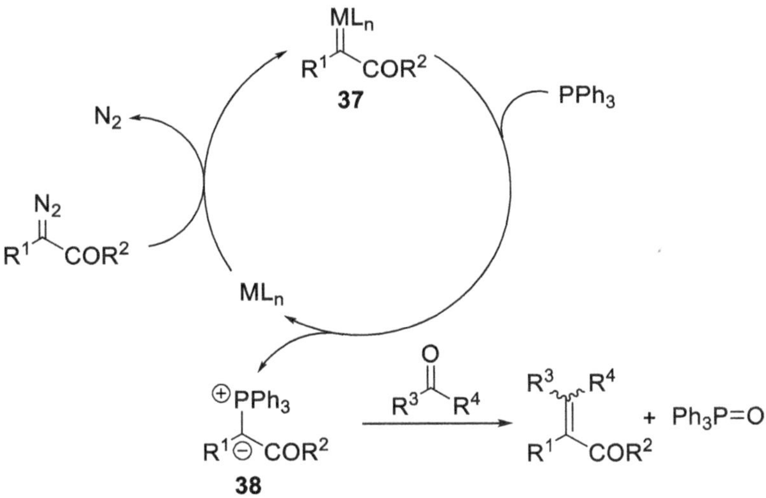

Scheme 8.17 Olefination of diazo compounds involving phosphorus ylide species.

In another possible reaction pathway, the metal carbene is first generated by the reaction of the metal complex with EDA. The phosphorus reagent attacks the metal carbene **37**, forming a phosphorus ylide species **38** that reacts with carbonyl derivatives by a Wittig reaction to afford the alkene products accompanied by phosphine oxide release (Scheme 8.17). In the reaction with iron and cobalt porphyrin, or Ru complexes as catalysts, this pathway is followed with the generation of the phosphorus ylide.[53–56,61–68,71] On the other hand, in the methylation of aldehyde by $TMSCHN_2$ with $ClRh(PPh_3)_3$ as catalyst, a different mechanism for the generation of the phosphorus ylide that does not involve a metal carbene is proposed.[58]

Tang and co-workers reported an olefination of ketenes with ethyl diazoacetate catalyzed by tetra(*p*-chlorophenyl)porphyrin iron chloride [Fe(TCP)Cl].[68] Moreover, when a chiral phosphine was used instead of Ph_3P, the olefination of ketone occurred, affording allenes **39** with high enantioselectivity (Scheme 8.18).

Another recent report showed that, with ruthenium-based olefin metathesis catalysts, α,β-unsaturated aldehydes could be olefinated with diazoacetates in the presence of triphenylphosphine. Based on this result, a cross-metathesis/olefination tandem transformation of terminal olefins into 1,3-dienoic esters **40** was developed by using a single Ru(II) carbene complex (Scheme 8.19).[56]

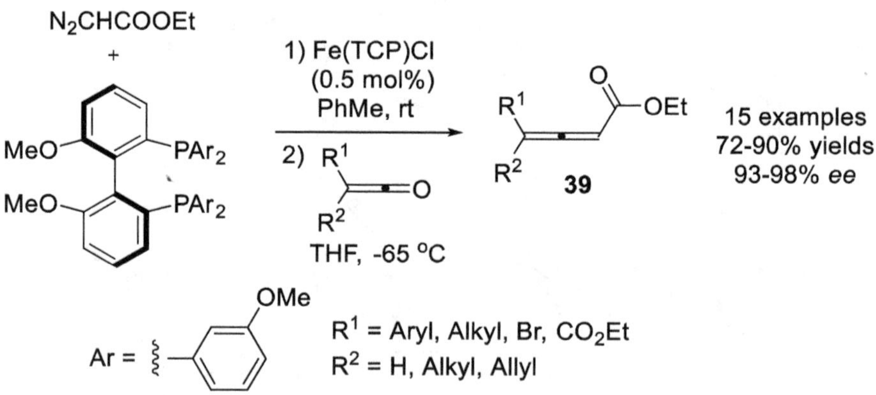

Scheme 8.18 Fe(Por)-catalyzed olefination of ketenes with EDA.

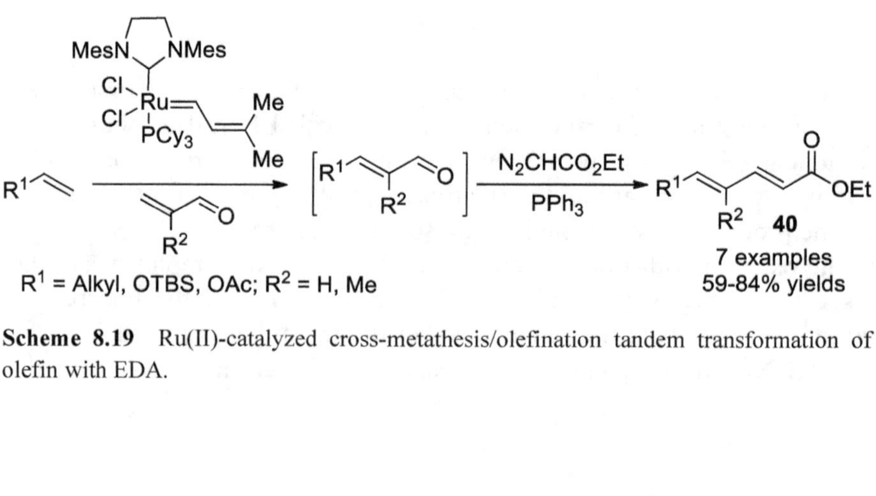

Scheme 8.19 Ru(II)-catalyzed cross-metathesis/olefination tandem transformation of olefin with EDA.

In the past decade, iron and copper complexes have been established as the most common transition-metal catalysts in phosphine-mediated olefinations. On the other hand, other transition-metal catalysts have also been explored, such as [(BINAP)Re(O)Cl_3],[72] hypervalent Sn(IV) porphyrins,[73] Ir(I) complexes[74] and polyethylene glycol (PEG)-immobilized Fe(II) porphyrin.[75] These metal catalysts have demonstrated efficient reactivity in catalyzing olefination of aldehydes with α-diazoester. This novel catalytic system enabled the preparation of diverse functionalized alkenes from a wide range of diazo compounds and carbonyl compounds. Especially, densely substituted vinyl phosphonates **41** could be obtained through Co(II)-porphyrin-catalyzed selective olefination (*E*/*Z* up to > 99:1) of acyl phosphonates with EDA in the presence of PPh_3 (Eq. (8.7)).[76]

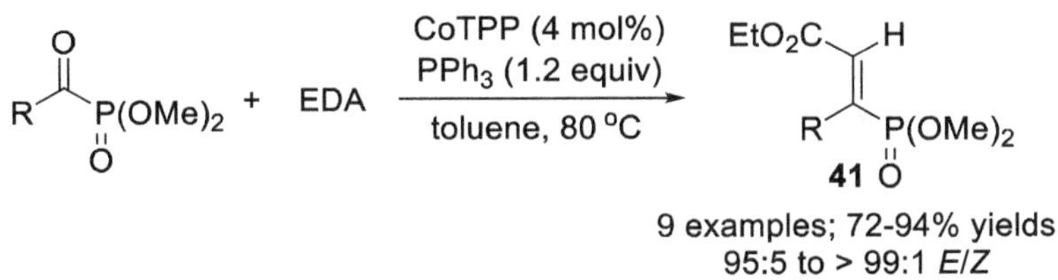

(8.7)

Rh(II)/Ce(III)-co-catalyzed olefination of ferrocenyl aldehydes with various diazoacetates was reported, providing a concise and efficient procedure to access the synthesis of alkenylferrocene derivatives.[77] Moreover, both Cu(I)[78] and Cu(II)[79] complexes were proved to be effective in catalyzing diazo olefination under non-basic conditions. Lebel and co-workers realized a one-pot process, providing *trans*-cyclopropyl compounds *via* Cu(I)-catalyzed olefination of aldehydes with sequential Pd(0)-catalyzed cyclopropanation with CH_2N_2 to afford GPR40 receptor agonists.[80] They extensively improved the olefination conditions by using readily available/recycle tetraarylphosphonium (TAP)-supported phosphorus ylides.[81]

Wei and co-workers reported Cu(II)-catalyzed olefination of aldehydes with *in situ* generated diazo species from *N*-tosylhydrazones analogue, as well as intermolecular metal-free phosphine-mediated diaza-Wittig reaction.[82] This transformation afforded either *trans*-alkenes **42** with moderate to good yields and selectivity, or unsymmetrical azines **43** with high efficiency (Scheme 8.20). Continuously, the same group and Lebel group developed the one-pot synthesis of *tri*-substituted alkenes from alcohols with *N*-tosylhydrazones or various diazo compounds. The reaction proceeds through Cu-catalyzed aerobic oxidation-olefination cascade.[83,84]

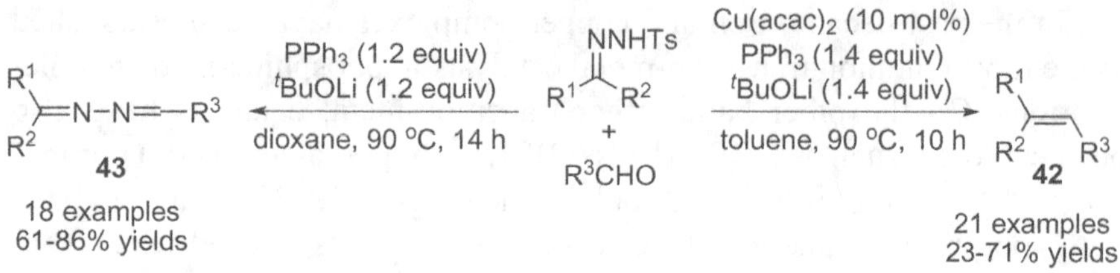

Scheme 8.20 Olefination and diaza-Wittig reaction of aldehydes with *N*-tosylhydrazones.

Tang and co-workers developed the Fe(Por)-catalyzed olefination by using PE-supported arsine.[85] This efficient protocol, which provided *di-*, *tri-* and *tetra*-substituted alkenes **44** with high yields, shows superior catalytic efficiency over a wide range of carbonyl substrates (Eq. (8.8)). By the addition of polymethylhydrosiloxane as an inexpensive reductant, the loading of PE-supported arsine could be reduced to 0.05 mol%.

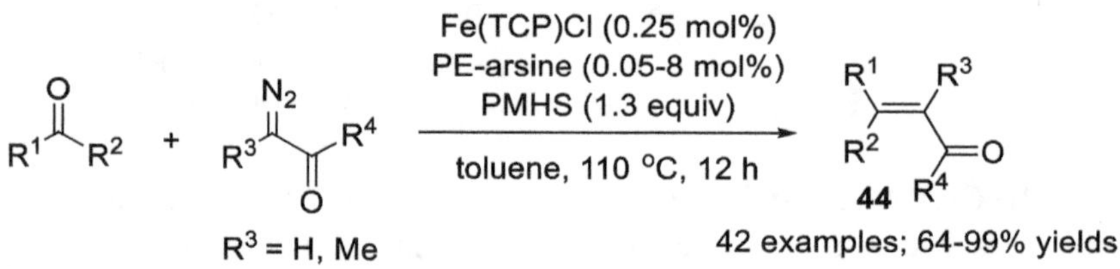

(8.8)

In addition, Tang and co-workers reported the multistep one-pot Fe-catalyzed cyclopentenones synthesis from alkenyl diazoketones **45** with acid chlorides.[86] This cascade sequence proceeded through catalytic Wittig olefination of *in situ* generated ketene **46** with subsequently acid-mediated Nazarov cyclization to afford cyclopentenone derivatives **47** without the isolation of allenes intermediate (Scheme 8.21).

Furthermore, the same group demonstrated a synthesis of *tetra*-substituted dienes and cyclopentadienes from *tri*-substituted alkenes with iron porphyrin carbene.[87] This intriguing reaction was initiated by cyclopropanation of diazo compounds to form **48**, followed by ring-opening process to generate vinyl phosphorus ylide intermediates **49/49′** (Scheme 8.22). Owing to the different configurations of the ylide intermediates, sequential either intra- or intermolecular Wittig reaction afforded cyclopentadienes or tetra-substituted dienes, respectively. Experimental as well as computational studies were carried out in order to gain insights into the mechanism of this unusual iron carbene transformation.[88]

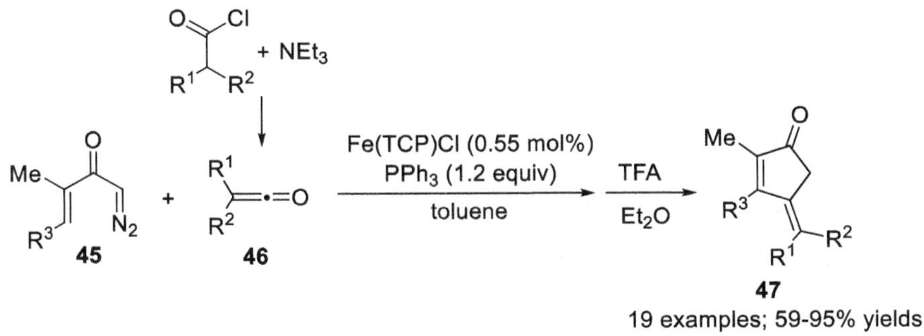

Scheme 8.21 One-pot Wittig/Nazarov cascade synthesis of cyclopentenone derivatives.

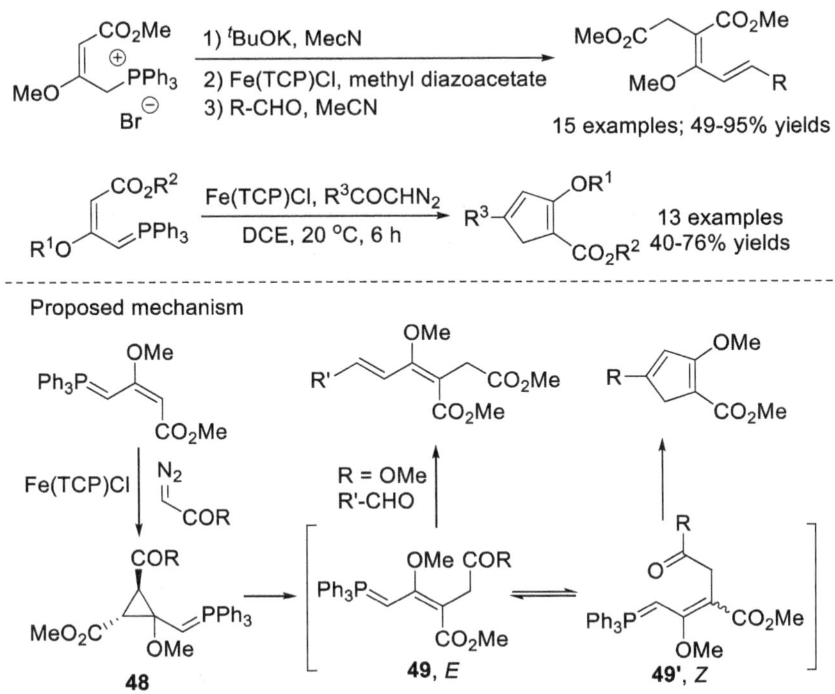

Scheme 8.22 Synthesis of *tetra*-substituted dienes and cyclopentadienes through tri-substituted alkenes with iron porphyrin carbenes.

Recently, Fe(IV)-corrole complexes (**50**) were investigated for catalyzing PPh_3-mediated aldehydes olefination with EDA.[89] This catalytic method enabled the high *trans*-selective olefination of aromatic aldehydes. Notably, the transformation from phosphazine to ylide **51**

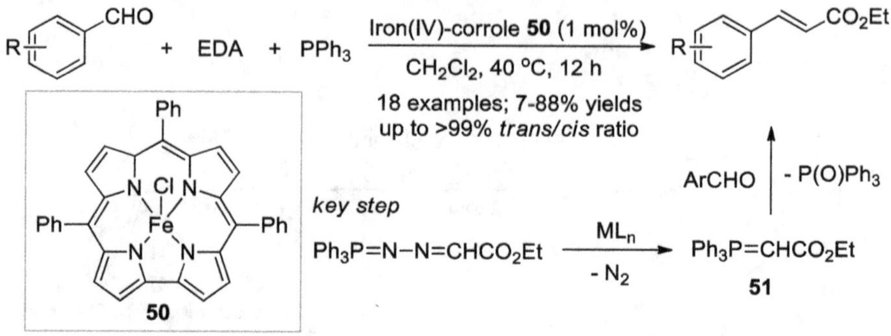

Scheme 8.23 Fe(IV)-corrole-catalyzed aldehydes olefination with EDA in the presence of PPh_3.

proved to be the key step in this Fe(IV)-corrole catalytic system (Scheme 8.23).

In summary, while olefination of diazo compounds is usually regarded as side reaction in transition-metal-catalyzed carbene transfer reactions, it is possible to develop useful synthetic methods based on this process. In particular, intramolecular olefination of bis-diazo compounds can avoid the problem of chemoselectivity of the reaction, making the carbene dimerization a straightforward method for the synthesis of unsaturated cyclic compounds. Finally, the olefination of diazo compounds through phosphorus ylide intermediates may be served as a complementary method for classic Wittig reaction.

References

1. Martin, V.; Blakey, S. *Tetrahedron Lett.* **2008**, *49*, 6800–6803.
2. Zhou, Y.; Trewyn, B. G.; Angelici, R. J.; Woo, L. K. *J. Am. Chem. Soc.* **2009**, *131*, 11734–11743.
3. Zhou, Y.; Angelici, R. J.; Woo, L. K. *Catal. Lett.* **2010**, *137*, 8–15.
4. Le Maux, P.; Nicolas, I.; Chevance, S.; Simonneaux, G. *Tetrahedron* **2010**, *66*, 4462–4468.
5. Basato, M.; Tubaro, C.; Biffis, A.; Bonato, M.; Buscemi, G.; Lighezzolo, F.; Lunardi, P.; Vianini, C.; Benetollo, F.; Del Zotto, A. *Chem. Eur. J.* **2009**, *15*, 1516–1526.
6. Dragutan, I.; Dragutan, V.; Verpoort, F. *Appl. Organometal. Chem.* **2014**, *28*, 211–215.

7. For an example for the study on *cis*/*trans* selectivity in Ru-catalyzed reaction of EDA, see: Graban, E.; Lemke, F. R. *Organometallics* **2002**, *21*, 3823–3826.
8. Guan, Z.; Zhang, L.-H.; Sinaÿ, P.; Zhang, Y. *J. Org. Chem.* **2012**, *77*, 8888–8895.
9. Zhu, C.; Falck, J. R. *Tetrahedron* **2012**, *68*, 9192–9199.
10. Hodgson, D. M.; Angrish, D. *Chem. Commun.* **2005**, 4902–4904.
11. Hodgson, D. M.; Angrish, D. *Chem. Eur. J.* **2007**, *13*, 3470–3479.
12. Li, G.-Y.; Che, C.-M. *Org. Lett.* **2004**, *6*, 1621–1623.
13. Rivilla, I.; Sameera, W. M. C.; Alvarez, E.; Díaz-Requejo, M. M.; Maseras, F.; Pérez, P. J. *Dalton Trans.* **2013**, *42*, 4132–4138.
14. Hansen, J. H.; Parr, B. T.; Pelphrey, P.; Jin, Q.; Autschbach, J.; Davies, H. M. L. *Angew. Chem. Int. Ed.* **2011**, *50*, 2544–2548.
15. Zhang, D.; Xu, G.; Ding, D.; Zhu, C.; Li, J.; Sun, J. *Angew. Chem. Int. Ed.* **2014**, *53*, 11070–11074.
16. Xia, Y.; Liu, Z.; Xiao, Q.; Qu, P.; Ge, R.; Zhang, Y.; Wang, J. *Angew. Chem. Int. Ed.* **2012**, *51*, 5714–5717.
17. Zhu, C.; Xu, G.; Liu, K.; Qiu, L.; Peng, S.; Sun, J. *Chem. Commun.* **2015**, *51*, 12768–12770.
18. Zhu, C.; Qiu, L.; Xu, G.; Li, J.; Sun, J. *Chem. Eur. J.* **2015**, *21*, 12871–12875.
19. Zhu, C.; Xu, G.; Ding, D.; Qiu, L.; Sun, J. *Org. Lett.* **2015**, *17*, 4244–4247.
20. González, J.; López, L. A.; Vicente, R. *Chem. Commun.* **2014**, *50*, 8536–8538.
21. Hu, F.; Yang, J.; Xia, Y.; Ma, C.; Xia, H.; Zhang, Y.; Wang, J. *Org. Chem. Front.* **2015**, *2*, 1450–1456.
22. Hu, M.; Ni, C.; Li, L.; Han, Y.; Hu, J. *J. Am. Chem. Soc.* **2015**, *137*, 14496–14501.
23. Zhang, Z.; Yu, W.; Wu, C.; Wang, C.; Zhang, Y.; Wang, J. *Angew. Chem. Int. Ed.* **2016**, *55*, 273–277.
24. Zheng, J.; Lin, J.-H.; Yu, L.-Y.; Wei, Y.; Zheng, X.; Xiao, J.-C. *Org. Lett.* **2015**, *17*, 6150–6153.
25. Blond, A.; Moumné, R.; Bégis, G.; Pasco, M.; Lecourt, T.; Micouin, L. *Tetrahedron Lett.* **2011**, *52*, 3201–3203.
26. Bott, T. M.; Atienza, B. J.; West, F. G. *RSC Adv.* **2014**, *4*, 31955–31959.
27. Wu, X.-P.; Su, Y.; Gu, P. *Org. Lett.* **2014**, *16*, 5339–5341.
28. Gu, P.; Wu, X.-P.; Su, Y.; Li, X.-Q.; Xue, P.; Li, R. *Synlett* **2014**, *25*, 535–538.
29. Mandler, M. D.; Truong, P. M.; Zavalij, P. Y.; Doyle, M. P. *Org. Lett.* **2014**, *16*, 740–743.
30. Ding, F.; Sun, Y.-G.; Verpoort, F.; Dragutan, V.; Dragutan, I. *J. Mol. Catal. A: Chem.* **2014**, *386*, 86–94.

31. Severa, L.; Vávra, J.; Kohoutová, A.; Čížková, M.; Šálová, T.; Hývl, J.; Šaman, D.; Pohl, R.; Adriaenssens, L.; Teplý, F. *Tetrahedron Lett.* **2009**, *50*, 4526–4528.
32. Le Paih, J.; Vovard-Le Bray, C.; Dérien, S.; Dixneuf, P. H. *J. Am. Chem. Soc.* **2010**, *132*, 7391–7397.
33. Vovard-Le Bray, C.; Dérien, S.; Dixneuf, P. H. *Angew. Chem. Int. Ed.* **2009**, *48*, 1439–1442.
34. Moulin, S.; Zhang, H.; Raju, S.; Bruneau, C.; Dérien, S. *Chem. Eur. J.* **2013**, *19*, 3292–3296.
35. Eckert, M.; Moulin, S.; Monnier, F.; Titanyuk, I. D.; Osipov, S. N.; Roisnel, T.; Dérien, S.; Dixneuf, P. H. *Chem. Eur. J.* **2011**, *17*, 9456–9462.
36. Vovard-Le Bray, C.; Klein, H.; Dixneuf, P. H.; Macé, A.; Berrée, F.; Carboni, B.; Dérien, S. *Adv. Synth. Catal.* **2012**, *354*, 1919–1925.
37. Qian, Y.; Shanahan, C. S.; Doyle, M. P. *Eur. J. Org. Chem.* **2013**, 6032–6037.
38. Xia, L.; Lee, Y. R. *Eur. J. Org. Chem.* **2014**, 3430–3442.
39. Jansone-Popova, S.; May, J. A. *J. Am. Chem. Soc.* **2012**, *134*, 17877–17880.
40. Jansone-Popova, S.; Le, P. Q.; May, J. A. *Tetrahedron* **2014**, *70*, 4118–4127.
41. Zheng, Y.; Mao, J.; Weng, Y.; Zhang, X.; Xu, X. *Org. Lett.* **2015**, *17*, 5638–5641.
42. Cambeiro, F.; López, S.; Varela, J. A.; Saá, C. *Angew. Chem. Int. Ed.* **2012**, *51*, 723–727.
43. Cambeiro, F.; López, S.; Varela, J. A.; Saá, C. *Angew. Chem. Int. Ed.* **2014**, *53*, 5959–5963.
44. González-Rodríguez, C.; Suárez, J. R.; Varela, J. A.; Saá, C. *Angew. Chem. Int. Ed.* **2015**, *54*, 2724–2728.
45. Herrmann, W. A.; Wang, M. *Angew. Chem., Int. Ed.* **1991**, *30*, 1641–1643.
46. Herrmann, W. A. In *Olefins from Aldehydes in Applied Homogeneous Catalysis with Organometallic Compounds*; 2nd ed.; Wiley VCH: Weinheim, 2002; Vol. 3, p. 1078.
47. Zhang, X.; Chen, P. *Chem. Eur. J.* **2003**, *9*, 1852–1859.
48. Harrison, R.; Mete, A.; Wilson, L. *Tetrahedron Lett.* **2003**, *44*, 6621–6624.
49. Santos, A. M.; Romão, C. C.; Kühn, F. E. *J. Am. Chem. Soc.* **2003**, *125*, 2414–2415.
50. Santos, A. M.; Pedro, F. M.; Yogalekar, A. A.; Lucas, I. S.; Romão, C. C.; Kühn, F. E. *Chem. Eur. J.* **2004**, *10*, 6313–6321.
51. Kühn, F. E.; Scherbaum, A.; Herrmann, W. A. *J. Organomet. Chem.* **2004**, *689*, 4149–4164.
52. Pedro, F. M.; Hirner, S.; Kühn, F. E. *Tetrahedron Lett.* **2005**, *46*, 7777–7779.
53. Lebel, H.; Paquet, V. *Organometallics* **2004**, *23*, 1187–1190.
54. Sun, W.; Yu, B.; Kühn, F. E. *Tetrahedron Lett.* **2006**, *47*, 1993–1996.

55. Pedro, F. M.; Santos, A. M.; Baratta, W.; Kühn, F. E. *Organometallics* **2007**, *26*, 302–309.
56. Murelli, R. P.; Snapper, M. L. *Org. Lett.* **2007**, *9*, 1749–1752.
57. Lebel, H.; Guay, D.; Paquet, V.; Huard, K. *Org. Lett.* **2004**, *6*, 3047–3050.
58. Lebel, H.; Paquet, V. *J. Am. Chem. Soc.* **2004**, *126*, 320–328.
59. Zhu, S.; Liao, Y.; Zhu, S. *Org. Lett.* **2004**, *6*, 377–380.
60. Nandra, G. S.; Pang, P. S.; Porter, M. J.; Elliott, J. M. *Org. Lett.* **2005**, *7*, 3453–3455.
61. Mirafzal, G. A.; Cheng, G.; Woo, L. K. *J. Am. Chem. Soc.* **2002**, *124*, 176–177.
62. Cheng, G.; Mirafzal, G. A.; Woo, L. K. *Organometallics* **2003**, *22*, 1468–1474.
63. Aggarwal, V. K.; Fulton, J. R.; Sheldon, C. G.; de Vicente, J. *J. Am. Chem. Soc.* **2003**, *125*, 6034–6035.
64. Chen, Y.; Huang, L.; Ranade, M. A.; Zhang, X. P. *J. Org. Chem.* **2003**, *68*, 3714–3717.
65. Chen, Y.; Huang, L.; Zhang, X. P. *Org. Lett.* **2003**, *5*, 2493–2496.
66. Chen, Y.; Huang, L.; Zhang, X. P. *J. Org. Chem.* **2003**, *68*, 5925–5929.
67. Sun, W.; Kühn, F. E. *Tetrahedron Lett.* **2004**, *45*, 7415–7418.
68. Li, C.-Y.; Wang, X.-B.; Sun, X.-L.; Tang, Y.; Zheng, J.-C.; Xu, Z.-H.; Zhou, Y.-G.; Dai, L.-X. *J. Am. Chem. Soc.* **2007**, *129*, 1494–1495.
69. Pang, W.; Zhu, S.; Jiang, H.; Zhu, S. *Tetrahedron* **2006**, *62*, 11760–11765.
70. Lebel, H.; Davi, M.; Díez-González, S.; Nolan, S. P. *J. Org. Chem.* **2007**, *72*, 144–149.
71. Lee, M.-Y.; Chen, Y.; Zhang, X. P. *Organometallics* **2003**, *22*, 4905–4909.
72. Strand, D.; Rein, T. *J. Organomet. Chem.* **2010**, *695*, 2220–2224.
73. Gharaati, S.; Moghadam, M.; Tangestaninejad, S.; Mirkhani, V.; Mohammadpoor-Baltork, I. *J. Organomet. Chem.* **2012**, *720*, 26-29.
74. Lebel, H.; Ladjel, C. *Organometallics* **2008**, *27*, 2676–2678.
75. Chinnusamy, T.; Rodionov, V.; Kühn, F. E.; Reiser, O. *Adv. Synth. Catal.* **2012**, *354*, 1827–1831.
76. Ertürk, E.; Demir, A. S. *Tetrahedron* **2008**, *64*, 7555–7560.
77. Chen, S.; Zhang, L.; Du, Y.; Li, B. *Synlett* **2012**, *23*, 943–947.
78. Lebel, H.; Davi, M. *Adv. Synth. Catal.* **2008**, *350*, 2352–2358.
79. Ho, P. E.; Tao, J.; Murphy, G. K. *Eur. J. Org. Chem.* **2013**, 6540–6544.
80. Davi, M.; Lebel, H. *Chem. Commun.* **2008**, 4974–4976.
81. Lebel, H.; Davi, M.; Roy, M.-N.; Zeghida, W.; Charette, A. B. *Synthesis* **2011**, 2275–2280.
82. Sha, Q.; Ling, Y.; Wang, W.; Wei, Y. *Adv. Synth. Catal.* **2013**, *355*, 2145–2150.
83. Sha, Q.; Wei, Y. *Synthesis* **2014**, *46*, 2353–2361.
84. Davi, M.; Lebel, H. *Org. Lett.* **2009**, *11*, 41–44.

85. Wang, P.; Liu, C.-R.; Sun, X.-L.; Chen, S.-S.; Li, J.-F.; Xie, Z.; Tang, Y. *Chem. Commun.* **2012**, *48*, 290–292.
86. Cao, P.; Sun, X.-L.; Zhu, B.-H.; Shen, Q.; Xie, Z.; Tang, Y. *Org. Lett.* **2009**, *11*, 3048–3051.
87. Wang, P.; Liao, S.; Wang, S. R.; Gao, R.-D.; Tang, Y. *Chem. Commun.* **2013**, *49*, 7436–7438.
88. Wang, P.; Ling, L.; Liao, S.-H.; Zhu, J.-B.; Wang, S. R.; Li, Y.-X.; Tang, Y. *Chem. Eur. J.* **2013**, *19*, 6766–6773.
89. Zou, H.-B.; Yang, H.; Liu, Z.-Y.; Mahmood, M. H. R.; Mei, G.-Q.; Liu, H.-Y.; Chang, C.-K. *Organometallics* **2015**, *34*, 2791–2795.

Chapter 9

Vinyl Diazoacetate Transformations

Vinyl diazoacetates have shown unique reactivity in transition-metal-catalyzed reactions, especially in regio- and enantioselective cycloadditions. Owing to the structure of these carbene precursors, the steric influences around the metal carbenic center hinder direct attack by nucleophiles, leading to preferential vinylogous attack. The electrophilic character on the vinylogous position is impacted by the ligated transition metal of the metal carbene. Besides, the substituents on the carbene center atom or double bond also play important roles in shaping the feasibility of diverse carbene reactions. In the past few years, metal-catalyzed transformations involving metalloenolcarbene species have been well explored and have shown wide applications in organic synthesis, particularly for the construction of various heterocyclic structures.[1,2] In this chapter, we will cover the recent advances in this area.

9.1 Vinylogous addition and formal insertion

As shown by the resonance structure of the vinyl metal carbene **2** that is generated from the corresponding vinyldiazoacetate **1**, in addition to the normal carbenic reactivity of the metal carbene, the vinyl metal carbene **2** also shows vinylogous nucleophilic (**3**) or electrophilic character (**4**) (Scheme 9.1). Thus, vinylogous addition reaction takes place with vinyldiazoacetate either serving as nucleophiles or electrophiles.

Davies and co-workers explored the Ag(I)-catalyzed reaction of vinyldiazoacetates **5**, demonstrating that the Ag(I) vinylcarbenes showed

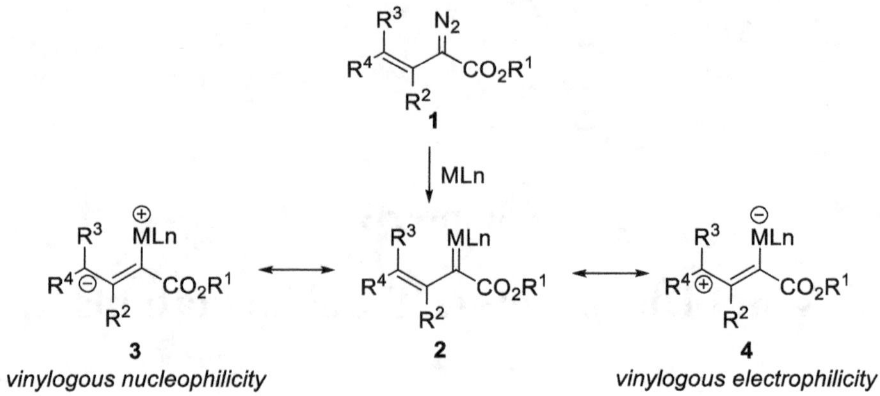

Scheme 9.1 Resonance structures of vinylcarbenes.

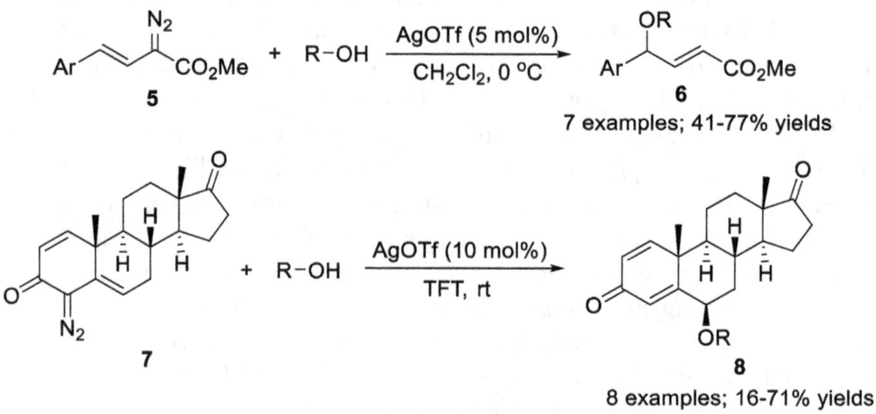

Scheme 9.2 Ag(I)-catalyzed vinylogous C–O bond formation of vinyldiazoacetates.

vinylogous reactivity as electrophiles leading to vinylogous O–H insertion products **6** (Scheme 9.2).[3] However, by using aniline as the substrate, the regioselectivity between vinylogous products and carbene center was much attenuated. The Ag(I)-catalyzed reaction was applied in late-stage vinylogous alkoxylation of pharmaceutically relevant substrates, as shown by the synthesis of aromatase inhibitors of C6 β-2-alkynyloxy derivatives of androsta-1,4-diene-3,17-dione **8**.[4]

The same group also demonstrated the Ag(I)-catalyzed vinylogous fluorination of vinyldiazoacetates, which afforded γ-fluoro-α,β-unsaturated carbonyls with high yields and excellent regioselectivity (Eq. (9.1)).[5]

The fluorination of farnesol and steroid derivatives demonstrated the successful application of this reaction in late-stage fluorination, which has potential utilization in the synthesis of PET radioligands.

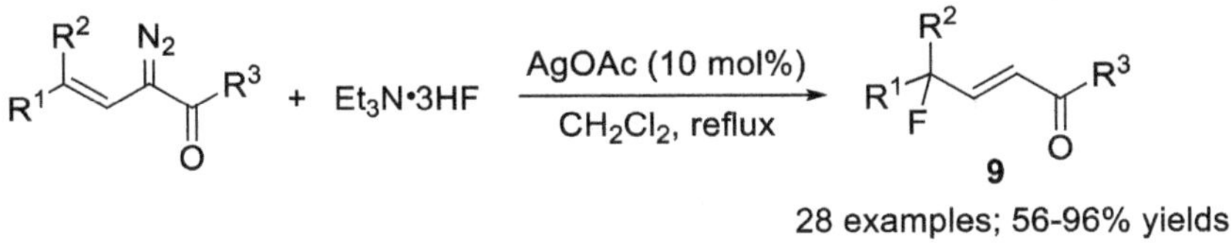

(9.1)

Additionally, Davies and co-workers reported the Rh(II)-catalyzed reaction of silyl enol ethers **10** with siloxyvinyldiazoacetates **11**, leading to the enantioselective formation of highly functionalized alkynoates **12** (Scheme 9.3).[6] It has been proposed that the reaction is initiated by the nucleophilic addition of silyl enol ether to vinylogous γ-position of the

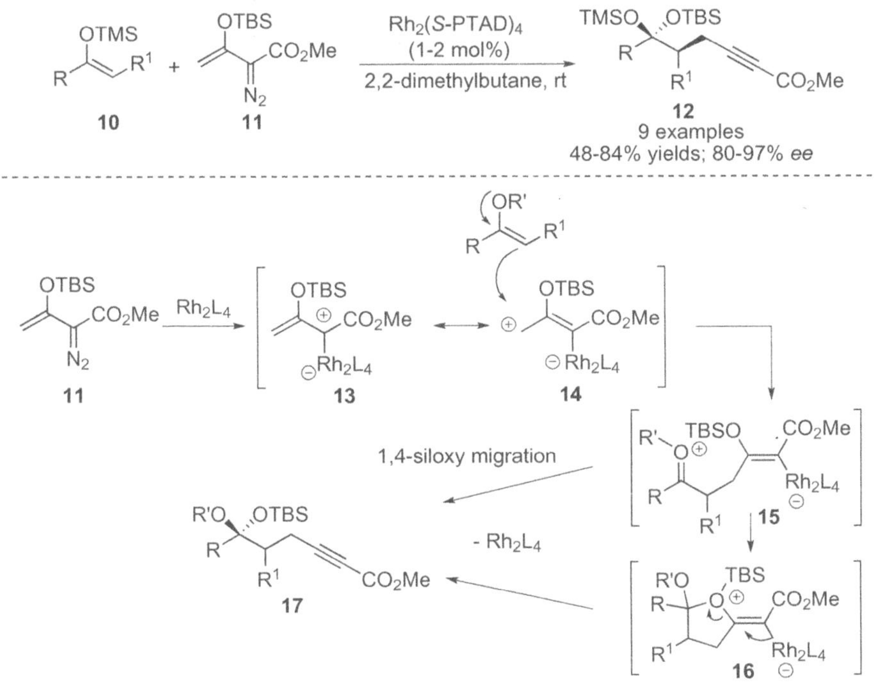

Scheme 9.3 Rh(II)-catalyzed synthesis of alkynoates from silyl enol ethers with siloxyvinyldiazoacetates.

β-siloxy Rh(II) vinylcarbene **14**, leading to the formation of zwitterionic intermediate **15**. Subsequently, direct 1,4-siloxy migration or stepwise addition to the oxocarbenium with sequential β-elimination delivers alkynoates **17**.

On the other hand, Doyle and co-workers developed Rh(II)-catalyzed cascade transformations from enoldiazoacetates **18** with arylhydrazones **19**, which enabled facile and efficient synthesis of highly functionalized 1,2,3,6-tetrahydropyridazines **20** with high diastereo- and excellent enantioselectivities.[7] Reaction mechanism has been assumed to be triggered by nucleophilic attack of the hydrazone on the vinylogous position of the Rh(II)-carbene to generate **24**, followed by subsequent Lewis acid-catalyzed intramolecular Mannich addition (Scheme 9.4).

Subsequently, Doyle and co-workers applied this cascade catalytic strategy to access regiospecific synthesis of pyrazole derivatives **26** by using donor–acceptor substituted hydrazones **25** (Scheme 9.5).[8] The reaction pathway involves the dinitrogen extrusion from enoldiazoacetates to generate Rh(II) carbene **27**, followed by nucleophilic attack by the carbon on the 1,3-dipole **28** to form **29** or **30**, depending on the substrate. Finally, one-pot Lewis acid-promoted cyclization and aromatization deliver the product **26**.

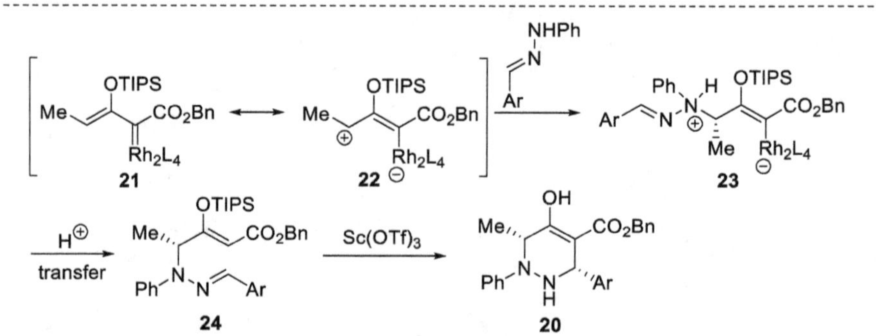

Scheme 9.4 Rh(II)-catalyzed cascade asymmetric synthesis of 1,2,3,6-tetrahydropyridazines.

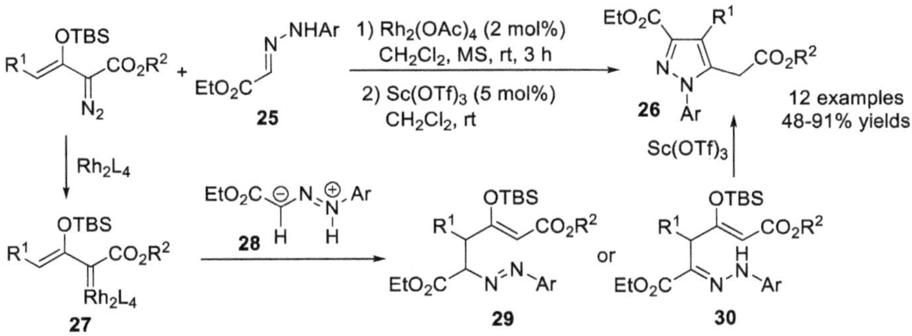

Scheme 9.5 Rh(II)-catalyzed synthesis of pyrazoles from enoldiazoacetates with hydrazones.

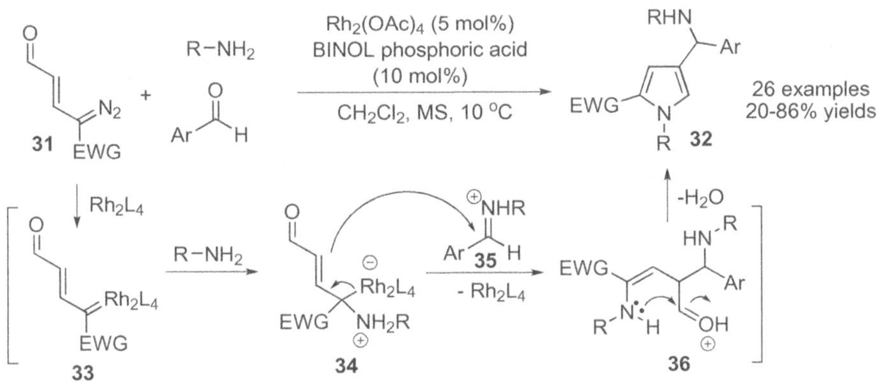

Scheme 9.6 Rh(II)/phosphoric acid-catalyzed multi-component reaction of enaldiazo compounds, arylamines and aryl aldehydes.

Katukojvala and co-workers developed a three-component reaction of enaldiazo compounds **31**, arylamines and aryl aldehydes in the presence of both Rh(II) catalyst and Brønsted acid, affording α-pyrrolyl-benzylamines **32** (Scheme 9.6).[9] The reaction is initiated by the formation of Rh(II) enolcarbene **33**, which is attacked by an amine to give intermediate **34**. Vinylogous Mannich addition of **34** to the iminium species **35** gave the intermediate **36**, followed by cyclization/dehydration cascade to deliver the pyrrole product **32**.

In addition to Rh(II)-catalyzed transformations, gold complexes are also found as efficient catalysts for the reactions of vinyldiazoacetates. Barluenga and López reported the Au(I)-catalyzed reaction of vinyldiazo

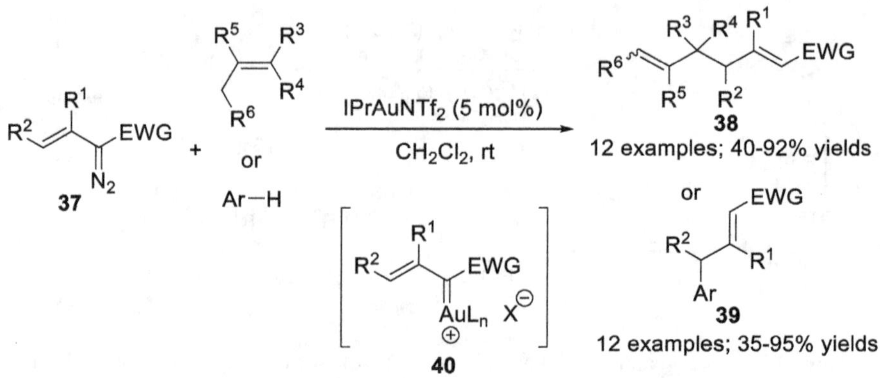

Scheme 9.7 Au(I)-catalyzed reaction of vinyldiazoacetates with arenes and alkenes.

compounds **37** with alkenes or arenes, in which regioselective C–C bond formations occurred to give **38** or **39** (Scheme 9.7).[10] In the reaction mechanism, a highly electrophilic alkenyl gold carbene **40** is proposed, which undergoes vinylogous nucleophilic attack by the arene or alkene.

Liu and co-workers developed an intriguing Au(I)-catalyzed three-component reaction of alkenyldiazo esters, nitriles and pyridine-derived oxides (Scheme 9.8).[11] The reaction is initiated by nucleophilic attack of nitrile on Au(I)-enolcarbene **43** to form intermediate **44**, followed by nucleophilic attack of the pyridine-based oxide to give **45**. The pyridinium ring of **45** is then attacked by iminyl nitrogen to produce azacyclic species **46**, which subsequently undergoes ring cleavage/protodeauration to deliver the final product **42**.

Barluenga and López reported a Cu(II)-catalyzed rearrangement of vinyldiazocarbonyl compounds in the presence of iodosylbenzene as the oxidant, leading to the formation of β-oxodiazo compounds **48** or the corresponding ketal-protected derivatives **49** (Scheme 9.9).[12] This intriguing process with the retention of diazo group occurred *via* oxidation to form cyclized oxonium species **51**, followed by an unusual 1,2-shift cascade of the diazoacetate moiety.

9.2 Intramolecular cyclization with diazo moiety

Intramolecular cyclization of vinyl diazoacetates or *in situ* generated vinyl diazo species allows the facile preparation of highly functionalized pyrazoles. Typical ring-closure pathway involves 1,5-electrocyclization of

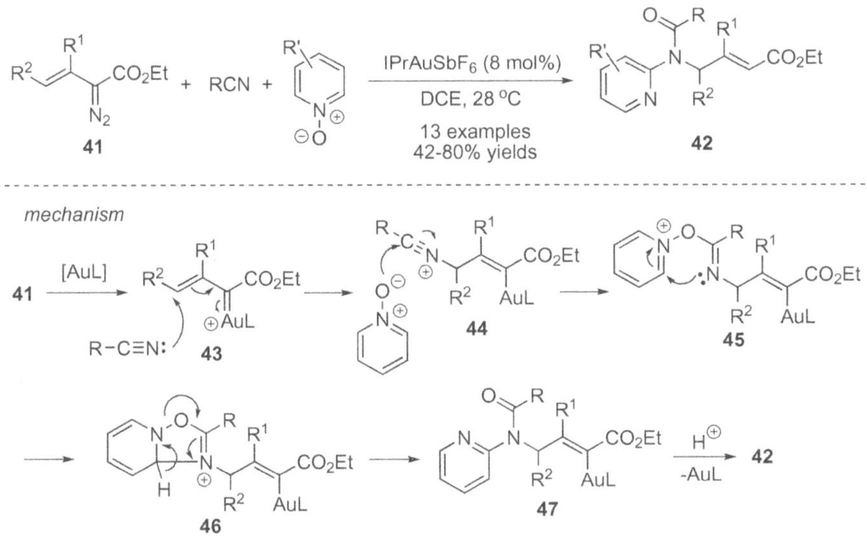

Scheme 9.8 Au(I)-catalyzed three-component oxidation of alkenyldiazo esters, nitriles and pyridine-based *N*-oxides.

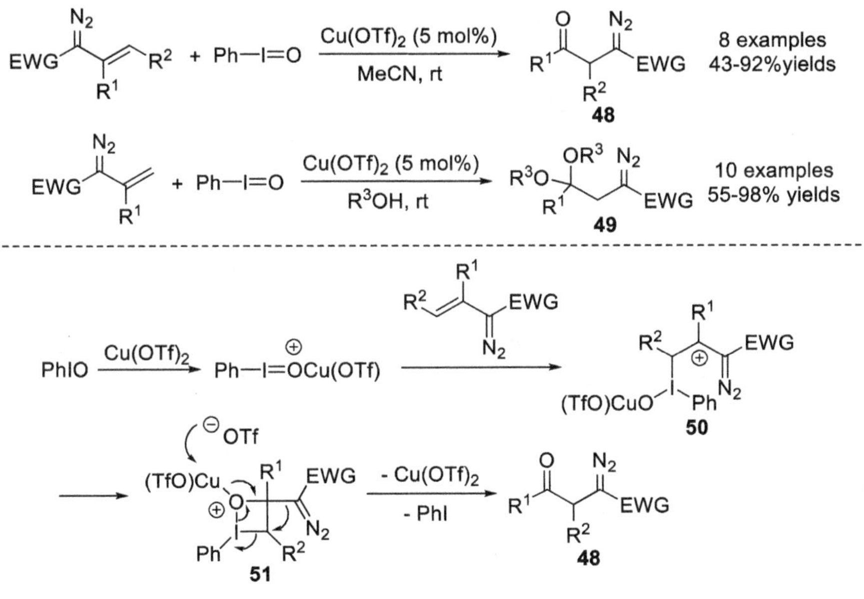

Scheme 9.9 Copper-catalyzed rearrangement of vinyldiazocarbonyl compounds under oxidative conditions.

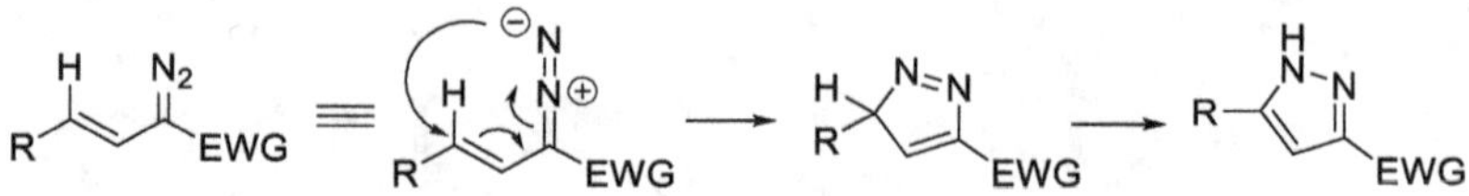

Scheme 9.10 1,5-Electrocyclization of vinyl diazo compounds.

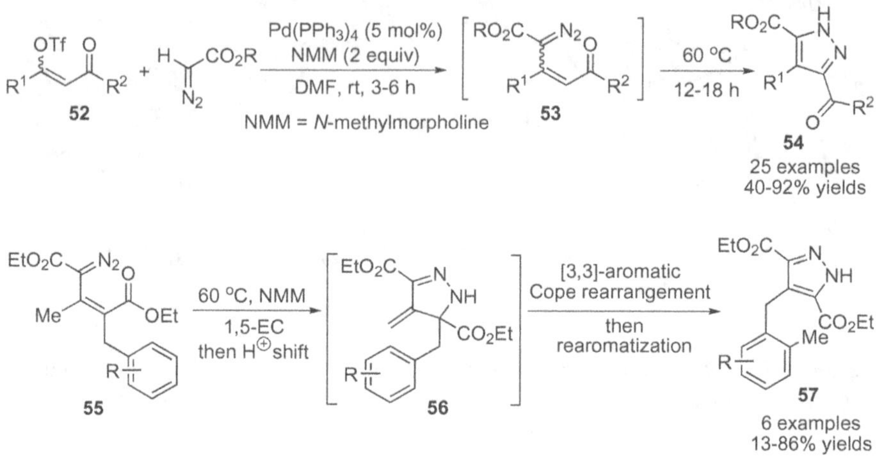

Scheme 9.11 Pd(0)-catalyzed coupling/electrocyclization and aromatic Cope rearrangement.

vinyl diazo compounds, followed by 1,5-migration (Scheme 9.10). For this type of reaction, the development of cascade and one-pot transformations have been focused on for achieving highly efficient synthesis of highly functionalized heterocyclic compounds.

For example, the integration of cross-coupling strategy with electrocyclization provided a novel route to access the poly-substituted pyrazoles. Frantz and co-workers reported the synthesis of 3,4,5-trisubstituted pyrazoles **54** through Pd-catalyzed tandem coupling/electrocyclization of enol triflates **52** and diazoacetates (Scheme 9.11).[13] This protocol is applicable to a wide range of substituted acyclic and cyclic enol triflates, affording the corresponding heterocycles. Moreover, the same group carried out both experimental and theoretical studies for gaining insights into the 6π electrocyclization/aromatic Cope rearrangement of vinyl diazoacetates **55**.[14]

Sun and co-workers reported a Co(II)-catalyzed cross-coupling reaction of aryldiazonium salts **58** with vinyl diazoacetates **59**, affording *N*-arylpyrazoles **60** with good yields (Scheme 9.12).[15] The reaction is

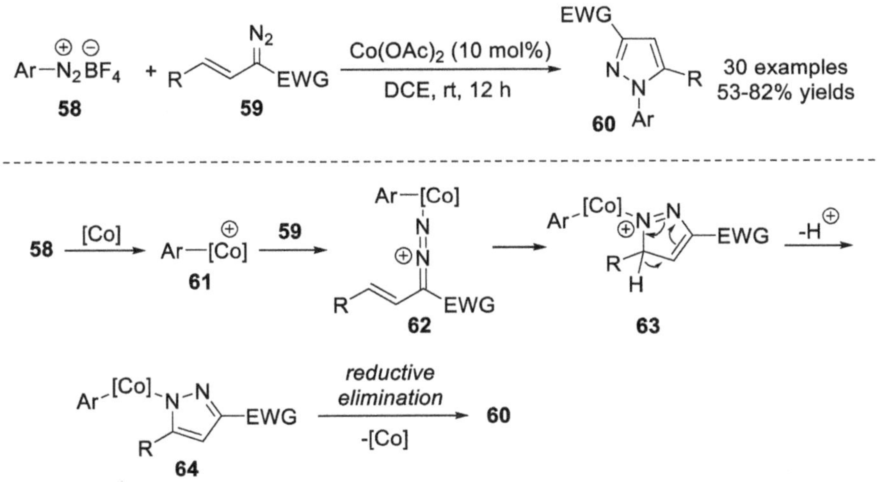

Scheme 9.12 Co(II)-catalyzed cross-coupling reaction of aryldiazonium salts with vinyl diazoacetates.

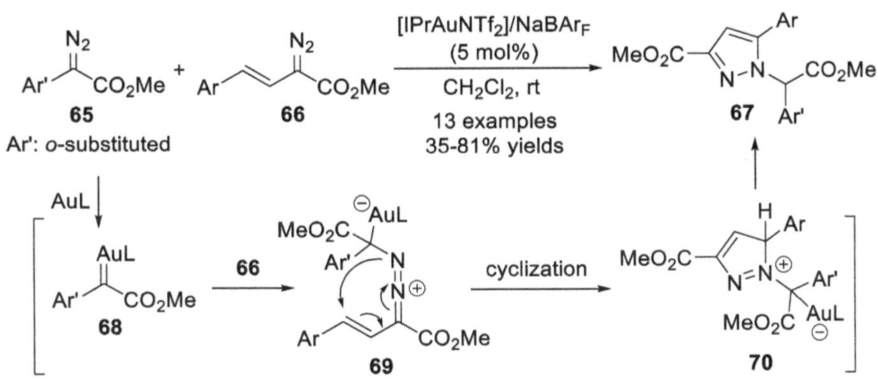

Scheme 9.13 Au(I)-catalyzed selective cross-coupling of diazo compounds to afford *N*-substituted pyrazoles.

initiated by oxidative addition to generate cationic aryl-cobalt complex **61**, followed by *N*-coordination to form diazo-cobalt species **62**. This complex undergoes 1,5-electrocyclization/tautomerization and deprotonation to generate intermediate **64**. Finally, reductive elimination furnished the final product **60**, along with the regeneration of cobalt catalyst.

Furthermore, they reported an Au(I)-catalyzed ligand-controlled selective cross-coupling reaction of aryldiazoacetates **65** with styryldiazoacetates **66**, affording *N*-substituted pyrazoles **67** (Scheme 9.13).[16]

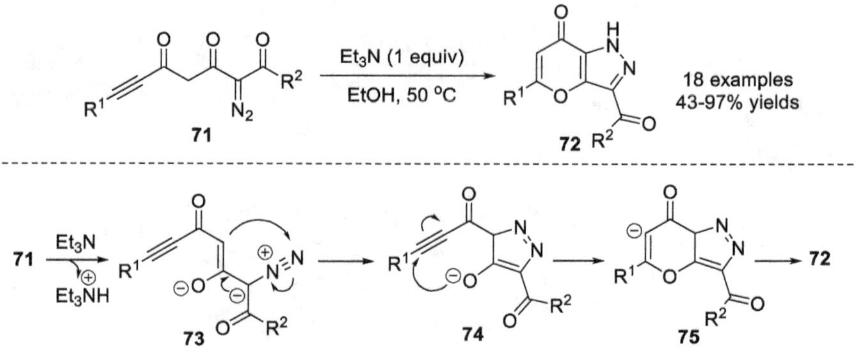

Scheme 9.14 Tandem cyclization of 2-diazo-3,5-dioxo-6-ynoates.

The reaction pathway involves the preferential formation of Au(I) carbene either from an aryldiazoacetate **65** or a styryldiazoacetate **66**, depending on the structure of the diazo substrates. The Au(I) carbene species serves as electrophile and undergoes nucleophilic attack by another diazo substrate, followed by cyclization cascade leading to the formation of the pyrazole product.

Deng and co-workers developed a base-mediated tandem cyclization of 2-diazo-3,5-dioxo-6-ynoates **71**, affording pyrano[3,2-*c*]pyrazol-7(1*H*)-one derivatives **72** in a straightforward manner (Scheme 9.14).[17] The proposed mechanism of this reaction involves a metal-free 6π-electrocyclic ring closure of the *in situ* generated vinyl diazo moiety of **73** to give intermediate **74**, followed by intramolecular Michael reaction to form **75** and finally protonation to deliver the final product **72**.

9.3 Intermolecular cyclization with metalloenolcarbenes

Metal enolcarbenes are highly effective annulation adducts for various cycloaddition reactions with unsaturated systems, such as 1,3-dipoles and dienes. These formal cycloaddition reactions provide convenient access to highly functionalized five-, six- and seven-membered-ring heterocyclic compounds. This section will summarize these cycloaddition methodologies involving metal enolcarbenes, especially for stereoselective synthesis of functionalized heterocyclic scaffolds.

9.3.1 *Formal* [3+2] *reaction*

Transition-metal-catalyzed formal [3+2] cycloadditions involving enolcarbenes provide expeditious methods for constructing five-membered heterocyclic structures. In 2010, Barluenga and co-workers reported a Cu(I)-catalyzed annulation of pyridines **77** with vinyl diazoacetates **76**, leading to the formation of indolizine derivatives **78** in a regioselective manner (Scheme 9.15).[18] This reaction represents the successful example of metal-catalyzed cyclization of vinyl diazo compounds with π-deficient heterocyclic system. Mechanistic rationale is proposed to involve nucleophilic attack of pyridine nitrogen on Cu(I)-enolcarbene **79**, followed by cyclization/oxidative aromatization sequence to generate indolizidine **78**.

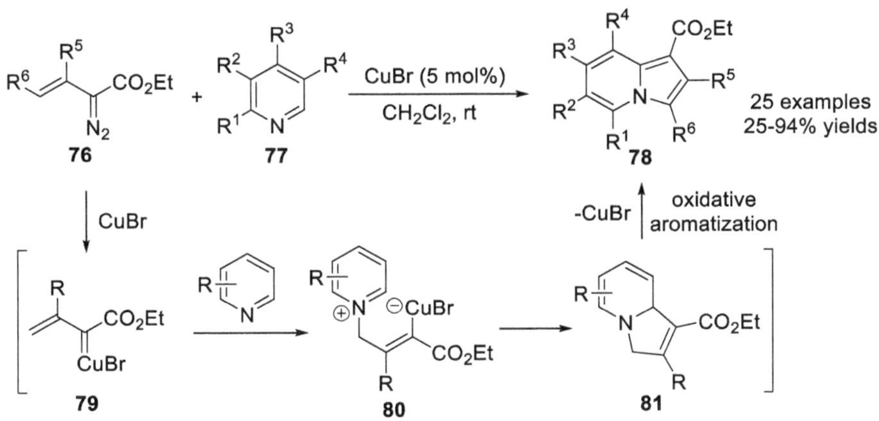

Scheme 9.15 Cu(I)-Catalyzed [3+2] cycloaddition with pyridine.

With Rh(II) and Au(I) catalysts, Davies and co-workers achieved a series of developments in catalytic [3+2] cycloaddition of vinyl diazo compounds. For example, they reported the Rh_2(*S*-DOSP)$_4$-catalyzed asymmetric cyclopentannulation of indolyl rings with (*E*)-arylvinyldiazoacetates (Scheme 9.16).[19] Depending on the substitution mode of indole derivatives, two distinct regio-isomers **82** and **83** could be obtained, respectively, with excellent enantioselectivity. It was proposed that the generation of *endo* or *exo* products was governed by the fact whether the ring closure occurred from the intermediate derived from *s-cis* (**84**) or *s-trans* (**85**) Rh(II)-enolcarbene.

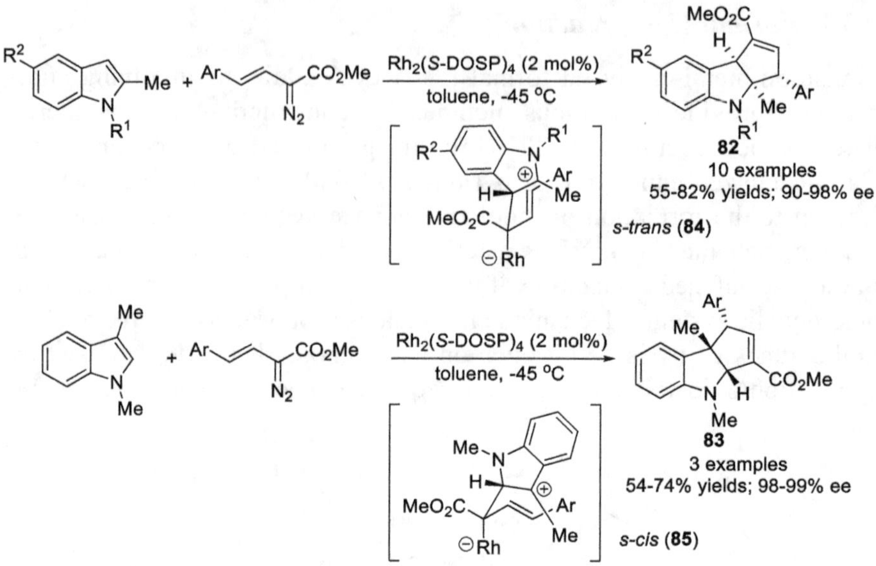

Scheme 9.16 Rh(II)-catalyzed asymmetric [3+2] annulation of indoles.

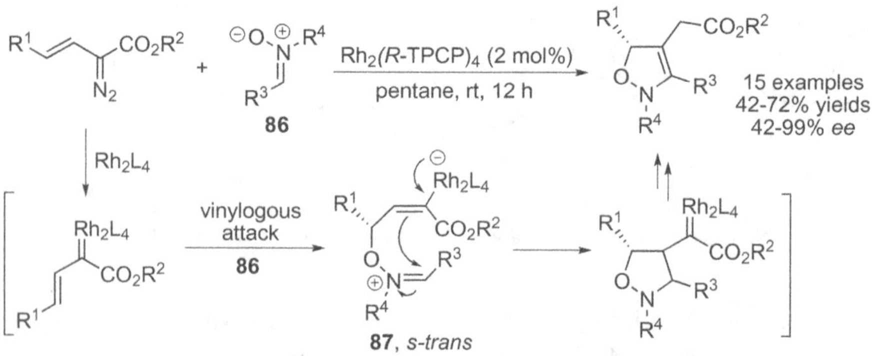

Scheme 9.17 Rh(II)-catalyzed asymmetric [3+2] cycloaddition of vinyldiazoacetates and nitrones.

In another example, the same group demonstrated Rh(II)-catalyzed enantioselective vinylogous addition of nitrones to terminally substituted vinyldiazoacetates for the efficient synthesis of 2,5-dihydroisoxazoles with high enantioselectivity (Scheme 9.17).[20] Notably, the bulky Rh(II) catalyst provided a preference to generate *s-trans* intermediate **87**, which was responsible for the selective [3+2] annulation.

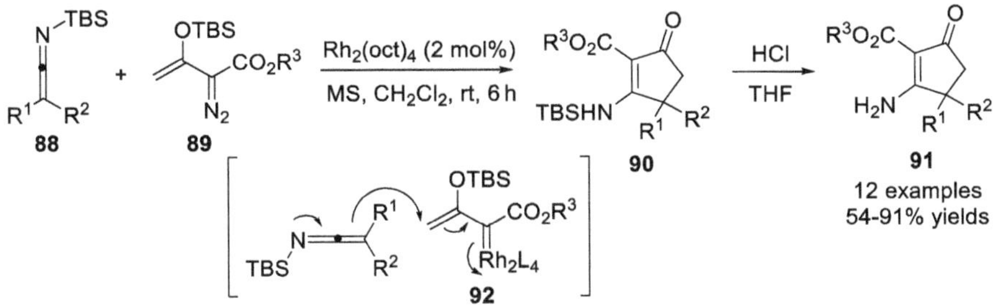

Scheme 9.18 Rh(II)-catalyzed [3+2] annulation of silylated ketene imines with enoldiazoacetates.

In addition to the chiral Rh(II)-catalyzed reaction, the same group also developed highly enantioselective [3+2] cycloaddition between vinyldiazoacetates and enol ethers catalyzed by binuclear gold complex activated by silver hexafluoroantimonate.[21] On the other hand, Doyle and co-workers developed a series of transformations involving catalytic [3+2] cycloaddition of silyl-protected enol diazoacetates.[22–24] For example, they reported a mild Rh(II)-catalyzed [3+2] annulation of silylated ketene imines **88** with enoldiazoacetates **89**, providing a synthetic method for the substituted 3-amino-2-cyclopentenones **91** (Scheme 9.18).[24] The reaction follows a pathway involving vinylogous electrophilic addition to vinyl rhodium carbene intermediate **92**.

9.3.2 *Formal* [3+3] *reaction*

In this type of cycloaddition, the metalloenolcarbenes generated from vinyldiazocarbonyl compounds serve as reactive dipolar species for the reactions with various 1,3-dipoles. Electron-donating groups such as siloxy group enhance the viability of electrophilic ring closure to the metal-bound carbon of the initial adduct from vinylogous addition, simultaneously inhibiting alternative [3+2] cycloaddition. The catalytic [3+3] cycloaddition of silyl-protected enoldiazoacetates has been extensively explored as a strategy for constructing various six-membered heterocyclic scaffolds, especially for stereoselective synthesis of functionalized heterocyclic compounds.[2]

Doyle and co-workers studied extensively the catalytic [3+3] cycloadditions of a variety of dipoles with electrophilic vinylcarbene intermediates. In 2011, they reported chiral Rh(II)-catalyzed asymmetric formal [3+3] cycloaddition of β-TBSO-substituted vinyldiazoacetates with nitrones **93**, producing 3,6-dihydro-1,2-oxazines with high yields

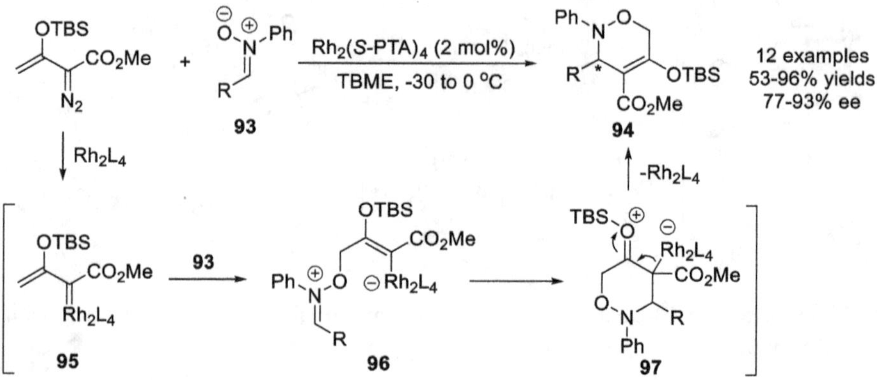

Scheme 9.19 Rh(II)-catalyzed asymmetric [3+3] cycloaddition of β-TBSO-substituted vinyldiazoacetates with nitrones.

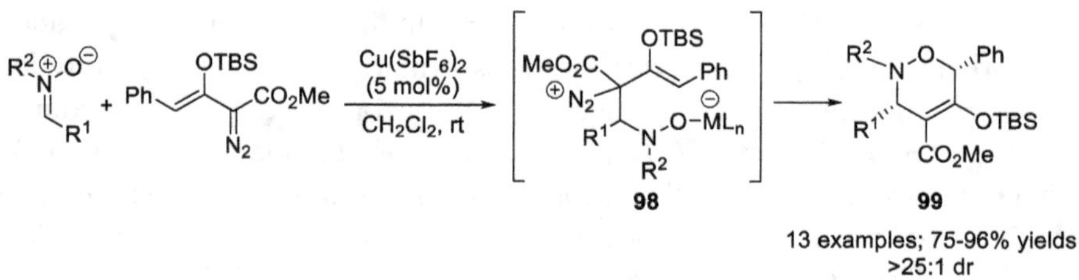

Scheme 9.20 Cu(II)-catalyzed formal [3+3] cycloaddition of enol diazoacetates with nitrones.

and enantioselectivities (Scheme 9.19).[25] Mechanistically, nucleophilic attack by the nitrone on the Rh(II) vinyl carbene **95** occurred at the vinylogous position, resulting in the formation of adduct **96**. Subsequently, intramolecular iminium ion addition took place in a stepwise or concerted manner, affording the intermediate **97** and then the [3+3] cycloadducts **94** along with the regeneration of the catalyst.

Doyle and co-workers also developed Cu(II)-catalyzed formal [3+3] cycloaddition of γ-aryl-enoldiazoacetates with nitrones, producing polysubstituted 1,2-oxazines **99** with high level of diastereoselective control (Scheme 9.20).[26] Instead of following metal carbene pathway, this transformation was proposed to occur through Lewis acid-catalyzed electrophilic addition of iminium ion by diazo carbon atom to form diazonium ion species **98**.

By using azomethine imines as the dipolar reactants, Doyle and co-workers developed a Rh(II)-catalyzed [3+3] cycloaddition of enoldiazoacetates, affording *N,N*-bicyclic pyrazolidinone derivatives with good regio- and excellent diastereoselectivity (Eq. (9.2)).[27] Reaction was initiated by selective nucleophilic attack of azomethine imine at the vinylogous position of metal carbene.

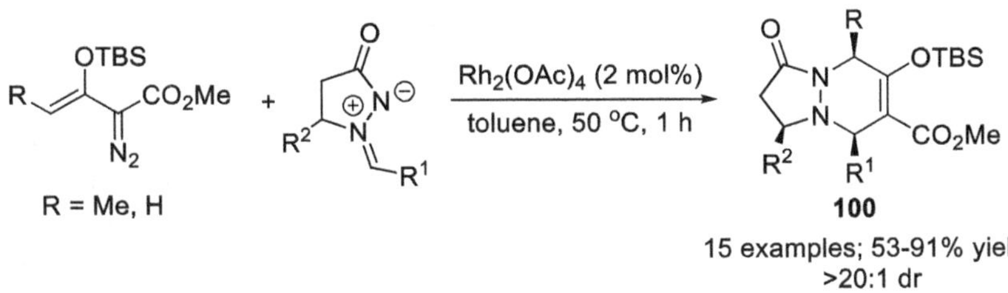

(9.2)

Moreover, asymmetric dearomatizing formal [3+3] cycloaddition that provided an effective access to bis/tricyclic 1,2,3,6-tetrahydropyridazine was achieved by chiral Rh(II)-catalyzed reaction of *N*-acyliminopyridinium ylides with enoldiazoacetates (Eq. (9.3)).[28] This cycloaddition was assumed to be triggered by vinylogous addition of Rh(II) enolcarbene with *N*-acyliminopyridinium ylide, followed by intramolecular asymmetric addition of pyridinium ylide to the catalyst-activated vinyl ether group.

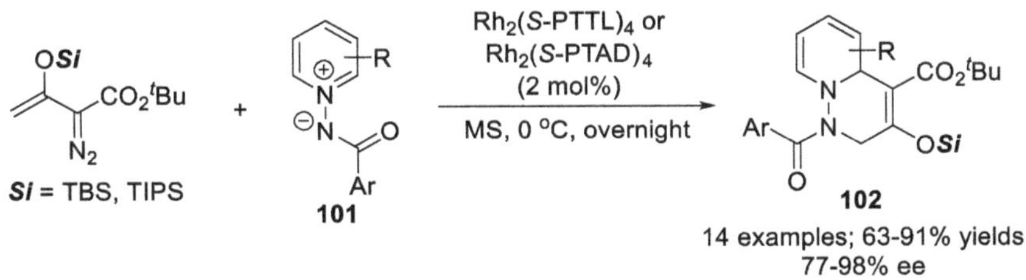

(9.3)

Moreover, Doyle and co-workers developed Rh(II)-catalyzed asymmetric dearomatizing formal [3+3] cycloaddition of isoquinolinium/pyridinium methylides with enoldiazoacetates, affording enantioenriched functionalized quinolizidines (Eq. (9.4)).[29] The indolizines were generated as the side-products, which were produced through diastereoselective [3+2] cycloaddition of methylides with donor–acceptor cyclopropene intermediates derived from rhodium carbene.

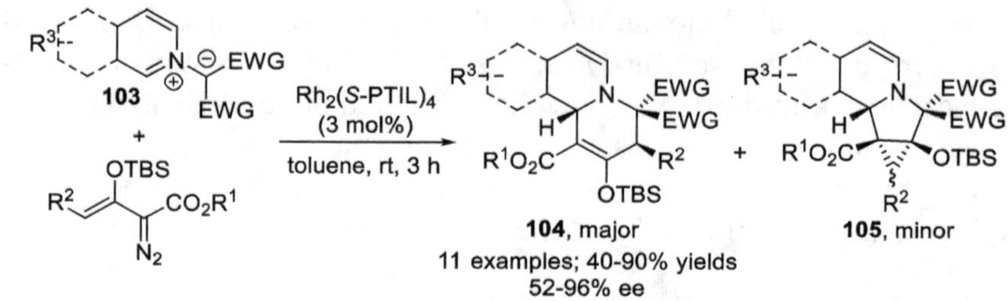

(9.4)

Recently, Doyle and co-workers reported a catalyst-dependent divergent addition of enoldiazoacetamides with stable 1,3-dipole of nitrones (Scheme 9.21).[30] In the presence of copper(I) tetrafluoroborate/bisoxazoline **107**, the [3+3] cycloadducts **108** were obtained with excellent yields and enantioselectivities. Alternatively, when the catalytic system was switched to acidic Cu(OTf), electrophilic vinylogous addition of enoldiazoacetamide to nitrone moiety occurred, affording Mannich addition product **109**.

On the other hand, Liu and co-workers developed an Au(I)/Ag(I)-catalyzed formal [3+3] cycloaddition between nitrosobenzenes **110** with alkenyl diazo esters, providing quinoline oxides **111** (Scheme 9.22).[31] Mechanistically, the reaction involves the attack of nitrosobenzene at

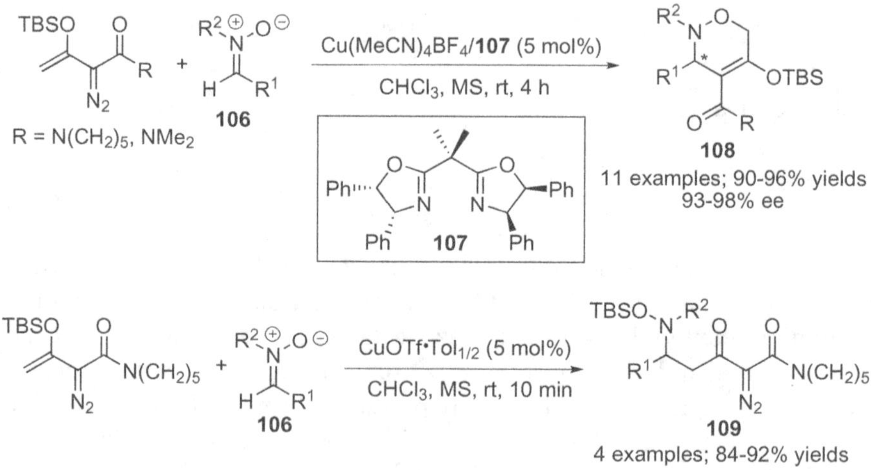

Scheme 9.21 Cu(I)-catalyzed divergent addition of enoldiazoacetamides with nitrones.

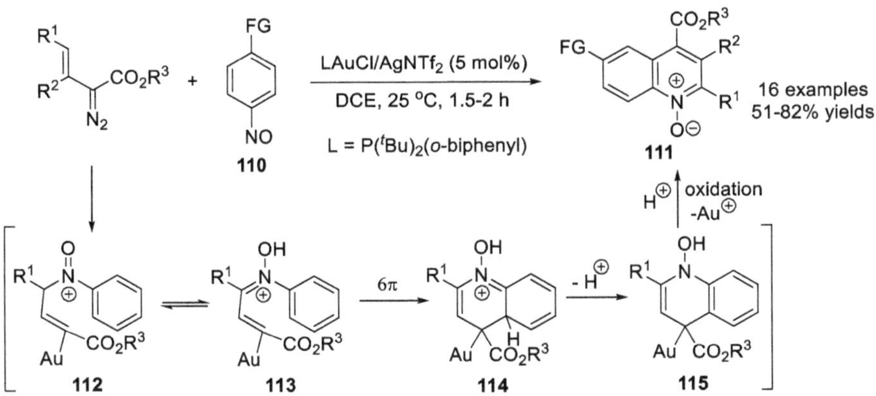

Scheme 9.22 Au(I)-catalyzed formal [3+3] cycloaddition of nitrosobenzene with alkenyl gold carbenes.

vinylogous position of the Au(I)-carbene, followed by oxime/nitroso tautomeric equilibrium to alkenyl iminium species **113**. Subsequently, **113** undergoes 6π-electrocyclization, deprotonation and oxidative aromatization to give quinoline oxides.

9.3.3 *Formal* [4+3] *reaction*

The formal [4+3] cycloaddition of vinyldiazoacetates and dienes represents a general and efficient method for the stereoselective construction of highly functionalized seven-membered rings. These transformations involve catalytic tandem cyclopropanation with the predominant formation of *cis*-1,2-divinyl compounds and subsequent Cope rearrangement. This type of tandem cyclopropanation/Cope rearrangement cascade has been extensively explored by Davies and co-workers.[32–34]

In 2008, Davies and co-workers reported an enantioselective synthesis of 6,5,7-tricyclic system derivatives through chiral Rh(II)-catalyzed asymmetric formal [4+3] cycloaddition of benzofuranyl diazoacetates with 1,3-dienes.[33] With the cyclopropanation/Cope rearrangement cascade followed by isomerization of intermediate **118**, a series of [4+3] cycloadducts **116** were obtained with high diastereo- (>94% de) and enantioselectivity (91–98% *ee*) (Scheme 9.23).

Moreover, Davies and co-workers reported an asymmetric Rh(II)-catalyzed [4+3] cycloaddition of vinyldiazoacetates and siloxy-substituted

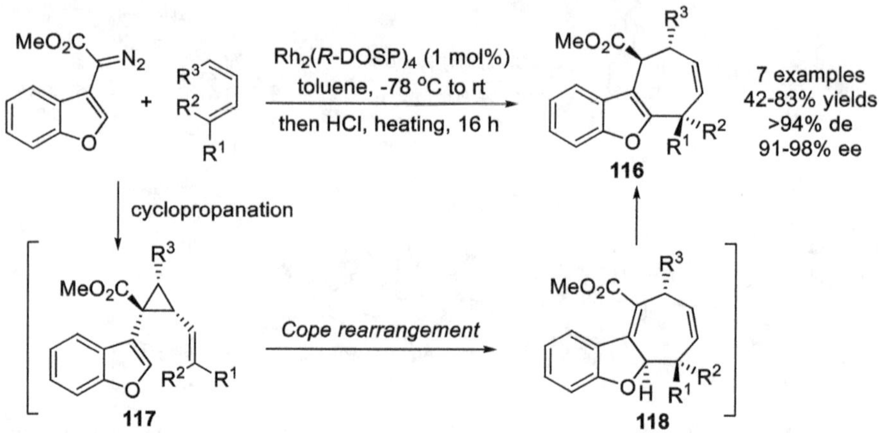

Scheme 9.23 Chiral Rh(II)-catalyzed asymmetric formal [4+3] cycloaddition between vinyldiazoacetates with 1,3-dienes.

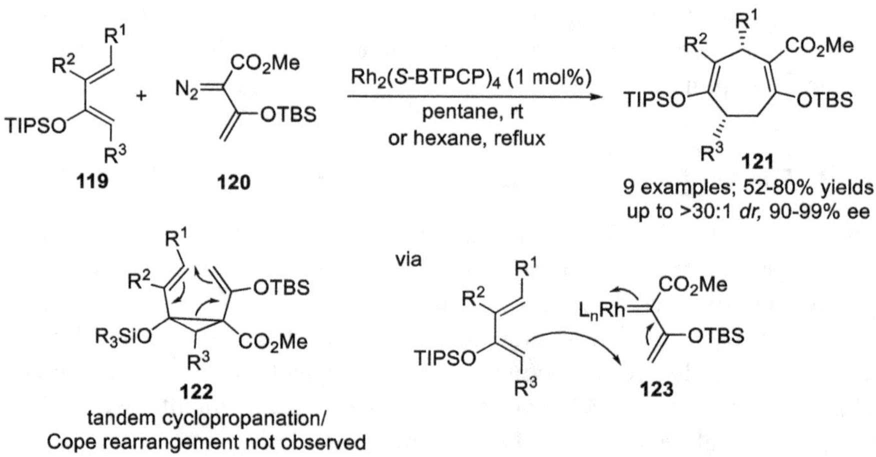

Scheme 9.24 Rh(II)-catalyzed [4+3] cycloaddition of vinyldiazoacetates and siloxy-substituted 1,3-dienes.

1,3-dienes.[34] Notably, this reaction afforded highly functionalized 1,4-cycloheptadienes **121** with reverse regioselectivity as compared with the conventional tandem cyclopropanation/Cope rearrangement pathway (Scheme 9.24). The reverse regioselectivity may be attributed to the steric hinderance of the bulky Rh(II) catalysts, which resulted in the preference for vinylogous reactivity. Thus, instead of cyclopropanation to generate

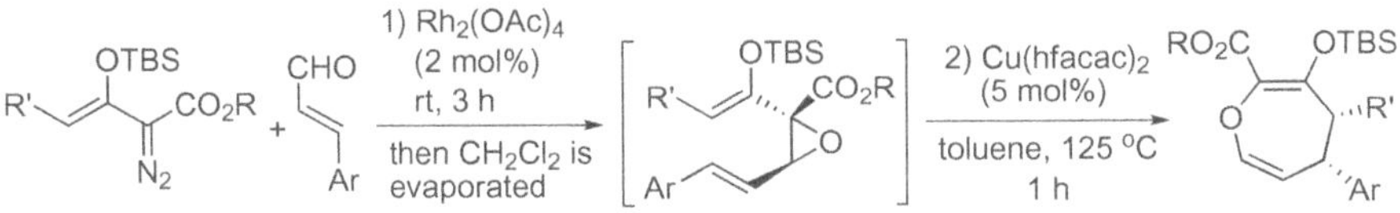

Scheme 9.25 Tandem Rh(II)- and Cu(II)-catalyzed epoxidation/Cope rearrangement.

122, this reaction is initiated by nucleophilic attack at vinylogous position of the Rh(II) carbene **123**, followed by cyclization in a stereoselective manner to afford **121**.

Furthermore, Doyle and co-workers developed a tandem reaction of siloxy-substituted vinyldiazoacetates with cinnamaldehydes catalyzed by Rh(II) and Cu(II) catalysts separately, affording 4,5-dihydrooxepinsin in high yields (81–95%).[35] This one-pot formal [4+3] cycloaddition proceeded through Rh(II)-catalyzed epoxidation to afford **124**, followed by Cope rearrangement to give **125** (Scheme 9.25).

9.3.4 *Other cyclization transformations*

In 2012, Barluenga and López reported a Cu-catalyzed formal [3+1] cycloaddition of vinyl diazoacetates with iminoiodinanes to afford functionalized 2-azetines (Scheme 9.26).[36] The reaction is proposed to proceed *via* Cu(I)-catalyzed nitrene aziridination of the alkenyl functionality to generate intermediate **130**, followed by metal carbene-type ring-expansion process through 1,2-amino migration. This may find potential application in synthesizing conformationally restricted α-amino acid.

Doyle and co-workers reported a regio- and diastereoselective Lewis acid-catalyzed formal [4+2] cycloaddition between imines **132** and donor–acceptor cyclopropenes **131**, which was *in situ* generated from silyl-protected enoldiazoacetates in the presence of Rh(II) complex (Scheme 9.27).[37] This reaction proceeded through catalytic Mannich addition/cyclization/aromatization cascade. The reaction afforded functionalized cyclopropane-fused tetrahydroquinoline **133** with good diastereoselectivities. **133** could be further converted into benzo[*b*]azepin-5-one **134** upon the treatment with TBAF.

Recently, Doyle and co-workers reported a Hg(II)-catalyzed intramolecular arene 1,4-addition of donor–acceptor cyclopropene carboxamides,

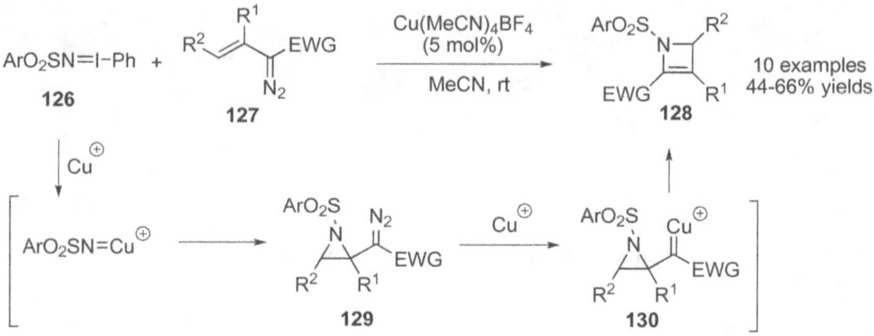

Scheme 9.26 Cu(I)-catalyzed formal [3+1] cycloaddition of vinyl diazoacetates with iminoiodinanes.

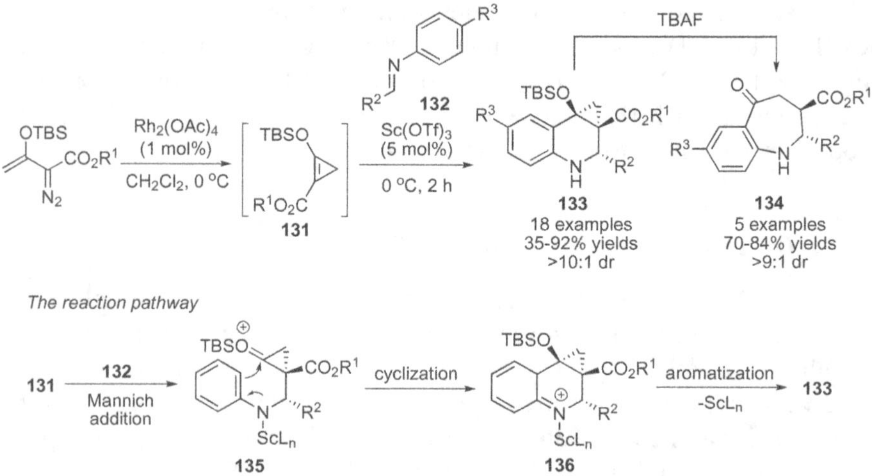

Scheme 9.27 Rh(II)-catalyzed formal [4+2] cycloaddition between the *in situ* generated donor–acceptor cyclopropenes and imines.

which were *in situ* generated from enoldiazoacetamides (Scheme 9.28).[38] The reaction is initiated by the π-complexation between $Hg(OTf)_2$ and cyclopropene **140**, resulting in the generation of coordinated Hg(II) species **141**. The bulky *tert*-butyl attachment to the amide nitrogen facilitates the proper orientation of the amide to enable nucleophilic attack of the aromatic ring on the activated cyclopropene. Subsequently, intramolecular nucleophilic addition to the *para*-position of the aromatic ring affords the desired cyclic compounds along with the release of Hg(II) catalyst.

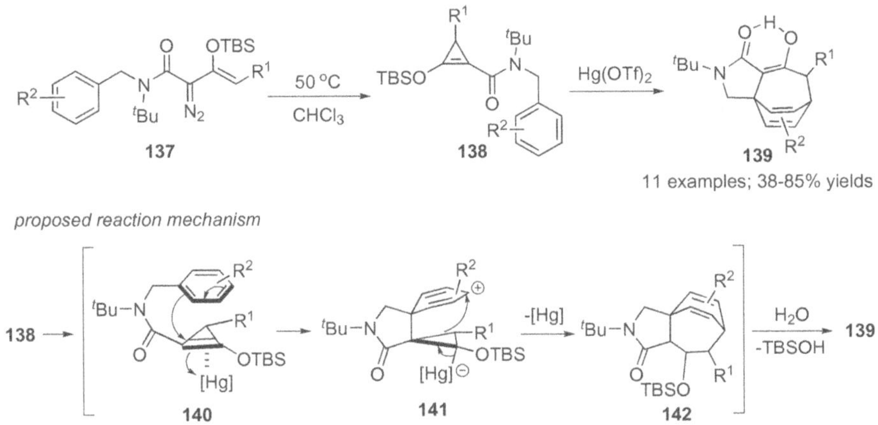

Scheme 9.28 Hg(II)-catalyzed intramolecular 1,4-addition of the cyclopropenes generated *in situ* from vinyl diazocarbonyl compounds.

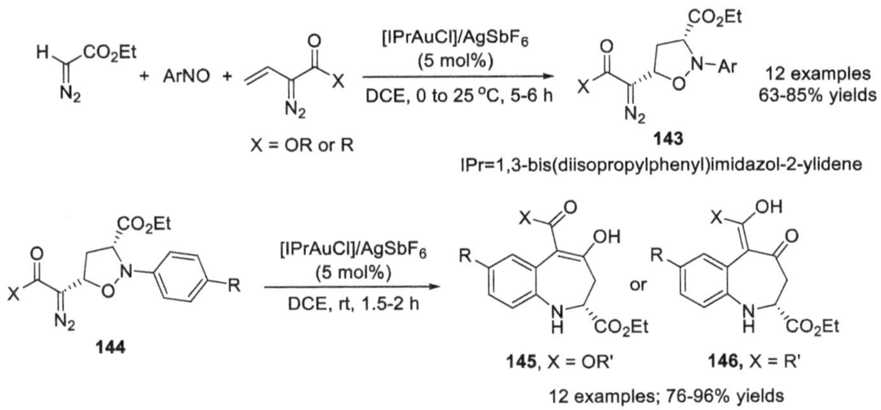

Scheme 9.29 Au(I)-catalyzed formal [3+2] cycloaddition and sequential rearrangement cascade.

Employing Au(I)/Ag(I) co-catalytic system, Liu and co-workers developed a formal [3+2] cycloaddition of EDA, nitrosoarenes and vinyldiazo carbonyl substrates to afford diazo-containing isoxazolidine derivatives **143** (Scheme 9.29).[39] Reaction mechanism involves *in situ* generation of nitrones as the electrophiles, which are subsequently attacked by vinyldiazoesters. Furthermore, by the treatment of the same catalyst, the isolated isoxazolidines **144** undergoes a 1,2-H shift/[3,3]

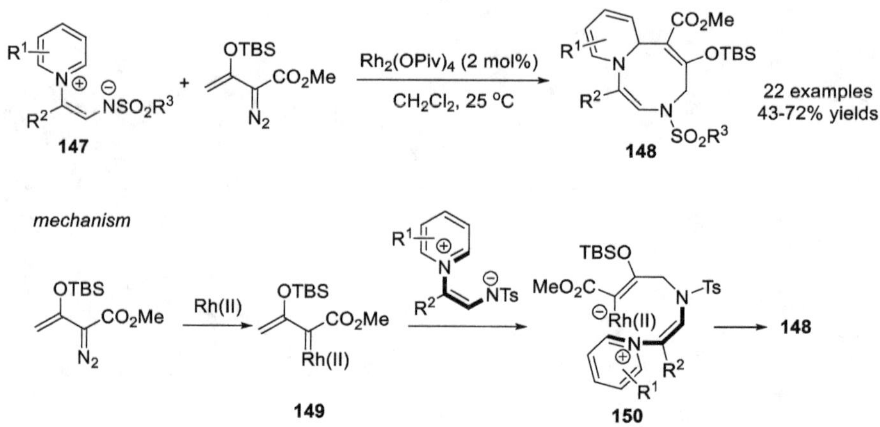

Scheme 9.30 Rh(II)-catalyzed formal [5+3] cycloaddition of pyridinium zwitterion 1,5-dipole equivalents.

rearrangement tandem process to yield benzo[*b*]azepine scaffold (**145** or **146**) with high yields. The reaction represents an application of Au-catalyzed cycloaddition strategy with vinyl diazoacetates in constructing heterocyclic compounds.

Recently, Yoo and co-workers developed an intermolecular Rh(II)-catalyzed [5+3] cycloaddition between isolable pyridinium zwitterion as the 1,5-dipole equivalents and enol diazoacetates, affording highly functionalized eight-membered heterocyclic scaffold **148** (Scheme 9.30).[40] This reaction is initiated by the nucleophilic attack of pyridinium zwitterion on the vinylogous position of Rh(II) enolcarbene **149**, followed by sequential intramolecular cyclization process of **150**.

References

1. Xu, X.; Doyle, M. P. *Aust. J. Chem.* **2014**, *67*, 365–373.
2. Xu, X.; Doyle, M. P. *Acc. Chem. Res.* **2014**, *47*, 1396–1405.
3. Hansen, J. H.; Davies, H. M. L. *Chem. Sci.* **2011**, *2*, 457–461.
4. Ghosh, D.; Lo, J.; Morton, D.; Valette, D.; Xi, J.; Griswold, J.; Hubbell, S.; Egbuta, C.; Jiang, W.; An, J.; Davies, H. M. L. *J. Med. Chem.* **2012**, *55*, 8464–8476.
5. Qin, C.; Davies, H. M. L. *Org. Lett.* **2013**, *15*, 6152–6154.
6. Valette, D.; Lian, Y.; Haydek, J. P.; Hardcastle, K. I.; Davies, H. M. L. *Angew. Chem. Int. Ed.* **2012**, *51*, 8636–8639.

7. Xu, X.; Zavalij, P. Y.; Doyle, M. P. *Angew. Chem. Int. Ed.* **2012**, *51*, 9829–9833.
8. Xu, X.; Zavalij, P. Y.; Hu, W.; Doyle, M. P. *J. Org. Chem.* **2013**, *78*, 1583–1588.
9. Dawande, S. G.; Kanchupalli, V.; Lad, B. S.; Rai, J.; Katukojvala, S. *Org. Lett.* **2014**, *16*, 3700–3703.
10. Barluenga, J.; Lonzi, G.; Tomás, M.; López, L. A. *Chem. Eur. J.* **2013**, *19*, 1573–1576.
11. Karad, S. N.; Liu, R.-S. *Angew. Chem. Int. Ed.* **2014**, *53*, 5444–5448.
12. Barluenga, J.; Lonzi, G.; Riesgo, L.; Tomás, M.; López, L. A. *J. Am. Chem. Soc.* **2011**, *133*, 18138–18141.
13. Babinski, D. J.; Aguilar, H. R.; Still, R.; Frantz, D. E. *J. Org. Chem.* **2011**, *76*, 5915–5923.
14. Babinski, D. J.; Bao, X.; El Arba, M.; Chen, B.; Hrovat, D. A.; Borden, W. T.; Frantz, D. E. *J. Am. Chem. Soc.* **2012**, *134*, 16139–16142.
15. Guo, H.; Zhang, D.; Zhu, C.; Li, J.; Xu, G.; Sun, J. *Org. Lett.* **2014**, *16*, 3110–3113.
16. Xu, G.; Zhu, C.; Gu, W.; Li, J.; Sun, J. *Angew. Chem. Int. Ed.* **2015**, *54*, 883–887.
17. Deng, G.; Wang, F.; Lu, S.; Cheng, B. *Org. Lett.* **2015**, *17*, 4651–4653.
18. Barluenga, J.; Lonzi, G.; Riesgo, L.; López, L. A.; Tomás, M. *J. Am. Chem. Soc.* **2010**, *132*, 13200–13202.
19. Lian, Y.; Davies, H. M. L. *J. Am. Chem. Soc.* **2010**, *132*, 440–441.
20. Qin, C.; Davies, H. M. L. *J. Am. Chem. Soc.* **2013**, *135*, 14516–14519.
21. Briones, J. F.; Davies, H. M. L. *J. Am. Chem. Soc.* **2013**, *135*, 13314–13317.
22. Xu, X.; Shabashov, D.; Zavalij, P. Y.; Doyle, M. P. *Org. Lett.* **2012**, *14*, 800–803.
23. Xu, X.; Shabashov, D.; Zavalij, P. Y.; Doyle, M. P. *J. Org. Chem.* **2012**, *77*, 5313–5317.
24. Xu, X.; Leszczynski, J. S.; Mason, S. M.; Zavalij, P. Y.; Doyle, M. P. *Chem. Commun.* **2014**, *50*, 2462–2464.
25. Wang, X.; Xu, X.; Zavalij, P. Y.; Doyle, M. P. *J. Am. Chem. Soc.* **2011**, *133*, 16402–16405.
26. Qian, Y.; Xu, X.; Wang, X.; Zavalij, P. J.; Hu, W.; Doyle, M. P. *Angew. Chem. Int. Ed.* **2012**, *51*, 5900–5903.
27. Qian, Y.; Zavalij, P. J.; Hu, W.; Doyle, M. P. *Org. Lett.* **2013**, *15*, 1564–1567.
28. Xu, X.; Zavalij, P. Y.; Doyle, M. P. *Angew. Chem. Int. Ed.* **2013**, *52*, 12664–12668.
29. Xu, X.; Zavalij, P. Y.; Doyle, M. P. *J. Am. Chem. Soc.* **2013**, *135*, 12439–12447.
30. Cheng, Q.-Q.; Yedoyan, J.; Arman, H.; Doyle, M. P. *J. Am. Chem. Soc.* **2016**, *138*, 44–47.

31. Pagar, V. V.; Jadhav, A. M.; Liu, R.-S. *J. Am. Chem. Soc.* **2011**, *133*, 20728–20731.
32. Davies, H. M. L.; Stafford, D. G.; Doan, B. D.; Houser, J. H. *J. Am. Chem. Soc.* **1998**, *120*, 3326–3331.
33. Olson, J. P.; Davies, H. M. L. *Org. Lett.* **2008**, *10*, 573–576. Corrections: *Org. Lett.* **2010**, *12*, 1144–1144.
34. Guzmán, P. E.; Lian, Y.; Davies, H. M. L. *Angew. Chem. Int. Ed.* **2014**, *53*, 13083–13087.
35. Xu, X.; Hu, W.-H.; Zavalij, P. Y.; Doyle, M. P. *Angew. Chem. Int. Ed.* **2011**, *50*, 11152–11155.
36. Barluenga, J.; Riesgo, L.; Lonzi, G.; Tomás, M.; López, L. A. *Chem. Eur. J.* **2012**, *18*, 9221–9224.
37. Truong, P. M.; Mandler, M. D.; Zavalij, P. Y.; Doyle, M. P. *Org. Lett.* **2013**, *15*, 3278–3281.
38. Deng, Y.; Jing, C.; Zavalij, P. Y.; Doyle, M. P. *Org. Lett.* **2015**, *17*, 4312–4315.
39. Pagar, V. V.; Liu, R.-S. *Angew. Chem. Int. Ed.* **2015**, *54*, 4923–4926.
40. Lee, D. J.; Ko, D.; Yoo, E. J. *Angew. Chem. Int. Ed.* **2015**, *54*, 13715–13718.

Chapter 10

Diazo Compounds as 1,3-Dipoles in [3+2] Cycloadditions

Due to the electronic character of the diazo group, the diazo compounds can serve as efficient 1,3-dipoles in [3+2] cycloadditions with various unsaturated compounds (Figure 10.1). These transformations have found practical applications in the synthesis of pyrazolines.[1–3] In this chapter, the recent advances are summarized according to the types of dipolarophiles.

10.1 Cycloaddition with alkenes

1,3-Dipolar [3+2] cycloaddition of diazo substrates with electron-deficient alkenes provides a straightforward access toward multisubstituted pyrazolines or pyrazoles. This type of cycloaddition has been extensively explored with the reaction of diazoalkanes/diazocarbonyl substrates and a wide range of alkene dipolarophiles, including substituted acrylate derivatives **1**,[4–7] Baylis–Hillman adducts **2**,[8] acrylonitriles **3**,[7] acrylamides **4**,[7] *trans*-dibenzyl glutaconate **5**,[9] α-sulfinyl-2-cyclopentenones **6** or 5,6-dihydropyran-2-one **7**,[10,11] 2-thio-3-chloroacrylamides,[12] conjugated dienes/α,β-unsaturated ketones **8** and **9**,[13] dimethyl 1-(formylamino)ethylene phosphonate **10**,[14] chlorins-derived alkenes **11**[15,16] and nitroalkenes **12**[17] (Figure 10.2).

It is worth mentioning that formal cyclopropanation product could be obtained from these [3+2] cycloaddition reactions, as shown in the reaction of 3,5-diaryl-1,2-thiaphospholes with 9-diazofluorene which

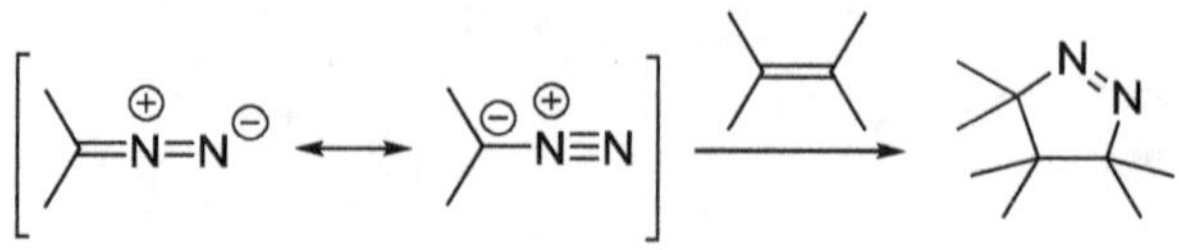

Figure 10.1 Diazo compounds as 1,3-dipoles in [3+2] cycloadditions.

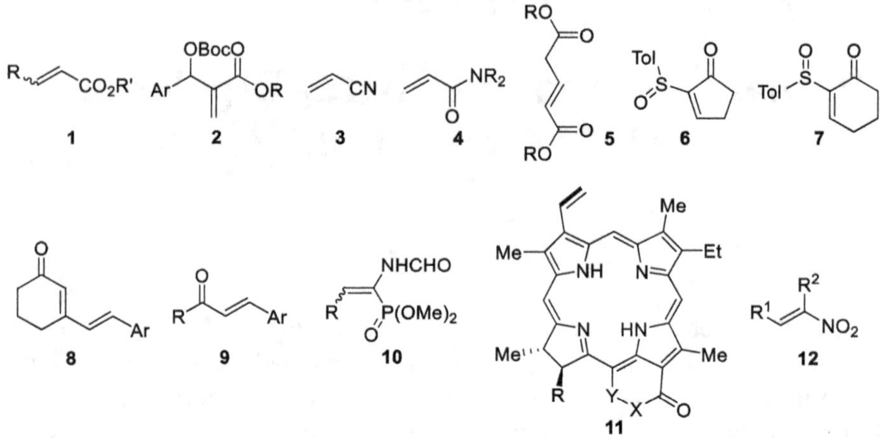

Figure 10.2 The alkenes as dipolarophiles in [3+2] cycloaddition.

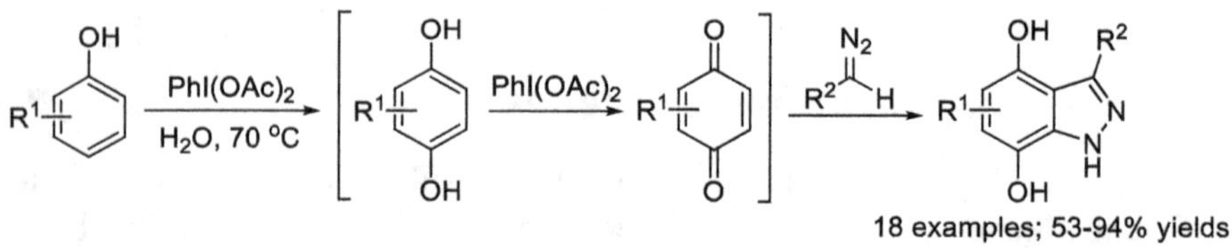

Scheme 10.1 1*H*-Indazole-4,7-diols from phenol and aryldiazomethane.

proceeded through a reaction sequence of thermal [3+2] cycloaddition/ring-contraction with nitrogen extrusion.[18]

A one-pot synthesis of 1*H*-indazole-4,7-diols derivatives was attained by hypervalent reagent-mediated 1,3-dipolar cycloaddition with diazo compounds in water (Scheme 10.1).[19] From readily available aryldiazomethane and phenols, various indazole derivatives were obtained in good yields under mild conditions. The oxidative cycloaddition procedure proceeded through the *in situ* generated dihydroxybenzene and benzoquinone, followed by [3+2] cycloaddition to deliver the cycloadducts.

Amine-promoted synthesis of substituted pyrazole-5-carboxylates was furnished by 1,3-dipolar cycloaddition of EDA with α-methylene carbonyl

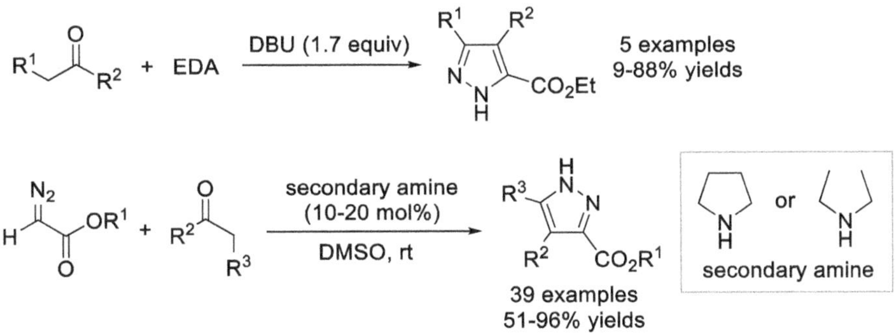

Scheme 10.2 Amine-mediated synthesis of pyrazoles from carbonyl compounds with diazoesters.

compounds. The reaction pathway involves DBU-promoted [3+2] cycloaddition of the diazoester with enolate, followed by spontaneous aromatization process through dehydration.[20] An amine-catalyzed 1,3-dipolar cycloaddition between diazoacetates and various carbonyl substrates, including ketones, aldehydes and 1,3-dicarbonyl compounds, was achieved, giving the substituted pyrazoles with high regioselectivity (Scheme 10.2).[21]

On the other hand, the structure diversity of the 1,3-dipoles, namely the diazo substrates, has also been explored. Thus, various diazo compounds, including 6-diazopenicillanates **13**,[22] diisopropyl diazomethylphosphonate **14** (with electron-deficient alkene),[23] lithium trimethylsilyldiazomethane **15** (with α,β-unsaturated cyclic enones),[24] were subjected to 1,3-dipolar cycloaddition, leading to the generation of functionalized pyrazoles. Moreover, three-component reactions, using the *in situ* generated diazo species from arylglyoxal monohydrates **16**[25] or aldehyde **17**[26] with *N*-tosylhydrazine in the presence of base, have been developed in one-pot process (Figure 10.3). Besides, cyclic ketone derived *N*-tosylhydrazones have also been subjected to the reaction with chalcones, affording spiropyrazolines with high selectivity and good yields.[27]

For example, Wu and co-workers reported the synthesis of polysubstituted pyrazoles **18** or **19** through tandem multicomponent reaction of arylglyoxal monohydrates **20**, *N*-tosylhydrazine and carbonyl compounds (Scheme 10.3).[28] By employing readily available substrates, the reaction afforded the products under simple operation conditions with good functional group tolerance. The proposed reaction pathway involves the [3+2] dipolar cycloaddition of the *in situ* generated enolates **22** with α-diazocarbonyl intermediates **21**, followed by dehydration and tautomerization.

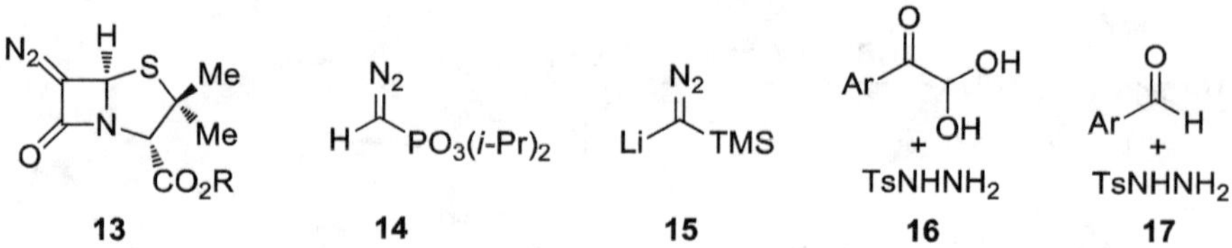

Figure 10.3 Diazo compounds used as 1,3-dipoles in [3+2] cycloaddition.

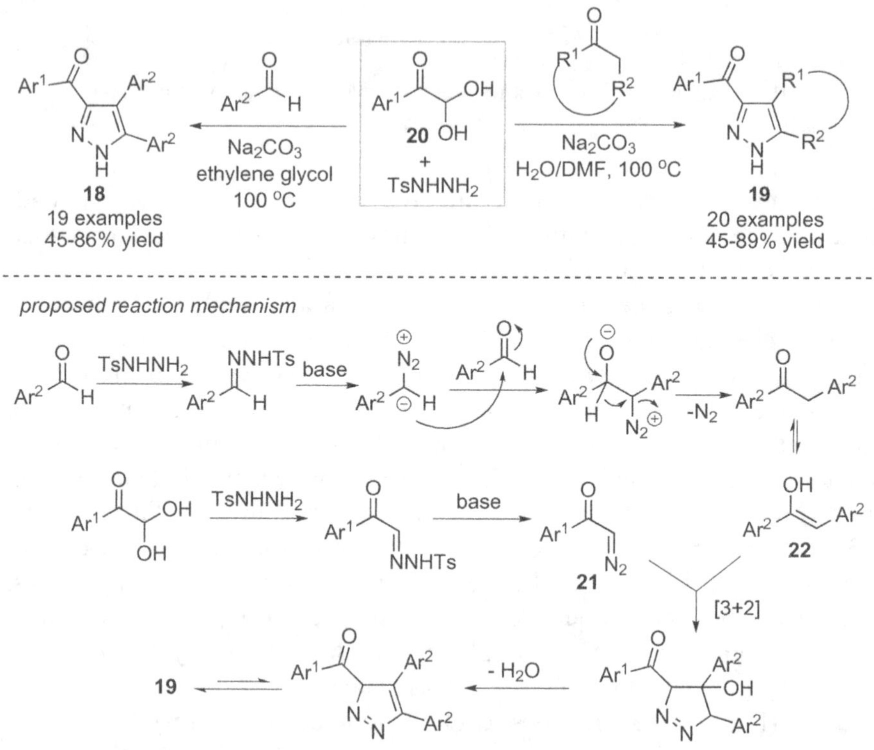

Scheme 10.3 Multicomponent procedure as convenient access to polysubstituted pyrazoles.

Intramolecular 1,3-dipolar cycloaddition of olefins with diazo substrate derived from *N*-tosylhydrazones has been established as an efficient method for the construction of a fused-ring system.[29–32] Taber and coworkers applied this strategy in the reaction of ω-alkenyl carbonyl substrates **23** with *N*-tosylhydrazine in the presence of base to generate bicyclic diazenes **24**, which were subsequently converted to cyclopropanes **25**

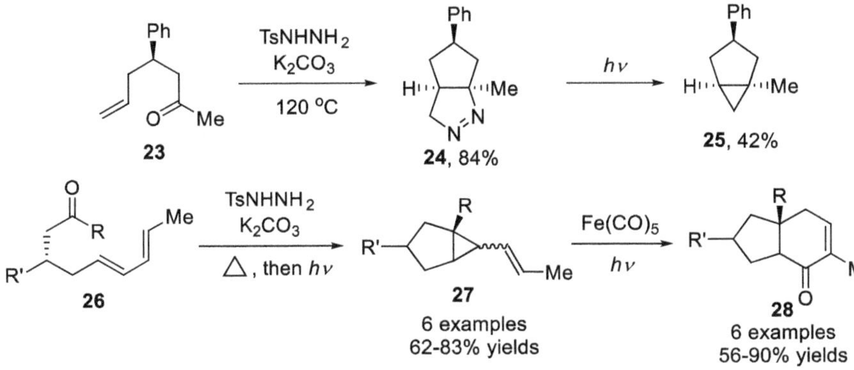

Scheme 10.4 1,3-Dipolar cycloaddition of diazo species derived from *N*-tosylhydrazones and sequential transformation to cyclopropanes.

by photo irradiation.[29] They also reported the construction of bicyclic scaffolds **27** from ω-dienyl ketones **26** through the *in situ* generation of diazo substrate and subsequent [3+2] dipolar cycloaddition. **27** was then followed by $Fe(CO)_5$-mediated cyclocarbonylation to afford **28**.[30] This transition-metal-free formal [1+4+1] protocol is a straightforward method for assembling fused ring system (Scheme 10.4).

The intramolecular 1,3-dipolar cycloaddition can serve as rapid entry to prepare pyrazoles-containing fused rings (Scheme 10.5).[31,32] Both one-pot and stepwise synthesis of benzofuropyrazoles **29** were realized from easily available salicylaldehyde substrates.[31] Besides, one-pot diastereoselective synthesis of tetrahydropyrrolo[3,4-*c*]pyrazoles derivatives **30** were developed by base-mediated intramolecular 1,3-dipolar cycloaddition from α-allylamino ketones.[32]

Mykhailiuk and co-workers explored the [3+2] cycloaddition of the *in situ* generated $CF_3CHN_2/C_2F_5CHN_2$ with electron-deficient alkenes, which afforded CF_3/C_2F_5-substituted pyrazolines or pyrazoles at low temperature.[33,34] The heterocycles containing fluorinated group have found potential applications as important precursors in drug discovery and medicinal research. However, at higher temperature, CF_3-substituted pyrazolines undergo ring contraction to generate cyclopropanes with simultaneous extrusion of N_2. In this case, various cyclopropanes containing CF_3 groups could be obtained, including 1-amino-2-(trifluoromethyl) cyclopropane-1-carboxylic acid,[35] 6-trifluoromethyl-3-azabicyclo[3.1.0] hexanes[36] and 3,3'-cyclopropyl spiro-oxindole derivatives (Scheme 10.6).[37]

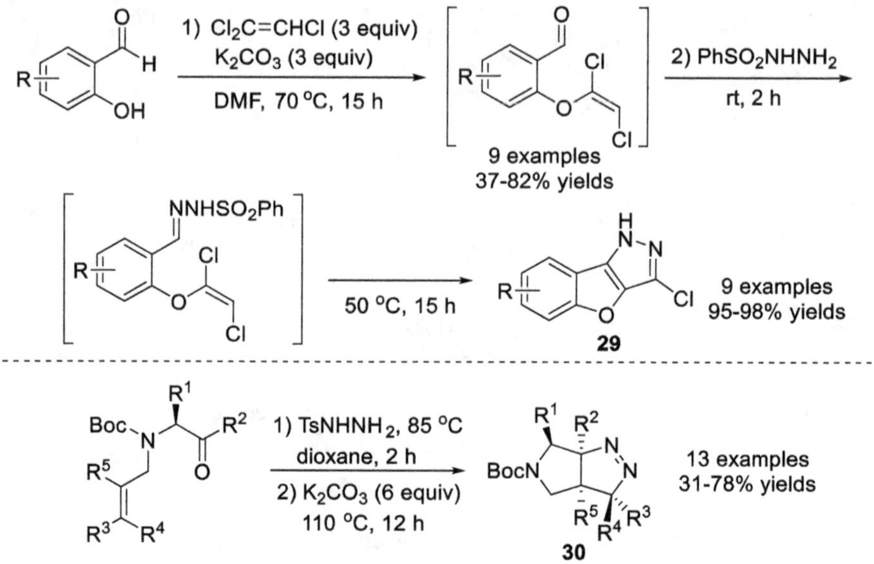

Scheme 10.5 Intramolecular 1,3-dipolar cycloaddition to synthesize substituted pyrazoles.

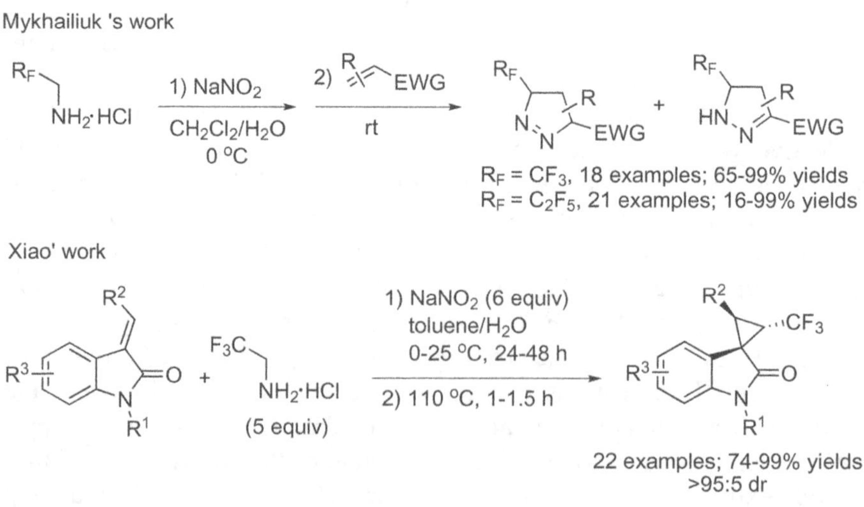

Scheme 10.6 1,3-Dipolar cycloaddition of the *in situ* generated fluorinated diazo intermediates and sequential ring contraction.

Namboothiri and co-workers reported the synthesis of functionalized pyrazoles through [3+2] cycloaddition of Bestmann–Ohira reagent[38,39,41] or α-diazo-β-ketosulfone[40,41] with electron-deficient alkenes. The base-promoted transformation proceeded through deacylation/[3+2] cycloaddition/elimination sequence. A wide range of sulfonyl pyrazoles and phosphonyl pyrazoles could be accessed by using various dipolarophiles, including nitroalkenes,[38,40] conjugated enones[39] and vinyl sulfones.[41]

10.2 Stereoselective cycloaddition with alkenes

To achieve stereoselectivity in this type of 1,3-dipolar cycloaddition, a chiral auxiliary approach was explored by using dipolarophiles bearing chiral substituents. García Ruano and co-workers utilized chiral vinyl sulfoxide as dipolarophile in 1,3-dipolar cycloadditions.[42–45] It was demonstrated that the chiral sulfinyl group was able to completely control the π-facial selectivity, affording the additional products in high diastereoselectivity. Interestingly, the diastereoselectivity could be inverted in the presence of Lewis acids, with $Yb(OTf)_3$ being the most efficient. Chelation of the $Yb(OTf)_3$ with carbonyl oxygen and sulfinyl oxygen was proposed to be responsible for the inversion of facial selectivity (Scheme 10.7).[45]

Asymmetric catalytic variant of this transformation remains a challenging problem, but progress has been made by Maruoka and co-workers. With catalytic chiral titanium BINOLate Lewis acids, (*S*)-BINOL/Ti(O*i*-Pr)$_4$ (2:1) and bis{((*S*)-binaphthoxy)-(isopropoxy)titanium}oxide,

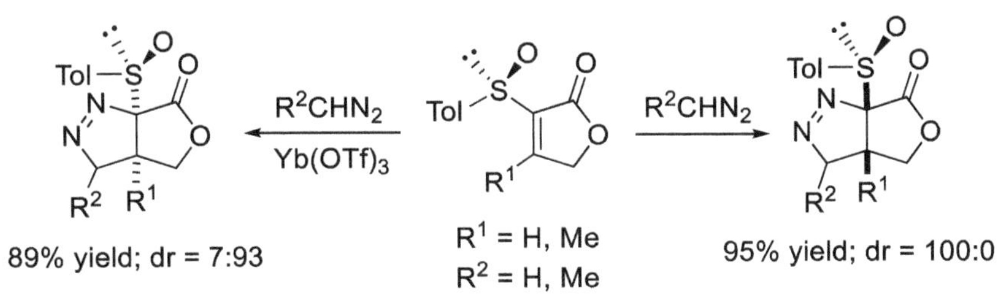

Scheme 10.7 Diastereoselective 1,3-dipolar cycloaddition with chiral vinyl sulfoxide.

high enantioselectivities were achieved in 1,3-dipolar cycloadditions between *t*-butyl diazoacetate and α-substituted acroleins (Eq. (10.1)).[46]

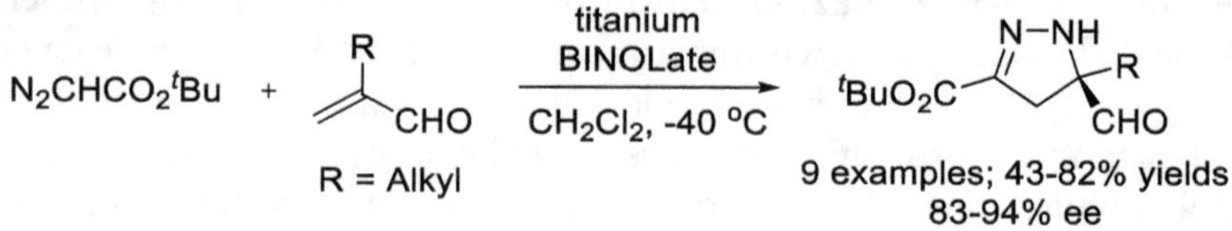

(10.1)

Subsequently, a series of asymmetric 1,3-dipolar [3+2] cycloadditions have been developed for the synthesis of polyfunctionalized chiral pyrazolines (Scheme 10.8).[47–50] For transition-metal-catalyzed reaction,

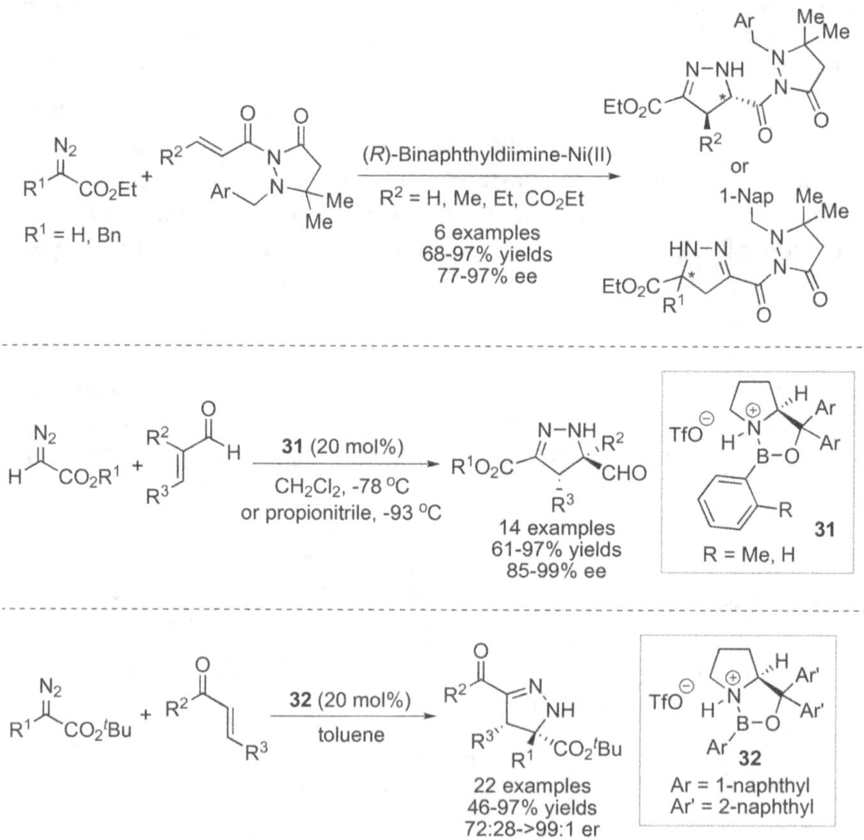

Scheme 10.8 Catalytic asymmetric 1,3-dipolar cycloaddition.

binaphthyldiimine-nickel(II) complex served as an excellent catalyst for asymmetric cycloaddition of diazoesters with 2-alkenoic acid derivatives with good to excellent enantioselectivity.[47] Alternatively, Ryu and co-workers explored boron-based chiral Lewis acids as the catalysts. They reported chiral oxazaborolidiniumion **31** catalyzed enantioselective 1,3-dipolar cycloaddition of alkyl diazoacetate with *α,β*-disubstituted acroleins,[48] as well as of oxazaborolidiniumion **32** catalyzed *α*-substituted diazoesters with enones.[49] Chiral 2-pyrazolines could be obtained in high enantioselectivities (up to 99% *ee*).

Peng and co-workers reported the asymmetric 1,3-dipolar cycloaddition of dimethyl (diazomethyl)phosphonate, the Seyferth–Gilbert reagent (SGR), with isatylidene malononitriles catalyzed by cinchona alkaloids **33** (Eq. (10.2)).[50] This enantioselective reaction could also be implemented in a three-component manner with isatin, malononitrile and SGR, which involved Knoevenagel condensation and sequential cycloaddition.

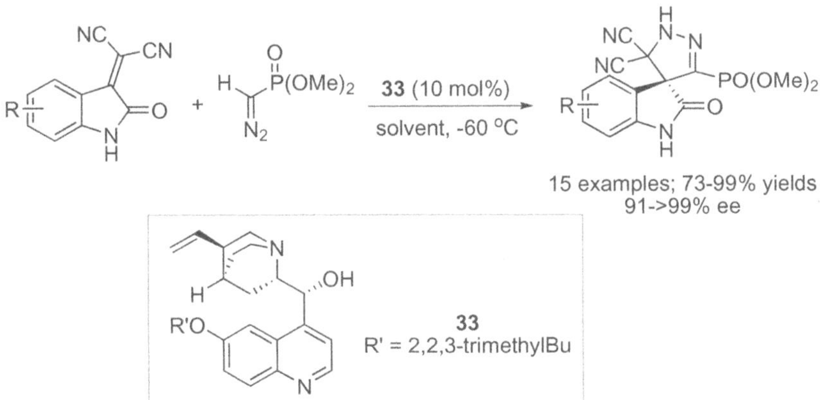

(10.2)

10.3 1,3-Dipolar cycloaddition of diazo compounds in multistep reactions

Multistep reactions involving 1,3-dipolar cycloaddition provide efficient synthesis of highly substituted pyrazolines[51] or pyrazoles[52–55] from readily available substrates. For example, various bi- and tricyclic Δ^1-pyrazolines **37** were obtained from 4-alkenyl carbonyls **34** with $TMSCHN_2$ through nucleophilic addition/1,3-Brook of **35**/intramolecular [3+2] cycloaddition sequence of **36** (Scheme 10.9).[51]

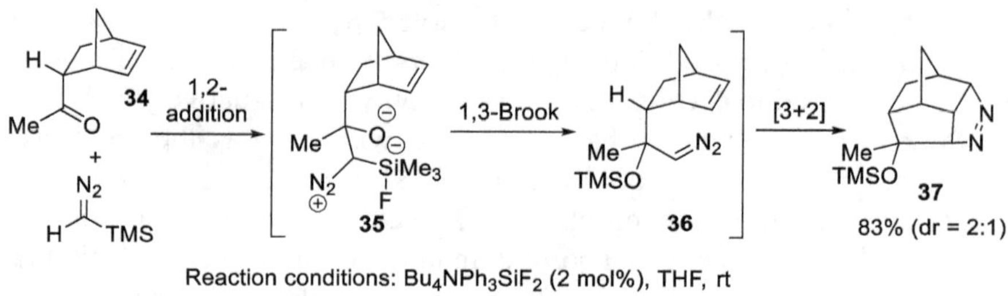

Scheme 10.9 Synthesis of substituted pyrazolines with $TMSCHN_2$.

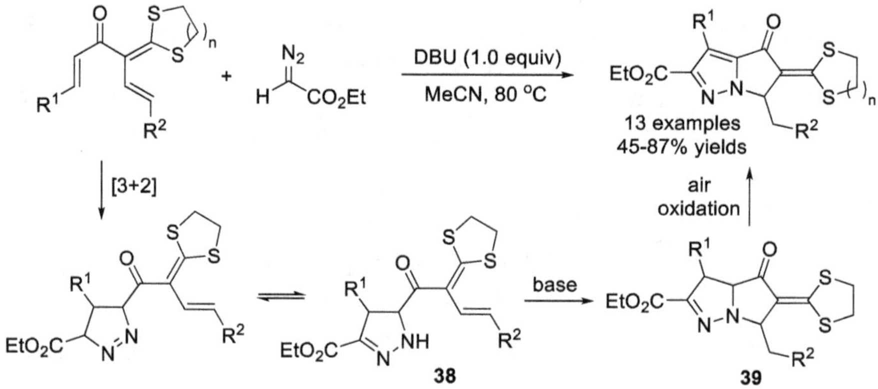

Scheme 10.10 Synthesis of fused pyrazoles through a 1,3-dipolar cycloaddition sequence.

For tandem synthesis of pyrazoles, Liu and co-workers reported the bicyclization protocol from EDA to achieve the synthesis of withasomnine alkaloids with high efficiency.[52] The reaction involves the 1,3-dipolar cycloaddition/aza-addition of **38**/oxidative aromatization sequential reactions of **39** (Scheme 10.10).

Smietana and co-workers developed a regioselective three-component reaction by using dimethyl (1-diazo-2-oxopropyl)phosphonate (Bestmann–Ohira reagent) **40** for the construction of phosphonyl pyrazole rings (Scheme 10.11). Initially, the one-pot protocol with aldehydes, cyanoacetic derivatives and Bestmann–Ohira reagent was developed, which proceeded *via* domino Knoevenagel condensation/formal 1,3-dipolar cycloaddition/elimination cascade.[53] Subsequently, they reported the synthesis of substituted 3-carbo-5-phosphonylpyrazoles from aldehydes, methyl ketone and **40** through Claisen–Schmidt/1,3-dipolar cycloaddition/oxidation sequence.[54]

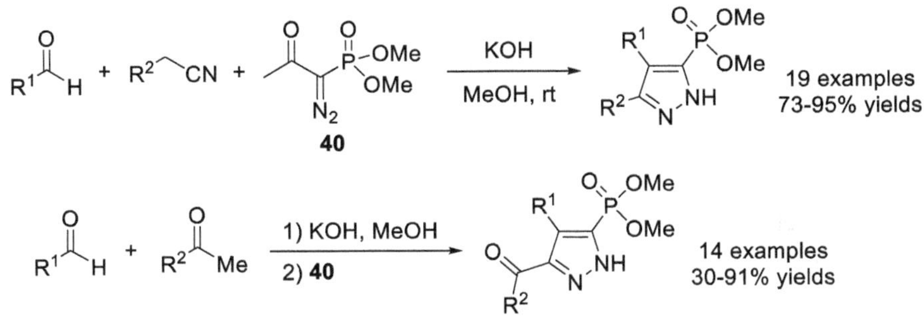

Scheme 10.11 Synthesis of phosphonyl pyrazoles from BOR in three-component reaction.

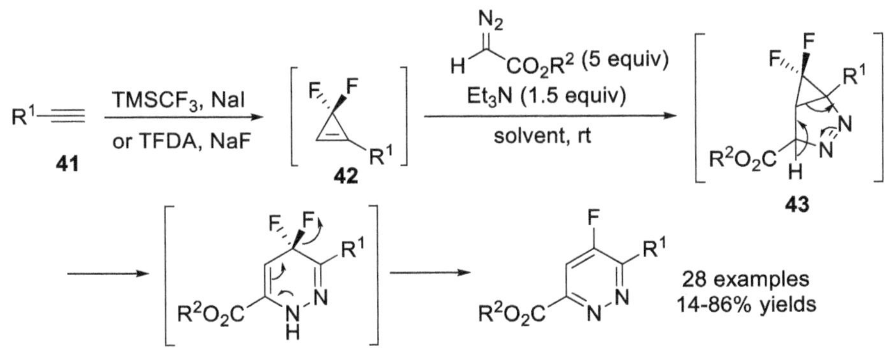

Scheme 10.12 Synthesis of 5-fluoropyridazines through a [2+1]/[3+2] cycloaddition cascade.

Moreover, densely functionalized vinyl pyrazoles have been prepared from domino reactions between one molecule of α,β-unsaturated aldehydes with two molecules of Bestmann–Ohira reagent **40**.[55] This reaction occurs through a cascade process of deacylative 1,3-dipolar cycloaddition, HWE homologation, 1,3-H shift and finally aromatization domino.

A three-component synthetic approach from terminal alkynes **41**, difluorocarbene precursors and diazoesters was realized, affording a wide range of highly functionalized 5-fluoropyridazines.[56] This reaction was initiated by cyclopropanation of difluorocarbene to generate *gem*-difluorocyclopropenes **42**, followed by [3+2] cycloaddition sequence to obtain cyclopropano[*c*]pyrazolines intermediate **43**. Sequential ring expansion and aromatization occurred, producing pyridazine derivatives along with the elimination of HF (Scheme 10.12).

10.4 Reaction with alkynes

Intermolecular 1,3-dipolar [3+2] cycloaddition of diazoalkane species with stable or *in situ* generated alkynes, especially electron-deficient derivatives, has been applied to the straightforward synthesis of pyrazoles. However, the reaction of less electron-rich α-diazocarbonyl compounds with electron-deficient alkynes usually requires activation by Lewis acids. Jiang and Li reported an intermolecular 1,3-dipolar cycloaddition of α-diazocarbonyl compounds with alkynes by $InCl_3$-catalyzed reaction in water (Eq. (10.3)).[57] The alkynes bearing carbonyl substituents reacted smoothly with diazoacetates to give the pyrazole products in good yields. However, phenyl acetylene failed to react under the same conditions. It is rationalized that $InCl_3$ activates the alkynes by coordinating the carbonyl group and thus lowers the LUMO of the alkyne moiety.

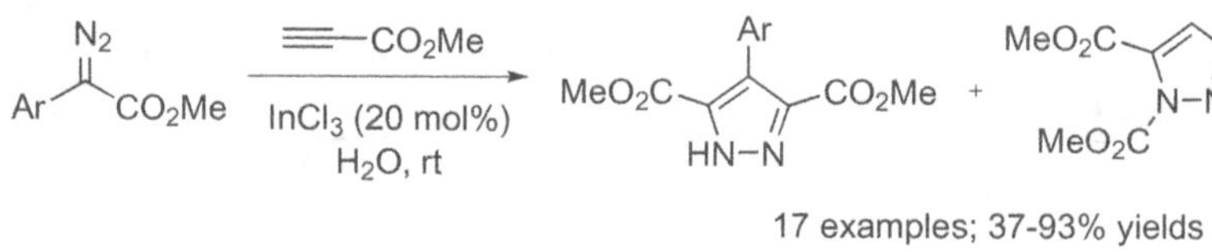

(10.3)

For simple alkyl or aryl alkynes, the reaction with α-diazocarbonyl compounds usually fails and Lewis acid activation cannot be applied in such cases. An alternative mode of activation by raising the HOMO level of the alkyne has been developed by Qi and Ready.[58] In their approach, the alkyne was converted into the Cu(I) acetylide, which was then reacted with α-diazoacetate. With this method, the cycloaddition occurred smoothly to afford the pyrazole derivatives in moderate to good yields (Eq. (10.4)). The cycloaddition involves the LUMO of the diazo compound, and the copper is considered as an electron-donating group to raise the HOMO of the alkyne. This is a rare example of an inverse-electron-demand cycloaddition reaction.

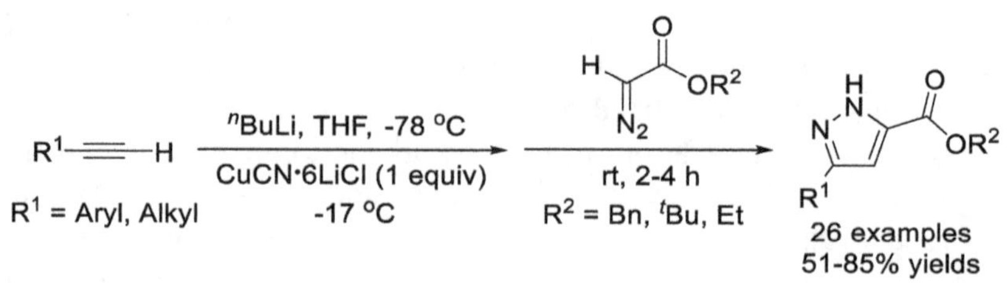

(10.4)

Other developments in the reactions of diazo compounds as 1,3-dipoles include the reactions of 1,3-dipolarophiles with trimethylsilyldiazomethane,[59] or with *in situ* generated aryldiazomethanes,[60] the reactions of conjugated nitroalkenes with the anion of diethyl 1-diazomethylphosphonate, generated *in situ* from diethyl 1-diazo-2-oxopropylphosphonate (Bestmann–Ohira reagent)[61] and a three-component reaction of *in situ* generated *N*-fluoropyridinium fluoride with isonitrile and diazo compound.[62]

Moreover, with α-diazocarbonyl compounds as the 1,3-dipolar reagents, a series of alkynes, including terminal alkynes,[63,64] cyclooctyne[65] and ynones (with Bestmann–Ohira reagent),[66] were employed as the substrates to synthesize substituted pyrazoles under Lewis acid[63] or base-promoted[66] conditions, as well as under catalyst-free and neat conditions.[67]

2,2,2-Trifluorodiazoethane has attracted considerable attention as a versatile reagent for introducing trifluoromethyl group in the products. Ma and co-workers reported the Ag(I)-mediated 1,3-dipolar [3+2] cycloaddition of terminal alkynes with *in situ* generated CF_3CHN_2.[68] This regioselective reaction is a mild and efficient method for the synthesis of 5-substituted 3-trifluoromethylpyrazole derivatives, which may be potentially useful in drug development (Eq. (10.5)).

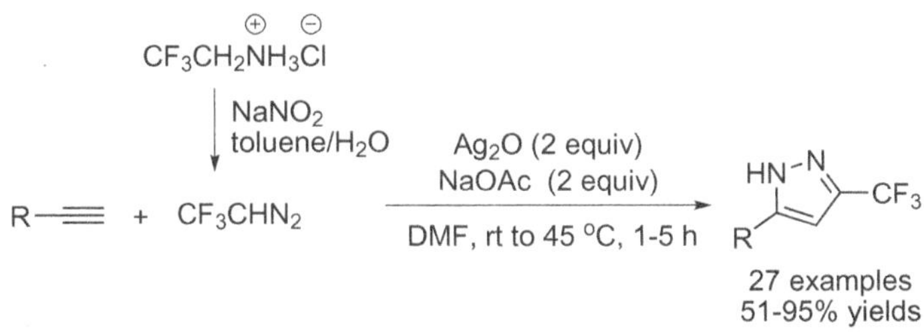

(10.5)

Mykhailiuk and co-workers reported the first [3+2] cycloaddition of *in situ* generated difluoromethyl diazomethane **44** with alkynes.[69] This one-pot procedure provides a convenient synthesis of difluoromethyl-substituted pyrazoles (Scheme 10.13). Moreover, practical procedure to access CF_2HCHN_2 species **44** was established, providing the opportunity for further application of **44** in other types of diazo transformations. The same group also reported the three-component synthesis of fluorinated pyrazoles from fluoroalkylamines ($RCF_2CH_2NH_2$), $NaNO_2$ and electron-deficient alkynes.[70]

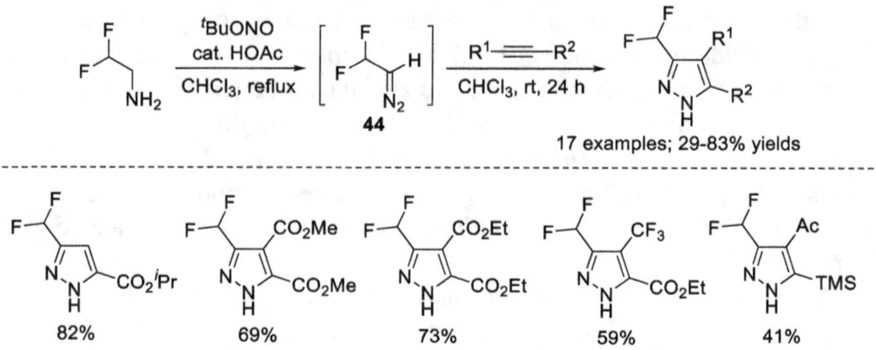

Scheme 10.13 [3+2] Cycloaddition of the *in situ* generated CF_2HCHN_2 with alkynes.

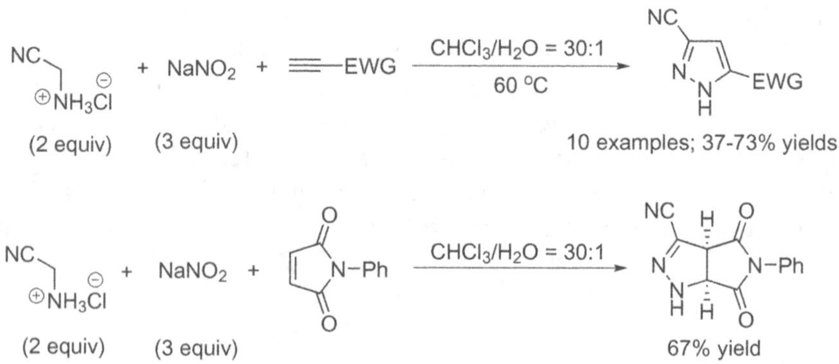

Scheme 10.14 1,3-Dipole cycloaddition by using *in situ* generated $NCCHN_2$.

The *in situ* generation method of unstable diazoacetonitrile (N_2CHCN) was applied in the 1,3-dipole cycloaddition by Mykhailiuk.[71] In the mixed solvent of chloroform/water, the 1,3-dipole cycloaddition of *in situ* generated N_2CHCN with electron-deficient alkynes could be carried out under safe operation conditions (Scheme 10.14). The potential of N_2CHCN was also demonstrated by the [3+2] cycloaddition with an alkene.

The [3+2] cycloaddition of benzyne intermediate derived from *ortho*-(trimethylsilyl)aryl triflates **45** in the presence of fluoride with diazo substrates was found highly efficient for the synthesis of substituted indazoles.[72] Jin and co-workers reported a [3+2] cycloaddition of various diazomethane derivatives with arynes generated from silyl aryl triflates by treatment with KF or CsF (Scheme 10.15).[73] Notably, when only 0.5 eq.

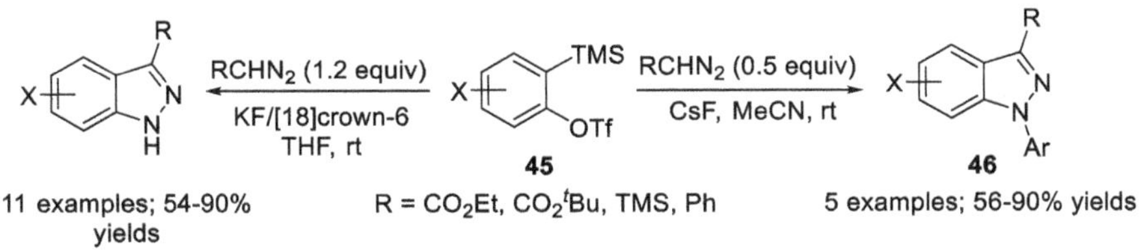

Scheme 10.15 Cycloaddition of diazomethane derivatives with arynes.

of the diazo compounds was used in this reaction, the 1,3-dipolar addition product further reacted with aryne to yield the product **46**.

Larock and co-workers reported the similar indazole synthesis by [3+2] cycloaddition of arynes with diazo compounds.[74] The reaction with ethyl diazoacetate afforded *N*-unsubstituted indazoles or 1-arylated indazoles through nucleophilic addition with excess of arynes, depending on the ratio of the reactants and the reaction conditions. On the other hand, the reaction with ethyl acyldiazoacetate afforded the 1-acyl indazoles exclusively. Hari and Aoyama developed a two-step synthesis of 3-hydroxymethyl indazoles with the [3+2] cycloaddition of benzyne with 2-diazo-(2-trimethylsilyl)ethanols **47**, which was generated from diazo(trimethylsilyl)-methylmagnesium bromide with various carbonyls.[75] Besides, Liu and co-workers developed a Ag(I)-catalyzed cycloaddition of arynes with diazocarbonyl compounds to synthesize 2-aryl-2*H*-indazoles, which were generated through subsequent nucleophilic addition of (1*H*-indazol-1-yl)silver **48** to benzyne (Scheme 10.16).[76]

Shi and co-workers applied the method of *in situ* generation of diazoalkanes from *N*-tosylhydrazones in the [3+2] cycloaddition with aryne.[77] Furthermore, they achieved the synthesis for both 3-substituted and 1,3-disubstituted 1*H*-indazoles from arynes and hydrazones, depending on substituents on the hydrazones.[78] The reaction of *N*-tosylhydrazones was assumed to proceed either *via* 1,3-dipolar cycloaddition of *in situ* generated diazo compounds **49**, or *via* annulation/elimination pathway. Alternatively, 1,3-disubstituted indazoles were formed in the case when the substituents on the nitrogen were aryl or alkyl groups. The reaction likely proceeds through a pathway of non-diazo annulation of **50**, followed by oxidation (Scheme 10.17).

Furthermore, the regioselective synthesis of pyrazoles was realized by [3+2] cycloaddition of alkynes with *N*-tosylhydrazones in the presence of base.[79–82] Valdés and co-workers reported a catalyst-free [3+2]

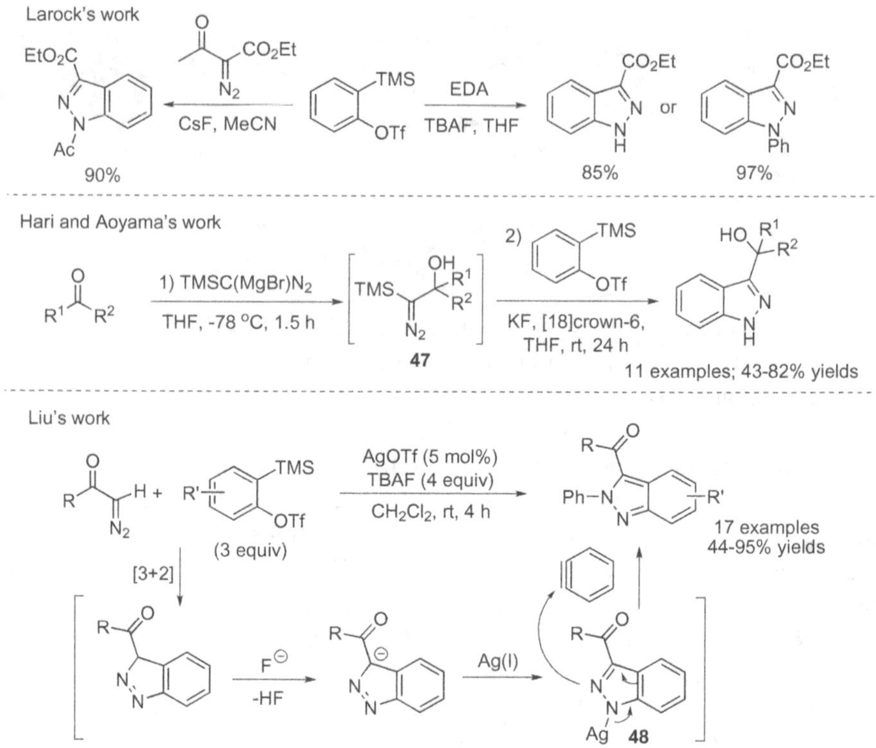

Scheme 10.16 Synthesis of indazoles by the reaction of diazo compounds with benzyne.

cycloaddition/[1,5] rearrangement cascade. The interesting observation is that the selectivity of [1,5] rearrangement is affected by the substituents of the *N*-tosylhydrazones. Thus, depending on the substrates, the cycloaddition intermediate **53** either gave pyrazoles **51**, or pyrazoles **52**, as the major products (Scheme 10.18).[79] When the reaction was applied to the *N*-tosylhydrazones **54** bearing stereogenic center, the [1,5] rearrangement occurred with retention of the configuration, affording chiral product **55**.[82]

1,3-Dipolar cycloaddition of *N*-tosylhydrazones with the *in situ* generated alkynyl bromides from *gem*-dibromoalkenes was demonstrated to afford 3,5-diaryl-4-bromopyrazoles.[80] Besides, Ley and co-workers reported the reaction of alkynes with cyclic *N*-tosylhydrazones **56** for the synthesis of spirocyclic pyrazoles **57**, which was further subjected

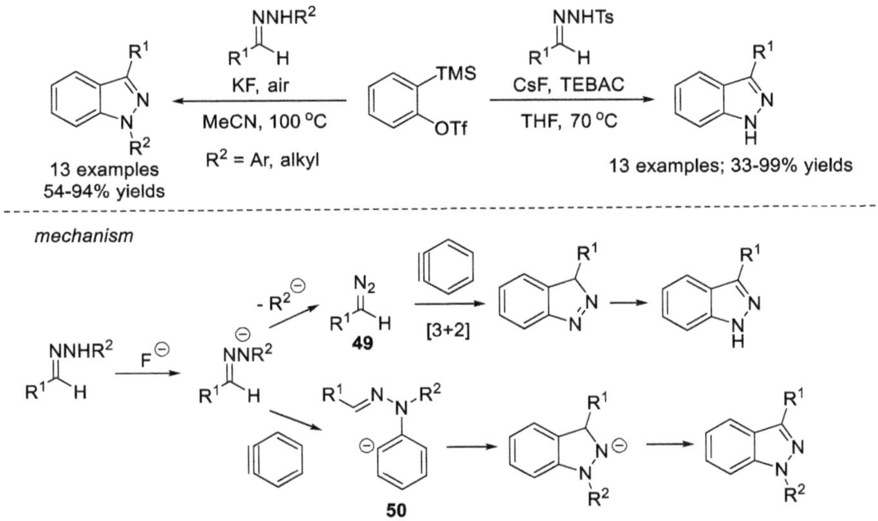

Scheme 10.17 Synthesis of 1*H*-indazoles by the reaction of hydrazones with arynes.

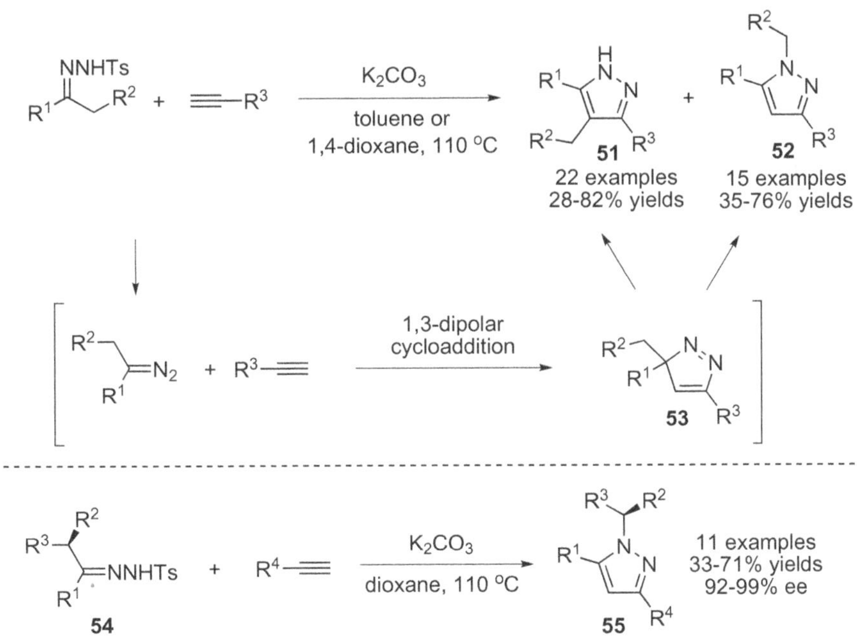

Scheme 10.18 Catalyst-free [3+2] cycloaddition/[1,5] rearrangement cascade.

to Lewis acid-mediated ring-expansion to afford the bicyclic pyrazoles **58** (Eq. (10.6)).[81]

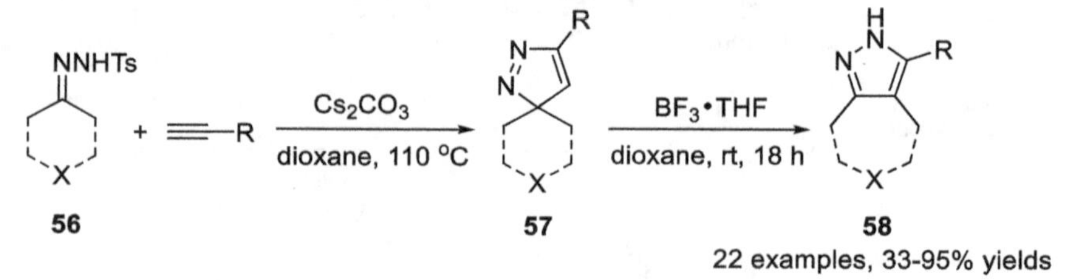

(10.6)

10.5 With allenes

1,3-Dipolar cycloaddition of diazo species with allenes has been realized in both intra- and intermolecular manner.[83–87] For the intramolecular reaction, a tandem approach toward fused triquinanes has been achieved with the substrates containing allene moiety and diazo precursor. This tandem cycloaddition protocol proceeded *via* 1,3-dipolar cycloaddition of the diazo intermediate **59**, generating trimethylenemethane diyl intermediate **60**, followed by sequential intramolecular diradical [3+2] cycloaddition to afford fused triquinanes **61** (Scheme 10.19).[83–85]

For intermolecular reaction, Shi and co-workers reported the 1,3-dipolar cycloaddition of diesters bearing vinylidenecyclopropane moiety with the

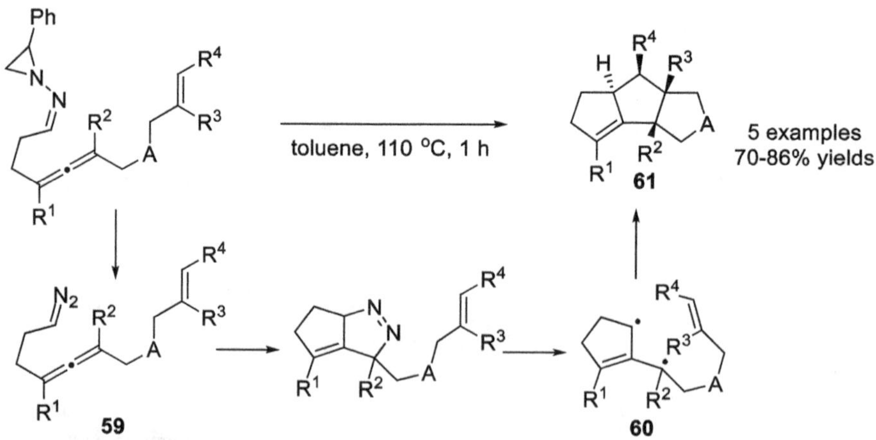

Scheme 10.19 Tandem cycloaddition strategy for the synthesis of fused triquinanes.

in situ generated diazo species (Eq. (10.7)).[86] The regioselectivity of the [3+2] cycloaddition is controlled by both steric interaction of the substituents and the atomic orbital coefficients between HOMO and LUMO.

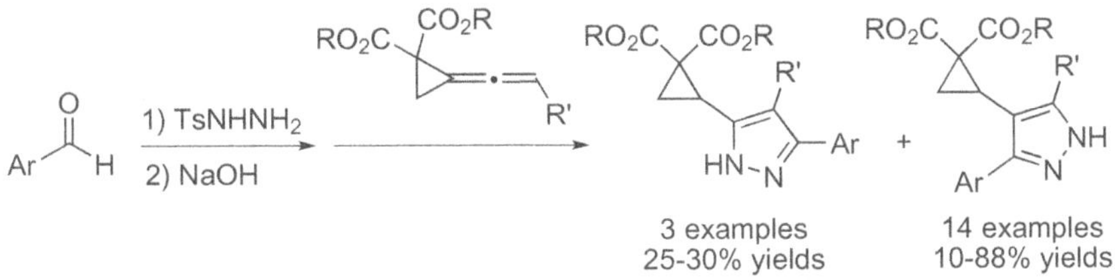

(10.7)

Besides, Ma and co-workers demonstrated the cycloaddition procedure of electron-deficient allenes **62** with CF_3CHN_2 under mild conditions (Scheme 10.20).[87] This regioselective synthetic approach afforded 5-(trifluoromethyl)pyrazolines **63** in good yields. When the reaction was carried out in the presence of NEt_3, 3-(trifluoromethyl)pyrazoles **64** could be obtained.

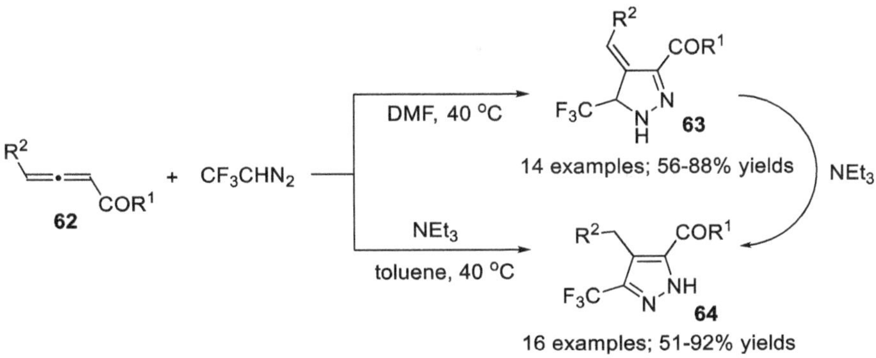

Scheme 10.20 Synthesis of CF_3-substituted pyrazolines and pyrazoles by the cycloaddition of allenes with CF_3CHN_2

10.6 With other dipolarophiles

Thiocarbonyl compounds **65** also serve as efficient dipolarophiles in [3+2] cycloaddition with diazo species to afford thiadiazolines **66**. From thiadiazolines **66**, alkenes **68** can be obtained through N_2 extrusion, generating

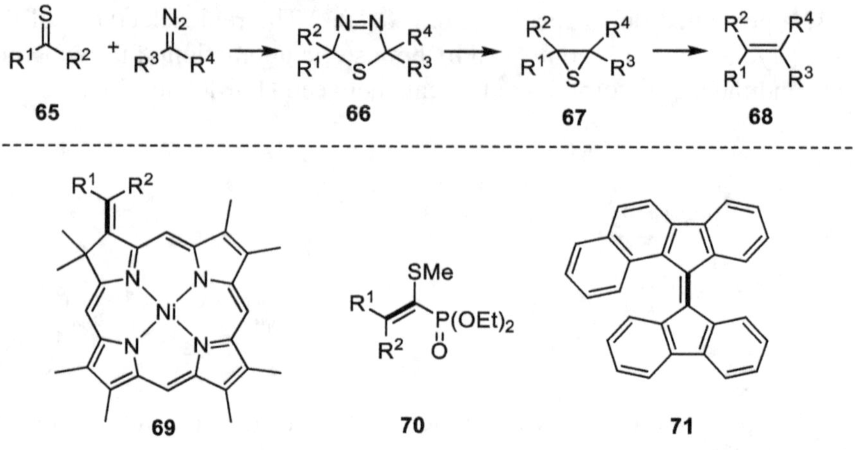

Scheme 10.21 Barton olefination by the reaction of thiocarbonyls and diazo compounds.

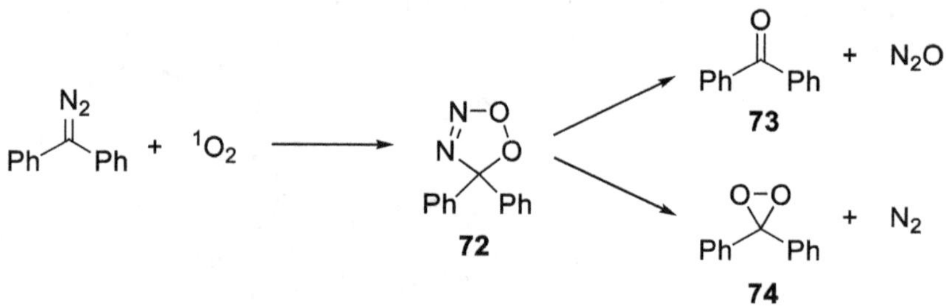

Scheme 10.22 1,3-Dipolar cycloaddition of diphenyldiazomethane with singlet oxygen.

thiirane **67** followed by further sulfur extrusion (Barton olefination).[88–90] Recently, this strategy has been applied to the synthesis of substituted chlorins **69**,[91] 2,2-disubstituted 1-(methylsulfanyl)vinyl phosphonates **70**[92] and 11-(9*H*-fluoren-9-ylidene)-11*H*-benzo[*a*]fluorene **71** (Scheme 10.21).[93]

An intriguing 1,3-dipolar cycloaddition between diphenyldiazomethane with singlet oxygen was investigated by time-resolved UV-vis and IR spectra.[94] It was observed that the elusive cycloadduct of 1,2,3,4-dioxadiazole **72** decomposed even at low concentration, resulting in a parallel generation of either benzophenone **73** and N_2O, or benzopheneone *O*-oxide **74** (Scheme 10.22).

Recently, Ma and co-workers reported a Ag(I)-catalyzed cycloaddition of isocyanides **75** as the dipolarophiles with CF_3CHN_2, diazoesters

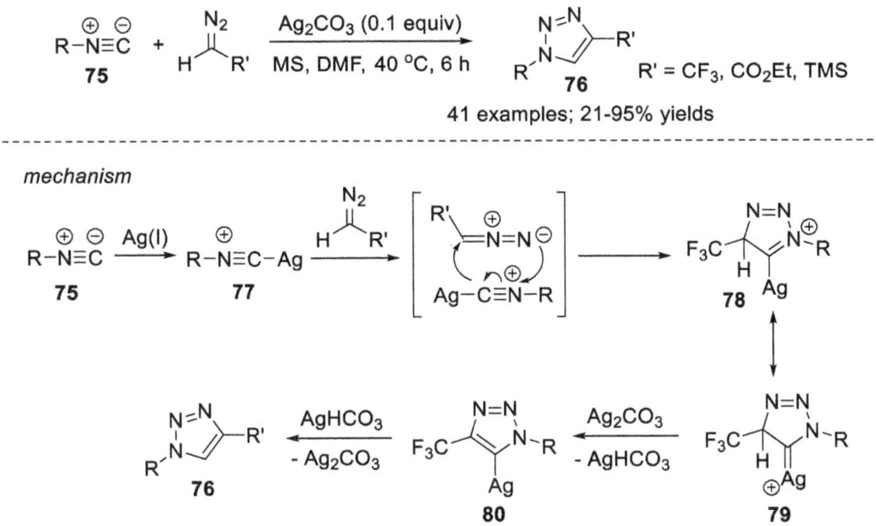

Scheme 10.23 Ag(I)-catalyzed 1,3-dipole cycloaddition of diazo compounds with isocyanides.

or $TMSCHN_2$.[95] This reaction allowed efficient assembly of 1,4-di-substituted 1,2,3-triazoles **76** with high regioselectivity under mild conditions (Scheme 10.23). The reaction was initiated by cycloaddition between silver isocyanide complex **77** with diazo substrate to generate intermediate **78** (in resonance with Ag(I)-carbene intermediate **79**), followed by deprotonation and sequential protonation to produce 1,2,3-triazoles **76**.

Besides, the cycloaddition reactions of nitriles with diazo compounds have been explored to afford triazoles. A recent example is shown by a synthesis of fused 1,2,3-triazoles from salicylaldehyde derivatives.[96] Both stepwise and one-pot procedure were successfully achieved through smooth cycloaddition of cyano group with the *in situ* generated aryl diazomethanes **84** (Scheme 10.24).

Alternatively, the synthesis of 5-aryl-1*H*-1,2,3-triazole-4-carboxylates could also be attained by the reaction of α-diazocarbonyl compounds with imines in the presence of base. Thus, the DBU-promoted reaction of *N*-PMP protected imine **86** afforded the fully substituted 1,2,3-triazoles **87** in good yield (Eq. (10.8)).[97] The reaction was proposed to be initiated by 1,3-dipolar cycloaddition with DBU-promoted, followed by aromatization sequence.

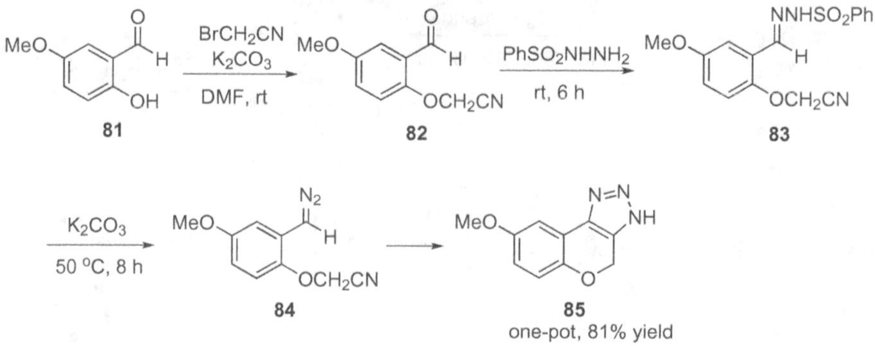

Scheme 10.24 Synthesis of fused 1,2,3-triazoles by intramolecular 1,3-dipolar cycloaddition of nitrile with diazo moiety.

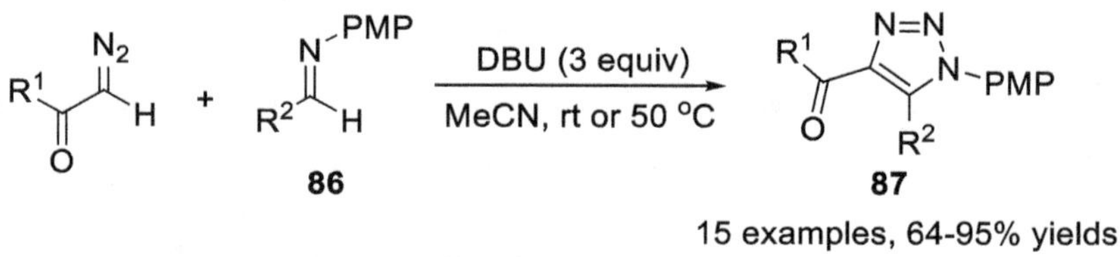

(10.8)

Diazonium salts have also been explored as the dipolarophiles. Ma and co-workers reported a regioselective [3+2] cycloaddition between aryldiazonium salts **88** with 2,2,2-trifluorodiazoethane, affording the 2-aryl-5-trifluoromethyltetrazoles **89** under mild conditions (Eq. (10.9)).[98] This cycloaddition protocol could also be performed in a one-pot diazotization/cycloaddition sequence directly from aniline derivatives.

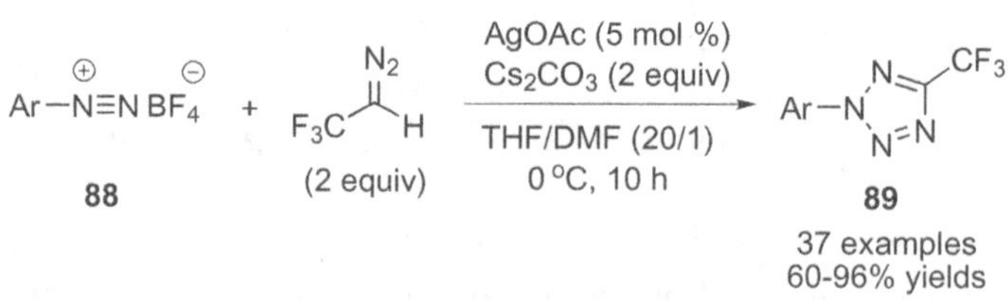

(10.9)

In summary, the diazo compounds represent a type of highly versatile 1,3-dipoles. Various dipolarophiles, including alkenes, alkynes, allenes, nitriles and isocyanates, have been explored in the [3+2] cycloaddition reactions with diazo compounds, providing a straightforward method for the synthesis of polyfunctionalized pyrazoles and pyrazolines.

References

1. Rodriguez, J. B. *Synthesis* **2014**, *46*, 1129–1142.
2. Dubrovskiy, A. V.; Markina, N. A.; Larock, R. C. *Org. Biomol. Chem.* **2013**, *11*, 191–218.
3. Hashimoto, T.; Maruoka, K. *Org. Biomol. Chem.* **2008**, *6*, 829–835.
4. Lozhkin, S. S.; Petrov, D. V.; Dokichev, V. A.; Tomilov, Y. V.; Nefedov, O. M. *Chem. Heterocycl. Comp.* **2009**, *45*, 937–941.
5. Novikov, R. A.; Platonov, D. N.; Dokichev, V. A.; Tomilov, Y. V.; Nefedov, O. M. *Russ. Chem. Bull. Int. Ed.* **2010**, *59*, 984–990.
6. Ovchinnikov, M. Y.; Yangirov, T. A.; Lobov, A. N.; Sultanova, R. M.; Khursan, S. L. *Int. J. Chem. Kinet.* **2013**, *45*, 499–507.
7. Krishna, P. R.; Sekhar, E. R.; Mongin, F. *Tetrahedron Lett.* **2008**, *49*, 6768–6772.
8. Sun, H.; Wang, X.; Zhan, M.; Liu, J.; Xie, Y. *Tetrahedron Lett.* **2013**, *54*, 3846–3850.
9. Wang, W.; Simovic, D. D.; Di, M.; Fieber, L.; Rein, K. S. *Bioorg. Med. Chem. Lett.* **2013**, *23*, 1949–1952.
10. García Ruano, J. L.; Alonso, M.; Cruz, D.; Fraile, A.; Martín, M. R.; Peromingo, M. T.; Tito, A.; Yuste, F. *Tetrahedron* **2008**, *64*, 10546–10551.
11. Cruz Cruz, D.; Yuste, F.; Martín, M. R.; Tito, A.; García Ruano, J. L. *J. Org. Chem.* **2009**, *74*, 3820–3826.
12. Kissane, M.; Lawrence, S. E.; Maguire, A. R. *Org. Biomol. Chem.* **2010**, *8*, 2735–2748.
13. Hamadi, N. B.; Msaddek, M. *C. R. Chimie* **2011**, *14*, 997–1001.
14. Goulioukina, N. S.; Makukhin, N. N.; Beletskaya, I. P. *Tetrahedron* **2011**, *67*, 9535–9540.
15. Liu, R.; Yin, J.; Li, J.; Wu, J.; Chen, G.; Jin, Y.; Wang, J. *Chin. J. Org. Chem.* **2012**, *32*, 544–551.
16. Yang, Z.; Wang, Z.; Xu, X.; Liu, Y.; Qi, C.; Wang, J. *Chin. J. Org. Chem.* **2012**, *32*, 2099–2107.
17. Xie, J.-W.; Wang, Z.; Yang, W.-J.; Kong, L.-C.; Xu, D.-C. *Org. Biomol. Chem.* **2009**, *7*, 4352–4354.
18. Maurer, S.; Jikyo, T.; Maas, G. *Eur. J. Org. Chem.* **2009**, 2195–2207.
19. Hou, Y.; Cai, C.; Yu, G. *Synlett* **2016**, *27*,773–776.
20. Gioiello, A.; Khamidullina, A.; Fulco, M. C.; Venturoni, F.; Zlotsky, S.; Pellicciari, R. *Tetrahedron Lett.* **2009**, *50*, 5978–5980.
21. Wang, L.; Huang, J.; Gong, X.; Wang, J. *Chem. Eur. J.* **2013**, *19*, 7555–7560.
22. Santos, B. S.; Nunes, S. C. C.; Pais, A. A. C. C.; Pinho e Melo, T. M. V. D. *Tetrahedron* **2012**, *68*, 3729–3737.
23. Marinozzi, M.; Tondi, S.; Marcelli, G.; Giorgi, G. *Tetrahedron* **2014**, *70*, 9485–9491.

24. Sun, C.; O'Connor, M. J.; Lee, D.; Wink, D. J.; Milligan, R. D. *Angew. Chem. Int. Ed.* **2014**, *53*, 3197–3200.
25. Shu, W.-M.; Ma, J.-R.; Zheng, K.-L.; Sun, H.-Y.; Wang, M.; Yang, Y.; Wu, A.-X. *Tetrahedron* **2014**, *70*, 9321–9329.
26. Zhang, G.; Ni, H.; Chen, W.; Shao, J.; Liu, H.; Chen, B.; Yu, Y. *Org. Lett.* **2013**, *15*, 5967–5969.
27. Wu, Q.-X.; Li, H.-J.; Wang, H.-S.; Zhang, Z.-G.; Wang, C.-C.; Wu, Y.-C. *Synlett* **2015**, *26*, 243–249.
28. Shu, W.-M.; Zheng, K.-L.; Ma, J.-R.; Sun, H.-Y.; Wang, M.; Wu, A.-X. *Org. Lett.* **2015**, *17*, 1914–1917.
29. Taber, D. F.; Guo, P. *J. Org. Chem.* **2008**, *73*, 9479–9481.
30. Taber, D. F.; Guo, P.; Guo, N. *J. Am. Chem. Soc.* **2010**, *132*, 11179–11182.
31. Sales, Z. S.; Mani, N. S. *J. Org. Chem.* **2009**, *74*, 891–894.
32. Barroso, R.; Escribano, M.; Cabal, M.-P.; Valdés, C. *Eur. J. Org. Chem.* **2014**, 1672–1683.
33. Slobodyanyuk, E. Y.; Artamonov, O. S.; Shishkin, O. V.; Mykhailiuk, P. K. *Eur. J. Org. Chem.* **2014**, 2487–2495.
34. Mykhailiuk, P. K. *Chem. Eur. J.* **2014**, *20*, 4942–4947.
35. Artamonov, O. S.; Mykhailiuk, P. K.; Voievoda, N. M.; Volochnyuk, D. M.; Komarov, I. V. *Synthesis* **2010**, 443–446.
36. Artamonov, O. S.; Slobodyanyuk, E. Y.; Shishkin, O. V.; Komarov, I. V.; Mykhailiuk, P. K. *Synthesis* **2013**, *45*, 225–230.
37. Li, T.-R.; Duan, S.-W.; Ding, W.; Liu, Y.-Y.; Chen, J.-R.; Lu, L.-Q.; Xiao, W.-J. *J. Org. Chem.* **2014**, *79*, 2296–2302.
38. Muruganantham, R.; Namboothiri, I. *J. Org. Chem.* **2010**, *75*, 2197–2205.
39. Verma, D.; Mobin, S.; Namboothiri, I. N. N. *J. Org. Chem.* **2011**, *76*, 4764–4770.
40. Kumar, R.; Namboothiri, I. N. N. *Org. Lett.* **2011**, *13*, 4016–4019.
41. Kumar, R.; Nair, D.; Namboothiri, I. N. N. *Tetrahedron* **2014**, *70*, 1794–1799.
42. García Ruano, J. L.; de Diego, S. A. A.; Martín, M. R.; Torrente, E.; Martín Castro, A. M. *Org. Lett.* **2004**, *6*, 4945–4948.
43. García Ruano, J. L.; Alonso, M.; Fraile, A.; Martin, R.; Peromingo, M. T.; Tito, A. *Phosphorus, Sulfur Silicon Relat. Elem.* **2005**, *180*, 1441–1442.
44. García Ruano, J. L.; González Gutiérrez, L.; Torrente, E.; Yuste, F.; Martín Castro, A. M. *Arkivoc* **2005**, *IX*, 146–158.
45. García Ruano, J. L.; Peromingo, M. T.; Alonso, M.; Fraile, A.; Martín, M. R.; Tito, A. *J. Org. Chem.* **2005**, *70*, 8942–8947.
46. Kano, T.; Hashimoto, T.; Maruoka, K. *J. Am. Chem. Soc.* **2006**, *128*, 2174–2175.
47. Suga, H.; Furihata, Y.; Sakamoto, A.; Itoh, K.; Okumura, Y.; Tsuchida, T.; Kakehi, A.; Baba, T. *J. Org. Chem.* **2011**, *76*, 7377–7387.
48. Gao, L.; Hwang, G.-S.; Lee, M. Y.; Ryu, D. H. *Chem. Commun.* **2009**, 5460–5462.

49. Lee, S. I.; Kim, K. E.; Hwang, G.-S.; Ryu, D. H. *Org. Biomol. Chem.* **2015**, *13*, 2745–2749.
50. Du, T.; Du, F.; Ning, Y.; Peng, Y. *Org. Lett.* **2015**, *17*, 1308–1311.
51. Liu, H.; O'Connor, M. J.; Sun, C.; Wink, D. J.; Lee, D. *Org. Lett.* **2013**, *15*, 2974–2977.
52. Wang, H.; Zhao, Y.-L.; Li, L.; Zhang, Z.-W.; Liu, Q. *Adv. Synth. Catal.* **2013**, *355*, 1540–1544.
53. Mohanan, K.; Martin, A. R.; Toupet, L.; Smietana, M.; Vasseur, J.-J. *Angew. Chem. Int. Ed.* **2010**, *49*, 3196–3199.
54. Martin, A. R.; Mohanan, K.; Toupet, L.; Vasseur, J.-J.; Smietana, M. *Eur. J. Org. Chem.* **2011**, 3184–3190.
55. Ahamad, S.; Gupta, A. K.; Kant, R.; Mohanan, K. *Org. Biomol. Chem.* **2015**, *13*, 1492–1499.
56. Tran, G.; Pardo, D. G.; Tsuchiya, T.; Hillebrand, S.; Vors, J.-P.; Cossy, J. *Org. Lett.* **2015**, *17*, 3414–3417.
57. Jiang, N.; Li, C.-J. *Chem. Commun.* **2004**, 394–395.
58. Qi, X.; Ready, J. M. *Angew. Chem. Int. Ed.* **2007**, *46*, 3242–3244.
59. Simovic, D.; Di, M.; Marks, V.; Chatfield, D. C.; Rein, K. S. *J. Org. Chem.* **2007**, *72*, 650–653.
60. Aggarwal, V. K.; de Vicente, J.; Bonnert, R. V. *J. Org. Chem.* **2003**, *68*, 5381–5383.
61. Muruganantham, R.; Mobin, S. M.; Namboothiri, I. N. N. *Org. Lett.* **2007**, *9*, 1125–1128.
62. Kiselyov, A. S. *Tetrahedron Lett.* **2006**, *47*, 2631–2634.
63. He, S.; Chen, L.; Niu, Y.-N.; Wu, L.-Y.; Liang, Y.-M. *Tetrahedron Lett.* **2009**, *50*, 2443–2445.
64. Cheung, K. M. J.; Reynisson, J.; McDonald, E. *Tetrahedron Lett.* **2010**, *51*, 5915–5918.
65. McGrath, N. A.; Raines, R. T. *Chem. Sci.* **2012**, *3*, 3237–3240.
66. Pramanik, M. M. D.; Kant, R.; Rastogi, N. *Tetrahedron* **2014**, *70*, 5214–5220.
67. Vuluga, D.; Legros, J.; Crousse, B.; Bonnet-Delpon, D. *Green Chem.* **2009**, *11*, 156–159.
68. Li, F.; Nie, J.; Sun, L.; Zheng, Y.; Ma, J.-A. *Angew. Chem. Int. Ed.* **2013**, *52*, 6255–6258.
69. Mykhailiuk, P. K. *Angew. Chem. Int. Ed.* **2015**, *54*, 6558–6561.
70. Mykhailiuk, P. K. *Org. Biomol. Chem.* **2015**, *13*, 3438–3445.
71. Mykhailiuk, P. K. *Eur. J. Org. Chem.* **2015**, 7235–7239.
72. Shoji, Y.; Hari, Y.; Aoyama, T. *Tetrahedron Lett.* **2004**, *45*, 1769–1771.
73. Jin, T.; Yamamoto, Y. *Angew. Chem. Int. Ed.* **2007**, *46*, 3323–3325.
74. Liu, Z.; Shi, F.; Martinez, P. D. G.; Raminelli, C.; Larock, R. C. *J. Org. Chem.* **2008**, *73*, 219–226.
75. Hari, Y.; Sone, R.; Aoyama, T. *Org. Biomol. Chem.* **2009**, *7*, 2804–2808.

76. Wang, C.-D.; Liu, R.-S. *Org. Biomol. Chem.* **2012**, *10*, 8948–8952.
77. Li, P.; Zhao, J.; Wu, C.; Larock, R. C.; Shi, F. *Org. Lett.* **2011**, *13*, 3340–3343.
78. Li, P.; Wu, C.; Zhao, J.; Rogness, D. C.; Shi, F. *J. Org. Chem.* **2012**, *77*, 3149–3158.
79. Pérez-Aguilar, M. C.; Valdés, C. *Angew. Chem. Int. Ed.* **2013**, *52*, 7219–7223.
80. Sha, Q.; Wei, Y. *Synthesis* **2013**, *45*, 413–420.
81. Merchant, R. R.; Allwood, D. M.; Blakemore, D. C.; Ley, S. V. *J. Org. Chem.* **2014**, *79*, 8800–8811.
82. Pérez-Aguilar, M. C.; Valdés, C. *Angew. Chem. Int. Ed.* **2015**, *54*, 13729–13733.
83. Kang, T.; Kim, W.-Y.; Yoon, Y.; Kim, B. G.; Lee, H.-Y. *J. Am. Chem. Soc.* **2011**, *133*, 18050–18053.
84. Qiao, Y.; Han, K.-L. *Org. Biomol. Chem.* **2014**, *12*, 1220–1231.
85. Lee, H.-Y. *Acc. Chem. Res.* **2015**, *48*, 2308–2319.
86. Wu, L.; Shi, M. *J. Org. Chem.* **2010**, *75*, 2296–2301.
87. Zhang, F.-G.; Wei, Y.; Yi, Y.-P.; Nie, J.; Ma, J.-A. *Org. Lett.* **2014**, *16*, 3122–3125.
88. Barton, D. H. R.; Guziec, F. S.; Shahak, I. *J. Chem. Soc., Perkin Trans. 1* **1974**, 1794–1799.
89. Back, T. G.; Barton, D. H. R.; Britten-Kelley, M. R.; Guziec, F. S. *J. Chem. Soc., Perkin Trans. 1* **1976**, 2079–2089.
90. Wang, Z. *Barton-Kellogg Olefination, in: Comprehensive Organic Name Reactions and Reagents*, Wiley, **2009**, vol. 1, chapter 56, pp. 249.
91. Dullweber, F.; Montforts, F.-P. *Synlett* **2008**, 3213–3215.
92. Mlostoń, G.; Urbaniak, K.; Linden, A.; Heimgartner, H. *Tetrahedron* **2009**, *65*, 8191–8198.
93. Assadi, N.; Pogodin, S.; Agranat, I. *Eur. J. Org. Chem.* **2011**, 6773–6780.
94. Torres-Alacan, J.; Sander, W. *J. Org. Chem.* **2008**, *73*, 7118–7123.
95. Wang, S.; Yang, L.-J.; Zeng, J.-L.; Zheng, Y.; Ma, J.-A. *Org. Chem. Front.* **2015**, *2*, 1468–1474.
96. Mani, N. S.; Fitzgerald, A. E. *J. Org. Chem.* **2014**, *79*, 8889–8894.
97. Chen, J.-H.; Liu, S.-R.; Chen, K. *Chem. Asian J.* **2010**, *5*, 328–333.
98. Chen, Z.; Fan, S.-Q.; Zheng, Y.; Ma, J.-A. *Chem. Commun.* **2015**, *51*, 16545–16548.

Chapter 11

Reactions with Diazo Compounds as Nucleophiles or Electrophiles

The neutral diazo compounds can serve as nucleophiles since the carbon attached to the dinitrogen is negatively polarized. Alternatively, acyldiazomethanes may be deprotonated to generate anions, which function as carbon nucleophiles (Figure 11.1).[1–5] The carbon-bearing diazo functionality can undergo nucleophilic addition to ketones, aldehydes or imines. They may also undergo Michael addition to α,β-unsaturated carbonyl compounds. These reactions represent unique approaches for the C–C bond formations, affording various products bearing diazo functionality. In this context, asymmetric catalysis using diazo compounds as nucleophiles has attracted particular attention in recent years and some highly enantioselective reactions have been developed.

The study by Mayr has indicated that diazomethane, phenyldiazomethane and trimethylsilyldiazomethane are comparable in nucleophilicity as ketene acetals, while the less reactive compounds, diphenyldiazomethane, ethyl diazoacetate and diazoacetone, are

Figure 11.1 Diazo compounds as nucleophiles.

comparable to typical silyl enol ethers and activated allylsilanes, and the nucleophilicity of diethyl diazomalonate corresponds to that of 1,1-dialkylethylenes or styrenes.[3]

The relative stability of the diazo group under neutral or basic conditions allows the generation of the enolate from β-oxo α-diazocarbonyl compounds. Thus, the α-diazo-β-keto compounds **1** can be enolized to generate enolates **2**, which function as nucleophiles (Eq. (11.1)). The most common procedure to enolize these diazo compounds is to convert them into boron or Ti(IV) enolates, or into silyl enol ethers. Similar to the common enolates, nucleophilic additions to C=O or C=N bonds are also possible with silyl enol ethers.

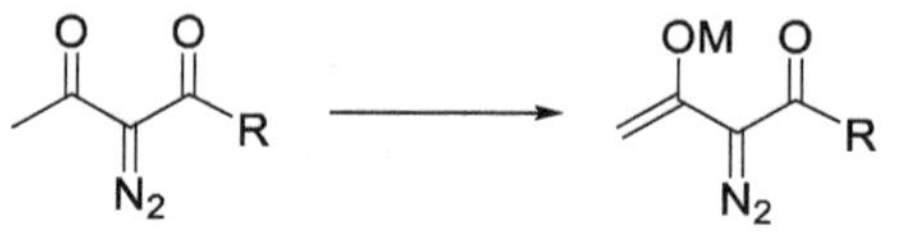

1 **2**, M = B, Ti, Si (11.1)

Although the sensitivity of the diazo functionality, in particular to acidic conditions, has limited the widespread use of the diazo group-based nucleophiles, some useful synthetic methodologies have been developed with diazo nucleophiles in recent years. In this section, we have categorized the recent developments according to the reaction partners and reaction types, which include the nucleophilic addition to carbonyl group or imine group; nucleophilic attack to boron compounds; Darzens reaction, aziridination, migration and other transformations.

While in most cases diazo compounds serve as nucleophiles (broadly speaking, this also includes the cases of diazo compounds as carbene precursors), diazo compounds can also be utilized as electrophiles to accept the attack by nucleophiles, owing to the partial positive charge on the nitrogen atom. These transformations are potentially useful in organic synthesis, especially for the construction of nitrogen-containing heterocycles. In this chapter, a series of reactions with diazo compounds as electrophiles will also be summarized according to the types of the reactions.

11.1 Aldol-type reaction

Aldol-type reaction between α-diazo compounds and aldehydes or ketones is an efficient method to access β-hydroxy α-carbonyl compounds. Usually, the reaction is conducted by the generation of anion from acyldiazomethane with strong base such as LDA at low temperature under rigorous anhydrous reaction conditions. Recent studies are focused on the development of mild and catalytic processes, and asymmetric catalysis. In 2008, Kantam and co-workers reported the preparation of silica-immobilized tetramethylguanidine as a sustainable catalyst, which was explored as the solid base in aldol reaction between aldehydes and EDA.[6] This catalyst could efficiently promote the aldol reaction in high yields and it could be quantitatively recovered from the reaction mixture (Eq. (11.2)).

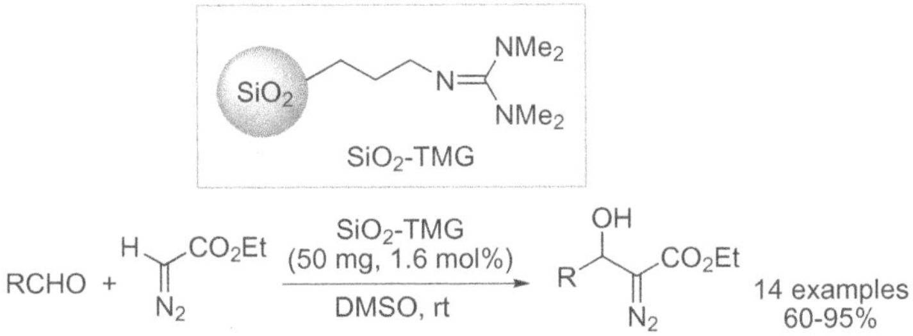

(11.2)

Subsequently, they reported the polymer-supported ionic liquid **3**, which could catalyze the corresponding aldol reaction in water.[7] The imidazonium salt catalyst was easily recovered by simple filtration. The recovered catalyst could be reused for five consecutive cycles without distinct loss of activity (Eq.(11.3)).

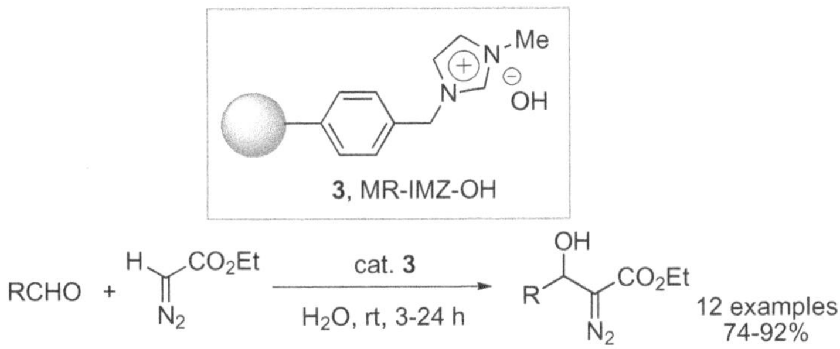

(11.3)

In addition to the base-promoted aldol reaction, benzoic acid catalyzed aldol-type condensation can occur at room temperature, which affords the β-hydroxy-α-diazocarbonyl products in excellent yields. When using sugar-derived aldehydes as the substrates, high yields and excellent diastereoselectivities were observed in this transformation.[8]

A DBU-catalyzed condensation of acyldiazomethanes with aldehydes or imines was achieved under mild conditions.[9,10] The reaction could also be carried out in water (Eq. (11.4)).[10] Moreover, phase-transfer catalyst quaternary ammonium salts have been introduced into this reaction.[11–13] The condensation can also be carried out under Lewis acid-catalyzed conditions.[14]

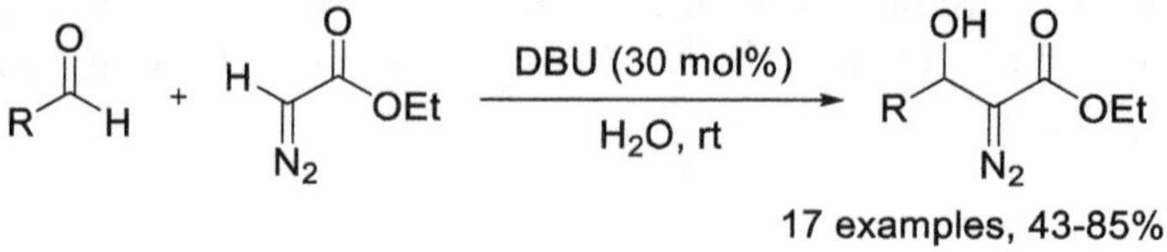

(11.4)

A further extension of this chemistry is the catalytic asymmetric aldol-type reactions of aldehydes with diazo esters. Wang and co-workers demonstrated the first catalytic asymmetric aldol-type reactions of aldehydes with diazo esters with moderate enantioselectivities by using chiral complex of (*S*)-6,6′-Br_2BINOL-Zr(O^tBu)$_4$ (Eq. (11.5)).[15]

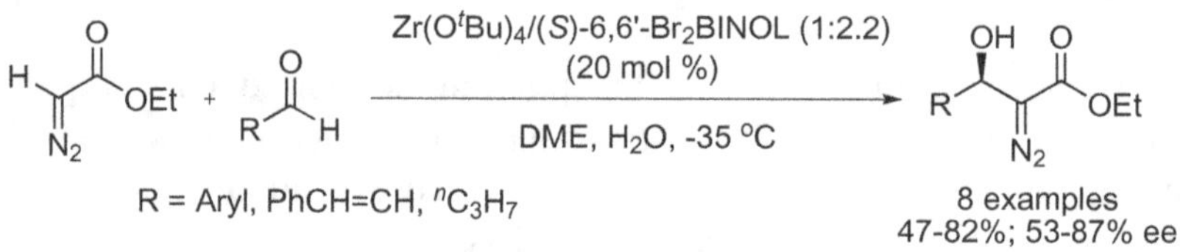

(11.5)

A similar level of enantioselectivity has been achieved by Nishida and co-workers in the condensation of aldehydes with *t*-butyl diazoacetate catalyzed by a phase-transfer catalyst, cinchonidinium salt **4** (Eq. (11.6)).[11,12]

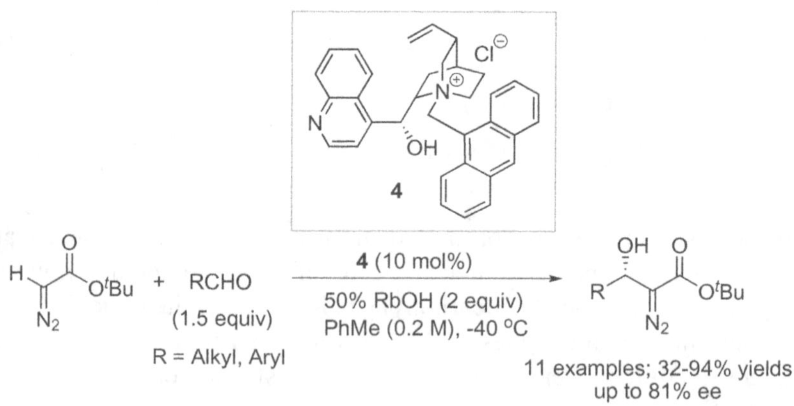

(11.6)

The Trost group subsequently achieved high enantioselectivity in this type of aldol reaction by using main group metal- or transition-metal-ProPhenol catalysts. In 2009, they reported Mg-catalyzed asymmetric aldol addition of EDA to aldehydes with up to 98% *ee*.[16,17] With this reaction, a series of β-hydroxy-α-diazoesters could be synthesized in good yields and with high enantioselectivities (Eq. (11.7)). The products could be further transformed into optically enriched β-amido-α-diazoesters, chiral β-amino acids,[18] as well as chiral 1,2-diols.[19]

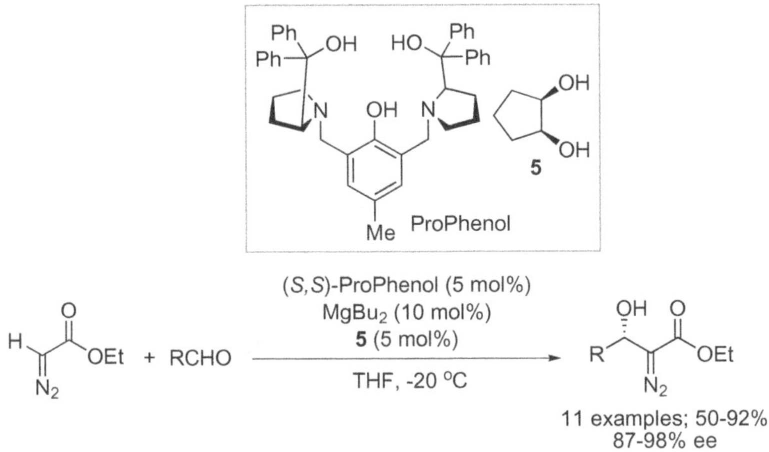

(11.7)

Subsequently, Feng and co-workers reported the asymmetric aldol-type reaction of EDA and aldehydes, affording the β-hydroxy-α-diazoesters in moderate yields and enantioselectivities.[20] They used titanium salt as the catalyst with BINOL **6** and cinchonine **7** as the chiral ligand (Eq. (11.8)).

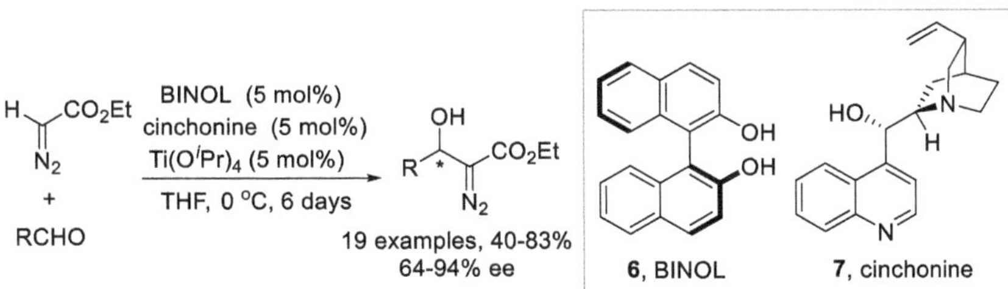

(11.8)

In 2009, the first catalytic asymmetric aldol-type condensation between EDA and ketones was described by Cozzi.[21] Norephedrine

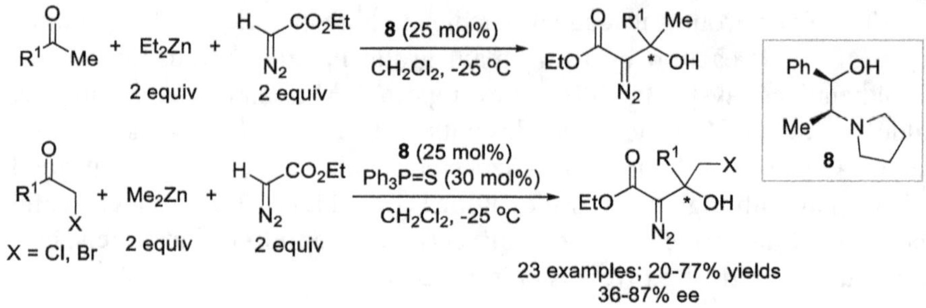

Scheme 11.1 Aldol-type reaction of EDA with ketones and α-haloketones.

derivatives **8** and dialkyl zinc were utilized as the promoters. The reaction with a series of ketones afforded moderate yields and enantioselectivities. Furthermore, the reaction gave good enantiomeric excesses with α-haloketones in the presence of $Ph_3P{=}S$ (Scheme 11.1).

An asymmetric catalytic system for α-ketoesters and diazoacetate esters was reported by Feng and co-workers in the same year.[22] Catalyzed by C_2-symmetric *N,N′*-dioxide (**9**)-Sc(III) complex, a series of tertiary alcohols with diazo group were obtained in good yields with excellent enantioselectivities (Eq. (11.9)).

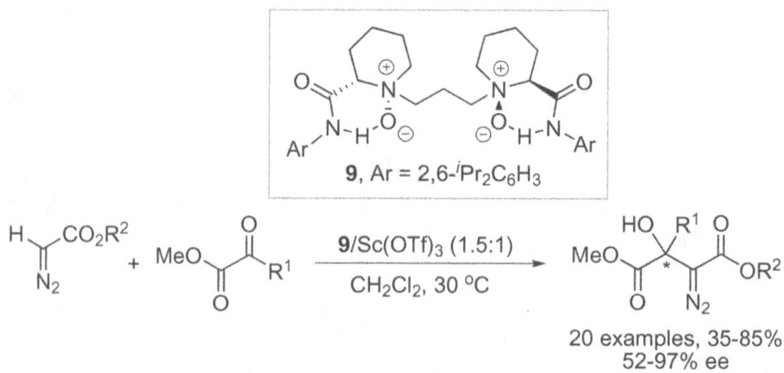

(11.9)

The aldol-type reaction between EDA and trifluoromethyl ketones **10** or **11** by using dimethyl zinc as the base has also been reported.[23] A series of trifluoromethyl-substituted aldol products were obtained in good yields. The substrates with electron-donating groups on the aromatic rings slow down the rate of this transformation, giving the products in relatively

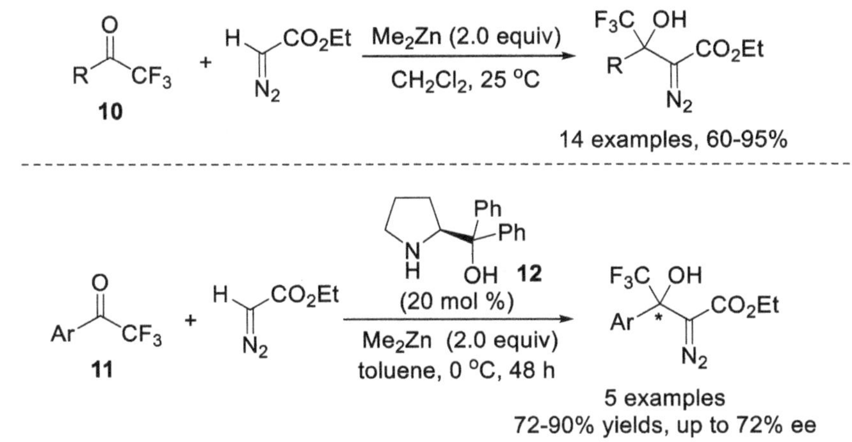

Scheme 11.2 Aldol-type reaction of trifluoromethyl ketones and EDA.

lower yields (Scheme 11.2). The corresponding asymmetric aldol reaction was explored with chiral amino alcohol **12**, but only moderate enantioselectivities were obtained.

11.2 Tandem reaction: Nucleophilic addition followed by [1,2]-migration

The β-hydroxy-α-diazocarbonyl compounds **13** obtained from nucleophilic addition of acyldiazomethane to aldehyde or ketone can be catalyzed by transition-metal complexes to give β-keto carbonyl compounds **14** through 1,2-migration (Scheme 11.3).

When the nucleophilic addition of diazo substrates to aldehydes or ketones is carried out under acidic conditions through the activation

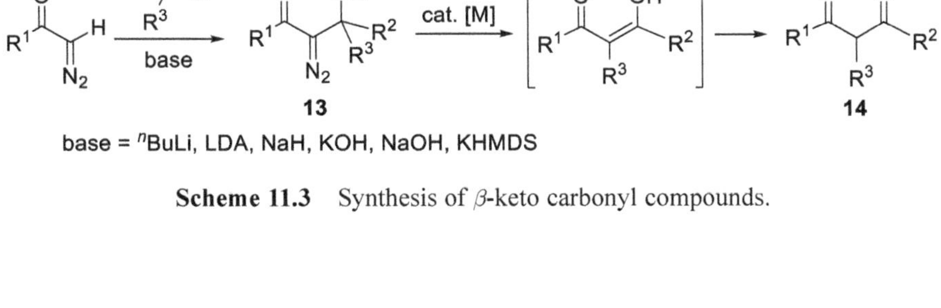

Scheme 11.3 Synthesis of β-keto carbonyl compounds.

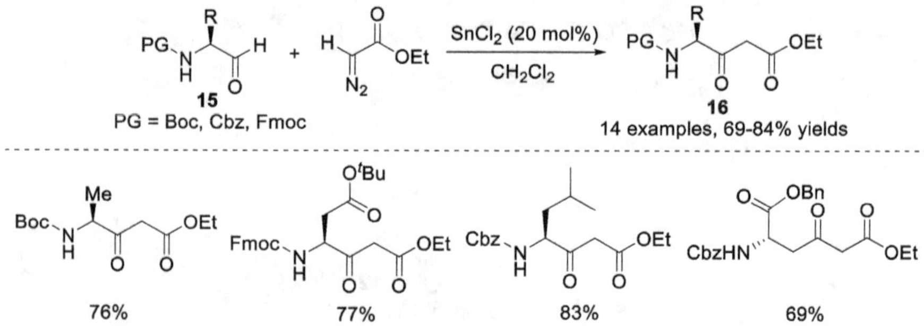

Scheme 11.4 $SnCl_2$-catalyzed synthesis of *N*-protected-γ-amino-β-ketoesters.

of carbonyl group, 1,3-dicarbonyl compounds are obtained as the final products through 1,2-H migration process. This transformation has been known for a long time and a recent application of this type of tandem reaction is the preparation of γ-amino-β-ketoesters **16** from amino aldehydes **15** and EDA (Scheme 11.4).[24]

Recent studies on this transformation have focused on the exploration of new catalysts. For example, the synthesis of α-substituted β-ketoesters from EDA and aldehydes improved by mesoporous silica was reported.[25] Another focus of the study is the stereoselectivity of the reaction. In this context, Maruoka and co-workers reported the Lewis acid catalyzed C–C bond formation between aldehydes and α-alkyl-α-diazocarbonyl compounds with chiral auxiliary group.[26] Upon screening a series of chiral auxiliaries, camphorsultam-derived substrates **17** provided good 1,2-H migration product **18** with high diastereoselectivity (Eq. (11.10)).

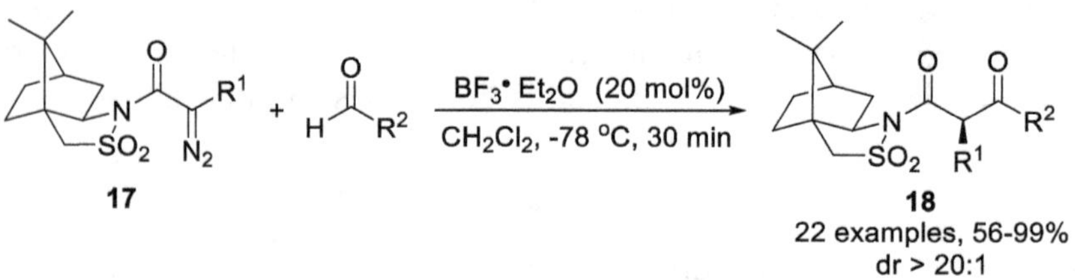

(11.10)

A two-directional homologation synthesis between diazoalkanes and formaldehyde is described by Kingsbury.[27] A series of symmetric ketones were obtained in good yields by Lewis acid-catalyzed reaction. The reaction was applied to the synthesis of (−)-dihydrocuscohygrine **19** (Scheme 11.5).

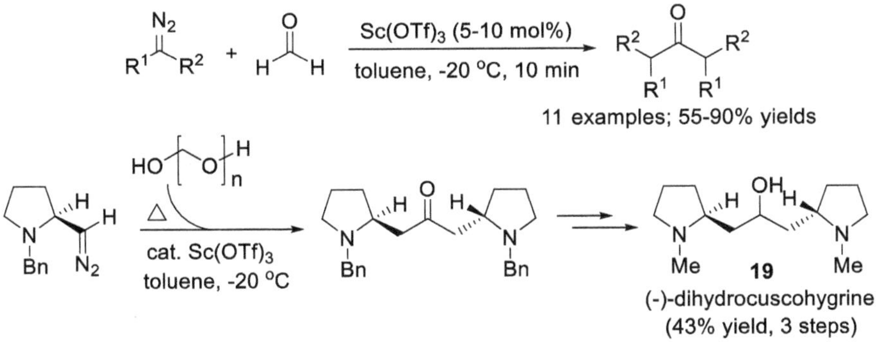

Scheme 11.5 Sc(III)-catalyzed synthesis of acyclic ketones of formaldehyde and diazoalkanes.

A series of β-ketonitriles were synthesized through BF_3-catalyzed nucleophilic addition between aromatic aldehydes with diazoacetonitrile **20**. From intermediate **21**, 1,2-H migration occurred exclusively and by-product *via* 1,2-Ar migration was not observed (Eq. (11.11)).[28]

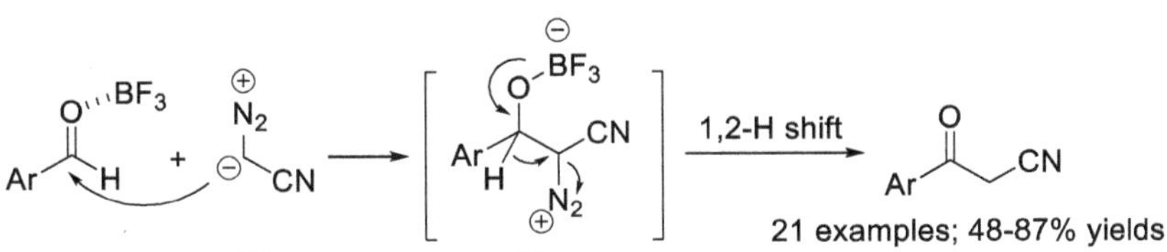

20 21

(11.11)

Ryu and Hwang reported Lewis acid-catalyzed reaction of diazo esters and cyclic enones.[29] The oxazaborolidinium ion **22** showed specific reactivity to afford formal C–H bond functionalization products in high yields (Scheme 11.6). The reaction mechanism is proposed to be initiated by 1,4-addition of unsaturated C=C bond to give **23**, followed by 1,2-H migration.

The 1,2-H shift of β-hydroxy-α-diazoesters could also be catalyzed by Pd(II) and Rh(II) catalysts.[30,31]

With *N*-tosylhydrazones as the precursors for non-stabilized diazoalkanes, the diazo nucleophilic addition/1,2-H shift reaction sequence was used for the synthesis of unsymmetrical ketones **25**.[32] By using excess amount of base, *N*-tosylhydrazones **24** derived from cyclic or open chain alkyl ketones were subjected to the reaction with aryl aldehydes, providing a wide scope of ketone derivatives with moderate to excellent yields (Scheme 11.7).

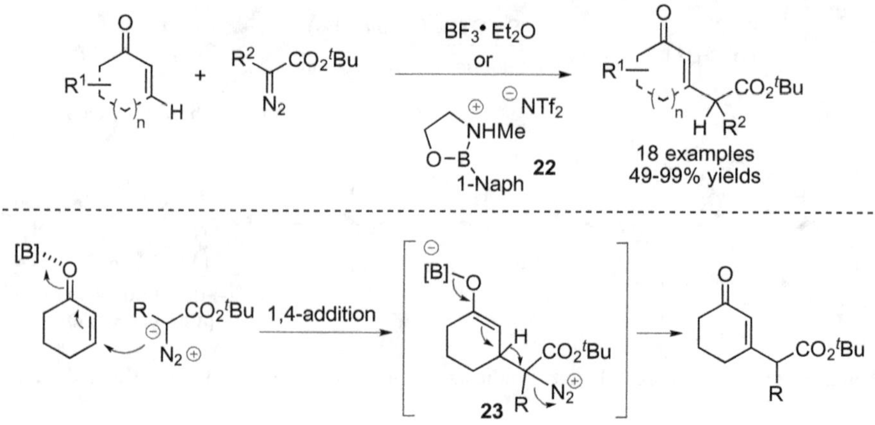

Scheme 11.6 Lewis acid-catalyzed reaction of cyclic enones with diazo esters.

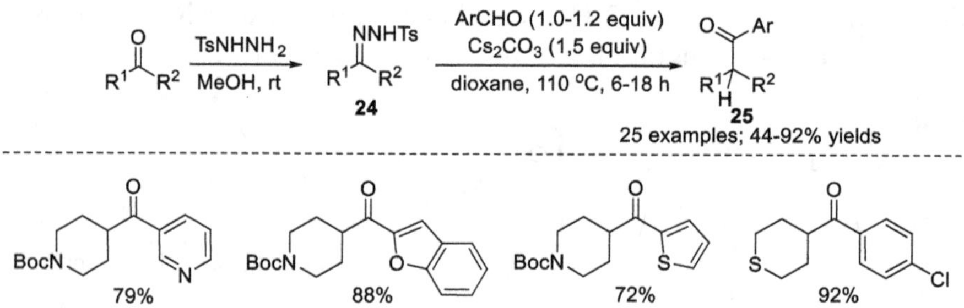

Scheme 11.7 Synthesis of ketones from aryl aldehydes and *N*-tosylhydrazones.

Notably, under certain conditions 1,2-Ar migration may override 1,2-H shift, which can be a unique strategy to generate C–C bonds. For example, Maruoka reported the construction of chiral quaternary center from aldehydes and aryldiazoacetate bearing chiral auxiliary group.[33] By using appropriate acid catalysts, 1,2-Ar migration could be optimized (Eq. (11.12)).

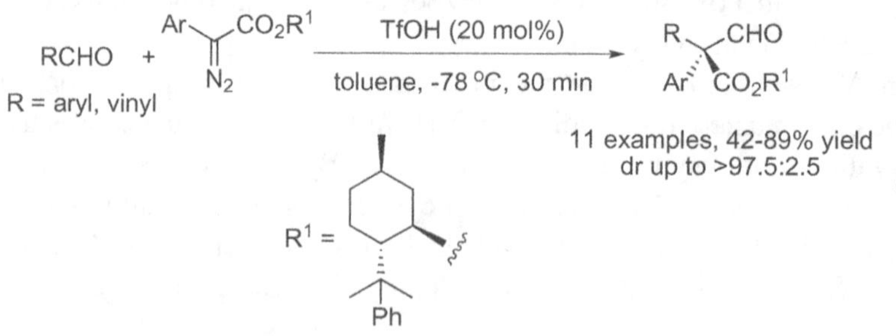

(11.12)

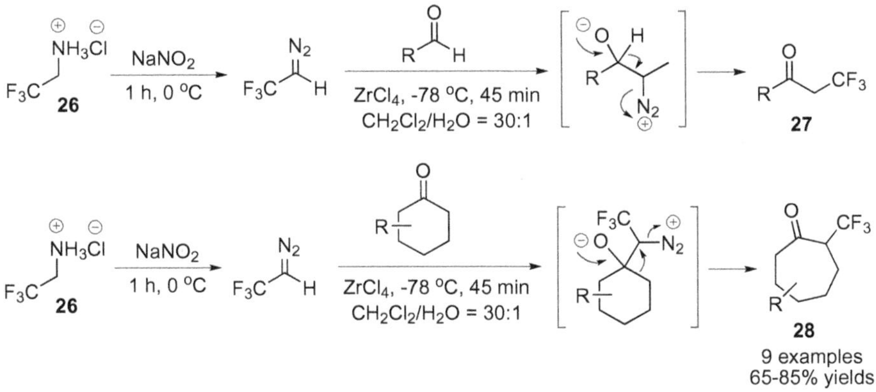

Scheme 11.8 Synthesis of α-CF_3 ketones from aldehydes or cyclic ketones with CF_3CHN_2.

The gold complex [IPrAu(NCMe)]BF_4-catalyzed reaction of benzaldehyde with EDA was reported to afford ethyl 3-hydroxy-2-phenyl-acrylate as the major product, which was generated *via* the 1,2-Ar migration.[34] Ethyl 3-oxo-3-phenylpropanoate was isolated as the side product from 1,2-H migration process. Moreover, for the diazo substrates, commercially available $TMSCHN_2$ was utilized in these types of reactions with oxazaborolidinium ion catalysts.[29,35]

Carreira and co-workers explored the C–C bond formation between aldehydes and CF_3CHN_2, generated *in situ* from trifluoroethylamine hydrochloride **26** and sodium nitrite.[36–38] The usefulness of CF_3CHN_2 as a synthon for introducing CF_3 group was demonstrated by the synthesis of CF_3-substituted ketones **27**.[36] The reaction with cyclic ketone derivatives gave ring-expansion α-CF_3 ketones **28** in good yields and with high diastereoselectivity (Scheme 11.8). For aliphatic aldehyde substrates, 1,2-H migration is favored over alkyl migration. In contrast, in the case of aromatic aldehydes the rate of aryl migration exceeds that of 1,2-H migration, followed by double homologation between CF_3CHN_2 with CF_3-substituted aldehydes.

Further application of CF_3CHN_2 was demonstrated by the synthesis of CF_3-substituted benzofuranols **30** from salicylaldehydes **29** (Scheme 11.9).[37] The reaction involves nucleophilic attack and 1,2-Ar migration, followed by Lewis acid-promoted cyclization. Carreira and co-workers also reported the homologative trifluoromethylation of acetals with CF_3CHN_2.[38]

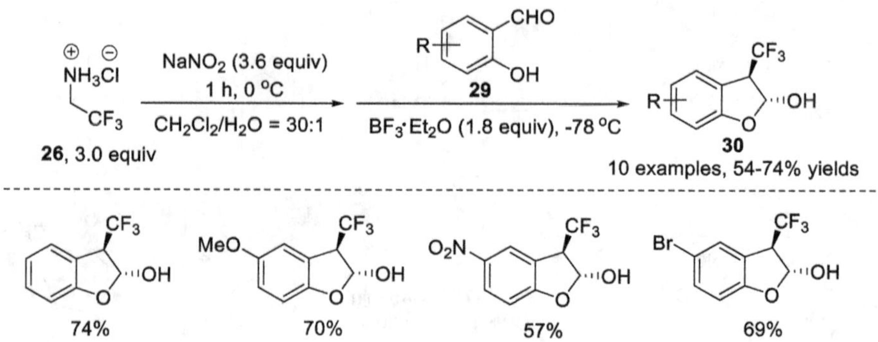

Scheme 11.9 Synthesis of CF_3-substituted benzofuranols from salicylaldehydes and CF_3CHN_2.

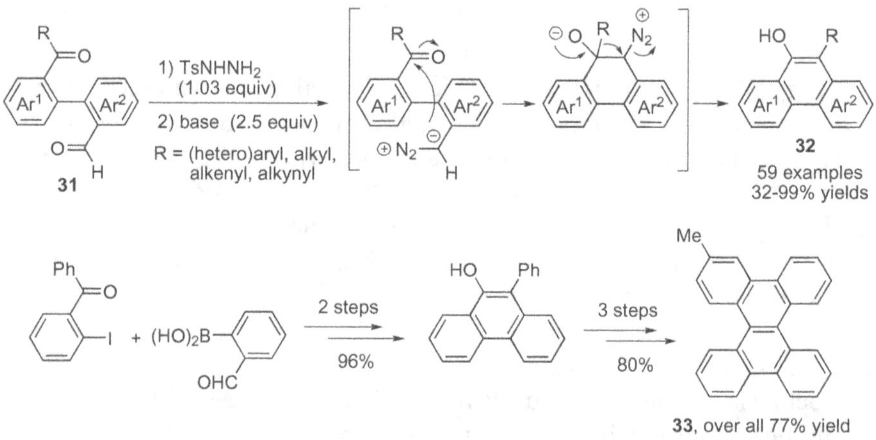

Scheme 11.10 Catalyst-free synthesis of phenanthrols and naphthols.

Intramolecular nucleophilic addition of diazo carbon to aldehydes is usually highly efficient. Wang and co-workers reported a method toward the synthesis of phenanthrols **32** and naphthols based on such process. With ketoaldehyde substrates **31**, the reaction starts from the selective formation of *N*-tosylhydrazone with the more reactive aldehyde carbonyl group. The *in situ* generated diazo moiety then attacks the ketone, followed by 1,2-R migration. The reaction shows a remarkable wide range of substrate scope and the yields are generally excellent (Scheme 11.10).[39] This protocol was applied in the synthesis of dibenzochrysene derivatives **33**.

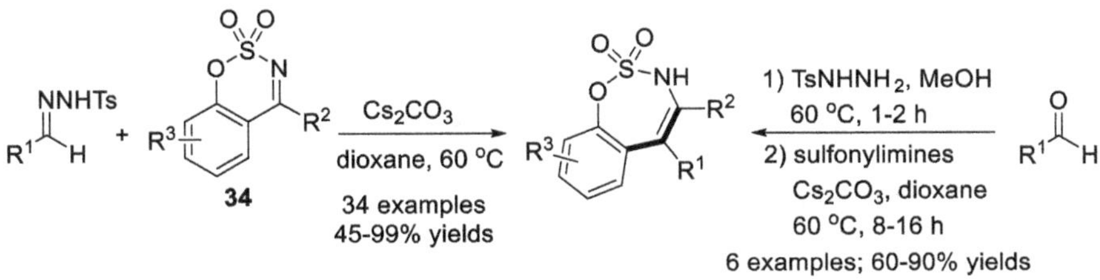

Scheme 11.11 Ring-expansion of six-membered sulfonylimines with *N*-tosylhydrazones.

Furthermore, a metal-free ring-expansion reaction of six-membered *N*-sulfonylimines **34** with *N*-tosylhydrazones derived from aldehydes has been recently reported by Liu and co-workers (Scheme 11.11).[40]

Catalytic asymmetric C–C bond formation between α-diazoesters and carbonyl compounds has attracted great attention in recent years. New advances were reported on the design and exploration of chiral ligands and Lewis acids.[41–48] In 2010, a significant breakthrough was made by Feng and co-workers by using only 0.05 mol% *N,N′*-dioxide **37** and **38** as the chiral ligands with scandium(III) as highly efficient Lewis acid catalyst in the reaction of diazo compounds with aldehydes.[41] This reaction, now recognized as Roskamp-Feng reaction, allows for the preparation of α-alkyl-β-ketoesters **34** in excellent yields and enantioselectivities. These types of ligands were further applied to the asymmetric homologation with α-diazoesters and α-ketoesters with $Y(OTf)_3$.[42] The corresponding products **35** with all-carbon quaternary center were obtained *via* 1,2-aryl/alkyl migration in good yields and with excellent enantioselectivities. In 2015, they further reported asymmetric intramolecular homologation of ketones with α-diazoesters, which resulted in the formation of chiral cyclic α-aryl/alkyl β-ketoesters **36** in high yields and enantioselectivities (Scheme 11.12).[43]

Ryu and co-workers explored the oxazaborolidinium ion **41** as the catalyst in the reaction of α-alkyl-α-diazoesters with aldehydes.[44,45] Interestingly, the migratory aptitude was found affected remarkably by the solvents: toluene favored 1,2-H shift to give **39**, while polar solvent propionitrile led to 1,2-aryl migration to give **40**. With chiral oxazaborolidinium ion catalyst, α-alkyl-β-ketoesters[44] and the aldehydes bearing all-carbon α-quaternary center[45] were obtained, respectively, in good yields and with excellent enantioselectivities (Scheme 11.13).

In connection to the previous studies,[29] catalytic asymmetric C–C bond formation of cyclic enones with α-diazoesters could be achieved by chiral oxazaborolidinium ion **43a** or **43b**.[46] β-Functionalized cyclic

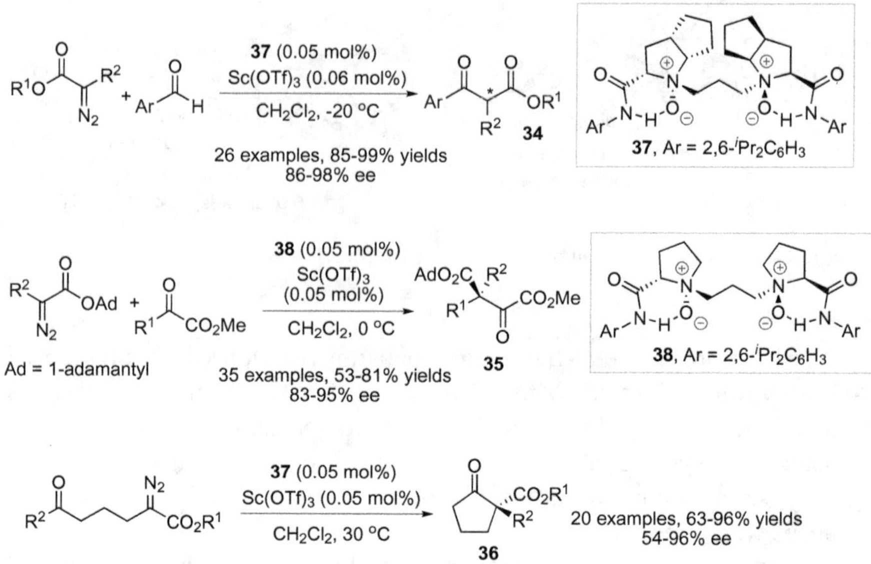

Scheme 11.12 Lewis acid/N,N'-dioxide-catalyzed C–C bond formation *via* 1,2-migration.

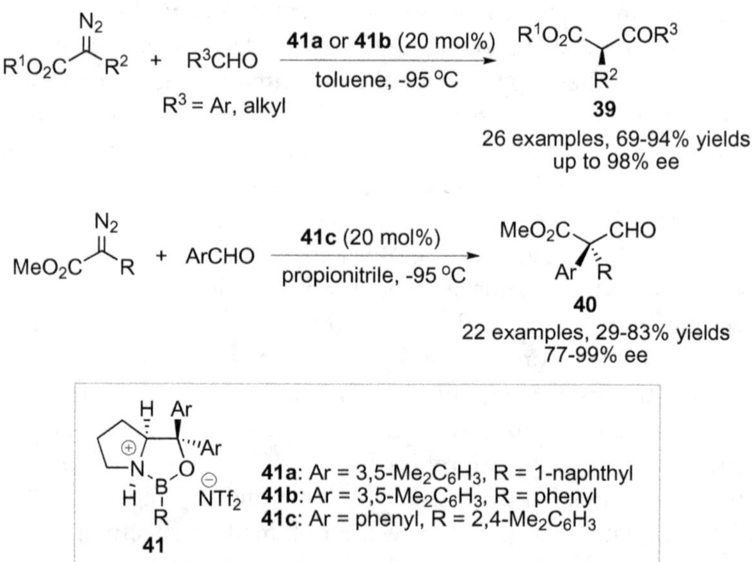

Scheme 11.13 Oxazaborolidinium ion-catalyzed asymmetric reaction between α-alkyl-α-diazoesters with aldehydes.

enones **42** were obtained in excellent yields and enantioselectivities (Eq. (11.13)). The method was applied to the stereoselective formal synthesis of (+)-epijuvabione.

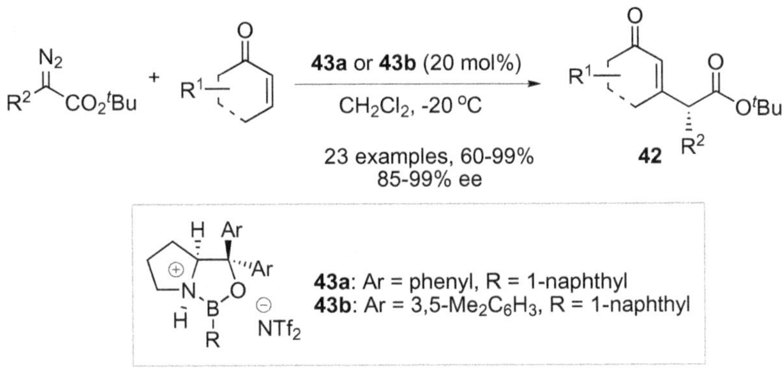

(11.13)

Moreover, Ryu and co-workers recently developed the catalytic asymmetric Roskamp reaction between aryldiazoalkanes with both aromatic and aliphatic aldehydes (Eq. (11.14)).[47] In the presence of Lewis acid catalyst **44**, this method provided an efficient route to prepare α-tertiary aryl ketones **45** in good yields (up to 94%) and with excellent enantioselectivities (up to 99% *ee*).

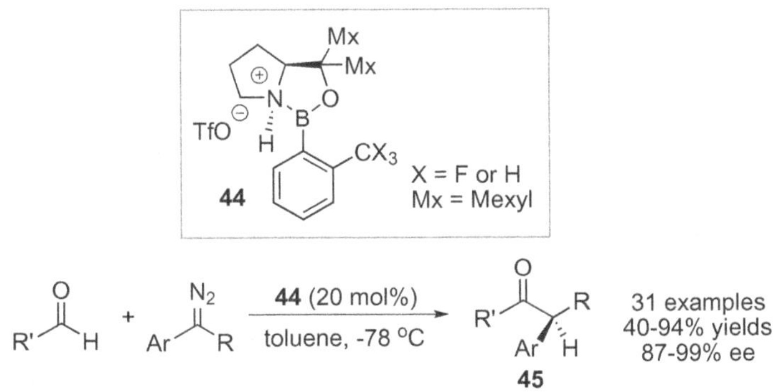

(11.14)

Besides, the same group also reported enantioselective reaction of α-aryl diazo Weinreb amides catalyzed by oxazaborolidinium ion **46**. The chiral β-ketoamide products were reduced *in situ* with $Zn(BH_4)_2$ to give *syn*-α-aryl-β-hydroxy Weinreb amides **47** (Eq. (11.15)).[48]

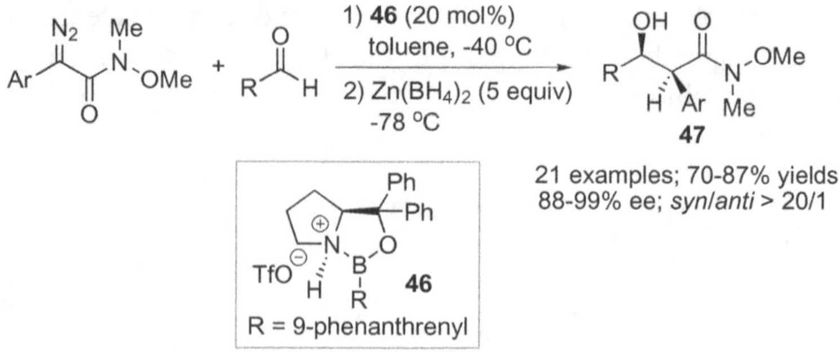

(11.15)

Homologation of cyclic ketones with diazo compounds is a useful strategy for ring-expansion of ketones. Kingsbury and co-workers reported the synthesis of a series of α-tertiary and α-quaternary cyclic ketones with diazomethane derivatives at room temperature.[49] Furthermore, they reported regioselective ring expansion of aryl cyclobutanones **48** with $TMSCHN_2$ catalyzed by scandium(III) acids.[50] The corresponding β-ketosilanes **49** or enolsilanes **50** were obtained with good regio- and diastereocontrol, which provided cyclopentanones **51** in high yields upon hydrolysis (Scheme 11.14).

Maruoka and co-workers reported the stereoselective construction of seven-membered cyclic ketones bearing all-carbon quaternary center from α-diazoesters with high diastereoselectivity.[51] Cyclic ketones derivatives and oxa-analogues were obtained in good yields, catalyzed by boron trifluoride at low temperature.

In addition to the previous synthetic method of ethyl 3-hydroxy-2(1*H*)-oxo-quinoline-4-carboxylate,[52] the one-pot protocol has been developed through the catalysis by NHC-dirhodium(II)/DBU.[53] The ring-expansion transformation from isatins with EDA afforded the products in good yields through nucleophilic addition and the subsequent 1,2-Ar migration.

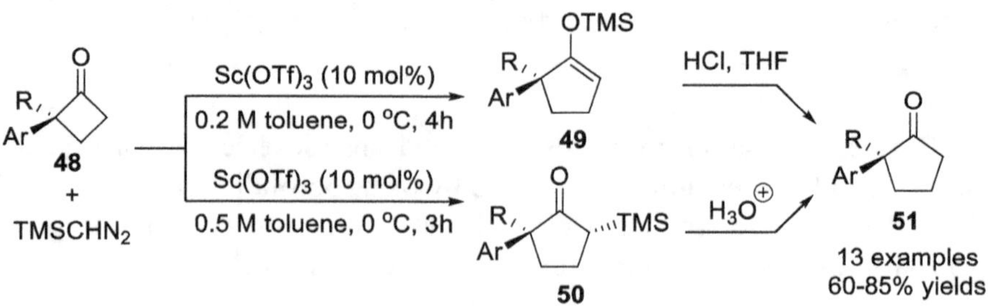

Scheme 11.14 Ring expansion of aryl cyclobutanones.

Cyclooctynes have shown wide applications in chemical biology because of their unique reactivity toward azides. Moreover, difluorinated cyclooctynes are particularly useful reagents in copper-free click chemistry for labeling azido-biomolecules. Bertozzi and co-workers reported a new approach for accessing difluorinated cyclooctynes **53** *via* the regioselective ring-expansion of di-fluorinated ketone **52** with $TMSCHN_2$ (Eq. (11.16)).[54] Notably, for the substrates of tri- and tetra-fluorinated cycloheptanones, oxirane formation was observed because the 1,2-migration of electron-deficient C–CF bond is not favored.

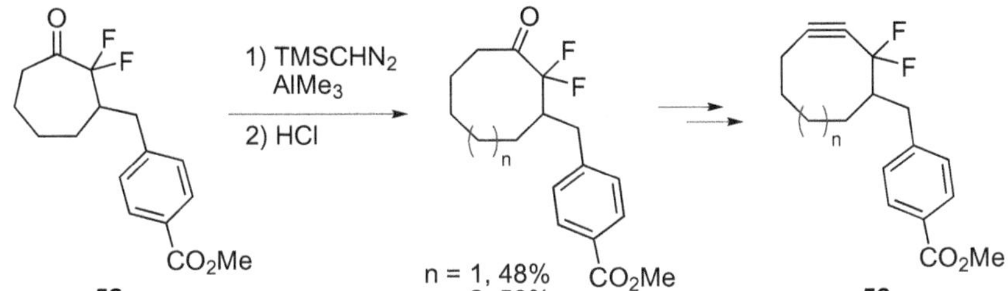

52 53

(11.16)

Catalytic enantioselective ring-expansion reactions from diazo compounds have been explored by the groups of Maruoka, Kingsbury and Feng. Various chiral Lewis acids and ligands are subjected to enhance the stereoselectivity. Maruoka first attempted the desymmetrizing asymmetric ring expansion by attaching a chiral auxiliary on diazo moiety.[55] Functionalized cycloheptanones were obtained in good yields and with excellent diastereoselectivities from *N*-α-diazoacetyl camphorsultam and cyclohexones. They have also succeeded in developing chiral aluminum Lewis acid-complex catalyzed desymmetrizing asymmetric ring expansion.[56] After screening the silyl substituent groups, 3,3′-bis(trimethylsilyl)-BINOL **54** was chosen as chiral ligand for this transformation. Seven-membered cyclic ketone derivatives bearing all-carbon quaternary center were obtained in good yields with high enantioselectivities (Scheme 11.15).

Moreover, they reported the desymmetrizing asymmetric ring-expansion reaction of α-hydroxy-α-diazoesters catalyzed by chiral dicarboxylic acid **55**.[57] Chiral cycloheptanones were obtained in moderate to good yields and with enantioselectivities, followed by decarboxylation (Scheme 11.16).

Kingsbury and co-workers reported the catalytic enantioselective homologation from diazoalkanes with $Sc(OTf)_3$ as the Lewis acid.[58] It was found that tris(oxazoline) **56** gave a better enantioselective control

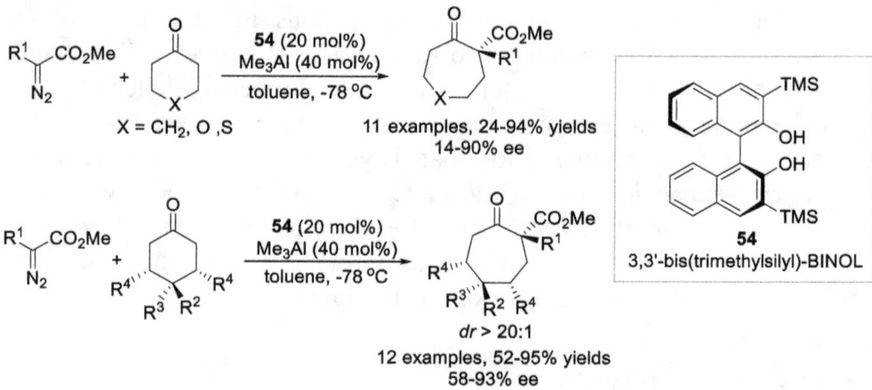

Scheme 11.15 Asymmetric ring expansion through desymmetrization catalyzed by chiral aluminum Lewis acid.

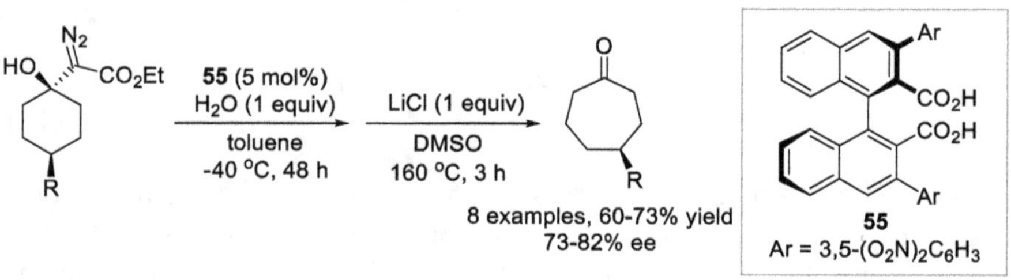

Scheme 11.16 Asymmetric ring expansion and decarboxylation catalyzed by chiral dicarboxylic acid.

than the bis(oxazoline) ligand (Eq. (11.17)). Notably, the catalyst loading of this reaction could be reduced to 0.5 mol% when the reaction was scaled up to 5 mmol.[59]

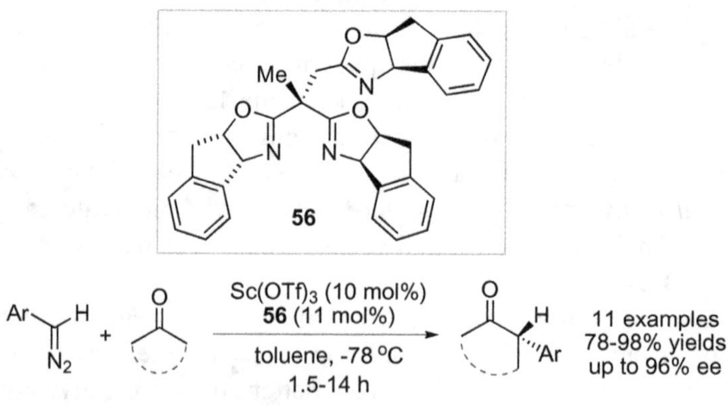

(11.17)

On the other hand, Feng and co-workers reported catalytic asymmetric ring-expansion reaction of isatins **57** with α-alkyl-α-diazoesters.[60] Highly functionalized C4-quaternary 2-quinolones **58** were obtained in high yields and enantioselectivities. The catalyst loading of the reaction could be reduced to 0.05–0.5 mol% of $Sc(OTf)_3$ with *N,N′*-dioxide ligand **59** (Eq. (11.18)).

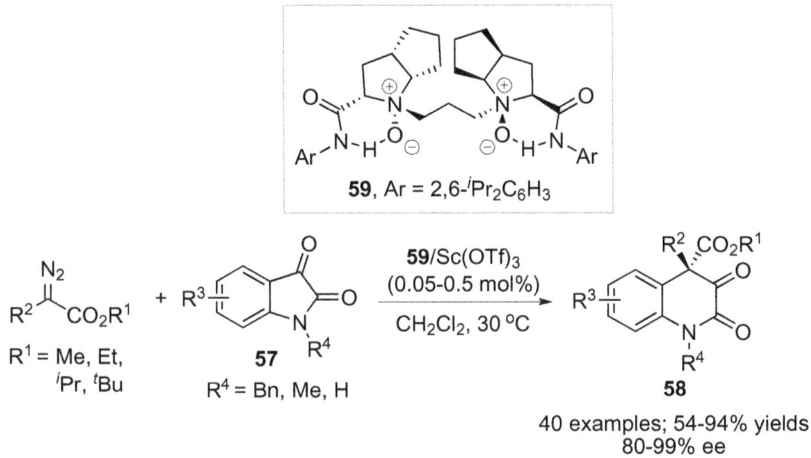

(11.18)

1,2-Migration is commonly encountered in transition-metal-catalyzed reactions of α-diazocarbonyl compounds. Although 1,2-migration is usually considered as a side reaction that competes with other typical metal carbene reactions such as X–H insertion and cyclopropanation, it has some useful applications in synthesis, e.g. Rh(II)-catalyzed reaction of β-hydroxy-α-diazocarbonyl compounds is a useful method to synthesize 1,3-dicarbonyl compounds.

Zhu and co-workers developed an approach to α-halo ketones **61** through halogenation/semipinacol rearrangement of α-diazo alcohols **60** in the presence of Lewis base.[61] The mechanism was proposed to involve electrophilic halogenation of diazo carbon, followed by 1,2-Ar migration of **62** (Scheme 11.17).

When there are different migrating groups adjacent to the carbene center, the question arises as to which group will migrate preferentially (Scheme 11.18). Previous experimental data have undoubtedly showed that 1,2-H migration is a highly favorable process. The migratory aptitude follows the order: hydrogen > phenyl > alkyl. The studies by Wang and co-workers demonstrated that the 1,2-migratory aptitude was significantly

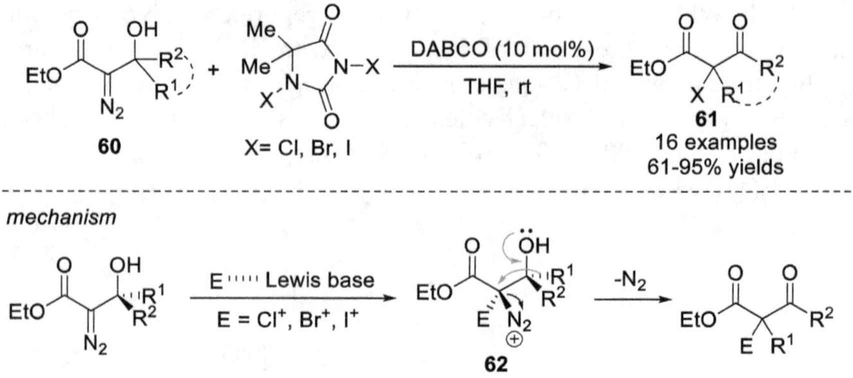

Scheme 11.17 Halogenation/semipinacol rearrangement of α-diazo alcohols.

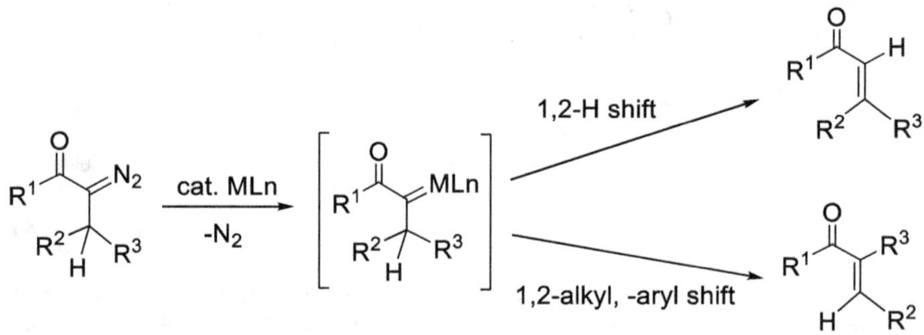

Scheme 11.18 Transition-metal-catalyzed 1,2-migration of diazocarbonyl compounds.

affected by the electronic and steric effects of the non-migrating substituents (the bystander groups).[62–67] The details of these studies are summarized in Chapter 7.

Gold-catalyzed 1,2-shift and related reaction of diazo compounds have been achieved by Liu and co-workers[68,69] They explored the Au(I)/ $AgNTf_2$-catalyzed Povarov reaction from vinyl α-diazoesters **63** or **65** with *N*-aryl imines. In the case of reaction with **65**, cycloadducts **66** containing diazo groups were obtained as major products with good yields.[68] Upon reaction optimization, they utilized HOTf as the Lewis acid to afford the cycloadducts **66** stereoselectively, followed by sequential gold-catalyzed ring expansion to generate seven-membered azacycles **67** (Scheme 11.19).

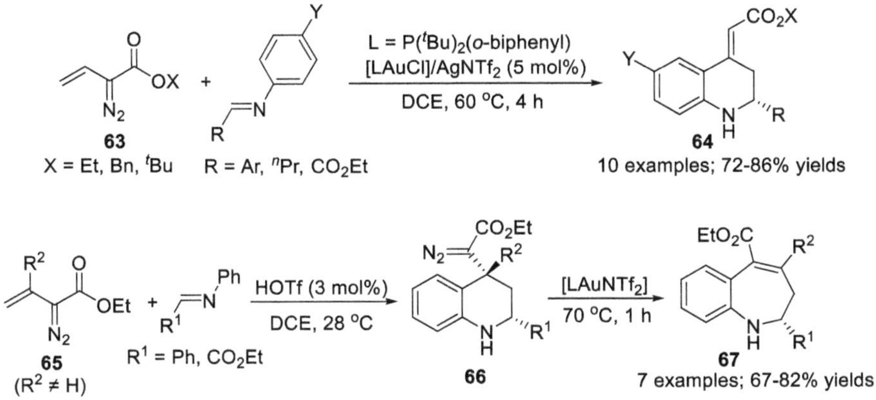

Scheme 11.19 Povarov reaction and ring expansions promoted by Au(I) catalysts.

Furthermore, Liu and co-workers reported Au(I)-catalyzed C–C bond formation between vinyldiazocarbonyl compounds **68** and acetals **69**.[69] Substituted alkyl 3,5-dimethoxy-5-pent-2-enoates were obtained in good yields and with excellent stereocontrol. The authors have proposed that the reaction proceeds through the formation of oxonium cation species **71**, followed by the formation of intermediate 1,3-dimethoxy-4-diazo ester **73** through Prins-type reaction. The final product is formed through 1,2-H shift of Au(I) carbene species **74** (Scheme 11.20).

11.3 Mannich-type reaction

Upon the activation of imines by Lewis acid, nucleophilic addition of diazo compounds to C=N bond leads to the formation of Mannich-type products. The addition of α-diazo imides to azomethines was promoted by using one equivalent of HOTf. A series of *syn*-glycolate Mannich products **76** were obtained in excellent isolated yields with high diastereoselectivities (Eq. (11.19)).[70]

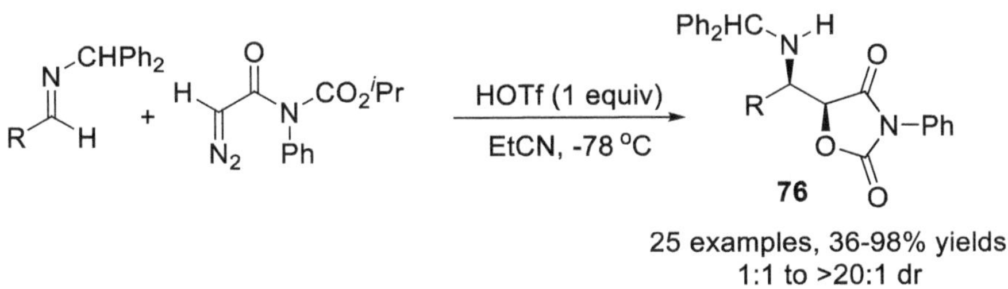

(11.19)

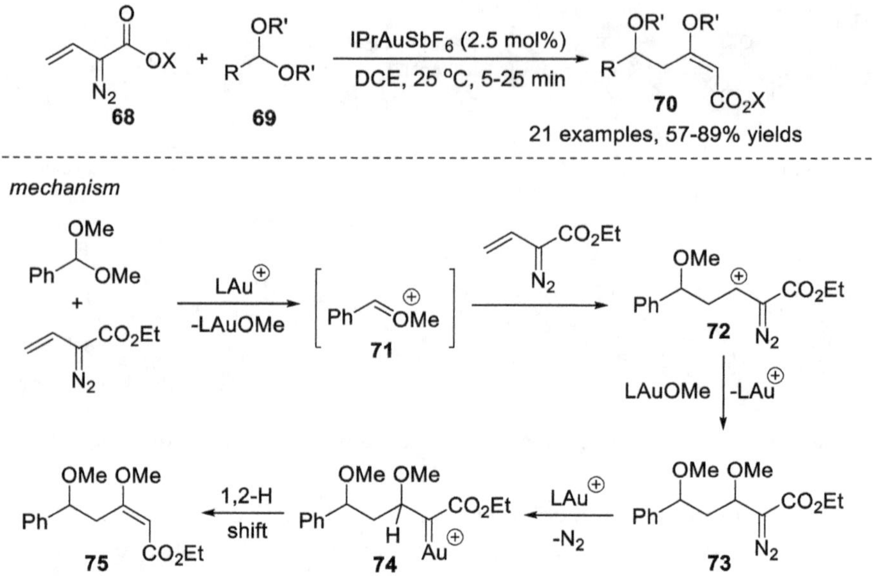

Scheme 11.20 Au(I)-catalyzed reaction of alkenyl diazocarbonyl compounds with acetals.

Furthermore, three-component Mannich reaction of 2-aminoazines, aromatic aldehydes and α-diazoesters has been developed with yttrium triflates as the catalyst. The obtained diazoesters **77** can be efficiently converted into a series of heterocycles (Scheme 11.21).[71]

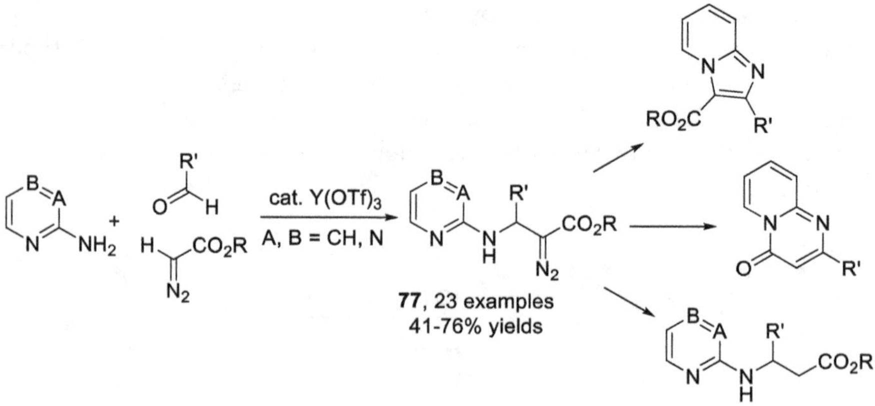

Scheme 11.21 Three-component coupling reaction to form β-amino-α-diazo compounds.

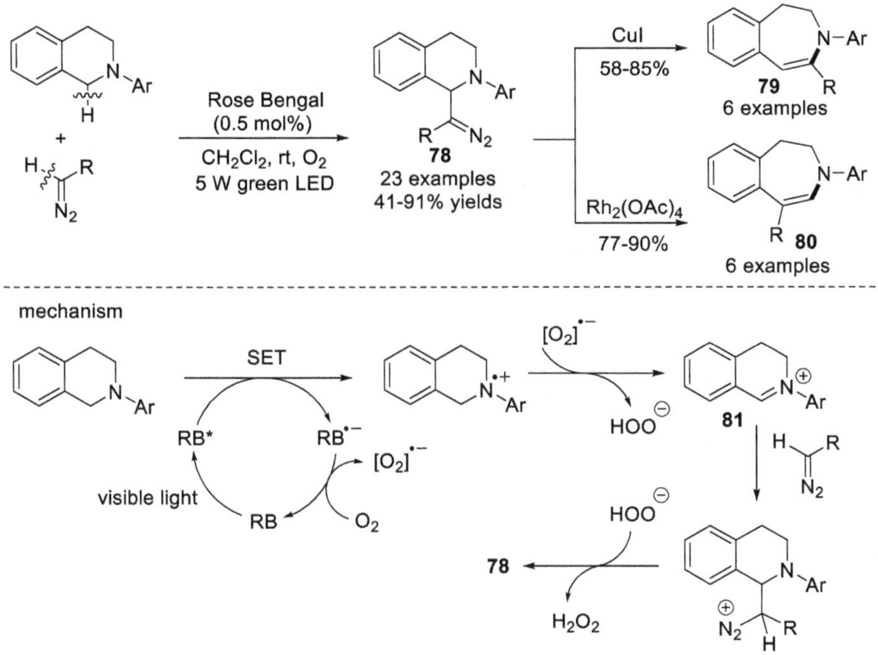

Scheme 11.22 Visible light-induced CDC reaction and catalytic ring expansion.

In 2013, Zhou and co-workers reported visible light-induced cross dehydrogenative coupling (CDC) reaction between amines and α-diazocarbonyl compounds.[72] This transformation is initiated by SET through the reductive quenching of the excited photo sensitizer, followed by hydrogen abstraction to afford iminium intermediate **81**. Nucleophilic addition of the diazo substrates to **81** followed by deprotonation affords β-amino-α-diazo adducts **78**. These products are then subjected to the catalyst-dependent ring-expansion reaction to afford **79** and **80** with high regioselectivity, respectively (Scheme 11.22).

Subsequently, the same group disclosed the Ag(I)-catalyzed three-component one-pot approach from 2-alkynylbenzaldehydes **82**, amines, and diazocarbonyl compounds employed as nucleophiles (Scheme 11.23).[73] At 100°C, 3-benzazepine skeleton **84** could be assembled based on a cascade reaction involving Ag(I)-catalyzed imine-yne cyclization/nucleophilic addition/1,2-aryl shift. On the other hand, at room temperature the reaction provided the diazo-containing dihydroisoquinolines **83** with high efficiency.

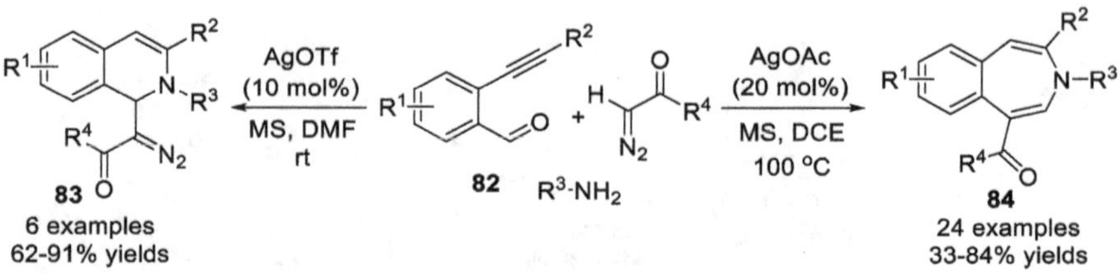

Scheme 11.23 Ag(I)-catalyzed three-component reaction to obtain dihydroisoquinolines or 3-benzazepines.

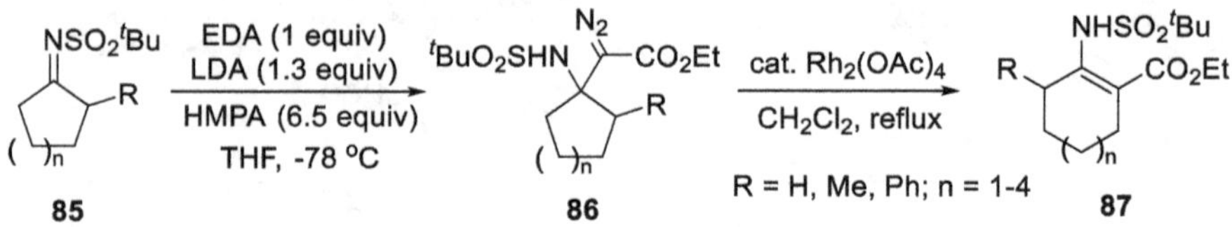

Scheme 11.24 Nucleophilic addition of ethyl diazoacetate to *N-tert*-butylsulfonyl cycloketimines.

Wang and co-workers demonstrated that the condensation of acyldiazomethanes could be carried out with imines.[74–77] Thus, the nucleophilic addition of ethyl diazoacetate derived anions to *N-tert*-butylsulfonyl cycloketimines **85** afforded diazo products **86**. The addition products **86** could be converted into cyclic enamines **87** *via* $Rh_2(OAc)_4$-catalyzed ring expansion (Scheme 11.24).[77] Besides, Lewis acid-catalyzed nucleophilic addition of diazoacetate to iminium ion has also been reported.[78,79]

In a further study, Wang and co-workers reported a stereoselective C=N addition of diazo group-containing anions with a chiral auxiliary approach (Scheme 11.25).[80] Thus, the anion derived from the diazoamide **88**, which bears Evans' chiral oxazolidinone auxiliary, was added to *N*-tosylimine with high diastereoselectivity. Under the optimized reaction conditions, high diastereoselectivity could be achieved for a series of imine substrates. The chiral auxiliary could be removed from the addition products **89** and the products were then converted into *syn*- or *anti*-α-hydroxy-β-amino acid derivatives (**90** or **91**).

The asymmetric catalytic version of this C=N bond addition was achieved by Terada and co-workers with a chiral binaphthol monophosphoric acid **92**,[81] and by Maruoka and co-workers by using a chiral

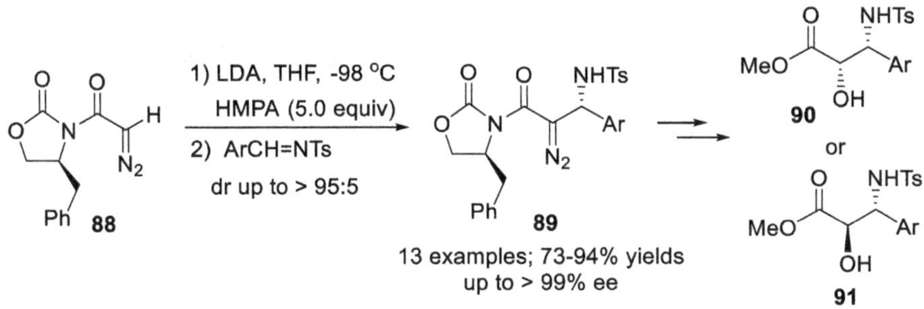

Scheme 11.25 Diastereoselective Mannich reaction by using Evans' chiral oxazolidinone auxiliary.

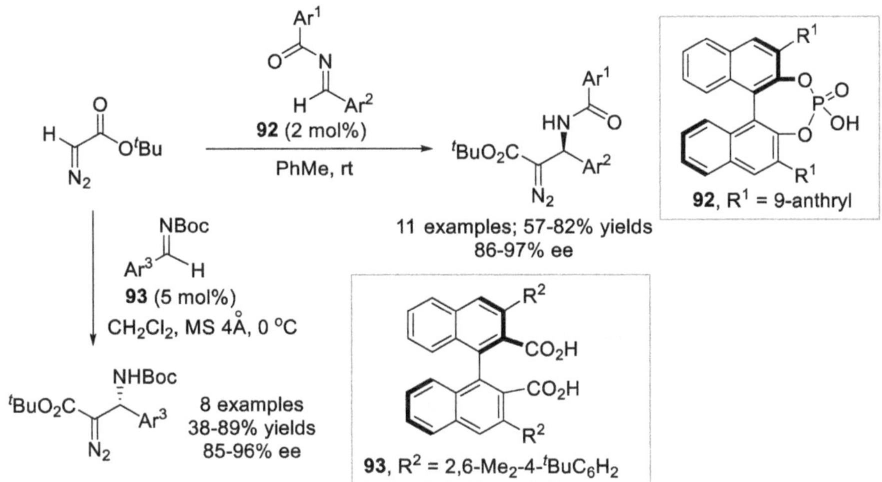

Scheme 11.26 Catalytic Asymmetric Mannich addition by using chiral binaphthol monophosphoric acid or chiral dicarboxylic acid.

dicarboxylic acid **93** (Scheme 11.26).[82] Moreover, the scope of this type of asymmetric catalysis was further expanded with other diazo substrates, including those bearing phosphonate and sulfone moieties.[83–87]

Maruoka and co-workers developed a three-component asymmetric Mannich reaction with acyclic azomethine imines, which was generated *in situ* by condensation of *N′*-benzylbenzoylhydrazide with aldehydes in the presence of axially chiral dicarboxylic acid **94**.[85] This protocol not only avoids the preparation of imines but also expands the substrate scope to alkyl, aryl and alkynyl aldehydes with excellent functional group tolerance (Eq. (11.20)).

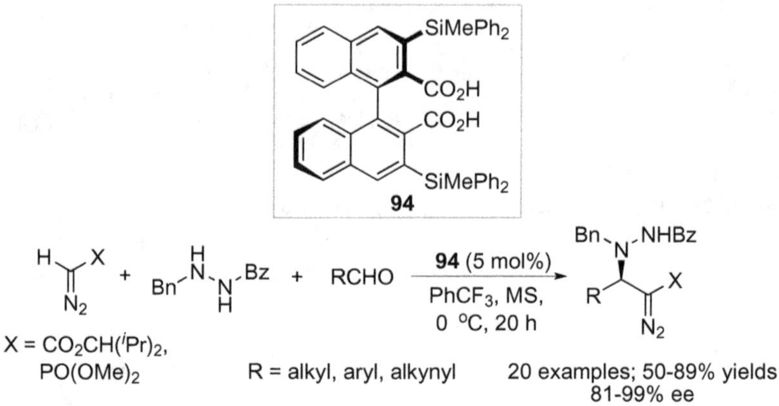

(11.20)

11.4 Darzens reaction

Although α-diazocarbonyl compounds undergo aldol-type reaction with aldehydes in most cases, the formation of epoxides (Darzens-type reaction) has also been observed. Gong and co-workers reported the first asymmetric catalytic Darzen-type reaction between diazoacetamides and aldehydes.[88] By using titanium/BINOL catalytic system, *cis*-glycidic amides **95** were obtained in good yields and with high enantioselectivity (Scheme 11.27).

Furthermore, they explored asymmetric Darzen reaction of diazoacetamides with alkyl or aryl aldehydes catalyzed by an air-stable chiral zirconium complex. In the presence of 3,3′-diiodobinaphthol as chiral Lewis acid, *cis*-glycidic amides were obtained in high yields with excellent enantioselectivities.[89] The titanium(IV)/(+)-pinanediol system was also reported to efficiently catalyze asymmetric Darzens reaction.[90]

Besides the diazoacetamides, α-aryl-α-diazoesters and aldehydes undergo Darzen-type reaction by using Ag(I)-*N*-heterocyclic carbene complex as the catalyst.[91] Catalytic active Ag(I) species were generated *in situ* from AgOTf with NHC-silver complex (IPrAgCl). A series of tri-substituted oxiranes **96** were prepared in moderate to good yields (Eq. (11.21)).

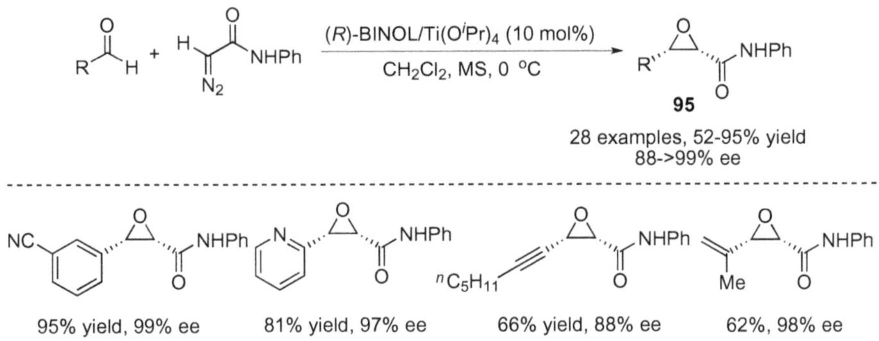

Scheme 11.27 Catalytic asymmetric Darzens-type reaction between diazoacetamides and aldehydes.

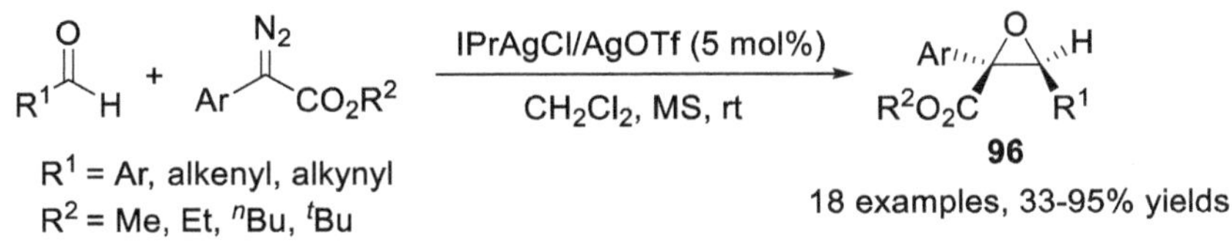

(11.21)

In addition to the synthesis of oxirane skeletons, the corresponding episulfide derivatives could also be obtained through nucleophilic addition of diazo substrate to C=S bonds. Feringa group applied this strategy to the synthesis of functionalized episulfide from diazofluorenone and thioketones.[92] The products were further utilized in the synthesis of molecules with light-driven rotary motors.

11.5 Acid-catalyzed aziridination

Similar to the Darzens-type reaction, the corresponding reaction of diazo compounds with aldimines affords aziridines **98**.[93–96] The formation of aziridines is frequently accompanied by Mannich-type by-products **99** or enamines **100**, generated from the intermediate **97** through protonation or 1,2-H shift (Scheme 11.28). The judicious choice of appropriate acid catalysts and *N*-protecting groups is crucial for the efficient aziridination.

Carreira and co-workers reported the synthesis of trifluoromethyl-substituted aziridines with *in situ* generated CF_3CHN_2 and *N*-PMP

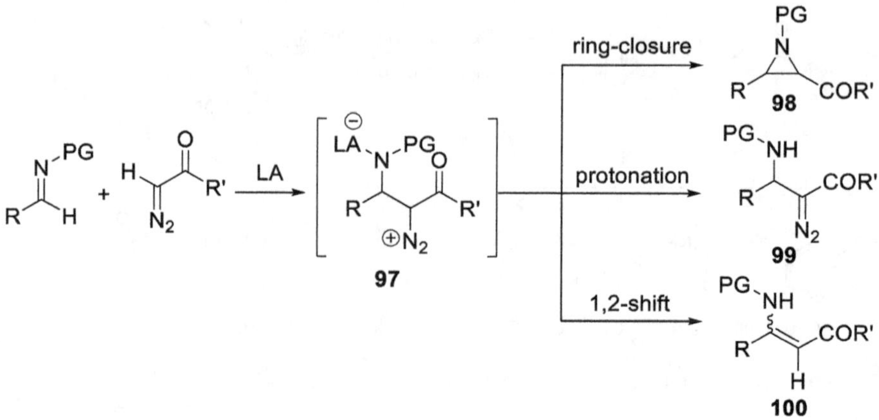

Scheme 11.28 Lewis acid-catalyzed reaction of α-diazocarbonyl compounds and aldimines.

glyoxal imines, with boron trifluoride as the catalyst.[97] The CF_3-substituted *N*-protected aziridines **101** were prepared in good yields with good *cis*-selectivity (Eq. (11.22)).

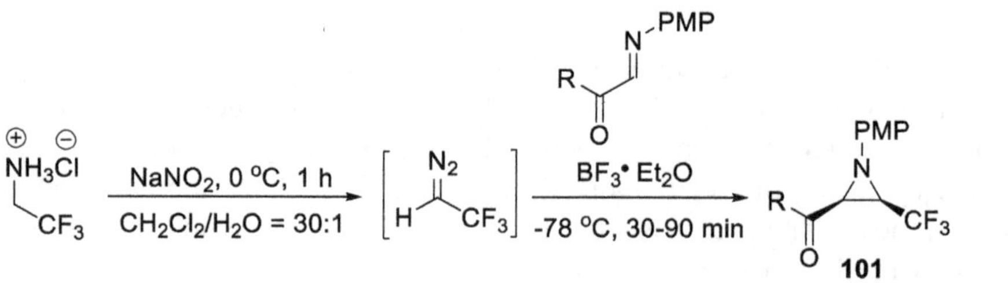

9 examples, 47-78% yield
cis:*trans* = 11:1-19:1

(11.22)

Stereoselective synthesis of aziridines were also explored with both chiral auxiliary and asymmetric catalysis strategies.[98–100] Maruoka and co-workers reported the synthesis of *tri*-substituted aziridines from α-substituted-α-diazocarbonyl compounds bearing camphorsultam as the auxiliary group and *N*-Boc/Cbz imines. Catalyzed by HOTf, a series of *trans*-tri-substituted aziridines **103** were prepared in good yields and with excellent diastereoselectivity.[98] The same reaction with *N*-α-diazoacetyl camphorsultam and *N*-Boc-α-ketimino esters also occurred with excellent diastereoselectivity. Furthermore, one-pot sequential aziridination/ring

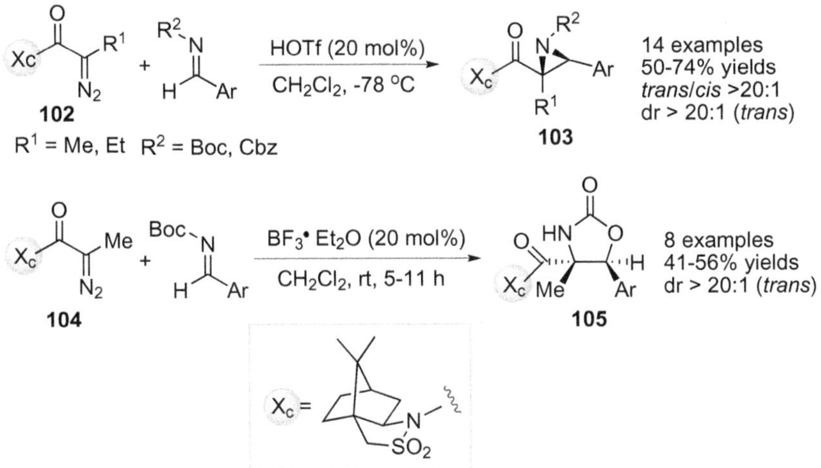

Scheme 11.29 Diastereoselective synthesis of aziridines with camphorsultam as the chiral auxiliary.

rearrangement was carried out to afford oxazolidinone derivatives **105**, which serve as the precursors for α,α-disubstituted β-hydroxy-α-amino acids (Scheme 11.29).[99]

Recently, boron trifluoride-catalyzed diastereoselective synthesis of *di*-substituted aziridines was reported with chiral auxiliary on imines moiety.[100] This reaction was carried out with α-diazocarbonyl compounds and *N*-(*O*-pivaloylated β-D-galactosyl)imines to give aziridines with high diastereoselectivity and moderate to good yields.

The corresponding catalytic asymmetric aziridination with imine and diazo compounds has been extensively explored by the groups of Wulff, Maruoka and others. New developments have been achieved on the design and synthesis of chiral Lewis or Brønsted acid for these types of reactions. Among these developments, Wulff and co-workers developed the efficient chiral catalysts by using borate $B(OPh)_3$ with biaryl phenyl ligands (*s*)-VANOL **106** and (*s*)-VAPOL **107** (Figure 11.2).[101–104] Detailed mechanistic studies indicated that complex **108** is the reactive species in the reaction system.

As shown in Scheme 11.30, the aziridination catalyzed by **106** or **107** borate catalyst afforded the products **109** with high enantioselectivity. Notably, the structure of the *N*-substituents of the imine had significant impact on the reaction. MEDAM was proved to be the best protecting

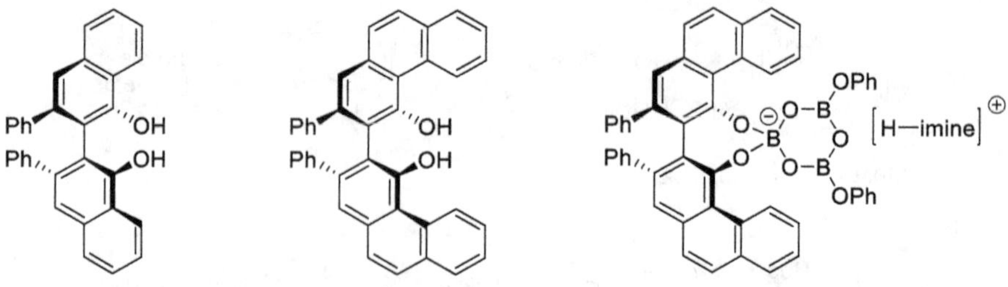

106, (*s*)-VANOL **107**, (*s*)-VAPOL **108**

Figure 11.2 Structure of (*s*)-VANOL and (*s*)-VAPOL.

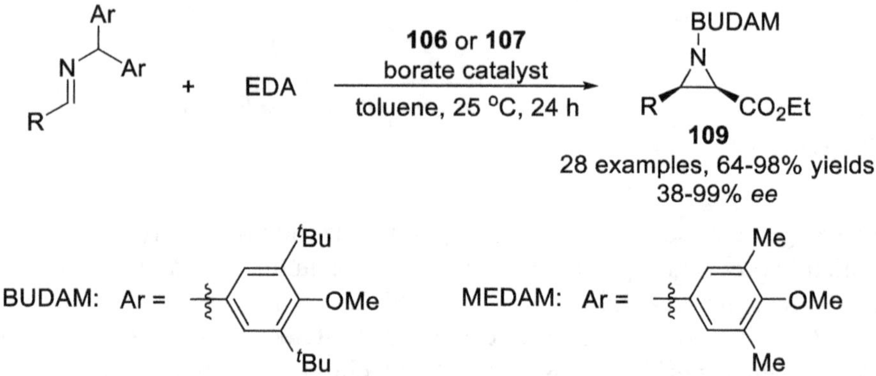

Scheme 11.30 Catalytic asymmetric aziridination with *N*-PG imines and EDA.

group for both aromatic and aliphatic imines. The aziridination products were obtained with high enantioselectivity and *cis*-selectivity (Scheme 11.30).[105] The mechanism of the reaction has been investigated by kinetic isotope effects and theoretical calculations, which reveal that the reversible formation of diazonium ion intermediate precedes the ring closure (the rate-limiting step).[106]

To demonstrate the practical usefulness of this method, the catalytic asymmetric synthesis of aziridinations has been subjected to gram scale experiment with high efficiency.[107] The aziridination products were hydrolyzed to give the aziridine-2-carboxylic acids, which were used in the asymmetric synthesis of β-lactams with oxalyl chloride.[108]

The substrate-controlled stereoselective synthesis of both *cis*- and *trans*-aziridines was then demonstrated with the same imines.[109] Compared with the

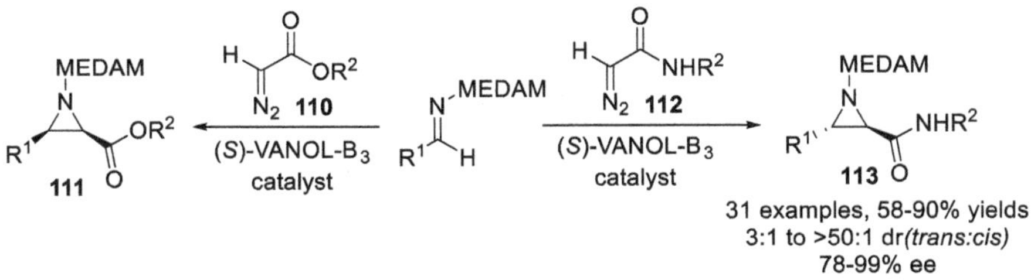

Scheme 11.31 Catalytic asymmetric synthesis of both *trans*- and *cis*-aziridines.

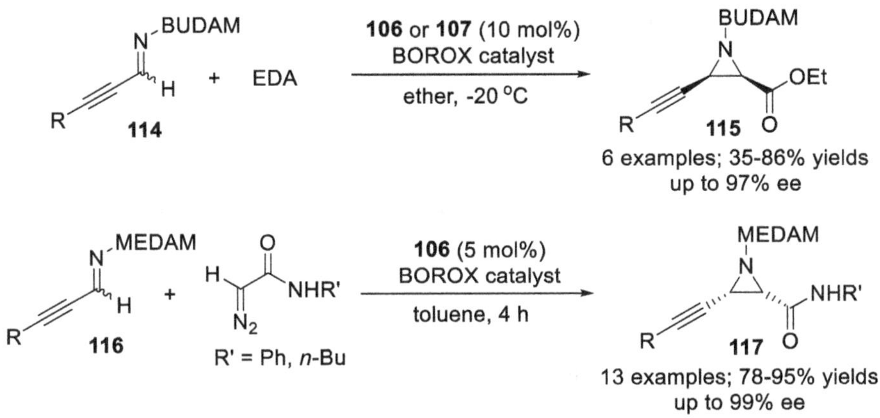

Scheme 11.32 Aziridination of alkynyl imines with diazoesters or diazoacetamides.

synthesis of *cis*-aziridines **111** from α-diazoesters **110**, this reaction with α-diazoacetamides **112** as the substrates afforded *trans*-aziridine carboxamides **113** with excellent enantioselectivity (Scheme 11.31). This interesting switch of diastereoselectivity is attributed to the hydrogen bonding between the amidic hydrogen of diazoacetamides with the (*s*)-VANOL-catalyst. This hypothesis can be corroborated with the formation of *cis*-isomer from tertiary diazoacetamide as substrates.[103] Moreover, the double stereodifferentiation in catalytic asymmetric aziridination with α-chiral amine-derived imines was studied with diazoesters and diazoacetamides.[110]

Very recently, Wulff and co-workers reported the aziridination of alkynyl imines **114** or **116** with diazoesters or diazoacetamides catalyzed by chiral boroxinate or BOROX catalyst.[111] Interestingly, the two diazo substrates afforded different enantiomers of the *cis*-aziridine products **115** and **117** from the same enantiomer of the catalyst (Scheme 11.32).

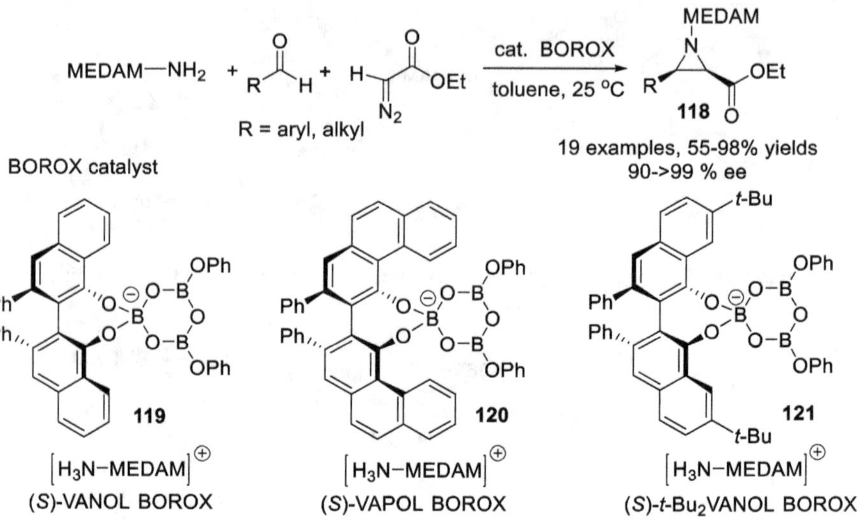

Scheme 11.33 Three-component reaction: Catalytic asymmetric aziridinations of aldehydes, amines and diazo compounds.

Since imines can be readily formed from the corresponding aliphatic aldehydes and amines, three-component catalytic asymmetric aziridinations from aldehydes, amines and diazo compounds were developed and high yields and enantioselectivities could be obtained.[112–115] For the substrates of aliphatic aldehydes, utilization of (*S*)-^{t}Bu$_2$VANOL BOROX **119** afforded higher yields than that using unsubstituted (*S*)-VANOL **106** BOROX and (*S*)-VAPOL **107** BOROX (Scheme 11.33). This method was then subjected to catalytic asymmetric synthesis of all four stereoisomers of sphinganine with hexadecanal as the starting material.[115]

To expand the substrates scope of the diazo compounds, tandem acylation and aziridination with TMSCHN$_2$ was developed, catalyzed by chiral polyborate Brønsted acid. The aziridines were obtained with high enantioselectivities by using aliphatic acid chlorides, TMSCHN$_2$ and imines as the substrates.[116] The asymmetric synthesis of *tri*-substituted aziridines was also achieved by the same group. Although *N*-MEDAM imines were inactive for the aziridination with α-diazo proprionate, the formation of *tri*-substituted aziridines could be achieved with *N*-Boc imines catalyzed by VANOL-catalyst.[117] *Tri*-substituted aziridines were prepared in moderate to good yields with excellent diastereo- and enantioselectivities from α-diazo propionate or α-diazo-*N*-acyloxazolidinones **122** (Scheme 11.34).

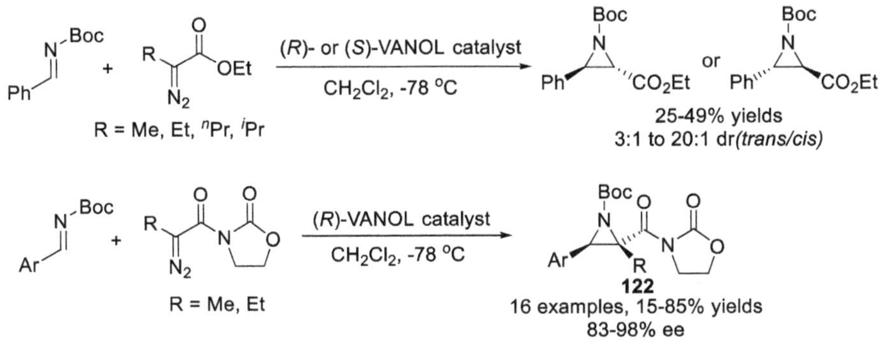

Scheme 11.34 Asymmetric synthesis of *tri*-substituted aziridines from *N*-Boc imines.

In addition to the chiral VANOL/VAPOL-polyborate anion system, the asymmetric aziridination with diazo compounds was also catalyzed by chiral phosphoric acids or dicarboxylic acids. In 2008, Maruoka and co-workers reported the synthesis of *trans*-selective asymmetric aziridination from α-diazoacetamides with *N*-Boc imines in moderate yields and with excellent enantioselectivities (Eq. (11.23)).[118] Similarly, Zhong and co-workers reported the efficient asymmetric approach to *trans*-aziridines **123** catalyzed by chiral phosphoric acid **124**.[119] The desired products were obtained with excellent yields (89–97%), diastereoselectivities (*trans/cis* >50:1) and enantioselectivities (88–98% *ee*).

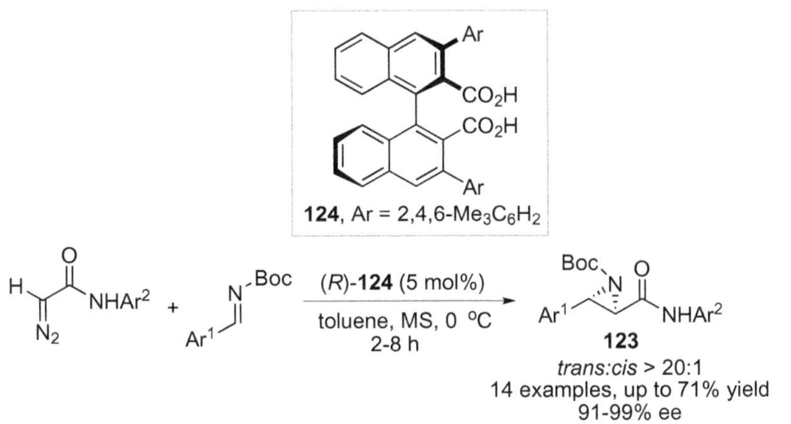

(11.23)

Furthermore, Maruoka and co-workers developed the chiral carboxylic acid catalyzed *trans*-selective asymmetric aziridination.[120] The key

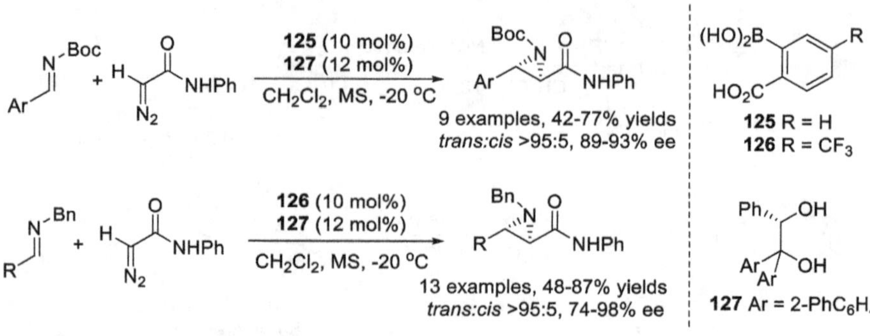

Scheme 11.35 Asymmetric aziridinations catalyzed by the boronate esters in situ assembled from chiral diol and boronic acids.

active species, the boronate ester, was assembled from chiral diol **127** and 2-boronobenzoic acid **125** or **126**, which successfully catalyzed *trans*-aziridination of *N*-Boc or *N*-benzyl imines in good yields and with high enantioselective control (Scheme 11.35).

Asymmetric *cis*-aziridination from EDA with *p*-anisidine and aryl glyoxal was catalyzed by chiral phosphoric acid.[121] This three-component synthesis has expanded the substrates scope by *in situ* generated aldimines. Moreover, multicomponent chiral Brønsted acid-catalyzed diastereo- and enantioselective synthesis of CF_3-substituted aziridines was reported from *in situ* generated CF_3CHN_2 and aldimines.[122] Competitive formation of triazoline derivatives composed the major side products.

Maruoka and co-workers also performed asymmetric synthesis of *tri*-substituted aziridines by using *N*-triflyl phosphoramide **128** as the chiral Brønsted acid.[123] Aryl aldehydes-derived *N*-Boc aldimines were reacted with α-alkyl α-diazocarbonyl compounds to prepare *trans*-aziridines in good yields and with good *ee* values. The reaction between α-ketimino esters and α-diazoacetyloxazolidinone were also carried out with good yields and enantioselectivities (Scheme 11.36).

11.6 Reaction through coordination with organoboron compounds

When the boron compounds serve as the electrophilic reagents, coordination of the nucleophilic diazo carbon atom with electron-deficient boron atom generates *tetra*-coordinated boron intermediate **129**. Subsequent

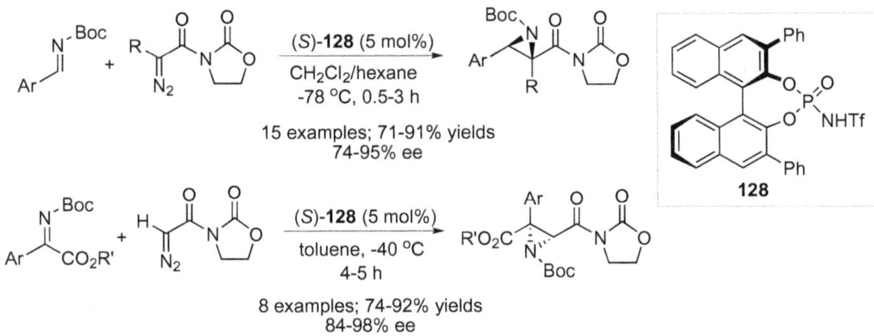

Scheme 11.36 *N*-Triflyl phosphoramide-catalyzed asymmetric synthesis of tri-substituted aziridines.

1,2-migration of one substituent from boronic moiety to diazo carbon center occurs with simultaneous N_2 extrusion (Eq. (11.24)). This process is a formal insertion of carbene into B–X bond, providing a unique approach toward the construction of C–B, C–C and C–Si bonds.[124,125]

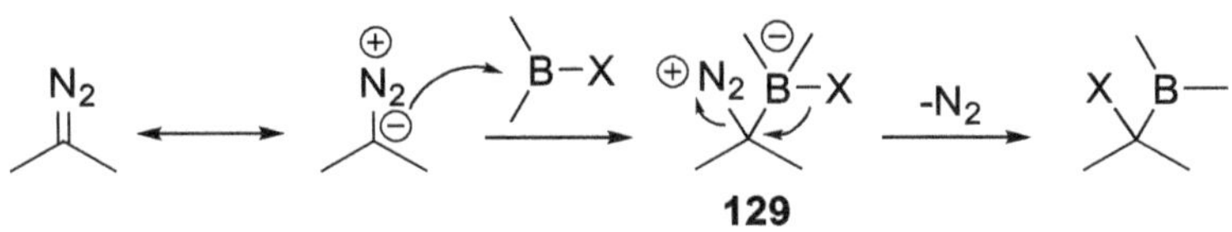

(11.24)

In 2009, Wang and co-workers reported the first transition-metal-free α-carbonyl C–C bond formation from α-diazocarbonyl compounds with boroxines, which are derived from the corresponding organoboronic acids by dehydration.[126] With diisopropylamine as the base, this transformation has been applied to aryl boroxines and alkenyl boroxines with diazoesters, diazoamides or diazoketones (Scheme 11.37).

Barluenga and Valdés in the same year reported the metal-free reductive cross-coupling reaction between *N*-tosylhydrazones and a series of boronic acids.[127] This transition-metal-free C–C bond forming reaction has shown excellent functional group tolerance, including the scope of electron-rich or deficient aromatic or alkyl boronic acids (Scheme 11.38). Furthermore, the *N*-tosylhydrazones could be generated *in situ* with high efficiency, making it possible to carry out the reaction in one-pot fashion directly starting from the carbonyl compounds with boronic acids.

Furthermore, Valdés and co-workers reported the same reaction with alkenylboronic acid **130** or **131** as the coupling partner.[128] This C–C bond

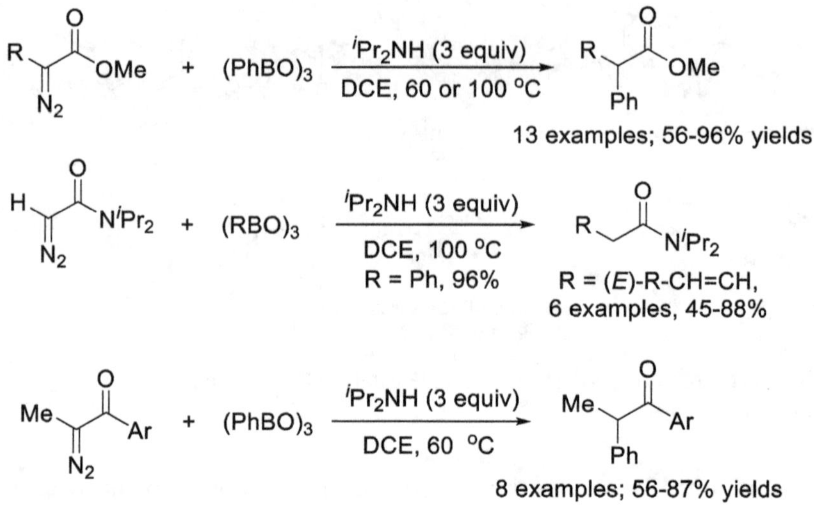

Scheme 11.37 Reactions of α-diazocarbonyls with aryl and alkenyl boroxines.

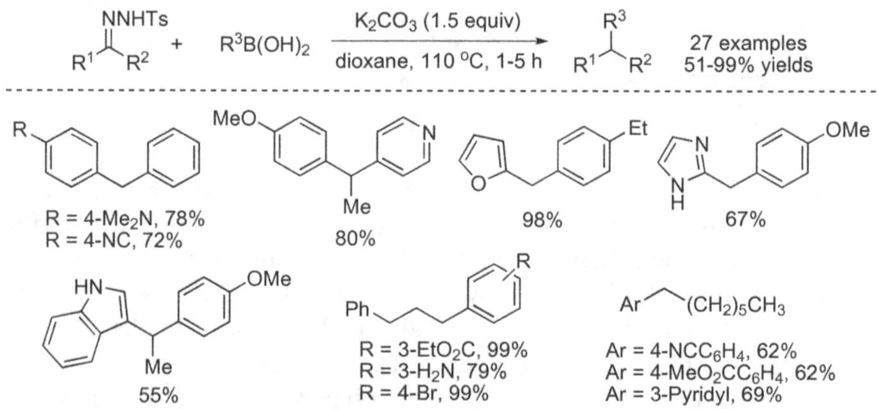

Scheme 11.38 Reductive coupling of *N*-tosylhydrazones with aryl or alkyl boronic acids.

forming reaction afforded two regio-isomers that differ in the position of double bond. When the styryl boronic acids and *N*-tosylhydrazones derived from ketones were subjected to this reaction, γ-protodeboronations products could be obtained exclusively. Otherwise, α-protodeboronations products were the major ones (Scheme 11.39).

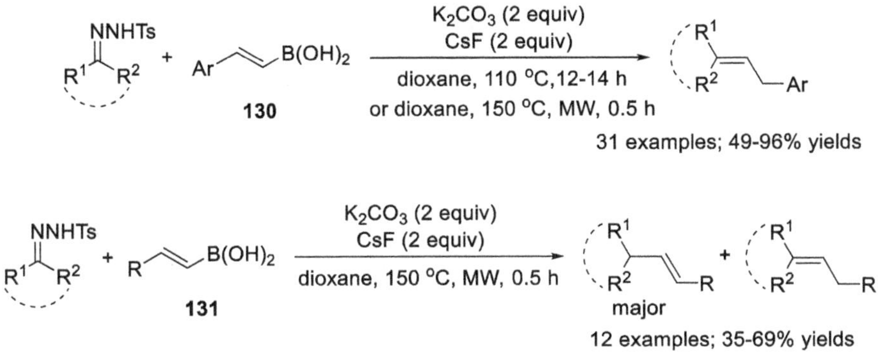

Scheme 11.39 Reductive coupling of alkenylboronic acids and *N*-tosylhydrazones.

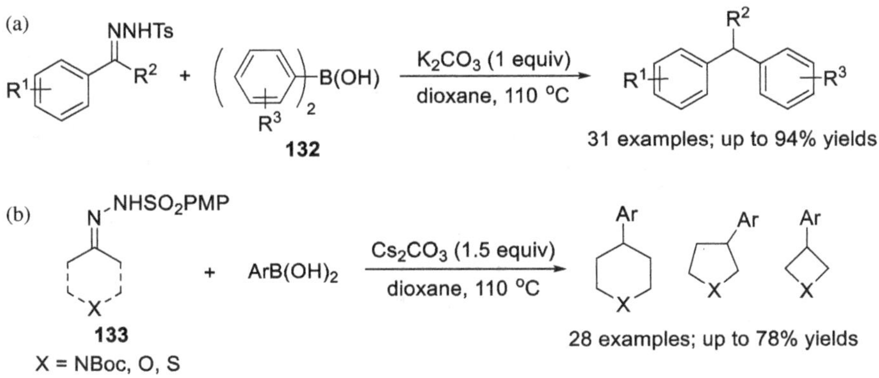

Scheme 11.40 Transition-metal-free reductive coupling reactions.

In the same year, Zou and co-workers demonstrated the transition-metal-free reductive coupling reaction of diarylboronic acids **132** with *N*-tosylhydrazones derived from aldehydes or ketones.[129] This procedure can serve as a complementary method for the synthesis of substituted diarylmethanes (Scheme 11.40(a)). Recently, Allwood and Ley reported the metal-free coupling reactions of saturated heterocyclic *p*-methoxyphenyl (PMP) sulfonylhydrazones **133** with aromatic boronic acids.[130] A range of aryl-substituted saturated heterocycles was prepared in moderate to good yields (Scheme 11.40(b)).

Notably, in these types of reactions the boron intermediate generated from the 1,2-shift of the *tetra*-coordinated boron complex can be trapped

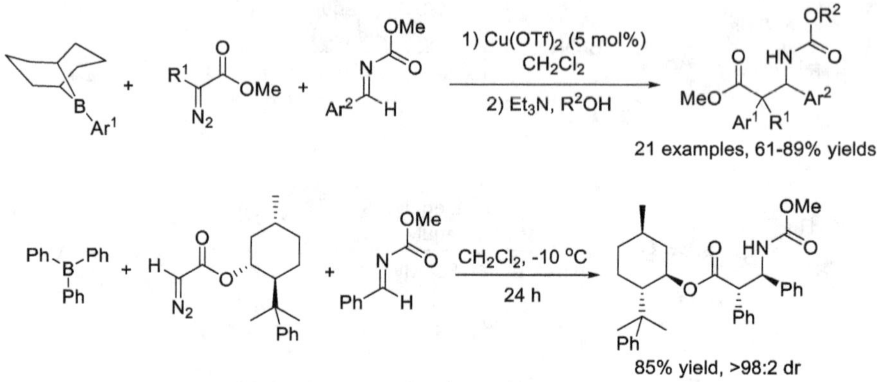

Scheme 11.41 Three-component reaction of diazo compounds, arylboranes and carbamoyl imines.

by an electrophile. In this context, a three-component reaction of α-diazocarbonyl compound, trialkylborane and aromatic aldehyde has been developed, providing an efficient approach toward the synthesis of β-ketoesters.[131] The reaction mechanism involves the generation of boron enolate, followed by trapping with aldehydes. Moreover, Schaus and co-workers also reported the three-component Mannich condensation from diazoesters, arylboranes and carbamoyl imines.[132] More recently, they expanded this reaction to the diastereoselective synthesis of β-amino carbonyl products under catalyst-free conditions.[133] Asymmetric variant of the reaction was conducted by using the (−)-phenylmenthol as chiral auxiliary (Scheme 11.41).

In addition to α-diazoesters, 2,2,2-trifluorodiazoethane (CF_3CHN_2), which is generated *in situ* from the commercially available starting materials, has demonstrated its applications in C–C bond formation by the reaction with organoboron reagents.[134–137] The groups of Molander and Wang independently reported the reaction by employing *in situ* generated CF_3CHN_2 intermediate with a series of organoboron reagents (Scheme 11.42).[134,135] In these cases, the diazo species represent useful building blocks for the introduction of either trifluoroethyl or *gem*-difluorovinyl groups to organic molecules.

Wang and co-workers further explored this strategy using *in situ* generated diazo compounds bearing other electron-withdrawing groups in the reaction with arylboronic acids. This transformation provides an effective

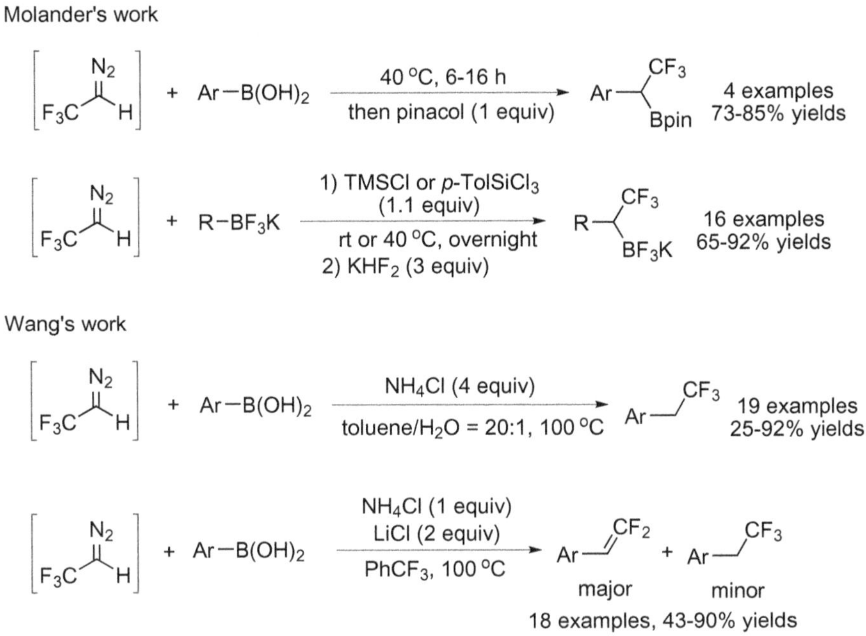

Scheme 11.42 Reaction of *in situ* generated CF_3CHN_2 with organoboron reagents.

access to the synthesis of α-aryl esters and nitriles from readily available substrates of α-aminoesters **134** and α-aminoacetonitriles **135** (Scheme 11.43).[136] The reaction could be carried out under essentially neutral conditions with excellent functional group tolerance.

More recently, Molander and co-workers reported the formation of bis(trifluoromethylated) alkyl boron compounds from CF_3CHN_2 with aryl or alkenyl boroxines.[137] The reaction is rationalized by employing the reaction of the 1:1 adduct with another CF_3CHN_2. The reaction showed high diastereoselectivity (Scheme 11.44).

In addition to the C–C bond formation through the coordination of diazo substrate with boron center, formal H–F bond insertion of α-diazo-β-ketoesters was also reported by treating diazo compounds with HBF_4.[138] Plausible mechanism involves nucleophilic attack by diazoesters to boron trifluoride. Fluoride migration generated boron enolate intermediate, which is followed by protonation with HF to afford the α-fluoro-β-ketoesters.

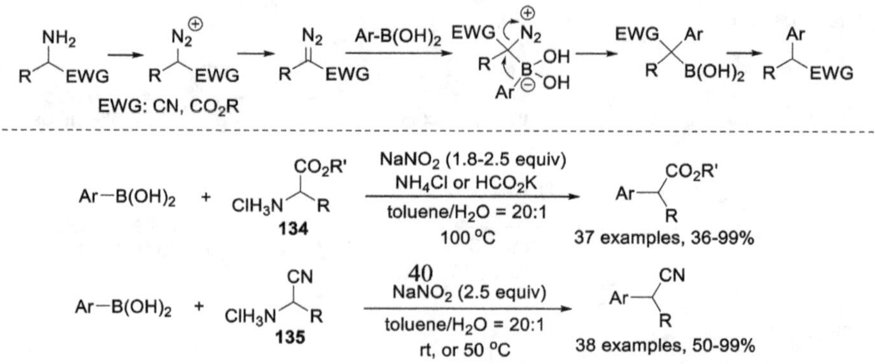

Scheme 11.43 Synthesis of α-aryl esters and nitriles by the reaction of arylboronic acids with α-aminoesters and α-aminoacetonitriles.

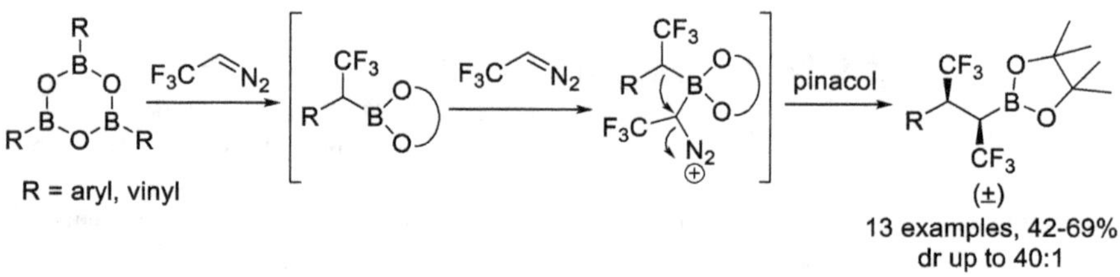

Scheme 11.44 Reaction of the *in situ* generated CF_3CHN_2 and aryl or alkenyl boroxines.

Moreover, the reaction of diazo compounds with organoboron compounds may be utilized in the synthesis of alkyl boron compounds. In this context, Wang and co-workers reported the transition-metal-free reaction of diboron pinacol ester with *N*-tosylhydrazones to afford benzyl-, alkyl- and allylboronates in good yields (Scheme 11.45).[139] The reaction has wide substrate scope with *N*-tosylhydrazones derived from both aldehydes and ketones. The key step in the mechanistic hypothesis is the 1,2-boron migration of **136** with subsequent protodeboronations or 1,2-H migration. In the case when proton is not available in the reaction system, protodeboronation will not occur, allowing the isolation of 1,1-diboronates. Based on this reaction, the same group furthermore developed a convenient method to access 1,1-diboronates **137** and 1-silyl-1-boron compounds.[140]

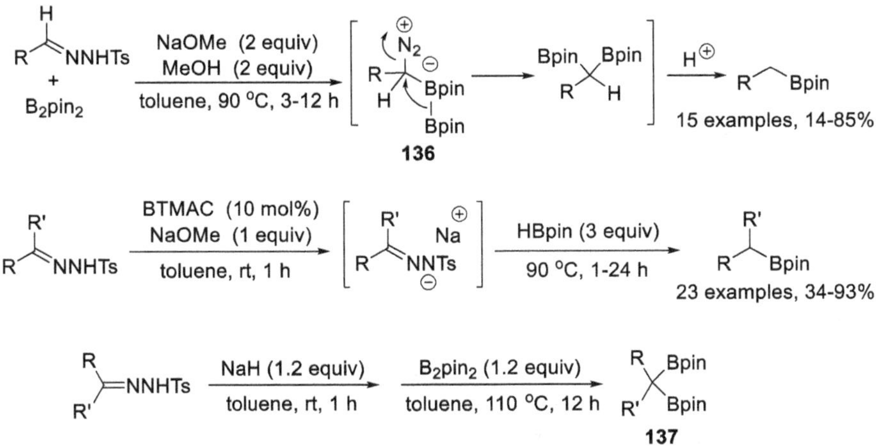

Scheme 11.45 Alkylboronates from the reaction of *N*-tosylhydrazones with B_2pin_2.

Furthermore, a one-pot transition-metal-free reaction between *in situ* generated diazo species **138** derived from carbonyl compounds with unsymmetrical diboron reagent was realized, affording unsymmetrical *gem*-diborylated products **139** with moderate to good yields (Eq. (11.25)).[141] Theoretical study indicated a concerted-asynchronous mechanism involving the formation of divergent C–B bonds. This method demonstrated the potential utility in the construction of a stereogenic center containing two differentiate diboron groups.

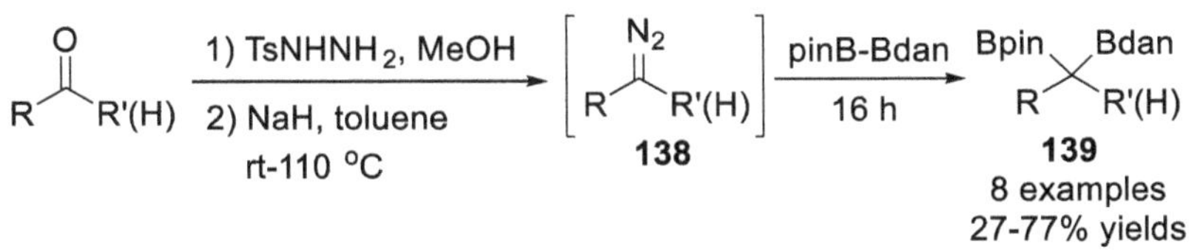

(11.25)

Recently, a selective Cu-catalyzed coupling reaction of allylboronic acids with α-diazoketones was demonstrated, affording branched allylic products **140** with high regioselectivity and moderate to good diastereoselectivity (Eq. (11.26)).[142] This process achieved the generation of a new $C(sp^3)$-$C(sp^3)$ bond with specific γ-selectivity. Although the mechanism is still unclear, this redox-free and mild reaction provided a complementary strategy for the selective synthesis of branched homoallyl carbonyl compounds.

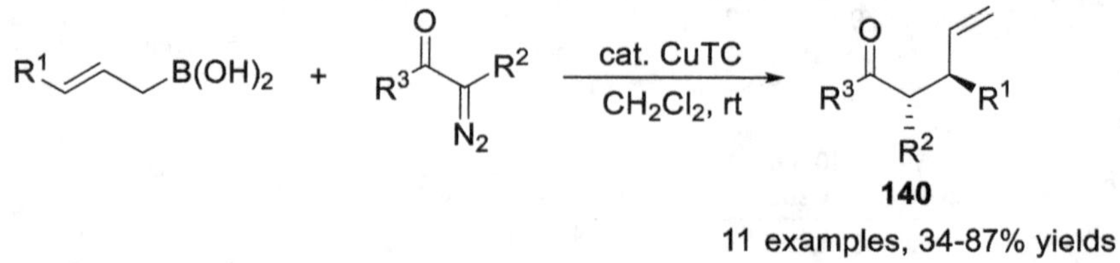

(11.26)

An intriguing ring-expansion procedure of boroles **141** with diazoalkanes has been recently realized. The reaction was affected by the steric bulk of the diazoalkane substrate. With bulky diazoalkanes, the ring-expansion product boracyclohexadienes **142** were obtained, while with less bulk diazoalkane, the reaction gave 1,2-azaborinines **143** (Scheme 11.46).[143] The results indicate an interesting chemoselectivity of the coordination of the diazo substrate to the electron-deficient boron (carbon versus nitrogen coordination).

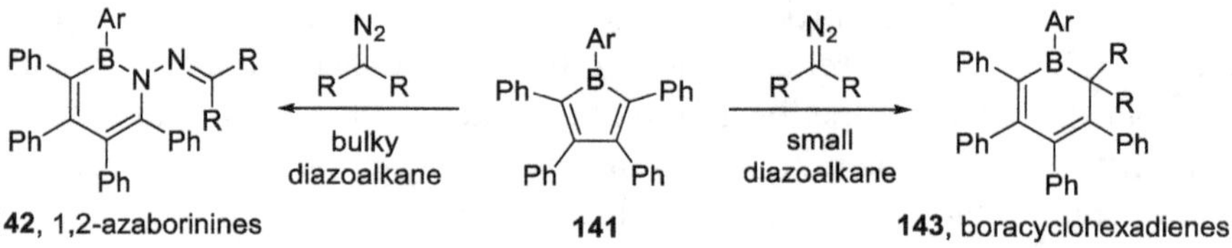

Scheme 11.46 Ring expansions of the boroles with diazo compounds.

11.7 Enoldiazo nucleophiles

The α-diazocarbonyl enolate intermediates, which can be readily generated from β-keto-α-diazoesters, have shown high reactivities in addition reactions with the retention of diazo functionality.[4] The reaction provides a useful access to the compounds of complexity while bearing diazo functionality. Besides, the addition adducts can be further used in the subsequent transformations.

It would be desirable if asymmetric catalysis can be achieved in the nucleophilic additions with α-diazocarbonyl enolate. In this context, it is conceivable that the Mukaiyama-aldol addition with silyl enol ethers may be similarly applied.[144,145] Doyle and co-workers explored such possibility with enol ether by using AgF/(*R*)-BINAP as the chiral catalyst. A moderately high level of enantioselectivity has been achieved (Eq. (11.27)).[145]

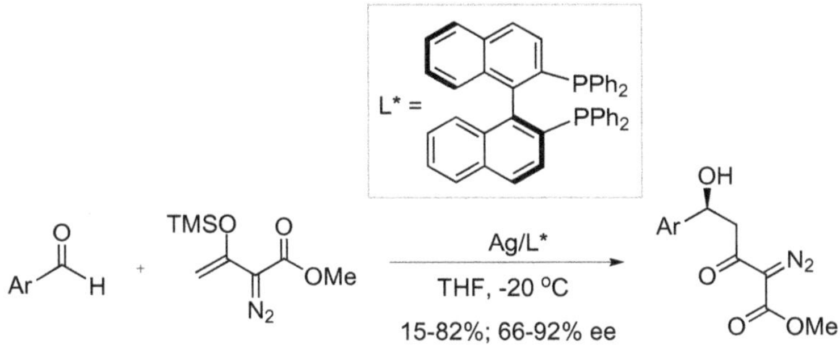

(11.27)

In a recent study, Doyle and co-workers reported an efficient catalytic Mukaiyama–Michael addition to α,β-unsaturated enones. Functionalized diazoacetoacetates **144** were obtained in good to excellent yields with mild Lewis acid Zn(II) triflate under low catalyst loading.[146] This catalytic system was also effective in the corresponding Mukaiyama-aldol reactions to afford diazo products **145**. Furthermore, asymmetric Mukaiyama–Michael addition was achieved in high enantioselectivities catalyzed by Cu(II) complex with chiral ^{t}Bu-BOX ligand.[147] Furthermore, the one-pot Mukaiyama-aldol reaction between diazoacetoacetate and aromatic or aliphatic aldehydes proceeded in good yields (Scheme 11.47).[148]

Furthermore, Doyle and co-workers reported the C–C bond formation from enoldiazoacetate nucleophiles with catalytically generated carbocations.[149] Benzhydryl, allylic and oxonium carbocations could be formed

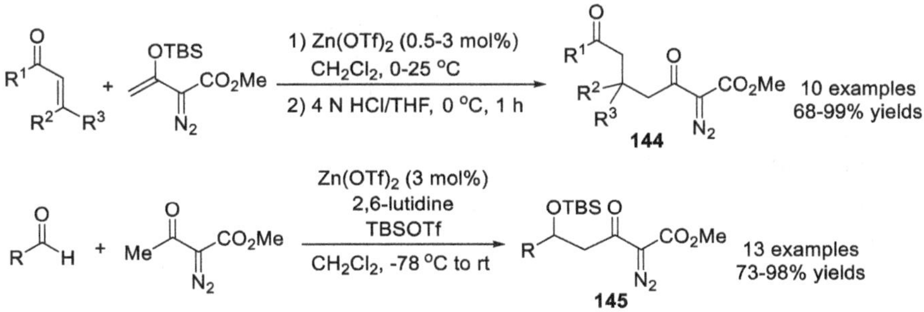

Scheme 11.47 Zn(II)-catalyzed Mukaiyama–Michael addition to α,β-unsaturated enones and Mukaiyama-aldol reactions with aldehydes.

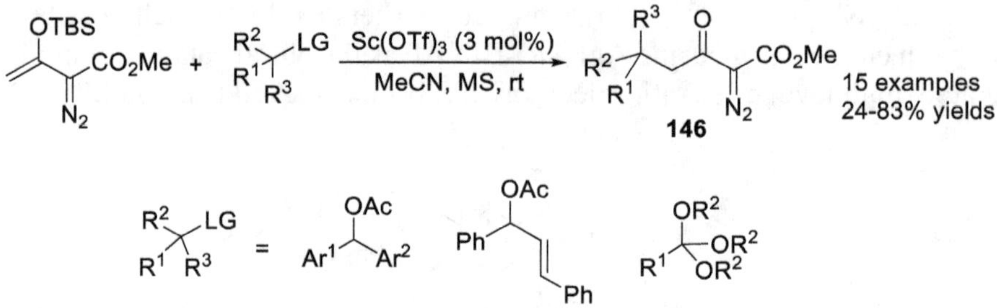

Scheme 11.48 Nucleophilic substitution reaction with enoldiazoacetates.

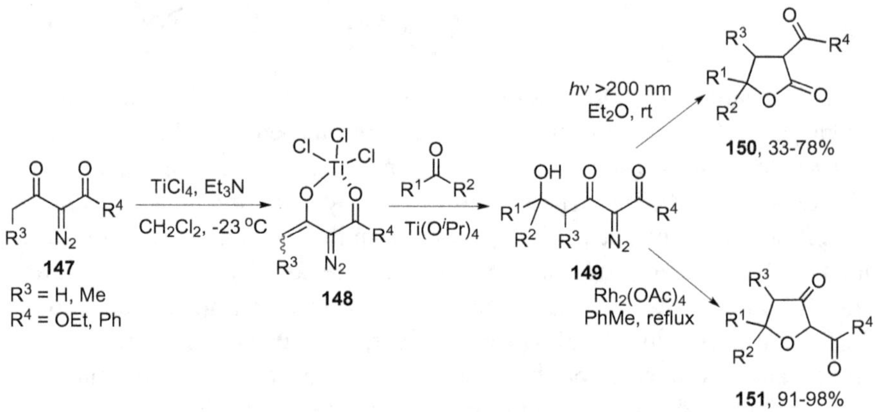

Scheme 11.49 Aldol-type addition followed by Wolff rearrangement or Rh(II)-catalyzed O–H insertion.

from a series of substrates with leaving groups (Scheme 11.48). The nucleophilic substitution products **146** bearing diazo moiety may be applied to the construction of natural product-like scaffolds.

The nucleophilic addition to C=O or C=N bonds by the enolates derived from β-keto-α-diazocarbonyl compounds and sequential cascade reactions can also be achieved similarly. Wang and co-workers reported the incorporation of these nucleophilic additions in the synthetic methods for heterocycles. As shown in Scheme 11.49, Ti(IV) enolates **148** derived from α-diazo-β-keto esters **147** reacted efficiently with aldehydes; however, the reaction became sluggish when ketones were used as the substrates. To activate the C=O group, $Ti(O^iPr)_4$ was used as Lewis acid in the reaction and the corresponding nucleophilic addition occurred

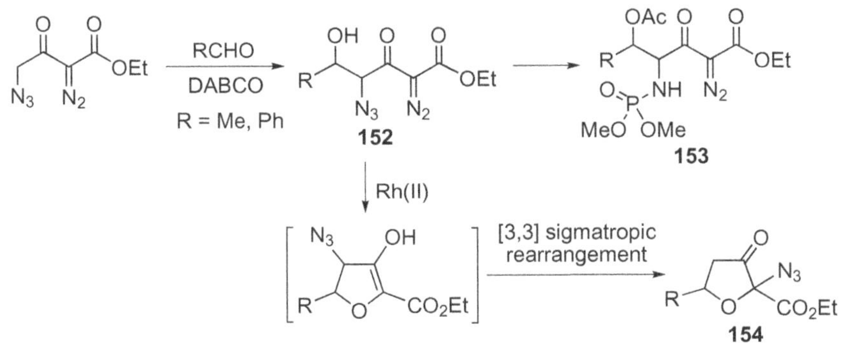

Scheme 11.50 DABCO-promoted aldol-type reactions of the azido-substituted diazo compounds with aldehydes.

smoothly with various ketone substrates. The addition products **149** were further converted into cyclic compounds through a photo-induced Wolff rearrangement to give **150**, or Rh(II)-catalyzed insertions to give **151**.[150,151] An analogous addition reaction was also carried out with *N*-tosylimines, and the addition products were similarly converted into γ-lactams or pyrroles.[152]

Padwa and co-workers reported the DABCO-promoted Mukaiyama-aldol-type reactions of the diazo dicarbonyl compounds bearing azido substituent with aldehydes. The products, γ-azido-α-diazo-δ-hydroxy-β-keto esters **152** can be further converted into products, such as **153** and **154** (Scheme 11.50).[153]

11.8 Miscellaneous transformations

11.8.1 *Transformations with electrophilic carbon or carbocations*

The reactions in which the diazo substrates were used as nucleophiles were further explored by expanding the electrophiles. In 2013, Zhu and co-workers reported the chiral Lewis base **156** catalyzed allylic alkylation of α-diazoesters with Morita–Baylis–Hillman carbonates as the electrophiles.[154] This C–C bond forming reaction has been applied in the S_N2'/S_N2' cascade transformation for the construction of highly functionalized diazo compounds in good yields with excellent enantioselectivities.

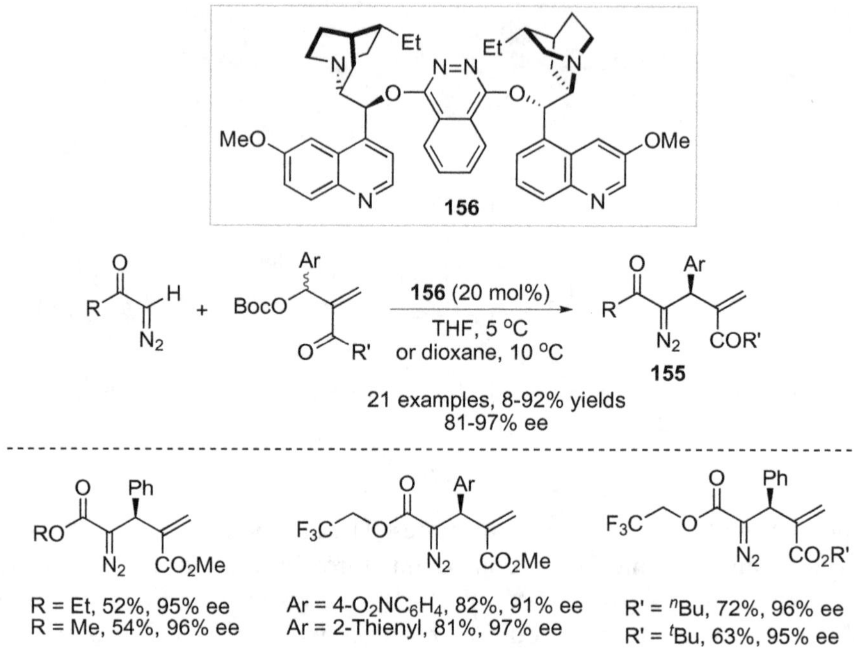

Scheme 11.51 Chiral Lewis base catalyzed allylic alkylation with carbonates.

The corresponding products could be used for the rapid synthesis of optically pure *N*-heterocycles (Scheme 11.51).

The reaction between diazo compounds with allene carbocation generated from propargyl alcohols **157** was used to access functionalized indenes **158** and furan-2-ones **159** (Scheme 11.52).[155] The reaction has been proposed to be initiated by the generation of allene carbocation intermediate *via* Meyer–Schuster rearrangement, followed by trapping of the carbocation by diazo carbon. The subsequent ring closure and other transformations led to the final products.

Furthermore, asymmetric synthesis of *2H*-chromene scaffolds was developed by the reaction of oxocarbenium precursors derived from *2H*-chromene acetals with EDA. With the catalytic system of Lewis acid and chiral phosphoric acid, the desired products were obtained in good yields.[156]

Recently, a Cu-catalyzed oxidative C–H bond functionalization and ring expansion of 9*H*-xanthene **160** or 9,10-dihydroacridine **161** with

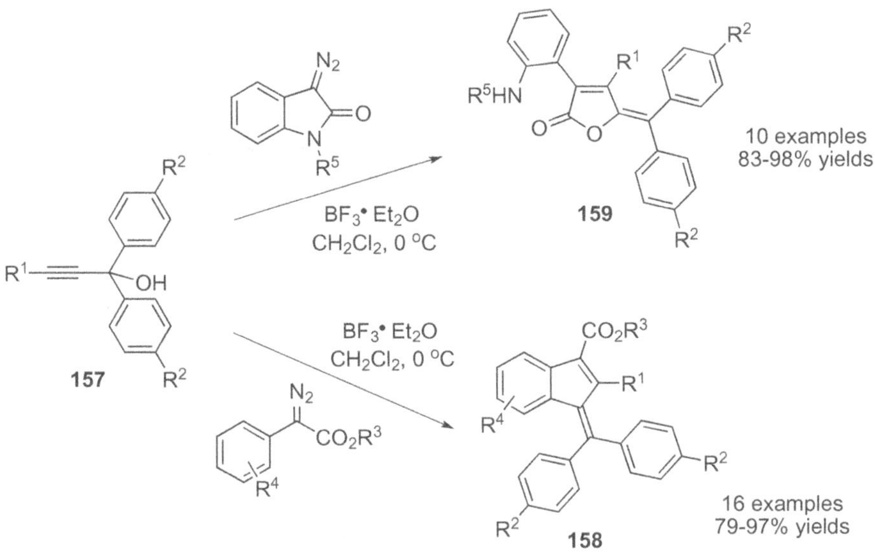

Scheme 11.52 Reaction of diazo compounds with propargyl alcohols.

$TMSCHN_2$ were reported to afford polycyclic dibenzoxepines **162** or dibenzoazepines **163** (Scheme 11.53).[157] Experimental as well as theoretical studies have suggested that the $Cu(bpy)_2$ complex is initially reduced *in situ* by diazoalkanes to generate Cu(I) species, which catalyzes the reductive O–O bond cleavage to form $PhCO_2\cdot$ radical and Cu(II)-carboxylate species **164**. Then, the substrate experiences two-electron radical oxidation to give carbocation **165** as the key intermediate, followed by nucleophilic attack of $TMSCHN_2$ and subsequent ring-expansion process. Finally, the desired tricycle products are obtained *via* elimination of silicon group with concomitant formation of trimethylsilylbenzoate.

11.8.2 *Generation of alkylidene carbene from $TMSCHN_2$ and carbonyl compounds*

A series of transformations with alkylidene carbene intermediates from $TMSC(Li)N_2$ have been explored by the group of Lee and others. The reaction is initiated by nucleophilic attack from diazo species to carbonyl

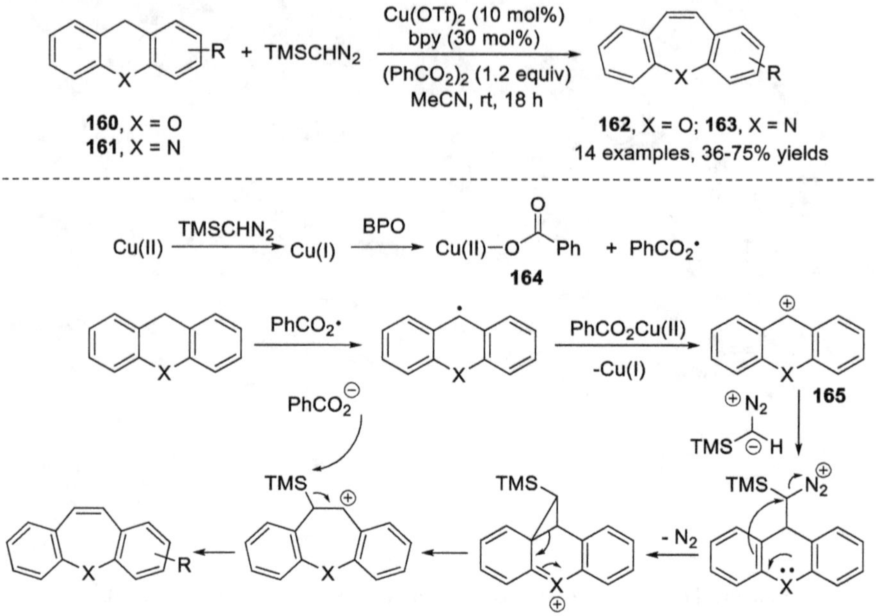

Scheme 11.53 Cu-catalyzed ring expansion with $TMSCHN_2$.

group to form diazo lithium adduct **166**. Then alkylidene carbene **168** is generated in two steps: 1,3-Brook rearrangement to form **167**, followed by extrusion of LiOTMS and N_2 (Eq. (11.28)).[158–160]

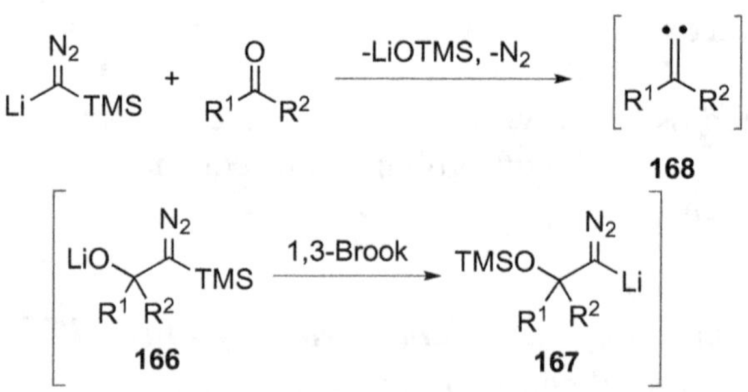

(11.28)

Lee and co-workers have explored the application of alkylidene carbene in organic synthesis (Scheme 11.54). Firstly, they investigated the reactivity of alkylidene carbenes in regioselective intramolecular

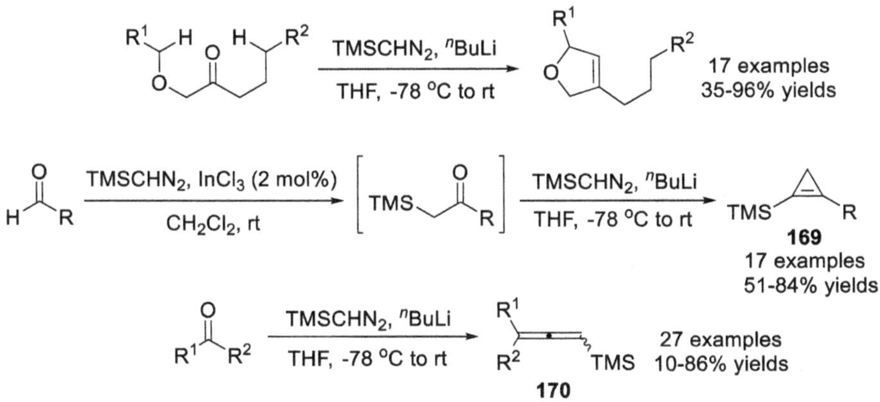

Scheme 11.54 Versatile reactivity of alkylidene carbenes generated from $TMSCHN_2$.

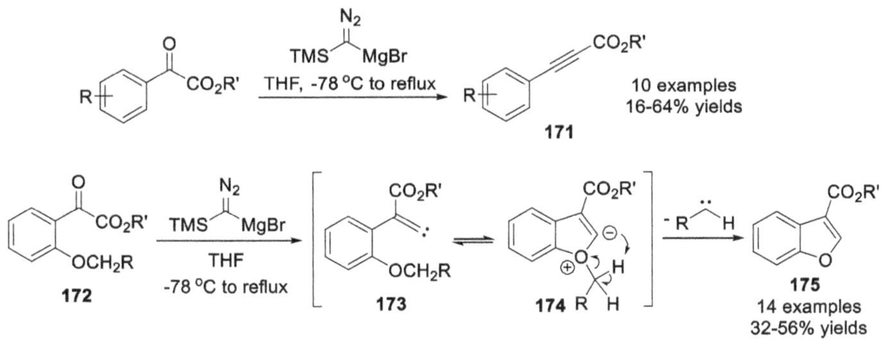

Scheme 11.55 Reaction of $TMSC(M)N_2$ with aryl(oxo)acetate derivatives.

C–H insertion and subsequent transformations.[161–164] The selectivity of competing C–H bond insertions was affected by stereoelectronic effect of different substituents.[161] Moreover, the synthesis of TMS-substituted cyclopropenes **169** was achieved *via* alkylidene carbene C–Si insertion.[162] Alkylidene carbene intermediates have also been subjected to access allenyl silanes **170** through intermolecular reaction with $TMSCHN_2$.[163]

Aoyama and co-workers explored the reaction of diazo(trimehtylsilyl) methyl magnesium bromide with aryl(oxo)acetates (Scheme 11.55).

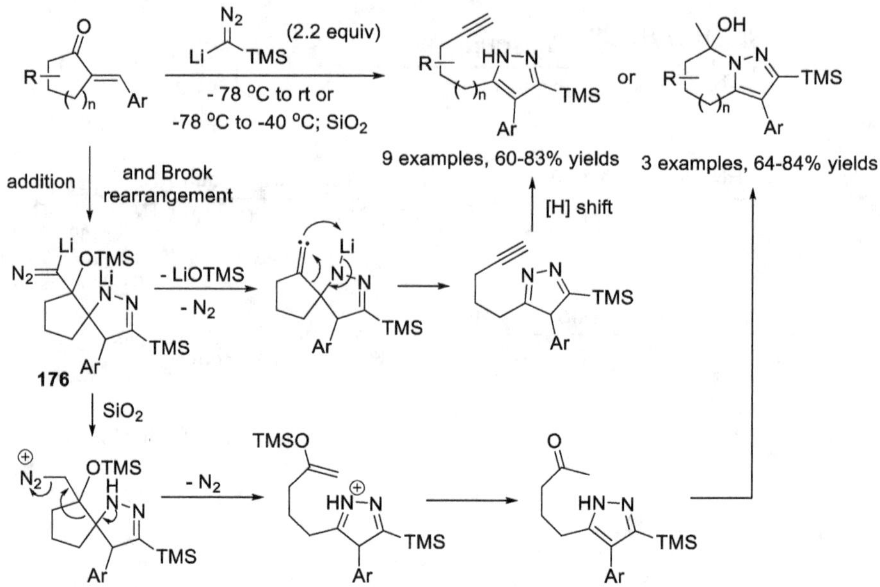

Scheme 11.56 Sequential 1,4-/1,2-addition of lithium(trimethylsilyl)diazomethane to α,β-unsaturated ketones.

Firstly, they developed the synthesis of arylpropiolates **171** through the formation of alkylidene carbene intermediates and sequential 1,2-migration process.[165] Furthermore, the substituted benzofuran-3-carboxylate derivatives **175** could be obtained through the reaction of diazo(trimehtylsilyl)methyl anion with *o*-alkoxyaryl(oxo)acetates **172**.[166] Plausible mechanism involves the generation of alkylidene carbene **173**, followed by oxonium ylide **174** formation *via* cyclization and elimination of alkylene.

Recently, Lee and co-workers reported a selective 1,4-/1,2-addition of lithium(trimethylsilyl)diazomethane with cyclic enones, affording various *N*-heterocyclic molecular frameworks (Scheme 11.56).[167] The reaction is proposed to be initiated by dipolar cycloaddition and Brook rearrangement to form **176**. Sequential transformations, such as Grob-type C–C fragmentation or alkylidene carbene mediated Li–N insertions, are possible depending on the ring size and the position of the double bond.

11.8.3 *Homologation by using diazophosphonates, CH_2N_2 and $TMSCHN_2$*

In addition to the abovementioned transformations, there are several other types of reactions that are related to diazo group-containing nucleophiles. The most notable is the addition of the anions derived from diazophosphonate to aldehydes, which is the key step in the synthesis of terminal alkynes from aldehydes (Scheme 11.57).

A valuable modification of the original procedure, which utilizes dimethyl diazomethylphosphonate **177**, is the use of more stable dimethyl-2-oxopropylphosphonate **178** (Figure 11.3).[168] More recently, the use of ROMP gel-supported ethyl1-diazo-2-oxopropylphosphonate **179** has been reported.[169] In these cases, the diazophosphonate anion is generated by base-promoted deacetylation. These diazo reagents have proven to be useful in natural product synthesis.[170–174]

Diazophosphonate reagents have shown wide applications in the homologation of aldehydes to the corresponding alkynes. Taber and co-workers reported an improved procedure, allowing this conversion to be carried out conveniently on a large scale under mild conditions.[175] The transformation can be subjected to the subsequent click reaction for a one-pot synthesis of 1,4-disubstituted 1,2,3-triazoles.[176] Furthermore, an

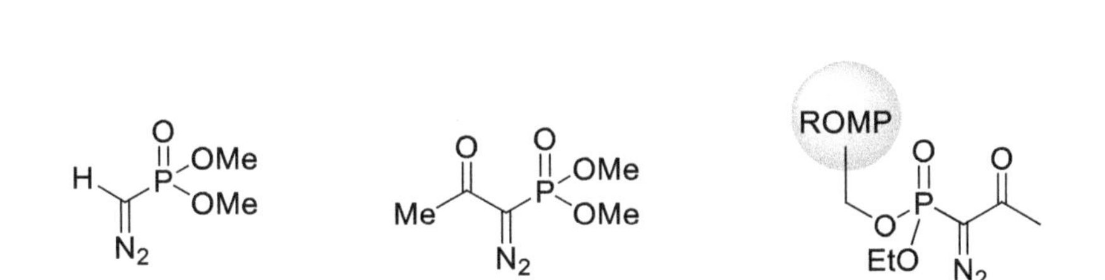

Scheme 11.57 Addition of the anion derived from diazophosphonate to aldehydes.

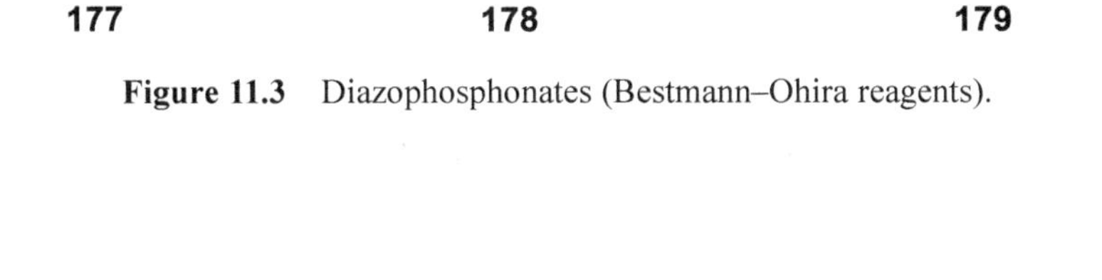

177 **178** **179**

Figure 11.3 Diazophosphonates (Bestmann–Ohira reagents).

improved method for *in situ* prepared Bestmann–Ohira reagent from sulfonyl azide was reported and utilized in scalable synthesis of terminal alkynes from aldehydes (Eq. (11.29)).[177]

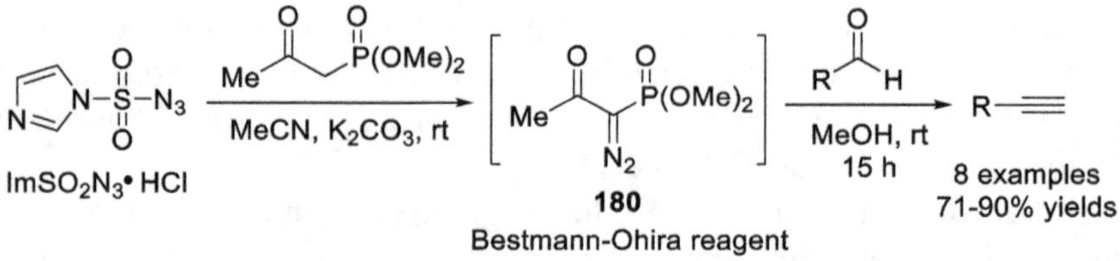

(11.29)

In 2014, a substituent-controlled nucleophilic substitution of Bestmann–Ohira reagent with benzyl or allylic bromides was reported by Rastogi for the synthesis of α-diazoalkylphosphonates **181** or vinylphosphonates **182** (Scheme 11.58).[178] The method can be served as a complementary strategy for the preparation of benzyl or alkyl diazophosphonate substrates.

More recently, the same group reported the synthesis of α-diazo-β-keto esters, phosphonates and sulfones *via* acylation of the corresponding diazomethyl anions with *N*-acylbenzotriazoles (Eq. (11.30)).[179] The reaction between dimethyl diazomethylphosphonate (DAMP or Seyferth–Gilbert reagent) with *N*-*o*-amino-acylbenzotriazoles represents a new transphosphorylation procedure for the synthesis of diazoacetyl phenylphosphoramidates.

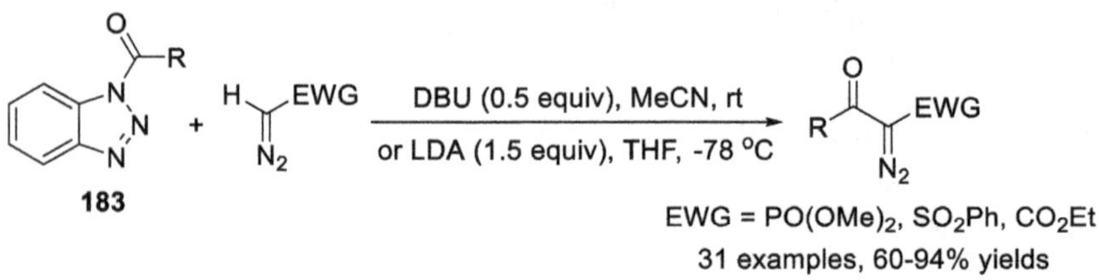

(11.30)

11.8.4 *Transformations between α-diazocarbonyl anion and electrophiles*

Scheidt and co-workers reported the base-promoted reaction of lithiated diazo esters with acylsilanes to afford β-ketoesters. The reaction

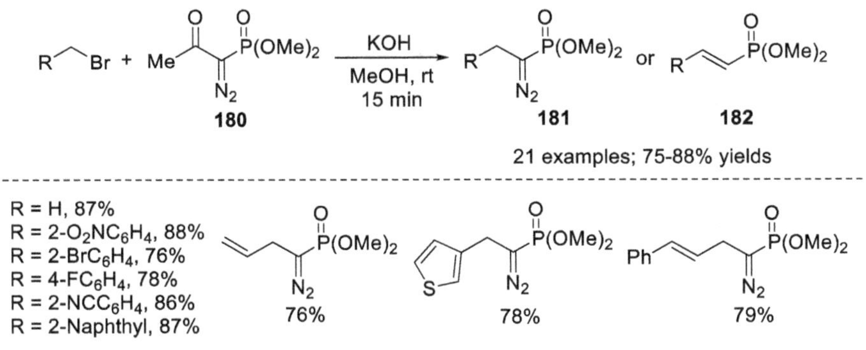

Scheme 11.58 Reaction of benzyl or alkyl bromide with Bestmann–Ohira reagent.

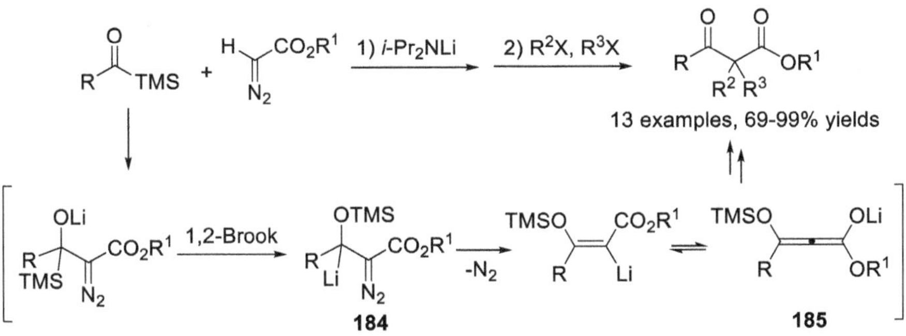

Scheme 11.59 Synthesis of β-ketoesters from lithiated diazo esters with acylsilanes.

mechanism is proposed to involve nucleophilic addition of the deprotonated diazo substrate to the acylsilane, followed by 1,2-Brook rearrangement to generate carbon anion intermediate **184**.[180] Upon elimination of N_2, β-silyloxy allenolate **185** is generated, which leads to the formation of β-ketoesters by sequential addition of electrophiles (Scheme 11.59).

Nucleophilic addition of diazo compounds to other electrophiles, such as carbonyl imidazolides,[181] nitrones,[182] azomethine imines[183] and 1,2-diaza-1,3-dienes,[184] has also been reported. Doyle and co-workers reported the Rh(II)-catalyzed formal [3+2+1]-cycloaddition reaction, starting from azomethine imines with two molecules of α-diazoketone.[183] Highly functionalized bicyclic pyrazolidinones could be prepared in moderate to good yields with high diastereoselectivities.

Favi and co-workers reported a formal [4+1] annulation of 1,2-diaza-1,3-diene **188**, generated *in situ* from α-halo *N*-acylhydrazones **186**,

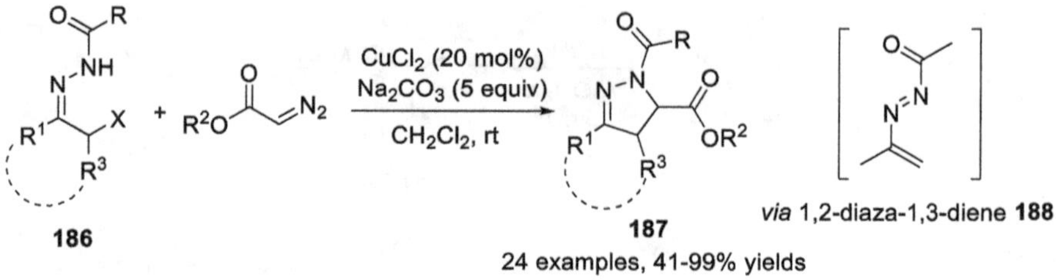

Scheme 11.60 Formal [4+1] annulation of α-halo *N*-acylhydrazones with α-diazoesters.

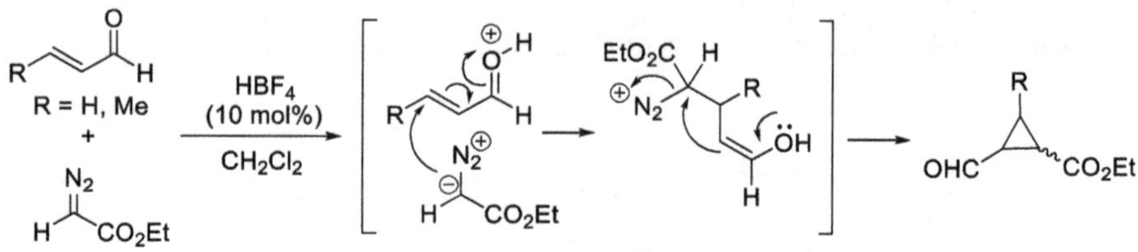

Scheme 11.61 HBF_4-catalyzed formal cyclopropanation of EDA with α,β-unsaturated aldehydes.

with α-diazoesters to give product **187** (Scheme 11.60).[184] The reaction mechanism is proposed to involve the diaza diene species **188** formation, and then 1,4-conjugate addition of diazoesters, which was followed by intramolecular annulation with displacement of N_2 to form dihydropyrazoles **187**.

Ethyl diazoacetate has also been reported to undergo a Michael addition-type reaction to add to the β carbon of α,β-unsaturated aldehydes. The reaction was catalyzed by acid and gave the cyclopropane derivatives as the final products (Scheme 11.61).[185]

11.8.5 *Diazo compound with terminal nitrogen as the nucleophile*

From the resonance structure of the diazo moiety, it is obvious that the terminal nitrogen atom also shows nucleophilicity. Although the addition of terminal nitrogen to carbon electrophile is less common, some examples are known in the literature. For example, the copper-catalyzed reaction of *2H*-azirines **189** and diazo compounds gave 2-substituted *2H*-1,2,3-triazoles (Scheme 11.62).[186] It has been proposed that the reaction involves the generation of vinyl nitrene intermediate **190**, which is

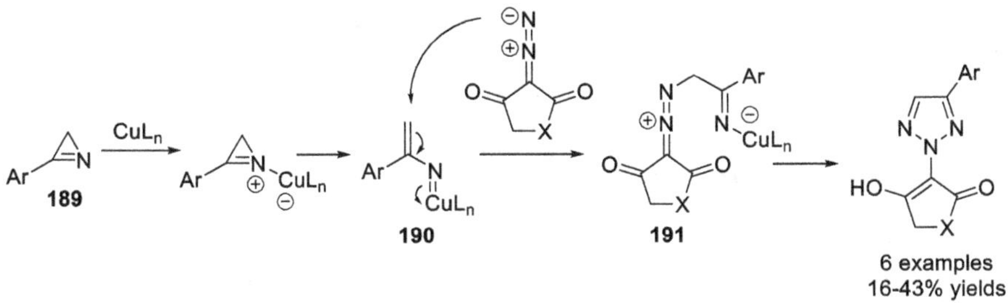

Scheme 11.62 Reaction of diazotetramic compounds and 2*H*-azirines.

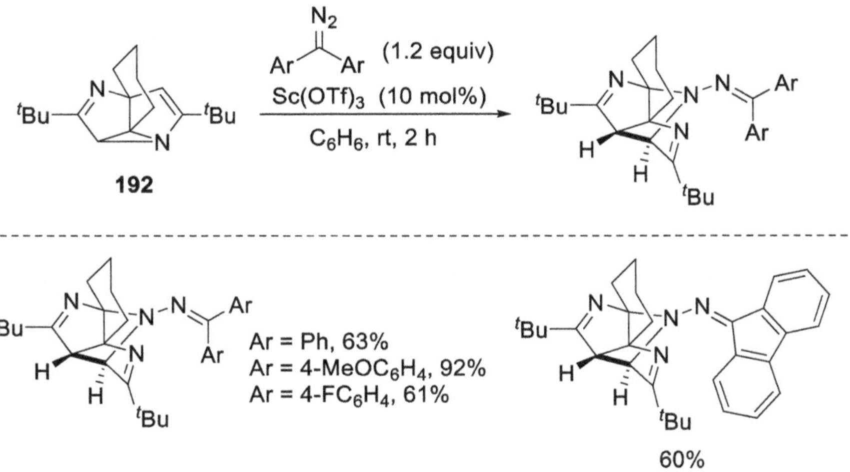

Scheme 11.63 Reaction of 2,6-diazasemibullvalenes with diaryl diazo compounds.

followed by nucleophilic attack by nitrogen atom of diazo species to form **191**, followed by ring closure to give the final products.

Scandium(III) triflate-catalyzed reaction of 2,6-diazasemibullvalenes **192** and diaryl diazo compounds was reported by Xi and co-workers (Scheme 11.63).[187] This transformation involves a tandem ring-opening/rearrangement/cycloaddition process, which is initiated by the nucleophilic attack of the terminal nitrogen atom of the diazo substrate.

11.9 Diazo compounds as electrophiles

Diazo compounds are amphoteric in terms of their electronic character. Although in most cases they behave as nucleophiles, there are also examples in which diazo compounds are used as electrophiles in the reaction. In this section, some of these reactions are summarized.

11.9.1 *Diaza-Wittig reaction*

Phosphorus-mediated diaza-Wittig reaction of diazo compounds has found applications in synthesizing functionalized pyridazines analogues, which are one of the privileged heterocyclic structures in drug design and biological research. This reaction proceeds *via* nucleophilic attack of phosphorus atom to terminal nitrogen to generate intermediate phosphazine **193**, followed by cyclization with simultaneous extrusion of phosphine oxide (Scheme 11.64). Recently, considerable attentions have been paid in expanding substrates scope and developing a new catalytic cascade system based on these types of transformations.

Intramolecular diaza-Wittig reaction of various diazo substrates, including substituted vinyldiazocarbonyl compounds **194**,[188,189] diazoamides **195**,[190] δ-oxo/hydroxy-α-diazo-β-ketoesters **196** and **197**[191–193] and α-aryl-α-diazoester **198**,[194] has been explored in the concise and convenient synthesis of polyfunctionalized pyridazines through the reaction of diazo moiety with the internal carbonyl group (Figure 11.4).

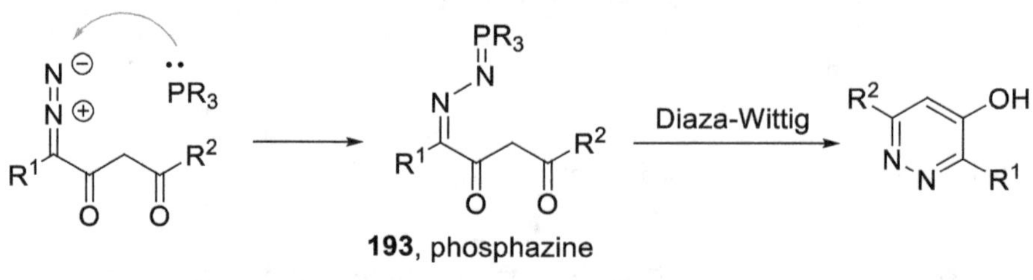

Scheme 11.64 Intramolecular diaza-Wittig reaction.

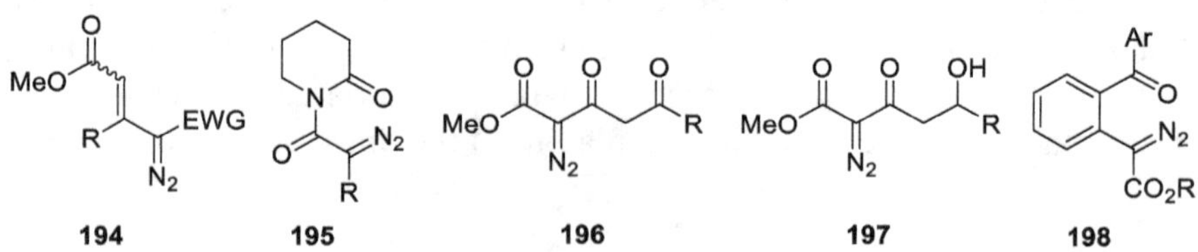

Figure 11.4 Substrates scope of intramolecular diaza-Wittig reaction.

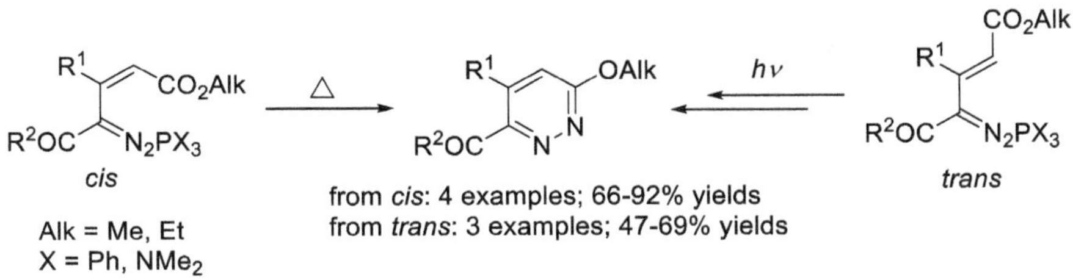

Scheme 11.65 The effect of configuration of vinyldiazo precursor in diaza-Wittig reaction.

Tandem oxidative-diaza-Wittig[192,193] as well as the sequential acyl-alkylation/diazo transfer/diaza-Wittig[194] strategy have been successfully carried out to access various annulation products from easily available substrates.

Nikolaev and co-workers investigated the impact of the configuration of vinyldiazo precursor in this reaction (Scheme 11.65).[195] As expected, the phosphazenes of vinyldiazocarbonyl compounds having *cis* stereochemistry readily produce pyridazines through diaza-Wittig process. For the diazo substrates having *trans* configuration, a *trans*-to-*cis* isomerization under UV irradiation is required for the intramolecular cyclization to occur.

Organocatalytic one-pot synthesis of polysubstituted pyridazines was achieved by intermolecular reaction between Morita–Baylis–Hillman carbonates **199** with diazo compounds under mild conditions.[196] The transformation starts with easily available starting materials and shows good functional group tolerance. The reaction pathway has been proposed to be initiated by allylic alkylation through base-promoted double S_N2' reaction to generate **200**, followed by phosphane-catalyzed intramolecular diaza-Wittig reaction to give **201** and aromatization process to deliver the pyridazines **202** (Scheme 11.66).

11.9.2 *Reaction with other nucleophilic reagents on terminal nitrogen*

In addition to the phosphine nucleophiles in diaza-Wittig reaction as described earlier, a series of other nucleophiles can be utilized in the nucleophilic addition to the terminal nitrogen of the diazo group, leading

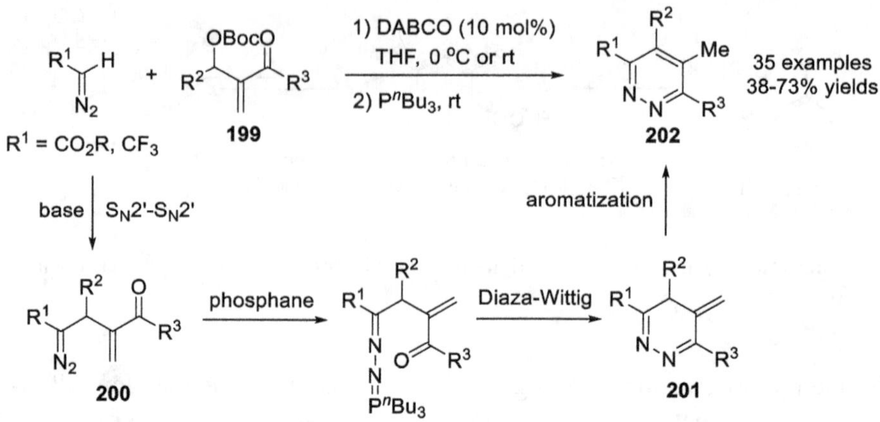

Scheme 11.66 One-pot synthesis of pyridazines from MBH carbonates with diazo compounds.

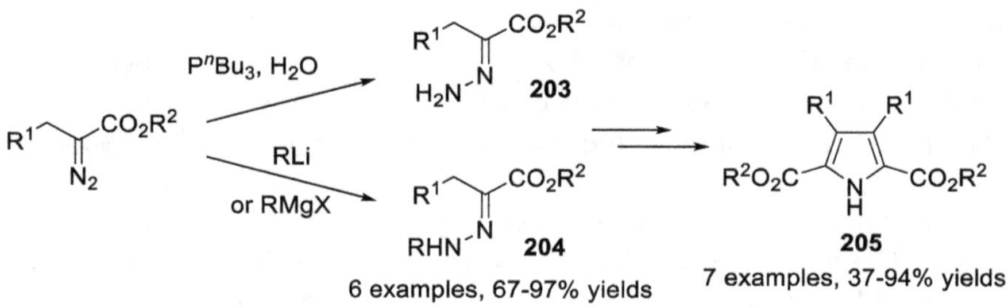

Scheme 11.67 Synthesis of hydrazones from the corresponding α-diazo esters.

to the development of a variety of unique transformations. Typically, hydrazones derivatives **203** and **204** could be directly obtained *via* nucleophilic addition of α-diazo esters to phosphines,[197] organolithium or Grignard reagents,[198,199] which can be viewed as a reductive process for the diazo compounds. The hydrazones obtained from α-diazo esters can be converted to polyfunctional pyrroles **205** in high yields (Scheme 11.67).[197,199]

Liu and co-workers developed a DBU-mediated C–N bond formation from active methylene compounds as carbon nucleophiles and α-diazocarbonyl compounds as *N*-terminal electrophiles

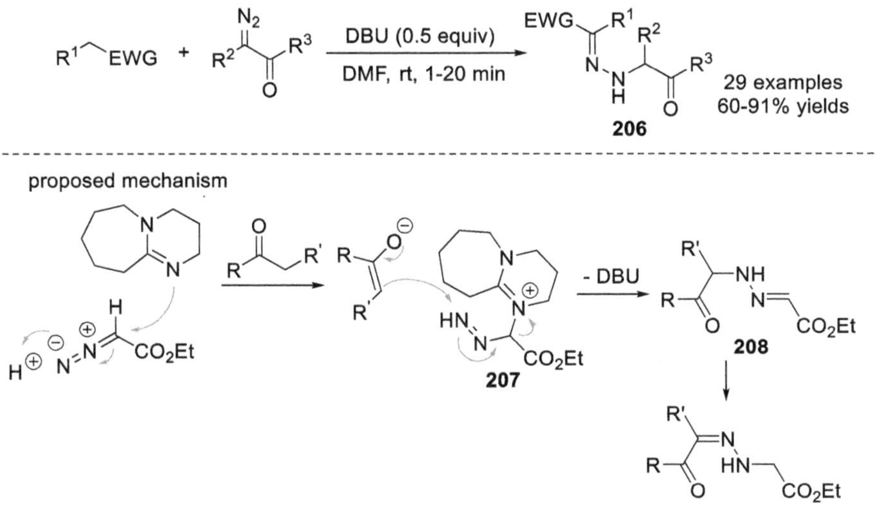

Scheme 11.68 DBU-mediated coupling reaction of α-diazocarbonyls as *N*-terminal electrophiles.

(Scheme 11.68).[200] This reaction allowed the straightforward synthesis of hydrazinyl acetate derivatives under ambient conditions. The products had potential applications in synthesizing various heterocycles, including 1,6-dihydro-1,2,4-triazin-5(4*H*)-ones and *N*-aryl 3-amino-5,6-dihydropyridazinones. In this reaction, DBU activated both α-diazocarbonyl substrates to form **207** and the active methylene component to facilitate the generation of **208** through enolization.

Xi and co-workers reported a nucleophilic addition of 1,4-dilithio-1,3-dienes species with diaryl diazomethanes, which provided an efficient synthetic approach for 1-imino-pyrroles.[201] Particularly, a zinc complex generating *via* transmetallation of dilithio intermediate was obtained and characterized, which provided insights into the mechanistic details of the reaction.

Furthermore, enantioselective C–N bond formation of the terminal nitrogen of diazo compounds under catalytic conditions was established by Feng and co-workers. Instead of Roskamp-type ring expansion, the products attributed to the asymmetric α-functionalization of ketones were obtained with α-diazoesters as the electrophiles in the presence of Feng's chiral *N,N′*-dioxide ligand **209**.[202] This α-hydrazonation reaction proceeded smoothly with high efficiency (up to 98% yields) and with excellent enantioselectivities (up to 99% *ee*) (Eq. (11.31)).

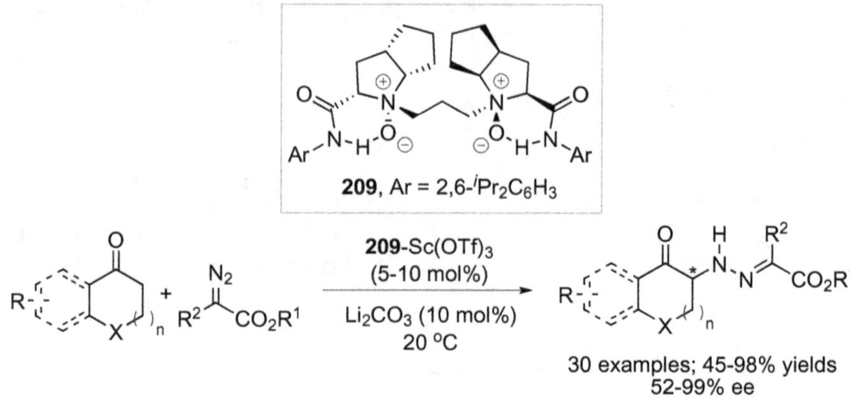

(11.31)

Gois and co-workers disclosed an NHC-catalyzed C–N bond formation of aldehydes with α-diazoesters, which could be a unique method for the synthesis of *N*-acylhydrazones.[203] The key intermediate in this C–N bond formation was acyl anion equivalent, namely Breslow intermediate **210**, which activates a nucleophilic attack on the terminal nitrogen of diazoesters (an umpolung process) (Eq. (11.32)).

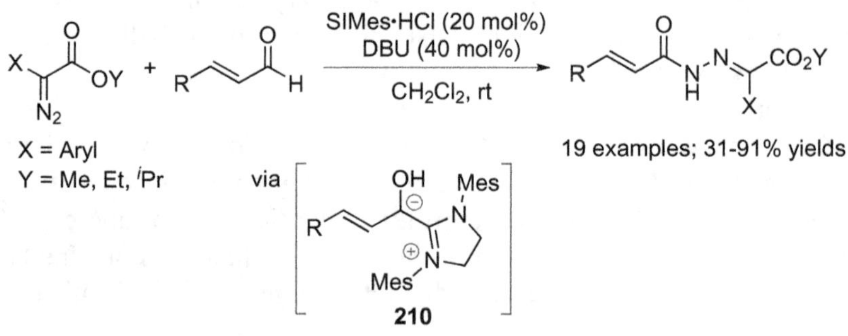

(11.32)

The Pd-catalyzed asymmetric [3+2] cycloaddition of 3-diazooxindoles as the dipolarophiles with vinyl cyclopropanes as the dipoles was achieved by Shi and co-workers, affording functionalized pyrazolidine analogues.[204] In addition, through the one-pot cycloaddition with maleimides, oxindole-fused spiropyrazolidine derivatives could be

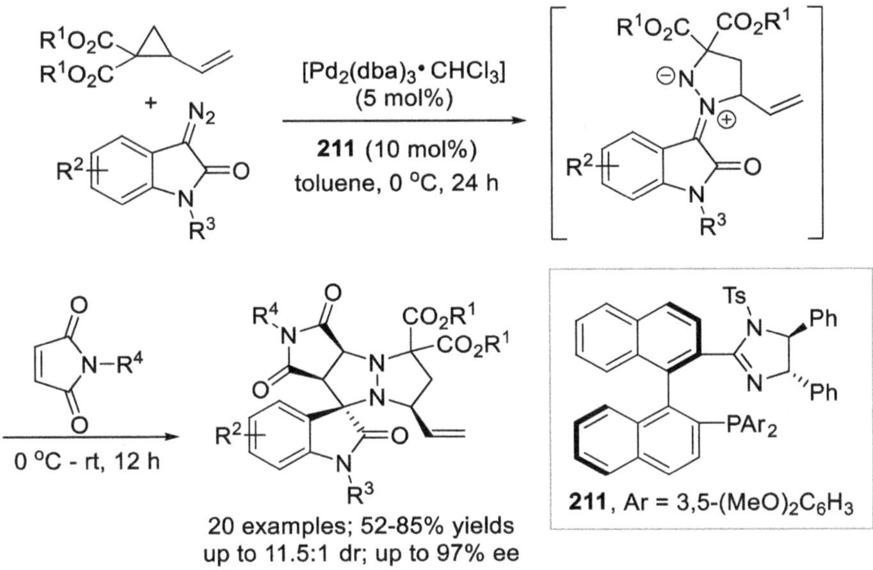

Scheme 11.69 Pd-catalyzed asymmetric tandem synthesis of oxindole-fused spiropyrazolidines.

obtained with high yields and good diastereo- and enantioselectivities (Scheme 11.69).

Recently, an Fe-catalyzed radical-type olefin hydroamination with diazocarbonyl compounds was reported to afford functionalized secondary and tertiary hydrazones (Scheme 11.70).[205] This mild hydroamination protocol was also implemented in an intramolecular manner. Mechanistically, this reaction is initiated by the generation of Fe(III) hydride species **212**, which adds to olefin to form intermediate **213** in a regioselective manner. Dissociation of **213** occurs to form Fe(II) species and alkyl radical **24**, which is trapped by diazo substrate to deliver radical adduct **215**. Sequential SET process between **215** and Fe(II) furnished anion **216**, which undergoes protonation and isomerization to afford the final product.

11.9.3 *Nucleophilic attack on the carbon bearing diazo group*

Ma and co-workers reported the Cu(I)-catalyzed coupling reaction of β-ketoacid with *in situ* generated trifluorodiazoethane (CF_3CHN_2)

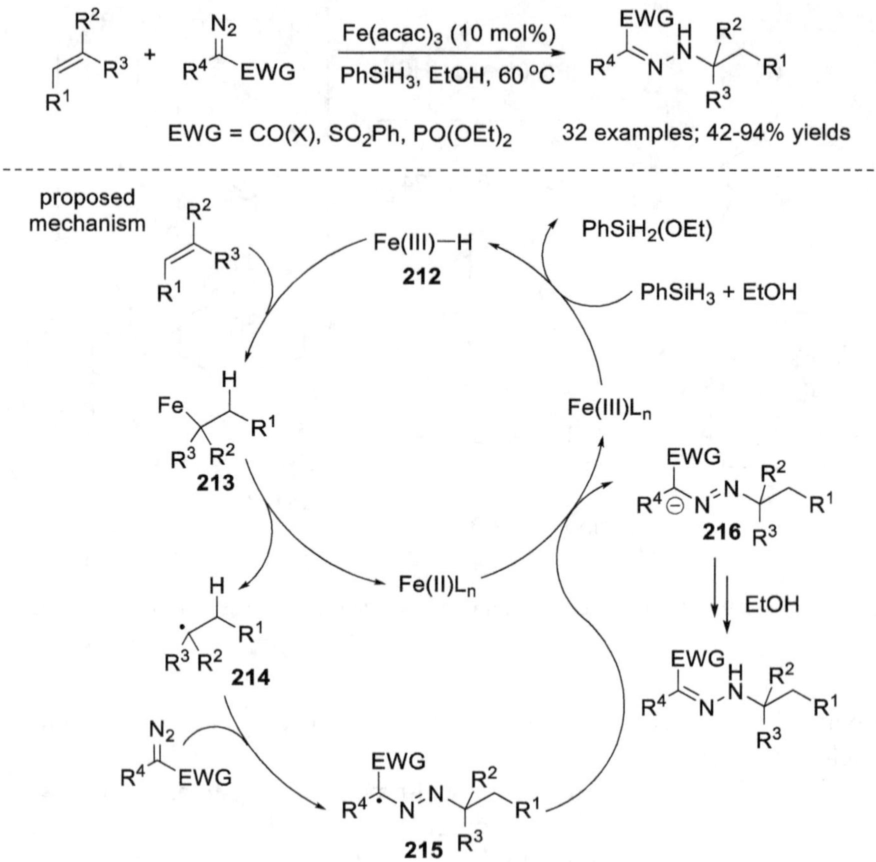

Scheme 11.70 Fe-catalyzed radical-type olefin hydroamination with diazo compounds.

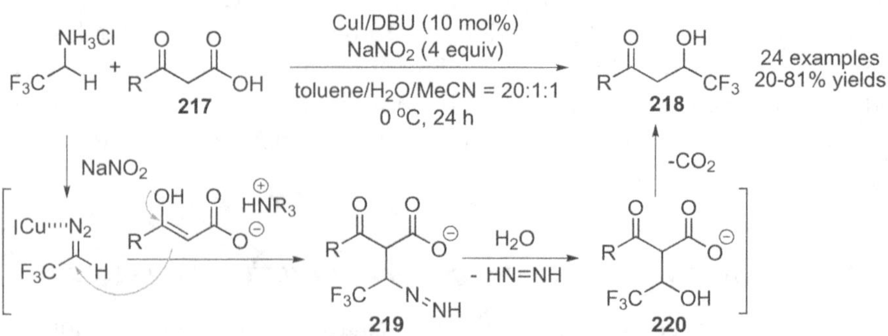

Scheme 11.71 Cu(I)-catalyzed one-pot reaction of β-ketoacids with CF_3CHN_2.

through denitrogenative-dehydrogenative-decarboxylative process.[206] This one-pot procedure afforded trifluoromethylated aldol adduct, with the formation of both C–C and C–O bond on the diazo carbon. Mechanistic hypothesis involves the nucleophilic attack on diazo species with enolate to give **219**. Subsequently, nucleophilic substitution with water to generate intermediate **220** was followed by decarboxylation to afford the final CF_3-substituted aldol product **218** (Scheme 11.71).

References

1. Padwa, A.; Austin, D. J. *Angew. Chem. Int. Ed. Engl.* **1994**, *33*, 1797–1815.
2. Zhao, Y.; Wang, J. *Synlett* **2005**, 2886–2892.
3. Bug, T.; Hartnagel, M.; Schlierf, C.; Mayr, H. *Chem. Eur. J.* **2003**, *9*, 4068–4076.
4. Zhang, Y.; Wang, J. *Chem. Commun.* **2009**, 5350–5361.
5. Hashimoto, T.; Maruoka, K. *Bull. Chem. Soc. Jpn.* **2013**, *86*, 1217–1230.
6. Likhar, P. R.; Roy, S.; Roy, M.; Subhas, M. S.; Kantam, M. L. *Synlett* **2008**, 1283–1286.
7. Likhar, P. R.; Roy, S.; Roy, M.; Subhas, M. S.; Kantam, M. L. *Catal. Commun.* **2009**, *10*, 728–731.
8. Krishna, P. R.; Prapurna, Y. L.; Alivelu, M. *Eur. J. Org. Chem.* **2011**, 5089–5095.
9. Jiang, N.; Wang, J. *Tetrahedron Lett.* **2002**, *43*, 1285–1287.
10. Xiao, F.; Liu, Y.; Wang, J. *Tetrahedron Lett.* **2007**, *48*, 1147–1149.
11. Arai, S.; Hasegawa, K.; Nishida, A. *Tetrahedron Lett.* **2004**, *45*, 1023–1026.
12. Hasegawa, K.; Arai, S.; Nishida, A. *Tetrahedron* **2006**, *62*, 1390–1401.
13. Varala, R.; Enugala, R.; Nuvula, S.; Adapa, S. R. *Tetrahedron Lett.* **2006**, *47*, 877–880.
14. Dudley, M. E.; Morshed, M. M.; Brennan, C. L.; Islam, M. S.; Ahmad, M. S.; Atuu, M.-R.; Branstetter, B.; Hossain, M. M. *J. Org. Chem.* **2004**, *69*, 7599–7608.
15. Yao, W.; Wang, J. *Org. Lett.* **2003**, *5*, 1527–1530.
16. Trost, B. M.; Malhotra, S.; Fried, B. A. *J. Am. Chem. Soc.* **2009**, *131*, 1674–1675.
17. For review, see: Trost, B. M.; Bartlett, M. J. *Acc. Chem. Res.* **2015**, *48*, 688–701.
18. Trost, B. M.; Malhotra, S.; Ellerbrock, P. *Org. Lett.* **2013**, *15*, 440–443.
19. Trost, B. M.; Malhotra, S.; Koschker, P.; Ellerbrock, P. *J. Am. Chem. Soc.* **2012**, *134*, 2075–2084.

20. Wang, W.; Shen, K.; Hu, X.; Wang, J.; Liu, X.; Feng, X. *Synlett* **2009**, 1655–1658.
21. Benfatti, F.; Yilmaz, S.; Cozzi, P. G. *Adv. Synth. Catal.* **2009**, *351*, 1763–1767.
22. Wang, F.; Liu, X.; Zhang, Y.; Lin, L.; Feng, X. *Chem. Commun.* **2009**, 7297–7299.
23. Cui, H.-F.; Wang, L.; Yang, L.-J.; Nie, J.; Zheng, Y.; Ma, J.-A. *Tetrahedron* **2011**, *67*, 8470–8476.
24. Bandyopadhyay, A.; Agrawal, N.; Mali, S. M.; Jadhav, S. V.; Gopi, H. N. *Org. Biomol. Chem.* **2010**, *8*, 4855–4860.
25. Murata, H.; Ishitani, H.; Iwamoto, M. *Tetrahedron Lett.* **2008**, *49*, 4788–4791.
26. Hashimoto, T.; Miyamoto, H.; Naganawa, Y.; Maruoka, K. *J. Am. Chem. Soc.* **2009**, *131*, 11280–11281.
27. Wommack, A. J.; Kingsbury, J. S. *J. Org. Chem.* **2013**, *78*, 10573–10587.
28. Yang, Z.; Son, K.-I.; Li, S.; Zhou, B.; Xu, J. *Eur. J. Org. Chem.* **2014**, 6380–6384.
29. Lee, S. I.; Kang, B. C.; Hwang, G.-S.; Ryu, D. H. *Org. Lett.* **2013**, *15*, 1428–1431.
30. Chen, S.; Yuan, F.; Zhao, H.; Li, B. *RSC Adv.* **2013**, *3*, 12616–12620.
31. Chen, S.; Du, Y.; Zhao, H.; Li, B. *RSC Adv.* **2014**, *4*, 12482–12485.
32. Allwood, D. M.; Blakemore, D. C.; Ley, S. V. *Org. Lett.* **2014**, *16*, 3064–3067.
33. Hashimoto, T.; Naganawa, Y.; Maruoka, K. *J. Am. Chem. Soc.* **2008**, *130*, 2434–2435.
34. Fructos, M. R.; Díaz-Requejo, M. M.; Pérez, P. J. *Chem. Commun.* **2009**, 5153–5155.
35. Kang, B. C.; Shim, S. Y.; Ryu, D. H. *Org. Lett.* **2014**, *16*, 2077–2079.
36. Morandi, B.; Carreira, E. M. *Angew. Chem. Int. Ed.* **2011**, *50*, 9085–9088.
37. Morandi, B.; Carreira, E. M. *Org. Lett.* **2011**, *13*, 5984–5985.
38. Hamilton, J. Y.; Morandi, B.; Carreira, E. M. *Synthesis* **2013**, *45*, 1857–1862.
39. Xia, Y; Qu, P.; Liu, Z.; Ge, R.; Xiao, Q.; Zhang, Y.; Wang, J. *Angew. Chem. Int. Ed.* **2013**, *52*, 2543–2546.
40. Xia, A.-J.; Kang, T.-R.; He, L.; Chen, L.-M.; Li, W.-T.; Yang, J.-L.; Liu, Q.-Z. *Angew. Chem. Int. Ed.* **2016**, *55*, 1441–1444.
41. Li, W.; Wang, J.; Hu, X.; Shen, K.; Wang, W.; Chu, Y.; Lin, L.; Liu, X.; Feng, X. *J. Am. Chem. Soc.* **2010**, *132*, 8532–8533.
42. Li, W.; Liu, X.; Tan, F.; Hao, X.; Zheng, J.; Lin, L.; Feng, X. *Angew. Chem. Int. Ed.* **2013**, *52*, 10883–10886.
43. Li, W.; Tan, F.; Hao, X.; Wang, G.; Tang, Y.; Liu, X.; Lin, L.; Feng, X. *Angew. Chem. Int. Ed.* **2015**, *54*, 1608–1611.
44. Gao, L.; Kang, B. C.; Hwang, G.-S.; Ryu, D. H. *Angew. Chem. Int. Ed.* **2012**, *51*, 8322–8325.

45. Gao, L.; Kang, B. C.; Ryu, D. H. *J. Am. Chem. Soc.* **2013**, *135*, 14556–14559.
46. Lee, S. I.; Hwang, G.-S.; Ryu, D. H. *J. Am. Chem. Soc.* **2013**, *135*, 7126–7129.
47. Kang, B. C.; Nam, D. G.; Hwang, G.-S.; Ryu, D. H. *Org. Lett.* **2015**, *17*, 4810–4813.
48. Shin, S. H.; Baek, E. H.; Hwang, G.-S.; Ryu, D. H. *Org. Lett.* **2015**, *17*, 4746–4749.
49. Moebius, D. C.; Kingsbury, J. S. *J. Am. Chem. Soc.* **2009**, *131*, 878–879.
50. Dabrowski, J. A.; Moebius, D. C.; Wommack, A. J.; Kornahrens, A. F.; Kingsbury, J. S. *Org. Lett.* **2010**, *12*, 3598–3601.
51. Hashimoto, T.; Naganawa, Y.; Maruoka, K. *J. Am. Chem. Soc.* **2009**, *131*, 6614–6617.
52. Gioiello, A.; Venturoni, F.; Marinozzi, M.; Natalini, B.; Pellicciari, R. *J. Org. Chem.* **2011**, *76*, 7431–7437.
53. Paterna, R.; André, V.; Duarte, M. T.; Veiros, L. F.; Candeias, N. R.; Gois, P. M. P. *Eur. J. Org. Chem.* **2013**, 6280–6290.
54. Sletten, E. M.; de Almeida, G.; Bertozzi, C. R. *Org. Lett.* **2014**, *16*, 1634–1637.
55. Hashimoto, T.; Naganawa, Y.; Maruoka, K. *Chem. Commun.* **2010**, *46*, 6810–6812.
56. Hashimoto, T.; Naganawa, Y.; Maruoka, K. *J. Am. Chem. Soc.* **2011**, *133*, 8834–8837.
57. Hashimoto, T.; Isobe, S.; Callens, C. K. A.; Maruoka, K. *Tetrahedron* **2012**, *68*, 7630–7635.
58. Rendina, V. L.; Moebius, D. C.; Kingsbury, J. S. *Org. Lett.* **2011**, *13*, 2004–2007.
59. Rendina, V. L.; Kaplan, H. Z.; Kingsbury, J. S. *Synthesis* **2012**, *44*, 686–693.
60. Li, W.; Liu, X.; Hao, X.; Cai, Y.; Lin, L.; Feng, X. *Angew. Chem. Int. Ed.* **2012**, *51*, 8644–8647.
61. Mao, H.; Tang, Z.; Hu, H.; Cheng, Y.; Zheng, W.-H.; Zhu, C. *Chem. Commun.* **2014**, *50*, 9773–9775.
62. Jiang, N.; Ma, Z.; Qu, Z.; Xing, X.; Xie, L.; Wang, J. *J. Org. Chem.* **2003**, *68*, 893–900.
63. Shi, W.; Jiang, N.; Zhang, S.; Wu, W.; Du, D.; Wang, J. *Org. Lett.* **2003**, *5*, 2243–2246.
64. Shi, W.; Xiao, F.; Wang, J. *J. Org. Chem.* **2005**, *70*, 4318–4322.
65. Xiao, F.; Wang, J. *J. Org. Chem.* **2006**, *71*, 5789–5791.
66. Xu, F.; Shi, W.; Wang, J. *J. Org. Chem.* **2005**, *70*, 4191–4194.
67. Xu, F.; Zhang, S.; Wu, X.; Liu, Y.; Shi, W.; Wang, J. *Org. Lett.* **2006**, *8*, 3207–3210.
68. Jadhav, A. M.; Pagar, V. V.; Liu, R.-S. *Angew. Chem. Int. Ed.* **2012**, *51*, 11809–11813.

69. Pagar, V. V.; Jadhav, A. M.; Liu, R.-S. *J. Org. Chem.* **2013**, *78*, 5711–5716.
70. Troyer, T. L.; Muchalski, H.; Johnston, J. N. *Chem. Commun.* **2009**, 6195–6197.
71. Gulevich, A. V.; Helan, V.; Wink, D. J.; Gevorgyan, V. *Org. Lett.* **2013**, *15*, 956–959.
72. Xiao, T.; Li, L.; Lin, G.; Mao, Z.-W.; Zhou, L. *Org. Lett.* **2014**, *16*, 4232–4235.
73. Xiao, T.; Peng, P.; Xie, Y.; Wang, Z.-Y.; Zhou, L. *Org. Lett.* **2015**, *17*, 4332-4335.
74. Jiang, N.; Ma, Z.; Qu, Z.; Xing, X.; Xie, L.; Wang, J. *J. Org. Chem.* **2003**, *68*, 893–900.
75. Zhao, Y.; Jiang, N.; Wang, J. *Tetrahedron Lett.* **2003**, *44*, 8339–8342.
76. Zhao, Y.; Jiang, N.; Chen, S.; Peng, C.; Zhang, X.; Zou, Y.; Zhang, S.; Wang, J. *Tetrahedron* **2005**, *61*, 6546–6552.
77. Chen, S.; Zhao, Y.; Wang, J. *Synthesis* **2006**, 1705–1710.
78. Cainelli, G.; Giacomini, D.; Galletti, P.; Quintavalla, A. *Eur. J. Org. Chem.* **2003**, 1765–1774.
79. Broccolo, F.; Cainelli, G.; Caltabiano, G.; Cocuzza, C. E. A.; Fortuna, C. G.; Galletti, P.; Giacomini, D.; Musumarra, G.; Musumeci, R.; Quintavalla, A. *J. Med. Chem.* **2006**, *49*, 2804–2811.
80. Zhao, Y.; Ma, Z.; Zhang, X.; Zou, Y.; Jin, X.; Wang, J. *Angew. Chem. Int. Ed.* **2004**, *43*, 5977–5980.
81. Uraguchi, D.; Sorimachi, K.; Terada, M. *J. Am. Chem. Soc.* **2005**, *127*, 9360–9361.
82. Hashimoto, T.; Maruoka, K. *J. Am. Chem. Soc.* **2007**, *129*, 10054–10055.
83. Hashimoto, T.; Maruoka, K. *Synthesis* **2008**, 3703–3706.
84. Hashimoto, T.; Kimura, H.; Nakatsu, H.; Maruoka, K. *J. Org. Chem.* **2011**, *76*, 6030–6037.
85. Hashimoto, T.; Kimura, H.; Kawamata, Y.; Maruoka, K. *Nat. Chem.* **2011**, *3*, 642–646.
86. Terada, M.; Kanomata, K. *Synlett* **2011**, 1255–1258.
87. Zhang, H.; Wen, X.; Gan, L.; Peng, Y. *Org. Lett.* **2012**, *14*, 2126–2129.
88. Liu, W.-J.; Lv, B.-D.; Gong, L.-Z. *Angew. Chem. Int. Ed.* **2009**, *48*, 6503–6506.
89. He, L.; Liu, W.-J.; Ren, L.; Lei, T.; Gong, L.-Z. *Adv. Synth. Catal.* **2010**, *352*, 1123–1127.
90. Liu, G.; Zhang, D.; Li, J.; Xu, G.; Sun, J. *Org. Biomol. Chem.* **2013**, *11*, 900–904.
91. Wang, Z.; Wen, J.; Bi, Q.-W.; Xu, X.-Q.; Shen, Z.-Q.; Li, X.-X.; Chen, Z. *Tetrahedron Lett.* **2014**, *55*, 2969–2972.
92. Pollard, M. M.; Wesenhagen, P. V.; Pijper, D.; Feringa, B. L. *Org. Biomol. Chem.* **2008**, *6*, 1605–1612.

93. Zhang, Y.; Lu, Z.; Wulff, W. D. *Synlett* **2009**, 2715–2739.
94. Degennaro, L.; Trinchera, P.; Luisi, R. *Chem. Rev.* **2014**, *114*, 7881–7929.
95. Johnston, J. N.; Muchalski, H.; Troyer, T. L. *Angew. Chem. Int. Ed.* **2010**, *49*, 2290–2298.
96. Sweeney, J. *Eur. J. Org. Chem.* **2009**, 4911–4919.
97. Künzi, S. A.; Morandi, B.; Carreira, E. M. *Org. Lett.* **2012**, *14*, 1900–1901.
98. Hashimoto, T.; Nakatsu, H.; Watanabe, S.; Maruoka, K. *Org. Lett.* **2010**, *12*, 1668–1671.
99. Hashimoto, T.; Nakatsu, H.; Yamamoto, K.; Watanabe, S.; Maruoka, K. *Chem. Asian J.* **2011**, *6*, 607–613.
100. Zhao, Y.; Wang, G.; Zhou, S.; Li, Z.; Meng, X. *Org. Biomol. Chem.* **2014**, *12*, 3362–3365.
101. Zhang, Y.; Desai, A.; Lu, Z.; Hu, G.; Ding, Z.; Wulff, W. D. *Chem. Eur. J.* **2008**, *14*, 3785–3803.
102. Hu, G.; Huang, L.; Huang, R. H.; Wulff, W. D. *J. Am. Chem. Soc.* **2009**, *131*, 15615–15617.
103. Vetticatt, M. J.; Desai, A. A.; Wulff, W. D. *J. Am. Chem. Soc.* **2010**, *132*, 13104–13107.
104. Hu, G.; Gupta, A. K.; Huang, R. H.; Mukherjee, M.; Wulff, W. D. *J. Am. Chem. Soc.* **2010**, *132*, 14669–14675.
105. Mukherjee, M.; Gupta, A. K.; Lu, Z.; Zhang, Y.; Wulff, W. D. *J. Org. Chem.* **2010**, *75*, 5643–5660.
106. Vetticatt, M. J.; Desai, A. A.; Wulff, W. D. *J. Org. Chem.* **2013**, *78*, 5142–5152.
107. Desai, A. A.; Ren, H.; Mukherjee, M.; Wulff, W. D. *Org. Process Res. Dev.* **2011**, *15*, 1108–1115.
108. Huang, L.; Zhao, W.; Staples, R. J.; Wulff, W. D. *Chem. Sci.* **2013**, *4*, 622–628.
109. Desai, A. A.; Wulff, W. D. *J. Am. Chem. Soc.* **2010**, *132*, 13100–13103.
110. Huang, L.; Zhang, Y.; Staples, R. J.; Huang, R. H.; Wulff, W. D. *Chem. Eur. J.* **2012**, *18*, 5302–5313.
111. Guan, Y.; López-Alberca, M. P.; Lu, Z.; Zhang, Y.; Desai, A. A.; Patwardhan, A. P.; Dai, Y.; Vetticatt, M. J.; Wulff, W. D. *Chem. Eur. J.* **2014**, *20*, 13894–13900.
112. Gupta, A. K.; Mukherjee, M.; Wulff, W. D. *Org. Lett.* **2011**, *13*, 5866–5869.
113. Gupta, A. K.; Mukherjee, M.; Hu, G.; Wulff, W. D. *J. Org. Chem.* **2012**, *77*, 7932–7944.
114. Guan, Y.; Ding, Z.; Wulff, W. D. *Chem. Eur. J.* **2013**, *19*, 15565–15571.
115. Mukherjee, M.; Zhou, Y.; Gupta, A. K.; Guan, Y.; Wulff, W. D. *Eur. J. Org. Chem.* **2014**, 1386–1390.
116. Ren, H.; Wulff, W. D. *Org. Lett.* **2010**, *12*, 4908–4911.
117. Huang, L.; Wulff, W. D. *J. Am. Chem. Soc.* **2011**, *133*, 8892–8895.

118. Hashimoto, T.; Uchiyama, N.; Maruoka, K. *J. Am. Chem. Soc.* **2008**, *130*, 14380–14381.
119. Zeng, X.; Zeng, X.; Xu, Z.; Lu, M.; Zhong, G. *Org. Lett.* **2009**, *11*, 3036–3039.
120. Hashimoto, T.; Gálvez, A. O.; Maruoka, K. *J. Am. Chem. Soc.* **2013**, *135*, 17667–17670.
121. Akiyama, T.; Suzuki, T.; Mori, K. *Org. Lett.* **2009**, *11*, 2445–2447.
122. Chai, Z.; Bouillon, J.-P.; Cahard, D. *Chem. Commun.* **2012**, *48*, 9471–9473.
123. Hashimoto, T.; Nakatsu, H.; Yamamoto, K.; Maruoka, K. *J. Am. Chem. Soc.* **2011**, *133*, 9730–9733.
124. Li, H.; Zhang, Y.; Wang, J. *Synthesis* **2013**, *45*, 3090–3098.
125. He, Z.; Zajdlik, A.; Yudin, A. K. *Dalton Trans.* **2014**, *43*, 11434–11451.
126. Peng, C.; Zhang, W.; Yan, G.; Wang, J. *Org. Lett.* **2009**, *11*, 1667–1670.
127. Barluenga, J.; Tomás-Gamasa, M.; Aznar, F.; Valdés, C. *Nat. Chem.* **2009**, *1*, 494–499.
128. Pérez-Aguilar, M. C.; Valdés, C. *Angew. Chem. Int. Ed.* **2012**, *51*, 5953–5957.
129. Li, X.; Feng, Y.; Lin, L.; Zou, G. *J. Org. Chem.* **2012**, *77*, 10991–10995.
130. Allwood, D. M.; Blakemore, D. C.; Brown, A. D.; Ley, S. V. *J. Org. Chem.* **2014**, *79*, 328–338.
131. Sanchez-Carmona, M. A.; Contreras-Cruz, D. A.; Miranda, L. D. *Org. Biomol. Chem.* **2011**, *9*, 6506–6508.
132. Luan, Y.; Schaus, S. E. *Org. Lett.* **2011**, *13*, 2510–2513.
133. Luan, Y.; Yu, J.; Zhang, X.; Schaus, S. E.; Wang, G. *J. Org. Chem.* **2014**, *79*, 4694–4698.
134. Argintaru, O. A.; Ryu, D.; Aron, I.; Molander, G. A. *Angew. Chem. Int. Ed.* **2013**, *52*, 13656–13660.
135. Wu, G.; Deng, Y.; Wu, C.; Wang, X.; Zhang, Y.; Wang, J. *Eur. J. Org. Chem.* **2014**, 4477–4481.
136. Wu, G.; Deng, Y.; Wu, C.; Zhang, Y.; Wang, J. *Angew. Chem. Int. Ed.* **2014**, *53*, 10510–10514.
137. Molander, G. A.; Ryu, D. *Angew. Chem. Int. Ed.* **2014**, *53*, 14181–14185.
138. Pasceri, R.; Bartrum, H. E.; Hayes, C. J.; Moody, C. J. *Chem. Commun.* **2012**, *48*, 12077–12079.
139. Li, H.; Wang, L.; Zhang, Y.; Wang, J. *Angew. Chem. Int. Ed.* **2012**, *51*, 2943–2946.
140. Li, H.; Shangguan, X.; Zhang, Z.; Huang, S.; Zhang, Y.; Wang, J. *Org. Lett.* **2014**, *16*, 448–451.
141. Cuenca, A. B.; Cid, J.; García-López, D.; Carbó, J. J.; Fernández, E. *Org. Biomol. Chem.* **2015**, *13*, 9659–9664. Correction: *Org. Biomol. Chem.* **2015**, *13*, 11772–11772.

142. Das, A.; Wang, D.; Belhomme, M.-C.; Szabó, K. J. *Org. Lett.* **2015**, *17*, 4754–4757.
143. Braunschweig, H.; Hupp, F.; Krummenacher, I.; Mailänder, L.; Rauch, F. *Chem. Eur. J.* **2015**, *21*, 17844–17849.
144. Doyle, M. P.; Kundu, K.; Russell, A. E. *Org. Lett.* **2005**, *7*, 5171–5174.
145. Kundu, K.; Doyle, M. P. *Tetrahedron: Asymmetry* **2006**, *17*, 574–577.
146. Liu, Y.; Zhang, Y.; Jee, N.; Doyle, M. P. *Org. Lett.* **2008**, *10*, 1605–1608.
147. Xu, X.; Hu, W.-H.; Doyle, M. P. *Angew. Chem. Int. Ed.* **2011**, *50*, 6392–6395.
148. Zhou, L.; Doyle, M. P. *Org. Lett.* **2010**, *12*, 796–799.
149. Liu, Y.; Liu, Y.; Shanahan, C. S.; Xu, X.; Doyle, M. P. *Org. Biomol. Chem.* **2014**, *12*, 5227–5234.
150. Deng, G.; Tian, X.; Wang, J. *Tetrahedron Lett.* **2003**, *44*, 587–589.
151. Liao, M.; Dong, S.; Deng, G.; Wang, J. *Tetrahedron Lett.* **2006**, *47*, 4537–4540.
152. Dong, C.; Deng, G.; Wang, J. *J. Org. Chem.* **2006**, *71*, 5560–5564.
153. Sá, M. M.; Silveira, G. P.; Bortoluzzi, A. J.; Padwa, A. *Tetrahedron* **2003**, *59*, 5441–5447.
154. Mao, H.; Lin, A.; Shi, Y.; Mao, Z.; Zhu, X.; Li, W.; Hu, H.; Cheng, Y.; Zhu, C. *Angew. Chem. Int. Ed.* **2013**, *52*, 6288–6292.
155. Muthusamy, S.; Sivaguru, M. *Org. Lett.* **2014**, *16*, 4248–4251.
156. Luan, Y.; Qi, Y.; Gao, H.; Ma, Q.; Schaus, S. E. *Eur. J. Org. Chem.* **2014**, 6868–6872.
157. Stopka, T.; Marzo, L.; Zurro, M.; Janich, S.; Würthwein, E.-U.; Daniliuc, C. G.; Alemán, J.; Mancheño, O. G. *Angew. Chem. Int. Ed.* **2015**, *54*, 5049–5053.
158. Miwa, K.; Aoyama, T.; Shioiri, T. *Synlett* **1994**, 461–462.
159. Sakai, A.; Aoyama, T.; Shioiri, T. *Tetrahedron* **1999**, *55*, 3687–3694.
160. Miyabe, R.; Shioiri, T.; Aoyama, T. *Heterocycles* **2002**, *57*, 1313–1318.
161. Yun, S. Y.; Zheng, J.-C.; Lee, D. *J. Am. Chem. Soc.* **2009**, *131*, 8413–8415.
162. Li, J.; Sun, C.; Lee, D. *J. Am. Chem. Soc.* **2010**, *132*, 6640–6641.
163. Zheng, J.-C.; Yun, S. Y.; Sun, C.; Lee, N.-K.; Lee, D. *J. Org. Chem.* **2011**, *76*, 1086–1099.
164. Zheng, J.-C.; Liu, H.; Lee, N.-K.; Lee, D. *Eur. J. Org. Chem.* **2014**, 506–510.
165. Hari, Y.; Date, K.; Kondo, R.; Aoyama, T. *Tetrahedron Lett.* **2008**, *49*, 4965–4967.
166. Hari, Y.; Kondo, R.; Date, K.; Aoyama, T. *Tetrahedron* **2009**, *65*, 8708–8713.
167. O'Connor, M. J.; Sun, C.; Guan, X.; Sabbasani, V. R.; Lee, D. *Angew. Chem. Int. Ed.* **2016**, *55*, 2222–2225.

168. Roth, G. J.; Bernd, L.; Müller, S. G.; Bestmann, H. J. *Synthesis* **2004**, 59–62.
169. Barrett, A. G. M.; Hopkins, B. T.; Love, A. C.; Tedeschi, L. *Org. Lett.* **2004**, *6*, 835–837.
170. Goundry, W. R. F.; Baldwin, J. E.; Lee, V. *Tetrahedron* **2003**, *59*, 1719–1729.
171. Xu, Z.; Peng, Y.; Ye, T. *Org. Lett.* **2003**, *5*, 2821–2824.
172. Marshall, J. A.; Schaaf, G. M. *J. Org. Chem.* **2003**, *68*, 7428–7432.
173. Maehr, H.; Uskokovic, M. R. *Eur. J. Org. Chem.* **2004**, 1703–1713.
174. Quesada, E.; Taylor, R. J. K. *Tetrahedron Lett.* **2005**, *46*, 6473–6476.
175. Taber, D. F.; Bai, S.; Guo, P.-F. *Tetrahedron Lett.* **2008**, *49*, 6904–6906.
176. Maisonneuve, S.; Xie, J. *Synlett* **2009**, 2977–2981.
177. Jepsen, T. H.; Kristensen, J. L. *J. Org. Chem.* **2014**, *79*, 9423–9426.
178. Pramanik, M. M. D.; Chaturvedi, A. K.; Rastogi, N. *Chem. Commun.* **2014**, *50*, 12896–12898.
179. Pramanik, M. M. D.; Rastogi, N. *Org. Biomol. Chem.* **2016**, *14*, 1239–1243.
180. Galliford, C. V.; Scheidt, K. A. *Chem. Commun.* **2008**, 1926–1928.
181. Saraireh, I. A. M. *Tetrahedron Lett.* **2012**, *53*, 2023–2025.
182. Kui, E. L.; Kanazawa, A.; Poisson, J.-F.; Py, S. *Tetrahedron Lett.* **2013**, *54*, 5103–5105.
183. Xu, X.; Xu, X.; Zavalij, P. Y.; Doyle, M. P. *Chem. Commun.* **2013**, *49*, 2762–2764.
184. Attanasi, O. A.; De Crescentini, L.; Favi, G.; Mantellini, F.; Mantenuto, S.; Nicolini, S. *J. Org. Chem.* **2014**, *79*, 8331–8338.
185. Branstetter, B.; Hossain, M. M. *Tetrahedron Lett.* **2006**, *47*, 221–223.
186. Rostovskii, N. V.; Novikov, M. S.; Khlebnikov, A. F.; Korneev, S. M.; Yufit, D. S. *Org. Biomol. Chem.* **2013**, *11*, 5535–5545.
187. Zhang, S.; Zhang, W.-X.; Xi, Z. *Angew. Chem. Int. Ed.* **2013**, *52*, 3485–3489.
188. Supurgibekov, M. B.; Hennig, L.; Schulze, B.; Nikolaev, V. A. *Russ. J. Org. Chem.* **2008**, *44*, 1840–1843.
189. Supurgibekov, M. B.; Zakharova, V. M.; Sieler, J.; Nikolaev, V. A. *Tetrahedron Lett.* **2011**, *52*, 341–345.
190. Muthusamy, S.; Srinivasan, P. *Tetrahedron Lett.* **2009**, *50*, 1331–1334.
191. Bel Abed, H.; Mammoliti, O.; Van Lommen, G.; Herdewijn, P. *Tetrahedron Lett.* **2012**, *53*, 6489–6491.
192. Bel Abed, H.; Mammoliti, O.; Bande, O.; Van Lommen, G.; Herdewijn, P. *J. Org. Chem.* **2013**, *78*, 7845–7858.
193. Bel Abed, H.; Mammoliti, O.; Bande, O.; Van Lommen, G.; Herdewijn, P. *Org. Biomol. Chem.* **2014**, *12*, 7159–7166.

194. Bel Abed, H.; Bande, O.; Mammoliti, O.; Van Lommen, G.; Herdewijn, P. *Tetrahedron Lett.* **2013**, *54*, 7056–7058.
195. Nikolaev, V. A.; Cantillo, D.; Kappe, C. O.; Medvedev, J. J.; Prakash, G. K. S.; Supurgibekov, M. B. *Chem. Eur. J.* **2016**, *22*, 174–184.
196. Mao, H.; Lin, A.; Tang, Z.; Hu, H.; Zhu, C.; Cheng, Y. *Chem. Eur. J.* **2014**, *20*, 2454–2458.
197. Yasui, E.; Wada, M.; Takamura, N. *Tetrahedron Lett.* **2009**, *50*, 4762–4765.
198. Yasui, E.; Wada, M.; Takamura, N. *Tetrahedron* **2009**, *65*, 461–468.
199. Yasui, E.; Wada, M.; Nagumo, S.; Takamura, N. *Tetrahedron* **2013**, *69*, 4325–4330.
200. Li, L.; Chen, J.-J.; Li, Y.-J.; Bu, X.-B.; Liu, Q.; Zhao, Y.-L. *Angew. Chem. Int. Ed.* **2015**, *54*, 12107–12111.
201. Zhan, M.; Zhang, S.; Zhang, W.-X.; Xi, Z. *Org. Lett.* **2013**, *15*, 4182–4185.
202. Li, W.; Liu, X.; Hao, X.; Hu, X.; Chu, Y.; Cao, W.; Qin, S.; Hu, C.; Lin, L.; Feng, X. *J. Am. Chem. Soc.* **2011**, *133*, 15268–15271.
203. Santos, F. M. F.; Rosa, J. N.; André, V.; Duarte, M. T.; Veiros, L. F.; Gois, P. M. P. *Org. Lett.* **2013**, *15*, 1760–1763.
204. Mei, L.-Y.; Tang, X.-Y.; Shi, M. *Chem. Eur. J.* **2014**, *20*, 13136–13142.
205. Zheng, J.; Qi, J.; Cui, S. *Org. Lett.* **2016**, *18*, 128–131.
206. Xiong, H.-Y.; Yang, Z.-Y.; Chen, Z.; Zeng, J.-L.; Nie, J.; Ma, J.-A. *Chem. Eur. J.* **2014**, *20*, 8325–8329.

Chapter 12

Transition-Metal-Catalyzed Cross-Coupling with Diazo Compounds

Transition-metal-catalyzed cross-coupling reactions of diazo compounds, as new types of coupling reactions that are distinct from the classic carbene transformations, have experienced rapid growth in the past decade. The common feature of these coupling reactions is that a carbene migratory insertion is involved as the key step for the formation of carbon–carbon bond. These types of coupling reactions have been proven to be general in terms of the coupling partners and the transition-metal catalysts. It is also possible to develop cascade transformations based on the carbene coupling process. Mechanistically, these reactions can be considered as incorporation of the carbene formation and migratory insertion steps into the traditional catalytic cycle of cross-coupling reactions (Scheme 12.1).

The key step in these coupling reactions is the carbene migratory insertion. This is comparable with the insertion of CO into metal–carbon bond, which is a fundamental step in transition metal-catalyzed carbonylations. The similar one-carbon unit insertion of carbene opens up new possibilities for the development of novel transition-metal-catalyzed C–C bond forming reactions. The carbene-based coupling has been proven to be general: various transition-metals including Pd, Cu, Rh, Ni, Co and Ir are effective catalysts; the scope of the reaction has also been extended to the substrates other than diazo compounds. This chapter will summarize the carbene coupling reactions and the related cascade transformations according to the type of transition-metal catalysts that have been employed. Since the fields have been extensively reviewed recently,[1–10]

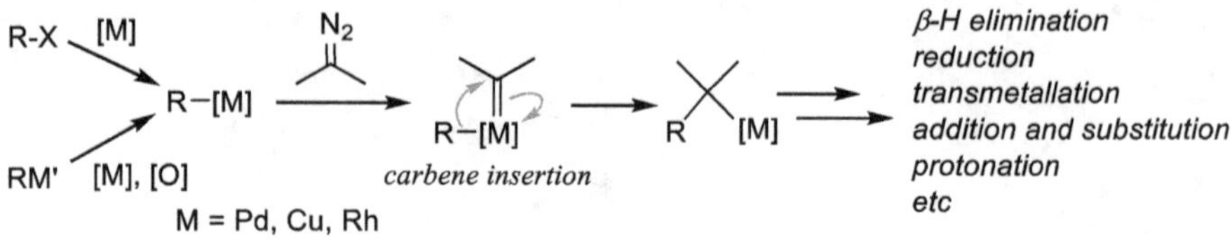

Scheme 12.1 General scheme of transition-metal-catalyzed coupling of diazo compounds involving migratory insertion.

this chapter will only present some representative examples to show the essence of these transformations.

12.1 Palladium-catalyzed coupling reaction

As the most prevalent transition-metal-catalyzed coupling of diazo compounds, the cross-coupling involving migratory insertion of palladium carbene species has been extensively explored and has demonstrated great potential in organic synthesis. The general outline of the coupling reactions is summarized in Scheme 12.2. First, Pd(II) intermediate **1** is generated either through transmetalation with other metal species or by oxidative addition of the organohalide analogues. Subsequent reaction with diazo compounds gave Pd(II) carbene **2** as the key intermediate, with the release of N_2. Subsequently the migratory insertion, in which the carbon ligand of the Pd(II) carbene migrates to the carbenic center, takes place to generate a new alkyl Pd(II) species **3**, followed by sequential *β*-H, *β*-C elimination or other cascade transformations to afford various products.

The first Pd-catalyzed cross-coupling with diazo compound was reported by Van Vranken in 2001.[11] However, the report did not attract much attention in the following years, presumably due to the limited scope of the reaction. In 2007, Barluenga and co-workers reported the palladium-catalyzed cross-coupling of aryl halides with *N*-tosylhydrazones (Eq. (12.1)).[12]

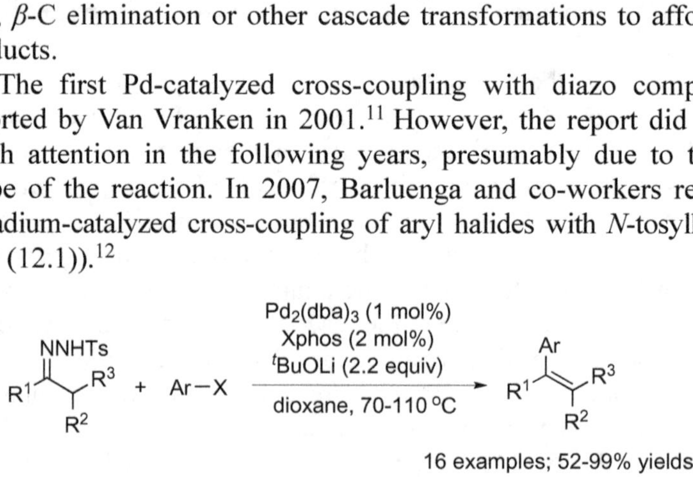

(12.1)

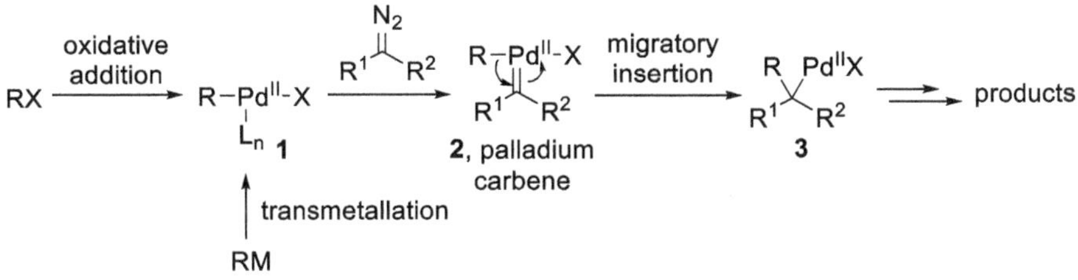

Scheme 12.2 Outline of Pd-catalyzed coupling reaction of diazo compounds involving migratory insertion.

In the following year, Wang and co-workers reported the palladium-catalyzed oxidative coupling of α-diazocarbonyl compounds with arylboronic acids (Eq. (12.2)).[13]

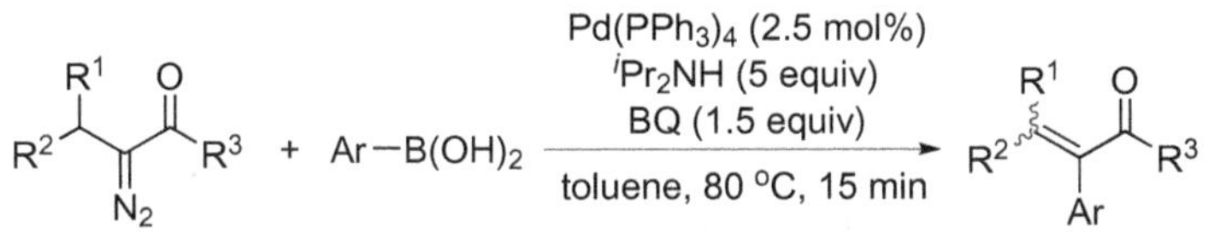

21 examples; 44-97% yields

(12.2)

In all the three coupling reactions, the formation of Pd(II) carbene and subsequent migratory insertion to generate alkyl palladium complexes are proposed as the key steps. Subsequent β-H elimination generates double bonds, along with the regeneration of Pd(0) species.

Afterward, Barluenga and Valdés launched a systematic study toward Pd-catalyzed cross-coupling reactions of aryl halides with *N*-tosylhydrazones derived from carbonyl compounds, providing efficient synthesis of various arylated alkenes.[14–20] They also utilized aryl nonaflates as the coupling partners in the Pd-catalyzed coupling with *N*-tosylhydrazones.[21,22]

An interesting recent development in this arena is the asymmetric catalysis reported by Gu and co-workers. With chiral pyrrolidine-based phosphoramidite ligand **7**, the Pd-catalyzed asymmetric synthesis of axially chiral 1-vinylnaphthalen-2-ylphosphine oxides from aryl bromides with *N*-tosylhydrazones derived from cyclic ketones could be achieved with high enantioselectivities (Scheme 12.3).[23] This reaction provides a unique method for the construction of axially chiral biaryl compounds with high enantioselectivities.

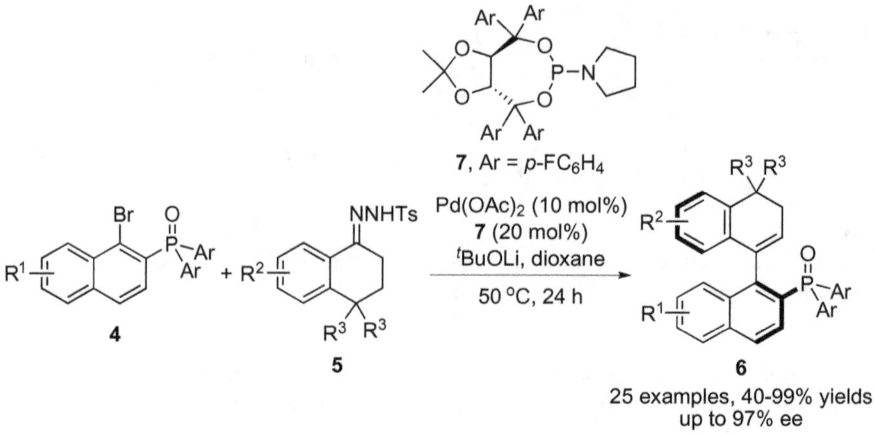

Scheme 12.3 Pd-Catalyzed enantioselective synthesis of axially chiral vinyl arenes.

Alami and co-workers have focused their research on the synthesis of 1,1-diarylethylene derivatives through Pd-catalyzed coupling reaction of polyoxygenated *N*-tosylhydrazones with aryl triflates.[24] They utilized the 1,1-diarylethylene scaffold in the synthesis of *iso*-combretastatins A analogues, which have been shown to have cytotoxic and antimitotic bioactivity.[25,26] Besides, 2-α-styrylpyridine derivatives, which exhibit antiproliferative and antimitotic activity, could be directly synthesized by the Pd-catalyzed coupling of *N*-tosylhydrazones with 2-halopyridines.[27] Alami and co-workers also reported the synthesis of *tetra*-substituted 1,1-diarylethylenes through the Pd-catalyzed coupling of aryl halides with sterically hindered *N*-tosylhydrazones.[28,29]

As a further extension, the aromatic C–F bond cleavage has been incorporated with the Pd-catalyzed carbene couplings. Pd-catalyzed cross-coupling reaction of aryl fluorides with *N*-tosylhydrazones **8** was realized to afford di- and tri-substituted olefins **9** (Scheme 12.4).[30] Mechanistically, the reaction was proposed to be initiated by electron-deficient group directed S_NAr-type process, which afforded the aryl-palladium(II) intermediate **10**.

Zhai and co-workers reported a synthesis of 2-arylphenols based on the Pd-catalyzed coupling of 6-diazocyclohexenone and aryl iodide, followed by aromatization sequence (Eq. (12.3)).[31]

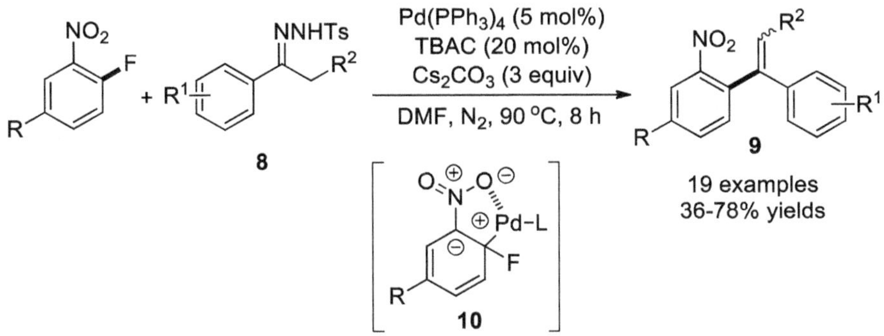

Scheme 12.4 Pd-catalyzed cross-coupling reaction of aryl fluorides with *N*-tosylhydrazones.

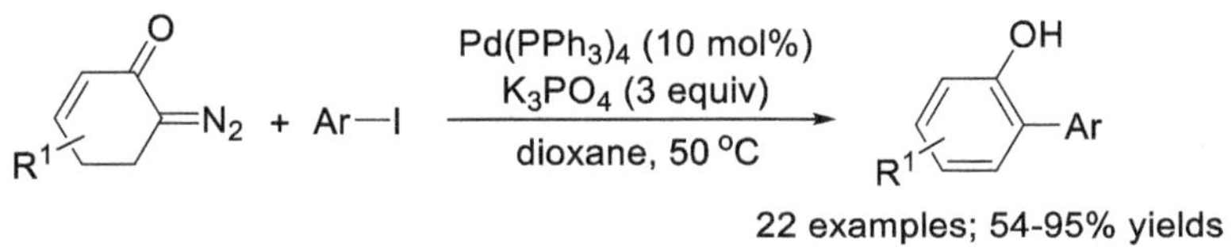

(12.3)

In the case when *β*-H is not available, the palladium intermediate generated from carbene migratory can be reduced to generate C–C single bond. For example, Wang and co-workers reported palladium-catalyzed reductive coupling of diaryl *N*-tosylhydrazones **11** and aryl bromide with ammonium formate as the reductant. The reaction provides useful access to functionalized triarylmethanes **12** (Eq. (12.4)).[32]

(12.4)

In addition to the coupling reaction with aryl halides, catalytic oxidative cross-coupling of arylboronic acids with diazo species was also developed. Mechanistically, the reaction involves the transmetalation between Pd(II) intermediate with boronic acid to generate aryl-Pd(II) species, as shown in Eq. (12.2).

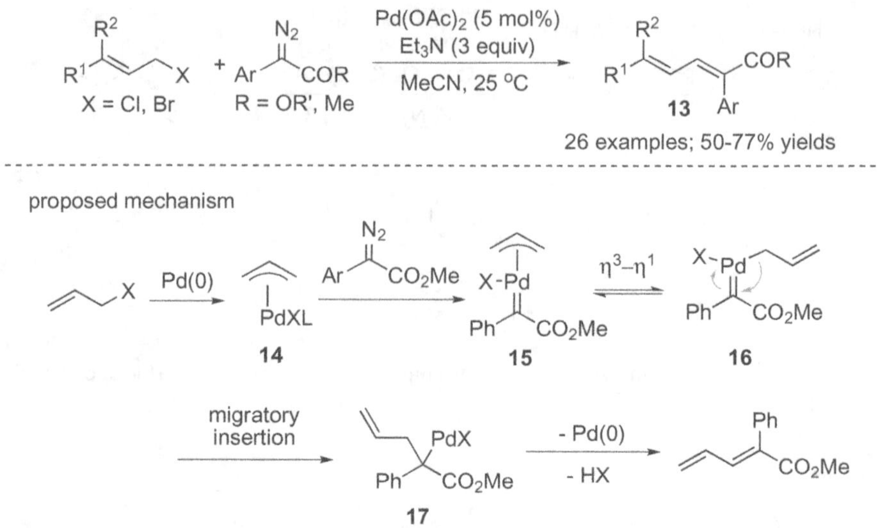

Scheme 12.5 Pd-catalyzed cross-coupling reaction of allyl halides with diazo compounds.

The cross-coupling with allyl halides was reported by Chen and Wang. The coupling reaction afforded 1,3-dienes **13** (Scheme 12.5).[33] The reaction was proposed to follow a pathway involving a π-allylic palladium carbene species, from which allyl migratory insertion occurs.

The cross-coupling involving π-allylic palladium carbene migratory insertion has been extended to allylic C–H bond olefinations by Gong and co-workers (Eq. (12.5)).[34] The cascade binary catalytic transformation with Pd(II)/Lewis acid has combined the allylic C–H activation with migratory insertion of allyl group.

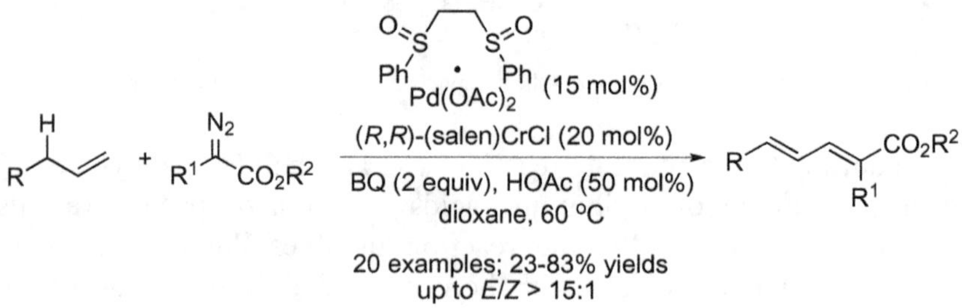

(12.5)

Furthermore, Wang and co-workers reported the Pd-catalyzed oxidative coupling between *N*-tosylhydrazones or diazo compounds and terminal alkynes **18**, affording conjugated enynes **19** (Eq. (12.6)).[35] The key steps in this transformation involve the migratory insertion of alkynyl group to metal carbene center, followed by β-H elimination to afford enynes.

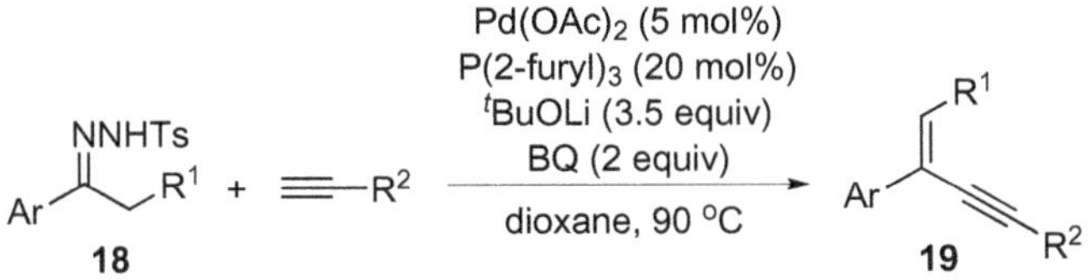

18 **19**

18 examples; 42-83% yields

(12.6)

The β-H elimination may be replaced by reductive elimination, thus forming two different C–C bonds in the carbonic carbon. For example, Liang and co-workers reported a Pd-catalyzed coupling reaction of aryl-diazoester **20** with allylic esters **21**, affording product **22** with the formation of two C–C bonds on the carbonic carbon (Scheme 12.6).[36]

Migratory insertion of other atoms, such as nitrogen and silicon, to carbene center has also been reported. For examples, Pd-catalyzed oxidative cross-coupling of *N*-tosylhydrazones with indoles was developed by Cui[37] and Alami,[38] providing a method for synthesizing *N*-vinylindoles

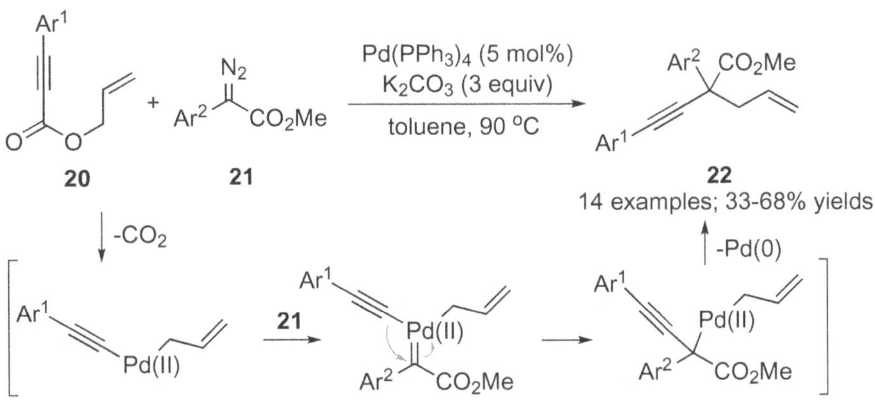

Scheme 12.6 Pd-Catalyzed coupling of diazo compounds with allylic esters.

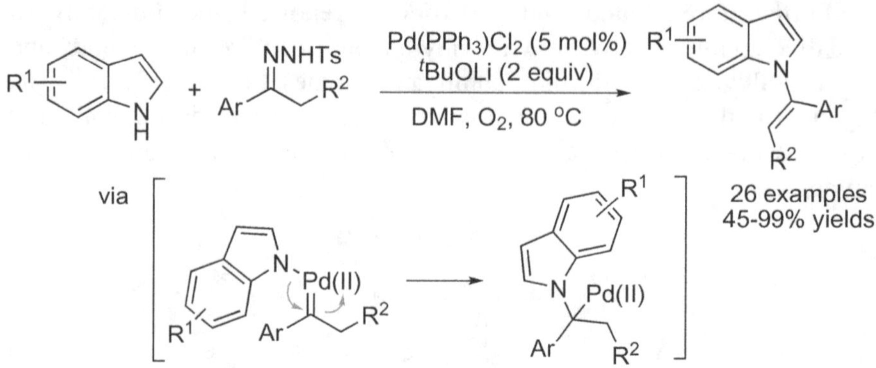

Scheme 12.7 Pd-catalyzed oxidative cross-coupling of *N*-tosylhydrazones with indoles.

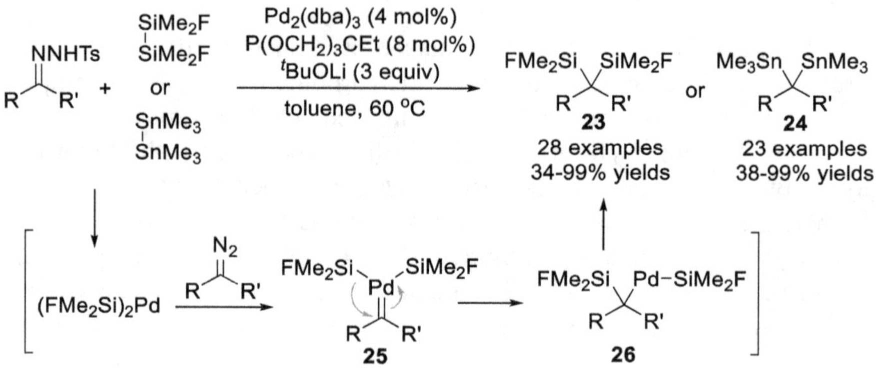

Scheme 12.8 Pd-catalyzed oxidative cross-coupling of *N*-tosylhydrazones with indoles.

as well as *N*-vinylazoles with good functional group tolerance (Scheme 12.7).

Wang and co-workers reported the Pd-catalyzed carbene insertion into Si–Si and Sn–Sn bonds, in which *N*-tosylhydrazones were used as the carbene precursors (Scheme 12.8).[39] This reaction afforded geminal bis(silane) **23** and geminal bis(stannane) **24** derivatives with high efficiency under mild conditions. This reaction was assumed to proceed *via* oxidation addition, migratory insertion of Pd carbene **25** into Pd–Si or Pd–Sn bond, and finally reductive elimination of alkyl palladium intermediate **26**.

Jiang and co-workers reported that Pd-catalyzed oxidative coupling of two molecules of *N*-tosylhydrazones afforded branched conjugated dienes **27**.[40] The proposed reaction mechanism involves diazo decomposition by alkenyl Pd species **28** to generate alkenyl-Pd carbene intermediate **29**.

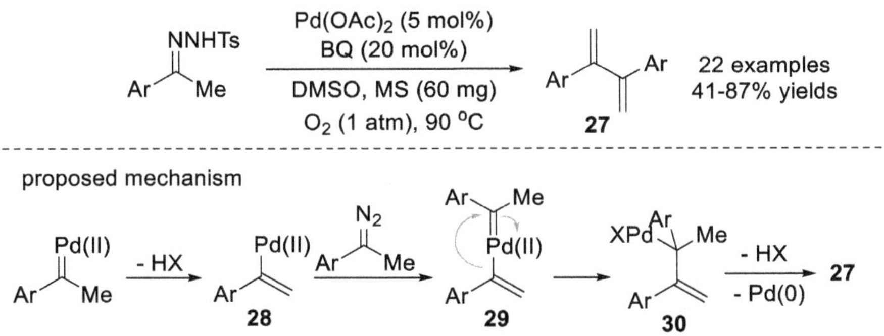

Scheme 12.9 Pd-catalyzed oxidative coupling of *N*-tosylhydrazones to give branched conjugated dienes.

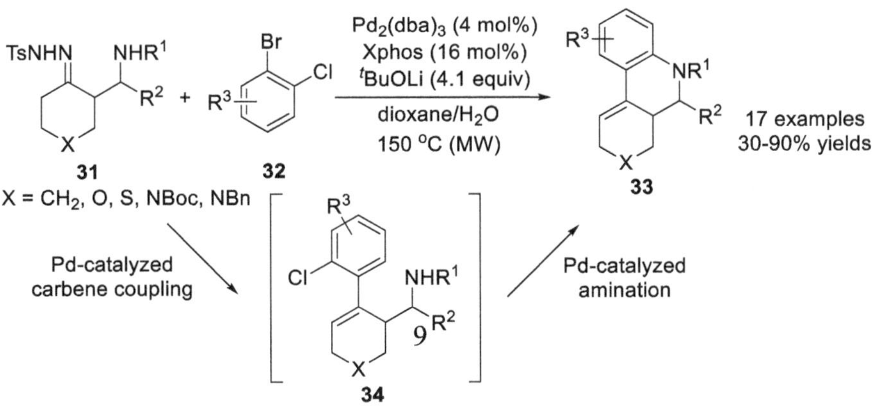

Scheme 12.10 Pd-catalyzed Auto-tandem transformation for C–C and C–N bond formation.

Migratory insertion of alkenyl group with subsequent β-H elimination gave rise to the formation of the final product (Scheme 12.9).

Pd-catalyzed carbene-based coupling reaction may be incorporated with other processes in which the same palladium catalyst is effective, namely a type of auto-tandem catalysis. For example, Barluenga and Valdés demonstrated the Pd-catalyzed tandem coupling of *N*-tosylhydrazones **31** with 1,2-dihalogenated arenes **32**.[41] This sequential C–C and C–N bond forming transformation provides a method for synthesizing quinoline derivatives **33** in an efficient manner. The reaction pathway is proposed following aryl group migratory insertion/β-H elimination and the Pd-catalyzed intramolecular amination of the intermediate **34** (Scheme 12.10).

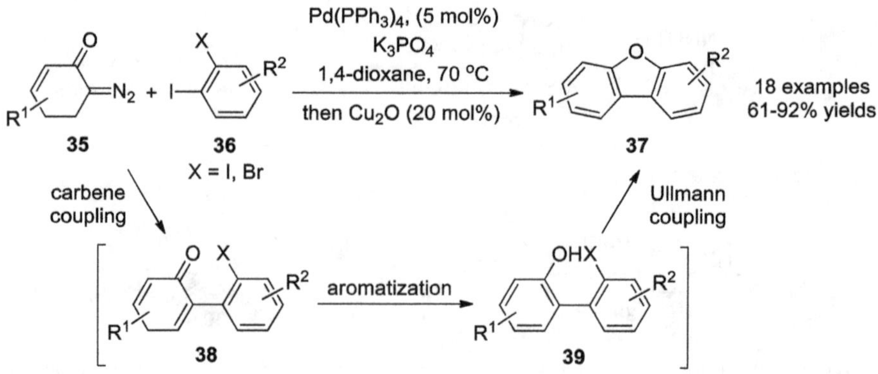

Scheme 12.11 Cascade of Pd-catalyzed carbene coupling/Cu-catalyzed Ullmann coupling.

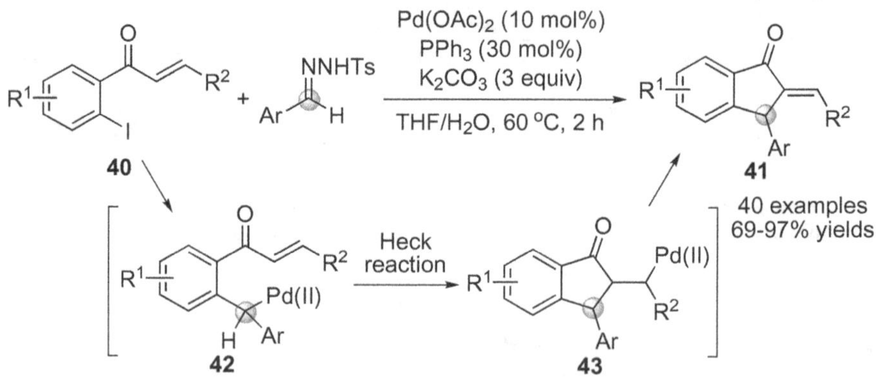

Scheme 12.12 Pd-catalyzed auto-tandem carbene coupling/Heck reaction.

In an example of cascade catalysis, Zhai and co-workers reported a concise method for the synthesis of functionalized dibenzofurans **37** through Pd-catalyzed coupling reaction of 6-diazo-2-cyclohexenone **35** with *ortho*-haloiodobenzenes **36** (Scheme 12.11).[42] This procedure proceeded *via* Pd-catalyzed carbene migratory insertion (**38**), aromatization (**39**) and finally Cu(I)-catalyzed Ullmann coupling.

Sekar and co-workers reported an auto-tandem catalytic reaction of carbene coupling and Heck reaction. The Pd-catalyzed reaction of the *N*-tosylhydrazones and 2'-iodochalcones **40** afforded 2-arylidene-3-aryl-1-indanones **41** (Scheme 12.12).[43] This reaction was proposed to proceed

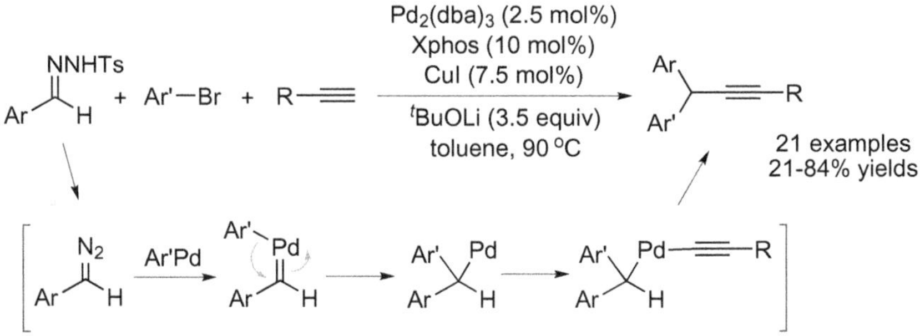

Scheme 12.13 Pd-catalyzed three-component coupling.

via migratory insertion of aryl group to generate **42**, followed by selective 5-*exo*-trig cyclization to give **43** and finally β-hydride elimination.

The organopalladium species generated from the Pd carbene migratory insertion may be devised to undergo transmetalation and reductive elimination, thus resulting in the formation of two different C–C bonds in a carbonic center. Such a transformation has been demonstrated by Wang and co-workers, who reported the Pd-catalyzed three-component cross-coupling reaction of aryl halides, *N*-tosylhydrazones and terminal alkynes (Scheme 12.13).[44] The undesirable Sonogashira coupling between the terminal alkyne and aryl bromide could be minimized by carefully tuning the reaction conditions.

12.2 Copper-catalyzed coupling reaction

While palladium catalysts have been mostly explored in carbene-based coupling reactions, Cu(I) complexes have also demonstrated high activity in catalyzing these types of cross-coupling reactions. In the reaction mechanism of Cu(I)-catalyzed coupling reactions, carbene migratory insertion is also involved in the C–C bond forming step. However, compared with the Pd-catalyzed reactions the formation of Cu(I) species prior to the Cu(I) carbene formation is generally through C–H bond metalation, rather than oxidative addition or transmetalation. Therefore, for the Cu(I)-catalyzed reaction the coupling substrates are different from the corresponding Pd-catalyzed reaction. Moreover, the organocopper species **46** generated from the Cu(I) carbene migratory insertion does not undergo β-H elimination, rather it readily undergoes protonation. Thus, a C–C

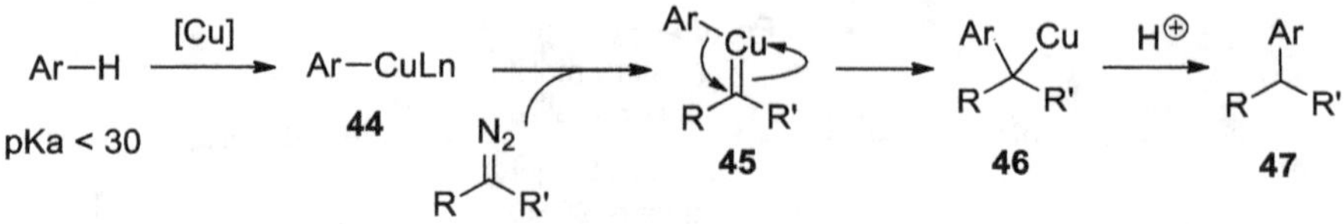

Scheme 12.14 The general scheme of the Cu(I)-catalyzed carbene coupling.

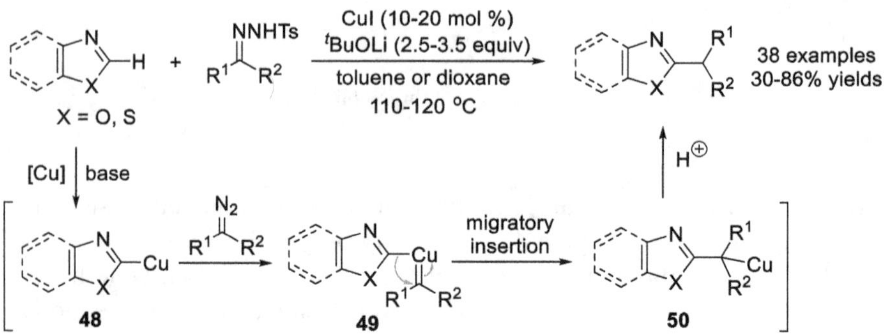

Scheme 12.15 Cu-catalyzed cross-coupling of *N*-tosylhydrazones with 1,3-azoles.

single bond is generated and the net transformation is a formal C–H bond insertion of carbene (Scheme 12.14).

Notably, such a type of formal C–H insertion is mechanistically different from the classic carbene C–H insertions (Chapter 5). For the classic C(sp^3)–H bond carbene insertions, the reaction typically follows a concerted mechanism; while for aromatic C(sp^2)–H bond formal insertion, Friedel–Crafts-type mechanism is involved. In both cases, electron-rich C–H bonds react preferentially, because the metal carbene in these reactions are electrophilic in character. On the contrary, in Cu(I)-catalyzed carbene coupling reaction with C–H bond (formally it is C–H bond insertion), electron-deficient C–H bond reacts preferentially because the C–H bond cleavage occurs in the step of C–H bond cuperation with the aid of base. This difference makes Cu(I)-catalyzed carbene coupling quite appealing in organic synthesis. In this section, some representative transformations based on the mechanism shown in Scheme 12.14 are presented.

In 2011, Wang and co-workers reported the first example of the Cu(I)-catalyzed carbene with C(sp^2)–H bond of 1,3-azoles with *N*-tosylhydrazones (Scheme 12.15).[45] The reaction afforded 2-benzylated

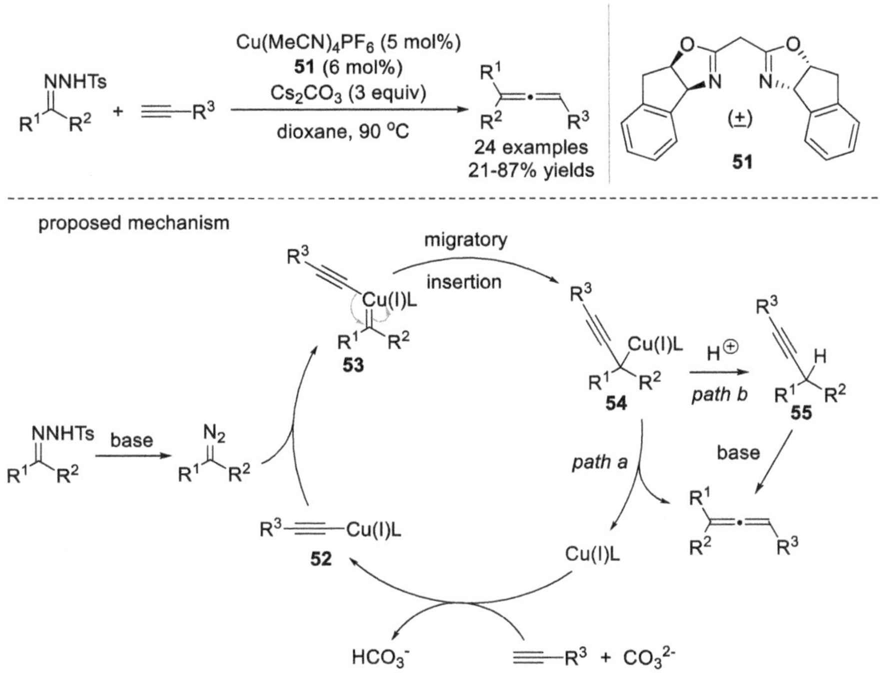

Scheme 12.16 Cu(I)-catalyzed coupling reaction of terminal alkynes with *N*-tosylhydrazones.

or 2-allylation products. Experimentally, it was verified that the reaction did not follow a direct Cu(I) carbene C–H insertion mechanism. The proposed reaction pathway involves C–H bond cuperation to form **48**, generation of Cu(I) carbene **49**, migratory insertion to give **50** and finally protonation to afford the formal C–H bond insertion product.

In the same year, Wang and co-workers reported the Cu(I)-catalyzed carbene coupling reaction of *N*-tosylhydrazones and terminal alkynes.[46] The reaction afforded *tri*-substituted allenes in good yields (Scheme 12.16). As the general scheme shown earlier, the mechanistic rationale for this coupling involves alkynyl migratory insertion of Cu(I) carbene **53**, followed by protonation on the distal triple bond of **54** to give the allene product (*path a*). However, it is also possible that alkyne **55** is first formed, which undergoes rearrangement under basic conditions to give allene (*path b*). Control experimental with an independently synthesized alkyne **55** suggested that *path a* is more likely.

Also, in the same year Fox and co-workers reported the Cu(I)-catalyzed coupling reaction of α-diazoesters with terminal alkynes to afford 2,4-disubstituted allenoates (Eq. (12.7)).[47] The ligand 3,6-di(pyridin-2-yl)-1,2,4,5-tetrazine **56** is crucial for the reaction. For the reaction mechanism, the authors proposed that the alkynoates were generated initially, which underwent isomerization to form the corresponding allenoate products in the presence of base.

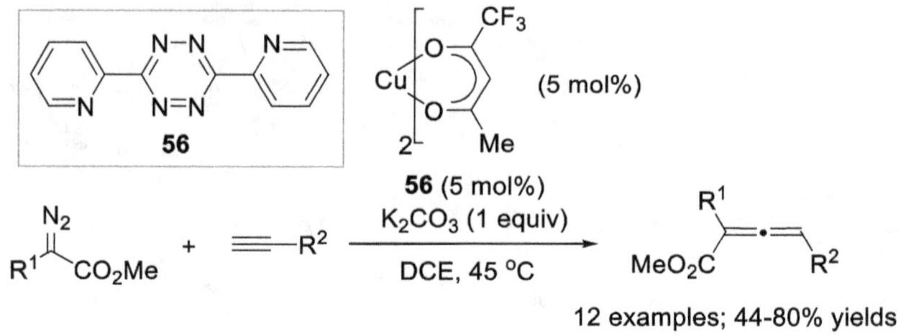

(12.7)

Subsequently, Feng and co-workers developed the catalytic asymmetric version of the reaction (Eq. (12.8)).[48] With guanidinium salts **57**/Cu(I) complex, the reaction produced chiral 2,4-disubstituted allenoates with good yields and enantioselectivities. Control experiments excluded the possibility of an asymmetric isomerization of alkynoates to allenes.

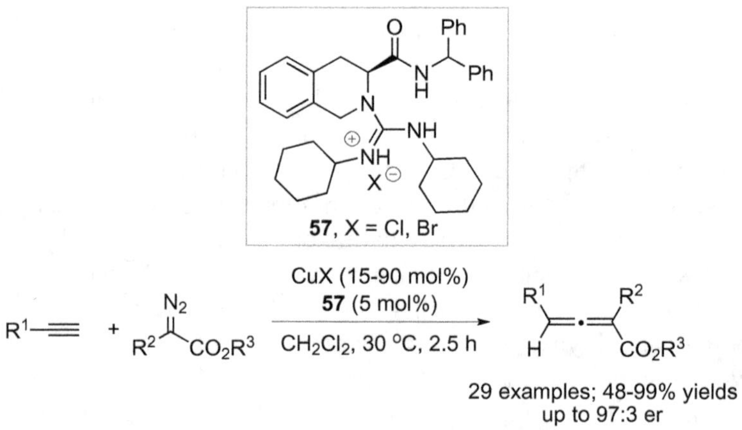

(12.8)

Interestingly, when silyl group-protected terminal alkynes were employed as the substrates in the Cu(I)-catalyzed coupling with

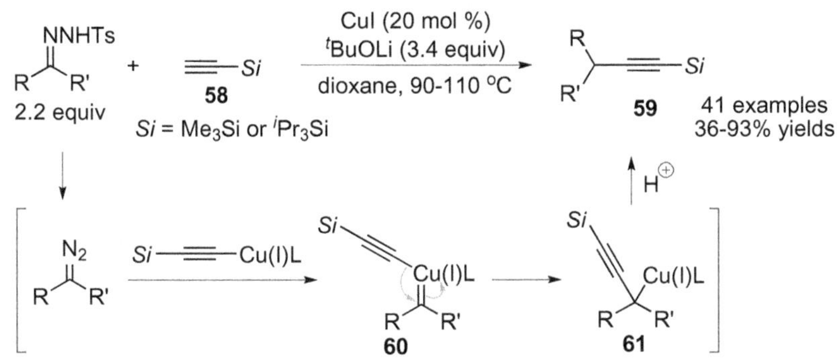

Scheme 12.17 Cu(I)-catalyzed cross-coupling reaction of *N*-tosylhydrazones with trialkylsilylethynes.

N-tosylhydrazone or diazo compounds, the final protonation occured at the sp^3 carbon where copper is bonded, affording the $C(sp^3)$–C(sp) bond coupling products.[49] Thus, with simple CuI as the catalyst, the coupling reaction of *N*-tosylhydrazones with trialkylsilylethyne **58** afforded the alkyne products in good yields (Scheme 12.17). The reaction shows wide substrate scope with various *N*-tosylhydrazones derived from aromatic and aliphatic ketones or aldehydes. Since the reaction afforded allene as a corresponding product when *tert*-butyl-substituted alkyne was employed as the substrate, the selectivity in the protonation is attributed to the electronic rather than steric reasons. Computational study has provided some insights into the reaction mechanism.[50]

Subsequently, Ma and co-workers reported the Cu(I)-catalyzed trifluoroethylation of terminal alkynes with *in situ* generated gaseous 2,2,2-trifluorodiazoethane **62**.[51] In the absence of additional base or ligand, trifluoroethylated alkynes **63** were obtained in good yields (Eq. (12.9)).

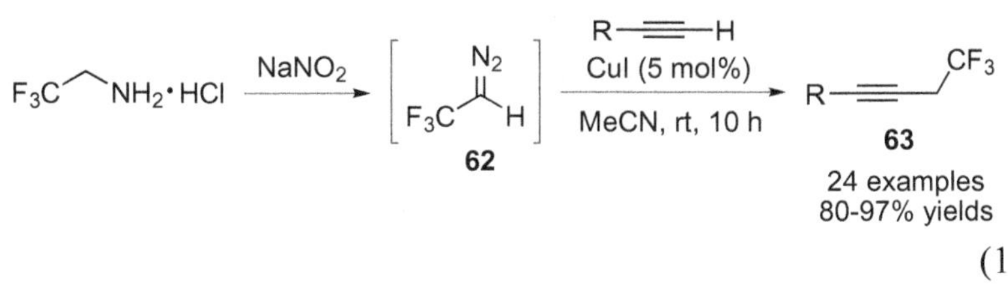

(12.9)

The copper species generated from Cu(I) carbene migratory insertion may undergo β-fluoride elimination, rather than protonation. Wang and co-workers demonstrated the Cu(I)-catalyzed cross-coupling reaction of

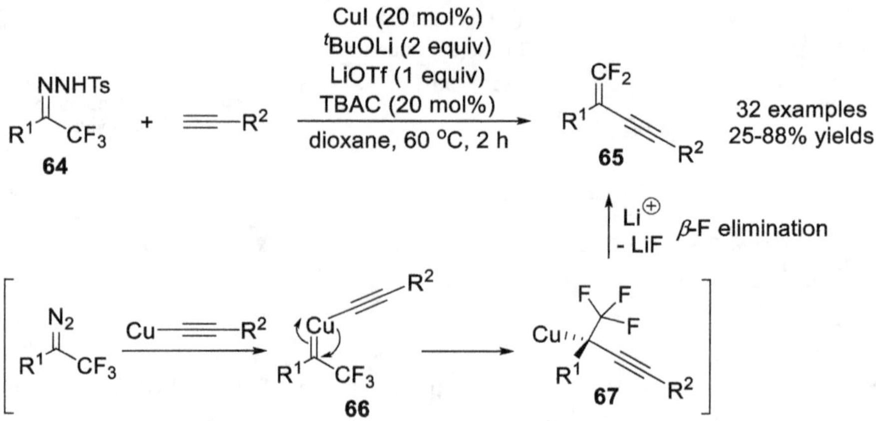

Scheme 12.18 Cu-Catalyzed synthesis of 1,1-difluoro-1,3-enynes from terminal alkynes with trifluoromethyl ketone *N*-tosylhydrazones.

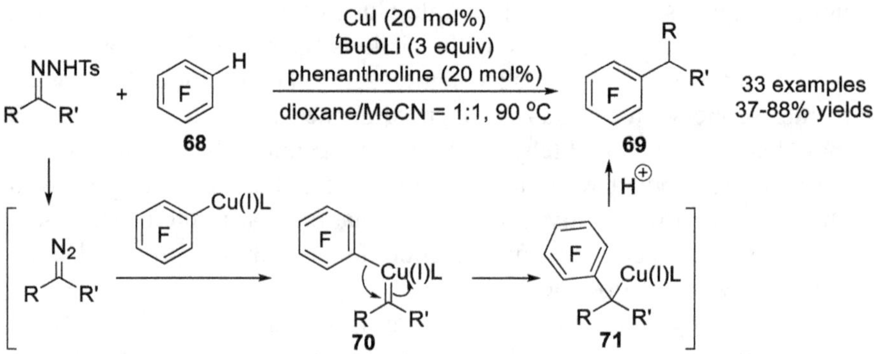

Scheme 12.19 Cu(I)-Catalyzed alkylation of polyfluoroarenes *via* C–H functionalization.

terminal alkynes with the *N*-tosylhydrazones **64**, derived from trifluoromethyl ketones, furnishing functionalized 1,1-difluoro-1,3-enyne derivatives **65** (Scheme 12.18).[52] The reaction pathway involves a β-fluoride elimination of intermediate **67**.

From the general reaction pathway shown in Scheme 12.14, it can be conceived that the substrates with similarly strong acidic C–H bonds, such as 1,2-azoles or terminal alkynes, are expected to undergo same type of carbene coupling with Cu(I) catalyst. In this context, Wang and co-workers reported a Cu(I)-catalyzed C–H alkylation of polyfluoroarenes **68** with *N*-tosylhydrazones or diazo compounds (Scheme 12.19).[53] Similar to the reactions shown previously, migratory insertion of the polyfluoroaryl

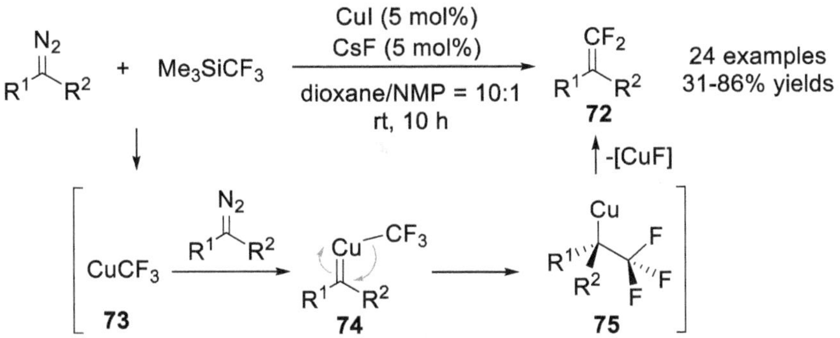

Scheme 12.20 Cu-catalyzed *gem*-difluoroolefination of diazo compounds with $TMSCF_3$.

copper carbene intermediate **70** and subsequent protonation of **71** are involved in the reaction pathway.

Cu(I)-catalyzed carbene coupling has been explored for the synthesis of organofluorine compounds. Hu and co-workers reported a Cu-catalyzed *gem*-difluoroolefination of diaryldiazomethanes with $TMSCF_3$ (Scheme 12.20).[54] This reaction enabled the efficient synthesis of structurally diverse 1,1-difluoroalkenes from commercially available substrates. In the proposed mechanism, carbene CF_3-migratory insertion of **74** takes place with sequential *β*-fluoride elimination to afford 1,1-difluoroolefins **72**. The reaction explored the intriguing reactivity of $CuCF_3$ species with the diazo substrates.

12.3 Rhodium-catalyzed coupling reaction

Rh(II) complexes, such as $Rh_2(OAc)_4$, have been extensively explored as efficient catalysts in catalytic carbene transfer reactions. The Rh(II) carbene in these reactions undergoes classic carbene transfer reactions, typically C–H bond insertions, cyclopropanations and ylide formations. On the contrary, the Rh(I)- or Rh(III)-catalyzed reactions of diazo compounds introduced in this section are completely different from the classic Rh(II)-catalyzed reactions. These transformations are outlined in Scheme 12.21. The organorhodium species *C*-Rh **76** can be generated through C–C bond cleavage, transmetalation or C–H bond cleavage. The *C*-Rh species further reacts with diazo substrate to generate Rh carbene **77**. The Rh carbene intermediate undergoes further migratory insertion to generate **78**, which may be followed by protonation or nucleophilic addition/substitution to afford the final products.

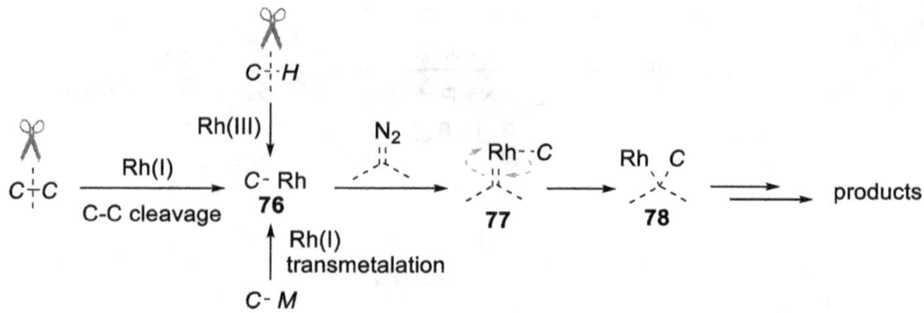

Scheme 12.21 Rh-catalyzed carbene coupling reactions

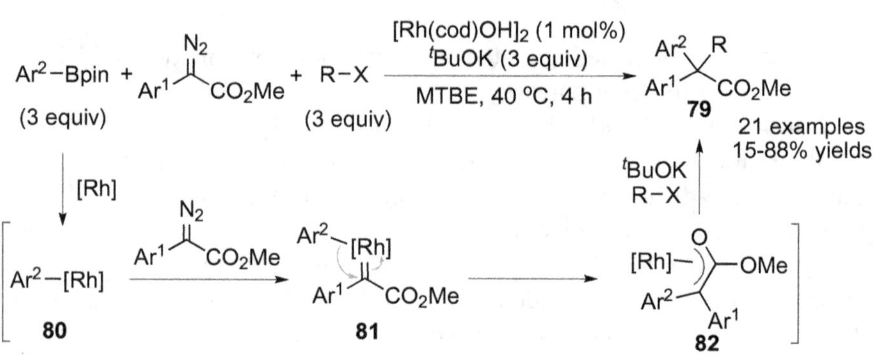

Scheme 12.22 Rh(I)-catalyzed coupling of arylboronates, *A*-aryldiazoacetates and alkyl halides.

In 2011, Yu and co-workers reported the first example of rhodium-catalyzed carbene reaction that follows the pathway shown earlier. This is a Rh(I)-catalyzed three-component coupling of arylboronates, α-aryldiazoacetates with alkyl halides, affording α,α-heterodiaryl carboxylic esters with quaternary carbon center (Scheme 12.22).[55] The proposed reaction pathway consists of migratory carbene insertion of **81** as the principal step to generate oxa-π-allylrhodium complex **82**, which underwent alkylation to form another C–C bond at the carbenic carbon.

Subsequently, Yu and co-workers explored the trapping of the Rh-enolate complex generated from Rh carbene migratory insertion with other electrophiles, such as *N*-chlorosuccinimide (NCS). As shown in Eq. (12.10), the Rh(III)-catalyzed three-component coupling of arylboronic acids, α-diazocarbonyl compounds with NCS afforded

α-aryl-α-chlorocarbonyl compounds.[56] The mechanism of this cascade arylation and chlorination reaction involves transmetalation, formation of Rh(III) carbene, migratory insertion to form Rh(III)-enolate complex, followed by nucleophilic displacement with NCS.

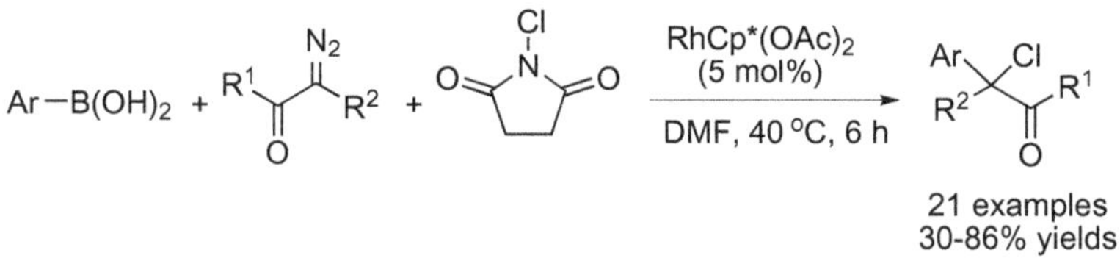

(12.10)

Wang and co-workers developed an Rh(I)-catalyzed cross-coupling of diazoesters with arylsiloxanes, providing a new method for constructing $C(sp^3)$–$C(sp^2)$ bonds.[57] This catalytic Hiyama-type reaction allowed the alternative synthesis of α-aryl esters with wide substrates scope (Eq. (12.11)). Mechanistically, the Rh(I)-carbene migratory insertion is proposed as the key step, with protonation as the terminating step in the catalytic cycle. Similar Stille-type cross-coupling of diazo compounds and organostannanes has also been developed (Eq. (12.12)).[58]

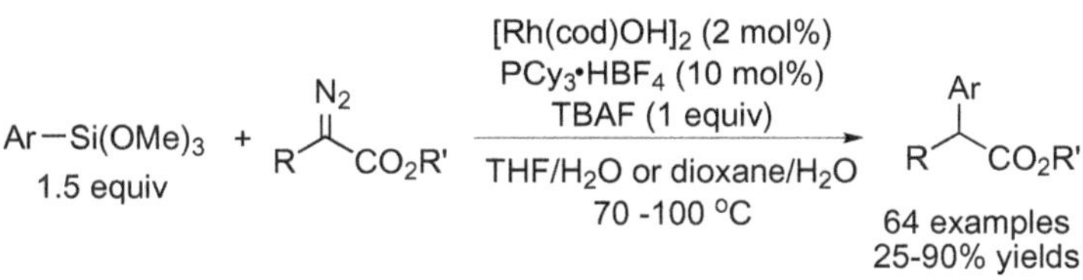

(12.11)

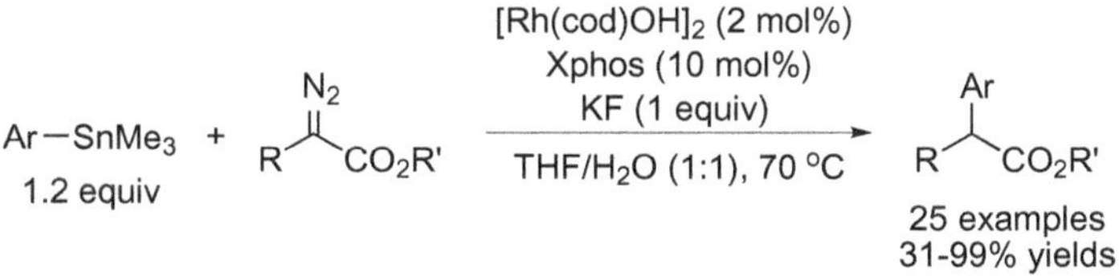

(12.12)

In addition to transmetalation, *C*-Rh species **76** can also be generated by C–C bond cleavage (Scheme 12.21), which may be further incorporated into the carbene coupling process. In 2014, Wang and co-workers developed the Rh(I)-catalyzed formal carbene insertion into C–C bond of benzocyclobutenols **83** with α-diazoesters as the carbene precursors (Scheme 12.23).[59] It has been proposed that the reaction is initiated by

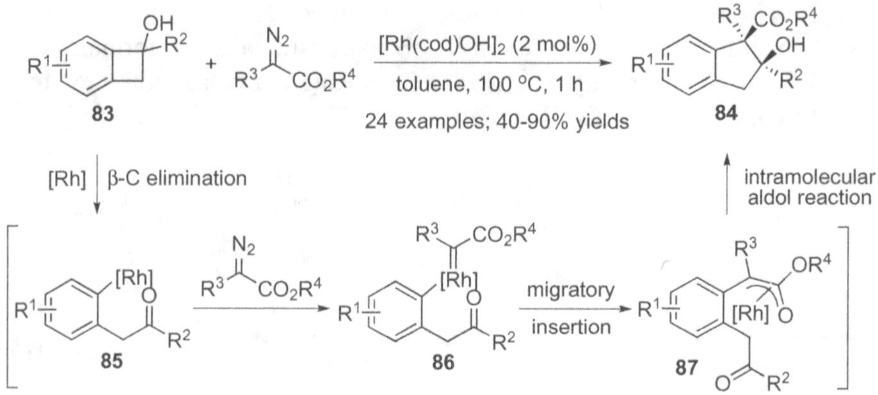

Scheme 12.23 Rh(I)-catalyzed formal carbene insertion into C–C bond of benzocyclobutenols with α-diazoesters.

selective C(sp^2)–C(sp^3) cleavage process to generate aryl-Rh(I) intermediate **85**. Subsequently, reaction with diazo substrate generates Rh(I) carbene **86**, followed by migratory insertion to afford oxa-π-allyl Rh(I) intermediate **87**. Finally, intramolecular aldol reaction followed by protonation occurred to deliver the indanol product **84**.

Meanwhile, Murakami and co-workers achieved Rh(I)-catalyzed asymmetric carbene insertion to C(sp^3)–C(sp^3) bond of cyclobutanols **88** with *N*-tosylhydrazones as the carbene precursor (Scheme 12.24).[60] With chiral phosphine ligands, cyclopentanol derivatives **89** and **89'** were obtained by ring expansion of cyclobutanols with excellent enantioselectivities (up to 99% *ee*).

In contrast to the Pd- and Cu-catalyzed carbene couplings, a unique feature of the Rh-catalyzed cross-coupling reaction involving migratory insertion is the combination with directing-group-assisted C–H bond activation strategy.[6] In 2012, Yu and co-workers reported the first Rh(III)-catalyzed intermolecular cross-coupling reaction of diazomalonates with aromatic C–H bond, assisted by divergent directing groups (Scheme 12.25).[61] Simple aromatic derivatives, such as acetophenone oximes **90**, 2-phenylpyridines **91**, benzylamines **92** and benzoic acids **93**, could be *ortho*-functionalized with high yields under mild conditions. The reaction mechanism has been proposed to be initiated by electrophilic C–H bond cleavage to form rhodacyclic intermediate **94**. From **94**, one possible pathway involves migratory insertion of

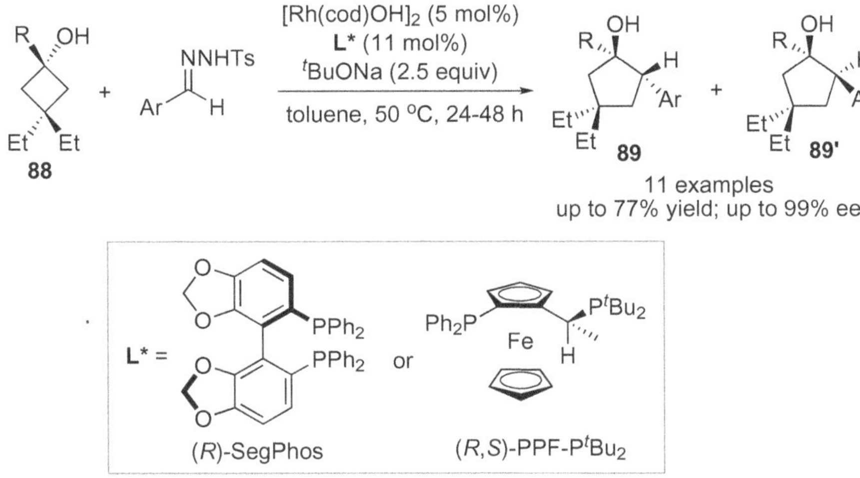

Scheme 12.24 Rh-catalyzed enantioselective carbene insertion of cyclobutanols with *N*-tosylhydrazones.

Rh-carbene **95**, followed by protonation to afford the product. The reaction may also follow intramolecular 1,2-aryl shift of **96** to generate the same intermediate **97**.

Following Yu's seminal work, various similar reactions were reported in the literature, in which various directing groups and the structure of diazo compounds were extensively explored. For example, Li and co-workers developed a Rh(III)-catalyzed C–H activation of phenacyl ammonium salts **98** assisted by an oxidizing C–N bond under redox-neutral condition, affording benzocyclopentanones **99** as the products.[62] DFT calculation revealed that C–H bond activation occurs through C-bound enolate-assisted concerted metalation-deprotonation. Upon coordination and dediaoniation of the α-diazoester substrate, the six-membered rhodacycle intermediate **102** is generated *via* carbene **101** migratory insertion of Rh-aryl bond, followed by α-elimination of Et_3N to form α-oxo Rh carbene species **103**. Subsequently, the second migratory insertion of **103** affords enolate Rh species **104**, which undergoes protonolysis to furnish the final product **99** (Scheme 12.26).

Generating the catalytic asymmetric version of the Rh-catalyzed C(sp^2)–H activation with carbene is a challenging task. Cramer and co-workers tackled this problem with their chiral cyclopentadienyl ligand. The induction of C_2-symmetric disubstituted Cp ligand with chiral

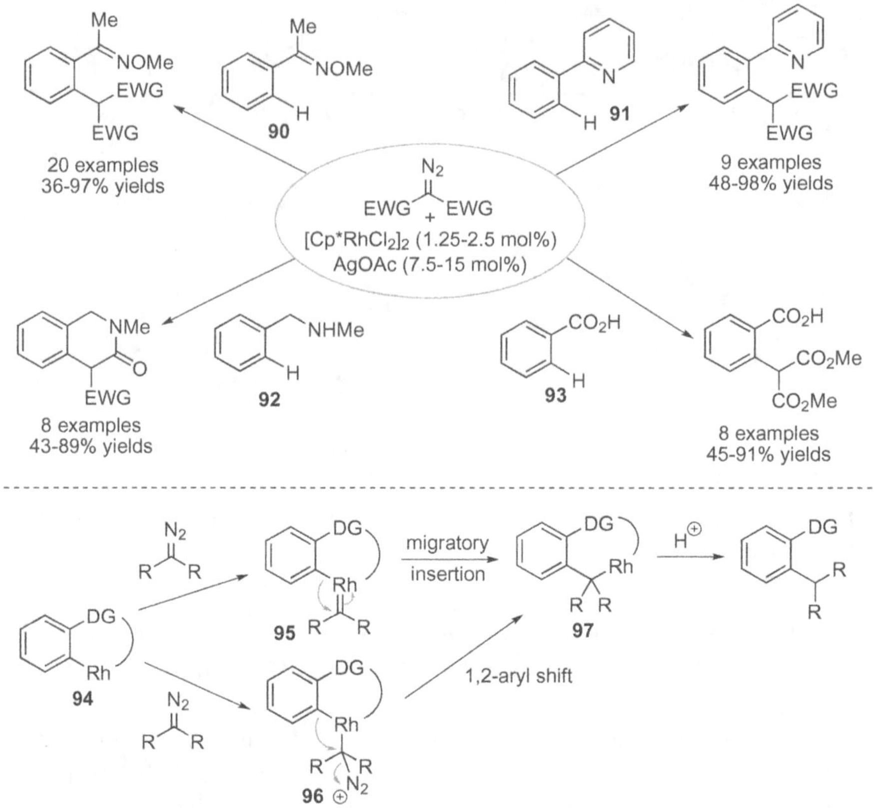

Scheme 12.25 Rh(II)-catalyzed aromatic C–H bond functionalization by α-diazomalonates.

backbone, as well as the employment of weak alkyl donor/acceptor diazo-carbonyl compounds **106**, allowed for achieving high level of enantioselectivity of isoindolone products **107** (Scheme 12.27).[63]

In addition to the aromatic C(sp^2)–H bond, the functionalization of C(sp^3)–H bond could also be incorporated with the carbene coupling reaction. Wang and co-workers reported a Rh(I)-catalyzed coupling of (quinolin-8-yl)methanone with *N*-tosylhydrazones, generating a C–C double bond through C(sp^3)–H functionalization (Scheme 12.28).[64] In the presence of base and Ag_2O, the Rh(I)-catalyzed reaction with *N*-tosylhydrazones with **108** afforded 3-aryl-1-(quinolin-8-yl)prop-2-en-1-one derivatives **109** in moderate yields. The reaction mechanism is

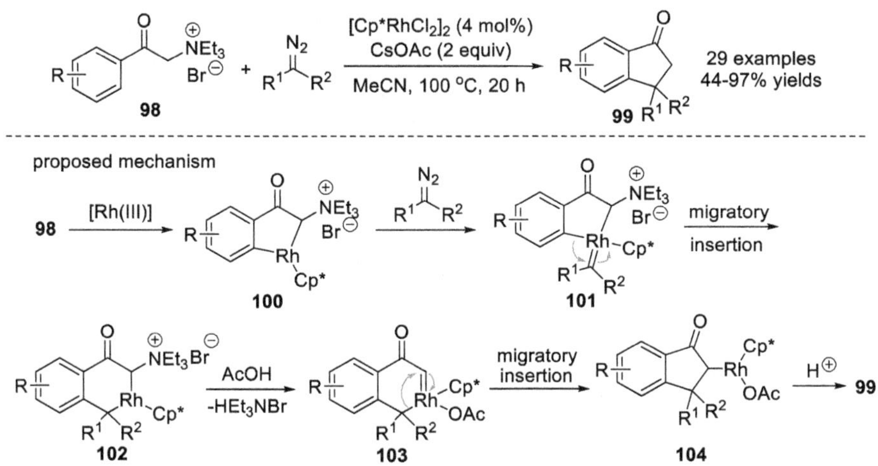

Scheme 12.26 Rh-catalyzed C–H activation of phenacyl ammonium salts with diazoesters.

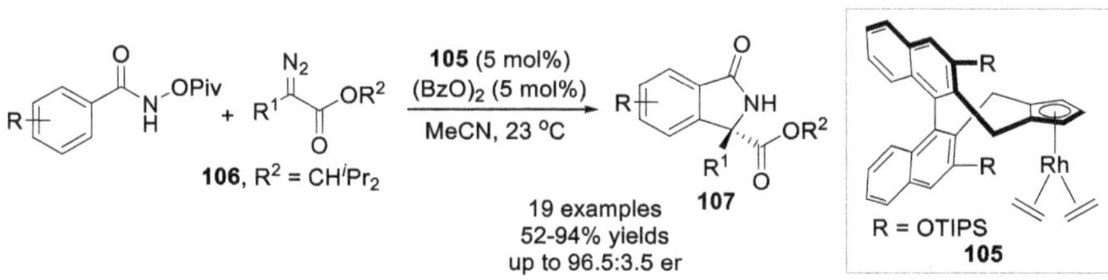

Scheme 12.27 Rh-catalyzed asymmetric coupling of benzamides with diazo compounds.

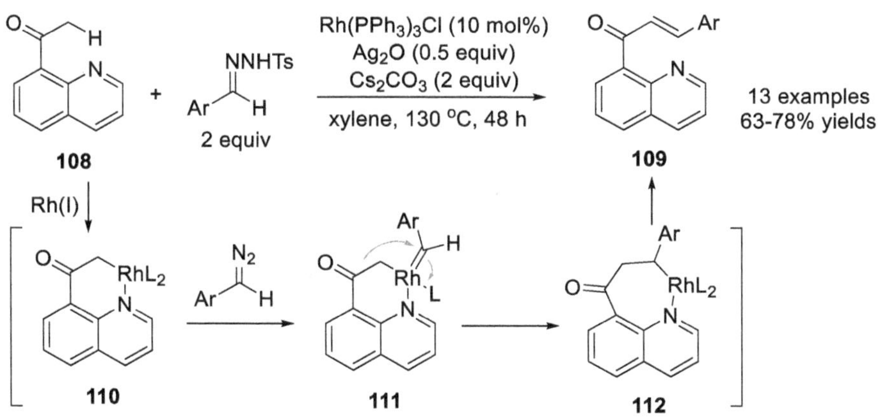

Scheme 12.28 Rh(I)-catalyzed C–H bond activation of (quinolin-8-yl)methanone with *N*-tosylhydrazones.

proposed to involve the directing group assisted C–H metalation to generate rhodacycle **110**, followed by carbene **111** formation and migratory insertion to form seven-membered rhodacycle **112**. Finally, β-H elimination delivers the olefin product **109**, with the formation of Rh-H species, which regenerates active Rh(I) catalyst in the presence of Ag_2O as the oxidant.

Cui and co-workers expanded the Rh(III)-catalyzed carbene C–H activation strategy to indoles and pyrroles assisted by a directing group. The reaction is an efficient route to fused heterocycles (Eq. (12.13)).[65]

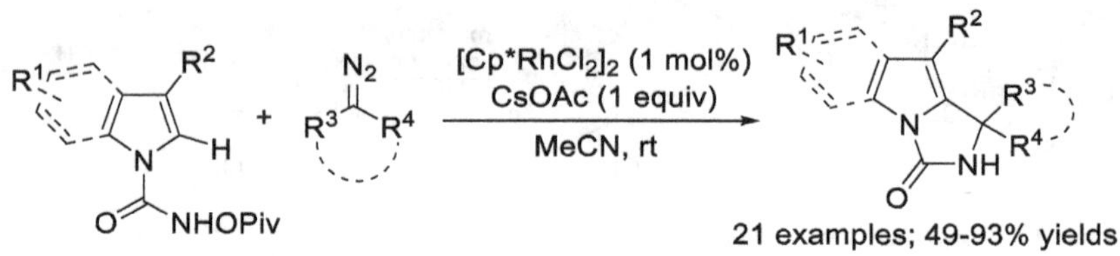

(12.13)

12.4 Other transition-metal-catalyzed carbene coupling reactions

Following the Cu(I)-catalyzed coupling of azoles with *N*-tosylhydrazones reported by Wang and co-workers (Scheme 12.15),[45] Miura and co-workers disclosed the direct 2-alkylation of azoles with *N*-tosylhydrazones by employing Ni(II) and Co(II) catalysis (Eqs. (12.14) and (12.15)).[66] While the reactions are similar, the Ni(II)- and Co(II)-catalyzed couplings expanded the scope of *N*-tosylhydrazones to those derived from aliphatic ketones, thus allowing the introduction of secondary alkyl groups into benzoxazole derivatives.

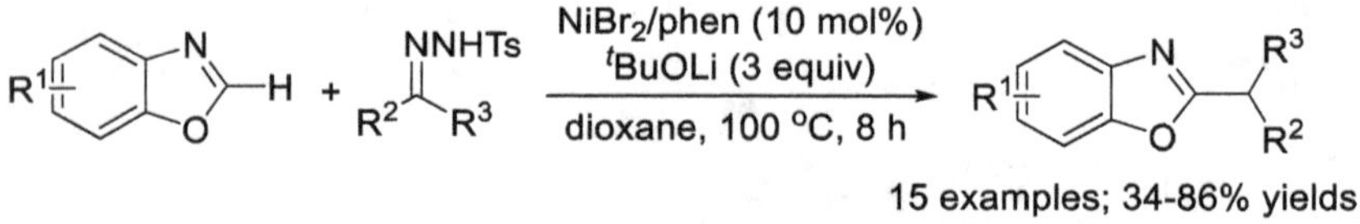

(12.14)

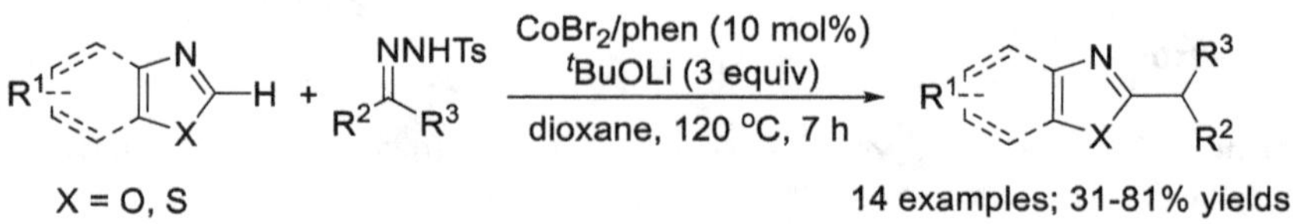

(12.15)

Iridium complexes have also proven to be effective catalysts in carbene couplings. Wang and co-workers reported an Ir(III)-catalyzed C–H functionalization of arenes with α-diazomalonates involving Ir(III)-carbene migratory insertion (Eq. (12.16)).[67]

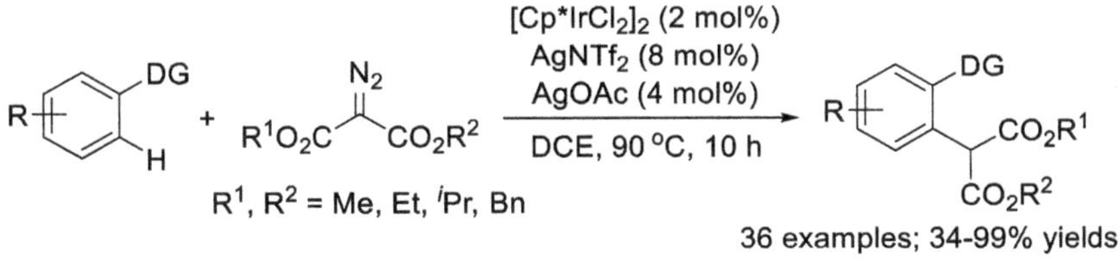

(12.16)

With pyridine as the directing group, Glorius and co-workers developed an efficient and scalable Co(III)-catalyzed C–H bond activation of aromatic and heteroaromatic compounds **113** with α-diazoesters (Scheme 12.29).[68] This reaction provides a straightforward access toward extended π-systems. The Co(III) complex plays multiple roles in the catalytic cycle: generation of cobaltacycle **115** through C–H bond cleavage, formation of Co(III) carbene **116** and as Lewis acid to assist nucleophilic addition to generate intermediate **119**.

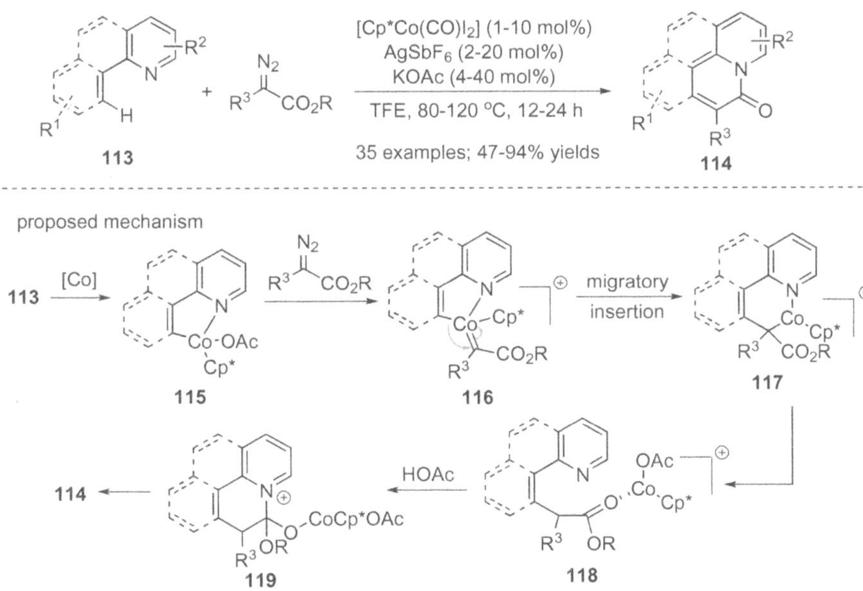

Scheme 12.29 Co(III)-catalyzed C–H bond activation/annulation with diazo compounds.

In summary, as a burgeoning area, the transition-metal-catalyzed cross-coupling reaction with diazo compounds involving migratory insertion has experienced remarkable development in the last decade. These novel catalytic carbene transformations represent landmark advances both for cross-coupling as well as for metal carbene chemistry. Again, diazo compounds play vital roles in these types of transformations. The most striking feature of these reactions is the generality and diversity of the transformations: the wide range of coupling partners, including organohalides, alkynes, organoboronic acids, organosilanes and C–H bonds, various metal catalysts and also different combinations. In particular, it is worth mentioning that the use of *N*-tosylhydrazones, the stable and readily available reagents, as the surrogates of diazo compounds makes this type of cross-coupling practically useful.

References

1. Zhang, Y.; Wang, J. *Eur. J. Org. Chem.* **2011**, 1015–1026.
2. Zhao, X.; Zhang, Y.; Wang, J. *Chem. Commun.* **2012**, *48*, 10162–10173.
3. Xiao, Q.; Zhang, Y.; Wang, J. *Acc. Chem. Res.* **2013**, *46*, 236–247.
4. Xia, Y.; Zhang, Y.; Wang, J. *ACS Catal.* **2013**, *3*, 2586–2598.
5. Liu, Z.; Wang, J. *J. Org. Chem.* **2013**, *78*, 10024–10030.
6. Hu, F.; Xia, Y.; Ma, C.; Zhang, Y.; Wang, J. *Chem. Commun.* **2015**, *51*, 7986–7995.
7. Barluenga, J.; Valdés, C. *Angew. Chem. Int. Ed.* **2011**, *50*, 7486–7500.
8. Shao, Z.; Zhang, H. *Chem. Soc. Rev.* **2012**, *41*, 560–572.
9. Xia, Y.; Qiu, D.; Wang, J. *Chem. Rev.* **2017**, *117*, 13810–13889.
10. Qiu, D.; Mo, F.; Zhang, Y.; Wang, J. *Adv. Organomet. Chem.* **2017**, *67*, 151–219.
11. Greenman, K. L.; Carter, D. S.; Van Vranken, D. L. *Tetrahedron* **2001**, *57*, 5219–5225.
12. Barluenga, J.; Moriel, P.; Valdés, C.; Aznar, F. *Angew. Chem. Int. Ed.* **2007**, *46*, 5587–5590.
13. Peng, C.; Wang, Y.; Wang, J. *J. Am. Chem. Soc.* **2008**, *130*, 1566–1567.
14. Barluenga, J.; Tomás-Gamasa, M.; Moriel, P.; Aznar, F.; Valdés, C. *Chem. Eur. J.* **2008**, *14*, 4792–4795.
15. Barluenga, J.; Escribano, M.; Moriel, P.; Aznar, F.; Valdés, C. *Chem. Eur. J.* **2009**, *15*, 13291–13294.
16. Barluenga, J.; Escribano, M.; Aznar, F.; Valdés, C. *Angew. Chem. Int. Ed.* **2010**, *49*, 6856–6859.
17. Barluenga, J.; Tomás-Gamasa, M.; Aznar, F.; Valdés, C. *Chem. Eur. J.* **2010**, *16*, 12801–12803.

18. Barluenga, J.; Tomás-Gamasa, M.; Aznar, F.; Valdés, C. *Adv. Synth. Catal.* **2010**, *352*, 3235–3240.
19. Florentino, L.; Aznar, F.; Valdés, C. *Chem. Eur. J.* **2013**, *19*, 10506–10510.
20. Jiménez-Aquino, A.; Vega, J. A.; Trabanco, A. A.; Valdés, C. *Adv. Synth. Catal.* **2014**, *356*, 1079–1084.
21. Barluenga, J.; Florentino, L.; Aznar, F.; Valdés, C. *Org. Lett.* **2011**, *13*, 510–513.
22. Florentino, L.; Aznar, F.; Valdés, C. *Org. Lett.* **2012**, *14*, 2323–2325.
23. Feng, J.; Li, B.; He, Y.; Gu, Z. *Angew. Chem. Int. Ed.* **2016**, *55*, 2186–2190.
24. Tréguier, B.; Hamze, A.; Provot, O.; Brion, J.-D.; Alami, M. *Tetrahedron Lett.* **2009**, *50*, 6549–6552.
25. Messaoudi, S.; Tréguier, B.; Hamze, A.; Provot, O.; Peyrat, J.-F.; De Losada, J. R.; Liu, J.-M.; Bignon, J.; Wdzieczak-Bakala, J.; Thoret, S.; Dubois, J.; Brion, J.-D.; Alami, M. *J. Med. Chem.* **2009**, *52*, 4538–4542.
26. Aziz, J.; Brachet, E.; Hamze, A.; Peyrat, J.-F.; Bernadat, G.; Morvan, E.; Bignon, J.; Wdzieczak-Bakala, J.; Desravines, D.; Dubois, J.; Tueni, M.; Yassine, A.; Brion, J.-D.; Alami, M. *Org. Biomol. Chem.* **2013**, *11*, 430–442.
27. Lawson, M.; Hamze, A.; Peyrat, J.-F.; Bignon, J.; Dubois, J.; Brion, J.-D.; Alami, M. *Org. Biomol. Chem.* **2013**, *11*, 3664–3673.
28. Brachet, E.; Hamze, A.; Peyrat, J.-F.; Brion, J.-D.; Alami, M. *Org. Lett.* **2010**, *12*, 4042–4045.
29. Roche, M.; Hamze, A.; Provot, O.; Brion, J.-D.; Alami, M. *J. Org. Chem.* **2013**, *78*, 445–454.
30. Luo, H.; Wu, G.; Xu, S.; Wang, K.; Wu, C.; Zhang, Y.; Wang, J. *Chem. Commun.* **2015**, *51*, 13321–13323.
31. Yang, K.; Zhang, J.; Li, Y.; Cheng, B.; Zhao, L.; Zhai, H. *Org. Lett.* **2013**, *15*, 808–811.
32. Xia, Y.; Hu, F.; Liu, Z.; Qu, P.; Ge, R.; Ma, C.; Zhang, Y.; Wang, J. *Org. Lett.* **2013**, *15*, 1784–1787.
33. Chen, S.; Wang, J. *Chem. Commun.* **2008**, 4198–4200.
34. Wang, P.-S.; Lin, H.-C.; Zhou, X.-L.; Gong, L.-Z. *Org. Lett.* **2014**, *16*, 3332–3335.
35. Zhou, L.; Ye, F.; Ma, J.; Zhang, Y.; Wang, J. *Angew. Chem. Int. Ed.* **2011**, *50*, 3510–3514.
36. Chen, Z.-S.; Duan, X.-H.; Zhou, P.-X.; Ali, S.; Luo, J.-Y.; Liang, Y.-M. *Angew. Chem. Int. Ed.* **2012**, *51*, 1370–1374.
37. Zeng, X.; Cheng, G.; Shen, J.; Cui, X. *Org. Lett.* **2013**, *15*, 3022–3025.
38. Roche, M.; Frison, G.; Brion, J.-D.; Provot, O.; Hamze, A.; Alami, M. *J. Org. Chem.* **2013**, *78*, 8485–8495.
39. Liu, Z.; Tan, H.; Fu, T.; Xia, Y.; Qiu, D.; Zhang, Y.; Wang, J. *J. Am. Chem. Soc.* **2015**, *137*, 12800–12803.

40. Jiang, H.; He, L.; Li, X.; Chen, H.; Wu, W.; Fu, W. *Chem. Commun.* **2013**, *49*, 9218–9220.
41. Barluenga, J.; Quiñones, N.; Cabal, M.-P.; Aznar, F.; Valdés, C. *Angew. Chem. Int. Ed.* **2011**, *50*, 2350–2353.
42. Zhao, H.; Yang, K.; Zheng, H.; Ding, R.; Yin, F.; Wang, N.; Li, Y.; Cheng, B.; Wang, H.; Zhai, H. *Org. Lett.* **2015**, *17*, 5744–5747.
43. Arunprasath, D.; Muthupandi, P.; Sekar, G. *Org. Lett.* **2015**, *17*, 5448–5451.
44. Zhou, L.; Ye, F.; Zhang, Y.; Wang, J. *J. Am. Chem. Soc.* **2010**, *132*, 13590–13591.
45. Zhao, X.; Wu, G.; Zhang, Y.; Wang, J. *J. Am. Chem. Soc.* **2011**, *133*, 3296–3299.
46. Xiao, Q.; Xia, Y.; Li, H.; Zhang, Y.; Wang, J. *Angew. Chem. Int. Ed.* **2011**, *50*, 1114–1117.
47. Hassink, M.; Liu, X.; Fox, J. M. *Org. Lett.* **2011**, *13*, 2388–2391.
48. Tang, Y.; Chen, Q.; Liu, X.; Wang, G.; Lin, L.; Feng, X. *Angew. Chem. Int. Ed.* **2015**, *54*, 9512–9516.
49. Ye, F.; Ma, X.; Xiao, Q.; Li, H.; Zhang, Y.; Wang, J. *J. Am. Chem. Soc.* **2012**, *134*, 5742–5745.
50. Wang, T.; Wang, M.; Fang, S.; Liu, J. *Organometallics* **2014**, *33*, 3941–3949.
51. Liu, C.-B.; Meng, W.; Li, F.; Wang, S.; Nie, J.; Ma, J.-A. *Angew. Chem. Int. Ed.* **2012**, *51*, 6227–6230.
52. Zhang, Z.; Zhou, Q.; Yu, W.; Li, T.; Wu, G.; Zhang, Y.; Wang, J. *Org. Lett.* **2015**, *17*, 2474–2477.
53. Xu, S.; Wu, G.; Ye, F.; Wang, X.; Li, H.; Zhao, X.; Zhang, Y.; Wang, J. *Angew. Chem. Int. Ed.* **2015**, *54*, 4669–4672.
54. Hu, M.; He, Z.; Gao, B.; Li, L.; Ni, C.; Hu, J. *J. Am. Chem. Soc.* **2013**, *135*, 17302–17305.
55. Tsoi, Y.-T.; Zhou, Z.; Yu, W.-Y. *Org. Lett.* **2011**, *13*, 5370–5373.
56. Ng, F.-N.; Lau, Y.-F.; Zhou, Z.; Yu, W.-Y. *Org. Lett.* **2015**, *17*, 1676–1679.
57. Xia, Y.; Liu, Z.; Feng, S.; Ye, F.; Zhang, Y.; Wang, J. *Org. Lett.* **2015**, *17*, 956–959.
58. Liu, Z.; Xia, Y.; Feng, S.; Wang, S.; Qiu, D.; Zhang, Y.; Wang, J. *Aust. J. Chem.* **2015**, *68*, 1379–1384.
59. Xia, Y.; Liu, Z.; Liu, Z.; Ge, R.; Ye, F.; Hossain, M. L.; Zhang, Y.; Wang, J. *J. Am. Chem. Soc.* **2014**, *136*, 3013–3015.
60. Yada, A.; Fujita, S.; Murakami, M. *J. Am. Chem. Soc.* **2014**, *136*, 7217–7220.
61. Chan, W.-W.; Lo, S.-F.; Zhou, Z.; Yu, W.-Y. *J. Am. Chem. Soc.* **2012**, *134*, 13565–13568.

62. Yu, S.; Liu, S.; Lan, Y.; Wan, B.; Li, X. *J. Am. Chem. Soc.* **2015**, *137*, 1623–1631.
63. Ye, B.; Cramer, N. *Angew. Chem. Int. Ed.* **2014**, *53*, 7896–7899.
64. Zhang, Y.; Wang, J.; Wang, J. *Synlett* **2013**, *24*, 1643–1648.
65. Zhang, Y.; Zheng, J.; Cui, S. *J. Org. Chem.* **2014**, *79*, 6490–6500.
66. Yao, T.; Hirano, K.; Satoh, T.; Miura, M. *Angew. Chem. Int. Ed.* **2012**, *51*, 775–779.
67. Xia, Y.; Liu, Z.; Feng, S.; Zhang, Y.; Wang, J. *J. Org. Chem.* **2015**, *80*, 223–236.
68. Zhao, D.; Kim, J. H.; Stegemann, L.; Strassert, C. A.; Glorius, F. *Angew. Chem. Int. Ed.* **2015**, *54*, 4508–4511.

Chapter 13

Polymerization with Diazo Compounds

13.1 Introduction

Polymerization with diazo compounds through carbene reactions will generate polymers with backbones consisting of one-carbon units, namely the polymers **1** that contain carbon–carbon single bond units (C1 polymerization) (Scheme 13.1). This is profoundly different from the traditional vinyl polymerization (C2 polymerization). An advantage of using diazo compounds as carbene precursors in polymerization is that the only by-product upon carbene generation is N_2.

Although polymerization of diazoalkanes has been known for a long time, the polymerization of the more stable α-diazocarbonyl compounds under transition-metal catalysis has only been explored recently. Since carbene polymerization has been recently reviewed by de Bruin and Ihara,[1–5] this chapter will focus on the most recent advances in this field.

13.2 Palladium-catalyzed polymerization of diazo compounds

Ihara and co-workers carried out extensive studies regarding the Pd(II)-catalyzed polymerization reactions with various functionalized diazo compounds as the monomers. The general mechanism of Pd-catalyzed polymerization, which includes initiation, propagation and termination processes (Scheme 13.2), is as follows.[1,2]

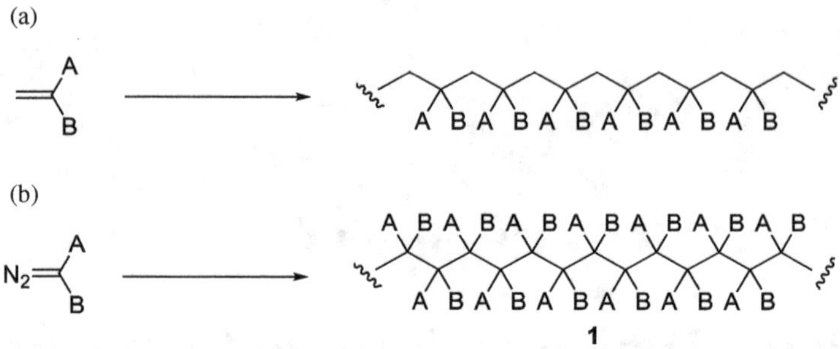

Scheme 13.1 Vinyl polymerization versus carbene polymerization.

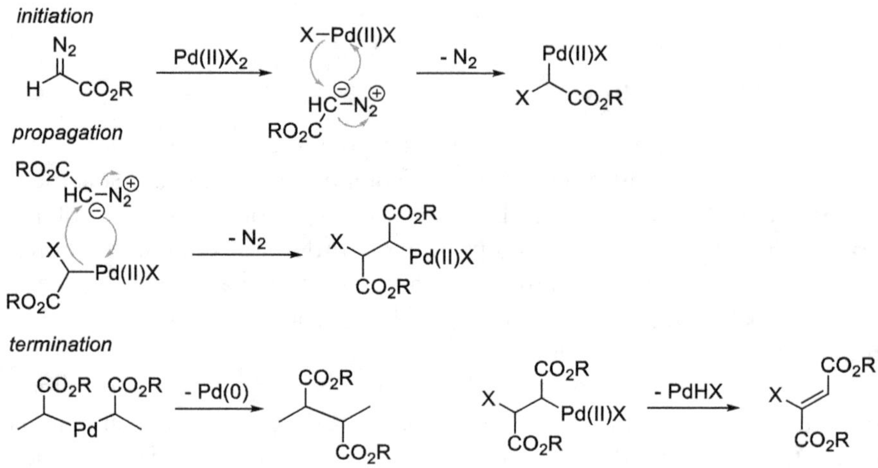

Scheme 13.2 Proposed mechanism for Pd-catalyzed polymerization of carbene.

In 2008, the same group reported the Pd-catalyzed polymerization of substituted diazoacetamides, affording polymers **2** bearing *N*-substituted carbamoylmethylene as the repeating unit with the incorporation of the azo group in the main chain.[6] Besides, copolymerization of diazoacetamides with diazoketone has also been implemented. The same group also realized the Pd-catalyzed copolymerization of cyclic diazoketones derived from cyclohexanone and 1-tetralone, as well as the homo- and copolymerization of *α,β*-unsaturated cyclic diazoketone.[7] These reactions provide polymers **3** and **4** containing unique repeating unit structures, where one of the six carbon atoms of the six-membered ring constitutes the C–C main chain framework (Scheme 13.3).

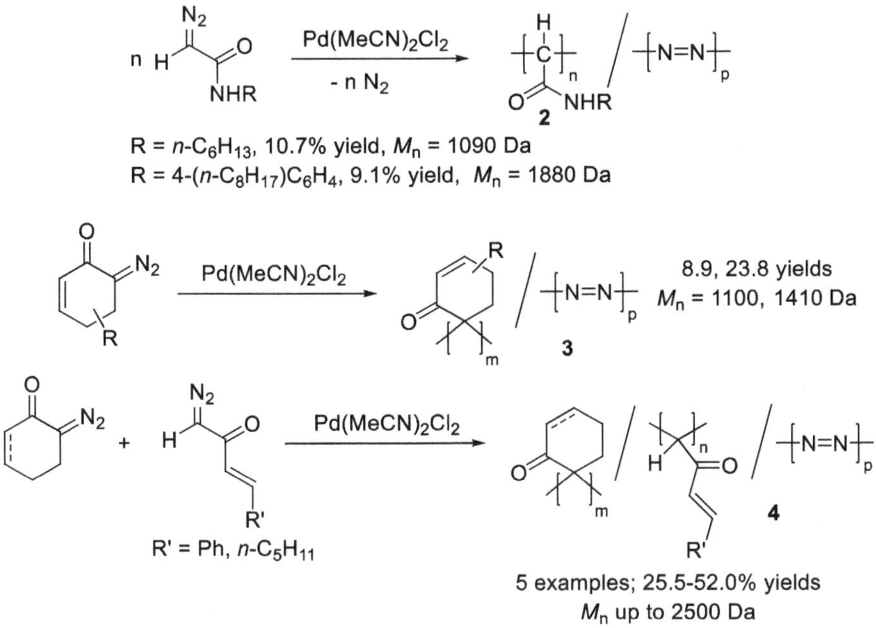

Scheme 13.3 Pd(II)-Catalyzed polymerization of diazoacetamides and cyclic diazoketones.

Subsequently, the same group reported a Pd(II)-catalyzed copolymerization of bifunctional diazocarbonyl compounds with monofunctional diazoketones, leading to the formation of new type of crosslinked polymers.[8] Higher density of substituents in each lattice in the network of polymers was achieved as compared with the analogous polymers obtained by conventional vinyl polymerization. Notably, Pd-catalyzed homo- and copolymerization of monofunctional diazoacetylferrocene were also examined, resulting in the formation of ferrocene-containing polymers (Scheme 13.4).

Other Pd(II) complexes have also been explored for promoting the polymerization of diazo compounds. In 2008, Ihara and Inoue employed the (NHC)Pd/borate initiating system for catalyzing polymerization of ethyl diazoacetate, affording high-molecular-weight polymers (M_n>20000 Da).[9] Based on the MALDI-TOF-MS experiments, a mechanism was proposed in which the reaction was initiated by cationic [(NHC)Pd(II) H]+ **5**, followed by insertion of EDA into Pd-H (initiation) and Pd-C (propagation), terminated by β-H elimination affording Pd-H species

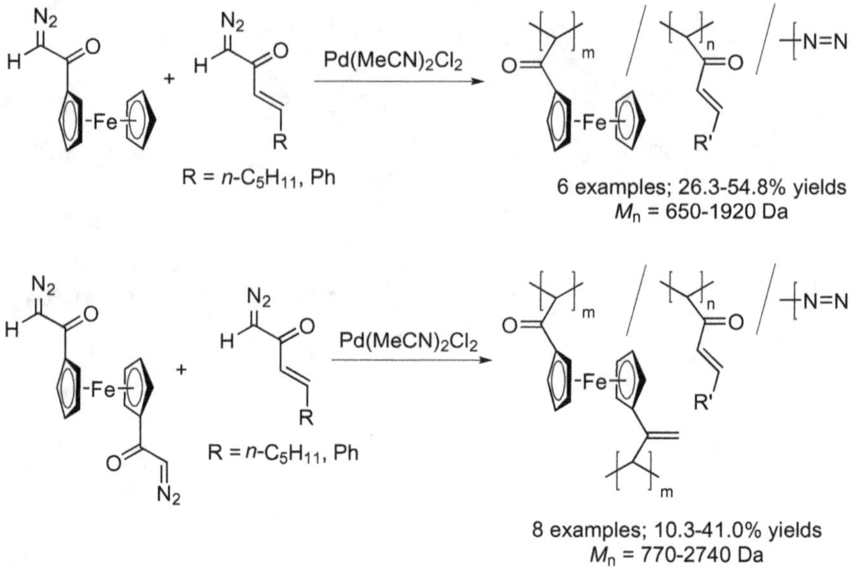

Scheme 13.4 Pd(II)-Catalyzed copolymerization of bifunctional diazocarbonyl compounds.

and C=C bond at chain ends. Furthermore, this catalytic system was efficiently applied for the polymerization of various alkyl diazoacetates with a variety of ester groups. Corresponding poly-(alkoxycarbonylmethylene) s were obtained with high molecular weight (M_n>10000 Da, R = *n*-hexyl, cyclohexyl, benzyl, and cholesteryl); however, 1-adamantyl diazoacetate afforded low-M_n polymers (M_n< 4000 Da), presumably due to the effect of steric hinderance (Scheme 13.5).[10]

Ihara and co-workers further explored π-allylPdCl-based systems for catalyzing the polymerization of diazo derivatives.[11] Compared with the previously reported reaction system with (NHC)Pd/borate, these catalysts show much higher initiation efficiencies, and polymerize the alkyl diazoacetates at low temperatures (−20–0 °C). MALDI-TOF-MS analyses provided the information for the chain-end structures of the corresponding polymers, thereby providing the information for the initiation and termination mechanisms. In 2013, the same group also realized the preparation of C–C main chain polymers having aromatic substituents on all the main chain carbons *via* Pd-mediated polymerization of aryl and arylmethyl diazoacetates.[12] These types of polymers demonstrated interesting photophysical properties (Scheme 13.6).

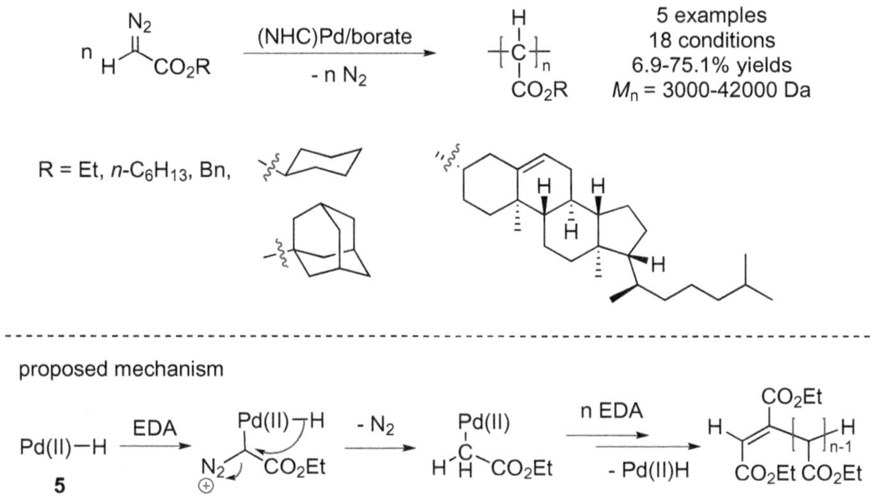

Scheme 13.5 Polymerization of alkyl diazoacetates catalyzed by (NHC)Pd/borate.

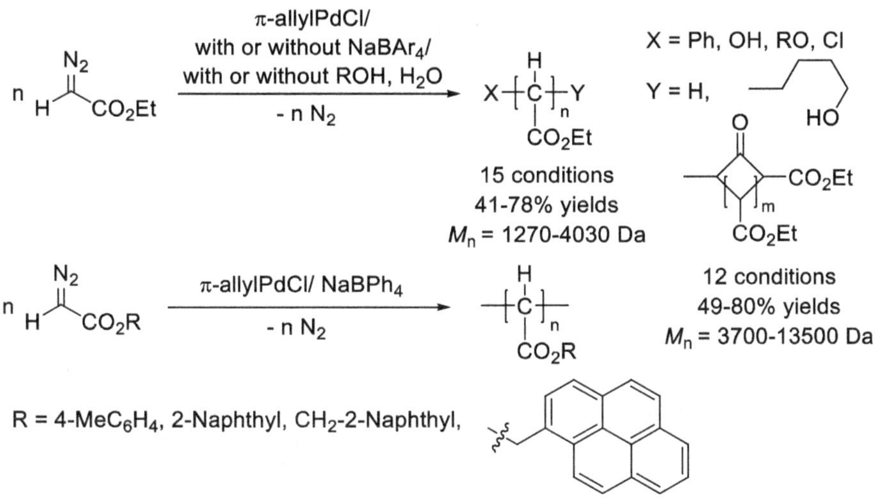

Scheme 13.6 Polymerization of diazoacetates catalyzed by π-allylPdCl.

Furthermore, Ihara and co-workers reported the polymerizations of hydroxy-containing diazoacetates, in which the reaction pathway was dependent on the initiation system (Scheme 13.7).[13] When initiated by π-allylPdCl complex, hydroxy-containing “poly(substituted methylene)s”

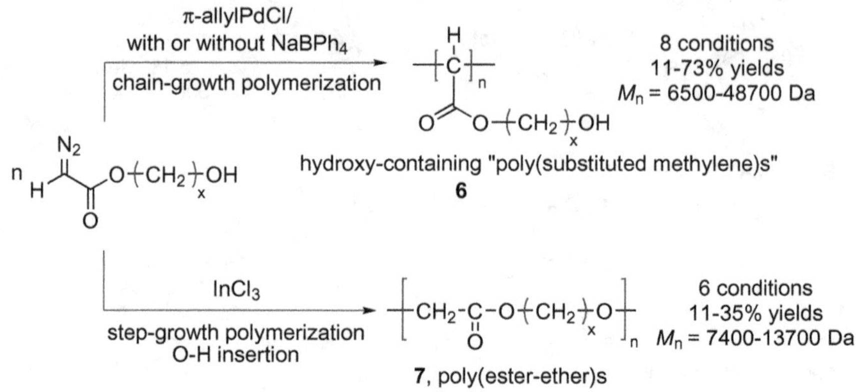

Scheme 13.7 Pd- and $InCl_3$-catalyzed polymerization of hydroxy-containing diazoacetates.

6 (C–C main chain polymers) were obtained through Pd-mediated chain-growth polymerization pathway. The obtained polymer had a slightly branched structure owing to chain transfer reaction with hydroxy group. However, the analoguous polymerization of silyl-protected substrates had a completely linear structure. On the other hand, oligomers **7** with a distinct repeating unit (ester–ether), where a new ether bond was generated through O–H insertion of diazocarbonyl groups into hydroxyl groups, were obtained by using $InCl_3$ as the catalyst through step-growth mechanism.

Alternatively, de Bruin and co-workers explored the copolymerization of olefins with carbene monomers in the presence of some well-defined Pd catalysts.[14] However, copolymer formation or even formation of the homopolymers was not observed, which indicated that olefin and carbene polymerization are incompatible. Detailed investigation concluded that low-valent Pd rather than Pd(II)-alkyl was the active species to mediate carbene polymerization of diazo compounds. Well-defined Pd(II)-alkyl species showed only a few insertions of carbene monomers, but longer oligomers (~20 carbene units) were formed from Pd(0) salts.

13.3 Rhodium-catalyzed polymerization of diazo compounds

Apart from Pd and Ru catalysts, rhodium complexes have also been explored as efficient catalysts for mediating polymerization of diazo compounds. A recent advance in this field is the report of a stereo-regular polymerization of ethyl diazoacetate, which gives a high-molecular-weight poly(ethyl 2-ylidene acetate) (up to 190 kDa) (Eq. (13.1)).[15,16] The reaction is catalyzed by a series of Rh(I) complexes, affording the polymers bearing unique chain structures.

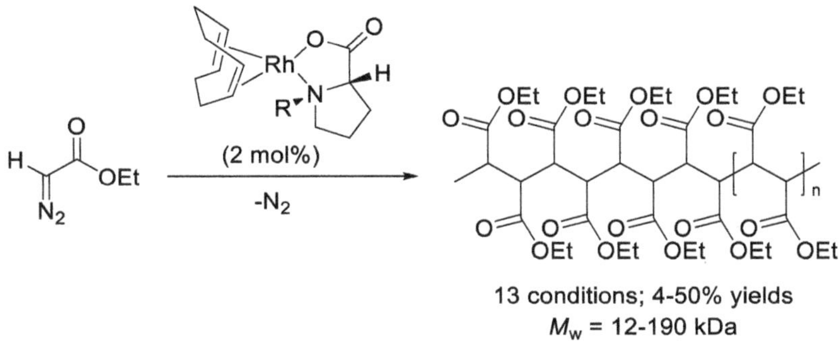

(13.1)

In 2007, de Bruin and co-workers reported their study on the ligand variation, kinetic investigations and the mechanism of an Rh(I)-mediated carbene polymerization. This carbene polymerization provides highly stereo-regular functionalized polymethylenes (PEA) of high molecular weight (Scheme 13.8).[17] Variation of the anionic *N*,*O*-ligand resulted in

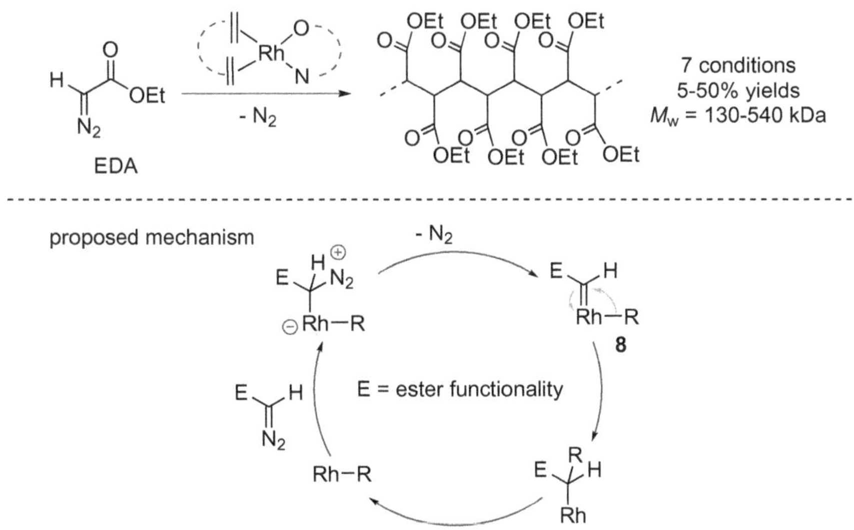

Scheme 13.8 Rh-Mediated polymerization of carbenes generated from EDA.

the change of the polymer yields but not on the polymer properties; whereas the variation of the diene moiety had a marked influence on the polymer properties, indicating that diene ligand remained bound to catalytically active species during the propagation steps. Kinetic study demonstrated that only a small fraction of Rh (1–5%) was involved in the catalysis of the polymerization. Based on DFT calculations, a chain-growth mechanism involving Rh(I)-carbene **8** migratory insertion into the growing alkyl chain was proposed.

In 2010, the same group reported the stereoselective polymerization of diazoesters catalyzed by [(*L*-prolinate)Rh(I)(Me_2cod)] **9**. With this C2-chiral and sterically more demanding diene ligand, highly syndiotactic carbon-chain polymers were obtained with the highest reported molecular weights and high yields (Eq. (13.2)).[18] This result demonstrated the importance of ligand design in synthesizing functionalized polymers with diverse property.

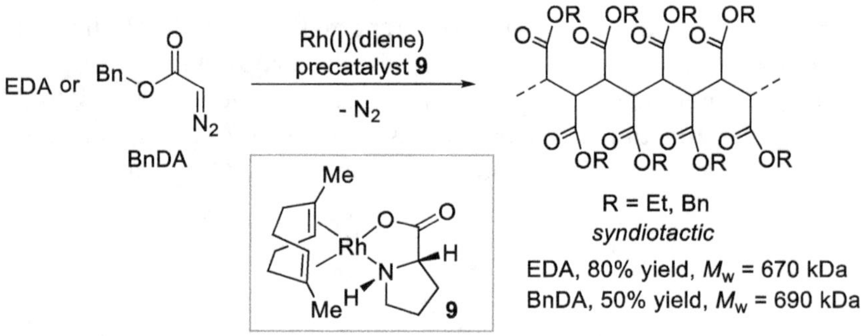

(13.2)

To gain insights into the catalytic stereo-regular carbene polymerization, de Bruin and co-workers utilized high-resolution ESI-MS to characterize the active rhodium species in the polymer-forming growing chain. The results showed that cationic [(allyl)Rh(III)(polymeryl)]+ **10** was the active intermediate (Scheme 13.9).[19] This cationic species is generated from oxygenated [(diene-O)Rh(III)] complex, which produces the polymer with higher yields than their non-oxidized [(diene)Rh(I)] precursors.

Furthermore, the details of the chain propagation, chain transfer and chain termination in the carbene polymerization catalyzed by cationic [(cycloocta-2,6-dien-1-yl)Rh(III)(alkyl)]$^+$ were investigated through DFT calculation.[20] Compared with the polymerization with neutral [(diene) Rh(I)(alkyl)] species, the chain propagation in this case is favored over

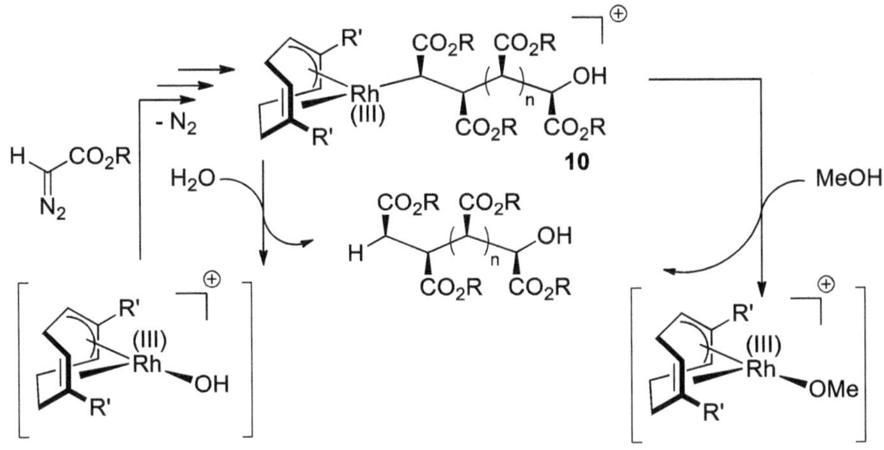

Scheme 13.9 Mechanism showing cationic species in carbene polymerization.

β-H elimination. The β-H elimination has been proposed to be responsible for the stereo-defects in Rh-mediated polymerization.[21] Thus, syndiotactic polymers with high molecular weight could be obtained by using cationic Rh(III) catalyst owing to chain-end-controlled propagation. The chain transfer involving alcohol-mediated protonolysis was also shown to be favored over β-H elimination as suggested by DFT calculation. The chain propagation from the intermediate with a stereo-error at the α-carbon of the growing chain was substantially slower as compared to the propagation from the syndiotactic intermediate (Scheme 13.10).

Detailed investigation on the Rh-mediated carbene polymerization was carried out through the end-group analysis, dilution-kinetic studies and the comparison of the activity of well-defined catalysts containing a preformed Rh–C bond.[22] It has been revealed that the chain is initiated by the insertion of a preformed carbene into an Rh-ligand or Rh-hydride bond, or by (internal or external) nucleophilic attack of water and/or alcohol on an Rh-carbene **11** moiety. Chain-termination takes place mainly by (nucleophilic) protonolysis involving water or alcohol; while β-H elimination only plays a minor role and is only observed in the formation of short oligomers. A chain-transfer mechanism involving nucleophilic attack on the alcohol (nucleophilic, σ-bond metathesis type, protonolysis) has been proposed as shown in Scheme 13.11.

Furthermore, de Bruin and co-workers utilized Rh(I) pre-catalysts to realize (co)polymerization of carbenes from diazoesters, affording high

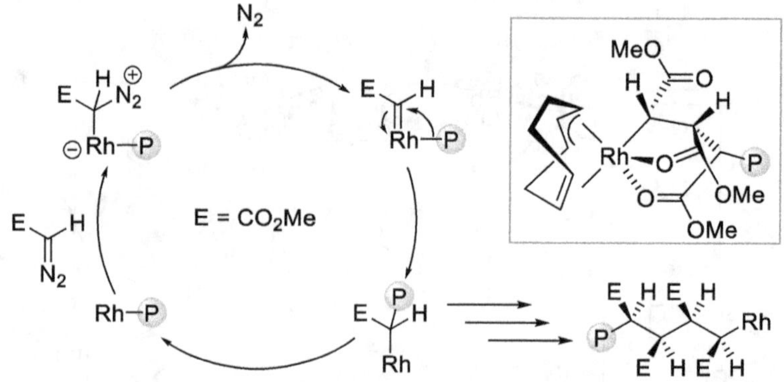

Scheme 13.10 Carbene polymerization catalyzed by [(cycloocta-2,6-dien-1-yl)Rh(III)(polymeryl)]$^+$.

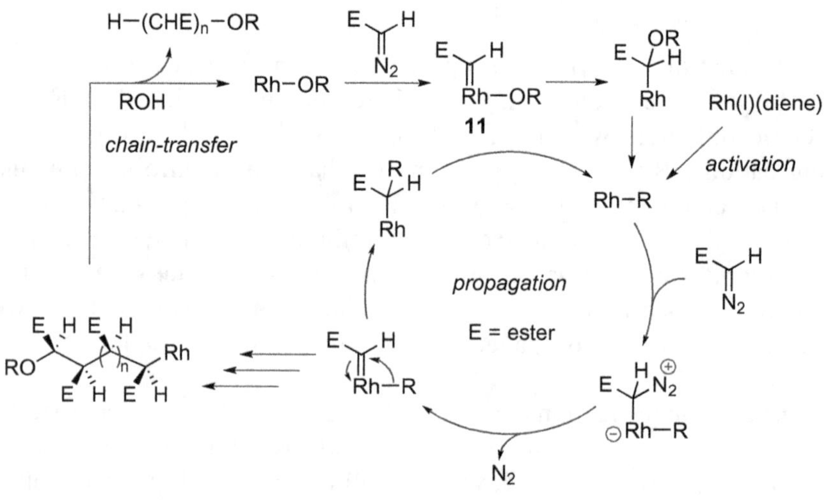

Scheme 13.11 Mechanism of alcohol-mediated chain transfer.

molecular weight (M_w of homopolymers up to 690 kDa and copolymers up to 430 kDa), and stereo-regular (co)polymers, contaminating a polar side group at every backbone carbon atom (Scheme 13.12).[23] The homopolymers poly(ethyl 2-ylideneacetate) and poly(benzyl 2-ylideneacetate) were analyzed by polarizing optical microscopy and revealed thermotropic and lyotric liquid crystallinity. On the other hand, the formation of block copolymers of carbenes was confirmed by NMR, differential

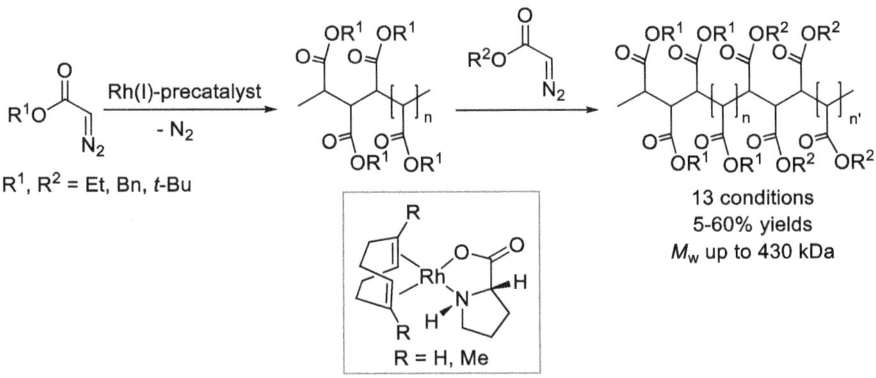

Scheme 13.12 Rh(I)-Mediated copolymerization of ethyl, benzyl and *tert*-butyl diazoacetate.

scanning calorimetry and size-exclusion chromatography. Thermal analysis indicated that the homopolymers and block copolymers were semicrystalline, but the random copolymers were completely amorphous. Besides, the chromatographic examination of the chemical composition and sequence distribution of copolymers from ethyl and benzyl diazoacetate was also implemented.[24]

The same group also realized the controlled synthesis of functional copolymers **12** with blocky microstructures through carbene copolymerization of diazomethane and diazoesters (Eq. (13.3)).[25] The reaction afforded high-M_w ethylene-acrylate type copolymers with a large variation in functional group content between 0% and 100%, which was tunable by changing the monomer feed ratio and the way of addition. The properties of this new class of polymers turned out to be highly dependent on the functional group content, and therefore they could be varied in the window between the properties of both homopolymers.

12, X_{ester} 0-100%
Blocky microstructure; stereoregularity

(13.3)

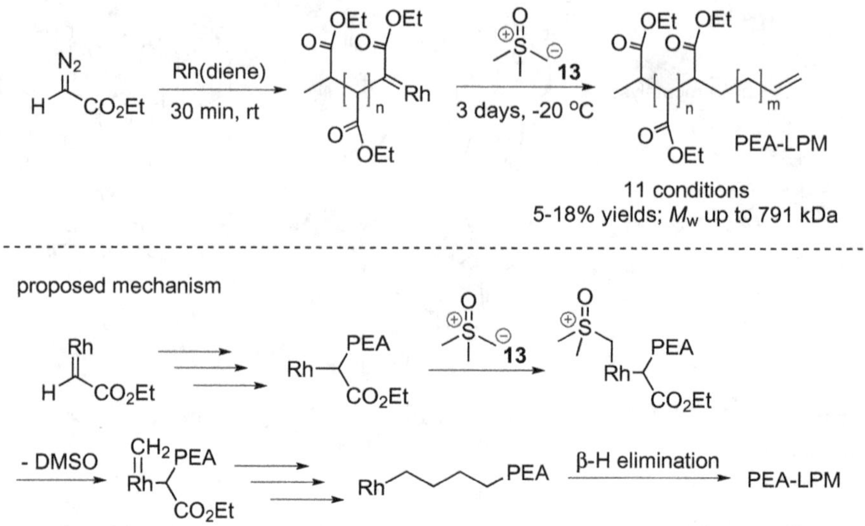

Scheme 13.13 Rh-catalyzed C1-copolymerization of diazoesters and sulfoxonium ylides.

It was observed that in the presence of Rh(III) catalyst, carbene copolymerization of diazomethane with EDA led to a significant increase in polymer yield and molecular weight as compared to the system using Rh(I) catalysts. Theoretical study toward chain propagation and termination steps in copolymerization mediated by the (cycloocta-2,5-dien-1-yl) Rh(III)(alkyl) species was conducted.[26] The results indicated that chain propagation *via* carbene insertion occurred with a much lower energy barrier in such a reaction system. Besides, the chain propagation leading to high-M_w polymers was found indeed favored over chain termination *via* β-hydride elimination.

Moreover, de Bruin and co-workers explored Rh-catalyzed carbene C1-copolymerization by using sulfur ylides as the carbene monomer precursors (Scheme 13.13).[27] Stereo-regular diblock copolymers from sulfoxonium ylides **13** could be generated with diazoesters as the respective carbene monomer precursors. High-M_w poly(methylene)-poly(estercarbene) copolymers were obtained with a diblock-syndiotactic microstructure with decent yields. These functionalized copolymers could act as blending agents to mix polyethylene or polymethylene with poly(ethyl 2-ylidene-acetate).

More recently, copolymerization of diazo compounds with olefins was reported (Eq. (13.4)).[28] The reaction provided a method for the controlled synthesis of high molecular-weight and highly functional polymers. Although Rh(III) species demonstrated high activity in homopolymerization of either olefins or diazo compounds, it turned out to be virtually inactive in this olefin–carbene copolymerization procedure. At higher ethene concentrations, copolymerization process slowed down as compared to the formation of homopolymers. Thus, the obtained copolymers exhibited a blocky microstructure, with the functional blocks being highly stereo-regular.

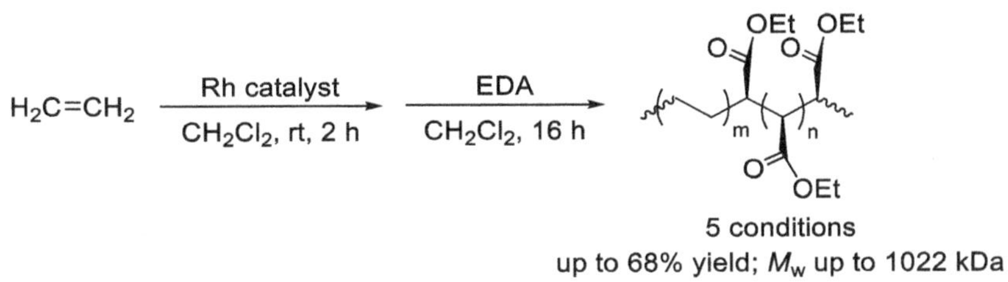

(13.4)

The polymers obtained through the carbene polymerization may have potential applications in materials science. In this context, de Bruin and co-workers demonstrated through a combination of experimental measurements and computational MM/MD modeling that poly(ethylidene acetate)(*st*-PEA) self-assembles with a fully sp^3-based backbone to form higher-order liquid-crystalline aggregates.[29] The results underscore the importance of the "tertiary structure" of synthetic polymers on their material properties.

On the other hand, traditional Rh(II)-catalyzed transformations, such as O–H insertion, have also been explored in the polymerizations. Ihara and Inoue reported the Rh(II)-catalyzed polycondensation of bis(diazocarbonyl) compounds with aromatic diols in cyclic ethereal solutions (Scheme 13.14).[30] Apart from the normal O–H insertion process, the polymers formed due to unexpected incorporation of the cyclic ethereal solvent were also obtained, but in low yields. The ring opening of the cyclic ether is initialized by the nucleophilic attack of the ethereal oxygen on the electron-deficient Rh(II) carbene center to form an Rh(II)-associated oxonium ylide, which undergoes nucleophilic attack by another ethereal oxygen.

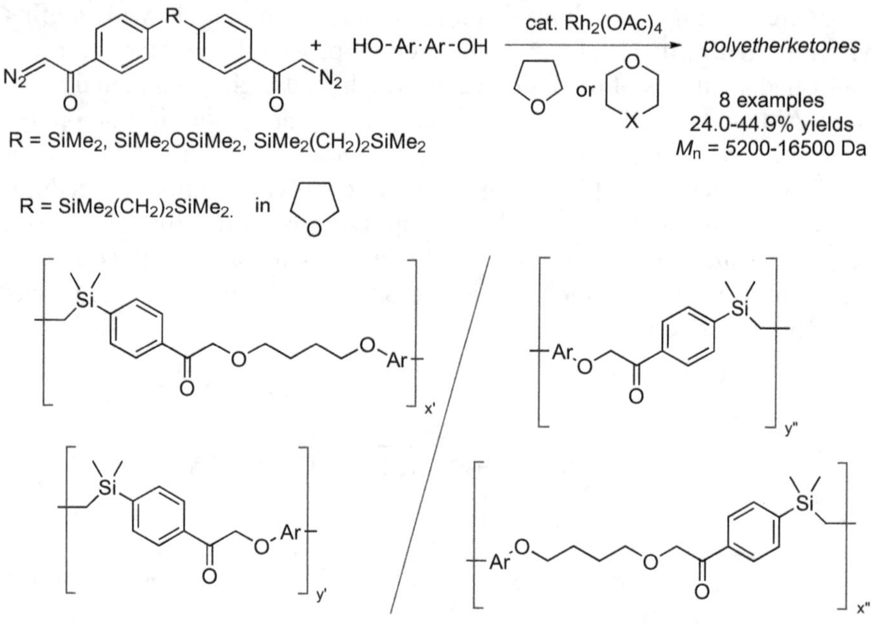

Scheme 13.14 Polycondensation of silyl-substituted bis(diazocarbonyl) compounds with aromatic diols and cyclic ethers.

In addition, Ihara and co-workers reported the synthesis of poly(ester ether ketone)s **14** through Rh(II)-mediated three-component polycondensation of bis(diazoketone), dicarboxylic acids with THF (Eq. (13.5)).[31] The advantage of this process is the use of aliphatic dicarboxylic acids as the monomers for polymerization. However, when THP was used as the solvent and co-monomer, normal insertion of carbene into the carboxyl O–H competed effectively with the insertion of THP with carbene.

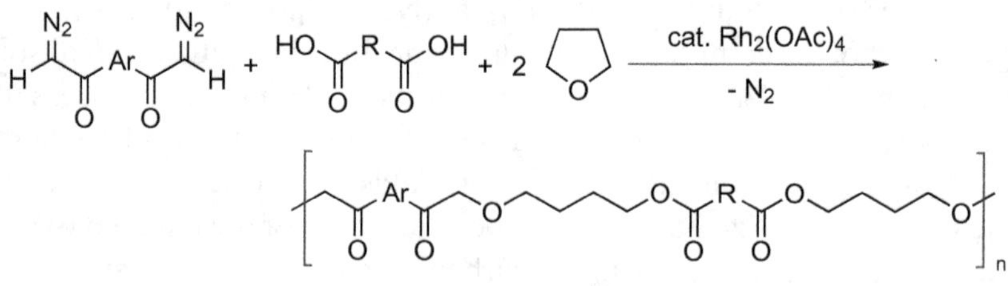

14, *new type of poly(ester ether ketone)s*

6 examples; 11 conditions; 17-47% yields
M_n = 7300-25300 Da

(13.5)

13.4 Polymerization of diazo compounds catalyzed by copper or ruthenium complexes

Liu and co-workers established another procedure for preparing C1/C1N2 copolymer through poly(imidazole-Cu)-mediated C1/C1N2 copolymerization of EDA, which involved Cu-based carbene radical and the generation of α-carbonyl diazomethane radical species (Scheme 13.15).[32] In the reaction mechanism, the diazo radical intermediate **15** was captured by radical trapping agent and was detected by room-temperature EPR spectroscopy.

Moreover, a series of bis-diazocarbonyl compounds were synthesized and subjected to Cu(II)-catalyzed de-nitrogen alkene polymerization, producing amorphous polymers with molecular weights of 2900–35400 Da and moderate yields (Scheme 13.16).[33] This catalytic intermolecular coupling protocol allowed efficient synthesis of unsaturated polyesters. Reaction pathway was proposed to involve the initial attack of the α-bisdiazo compound at the metal center to generate Cu-carbene species **16**. The polymer chain propagates in the presence of excess of Cu(II) catalyst and bis-diazo compounds. Finally, cyclic polymers **17** or linear polymers **18** were formed, respectively.

Ihara and co-workers realized the preparation of a variety of aliphatic and aromatic unsaturated polyesters through catalytic single-component polycondensation of bis(diazoacetate)s under mild reaction conditions (Scheme 13.17).[34] With the employment of Grubbs 2nd catalyst, well-defined unsaturated polyesters of high *cis*-C=C bond configuration were obtained selectively. Other transition metal catalysts, including Grubbs 1st catalyst, $Rh_2(OAc)_4$ or $Cu(acac)_2$, also mediated this polycondensation to produce unsaturated polyesters, but with low stereocontrol. Besides, the *cis*-C=C bond could be isomerized quantitatively into *trans*-configuration

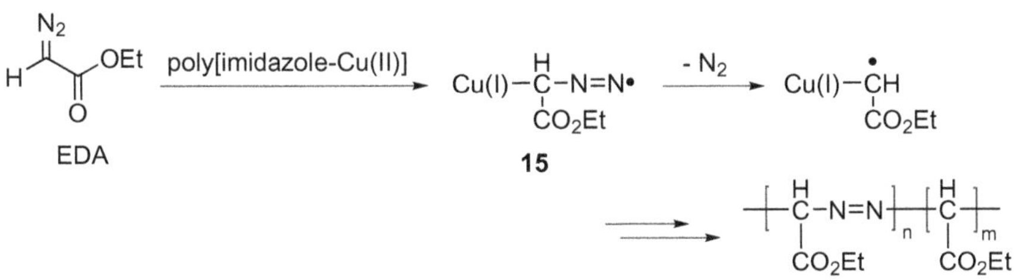

Scheme 13.15 Poly(imidazole-Cu)-mediated C1/C1N2 copolymerization of EDA.

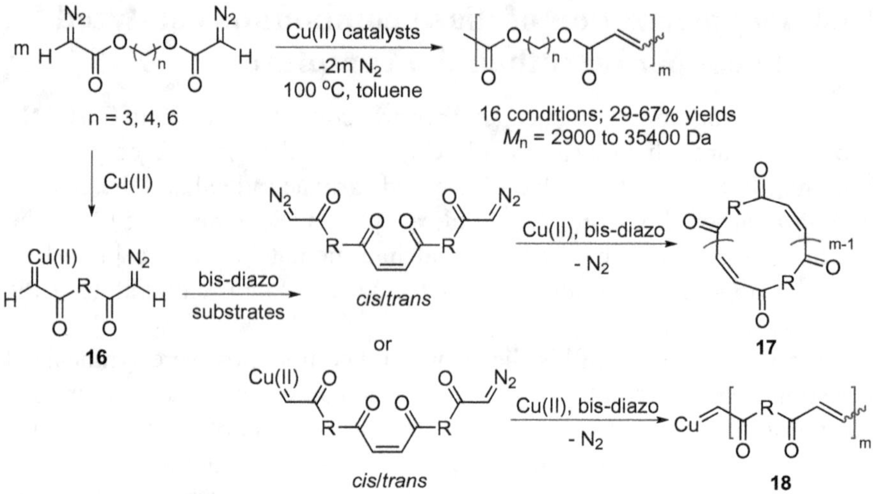

Scheme 13.16 Cu(II)-catalyzed de-nitrogenative alkene polymerization of bis-diazo compounds.

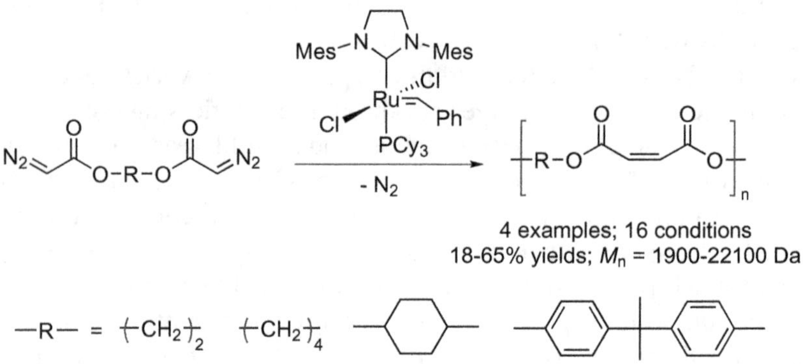

Scheme 13.17 Ru-catalyzed polycondensation of bis(diazoacetate)s leading to unsaturated polyesters.

with a catalytic amount of diethylamine, resulting in the formation of a well-defined polymer with only *trans*-C=C bond.

13.5 Polymerization without carbene formation

Apart from catalytic carbene polymerization of various diazo compounds, other types of polymerization strategies have also been explored to access

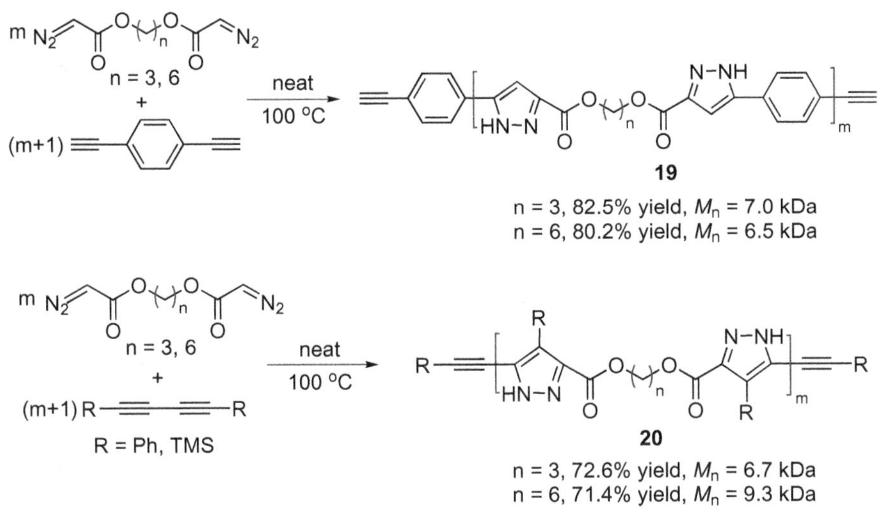

Scheme 13.18 Preparation of polymers *via* 1,3-diploar cycloaddition of bis-diazo compounds with bis-alkynes.

the functionalized polymers. In particular, Shea and co-workers have investigated the polymerization of ethyl diazoacetate with BH_3. The polymerization proceeds with the formation of homologated C-boron enolate.[35]

In 2014, Liu and co-workers reported the one-pot synthesis of polypyrazoles **19** and **20** through non-catalytic 1,3-dipolar [3+2] cycloaddition of bis-diazocarbonyl compounds with bisalkynes (Scheme 13.18).[36] Under thermal conditions, the substituted polypyrazoles were obtained with M_w in the range of 6500–9300 Da and 71–83% yields. The self-assembled porous polypyrazole-copper complex (pore diameters 1–2 μm) could be prepared by the coordination of the polypyrazole with copper sulfate, which showed excellent catalytic activity in click chemistry and could be served as a reusable catalyst.

The synthesis of polymers with the retention of diazo moiety was developed through radical polymerization procedure. In 2008, Kim and co-workers reported radical polymerization of 3-methacryloxypropyl trimethoxysilane (MPMS), 2-(2-diazo-3-oxobutyryloxy)ethyl methacrylate (DOBEMA) and γ-butyrolactonyl methacrylate (GBLMA) (Scheme 13.19).[37] The reaction afforded poly(DOBEMA-*co*-MPMS-*co*-GBLMA) **21** under mild conditions. The diazo group-containing polymers could be served as silicon-containing bilayer resist based on a

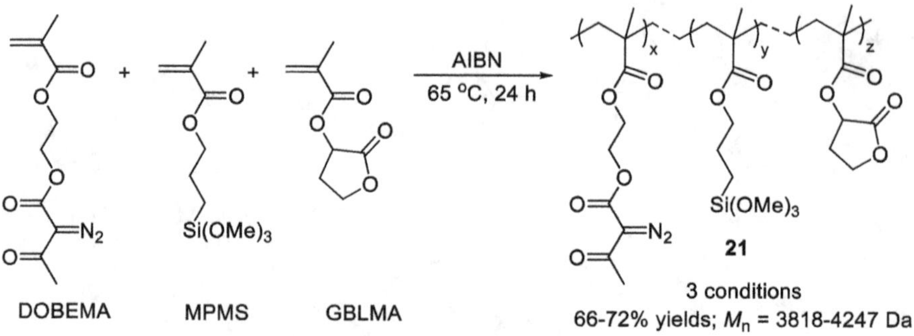

Scheme 13.19 Copolymerization of MPMS, DOBEMA and GBLMA.

non-chemically amplified resist system. The initial lithographic evaluation of the resist indicated the potential of this system for the next generation resists.

In the same year, Kim and co-workers further reported the synthesis of water-processable photopolymers through copolymerization of DOBEMA, hydroxyethyl methacrylate (HEMA), methacryllic acid (MAA) and sodium 4-vinylbenzenesulfonate (SVBS).[38] Two kinds of copolymers, namely poly(DOBEMA-*co*-HEMA-*co*-SVBS) **22** for negative-tone resist and poly(DOBEMA-*co*-MAA) **23** for positive-tone resist, were respectively prepared and their photo-lithographic properties were investigated (Scheme 13.20). The initial lithographic studies indicated that both of these negative and positive systems showed feasibility as water-processable polymers.

Subsequently, the same group developed radical copolymerization of two monomers with diazo moiety, cholic acid 3-diazo-3-ethoxycarbonyl-2-oxo-propyl ester methacrylate (CDEOPE-MA) and 2-diazo-6-hydroxy-3-oxo-hexanoic acid ethyl ester methacrylate (DHOHEE-MA) with γ-butyrolacton-2-yl methacrylate.[39] After evaluating lithographic properties of the polymers, the introduction of cholate-monomer (CDEOPE-MA) seemed to enhance the film formability as well as film stability in aqueous medium (Figure 13.1).

In 2012, Liu and Zou reported the synthesis of photoactive polymer **24** containing aryl substituted diazoketo moiety through radical copolymerization of methyl methacrylate (MMA), 2-hydroxyethylmethacrylate (HEMA) and 2-(2-diazo-3-oxo-3-(4-dimethylamino- phenyl)-propionyloxy)-ethyl methacrylate (Scheme 13.21).[40] Upon UV irradiation,

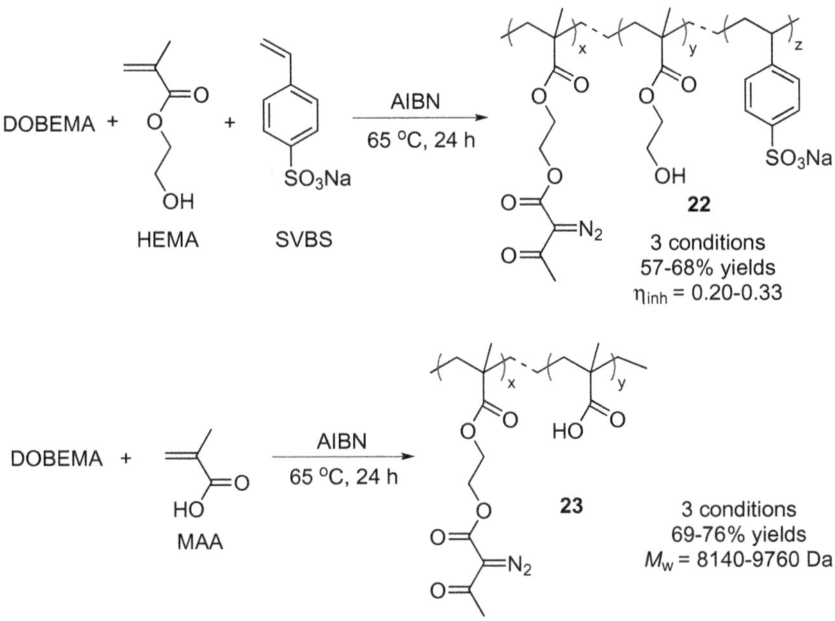

Scheme 13.20 Copolymerization for the synthesis of poly(DOBEMA-*co*-HEMA-*co*-SVBS) and poly(DOBEMA-*co*-MAA).

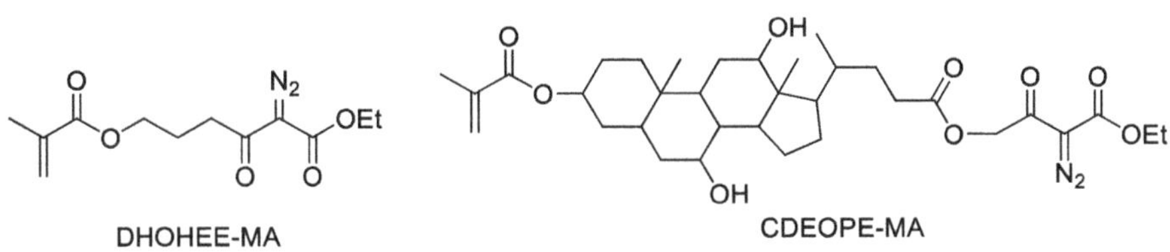

Figure 13.1 Structure of DHOHEE-MA and CDEOPE-MA.

the diazoketo group of the poly(diazo monomer-*co*-MMA-*co*-HEMA) underwent Wolff rearrangement, affording carboxylic acid moieties upon the reaction of the ketene intermediate with water. This type of photoactive polymer shows potential as a positive photoresist in TFT-LCD and IC discrete devices fabrication processes.

In 2017, Rühe and co-workers reported the thermal or photochemical copolymerization of diazomalonates with olefins. Subsequent decomposition of the diazo moiety in these copolymers occurs to form a carbene species, which induces C–H insertion cross-linking reactions. The

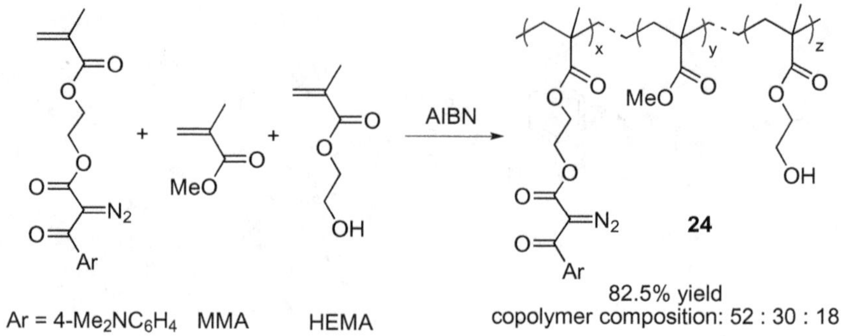

Scheme 13.21 Copolymers of 2-(2-diazo-3-oxo-3-(4-dimethylaminophenyl)propionyloxy)-ethyl methacrylate, MMA and HEMA.

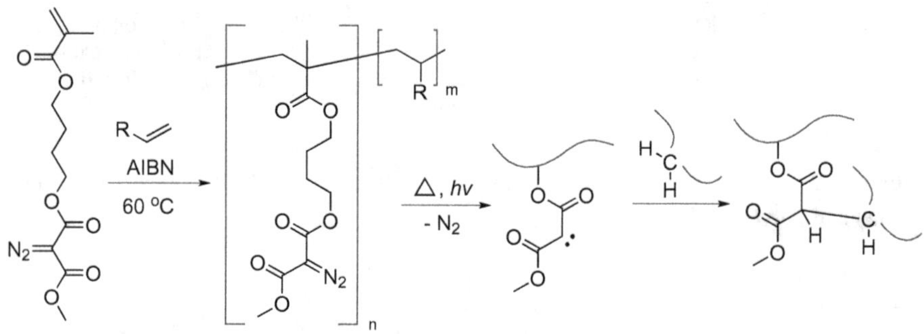

Scheme 13.22 Synthesis and copolymerization of diazomalonic ester and C–H insertion crosslinking reactions.

reaction can be used for the generation of a wide range of polymer layers and microstructures on a abroad spectrum of surfaces (Scheme 13.22).[41]

13.6 Polymerization of diazo compounds in the absence of transition-metal catalysts

As discussed earlier, Ihara and co-workers investigated the palladium-mediated oligomerization of diazoacetates and related diazocarbonyl compounds. However, the viscous oils obtained with palladium-catalyzed polymerization with alkyl diazoacetates generally have low molecular weight and low selectivity (average degree of polymerization up to about

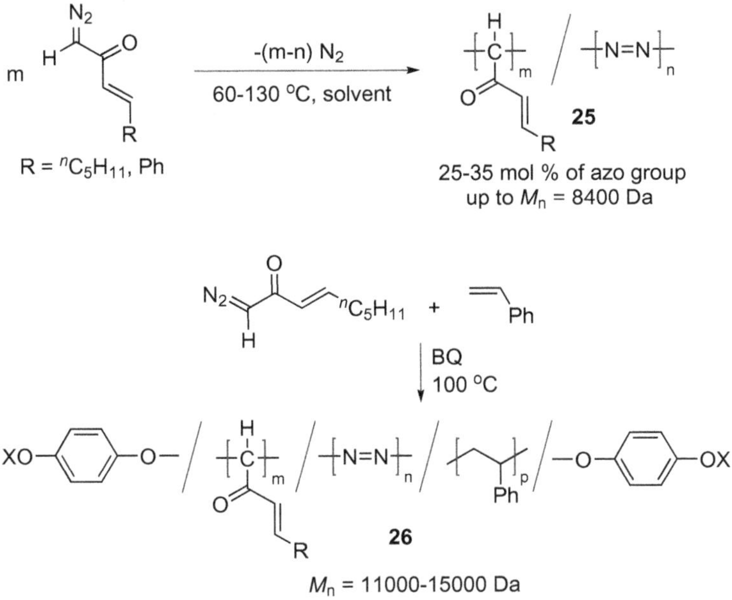

Scheme 13.23 Thermally induced polymerization of diazoketones and copolymerization with styrene in the presence of BQ.

100).[42–45] Oligomerization of alkyl diazoacetates with copper powder was initially reported by Liu and co-workers.[46] Ihara and co-workers demonstrated that the thermally induced polymerization of substituted diazoketones afforded polymers **25** with high azo content (25–35 mol%) and acyl groups in their main chains (Scheme 13.23).[47] Besides, the copolymerization of diazoketones with styrene in the presence of benzoquinone resulted in the generation of copolymers **26** (M_n=11000–15000 Da), which contained acylmethylene, azo group and repeating unit from styrene in their main chains. Polymerization mechanism was proposed to proceed *via* radical propagating chain process.

In addition to the polymerization with thermolysis of diazo compounds, microwave-irradiated polymerization of cyclic ester with α-diazoacetate in absence of any metal catalyst was reported by Liu and co-workers (Scheme 13.24).[48] This strategy in which ring opening and carbene polymerization are involved enabled the one-pot preparation of copolymers of polyester and polyolefin. Under optimized conditions, the copolymer with M_n 36.1 kDa and PDI of 1.86 was obtained. The same

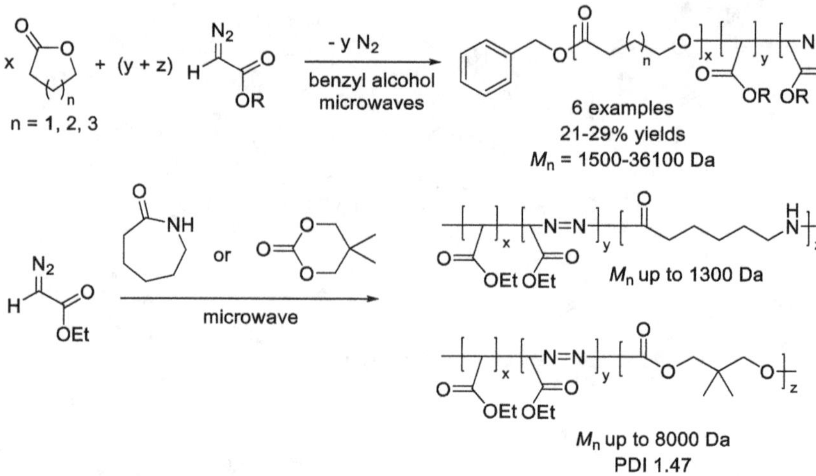

Scheme 13.24 Microwave-irradiated copolymerization of diazoacetate.

group also expanded this microwave-assisted one-pot copolymerization procedure to other cyclicmonomers (ε-caprolactam and 2,2-dimethyltrimethylene carbonate) with EDA.[49] Elemental analysis as well as the FT-IR spectroscopy data indicated the incorporation of the azo group in the obtained copolymers.

In summary, polymerization with diazo compounds has attracted significant attention in the past decades, although it is only a minor subdivision in the rich chemistry of diazo compounds. The classic transition-metal-catalyzed carbene transfer reactions, such as insertions, cyclopropanations and ylide reactions, have not been extensively explored in polymerization. For the polymerization based on metal carbene migratory insertion, namely the C1 polymerization, it is expected that living polymerization is possible and thus the molecular weight, distribution and controlled copolymerization are conceivable. Owing to the diverse reactivity of diazo compounds, the application of diazo compounds in polymerization synthesis is expected to boom in the coming years.

References

1. Ihara, E. *J. Synth. Org. Chem. Jpn.* **2008**, *66*, 358–367.
2. Ihara, E. *Adv. Polym. Sci.* **2010**, *231*, 191–231.

3. Jellema, E.; Jongerius, A. L.; Reek, J. N. H.; de Bruin, B. *Chem. Soc. Rev.* **2010**, *39*, 1706–1723.
4. Franssen, N. M. G.; Walters, A. J. C.; Reek, J. N. H.; de Bruin, B. *Catal. Sci. Technol.* **2011**, *1*, 153–165.
5. Franssen, N. M. G.; Reek, J. N. H.; de Bruin, B. *Chem. Soc. Rev.* **2013**, *42*, 5809–5832.
6. Ihara, E.; Hiraren, T.; Itoh, T.; Inoue, K. *Polym. J.* **2008**, *40*, 1094–1098.
7. Ihara, E.; Hiraren, T.; Itoh, T.; Inoue, K. *J. Polym. Sci. Part A: Polym. Chem.* **2008**, *46*, 1638–1648.
8. Ihara, E.; Goto, Y.; Itoh, T.; Inoue, K. *Polym. J.* **2009**, *41*, 1117–1123.
9. Ihara, E.; Ishiguro, Y.; Yoshida, N.; Hiraren, T.; Itoh, T.; Inoue, K. *Macromolecules* **2009**, *42*, 8608–8610.
10. Ihara, E.; Takahashi, H.; Akazawa, M.; Itoh, T.; Inoue, K. *Macromolecules* **2011**, *44*, 3287–3292.
11. Ihara, E.; Akazawa, M.; Itoh, T.; Fujii, M.; Yamashita, K.; Inoue, K.; Itoh, T.; Shimomoto, H. *Macromolecules* **2012**, *45*, 6869–6877.
12. Ihara, E.; Okada, R.; Sogai, T.; Asano, T.; Kida, M.; Inoue, K.; Itoh, T.; Shimomoto, H.; Ishibashi, Y.; Asahi, T. *J. Polym. Sci. Part A: Polym. Chem.* **2013**, *51*, 1020–1023.
13. Shimomoto, H.; Itoh, E.; Itoh, T.; Ihara, E.; Hoshikawa, N.; Hasegawa, N. *Macromolecules* **2014**, *47*, 4169–4177.
14. Franssen, N. M. G.; Reek, J. N. H.; de Bruin, B. *Polym. Chem.* **2011**, *2*, 422–431.
15. Noels, A. F. *Angew. Chem. Int. Ed.* **2007**, *46*, 1208–1210.
16. Hetterscheid, D. G. H.; Hendriksen, C.; Dzik, W. I.; Smits, J. M. M.; Eck, E. R. H. v.; Rowan, A. E.; Busico, V.; Vacatello, M.; Castelli, V. V. A.; Segre, A.; Jellema, E.; Bloemberg, T. G.; de Bruin, B. *J. Am. Chem. Soc.* **2006**, *128*, 9746–9752.
17. Jellema, E.; Budzelaar, P. H. M.; Reek, J. N. H.; de Bruin, B. *J. Am. Chem. Soc.* **2007**, *129*, 11631–11641.
18. Jellema, E.; Jongerius, A. L.; Walters, A. J. C.; Smits, J. M. M.; Reek, J. N. H.; de Bruin, B. *Organometallics* **2010**, *29*, 2823–2826.
19. Walters, A. J. C.; Troeppner, O.; Ivanović-Burmazović, I.; Tejel, C.; del Río, M. P.; Reek, J. N. H.; de Bruin, B. *Angew. Chem. Int. Ed.* **2012**, *51*, 5157–5161.
20. Walters, A. J. C.; Reek, J. N. H.; de Bruin, B. *ACS Catal.* **2014**, *4*, 1376–1389.
21. Finger, M.; Reek, J. N. H.; de Bruin, B. *Organometallics* **2011**, *30*, 1094–1101.
22. Walters, A. J. C.; Jellema, E.; Finger, M.; Aarnoutse, P.; Smits, J. M. M.; Reek, J. N. H.; de Bruin, B. *ACS Catal.* **2012**, *2*, 246–260.

23. Jellema, E.; Jongerius, A. L.; van Ekenstein, G. A.; Mookhoek, S. D.; Dingemans, T. J.; Reingruber, E. M.; Chojnacka, A.; Schoenmakers, P. J.; Sprenkels, R.; van Eck, E. R. H.; Reek, J. N. H.; de Bruin, B. *Macromolecules* **2010**, *43*, 8892–8903.
24. Reingruber, E. M.; Chojnacka, A.; Jellema, E.; de Bruin, B.; Buchberger, W.; Schoenmakers, P. J. *J. Chromatogr. A* **2012**, *1255*, 259–266.
25. Franssen, N. M. G.; Remerie, K.; Macko, T.; Reek, J. N. H.; de Bruin, B. *Macromolecules* **2012**, *45*, 3711–3721.
26. Franssen, N. M. G.; Finger, M.; Reek, J. N. H.; de Bruin, B. *Dalton Trans.* **2013**, *42*, 4139–4152.
27. Suarez, A. I. O.; del Río, M. P.; Remerie, K.; Reek, J. N. H.; de Bruin, B. *ACS Catal.* **2012**, *2*, 2046–2059.
28. Franssen, N. M. G.; Reek, J. N. H.; de Bruin, B. *Dalton Trans.* **2013**, *42*, 9058–9068.
29. Franssen, N. M. G.; Ensing, B.; Hegde, M.; Dingemans, T. J.; Norder, B.; Picken, S. J.; Alberda van Ekenstein, G. O. R.; van Eck, E. R. H.; Elemans, J. A. A. W.; Vis, M.; Reek, J. N. H.; de Bruin, B. *Chem. Eur. J.* **2013**, *19*, 11577–11589.
30. Ihara, E.; Saiki, K.; Goto, Y.; Itoh, T.; Inoue, K. *Macromolecules* **2010**, *43*, 4589–4598.
31. Ihara, E.; Hara, Y.; Itoh, T.; Inoue, K. *Macromolecules* **2011**, *44*, 5955–5960.
32. Li, F.; Xiao, L.; Li, Y.; Chen, C.; Liu, L. *Chem. Commun.* **2015**, *51*, 11964–11967.
33. Xiao, L.; Li, Y.; Liao, L.; Liu, L. *New J. Chem.* **2013**, *37*, 1874–1877.
34. Shimomoto, H.; Hara, Y.; Itoh, T.; Ihara, E. *Macromolecules* **2013**, *46*, 5483–5487.
35. Bai, J.; Burke, L. D.; Shea, K. J. *J. Am. Chem. Soc.* **2007**, *129*, 4981–4991.
36. Xiao, L.; Cai, S.; Liu, Q.; Liao, L.; Guo, X.; Li, Y.; Jia, X.; Li, F.; Liu, L. *Polym. Chem.* **2014**, *5*, 607–613.
37. Park, J. Y.; Kim, J.-B. *Chem. Lett.* **2008**, *37*, 920–921.
38. Park, J. Y.; Yun, J. M.; Kim, J.-B. *J. Polym. Sci. Part. A: Polym. Chem.* **2008**, *46*, 7534–7542.
39. Ganesan, R.; Kim, J.-B. *Microelectron. Eng.* **2011**, *88*, 93–98.
40. Liu, L.; Zou, Y. *J. Appl. Polym. Sci.* **2012**, *123*, 554–561.
41. Kotrade, P. F.; Rühe, J. *Angew. Chem. Int. Ed.* **2017**, *56*, 14405–14410.
42. Ihara, E.; Haida, N.; Iio, M.; Inoue, K. *Macromolecules* **2003**, *36*, 36–41.
43. Ihara, E.; Fujioka, M.; Haida, N.; Itoh, T.; Inoue, K. *Macromolecules* **2005**, *38*, 2101–2108.
44. Ihara, E.; Nakada, A.; Itoh, T.; Inoue, K. *Macromolecules* **2006**, *39*, 6440–6444.

45. Ihara, E.; Kida, M.; Fujioka, M.; Haida, N.; Itoh, T.; Inoue, K. *J. Polym. Sci. Part A: Polym. Chem.* **2007**, *45*, 1536–1545.
46. Liu, L.; Song, Y.; Li, H. *Polym. Int.* **2002**, *51*, 1047–1049.
47. Ihara, E.; Nishida, H.; Fujii, M.; Itoh, T.; Inoue, K. *Polym. Bull.* **2011**, *66*, 3–15.
48. Xiao, L.; Liao, L.; Guo, X.; Liu, L. *Macromol. Chem. Phys.* **2013**, *214*, 2500–2506.
49. Xiao, L.; Li, Y.; Jia, X.; Liao, L.; Liu, L. *Polym. Int.* **2014**, *63*, 1154–1158.

Chapter 14

Diazo Compounds in Flow Chemistry

In the last decades, the continuous-flow reactors with channel dimensions in the micro- or millimeter region have demonstrated potential widespread applications in organic synthesis. This technology has now been recognized as an efficient tool for the safe manufacturing of active pharmaceutical ingredients, when hazardous reactions or extreme process windows are necessarily employed.[1–6] On the other hand, diazo compounds are a series of extremely important and versatile substrates or key intermediates in organic synthesis and other fields. As shown in the previous chapters, these compounds widely serve as carbene and metal carbene precursors. However, diazo compounds are generally unstable, toxic and explosive, which limits their synthetic utility both in the laboratory and in the industry. Thus, continuous-flow processes are particularly appealing when diazo compounds are used as substrates in organic synthesis.[7–9] In this chapter, we will highlight the recent advances and synthetic utilities of diazo compounds in flow chemistry.

14.1 Continuous flow reaction of diazomethane and $TMSCHN_2$

Diazomethane (CH_2N_2, **1**) (bp −23°C) has been widely utilized as a versatile C1 building block in organic synthesis. However, diazomethane is notorious for its explosive and toxic nature. Continuous-flow technology with microreactors provides a safety procedure for the production and *in situ* transformation of CH_2N_2. In 1998, Aerojet corporation reported the

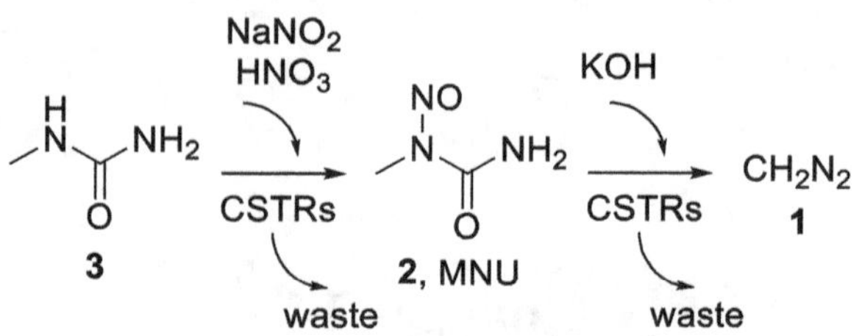

Scheme 14.1 Aerojet continuous flow technique for CH_2N_2 production.

application of continuous stirred tank reactor (CSTR) in producing CH_2N_2 from *N*-methyl-*N*-nitroso urea **2** (MNU) with little or no explosive hazard by using *N*-methyl-*N*-nitroso amine **3** as the starting material (Scheme 14.1).[10] In 2002, Proctor and Warr described the development of a safe process for the industrial generation of CH_2N_2 **1** with *N*-methyl-*N*-nitroso-*p*-toluene sulphonamide (Diazald, **4**) as the precursor. In this procedure, **1** is produced and consumed in a continuous process capable of producing between 50 and 60 tons per year.[11]

The groups of Löbbecke[12] and Stark[13] attempted the methylation of benzoic acid by using CH_2N_2 **1** produced in microreactors. However, the substrate scope is limited due to the strong basic KOH solution condition that is required for producing diazomethane. In 2011, Kim and co-workers reported on the poly(dimethylsiloxane) (PDMS) dual-channel microreactor to integrate generation, separation and reaction of diazomethane in continuous-flow system with high efficiency (Scheme 14.2).[14] The microreactor consisted of two parallel channels separated by a gas-permeable, 45 μm thick PDMS membrane. The base-induced decomposition of Diazald occurred in the lower channel, affording diazomethane which diffused through the membrane to the upper channel. Separated CH_2N_2 **1** was sequentially subjected to different reactions to afford the corresponding products in excellent yields with throughputs of 0.58–2.88 mmol/day.

Later, Kappe and co-workers applied this membrane separation concept in the design of the tube-in-tube reactor. This apparatus allows the laboratory-scale generation of anhydrous CH_2N_2 with 3.6 mmol/h.[15] The inner tube of this device was made of hydrophobic, gas-permeable TeflonAF-2400 with an internal volume of 2 mL. CH_2N_2 **1** was produced in the inner tube from a methanolic solution of Diazald **4** in the presence of base, which then diffused through the permeable membrane into the outer chamber and was subjected to 1,3-dipole cycloaddition, cyclopropanation, esterification and other reactions (Scheme 14.3).

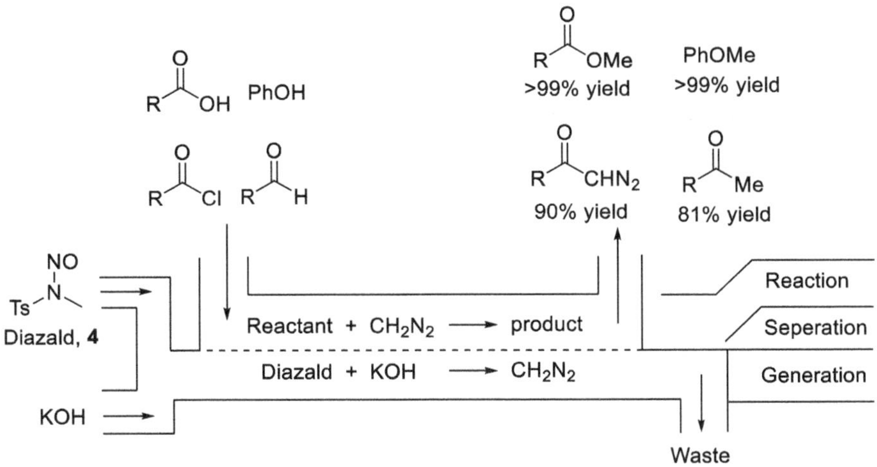

Scheme 14.2 A dual-channel microreactor for continuous generation, separation and reaction of CH_2N_2.

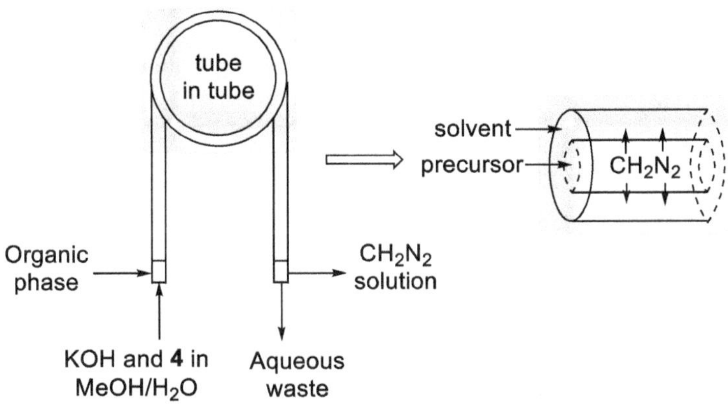

Scheme 14.3 Principle and schematic view of the tube-in-tube reactor.

Furthermore, the same group utilized this continuous-flow setup to achieve multistep synthetic procedure of α-chloro ketones **8** from *N*-protected α-amino acid **5**, which serve as important chiral building blocks for HIV protease inhibitors (Scheme 14.4(a)).[16] The mixed anhydride **6** was formed in the first residence loop at room temperature and passed through the outer chamber of this tube-in-tube reactor. The reaction of mixed anhydride and CH_2N_2 in the outer chamber resulted in the generation of α-diazoketone **7**, which was eventually converted to the

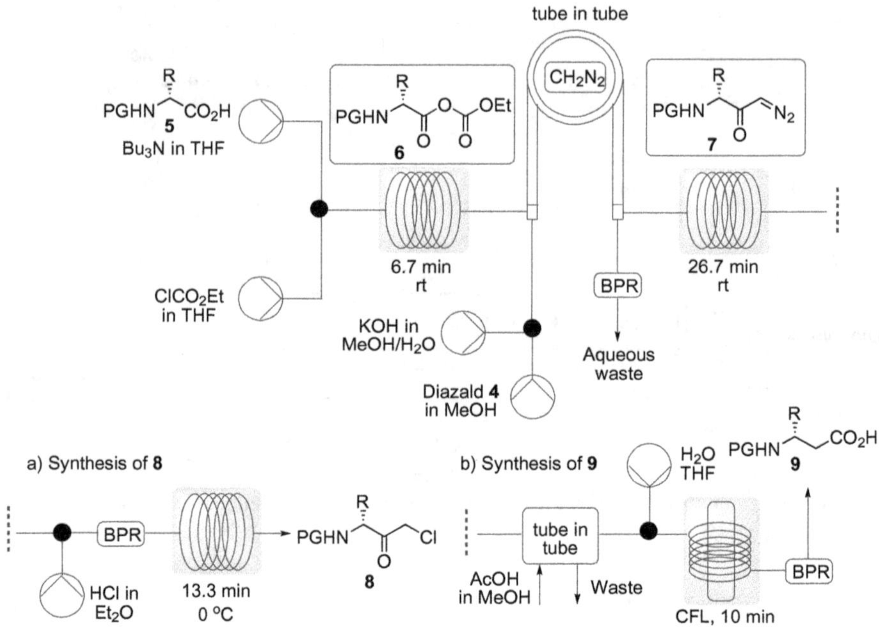

Scheme 14.4 Continuous-flow synthesis of α-halo ketones through the generation of anhydrous CH_2N_2.

α-chloro ketone **8** with 87% yield after the continuous-flow three-step sequence. This fully continuous process allowed the synthesis of 1.84 g of enantiopure α-chloro ketones **8** in 4.5 h.

In 2014, the same group reported a modified fully continuous four-step process, involving Arndt–Eistert homologation approach (Scheme 14.4(b)).[17] This process allowed a concise synthesis of β-amino acids **9** in around 50% overall yields from the corresponding protected α-amino acids **5**. In this procedure, the excess amount of the diazomethane was immediately removed from the stream with a second gas-selective membrane and it was immediately quenched with an alcoholic solution of acetic acid. Subsequently, Wolff rearrangement was performed either photochemically in a photoreactor with compact fluorescent light (CFL), or alternatively with the reaction cartridges packed with silver oxide. Thus, the on-demand generation-purification-consumption of CH_2N_2 eliminates possible human exposure to this hazardous reagent and reduces the risk of explosive decomposition.

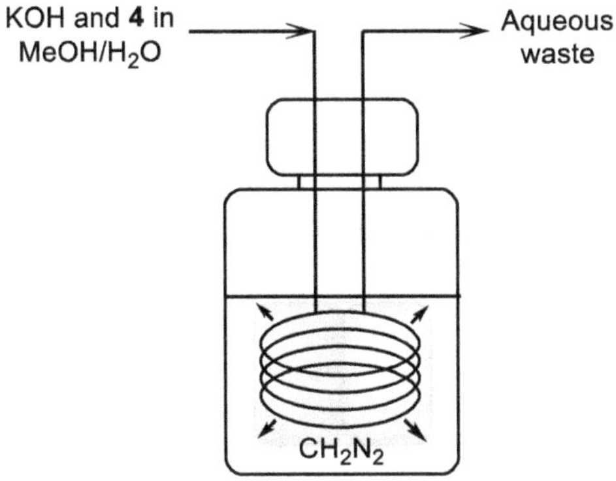

Scheme 14.5 Schematic view of the tube-in-flask reactor.

Based on the tube-in-tube setup, Kappe and Gutmann recently developed a simple and robust semi-batch apparatus tube-in-flask reactor (Scheme 14.5).[18–20] Diazald was placed within a semi-permeable Teflon AF-2400 tubing and then the generated diazomethane was selectively separated from the tubing into a solvent-and-substrate-filled flask. In this process, the step of generating CH_2N_2 is a flow reaction, while the reaction between CH_2N_2 and the substrate is a batch reaction. The reactor has been employed for the methylation of carboxylic acids, the synthesis of α-chloro ketones[19] and the preparation of CH_2N_2 solution at a rate of 42.6 mmol/h.

In 2012, Maggini and Woehl reported the base-induced decomposition of *N*-methyl-*N*-nitroso urea under continuous-flow conditions.[21] By using Corning Advanced-Flow Reactors, this continuous-flow reaction enables the production of diazomethane at a rate of 19 mol/d, at a total flow rate of 53 mL/min. However, excess benzoic acid was used to neutralize the excess base. In 2017, Lehmann developed a simple and robust continuous-flow two-step process, which started from non-hazardous *N*-methyl urea, and generates and consumes MNU **2** and diazomethane **1** in-line with productivity up to 117 mmol/h (Scheme 14.6).[22] The closed reactor system combined with small reaction volumes in a continuous flow eliminates human exposure and minimizes the risk of an explosive decomposition.

Serving as a less hazardous and less explosive alternative of CH_2N_2, trimethylsilyldiazomethane ($TMSCHN_2$, **10**, bp 96°C) also demonstrated

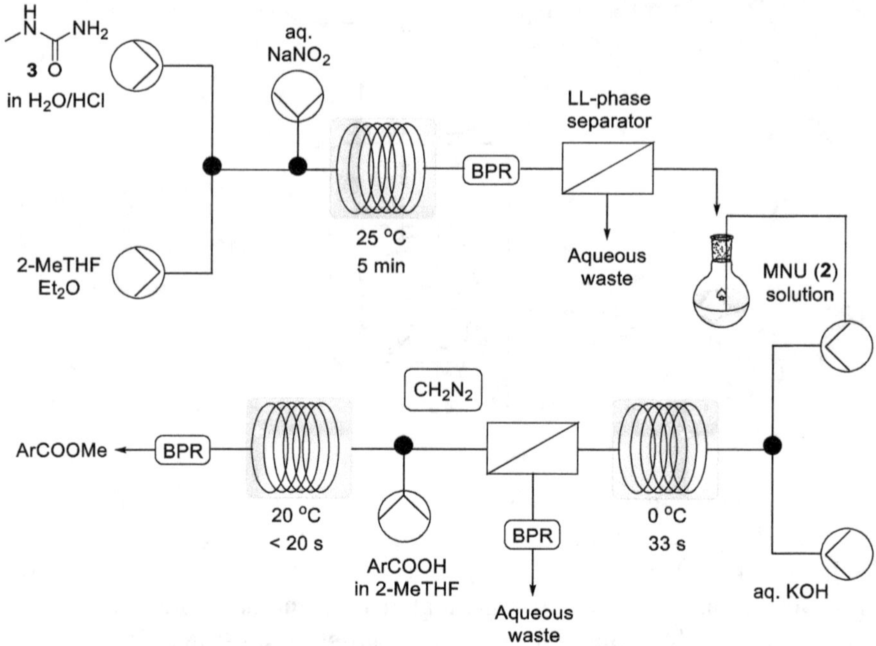

Scheme 14.6 In-line generation of MNU and CH_2N_2.

considerable utilities in various diazo transformations. In 2009, Liotta was the first to use $TMSCHN_2$ **10** in the continuous-flow synthesis of pharmaceutical intermediates.[23] In 2011, Ley and co-workers established a safe continuous-flow procedure for the synthesis of aliphatic and aromatic diazoketones from acyl chloride with $TMSCHN_2$ **10**. They also illustrated the telescoped flow sequence involving diazoketone formation followed by Cu(II)-catalyzed quinoxaline **12** synthesis from *o*-phenylenediamine **11** without isolation of the potentially explosive diazo intermediate (Scheme 14.7).[24] With the employment of efficient in-line purification system particularly notable for the polystyrene-supported scavengers, the desired heterocyclic compounds **12** were obtained in good yields.

Moreover, Fuse and Tanaka reported the Arndt–Eistert synthesis from **10** with α-aryl carboxylic acid, affording α-aryl esters and amides.[25,26] The sequence of acid chloride formation-nucleophilic acyl substitution-Wolff rearrangement-nucleophilic addition was performed entirely in a microflow system without isolating any intermediates.

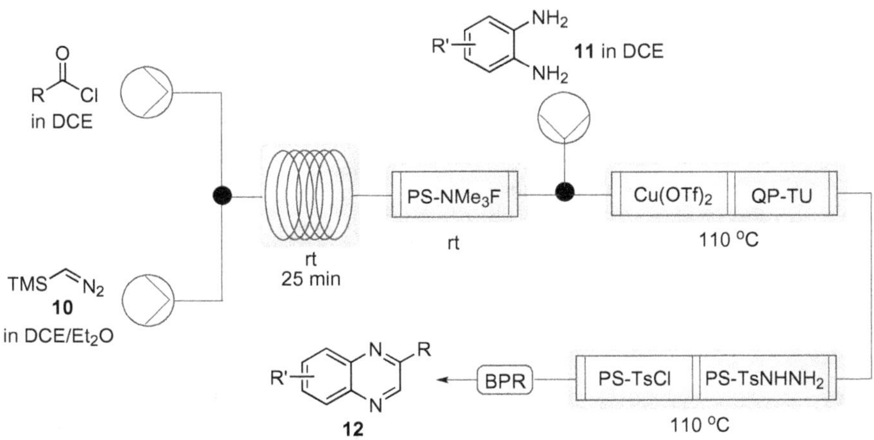

Scheme 14.7 Flow synthesis of quinoxaline derivatives.

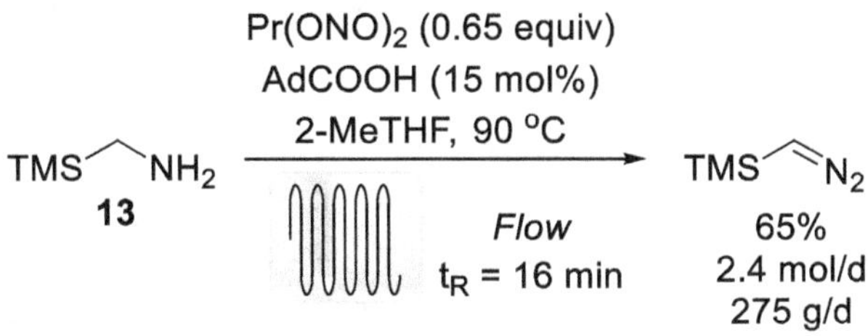

Scheme 14.8 Continuous-flow synthesis of $TMSCHN_2$ *via* diazotization.

In 2017, Lebel and co-workers reported the synthesis of $TMSCHN_2$ **10** by diazotization of $TMSCH_2NH_2$ **13** under batch and continuous-flow condition.[27] With the employment of dinitrite reagent $Pr(ONO)_2$ and sterically hindered 1-adamantanecarboxylic acid $AdCO_2H$, this procedure affords a daily production of 2.4 mol of $TMSCHN_2$ **10** (Scheme 14.8). Other silylated methylamines were also successfully applied to the same reaction conditions to furnish various silicon-bearing diazomethane reagents.

14.2 Continuous flow reaction of alkyl diazo compounds

Compared to diazomethane, other alkyl diazo compounds are even more difficult to apply in organic synthesis due to their instability and related

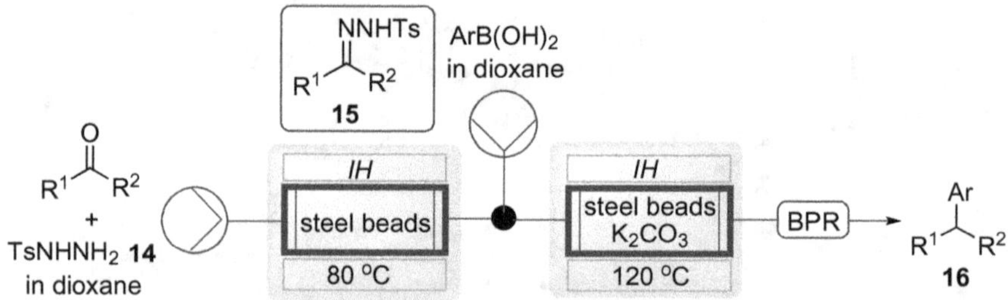

Scheme 14.9 Coupling reaction of *N*-tosylhydrazone with aryl boronic acids.

safety issues. Typically, aryl diazo compounds are slightly more stable than the corresponding alkyl diazo compounds due to the conjugation effect of aromatic rings. However, they are still easily decomposed. Continuous-flow chemistry affords potential solutions for *in situ* generation and the subsequent reactions of these diazo compounds. In 2009, Barluenga and co-workers reported the metal-free coupling reactions of boronic acids with *N*-tosylhydrazones in the presence of base.[28] This process was achieved by Kirschning group with continuous-flow reaction system (Scheme 14.9).[29] The mixed solution of ketones and *p*-toluenesulfonyl hydrazide **14** was first heated by electromagnetic induction of a fixed bed material based on steel beads, affording *N*-tosylhydrazones **15**. Then the solution was subjected to the second electromagnetic induction heating setup to react with aryl boronic acids, affording the final coupling products **16**.

In 2017, Charette and co-workers developed a continuous-flow synthesis and purification of aryldiazomethanes by using steric hindered mesitylsulfonylhydrazones to reduce the temperature (Scheme 14.10).[30] Upon washing with ammonium bicarbonate solution to remove the excess base and sulfinate by-products, the pure aryldiazomethane solution or diaryl diazomethane solution **18** was obtained by the liquid–liquid separation device. The resulting diazo compounds solution could be directly used for $Sc(OTf)_3$-catalyzed cyclic ketone ring-expansion reaction, Cu-catalyzed aldehyde epoxidation and ZnI_2-catalyzed Simmons–Smith cyclopropanation with good efficiency.

In 2011, Maggini and co-workers reported the *in situ* generation of aryl diazo compounds by using NiO_2 as the oxidant under continuous-flow condition.[31] This method was applied in the synthesis of a photoelectrically active methanofullerene. In 2015, Ley and co-workers developed

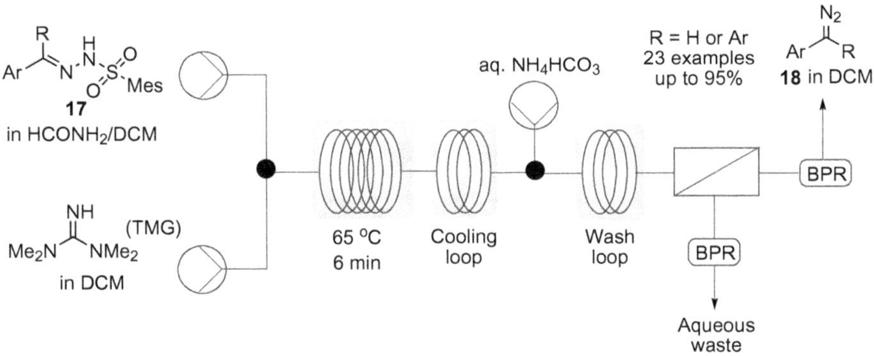

Scheme 14.10 Flow synthesis and purification of aryldiazomethane *via* hydrazone fragmentation.

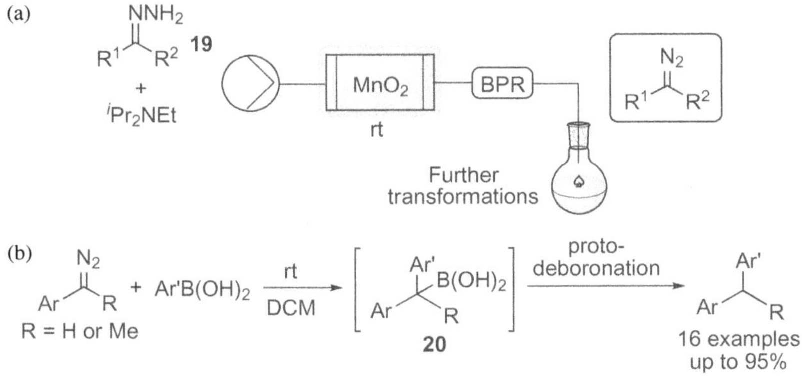

Scheme 14.11 Diazo compound formation by hydrazone oxidation and subsequent coupling.

the continuous-flow reaction system by using MnO_2 as the oxidant. By means of in-line IR, they found that the addition of a small amount of sacrificial hydrazone and Hünig's base could effectively reduce the by-products (Scheme 14.11(a)). They realized the C(sp^3)–C(sp^2) cross-coupling reaction with arylboronic acids under mild and simple reaction conditions, affording the alkylated products in good to excellent yields (Scheme 14.11(b)).[32] Mechanistically, the reaction is initialized by the nucleophilic attack of the diazo compound to the boronic, followed by a rapid 1,2-carbon migration to form the benzyl boronic species **20**.

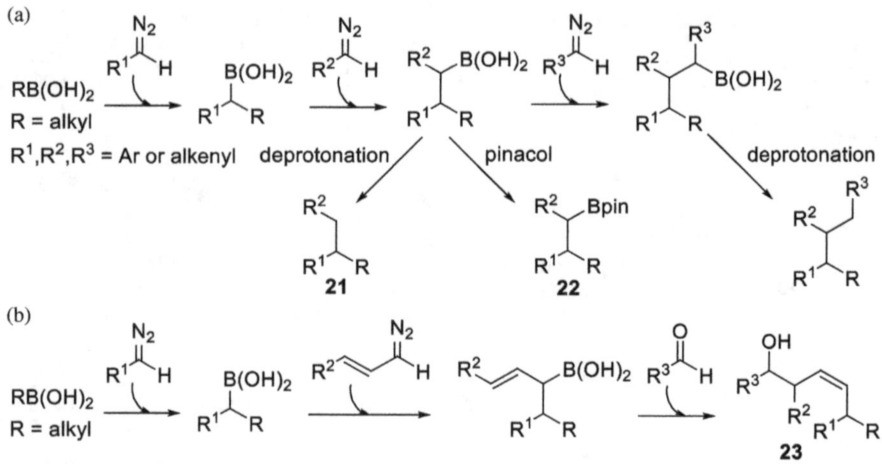

Scheme 14.12 Iterative C–C bond formation reaction of diazo compounds.

Then intermediate **20** undergoes protodeboronation to give the coupling products.

In 2016, Ley group reported the iterative C–C bond formation reactions through the reaction of flow-generated diazo compounds with boronic acids (Scheme 14.12).[33] The benzyl boronic acid intermediate generated from aryl diazo compounds with alkyl boronic acids continuously reacted with another diazo molecule. The resulting newly formed boronic acid species could undergo proto-deboronation to afford the coupling product **21**, or to afford boronic acid ester **22**. This method realized the formation of up to three C–C bonds in a sequence to generate a range of new chemical structures (Scheme 14.12(a)). By employing vinyl diazo compounds as the substrates, this iterative C–C bond formation reaction could afford homoallylic alcohols **23** through the final trapping of a reactive boronic acid species with an aldehyde (Scheme 14.12(b)).

Moreover, Ley and co-workers recently utilized a room temperature continuous-flow setup to realize non-catalytic cyclopropanation of electron-poor olefins with unstable diazo species, which were generated by oxidation in flow condition (Scheme 14.13(a)),[34] and Cu-catalyzed coupling reaction with terminal alkynes (Scheme 14.13(b)).[35] The latter reaction provided a practical method for the synthesis of *di-* and *tri-*substituted allenes under room temperature conditions. In 2017, Ley and co-workers also developed a Cu-catalyzed asymmetric coupling of flow-generated diazo compounds and propargylated amines (Scheme 14.13(c)).[36]

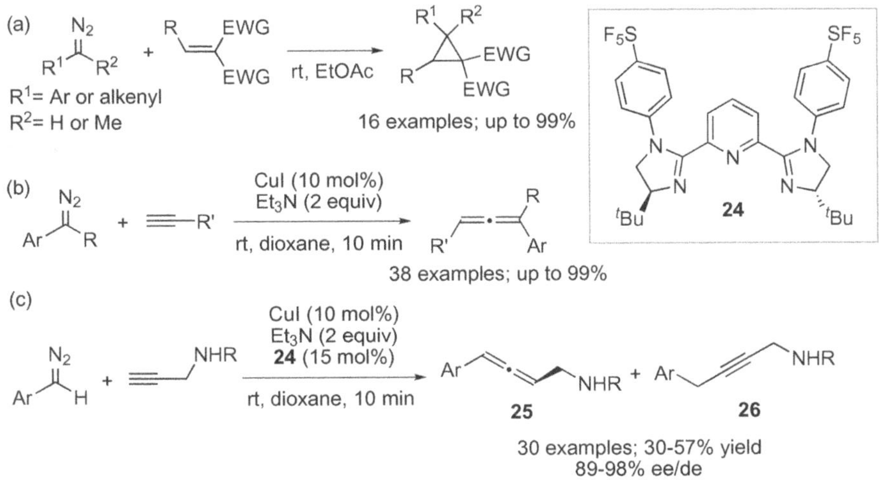

Scheme 14.13 Cyclopropanation and Cu(I)-catalyzed allene synthesis with continuous-flow setup.

By employing a pyridine bis(imidazoline) tridentate ligand **24**, this method achieved the asymmetric synthesis of disubstituted allenes **25** with high enantioselectivity (89–98% *de*/*ee*).

Compared with semi-stable aryl diazo compounds, the synthesis and application of unstable alkyl diazo compounds are more challenging. In 1989, Warkentin reported the photolysis of 1,3,4-oxadiazolines **27** under UV light irradiation, affording the corresponding diazoalkanes.[37] In 2017, Ley and co-workers further reported the preparation of a series of stable oxadiazolines, which can be utilized in continuous-flow generation of alkyl diazo compounds.[38] The C(sp^2)–C(sp^3) bond coupling reaction was realized through the reaction of aryl boronic acids with *in situ* generated diazo species. The alkyl boronic acid intermediate **28** underwent subsequent protodeboronation or oxidation process, affording the coupling products **29** and tertiary alcohols **30** (Scheme 14.14).

14.3 Continuous flow reaction of *α*-diazocarbonyl compounds

In 2013, Rutjes and co-workers reported the synthesis of EDA in a safe manner through in-line coupling of microreactor synthesis and separation

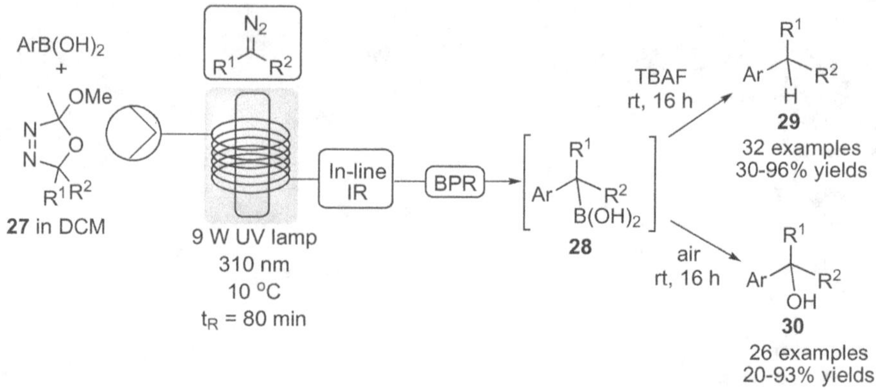

Scheme 14.14 Flow synthesis and transformation of alkyl-substituted diazo compounds.

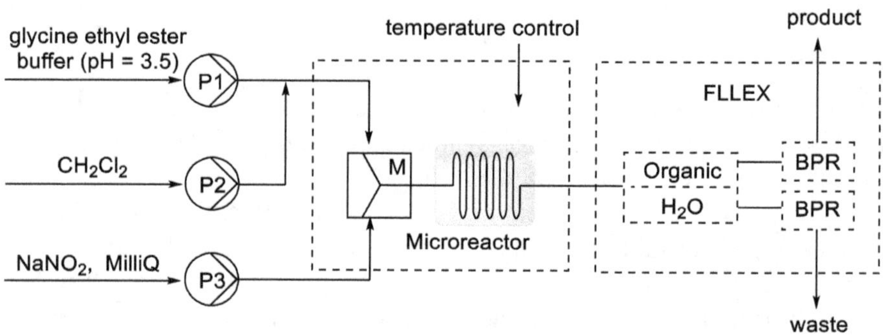

Scheme 14.15 EDA synthesis in flow by using FLLEX directly coupled to the microreactor.

technology (Scheme 14.15).[39] From cheap and commercially available substrates of glycine ethyl ester hydrochloride **31**, sodium nitrite and dichloromethane, approximately 20 g/day of EDA was produced using a microreactor with an internal volume of 100 μL. Besides, the combination of plug-and-play microreactor setup with temperature control system and Flow-Liquid-Liquid-Extraction module (FLLEX) was established to perform a direct in-line extraction of the desired product.

Kim[40] and Wirth[41] reported the aldol addition reaction of EDA with aldehyde under weak basic conditions in the continuous-flow system,

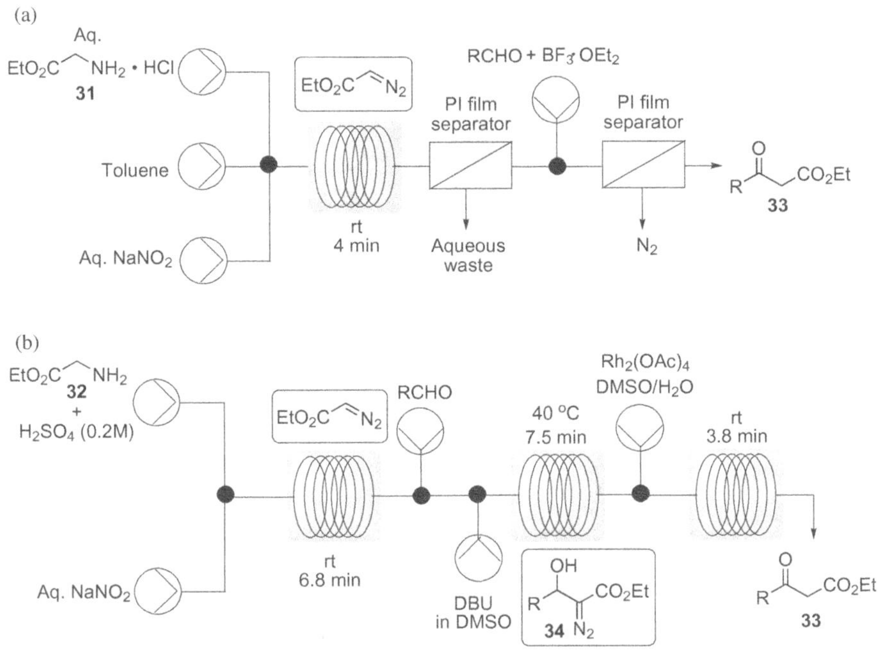

Scheme 14.16 Flow synthesis and transformation of EDA.

affording β-hydroxy-α-diazoesters **34**. In 2014, Kim utilized a novel integrated microfluidic system, including droplet technique for liquid–liquid separation and gas–liquid separation systems, a dual channel membrane technique and capillary microreactor, to allow the synthesis of relatively hazardous and explosive EDA in a safer and efficient manner.[40] They also explored the $BF_3{\cdot}OEt_2$ catalyzed Roskamp reaction of EDA with aldehydes (Scheme 14.16(a)). Wirth and co-workers studied the possibility of a sequential three-step reaction in series in a continuous-flow system.[41] They introduced $Rh_2(OAc)_4$ solution in the system to catalyze 1,2-hydrogen migration of β-hydroxy-α-diazoester **34** to afford the same β-ketoester product **33** (Scheme 14.16(b)).

While the addition of EDA to aldehydes can be performed using organic bases, the addition of EDA to other carbonyl electrophiles requires the use of strong base, such as lithium diisopropylamide (LDA). The generated ethyl lithiodiazoacetate is highly reactive and decomposes rapidly, even at low temperatures. Wirth and co-workers developed an efficient protocol by using continuous-flow technology for the *in situ* generation

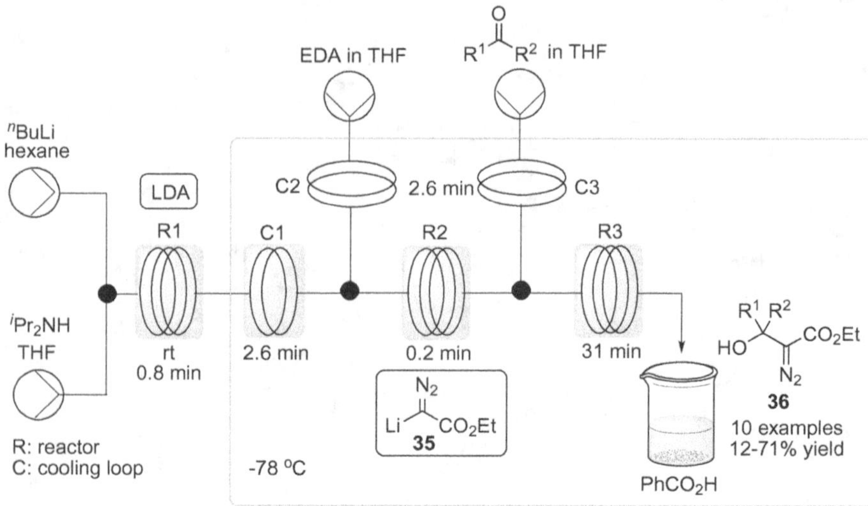

Scheme 14.17 Application of LDA/EDA in flow system.

and the direct use of ethyl lithiodiazoacetate from EDA (Scheme 14.17).[42] The LDA generated *in situ* in the flow system is cooled by the coil and reacted with EDA at low temperature. Then it reacted with ketone or lactone to afford the addition product **36**. The extremely short reaction time between EDA and LDA is compatible to the reactivity of the unstable intermediate **35**. The reaction demonstrates that the continuous-flow process easily deals with the unstable metal organic intermediates.

One of the advantages of continuous-flow chemistry is the ease with which multiphase reactions can be achieved, including heterogeneous catalytic reactions. Thus, the catalyst can be immobilized on the polymer materials at first, and then placed into a packed bed reactor. Martínez-Merino and co-workers reported the preparation of polymer-immobilized bisoxazoline-copper (BOX-Cu) and copper-pyridineoxazoline (pyox-Cu) complexes, and utilized these catalytic systems in the catalytic cyclopropanation reaction of EDA with styrene under batch and flow conditions.[43,44] Interestingly, the monolithic systems under flow conditions led to an increase in enantioselectivities of ca. 20% relative to those of the homogenous batch process. The use of $scCO_2$ as a solvent creates a much more efficient and environmentally friendly process. Besides, Caselli and co-workers developed the continuous-flow heterogeneous asymmetric cyclopropanations, catalyzed by supported hydrogen-bonded (SHB)

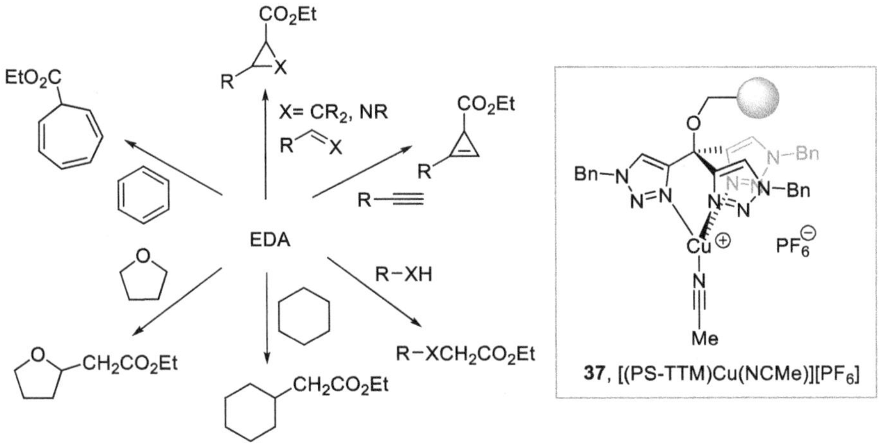

Scheme 14.18 Application of heterogeneous Cu(II) catalyst.

chiral Cu(I) complexes of pyridine containing tetraazamacrocyclic ligands (Pc-L*-Cu) using CO_2 as a transport vector.[45]

In 2015, Pérez and co-workers reported the synthesis of polystyrene-linked tris(triazolyl)methane copper(I) cationic species **37** as the heterogeneous catalysts, used for various carbene transfer reactions of EDA with an array of substrates under batch and continuous-flow conditions.[46] Not only the X–H insertion or C–H functionalization process but also cyclopropenation and Buchner reaction took place with high efficiency and with the recyclability and reuse of the catalyst (Scheme 14.18). In the continuous-flow system, the high and constant turnover frequency of this heterogenized cationic catalyst was quite remarkable (after 48 h of operation, complete conversion through O–H insertion with ethanol can be achieved with a residence time of 1 min). The chemical suitability allowed subsequent carbene transformations with a simple and affordable experimental setup.

Furthermore, the synthesis of heterocyclic compounds by using diazo components as nucleophiles was reported by using continuous-flow setup. In 2010, Moody and Hayes achieved the synthesis of 2,4,6-trisubstituted pyrimidines **39** through Lewis acid-catalyzed Roskamp reaction of EDA with aldehydes in a flow system.[47] The corresponding *in situ* generated β-keto esters **33** in flow were subjected to sequential condensation procedure with amidine hydrochlorides **38**, affording a variety of 2,6-substituted pyrimidin-4-ols **39** with moderate to good yields (Scheme 14.19). This

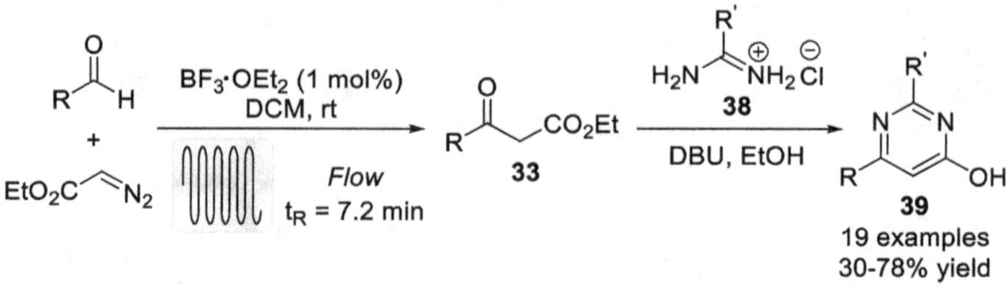

Scheme 14.19 Synthesis of substituted pyrimidine using continuous-flow setup.

reaction demonstrates the application of two toxic agents (EDA and BF_3•OEt_2) in the same flow system. These two substrates are also used in the ring-expansion reaction of piperidone, reported by Zhang and co-workers.[48]

In addition to EDA, other commonly used α-carbonyl diazo compounds are generally classified into donor/acceptor diazo compounds (such as aryl diazo esters), and acceptor/acceptor diazo compounds (such as α-diazo diester compounds). However, the potential explosion risk of diazo compounds and the use of organic azide substrates in the preparation process involve safety hazards, so it has certain risks in large-scale application. Thus, flow chemistry procedures are particularly useful in circumventing such problems when more general diazo compounds are used in large scale.

Moody and Hayes developed the continuous-flow Bamford–Stevens reaction of arylsulfonyl hydrazones **40** to prepare α-diazoesters **41** (Scheme 14.20).[49,50] The CH_2Cl_2 solution of arylsulfonyl hydrazones **40** and Et_3N were respectively introduced into the convection flow coil reactor, and reacted under heating conditions. After removing the ammonium salt through a silica gel column mixed with Et_3N, the solution of **41** was obtained, which was subjected to Rh(II)-catalyzed O–H, N–H, S–H insertion and Cu(I)-catalyzed P–H insertion without further purification (*path a*). In addition, if the purification column in the flow system was removed, the residual sulfinate ion in the reaction solution could directly react with α-diazoesters **41** under $Rh_2(OAc)_4$ catalysis, affording the sulfone product **42** (*path b*).

In 2015, Moody and co-workers reported a continuous-flow generation of diazo compounds, including diazoamides and diazophosphonates, by using a recyclable polymer-supported oxidant of solid-phase *N*-iodosulfonamide.[51] The solution of the hydrazone **43** in CH_2Cl_2 was

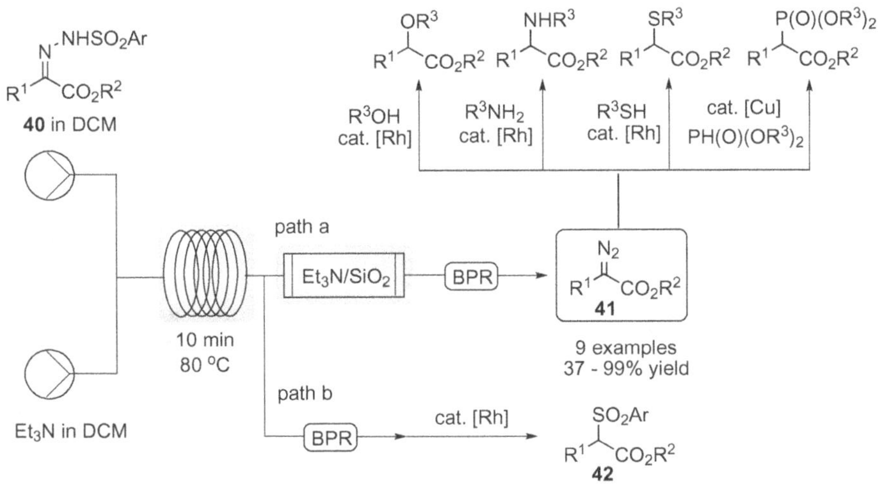

Scheme 14.20 Flow synthesis with diazoesters *via* Bamford–Stevens reaction.

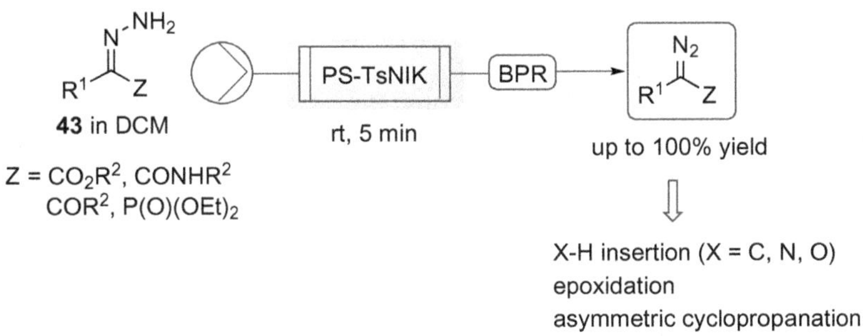

Scheme 14.21 Flow synthesis of diazo compounds using supported oxidant.

pumped through the column reactor filled with PS-TsNIK to generate the corresponding diazo compounds with a retention time of 5 min. Furthermore, the diazo compounds are subjected to sequential Rh(II)-catalyzed X–H insertion and epoxidation, providing heteroatom alkylated products and epoxides with only water, KI and N_2 as the by-products (Scheme 14.21). In 2017, the same group used the α-diazoesters produced by this method in the intramolecular C–H bond insertion reaction, and achieved the synthesis of a series of spiro β- or γ-lactones.[52] Moreover, Davies utilized this method in the enantioselective Rh(II)-catalyzed

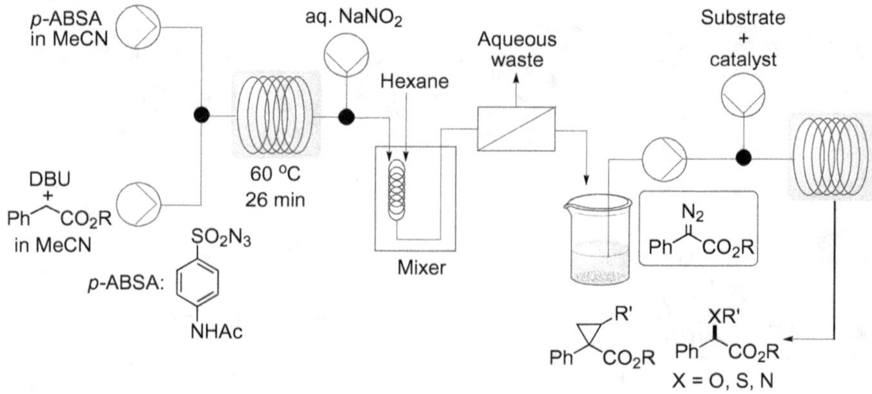

Scheme 14.22 Diazo transfer reaction in flow.

cyclopropanation and intermolecular C–H bond functionalization with high level of *ee*.[53] Since trace water and iodine in the solution containing diazo compound interfere in the subsequent reaction, the protocol was modified by adding anhydrous sodium thiosulfate at the bottom of the molecular sieves cartridge.

Compared with the batch reactions, the continuous-flow diazo transfer reactions are also an effective strategy to reduce the safety risk of organic azide compounds. In 2015, Wirth and co-workers developed a continuous process for the rapid and safe generation of α-aryl-α-diazoesters as donor/acceptor carbene precursors through diazo transfer reaction (Scheme 14.22).[54] The safety profile of the first step was determined by in-depth kinetic study by using in-line IR spectroscopy, which provided quantitative information on diazo formation and consumption. After elective liquid/liquid extraction, separated aryl diazoacetate could be obtained with high purity, which was subjected to the subsequent catalytic carbene transformations. Finally, a series of functionalized α-substituted phenylacetates and cyclopropanes were obtained in good yields. This process has been utilized in the continuous-flow synthesis of milnacipran analogues, which have been used for the treatment of depression and fibromyalgia.[55]

In 2016, Maguire and Collins developed the continuous-flow diazo transfer process by using heat and shock sensitive tosyl azide as the diazo transfer reagent.[56] They achieved the flow synthesis of cyclic α-diazo-β-keto sulfoxides **44** (Scheme 14.23).[57,58] Due to the overlap of the sulfinyl

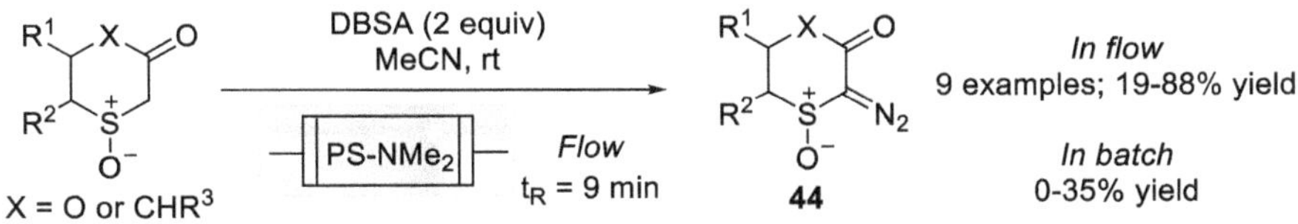

Scheme 14.23 Synthesis of α-diazo-β-carbonylsulfoxide in flow.

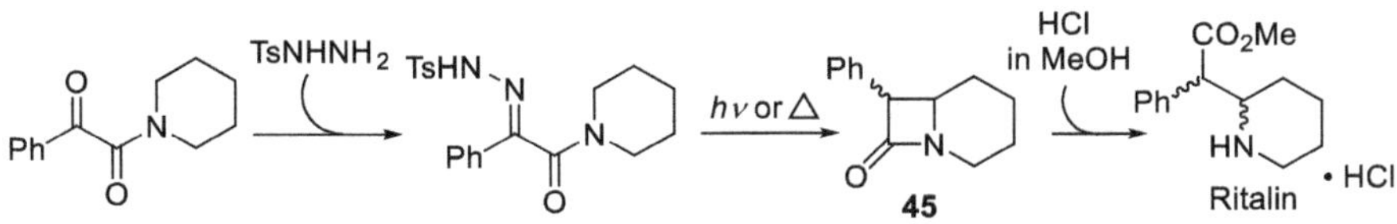

Scheme 14.24 Synthesis of Ritalin in flow.

lone pair with the anti-bonding orbital of the diazo moiety, the stability of α-diazo-β-keto sulfoxides is lower than the corresponding α-diazocarbonyl compounds. Only the cyclic α-diazo-β-carbonyl sulfoxide can be isolated. The yield of these compounds is about 30% under batch condition because of the decomposition during the diazo transformation process. In this case, the high yielding reactions are enabled by flowing through a bed of polystyrene-supported base (PS-DBU or PS-NMe_2) with highly controlled residence times.

In 2017, Monbaliu and co-workers reported the carbene intermolecular C–H bond insertion applied in the continuous-flow synthesis of Ritalin, which served as the most widely prescribed stimulant medication to treat attention deficit hyperactivity disorder (ADHD) and narcolepsy (Scheme 14.24).[59] The daily productivity of the key intermediate β-lactam **45** could reach 4.25 kg.

Furthermore, due to the operational simplicity and the easy extension of the continuous-flow system for the photochemical reaction, the tandem flow reaction based on photo-induced Wolff rearrangement has attracted great attention.[60] In 2010, Konopelski and co-workers reported the photo-induced Wolff rearrangement under continuous-flow system (Scheme 14.25).[61] The α-diazo-*N*-methoxy-*N*-methyl-β-ketoamides **46** underwent Wolff rearrangement to afford ketene intermediate **47**, followed by intramolecular ring closure to give β-lactams **48**. The employment of 100 W CFL could significantly increase the ratio of *trans*-**48a** as the final product. The use of the Weinreb amide substituent permitted regioselective control on the C3 position.

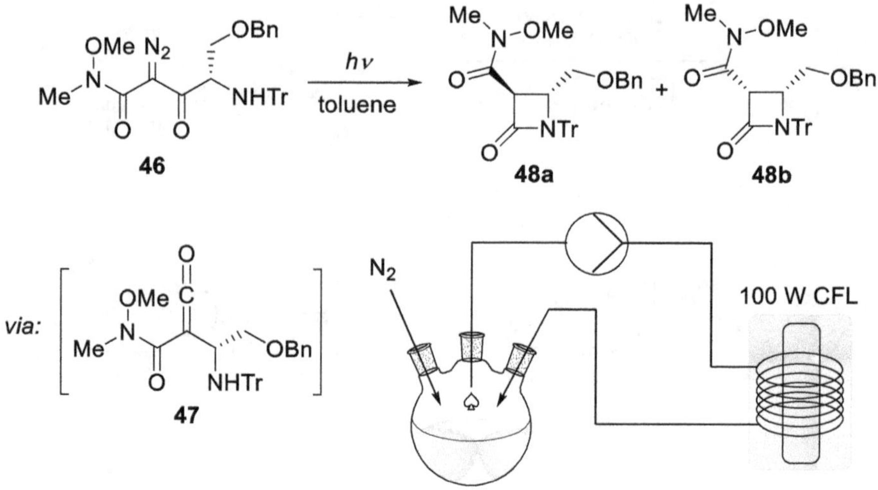

Scheme 14.25 Synthesis of β-lactams in flow *via* Wolff rearrangement.

In 2013, Danheiser and co-workers reported the batch and flow photochemical benzannulations of α-diazoketones **49** with ynamides **50** (Scheme 14.26(a)).[62] Wolff rearrangement of **49** occurs to generate ketene intermediate **51**, followed by [2+2] cycloaddition with **50**. The resulting vinylcyclobutenone **52** undergoes 4π electrocyclic cleavage and sequential 6π electrocyclic ring closure and tautomerization, affording polysubstituted phenol **54**. In 2017, Xiao and co-workers reported the Pd-catalyzed asymmetric [4+2] cycloaddition of benzoxazinanones **55** with ketene intermediate, which was similarly generated through photo-induced Wolff rearrangement (Scheme 14.26(b)).[63] This procedure successfully implemented a gram-scale flow synthesis of product **57**.

In 2015, a photo-induced multicomponent Passerini-type reaction of α-diazoketones, carboxylic acids with isocyanides, was efficiently performed under continuous-flow conditions (Scheme 14.27).[64] As a sustainable and green procedure toward high atom economy, this method allowed for the synthesis of captodative olefins, which acted as versatile synthons in organic synthesis, with high level of stereoselectivity and productivity. In this process, it is advantageous to form a *Z*-type olefin, however, photo-induced olefin isomerization process may make the reaction less selective. The presence of *trans*-piperylene, which is normally used as quencher of triplet excited states, can suppress the formation of

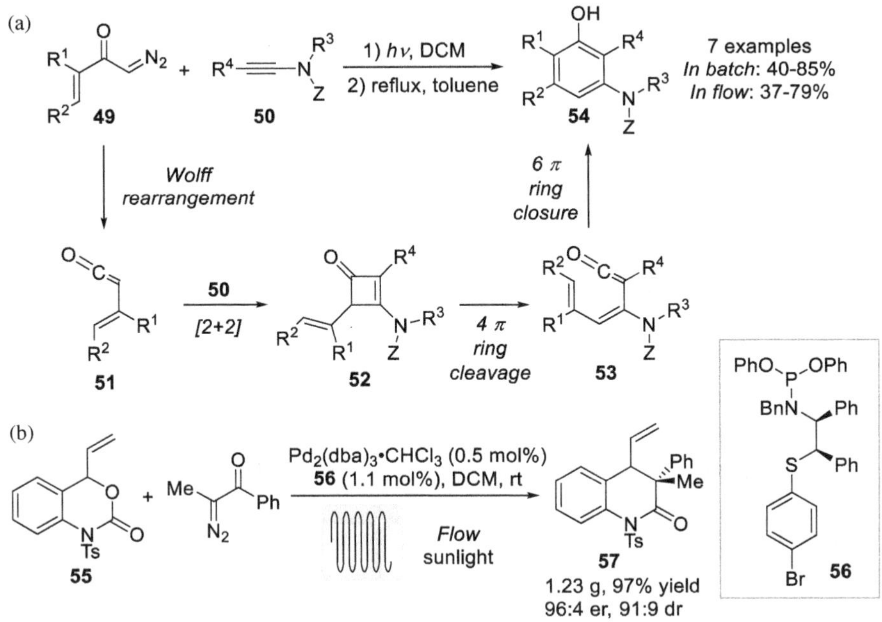

Scheme 14.26 Cycloaddition reactions of ketene intermediates under flow conditions.

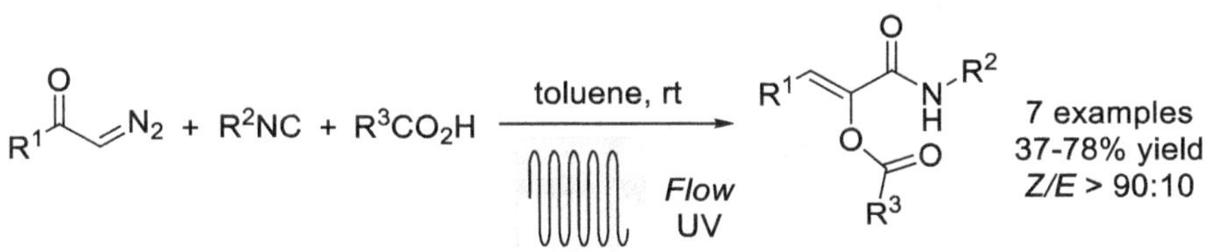

Scheme 14.27 Passerini-like three-component reaction in flow.

E-captodative olefin formed through the photo-induced isomerization.[65] In the flow system, the product will quickly leave the reaction system upon the completion of the reaction, which effectively inhibits the formation of *E*-type products. This example showcases the advantages of continuous-flow reaction in controlling the selectivity by avoiding side reactions.

In 2016, Ley and co-workers reported the microwave-induced Wolff–Staudinger reaction of α-diazoketones with imines under continuous-flow system (Scheme 14.28).[66] The ketenes generated by thermal reaction of

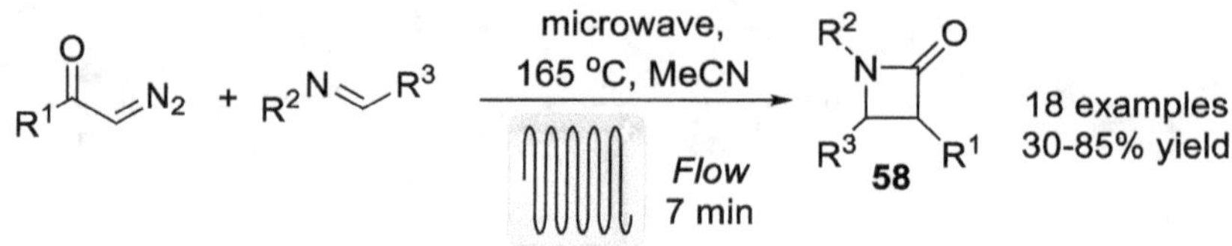

Scheme 14.28 Wolf–Staudinger cascade reaction in flow.

α-diazoketones undergo [2+2] cycloaddition with imines, affording *trans*-configured β-lactams **58**. In this procedure, the small size of the flow reactor ensures the efficiency of the microwave radiation.

The α-diazocarbonyl compounds are also utilized in heterogeneous catalytic asymmetric reactions. In 2011, Hashimoto and coworkers reported the synthesis of immobilized [Rh_2(*S*-PTTL)(*S*-TCPTTL)$_3$] complex (**60**) by three-component copolymerization of Rh(II)-complex-containing monomer, 2-(trifluoromethyl)styrene, with flexible cross-linker.[67] This heterogeneous complex enabled the catalytic asymmetric carbonyl ylide [3+2] cycloaddition of 2-diazo-3,6-diketoesters with styrene or phenylacetylene under continuous-flow condition, affording cycloadducts with high yields and high level of asymmetric induction (up to 99% *ee*) (Scheme 14.29).

The high pressure in the traditional packed-bed reactor is a crucial challenge when employing micro-reactors. In 2015, Davies and Jones utilized flexible composite polymer/oxide hollow fibers as the flow reactors, and immobilized [Rh_2(*S*-DOSP)$_4$] analogues on the silica faces.[68] When the reactant solution enters into the hollow fiber, another stream flows perpendicularly to the porous tube wall, causing the reactant solution to flow out from the other side through the tube wall. During this procedure, the reactants interact with the catalyst in the tube wall to make the reaction occur (Scheme 14.30). This method has been utilized in asymmetric cyclopropanation and C–H functionalization of α-diazocarbonyl compounds with high yields and high level of *ee* value.

In 2012, Moody and Hayes demonstrated the nucleophilic fluorination of α-diazo-β-ketoesters **61** with tetrafluoroboric acid either using conventional batch condition or in flow.[69] The resulting fluorinated products **62** demonstrated wide applications in synthesizing fluoro-containing heterocycles, as illustrated by the direct formation of fluoro-pyrimidines-pyrazoles and coumarins in one step (Scheme 14.31).

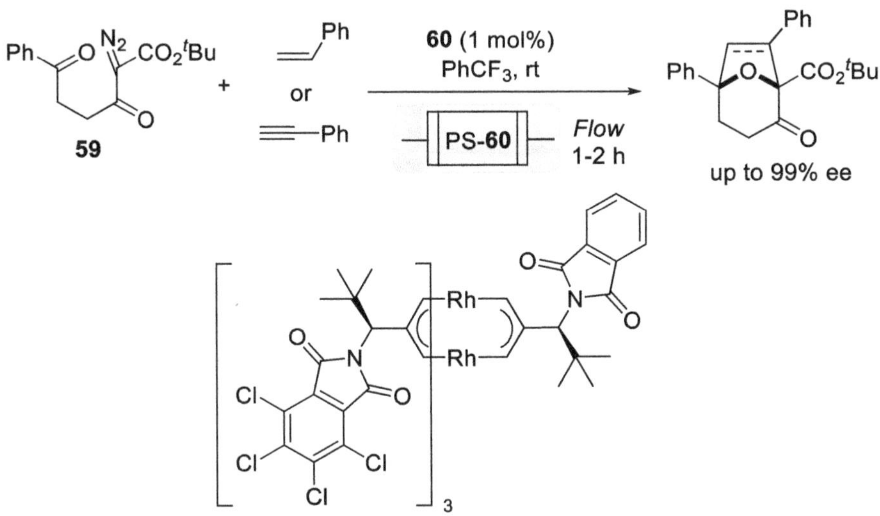

[Rh_2(*S*-PTTL)(*S*-TCPTTL)$_3$] **60**

Scheme 14.29 Carbonyl ylide [3+2] cycloaddition reaction under heterogeneous catalysis.

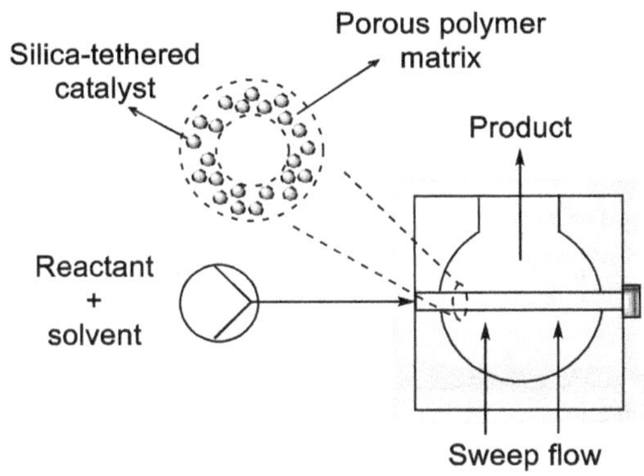

Scheme 14.30 Schematic view of the hollow fiber flow reactor.

In 2017, Beeler and co-workers reported a regioselective and enantioselective intermolecular Buchner ring expansion under continuous-flow conditions (Scheme 14.32).[70] The reaction of ethyl diazoacetate with arenes affords cycloheptatrienes in good yields and with excellent regioselectivities. In addition, by employing [Rh_2(*R*-PTAD)$_4$] as the chiral

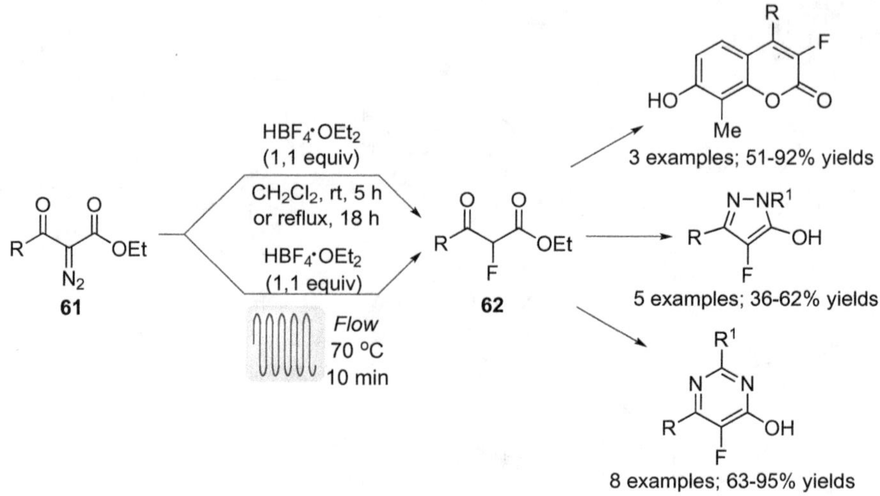

Scheme 14.31 Nucleophilic fluorination of α-diazo-β-ketoester.

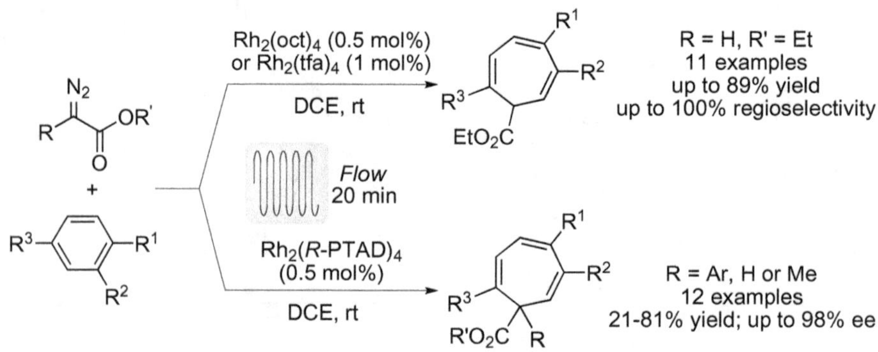

Scheme 14.32 Selective Buchner ring expansion in flow.

catalyst, this method affords cycloheptatrienes with an all-carbon quaternary center with high enantioselectivity.

14.4 Continuous-flow reaction of fluorinated alkyl diazo compounds

In recent years, the synthesis of fluorine-containing organic compounds has attracted great attention. The introduction of fluorine-containing

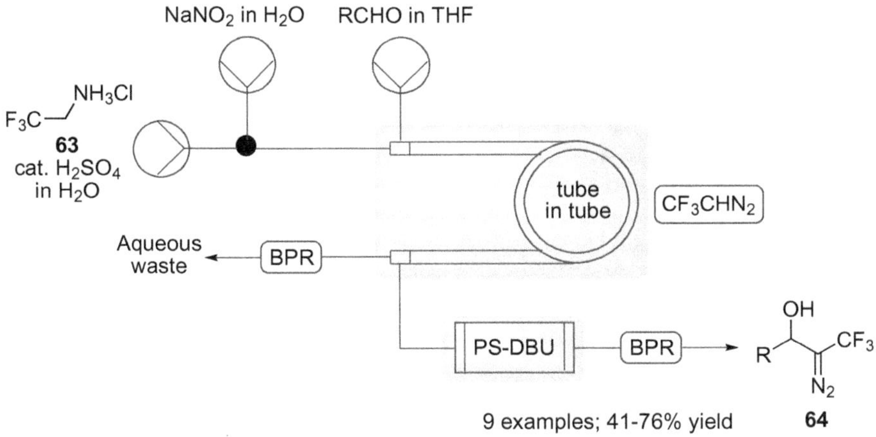

Scheme 14.33 Generation and reaction of CF_3CHN_2 in tube-in-tube reactor.

groups can significantly improve the biochemical activity of organic molecules. The application of fluorine-containing alkyl diazo compounds in flow system has been developed since 2016. Kappe and Pieber realized the *in situ* generation and transformation of trifluoromethyl diazomethane in tube-in-tube flow reactor (Scheme 14.33).[71] The aqueous solution of trifluoroethylamine hydrochloride **63** was mixed with $NaNO_2$ in the inner tube made out of Teflon AF-2400. Simultaneously, dry THF was pumped through the outer tube. The CF_3CHN_2 solution entered into a column reactor equipped with a supported DBU. Finally, trifluoromethylated α-hydroxy diazo compounds **64** could be obtained through base-promoted aldol addition.

By using *tert*-butyl nitrite as the diazotization reagent, Koenigs realized the continuous-flow synthesis of a series of fluoroalkyl-substituted diazo substrates, including CF_3CHN_2, CF_2HCHN_2 and other diazo compounds (Scheme 14.34(a)).[72] The resulting solution of diazo compounds was directly subjected to the reaction flask containing a solution of olefin or alkyne, followed by [3+2] cycloaddition reaction under room temperature. The process afforded the trifluoromethylated pyrazoles (**65**) and pyrazolines (**66**). Moreover, the same group studied the differences between CF_2HCHN_2 and CF_3CHN_2 in batch and flow conditions by using vinyl sulfones **67** as the substrates.[73] The CF_3CHN_2 generated by using $NaNO_2$ solution as the diazotization reagent was directly subjected to batch [3+2] cycloaddition reaction to proceed with high yields. However, CF_2HCHN_2 generated by this method could not participate in

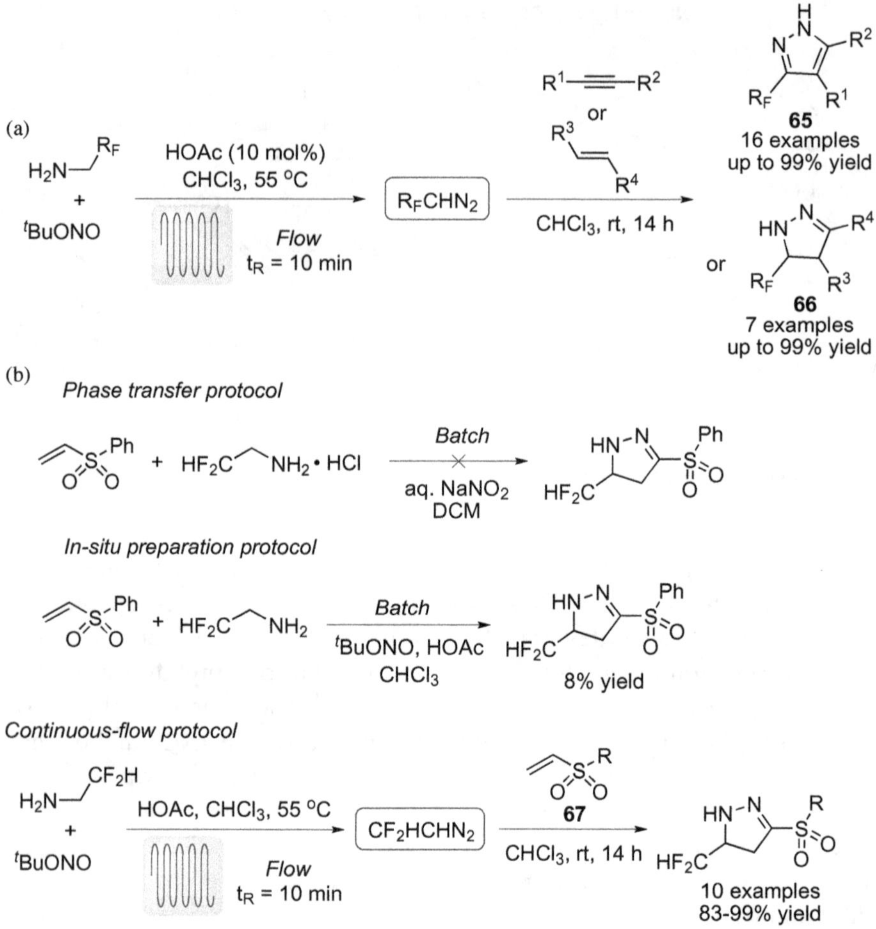

Scheme 14.34 [3+2] cycloaddition reactions of fluoroalkyl-substituted diazo compounds.

the batch [3+2] cycloaddition reaction. When employing *tert*-butyl nitrite as the diazotization reagent, the yield was also low under batch conditions (Scheme 14.34(b)). It was reasoned that the instability of CF_2HCHN_2 under batch conditions caused this difference. The similar instability is also reflected in the Rh(II)-catalyzed cyclopropanation of CF_2HCHN_2 with olefins reported by the same group.[74] These examples demonstrate the advantages of the flow system in handling unstable intermediates.

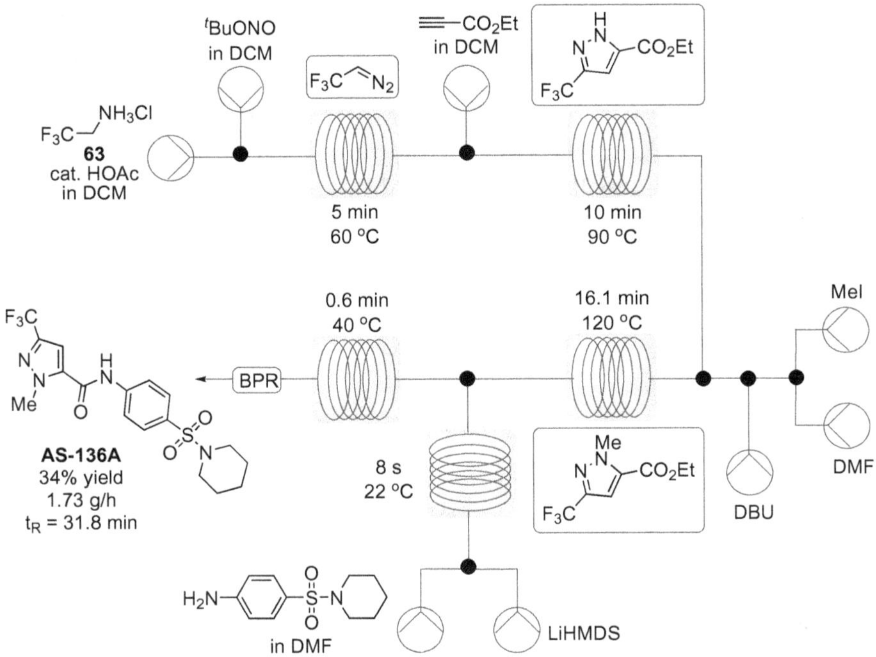

Scheme 14.35 Synthesis of AS-136A in flow.

Recently, Jamison and co-workers reported a unified continuous-flow assembly-line synthesis of biologically active fluoroalkyl substituted pyrazoles and pyrazolines.[75] The key step in this process is the [3+2] cycloaddition of fluorinated diazo compounds with alkenes or alkynes. By operating a back pressure of 325 psi allowing CH_2Cl_2 to be superheated without vaporization, this reaction was successfully implemented under high temperature (90–132°C). The pyrazole core was then sequentially modified through additional reactor modules involving *N*-alkylation and arylation, deprotection and amidation to achieve molecular diversity. To demonstrate the capabilities of this system, AS-136A was synthesized in 75% yield using three sets of individual modules, or through a telescoped continuous-flow multistep process yielding 1.73 g/h (34% yield) of the isolated product (Scheme 14.35).

In conclusion, as an emerging reaction technology, continuous-flow reaction has been widely used in various fields. The advantages of this technology in the treatment of harmful or unstable compounds make it suitable for the reaction of various diazo compounds. Some semi-stable or unstable diazo compounds can be utilized in the safe and efficient conversion under continuous-flow conditions.[76] It is obvious that this strategy greatly expands the applications of diazo compounds. On the basis of ensuring safety, the flow method can be easily applied to scale-up experiments, which provides the possibility for industrial application of diazo compounds. In addition, the advantages of flow chemistry in the treatment of reactions with high temperature and pressure, photochemical reactions, fast reactions, heterogeneous catalytic reactions and multistep reactions, also allow chemists to discover new reaction patterns of diazo compounds that may be relevant to organic synthesis.

References

1. Noël, T.; Buchwald, S. L. *Chem. Soc. Rev.* **2011**, *40*, 5010–5029.
2. Pastre, J. C.; Browne, D. L.; Ley, S. V. *Chem. Soc. Rev.* **2013**, *42*, 8849–8869.
3. Gutmann, B.; Cantillo, D.; Kappe, C. O. *Angew. Chem. Int. Ed.* **2015**, *54*, 6688–6728.
4. Cambié, D.; Bottecchia, C.; Straathof, N. J. W.; Hessel, V.; Noël, T. *Chem. Rev.* **2016**, *116*, 10276–10341.
5. Britton, J.; Raston, C. L. *Chem. Soc. Rev.* **2017**, *46*, 1250–1271.
6. Plutschack, M. B.; Pieber, B.; Gilmore, K.; Seeberger, P. H. *Chem. Rev.* **2017**, *117*, 11796–11893.
7. Müller, S. T. R.; Wirth, T. *ChemSusChem* **2015**, *8*, 245–250.
8. Deadman, B. J.; Collins, S. G.; Maguire, A. R. *Chem. Eur. J.* **2015**, *21*, 2298–2308.
9. Movsisyan, M.; Delbeke, E. I. P.; Berton, J. K. E. T.; Battilocchio, C.; Ley, S. V.; Stevens, C. V. *Chem. Soc. Rev.* **2016**, *45*, 4892–4928.
10. Archibald, T. G.; Barnard, J. C.; Reese, H. F. US Patent *5854405*, **1998**.
11. Proctor, L. D.; Warr, A. J. *Org. Process Res. Dev.* **2002**, *6*, 884–892.
12. Ferstl, W. F.; Schwarzer, M. S.; Loebbecke, S. L. *Chem. Ing. Tech.* **2004**, *76*, 1326–1327.
13. Struempel, M.; Ondruschka, B.; Daute, R.; Stark, A. *Green Chem.* **2008**, *10*, 41–43.
14. Maurya, R. A.; Park, C. P.; Lee, J. H.; Kim, D.-P. *Angew. Chem. Int. Ed.* **2011**, *50*, 5952–5955.
15. Mastronardi, F.; Gutmann, B.; Kappe, C. O. *Org. Lett.* **2013**, *15*, 5590–5593.
16. Pinho, V. D.; Gutmann, B.; Miranda, L. S. M.; de Souza, R. O. M. A.; Kappe, C. O. *J. Org. Chem.* **2014**, *79*, 1555–1562.
17. Pinho, V. D.; Gutmann, B.; Kappe, C. O. *RSC Adv.* **2014**, *4*, 37419–37422.

18. Dallinger, D.; Pinho, V. D.; Gutmann, B.; Kappe, C. O. *J. Org. Chem.* **2016**, *81*, 5814–5823.
19. Garbarino, S.; Guerra, J.; Poechlauer, P.; Gutmann, B.; Kappe, C. O. *J. Flow Chem.* **2016**, *6*, 211–217.
20. Dallinger, D.; Kappe, C. O. *Nat. Protoc.* **2017**, *12*, 2138–2147.
21. Rossi, E.; Woehl, P.; Maggini, M. *Org. Process Res. Dev.* **2012**, *16*, 1146–1149.
22. Lehmann, H. *Green Chem.* **2017**, *19*, 1449–1453.
23. Pollet, P.; Cope, E. D.; Kassner, M. K.; Charney, R.; Terett, S. H.; Richman, K. W.; Dubay, W.; Stringer, J.; Eckert, C. A.; Liotta, C. L. *Ind. Eng. Chem. Res.* **2009**, *48*, 7032–7036.
24. Martin, L. J.; Marzinzik, A. L.; Ley, S. V.; Baxendale, I. R. *Org. Lett.* **2011**, *13*, 320–323.
25. Mifune, Y.; Fuse, S.; Tanaka, H. *J. Flow Chem.* **2014**, *4*, 173–179.
26. Fuse, S.; Otake, Y.; Mifune, Y.; Tanaka, H. *Aust. J. Chem.* **2015**, 68, 1657–1661.
27. Audubert, C.; Gamboa Marin, O. J.; Lebel, H. *Angew. Chem. Int. Ed.* **2017**, *56*, 6294–6297.
28. Barluenga, J.; Tomás-Gamasa, M.; Aznar, F.; Valdés, C. *Nat. Chem.* **2009**, *1*, 494–499.
29. Kupracz, L.; Kirschning, A. *J. Flow Chem.* **2013**, *3*, 11–16.
30. Lévesque, É.; Laporte, S. T.; Charette, A. B. *Angew. Chem. Int. Ed.* **2017**, *56*, 837–841.
31. Rossi, E.; Carofiglio, T.; Venturi, A.; Ndobe, A.; Muccini, M.; Maggini, M. *Energy Environ. Sci.* **2011**, *4*, 725–727.
32. Tran, D. N.; Battilocchio, C.; Lou, S.-B.; Hawkins, J. M.; Ley, S. V. *Chem. Sci.* **2015**, *6*, 1120–1125.
33. Battilocchio, C.; Feist, F.; Hafner, A.; Simon, M.; Tran, D. N.; Allwood, D. M.; Blakemore, D. C.; Ley, S. V. *Nat. Chem.* **2016**, *8*, 360–367.
34. Roda, N. M.; Tran, D. N.; Battilocchio, C.; Labes, R.; Ingham, R. J.; Hawkins, J. M.; Ley, S. V. *Org. Biomol. Chem.* **2015**, *13*, 2550–2554.
35. Poh, J.-S.; Tran, D. N.; Battilocchio, C.; Hawkins, J. M.; Ley, S. V. *Angew. Chem. Int. Ed.* **2015**, *54*, 7920–7923.
36. Poh, J.-S.; Makai, S.; von Keutz, T.; Tran, D. N.; Battilocchio, C.; Pasau, P.; Ley, S. V. *Angew. Chem. Int. Ed.* **2017**, *56*, 1864–1868.
37. Majchrzak, M. W.; Bekhazi, M.; Tse-Sheepy, I.; Warkentin, J. *J. Org. Chem.* **1989**, *54*, 1842–1845.
38. Greb, A.; Poh, J.-S.; Greed, S.; Battilocchio, C.; Pasau, P.; Blakemore, D. C.; Ley, S. V. *Angew. Chem. Int. Ed.* **2017**, *56*, 16602–16605.
39. Delville, M. M. E.; van Hest, J. C. M.; Rutjes, F. P. J. T. *Beilstein J. Org. Chem.* **2013**, *9*, 1813–1818.
40. Maurya, R. A.; Min, K.-I.; Kim, D.-P. *Green Chem.* **2014**, *16*, 116–120.

41. Müller, S. T. R.; Smith, D.; Hellier, P.; Wirth, T. *Synlett* **2014**, *25*, 871–875.
42. Müller, S. T. R.; Hokamp, T.; Ehrmann, S.; Hellier, P.; Wirth, T. *Chem. Eur. J.* **2016**, *22*, 11940–11942.
43. Burguete, M. I.; Cornejo, A.; García-Verdugo, E.; García, J.; Gil, M. J.; Luis, S. V.; Martínez-Merino, V.; Mayoral, J. A.; Sokolova, M. *Green Chem.* **2007**, *9*, 1091–1096.
44. Aranda, C.; Cornejo, A.; Fraile, J. M.; García-Verdugo, E.; Gil, M. J.; Luis, S. V.; Mayoral, J. A.; Martinez-Merino, V.; Ochoa, Z. *Green Chem.* **2011**, *13*, 983–990.
45. Castano, B.; Gallo, E.; Cole-Hamilton, D. J.; Santo, V. D.; Psaro, R.; Caselli, A. *Green Chem.* **2014**, *16*, 3202–3209.
46. Maestre, L.; Ozkal, E.; Ayats, C.; Beltrán, Á.; Díaz-Requejo, M. M.; Pérez, P. J.; Pericàs, M. A. *Chem. Sci.* **2015**, *6*, 1510–1515.
47. Bartrum, H. E.; Blakemore, D. C.; Moody, C. J.; Hayes, C. J. *J. Org. Chem.* **2010**, *75*, 8674–8676.
48. Zhang, X.; Stefanick, S.; Villani, F. J. *Org. Process Res. Dev.* **2004**, *8*, 455–460.
49. Bartrum, H. E.; Blakemore, D. C.; Moody, C. J.; Hayes, C. J. *Chem. Eur. J.* **2011**, *17*, 9586–9589.
50. Bartrum, H. E.; Blakemore, D. C.; Hayes, C. J.; Moody, C. J. *Tetrahedron* **2013**, *69*, 2276–2282.
51. Nicolle, S. M.; Hayes, C. J.; Moody, C. J. *Chem. Eur. J.* **2015**, *21*, 4576–4579.
52. Nicolle, S. M.; Nortcliffe, A.; Bartrum, H. E.; Lewis, W.; Hayes, C. J.; Moody, C. J. *Chem. Eur. J.* **2017**, *23*, 13623–13627.
53. Rackl, D.; Yoo, C.-J.; Jones, C. W.; Davies, H. M. L. *Org. Lett.* **2017**, *19*, 3055–3058.
54. Müller, S. T. R.; Murat, A.; Maillos, D.; Lesimple, P.; Hellier, P.; Wirth, T. *Chem. Eur. J.* **2015**, *21*,7016–7020.
55. Müller, S. T. R.; Murat, A.; Hellier, P.; Wirth, T. *Org. Process Res. Dev.* **2016**, *20*, 495–502.
56. Deadman, B. J.; O'Mahony, R. M.; Lynch, D.; Crowley, D. C.; Collins, S. G.; Maguire, A. R. *Org. Biomol. Chem.* **2016**, *14*, 3423–3431.
57. McCaw, P. G.; Deadman, B. J.; Maguire, A. R.; Collins, S. G. *J. Flow Chem.* **2016**, *6*, 226–233.
58. McCaw, P. G.; Buckley, N. M.; Eccles, K. S.; Lawrence, S. E.; Maguire, A. R.; Collins, S. G. *J. Org. Chem.* **2017**, *82*, 3666–3679.
59. Gérardy, R.; Winter, M.; Vizza, A.; Monbaliu, J.-C. M. *React. Chem. Eng.* **2017**, *2*, 149–158.
60. Fuse, S.; Otake, Y.; Nakamura, H. *Eur. J. Org. Chem.* **2017**, 6466–6473.
61. Vaske, Y. S. M.; Mahoney, M. E.; Konopelski, J. P.; Rogow, D. L.; McDonald, W. J. *J. Am. Chem. Soc.* **2010**, *132*, 11379–11385.

62. Willumstad, T. P.; Haze, O.; Mak, X. Y.; Lam, T. Y.; Wang, Y.-P.; Danheiser, R. L. *J. Org. Chem.* **2013**, *78*, 11450–11469.
63. Li, M.-M.; Wei, Y.; Liu, J.; Chen, H.-W.; Lu, L.-Q.; Xiao, W.-J. *J. Am. Chem. Soc.* **2017**, *139*, 14707–14713.
64. Garbarino, S.; Protti, S.; Basso, A. *Synthesis* **2015**, *47*, 2385–2390.
65. Basso, A.; Banfi, L.; Garbarino, S.; Riva, R. *Angew. Chem. Int. Ed.* **2013**, *52*, 2096–2099.
66. Musio, B.; Mariani, F.; Śliwiński, E. P.; Kabeshov, M. A.; Odajima, H.; Ley, S. V. *Synthesis* **2016**, *48*, 3515–3526.
67. Takeda, K.; Oohara, T.; Shimada, N.; Nambu, H.; Hashimoto, S. *Chem. Eur. J.* **2011**, *17*, 13992–13998.
68. Moschetta, E. G.; Negretti, S.; Chepiga, K. M.; Brunelli, N. A.; Labreche, Y.; Feng, Y.; Rezaei, F.; Lively, R. P.; Koros, W. J.; Davies, H. M. L.; Jones, C. W. *Angew. Chem. Int. Ed.* **2015**, *54*, 6470–6474.
69. Pasceri, R.; Bartrum, H. E.; Hayes, C. J.; Moody, C. J. *Chem. Commun.* **2012**, *48*, 12077–12079.
70. Fleming, G. S.; Beeler, A. B. *Org. Lett.* **2017**, *19*, 5268–5271.
71. Pieber, B.; Kappe, C. O. *Org. Lett.* **2016**, *18*, 1076–1079.
72. Mertens, L.; Hock, K. J.; Koenigs, R. M. *Chem. Eur. J.* **2016**, *22*, 9542–9545.
73. Hock, K. J.; Mertens, L.; Metze, F. K.; Schmittmann, C.; Koenigs, R. M. *Green Chem.* **2017**, *19*, 905–909.
74. Hock, K. J.; Mertens, L.; Koenigs, R. M. *Chem. Commun.* **2016**, *52*, 13783–13786.
75. Britton, J.; Jamison, T. F. *Angew. Chem. Int. Ed.* **2017**, *56*, 8823–8827.
76. For the recent review, see: Gao, Y.; Wang, J. *Chin. J. Org. Chem.* **2018**, *38*, 1275–1291.

Chapter 15

Application of Diazo Compounds in the Synthesis of Natural Products

Due to their diverse reactivities, the diazo compounds have been widely explored as efficient synthons in the synthesis of natural products, biologically relevant compounds and other functional molecules. In particular, transition-metal-catalyzed catalytic carbene transfer reactions have proven to be highly useful for the construction of complex scaffolds in a concise and efficient manner, and in many cases with high stereochemical control. Although the application of diazo compounds in natural product synthesis has been mentioned in some reviews, there are only few articles that collectively summarize this topic.[1,2] In this chapter, selected recent examples of the synthesis of natural products or other functional compounds utilizing the transformations of diazo compounds will be introduced, classified according to the corresponding type of carbene transfer reaction used in the synthesis.

15.1 Diazo compounds as nucleophiles

In 2009, Kocienski and co-workers realized the total synthesis of ionomycin calcium(II) complex in 33 steps (Scheme 15.1).[3] The synthesis involves the diazomethane homologation with *in situ* generated acyl chloride to form **2**, followed by LDA mediated nucleophilic addition of diazo anion to aldehyde **3** to form α-diazo-β-hydroxyketone **4**. Subsequently, the Rh(II)-catalyzed carbene 1,2-H migration affords β-diketone **5**. Finally, removal of protecting groups from **5**, followed by hydrolyzation

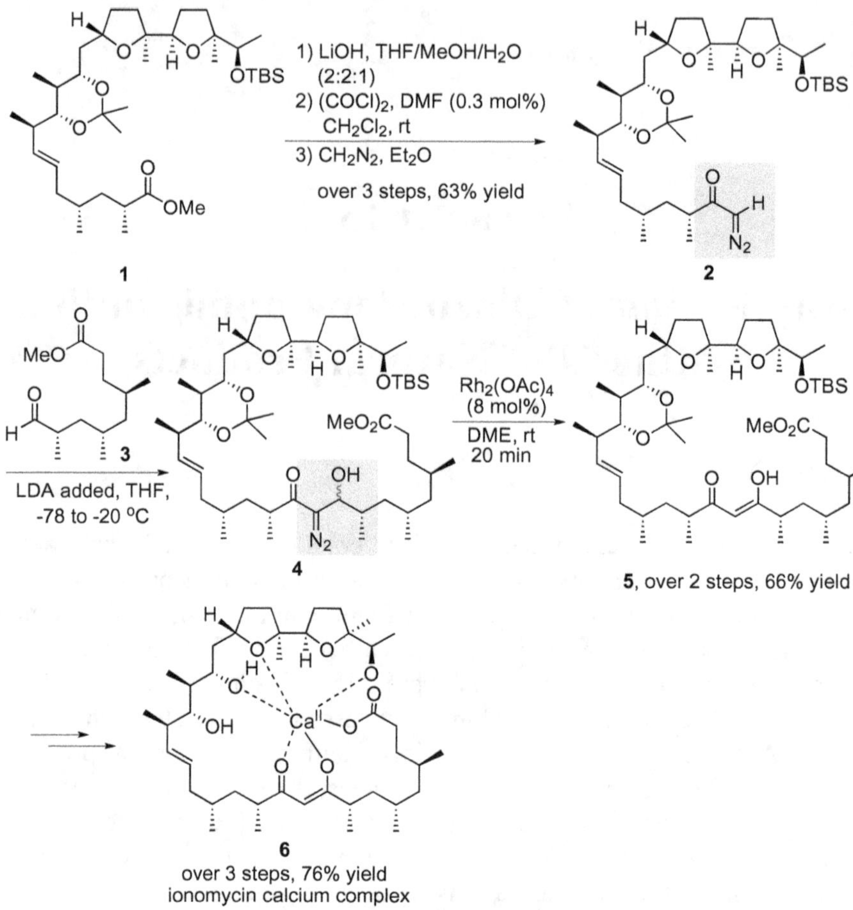

Scheme 15.1 Synthesis of ionomycin calcium(II) complex.

and reaction with calcium chloride, affords the target product **6** as a crystalline calcium salt.

In 2012, Tang and co-workers applied Ag-mediated ring-expansion strategy to the total synthesis of pipercyclobutanamide A **9** and piperchabamide G **10**.[4,5] The key step of the synthesis was the ring expansion of a cyclopropyl Ag(I) carbene derived from diazoester **7**, affording cyclobutenoate **8** with excellent yield, and high regio- and stereoselectivities (Scheme 15.2). With this key intermediate, pipercyclobutanamide A **9** and piperchabamide G **10** could be synthesized through sequential

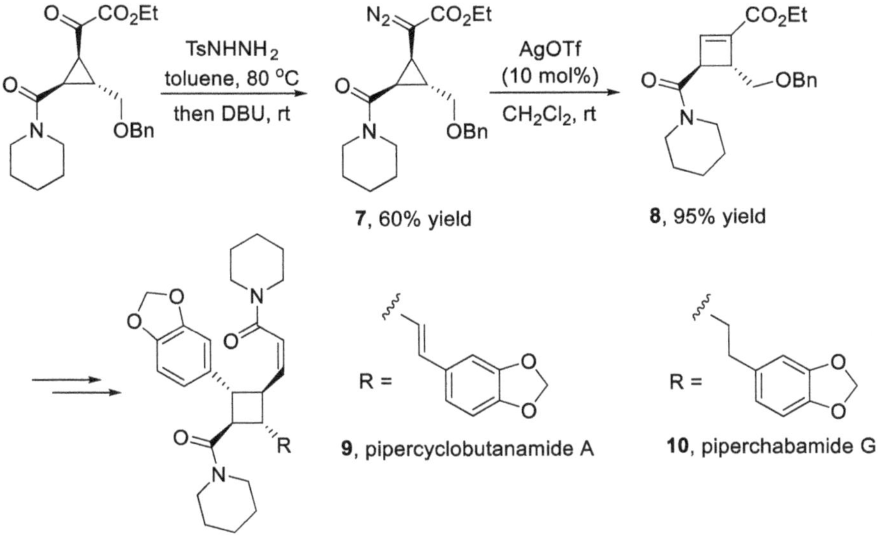

Scheme 15.2 Total synthesis of pipercyclobutanamide A and piperchabamide G.

transformations. The success demonstrates the usefulness of metal carbene migration in the construction of small ring systems, which exist in many biologically relevant natural or non-natural products.

15.2 Insertions with diazo compounds

15.2.1 *X–H insertions*

Carbene N–H insertion reactions as an efficient tool for the construction of carbon–nitrogen bond have been extensively utilized in organic synthesis. In 2008, Moody and co-workers reported the total synthesis of an unusual bis-oxazole natural product siphonazole **14**, and its *O*-methyl derivative **15** (Scheme 15.3).[6] This synthesis used the Rh(II)-catalyzed reaction of β-ketoester **11** twice. First, Rh(II)-catalyzed reaction of **11** with α-iodoacetonitrile generated oxazole intermediate **12**. Then amide intermediate **13** was prepared by Rh(II)-catalyzed carbene N–H insertion with β-ketoester **11**, which was followed by cyclization, alkylation with **12** and amidation to afford the siphonazole **14** or its *O*-methyl derivative **15**.

Catalytic asymmetric carbene Si–H insertion was previously studied by Panek and co-workers.[7] They applied this unique reaction in

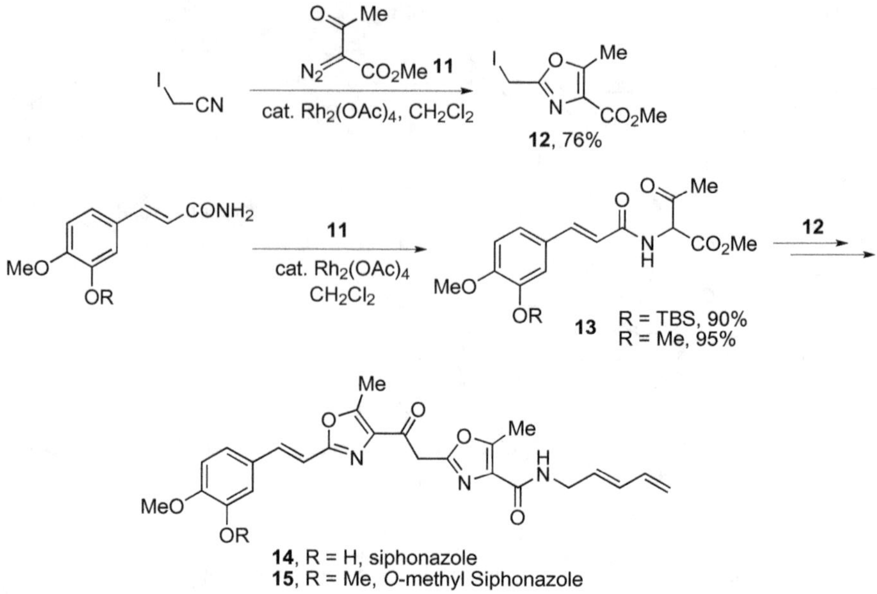

Scheme 15.3 Total synthesis of siphonazole and its *O*-methyl derivative.

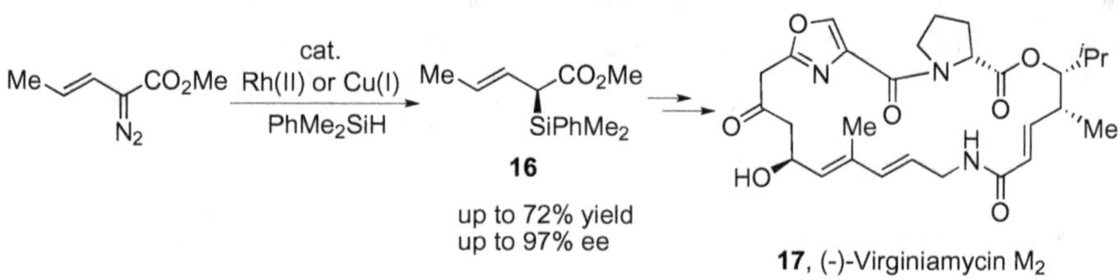

Scheme 15.4 Enantioselective carbene Si–H insertion for the total synthesis of (−)-virginiamycin M_2.

the enantioselective synthesis of antibiotic (−)-virginiamycin M_2 **17** (Scheme 15.4).[8] In this synthesis, the key intermediate **16** was obtained by highly enantioselective Rh(II)- or Cu(I)-catalyzed carbene insertion to Si–H bond.

In 2010, Yang and Li reported the first total synthesis of (±)-maoecrystal V **21**, which involved selective Rh(II)-catalyzed O–H insertion of diazo compounds as the key step (Scheme 15.5).[9,10] The diazo moiety was introduced in two steps from alcohol **18** to afford **19** in 69% yield.

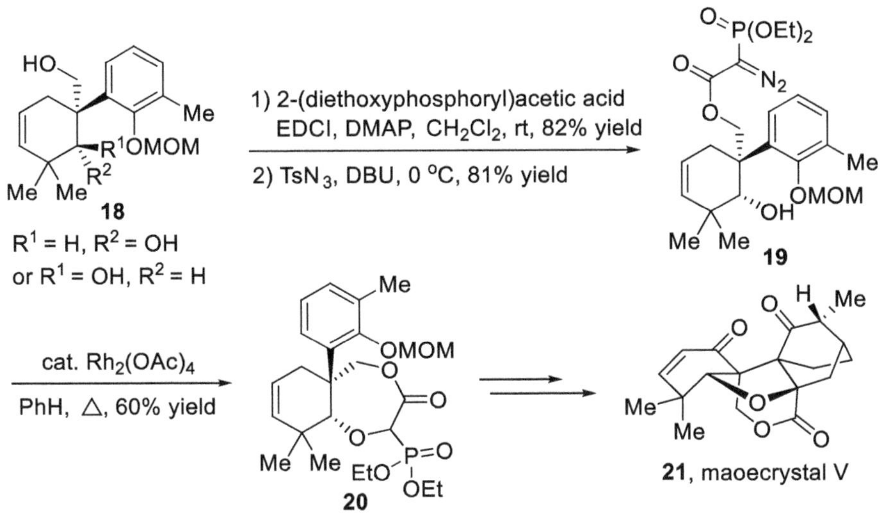

Scheme 15.5 Application of intermolecular Rh(II)-catalyzed O–H insertion in the total synthesis of maoecrystal V.

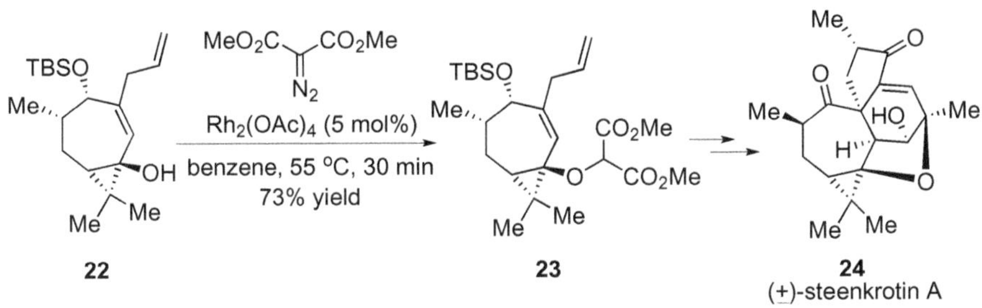

Scheme 15.6 Application of intermolecular Rh(II)-catalyzed O–H insertion for the total synthesis of (±)-steenkrotin A.

Subsequently, intramolecular Rh(II)-catalyzed carbene insertion to O–H bond afforded the corresponding product **20**, which could be converted into maoecrystal V **21.**

In another example of the application of carbene O–H insertion in natural product synthesis, Ding and co-workers reported the first diastereoselective total synthesis of diterpenoid (±)-steenkrotin A **24** (Scheme 15.6).[11] The synthesis involves the preparation of intermediate **23** through intermolecular Rh(II)-catalyzed O–H insertion of alcohol **22**

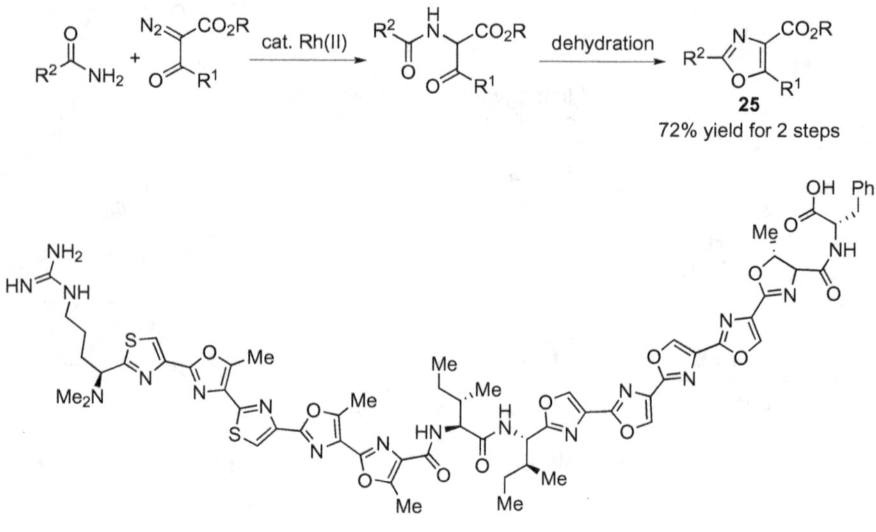

Scheme 15.7 Application of Rh(II)-catalyzed carbene N–H insertion in the total synthesis of plantazolicin A.

with readily available diazomalonic ester, followed by intramolecular carbonyl-ene reaction to construct the tetrahydrofuran subunit.

On the other hand, Moody and co-workers demonstrated the application of carbene N–H insertion in the total synthesis of the post-translationally modified polyazole peptide plantazolicin A **26**, which served as a novel soil bacterium metabolite.[12] The key structural motifs, the oxazole rings **25**, could be readily accessed by a two-step process involving carbene N–H insertion and dehydration from readily available amides and α-diazo β-ketoester (Scheme 15.7). Some of the seven oxazole rings in this natural product were assembled by this Rh(II)-catalyzed N–H insertion/dehydration strategy.

15.2.2 *C–H insertions*

As a unique method for C–H bond functionalization, both intra- and intermolecular carbene C–H insertion reactions of diazo compounds have found widespread applications in the synthesis of natural products and other functional molecules. In 2008, Taber and co-workers achieved the

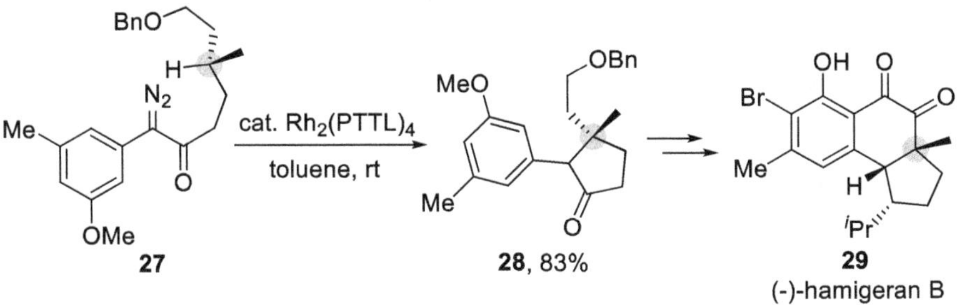

27 28, 83% 29 (-)-hamigeran B

Scheme 15.8 Application of intramolecular C–H insertion in the synthesis of (−)-hamigeran B

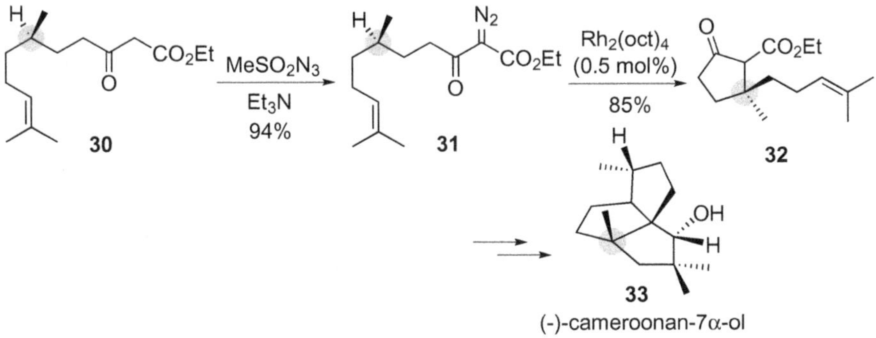

30 31 32 33 (-)-cameroonan-7α-ol

Scheme 15.9 Synthesis of (−)-cameroonan-7α-ol.

synthesis of (−)-hamigeran B **29** based on an Rh(II)-catalyzed intramolecular carbene C–H insertion of substituted α-aryl-α-diazoketone **27** (Scheme 15.8).[13] The Rh(II)-catalyzed C–H bond insertion proceed with the retention of the chirality of the stereogenic center, affording the key intermediate **28**.

Another example of the application of stereospecific carbene C–H bond insertion has been demonstrated by the same group in the first total synthesis of tricyclic sesquiterpene (−)-Cameroonan-7α-ol **33** from (+)-citronellal (Scheme 15.9).[14] In this synthesis, the Rh(II)-catalyzed reaction of diazo substrate **31** occurred with retention of chiral center of the target C–H bond to afford cyclic product **32**, demonstrating the power of the insertion reaction in the generation of the all-carbon quaternary chiral center.

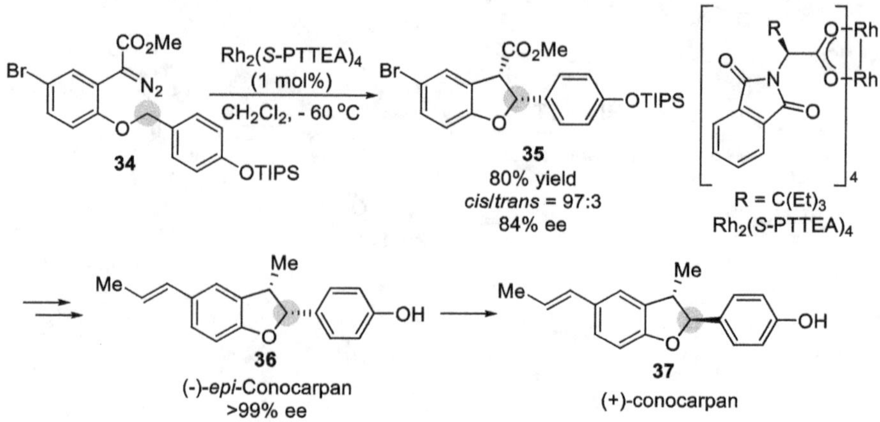

Scheme 15.10 Application of asymmetric catalytic carbene C–H bond insertion in the synthesis of (−)-*epi*-conocarpan and (+)-conocarpan

Asymmetric catalytic carbene C–H bond insertion has also been incorporated in the natural product synthesis. Hashimoto and co-workers utilized their chiral Rh(II) catalysts to achieve a series of asymmetric carbene C–H functionalizations. In 2009, they reported the asymmetric synthesis of neolignan natural products (−)-*epi*-conocarpan **36** and (+)-conocarpan **37** with enantio- and diastereoselective intramolecular carbene C–H insertions (Scheme 15.10).[15] The Rh_2(*S*-PTTEA)$_4$ complex, which incorporated *N*-phthaloyl-(*S*)-triethylalaninate as the chiral bridging ligand, exhibited exceptionally high diastereo- and enantioselectivities for catalyzing C–H insertion of 5-bromoaryldiazoacetate **34**, affording the key intermediate **35** effectively.

In 2008, Kan and co-workers reported the total synthesis of (−)-serotobenine **40** based on Davies's Rh_2(*S*-DOSP)$_4$-catalyzed C–H bond insertion of diazo substrate **38**, which bears a chiral ester moiety (Scheme 15.11).[16] The asymmetric C–H insertion afforded the desired indole-fused dihydrobenzofuran **39** with excellent yield and high stereoselectivity.

Yu and co-workers have combined the Rh(II)-catalyzed asymmetric carbene C(sp^3)–H bond insertion and Pd(II)-catalyzed aromatic C(sp^2)–H olefination in the total synthesis of (+)-Lithospermic acid **44** (Scheme 15.12).[17] Herein Davies's Rh_2(*S*-DOSP)$_4$ has also proven an efficient catalyst in promoting the highly stereoselective C(sp^3)–H bond

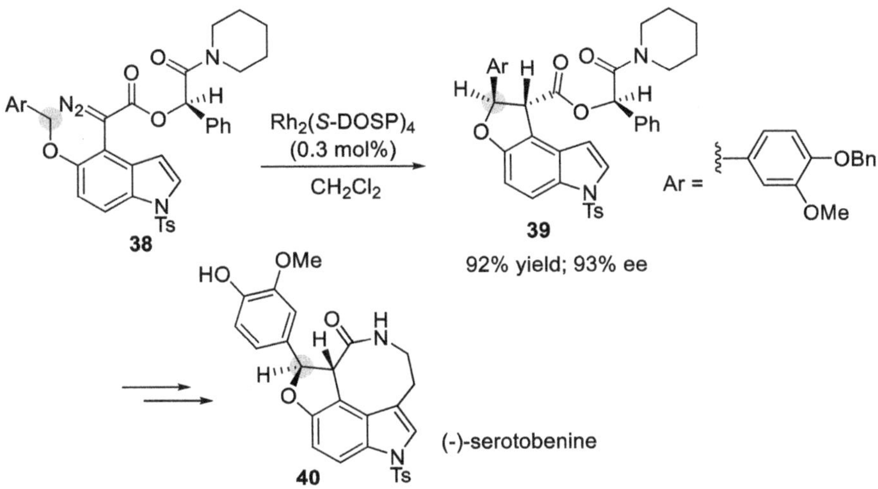

Scheme 15.11 Total synthesis of (−)-serotobenine.

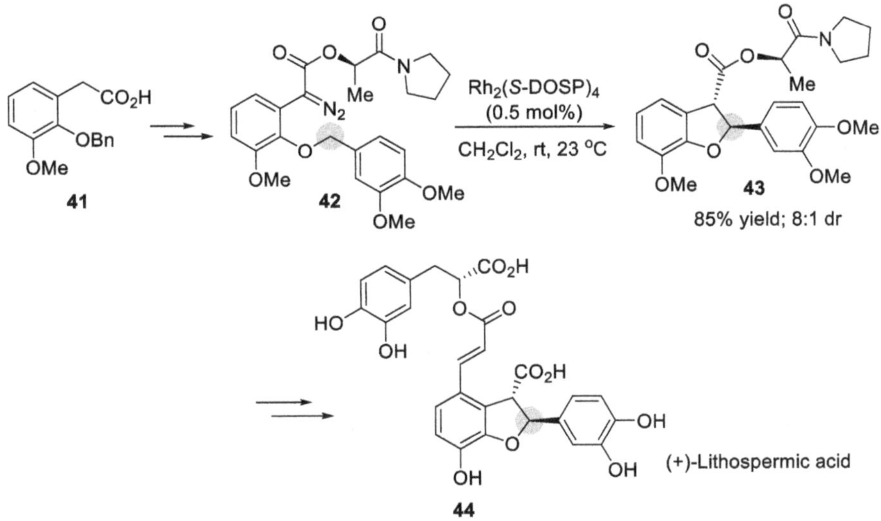

Scheme 15.12 Total synthesis of (+)-lithospermic acid *via* Rh(II)-catalyzed C–H insertion and late-stage Pd(II)-catalyzed C–H olefination.

insertion, affording the dihydrobenzofuran **43** with high diastereoselectivity. The total synthesis of (+)-lithospermic acid **44** was achieved in 12 steps and 11% overall yield from *o*-eugenol.

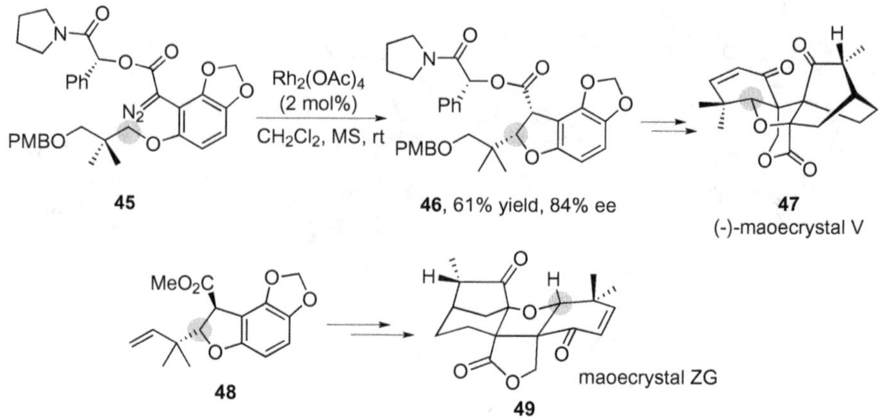

Scheme 15.13 Enantioselective synthesis of (−)-maoecrystal V *via* chiral auxiliary-assisted asymmetric C–H insertion.

Zakarian and co-workers completed an enantioselective total synthesis of maoecrystal V **47** through chiral auxiliary-assisted Rh(II)-catalyzed donor/acceptor carbene C–H insertion as the key step for determining the chiral centers (Scheme 15.13).[18] After the optimization of auxiliaries, the C–H bond insertion reaction led to the construction of the 2,3-dihydrobenzofuran core structure of **46** with 84% *ee*. Moreover, the isomer of maoecrystal V, named as maoecrystal ZG **49**, was also synthesized by a similar approach.

Apart from the construction of oxa-heterocycles, the aza-heterocyclic core structure was also constructed in a stereoselective manner through chiral Rh(II)-catalyzed carbene insertion to C–H bonds.[19,20] From the common chiral non-racemic bicyclic lactam lactone prepared by Rh(II)-catalyzed C–H insertion of substituted diazoacetate, Wee and co-workers completed stereoselective synthesis of a series of 2,6-disubstituted 3-hydroxypiperidine (**52**, **53**) and *cis*-decahydroquinoline alkaloids (**54**, **55**) from a common intermediate **51** (Scheme 15.14).[20] The chiral non-racemic intermediate **51** could be synthesized through enantioselective intramolecular C–H bond insertion of **50** with chiral Rh(II) catalyst $Rh_2(4S\text{-MPPIM})_4$.

In 2014, Kan and co-workers completed the enantioselective synthesis of SB-203207 **59**.[21] The noteworthy feature of this synthesis was the Lewis acid-catalyzed Roskamp reaction of diazo substrate bearing a chiral

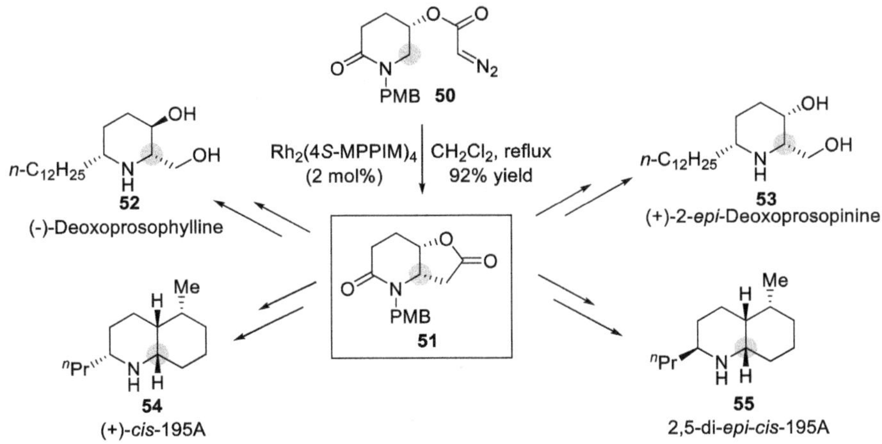

Scheme 15.14 Enantioselective syntheses of alkaloids from a chiral non-racemic bicyclic lactam lactone.

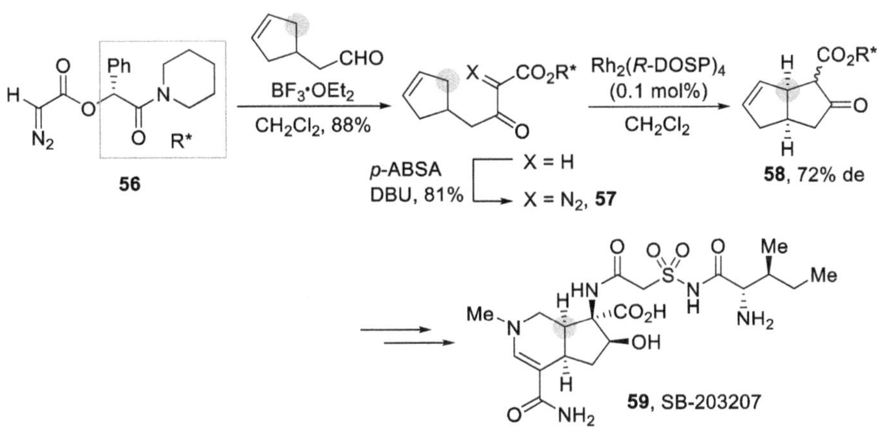

Scheme 15.15 Application of Roskamp reaction and Rh(II)-catalyzed C–H insertion in the total synthesis of SB-203207.

auxiliary **56** with aldehyde. This was followed by a diazo transfer reaction and subsequent diastereoselective Rh(II)-catalyzed C–H insertion of diazoester **57** to afford biscyclic product **58** through desymmetrization (Scheme 15.15).

Oxidized phospholipids show pro-inflammatory biological activity associated with diseases such as rheumatoid arthritis, emphysema and

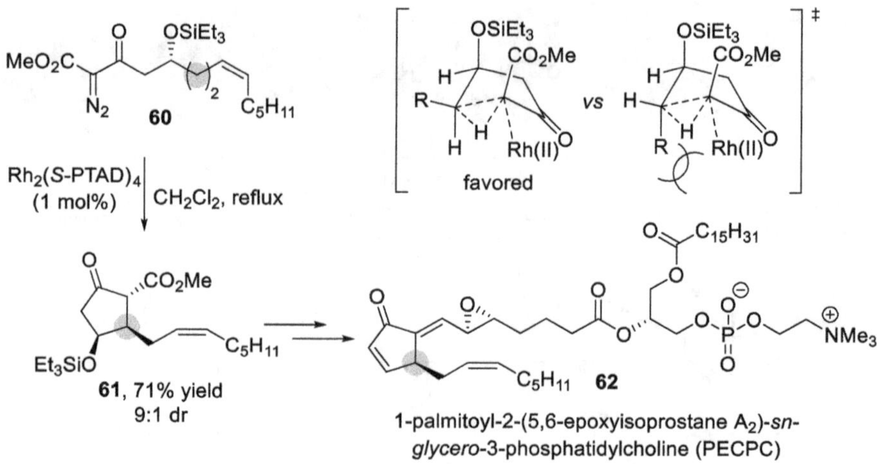

Scheme 15.16 Synthesis of epoxyisoprostane through stereoselective Rh(II)-catalyzed C–H insertion.

atherosclerosis. Carreira and co-workers established a general synthetic route to a class of cyclized oxidized phospholipids, including PECPC (5.4% overall yield, 11 steps), PEIPC (1.8% overall yield, 13 steps) and the corresponding carboxylic acids EC and EI.[22] The key step for constructing cyclopentenone scaffold **61** involved the chiral Rh-catalyzed C–H insertion in a diastereo- and regioselective manner (Scheme 15.16). Analysis of putative competing transition states indicated that the better stereocontrol was attributed to steric hinderance between the sterically demanding Rh(II)-center and the R group.

The spirocyclic scaffolds can also be constructed with Rh(II)-catalyzed carbene C–H bond insertion. Wood and Villalobos demonstrated such a strategy through model study toward the total synthesis of syringolides **63** with different ring sizes (Scheme 15.17).[23] From various substituted aryl diazoacetates, vinyl diazoacetates and diazoacetates, highly stereospecific synthesis of functionalized spirolactones could be achieved by utilizing this strategy.

From the substrate of iodoallylic phosphate, Tu and co-workers completed an efficient total synthesis of (+)-przewalskin B **66**, which exhibited modest anti-HIV-1 activity.[24] The key step in this synthesis includes the Rh(II)-catalyzed intramolecular carbene insertion into tertiary C–H bond, leading to the construction of spirocyclic core structure (Scheme 15.18). It

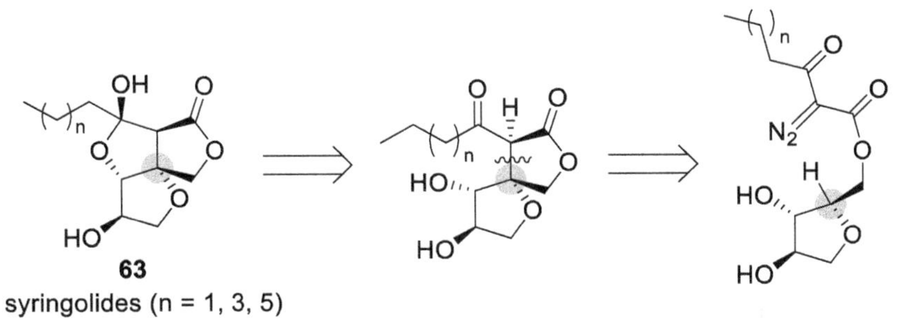

Scheme 15.17 Retro-synthetic analysis toward the total synthesis of syringolides.

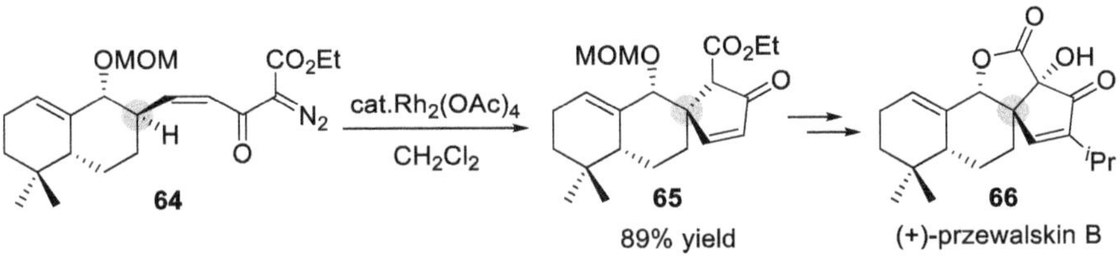

Scheme 15.18 Total synthesis of (+)-przewalskin B *via* Rh(II)-catalyzed tertiary C–H bond insertion.

is also noteworthy that a *cis*-configured C=C double bond is located between the diazo moiety and the target C–H bond.

Intermolecular carbene C–H bond insertion has also been applied in the synthesis of natural products. Kan and co-workers reported the enantioselective total syntheses of (+)-methoxyphenylkainic acid and (+)-phenylkainic acid **70** with Rh(II)-catalyzed intermolecular C–H insertion of diazophenylacetic acid derivatives **67** with 1,4-cyclohexadiene **68** as the key step (Scheme 15.19).[25] With the employment of (*R*)-2-hydroxy-2-phenyl-1-(piperidin-1-yl)ethanone as the chiral auxiliary, the corresponding insertion product **69** was obtained in good yields and with high level of diastereoselectivity.

Non-carbonyl-stabilized metal carbene can be derived from *N*-aziridinylimines, which may undergo selective C–H bond insertion to afford cyclic compounds. The application of this strategy was demonstrated by the total synthesis of the racemic natural product cryptaustoline **73** (Scheme 15.20).[26] A series of *N*-aziridinylimines **71** underwent intramolecular C–H insertion under catalysis with $Rh_2(5S\text{-MEPY})_4$ to afford

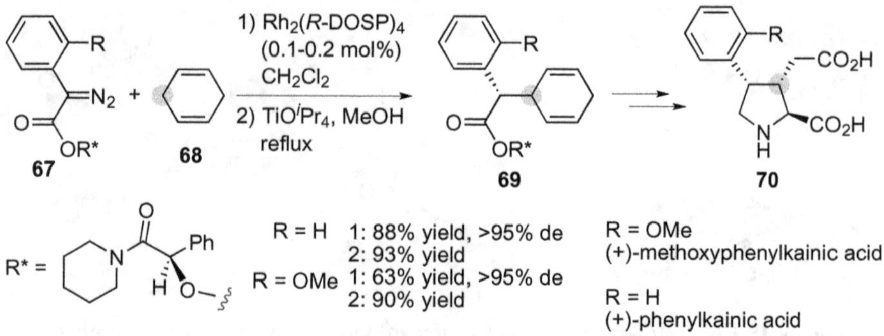

Scheme 15.19 Synthesis of (+)-methoxyphenylkainic acid and (+)-phenylkainic acid *via* intramolecular carbene C–H insertion.

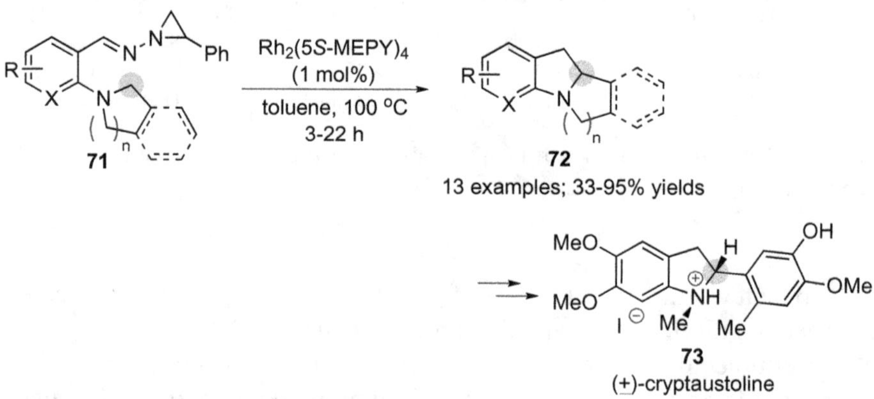

Scheme 15.20 Synthesis of *N*-fused indolines *via* Rh(II)-catalyzed carbene insertion from *N*-aziridinylimines.

the cyclic products **72**, one of which was further converted to cryptaustoline **73**.

15.3 Cyclopropanation with diazo compounds

Intramolecular cyclopropanation reaction of diazo derivatives has been proven to be of significant value in the synthesis of natural products. In 2009, Taber and co-workers completed the enantioselective synthesis of (+)-coronafacic acid **77** from diazo menthyl ester **74**.[27] In the first step, the

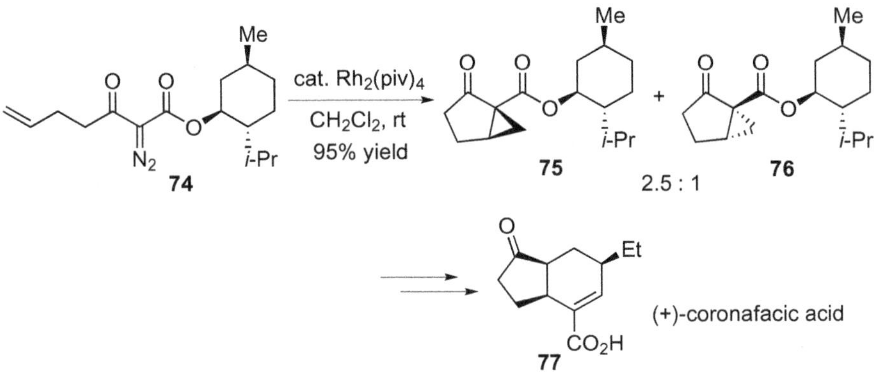

Scheme 15.21 Synthesis of (+)-coronafacic acid *via* Rh(II)-catalyzed cyclopropanation of α-diazoester.

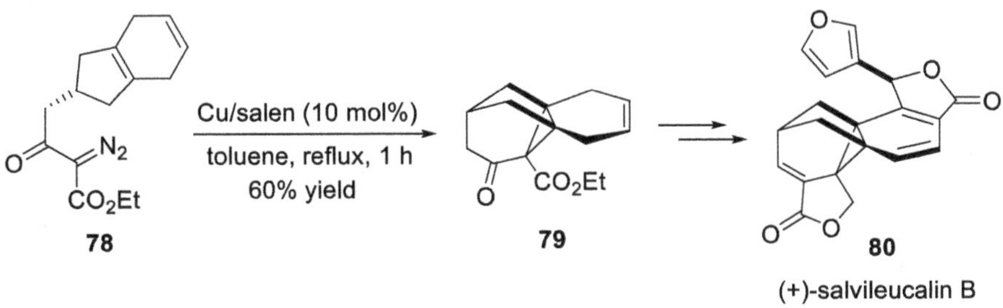

Scheme 15.22 Construction of the pentacyclic core of (+)-salvileucalin B with copper-catalyzed cyclopropanation.

two stereogenic centers were constructed by Rh(II)-catalyzed cyclopropanation with moderate diastereoselectivity (2.5:1) (Scheme 15.21). With the major isomer **75**, sequential $Fe(CO)_5$-mediated cyclocarbonylation afforded the bicyclic enone skeleton, which was then hydrogenated and carried on to the final product **77**.

In 2014, Taber and co-workers reported a concise construction of the prochiral pentacyclic core of (+)-salvileucalin B **80**.[28] The key step in the synthesis was the Cu(II)/salicylideneimine-catalyzed intramolecular cyclopropanation of a symmetrical indane-derived α-diazo-β-keto ester **78**, affording the fused polycyclic neoclerodane core **79** in 60% yield (Scheme 15.22).

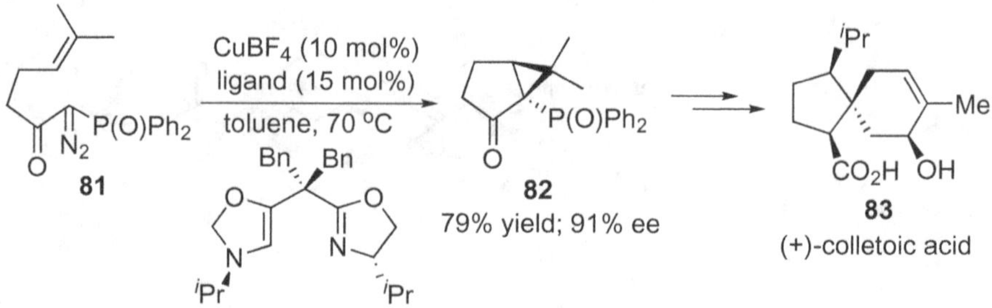

Scheme 15.23 Total synthesis of (+)-colletoic acid *via* Cu(I)-catalyzed intramolecular asymmetric cyclopropanation.

Catalytic asymmetric cyclopropanation has been well developed, which provides reliable method for natural product synthesis. For example, Nakada and co-workers realized the first total synthesis of (+)-colletoic acid **83**, which served as a potent naturally occurring 11*β*-hydroxysteroid dehydrogenase type 1 (11*β*-HSD1) inhibitor.[29] This enantioselective synthetic approach featured an asymmetric Cu(I)-catalyzed intramolecular cyclopropanation of *α*-diazo-*β*-ketodiphenyl phosphine oxide **81** with good yield and high level of enantioselectivity (Scheme 15.23). Further incorporation of 9 reaction steps including 5 highly stereoselective reactions allowed the synthesis of (+)-colletoic acid **83** in a highly efficient manner.

The cyclopropane moiety constructed by cyclopropanation may be further converted to various structures through ring-opening process, thus significantly expanding the synthetic application of cyclopropanation in natural product synthesis. Qin and co-workers have applied such approaches to the total synthesis of indole-derived alkaloids.[30–33] In 2008, they completed the total synthesis of (±)-minfiensine **88** with 4% overall yield in 12 steps.[30] In this synthesis, an efficient strategy for assembling the tetracyclic skeleton of indole alkaloid scaffold **87** was achieved by a three-step one-pot cascade, which consisted of Cu(I)-catalyzed intramolecular cyclopropanation of **84**, followed by ring-opening and ring-closure sequences (Scheme 15.24).

With similar approach, the same group accomplished the total synthesis of (±)-vincorine **91** as the members of *akuammiline* alkaloids with approximately 1% overall yield in 31 steps.[31] This achievement benefited from the previous strategy of concisely assembling the key tetracyclic core through three-step one-pot cascade from *α*-diazo-*β*-keto ester **89** (Scheme 15.25).

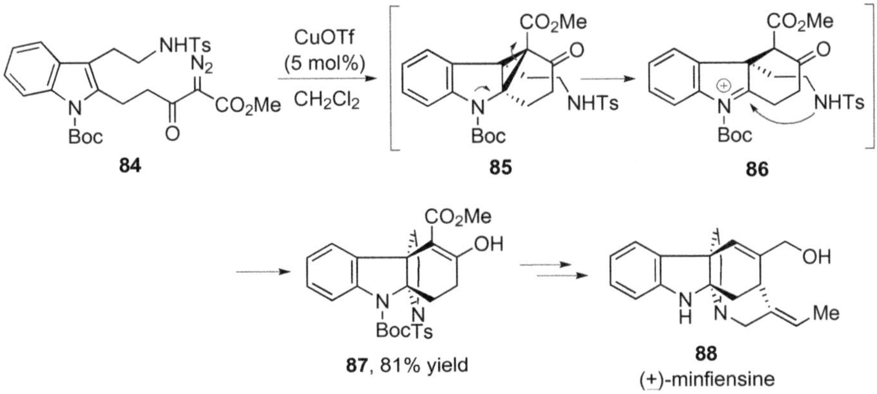

Scheme 15.24 Total synthesis of (±)-minfiensine *via* Cu(I)-catalyzed intramolecular cyclopropanation and ring-opening/closure sequences.

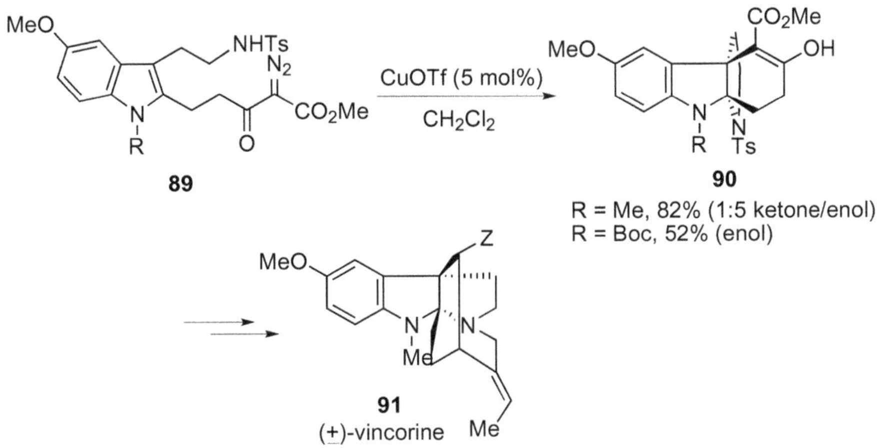

Scheme 15.25 Total synthesis of (±)-vincorine.

In 2009, they completed the total synthesis of potent *anti*-MDR alkaloids (−)-ardeemin **96** with approximately 2% overall yield in 17 steps.[32] The notable feature in this synthesis was the Cu(II)-catalyzed three-step one-pot cascade, which proceeded *via* carbene intermolecular cyclopropanation of indoles **92** with EDA to afford **93**, ring-opening to generate **94**, followed by ring-closure sequence to give the chiral 3-substituted hexahydropyrrolo[2,3-*b*]indole **95** with good yield (Scheme 15.26).

Recently, they further achieved the total synthesis of bioactive *Kopsia* alkaloids (−)-lundurine A **100**, which is the only natural product so far

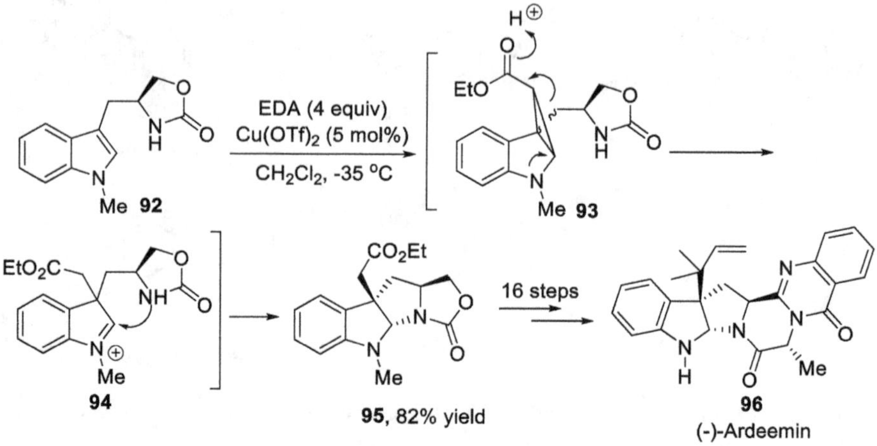

Scheme 15.26 Total synthesis of (−)-ardeemin involving Cu(II)-catalyzed intermolecular cyclopropanation and ring-opening cascade.

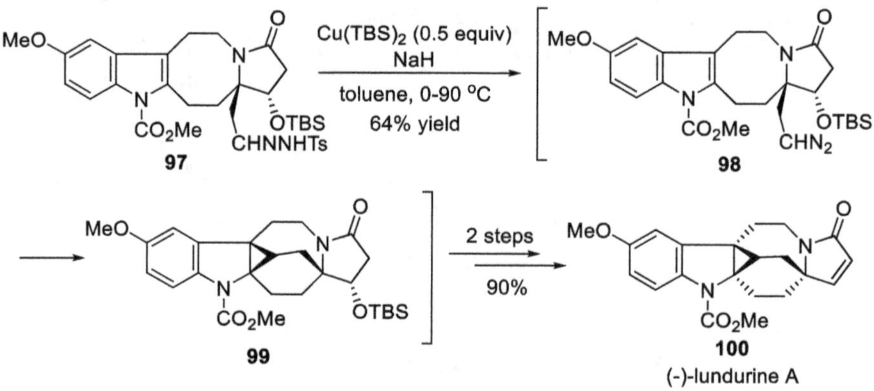

Scheme 15.27 Total synthesis of (−)-Lundurine A *via* Bamford–Stevens diazotization and copper-catalyzed cyclopropanation cascade.

known to contain indolylcyclopropane skeleton.[33] The initial attempt toward the construction of cyclopropane moiety with carbene cyclopropanation resulted in the undesired C–H insertion products from α-diazocarboxylates or -cyanides. Fortunately, with the diazoalkane substrate **98**, which is generated from Bamford–Stevens diazotization, the copper-catalyzed cyclopropanation occurs smoothly to afford the desired intermediate **99** in good yield (Scheme 15.27).

The highly reactive carbene can undergo cyclopropanation with arenes, which serves a valuable method for dearomatization of aromatic compounds. Reisman and Maguire demonstrated the significant utility of intramolecular arene cyclopropanation in the total synthesis of natural products. In 2010, Reisman reported the construction of the core structure for the synthesis of the cytotoxic natural product salvileucalin B **103**.[34] From α-diazo-β-keto cyanide **101**, the microwave-irradiated Cu(II)-catalyzed arene cyclopropanation was carried out to afford the central norcaradiene scaffold **102** containing fully substituted cyclopropanes in 64% yield (Scheme 15.28). Thus, the work laid the foundation toward the enantioselective total synthesis of salvileucalin B **103**.

By using this Cu(II)-catalyzed arene cyclopropanation strategy, the same group accomplished the first enantioselective total synthesis of diterpenoid natural product (+)-salvileucalin B **103** in 18 steps from prop-2-yn-1-ol derivatives (Scheme 15.29).[35]

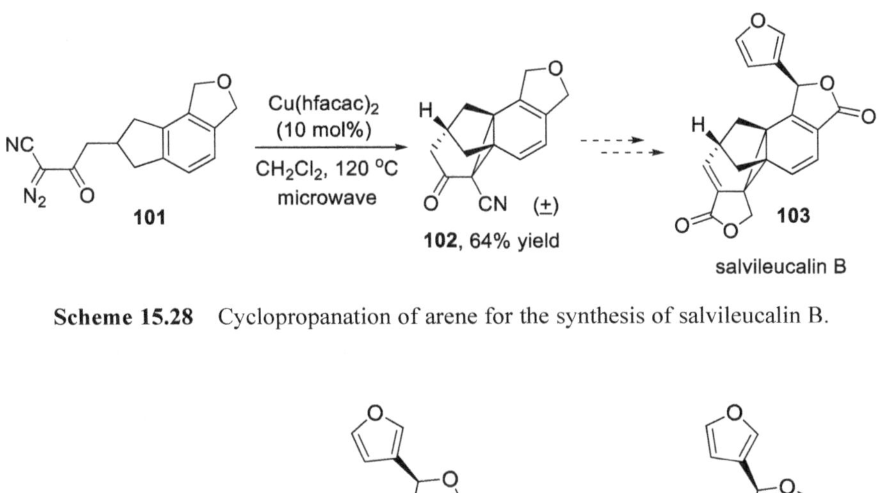

Scheme 15.28 Cyclopropanation of arene for the synthesis of salvileucalin B.

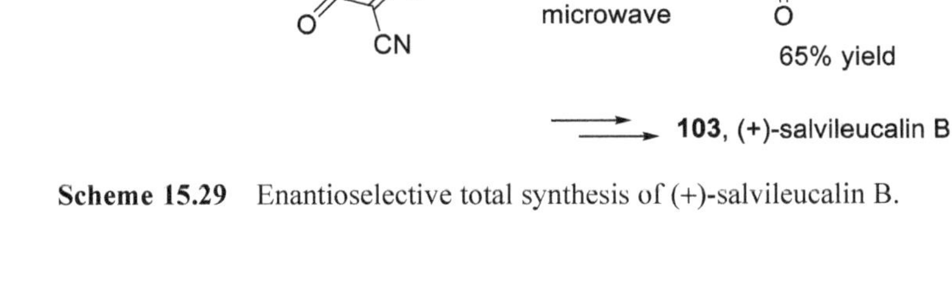

Scheme 15.29 Enantioselective total synthesis of (+)-salvileucalin B.

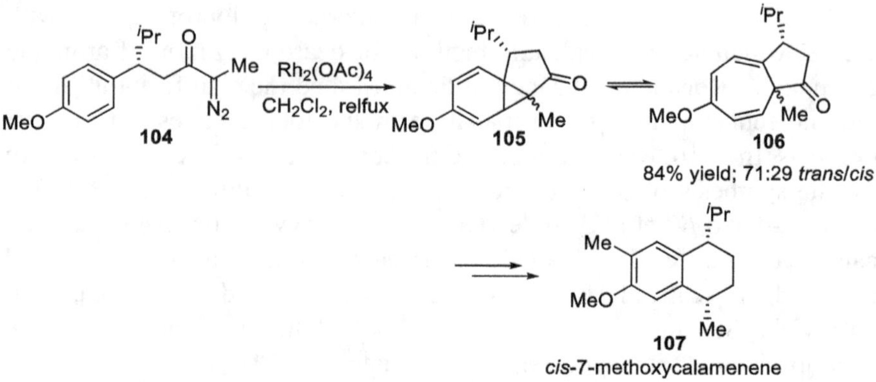

Scheme 15.30 Asymmetric synthesis of *cis*-7-methoxycalamenene *via* Buchner reaction of α-diazoketone.

Maguire and co-workers reported the asymmetric synthesis of *cis*-7-methoxycalamenene by using Rh(II)-catalyzed intramolecular Buchner cyclization of α-diazoketone (Scheme 15.30).[36] The desired equilibrating azulenone derivative **106** could be obtained in 84% yield with 71:29 dr, followed by reduction and acid-promoted rearrangement to afford the dihydro-naphthalene with the 6,6-membered bicyclic ring system.

The intramolecular cyclopropanation followed by ring-opening sequence has also been applied in the synthesis of natural products. In 2013, Tu and co-workers established a novel strategy toward the divergent synthesis of fawcettimine class alkaloids. They completed the first asymmetric total synthesis of (−)-lycojaponicumin C **114**, and an improved total synthesis of (−)-8-deoxyserratinine **115**, (+)-fawcettimine **116** and (+)-fawcettidine **117**.[37] The one-pot procedure, which involved the Mukaiyama–Michael addition with vinyl diazoacetate to give diazo intermediate **109**, copper-catalyzed intramolecular carbene cyclization/ring-opening process and decarboxylation, afforded the key intermediate the dione **113** on gram scale (Scheme 15.31). Mechanistically, the generation of bicyclic structure of **112** from **109** may involve the cyclopropane intermediate **110** (path *a*) or the zwitterionic intermediate **111** (path *b*).

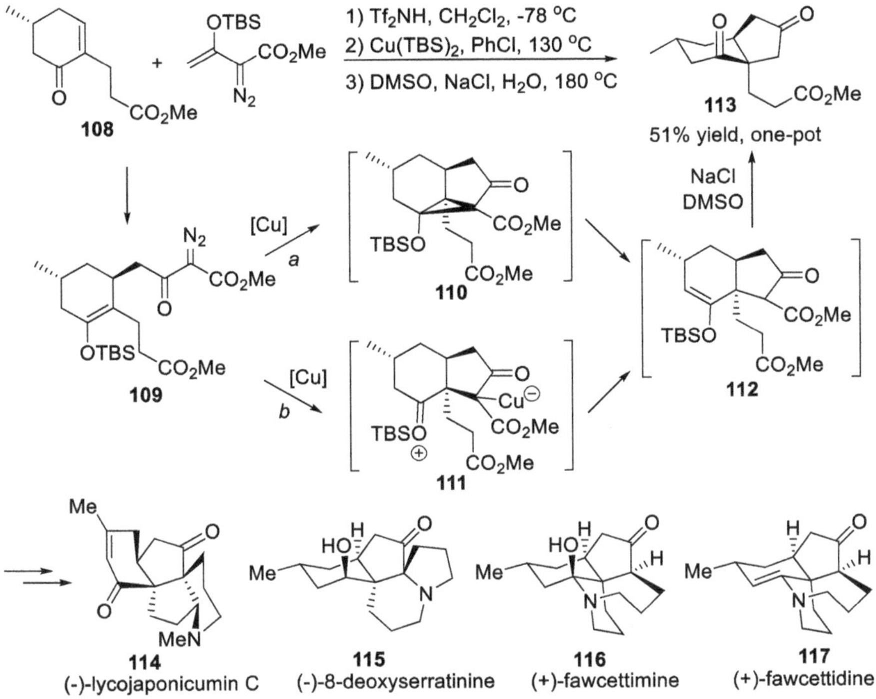

Scheme 15.31 Copper-catalyzed cyclopropanation/ring-opening cascade in the divergent synthesis of *lycopodium* alkaloids.

15.4 Ylide formation and related reactions with diazo compounds

15.4.1 [1,2]-*Sigmatropic rearrangement*

[1,2]-Shift reaction through the ylides generated from carbene, namely the Stevens rearrangement, has also been explored as the key step in natural product synthesis. In 2011, West and co-workers demonstrated an expedient and stereoselective synthetic route to fused 5-7-6 tricyclic core skeleton in the class of tigliane and daphnane diterpenes natural products.[38] From the substrates of α-diazoketones and diazoketoesters **118**, the highly efficient and stereospecific copper-catalyzed reaction to generate oxonium ylide **119**, underwent Stevens rearrangement to afford **120**, in which

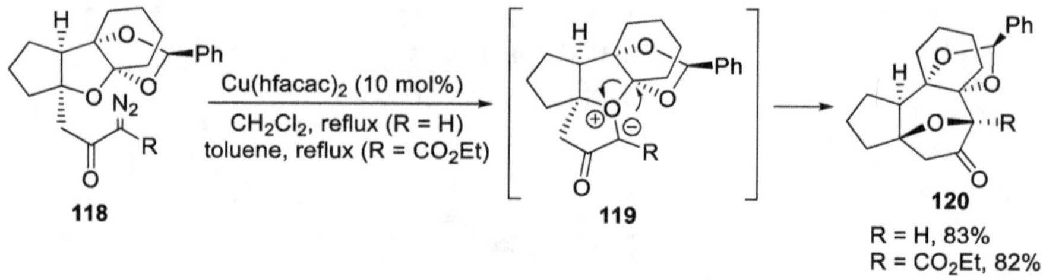

Scheme 15.32 Construction of tigliane and daphnane structures *via* oxonium ylide Stevens rearrangement.

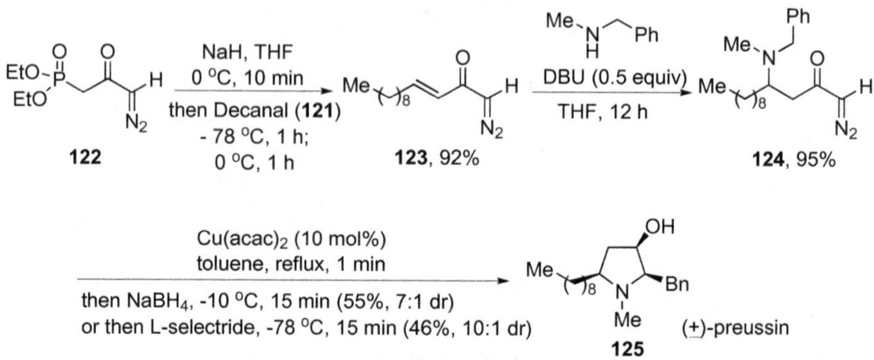

Scheme 15.33 A concise synthesis of (±)-preussin.

the seven-membered ring system with oxygenation at two key bridgehead positions was implemented (Scheme 15.32). This reaction laid the foundation for the total synthesis of tigliane and daphnane derivatives.

As an example of ammonium ylide Stevens rearrangement, Burtoloso and co-workers reported a concise and stereoselective synthesis of preussin alkaloid **125** by using readily available decanal and diethyl 3-diazo-2-oxopropylphosphonate **122** as the substrates.[39] The synthesis involved the olefination and aza-Michael reaction of the diazo substrate **122**, followed by highly stereoselective Cu(II)-catalyzed ammonium ylide Stevens rearrangement/reduction sequence. The racemic preussin **125** was obtained in 40% overall yield with good diastereoselectivity (Scheme 15.33). Additionally, the preparation of various diazoketones as key intermediates enabled the efficient synthesis of preussin analogues as well as other *all-cis*-substituted pyrrolidines.

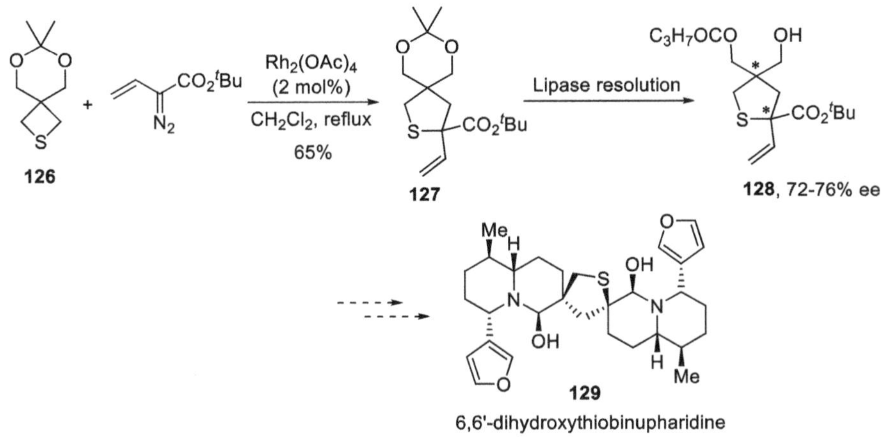

Scheme 15.34 Rh(II)-catalyzed synthesis of the thiolane subunit *via* Stevens rearrangement.

Zakarian and co-workers reported a stereodivergent approach to the central thiolane subunit core for the total synthesis of *nuphar* sesquiterpene thioalkaloids, including thiobinupharidine, 6-hydroxythiobinupharidine, 6,6'-dihydroxythiobinupharidine **129** and others.[40] The notable feature of this synthesis is the Rh(II)-catalyzed sulfur ylide formation from **126** and subsequent Stevens rearrangement to afford **127**, followed by the incorporation with enzyme-mediated optical resolution (Scheme 15.34). All four stereoisomers of thiolane **128** could be obtained feasibly.

15.4.2 [2,3]-*Sigmatropic rearrangement*

[2,3]-Sigmatropic rearrangement of ylides represents a type of unique bond reorganization in organic chemistry. This type of reaction has also been explored in natural product synthesis. In particular, the oxonium ylide generation followed by sequential [2,3]-sigmatropic rearrangement has demonstrated great synthetic value in constructing oxa-heterocyclic compounds for synthesis of natural products.

In 2008, Hodgson and co-workers reported an efficient synthesis of *rac*-hyperolactone C **135**, which acted as the anti-HIV agent and active species in biochemical research.[41] The key steps of the synthesis involved the consecutive Ru-catalyzed alkene cross-metathesis from

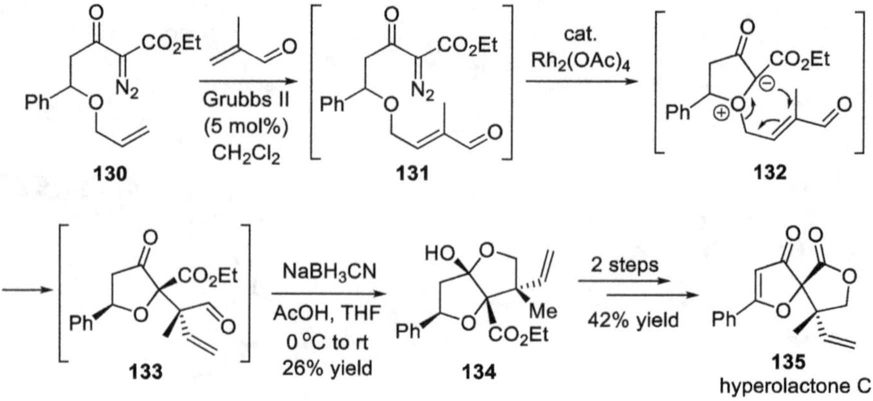

Scheme 15.35 Synthesis of hyperolactone C *via* cross-metathesis/oxonium ylide [2,3]-sigmatropic rearrangement cascade.

α-diazo-β-ketoesters **130**, bearing an allylic tether, with sequential Rh(II)-catalyzed oxonium ylide **132** generation and [2,3]-sigmatropic rearrangement to generate aldehyde **133**, which was further subjected to *in situ* reduction to give **134** (Scheme 15.35). Furthermore, they achieved the asymmetric total synthesis of anti-HIV spirolactone(-)-hyperolactone C, starting from readily available (*S*)-styrene oxide in 10 steps.[42]

Clark and co-workers utilized this conventional catalytic oxonium formation/[2,3]-sigmatropic rearrangement strategy in the synthesis of various natural products.[43–47] In 2008, they reported a stereoselective methodology toward the construction of fully functionalized tricyclic core of the plant-derived sesquiterpene natural product neoliacinic acid **141**.[43] The highly efficient assembly for the pivotal bicyclic ether-bridged scaffold was realized with the Rh(II)-catalyzed intramolecular diastereoselective C–H insertion of α-diazoketone **136** to form tetrahydrofuran ring of **137**, followed by Cu(II)-catalyzed oxonium ylide formation/[2,3]-sigmatropic rearrangement to afford desired oxabicyclic intermediate **139** (Scheme 15.36). Subsequently, **139** was converted to diol intermediate **140**, which possessed all the three rings and all six of the stereogenic centers for the synthesis of neoliacinic acid **141**.

In 2010, the same group reported the asymmetric total synthesis of (−)-cladiella-6,11-dien-3-ol **144** and the other two cladiellin natural products, including (−)-cladiell-11-ene-3,6,7-triol and (−)-3-acetoxycladiella-6,11-diene.[44] The entire family of this type of marine diterpenes could be

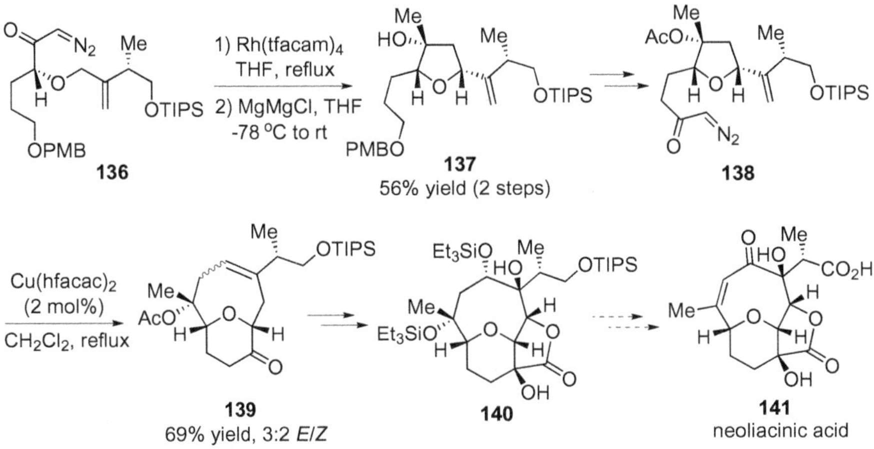

Scheme 15.36 Synthesis of tricyclic core of neoliacinic acid.

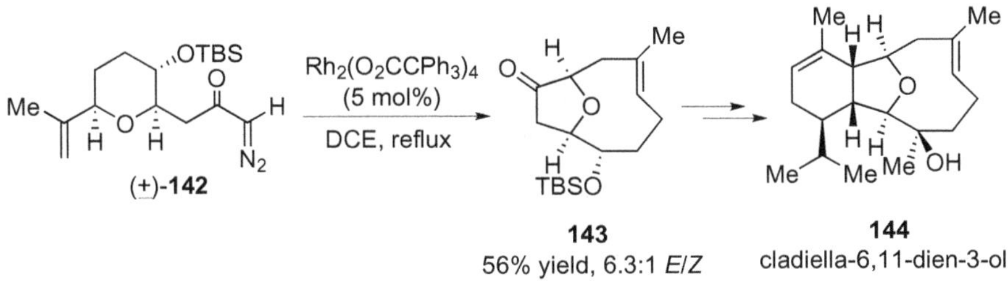

Scheme 15.37 Synthesis of cladiella-6,11-dien-3-ol.

synthesized from a common intermediate (*E*)-bridged bicyclic ether **143**, which was prepared by a Rh(II)-catalyzed stereoselective [2,3]-sigmatropic rearrangement (Scheme 15.37). After optimizing appropriate catalysts and solvents, this catalyst-dependent reaction was tuned to afford predominantly desired (*E*)-cyclic ether **143** with sterically hindered Rh(II) catalyst.

As an example, the first total synthesis of purported structure of sclerophytin F **148**, belonging to natural product members, was completed in 24 steps.[45] The synthesis relied on the Cu(II)-catalyzed ylide-type rearrangement to generate oxabicyclo[6.2.1]-5-undecen-9-one scaffold in high yield but with low diastereoselectivity (Scheme 15.38). Fortunately, the undesired *E*-bridged bicyclic ethers **146** could be transferred into the corresponding *Z*-ketone **147** in the presence of VAZO and a

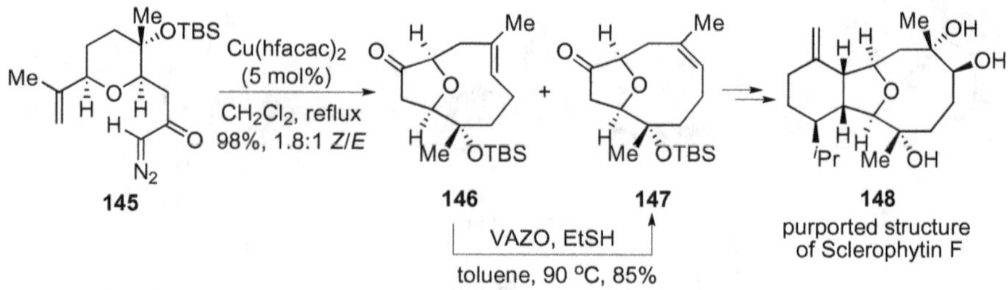

Scheme 15.38 Synthesis of purported structure of sclerophytin F.

sub-stoichiometric amount of ethanethiol, which allowed the synthesis of *Z*-bicyclic ketone with 83% overall yield in two steps. Recently, all the three diastereomeric (at C-6 and C-7) triols of the proposed structure of sclerophytin F were synthesized.[46] Re-analysis with NMR spectroscopy of these three diastereoisomers was implemented to allow the assignment of the structures of the natural products.

Furthermore, they focused their effort on the total synthesis of amphidinolides, which were macrolides isolated from marine natural products produced by dinoflagellates of the genus *amphidinium*. In 2013, they completed the total synthesis of amphidinolides T1 (**151**), T3 (**152**) and T4 (**153**).[47] Thus, from a readily available chiral alcohol, the overall yields of **151**, **152** and **153** were 6.9%, 5.9% and 5.5%, respectively. The crucial tetrahydrofuran subunit of these products was assembled by the oxonium ylide generated from copper carbene derived from α-diazoketone **149** with allylic moiety, followed by sequential [2,3]-sigmatropic rearrangement to generate key intermediate **150** with high yield (Scheme 15.39).

In 2013, Fujiwara reported the Rh(II)-catalyzed oxonium ylide formation/[2,3] rearrangement of substituted α-diazo-β-keto ester derived from D-glucose, allowing the concise preparation of tetrahydrofuran-3-one **155** as a single diastereomer with high yield and excellent stereoselectivity (Scheme 15.40).[48] This strategy was applied to the synthesis of 2-*epi*-cinatrin C_1 dimethyl ester **156** as the key step, and it also demonstrated potential utility in the synthesis of cinatrin derivatives.

15.4.3 [3+2] *Cycloaddition of the ylides*

[3+2] Cycloaddition of the ylides generated from carbene and ketone/aldehyde or imine is a powerful method for the construction of

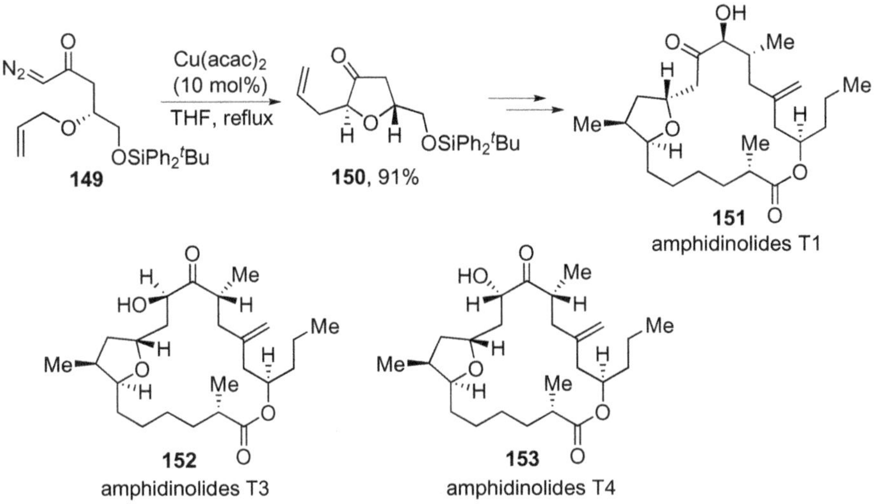

Scheme 15.39 Synthesis of amphidinolides T1, T3 and T4.

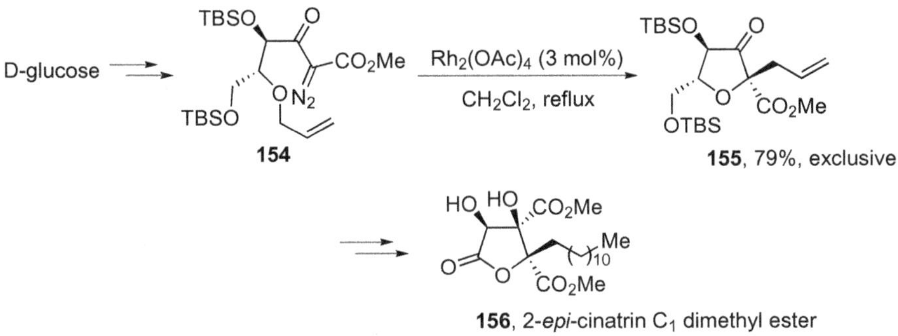

Scheme 15.40 Synthesis of 2-*epi*-cinatrin C_1 dimethyl ester.

heterocyclic structure. Padwa and co-workers have extensively explored this strategy and they completed the total synthesis of a series of natural products by using Rh(II)-catalyzed cyclization/cycloaddition sequence involving carbonyl ylide species.[49–53] In 2008, they reported a concise construction of the icetexane core through Rh(II)-catalyzed intramolecular cycloaddition of carbonyl ylide with a tethered double bond as the crucial step (Scheme 15.41).[49] The functionalized 6-7-6 fused tricyclic framework was established with good yields.

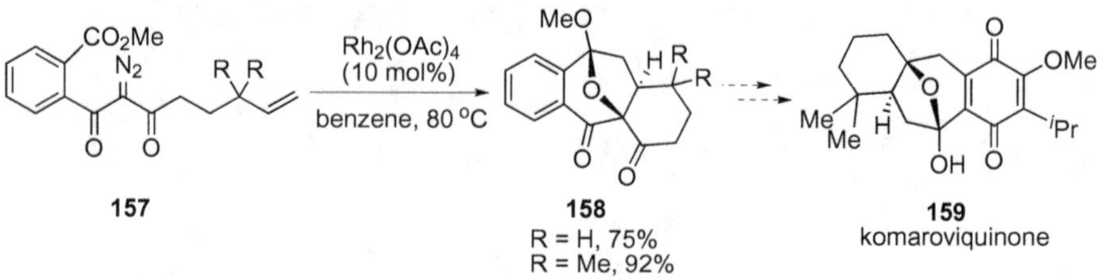

Scheme 15.41 Construction of the icetexane core of komaroviquinone.

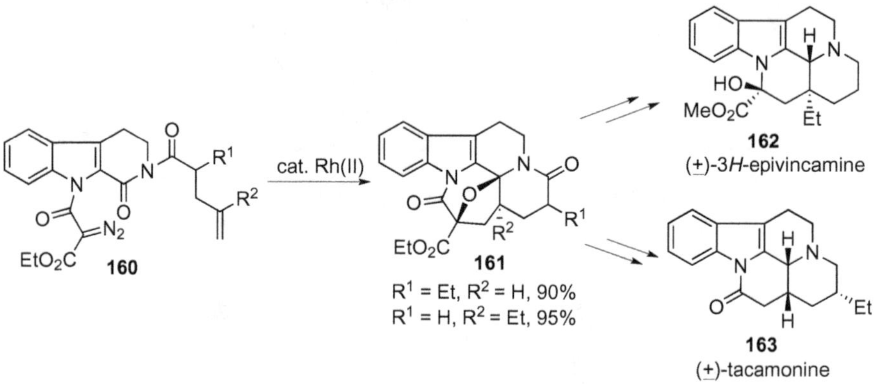

Scheme 15.42 Rh(II)-catalyzed cyclization/cycloaddition to access *vinca* and *tacaman* alkaloids.

Padwa and co-workers also completed the total synthesis of *vinca* and *tacaman* class of indole alkaloids, including (±)-3*H*-epivincamine **162**, (±)-tacamonine **163** and (±)-apotacamine.[50] The key step in these racemic synthetic routes was the Rh(II)-catalyzed carbonyl ylide formation/intramolecular cycloaddition cascade of indole-derived α-diazo-β-keto esters **160**, delivering the desired pentacyclic core of the natural product with excellent yields (Scheme 15.42).

The same group utilized the Rh(II)-catalyzed carbonyl ylide intramolecular 1,3-dipolar cycloaddition cascade of indolyl-substituted α-diazo imide in the total synthesis of pentacyclic alkaloid (±)-aspidophytine **166** (Scheme 15.43).[51,52] This reaction exemplified the utility of indolyl double bond as the two-carbon unit in the [3+2] cycloaddition in the stereocontrolled installation of the functionalized natural products scaffold.

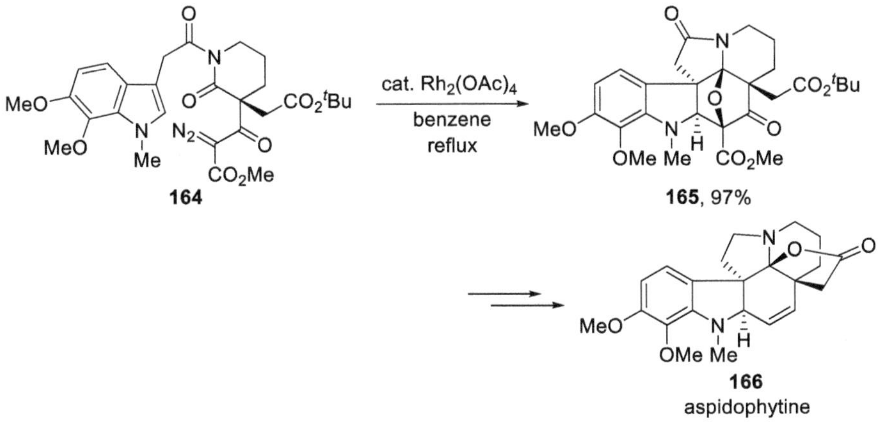

Scheme 15.43 Synthesis of (±)-aspidophytine.

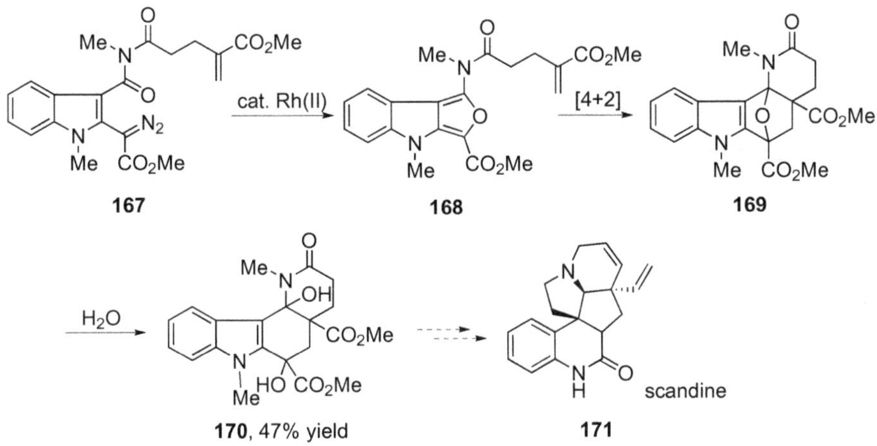

Scheme 15.44 Intramolecular cyclization/[4+2] cycloaddition of furo[3,4-*b*]indoles for the synthesis of scandine.

Furthermore, Padwa and co-workers described an intramolecular Rh(II)-catalyzed cyclization of α-diazo substrate **167** to form a reactive benzo[*c*]furan intermediate **168**, followed by tandem [4+2] intramolecular cycloaddition across the tethered alkene moiety, assembling a heavily functionalized azapolycyclic ring scaffold **169** (Scheme 15.44).[53]

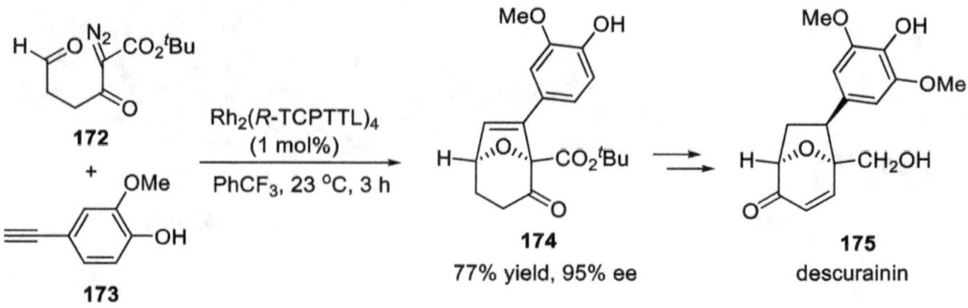

Scheme 15.45 Asymmetric synthesis of descurainin *via* carbonyl ylide cycloaddition.

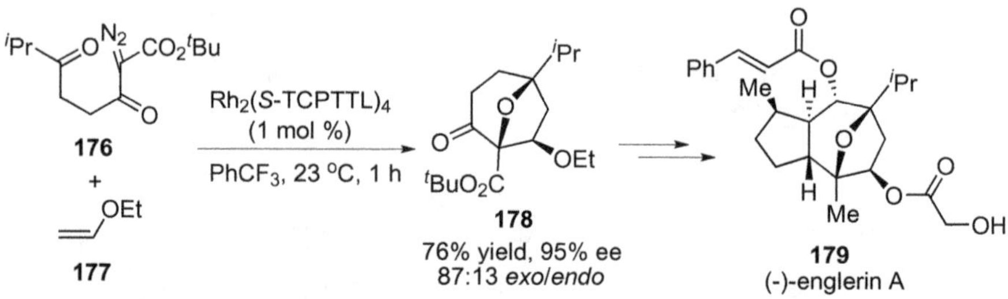

Scheme 15.46 Asymmetric total synthesis of (−)-englerin A.

Hashimoto and co-workers have explored catalytic asymmetric carbonyl ylide cycloadditions in synthesis of natural products.[54,55] For example, they completed the asymmetric synthesis of descurainin **175** by the incorporation of asymmetric 1,3-dipolar cycloaddition, stereoselective alkene hydrogenation and oxidation with Fremy's salt and a regioselective demethylation with $NbCl_5$.[54] In this synthesis, catalytic enantioselective intermolecular cycloaddition of carbonyl ylide derived from *tert*-butyl 2-diazo-5-formyl-3-oxopetanoate **172** with 4-hydroxy-3-methoxyphenylacetylene **173** was implemented by using chiral Rh(II) catalyst, giving rise to 8-oxabicyclo[3.2.1]octane skeleton **174** with high yield and excellent stereocontrol (Scheme 15.45).

In another example, Hashimoto and co-workers completed the asymmetric total synthesis of the guaiane sesquiterpene (−)-englerin A **179** (25 steps, 5.2% overall yield), which is a potent and selective inhibitor of the growth of renal cancer cell lines.[55] The key feature of this approach was the chiral Rh(II)-catalyzed high efficient diastereo- and enantioselective carbonyl ylide cycloadditions by using ethyl vinyl ether **177** as the dipolarophile (Scheme 15.46). The substituted oxabicyclo[3.2.1]octane

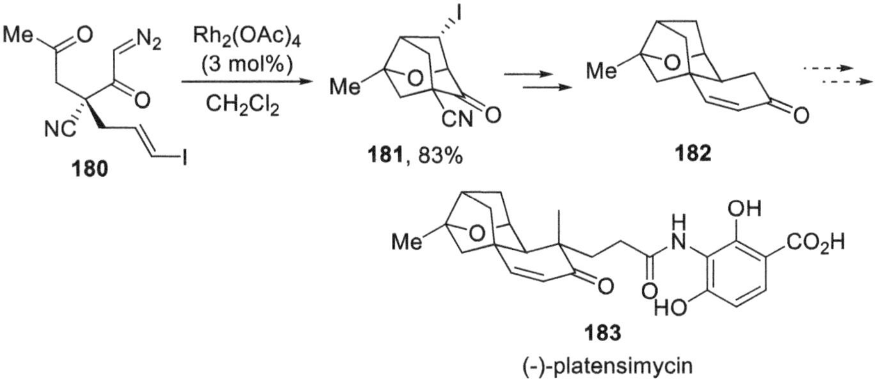

Scheme 15.47 Skeleton construction of platensimycin *via* carbonyl ylide cycloaddition.

skeleton of **178**, with concomitant introduction of the ethoxy substituent on the *exo*-face, was constructed with good yield as well as high level of diastereo- and enantioselective control.

A more complicated ring system could be constructed in a straightforward manner by installing the dipolarophile moiety intramolecularly. With such an approach, Lee and co-workers completed a concise formal synthesis of (−)-platensimycin derivatives.[56] From readily available substrate of isopropyl cyanoacetate, this system used an Rh(II)-catalyzed intramolecular carbonyl cycloaddition of α-diazoketone with an alkene tether **180** for the construction of tetracyclic core of **183** (Scheme 15.47).

Compared to the carbonyl ylides, the corresponding azomethine ylides are less explored in the synthesis of natural products. In 2012, Martin and co-workers reported the construction of the aza-polycyclic scaffold from commercially available substrate of 2-deoxy-D-ribose for the enantioselective total synthesis of didehydrostemofoline and isodidehydrostemofoline **186**.[57] In this synthesis, the diastereo- and regioselective Rh(II)-catalyzed dipolar cycloaddition of azomethine ylide derived from diazo imine played a curial role for the selective assembly of the key tricyclic intermediate **185** (Scheme 15.48).

15.4.4 *Other ylide-related transformations*

Some miscellaneous ylide-related transformations have been applied in the synthesis of natural products. For example, Xie and co-workers reported the total synthesis of bioactive natural products hyperolactone C

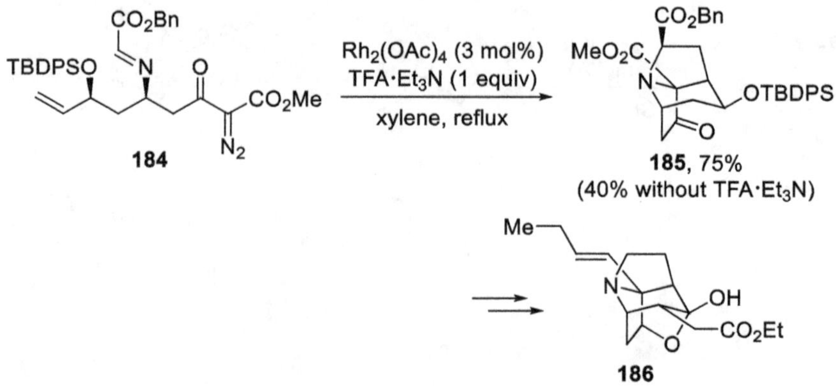

Scheme 15.48 Construction of azapolycyclic core toward the formal total synthesis of (iso)didehydrostemofoline.

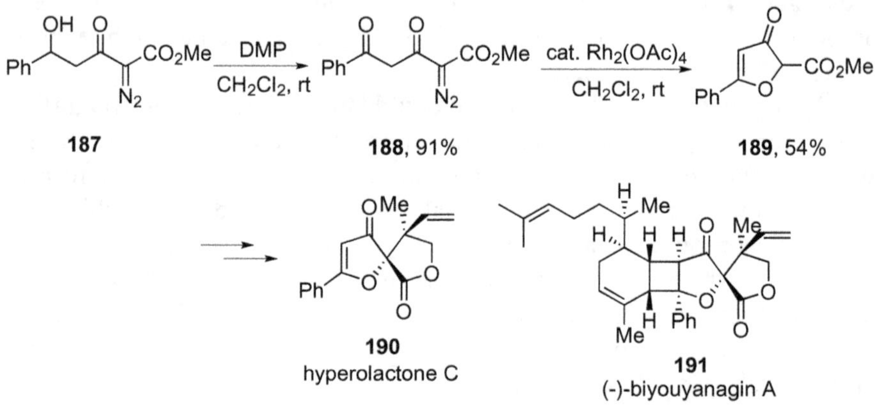

Scheme 15.49 Total synthesis of hyperolactone C and (−)-biyouyanagin A.

190 and (−)-biyouyanagin A **191**, which contain two vicinal quaternary carbon centers.[58] In their synthesis, the diazo substrate of α-diazo-β-keto-δ-hydroxyester **187**, prepared by the diazo transfer, was oxidized to give α-diazo-β,δ-ketoester **188**. Then, Rh(II)-catalyzed intramolecular carbonyl ylide generation with sequential hydrogen migration process afforded the cyclic β-ketoester **189** as the key intermediate (Scheme 15.49).

In another example, Bode and co-workers established the synthetic route for the bisallyl bullvalone scaffold, which served as the precursor

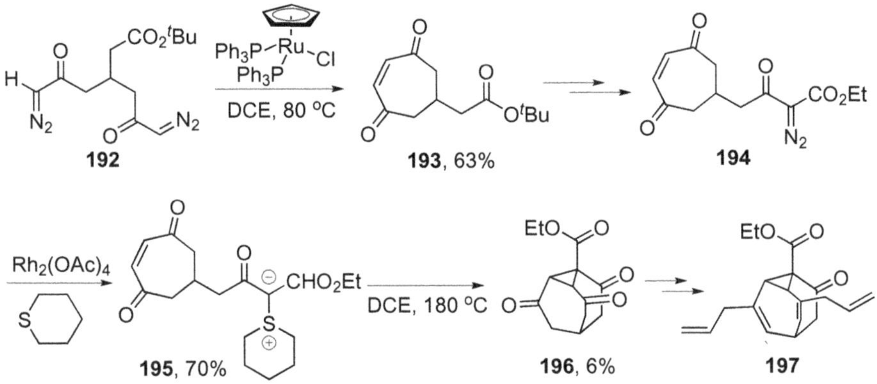

Scheme 15.50 Attempted synthesis of bisallyl bullvalene.

for the synthesis of functionalized bullvalenes, including bisporphyrin bullvalene and bisallyl bullvalene.[59] The first step in this synthesis was the Ru-catalyzed intramolecular dimerization of bisdiazoketone **192**, affording the desired enedione intermediate **193** with moderate yield. **193** was then converted into diazo substrate **194**. Subsequently, sequential formal intramolecular cyclopropanation was implemented with stoichiometric amounts of pentamethylene sulfide and catalytic $Rh_2(OAc)_4$, resulting in the generation of triketone **196**, albeit in very low yields (Scheme 15.50).

15.5 Wolff rearrangement

Wolff rearrangement, as a classic carbene reaction, has long been utilized in the synthesis of natural products. In 2010, the Wang group established the enantioselective synthesis of condensed and transannular ring structures containing pyrrolidine moiety, which relied on the highly diastereoselective addition of lithium enolate of α-diazoacetoacetate to chiral *N*-sulfinylimines, followed by photo-induced Wolff rearrangement and catalytic ring-closing metathesis (Scheme 15.51).[60] Heterocyclic skeletons as well as natural product (*R*)-pyrolam A **203** were synthesized following this strategy.

Burtoloso and co-workers reported a concise total synthesis of (−)-indolizidine 167B as well as two formal synthesis of (−)-indolizidine 167B.[61] The key steps in this synthesis were Horner–Wadsworth–Emmons

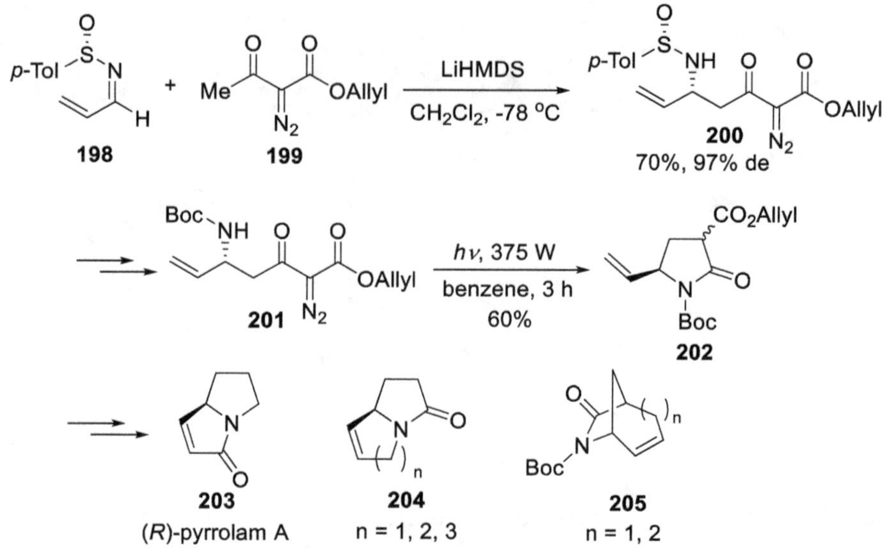

Scheme 15.51 Synthesis of condensed and transannular ring skeletons with pyrrolidine moiety.

olefination of a chiral *N*-Cbz-prolinal **207** with 3-diazo-2-oxopropylphosphonate **206**, followed by photochemical Wolff rearrangement of *α,β*-unsaturated diazoketone **208** at room temperature, giving rise to the Arndt–Eistert homologated product **210** with high yields (Scheme 15.52). Furthermore, the key intermediate **210** has been utilized for the synthesis of a series of indolizidines derivatives, including 1-deoxy-8,8*a*-di-epi-castanospermine **212**, 1,6-dideoxy-epi-castanospermine **213** and pumiliotoxin 251D **214**.[62]

Arndt–Eistert homologation process was applied for the construction of core structures of some complicated natural products.[63,64] For example, a racemic synthetic approach toward the core structure of polycyclic alkaloid palhinine A **218** was reported.[64] The key steps involved a domino Michael reaction to build bicyclo[2.2.2]octane motif, followed by Arndt–Eistert homologation and intramolecular aldol reaction to construct partly substituted isotwistane core structure (Scheme 15.53).

One of the unique applications of Wolff rearrangement is the ring contraction. A recent example is shown by a synthetic approach toward the cyclopent[g]isoquinoline building block, which constitutes the D/E/F ring system of fredericamycin A **221**.[65] In this synthetic strategy, the

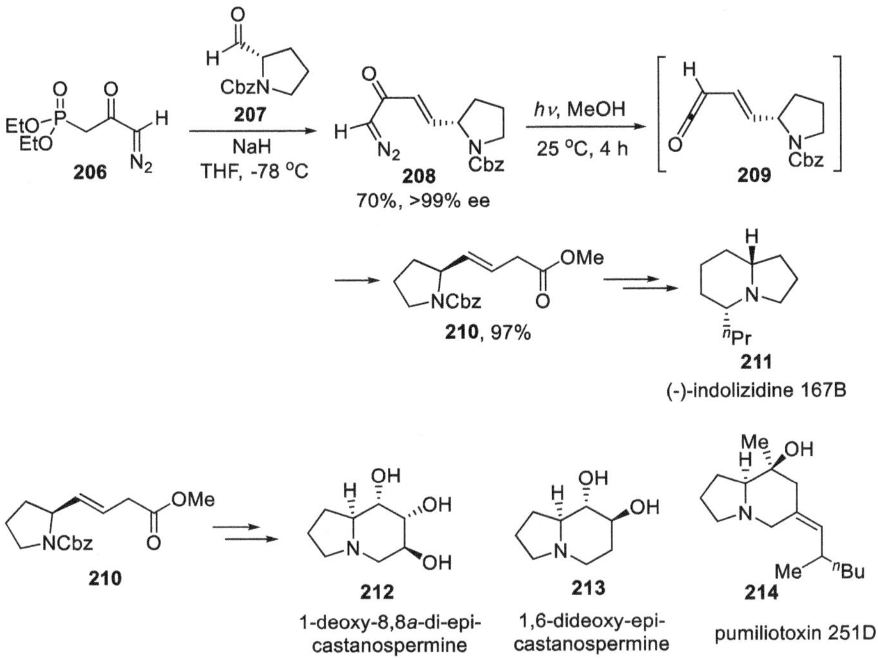

Scheme 15.52 Total synthesis of (−)-indolizidine 167B and related indolizidine derivatives *via* Wolff rearrangement of diazoketone.

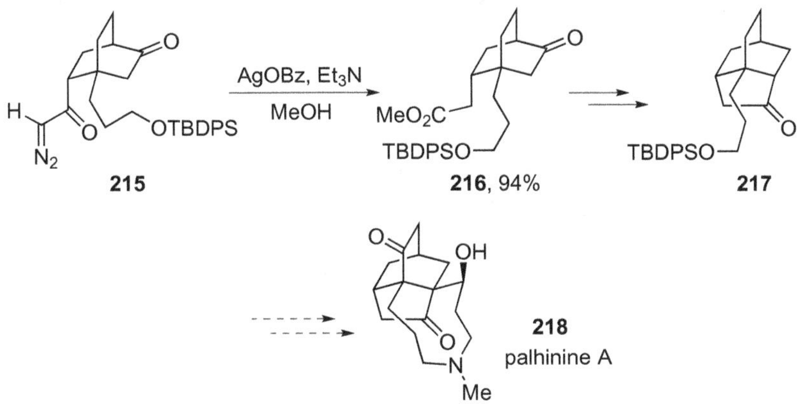

Scheme 15.53 Construction of core structure of polycyclic alkaloid palhinine A.

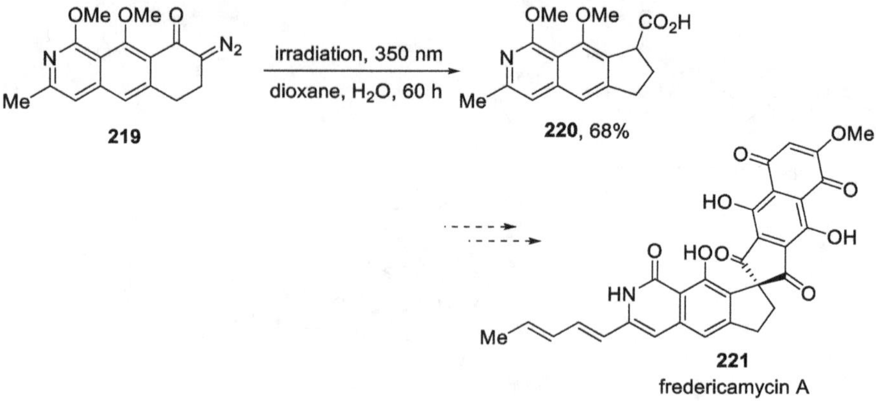

Scheme 15.54 Construction of D/E/F ring system of fredericamycin A.

photo-induced Wolff rearrangement of cyclic diazoketone **219** afforded the desired cyclopent[*g*]isoquinoline carboxylic acid **220** in 27% yield (Scheme 15.54).

15.6 The diazo moiety as 1,3-dipole in [3+2] cycloaddition

In addition to carbene-based transformations, the diazo group can directly react as 1,3-dipole in [3+2] cycloaddition. For example, the synthesis of various Δ^2-pyrazolines containing α,α-disubstituted α-amino carbon centers was achieved through 1,3-dipole [3+2] cycloaddition of lithium(trimethylsilyl)diazomethane with α,β-unsaturated esters **222** (Scheme 15.55).[66] Mechanistically, the generation of desired bicyclic pyrazolines scaffold **224** involved cycloaddition to give intermediate **223** with sequential intramolecular nucleophilic substitution cascade. The synthetic value of this versatile cycloaddition-based strategy was demonstrated by the total synthesis of amathaspiramide C **225** and the formal synthesis of amathaspiramide class of natural products.

Lee and co-workers established a novel synthetic methodology toward the synthesis of natural products through 1,3-dipole cycloaddition of diazo compounds with allenes, followed by tandem [3+2] cycloaddition *via* a trimethylenemethane (TMM) diyl intermediate.[67–70] In 2012, they accomplished the total synthesis of ceratopicanol **230** in 11 steps through tandem 1,3-dipole cycloaddition of aziridinylimine (as diazo

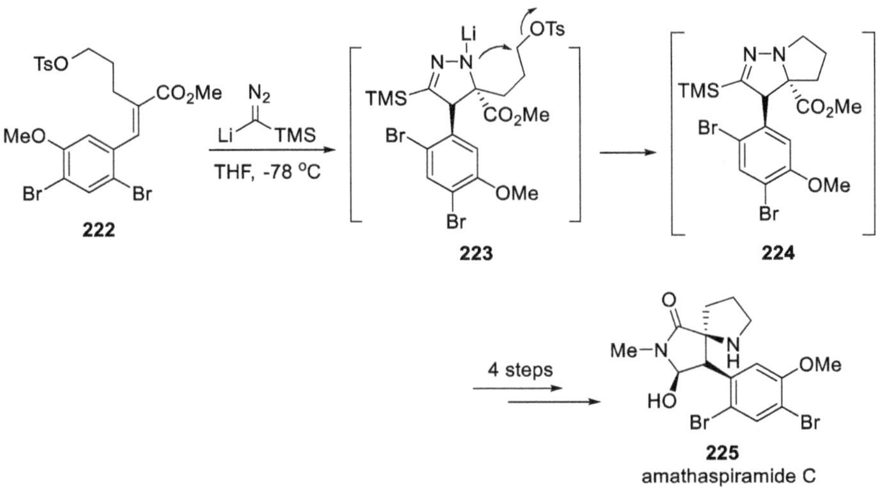

Scheme 15.55 Construction of bicyclic pyrazolines and total synthesis of amathaspiramide C.

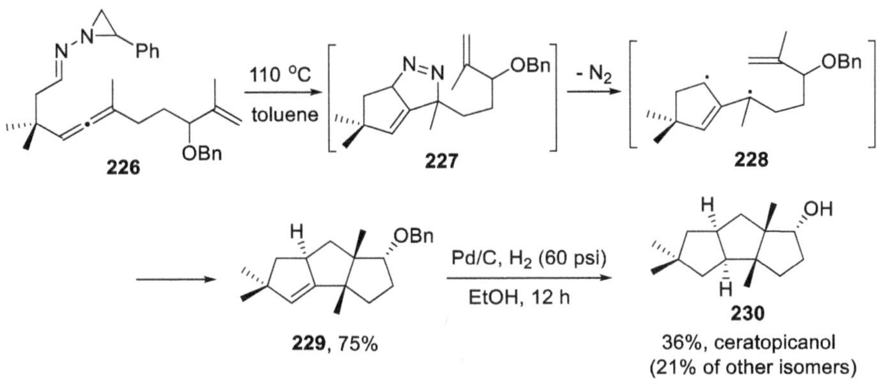

Scheme 15.56 Total synthesis of ceratopicanol *via* tandem dipole cycloaddition with [3+2] cycloaddition of TMM diyl.

precursor) with allenes, followed by TMM diyl **228** mediated intramolecular [2+3] cycloaddition sequence (Scheme 15.56).[67] From a linear substrate, this strategy constructed the desired polycyclic scaffold containing consecutive quaternary centers in one step, as well as with high level of diastereoselectivity. Thus, the method provided a concise route to access the natural products containing fused triquinanes moiety.

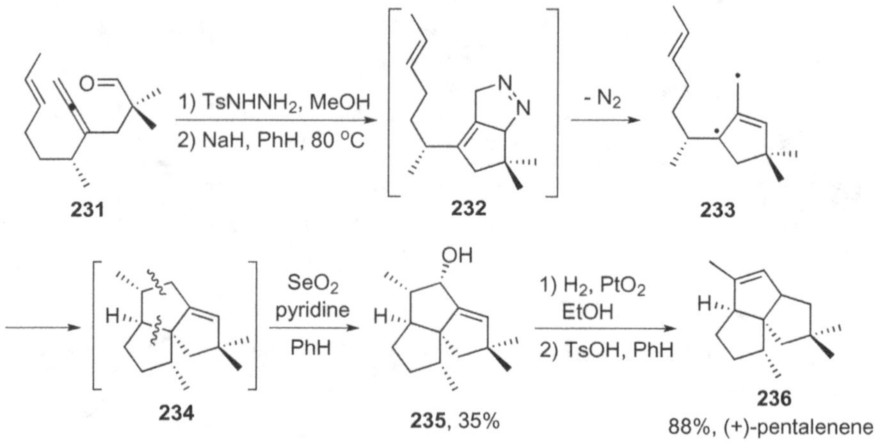

Scheme 15.57 Total synthesis of (+)-pentalenene through tandem cycloaddition.

In addition, Lee and co-workers utilized this tandem cycloaddition strategy to achieve the stereoselective total synthesis of (+)-pentalenene **236** with 6.8% overall yield in 14 steps.[68] In the synthesis, the corresponding allenyl diazo substrate was prepared *in situ* from *N*-tosylhydrazone of the aldehyde **231**, which was readily available as the starting material of (+)-citronellal. Thus, with this strategy, the angularly fused triquinane core structure could be established from the acyclic substrate with high efficiency (Scheme 15.57).

In 2014, the same group successfully disclosed the first total synthesis of (−)-crinipellin A **241** from 2-methyl-2-cyclopenten-1-one.[69] By using *N*-tosylhydrazone **237** as the diazo precursor, the TMM diyl **239** mediated tandem [3+2] cycloaddition allowed the efficient assembly of unique tetraquinane core **240** with high yields (Scheme 15.58). The absolute stereochemistry of crinipellin A was confirmed through this synthetic approach.

With a similar approach, the total synthesis of the proposed structure of "panaginsene" was completed.[70] However, the spectroscopic data of the synthesized "panaginsene" **246** did not match that of the natural product. Furthermore, the success of stereoselective total synthesis of another stereoisomer indicated that the actual structure of natural product panaginsene was the 11-*epi* stereoisomer of originally assigned "panaginsene" (Scheme 15.59).

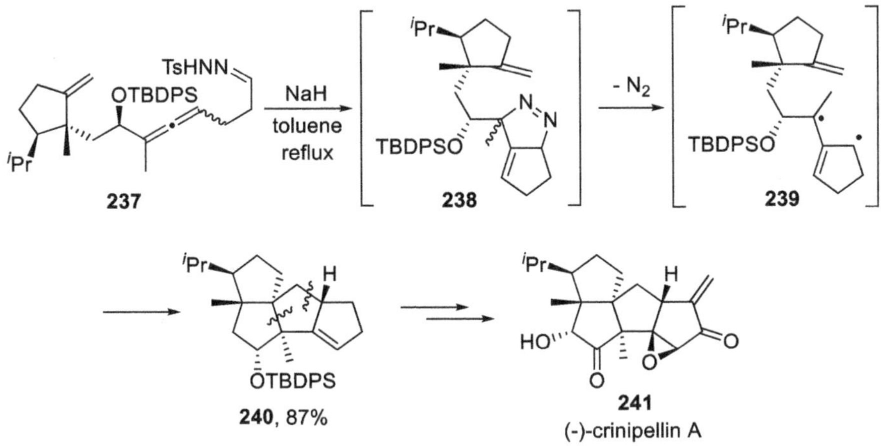

Scheme 15.58 Total synthesis of (−)-crinipellin A *via* tandem cycloaddition.

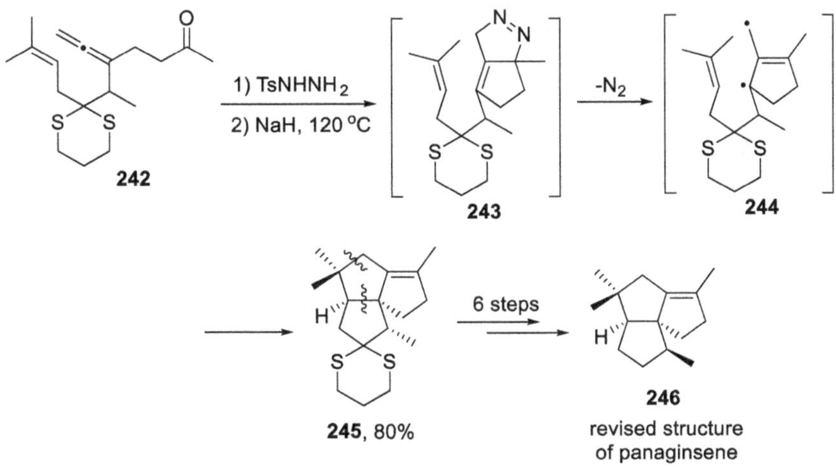

Scheme 15.59 Total synthesis of panaginsene with structural revision.

15.7 Vinyl diazoacetate transformations

The transformations involving vinyl diazoacetate derivatives and the corresponding vinyl metal carbene species have found applications in the construction of natural products through a cyclopropanation/Cope

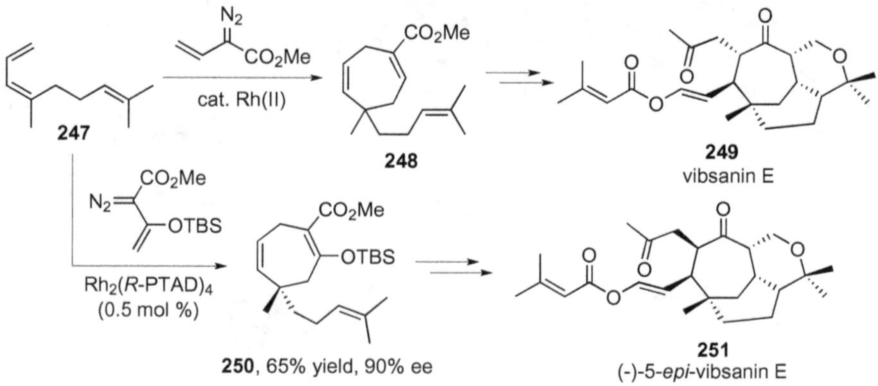

Scheme 15.60 Total synthesis of vibsanin E and (−)-5-*epi*-vibsanin E.

rearrangement cascade, especially for the construction of the functionalized seven-membered ring and heterocyclic scaffold. In 2009, Davies and Williams reported the total synthesis of diterpene racemic vibsanin E **249**, which consisted of a compact 3-oxatricyclo[6.3.2.$0^{5,10}$]tridecane core.[71] The key step in this 14-step synthesis was the Rh(II)-catalyzed formal [4+3] cycloaddition strategy to generate the desired all-carbon seven-membered ring scaffold **248**.

Simultaneously, they reported the asymmetric total synthesis of (−)-5-*epi*-Vibsanin E **251**, which involved chiral rhodium-catalyzed [4+3] cycloaddition between a vinylcarbenoid and a diene (Scheme 15.60).[72] Based on this cyclopropanation/Cope rearrangement cascade, the quaternary stereogenic center of **250** was constructed with good yields and high enantioselective control.

With a similar approach, the total synthesis of norhalichondrin B **257**, which served as a member of halichondrin family of marine polyethers, was achieved in 37 steps from β-furylethanol.[73] In this synthesis, one of the key steps involves the Rh(II)-catalyzed cyclopropanation/Cope rearrangement of vinyl diazoacetate **253** with furan to afford oxabicyclo[3.2.1]octane scaffold **254** (Scheme 15.61). Another key step included the Cu(II)-catalyzed oxonium ylide formation/[2,3] rearrangement to construct fragment **256**. Moreover, the reaction of dimethyl(diazomethyl)phosphonate was also used in this synthesis as a late-stage transformation. Thus, the utility of diazo compounds in synthesis was exemplified by this success for the total synthesis of polyether natural products.

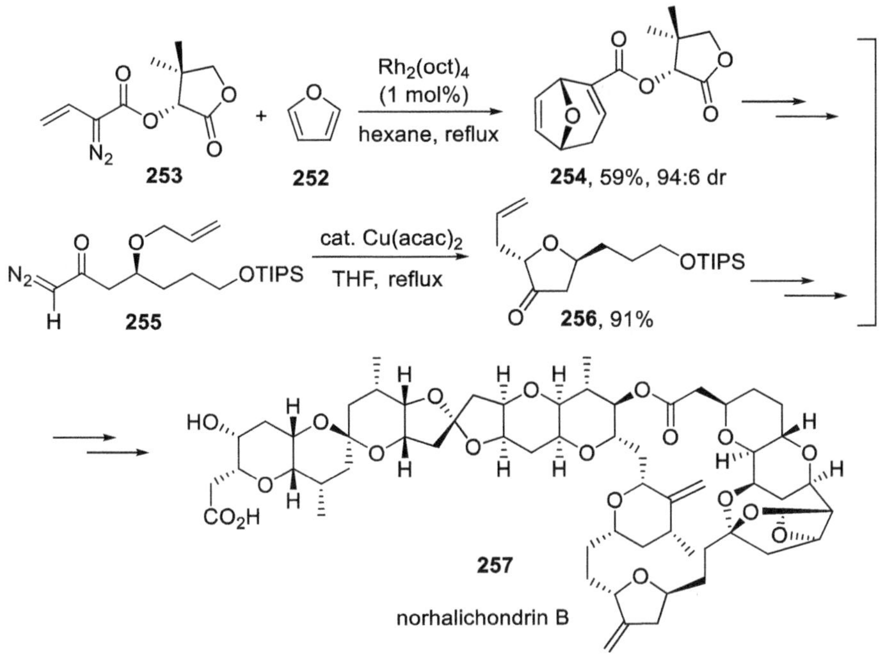

Scheme 15.61 Total synthesis of norhalichondrin B: application of Rh(II)-catalyzed cycloaddition and Cu(II)-catalyzed ylide rearrangement.

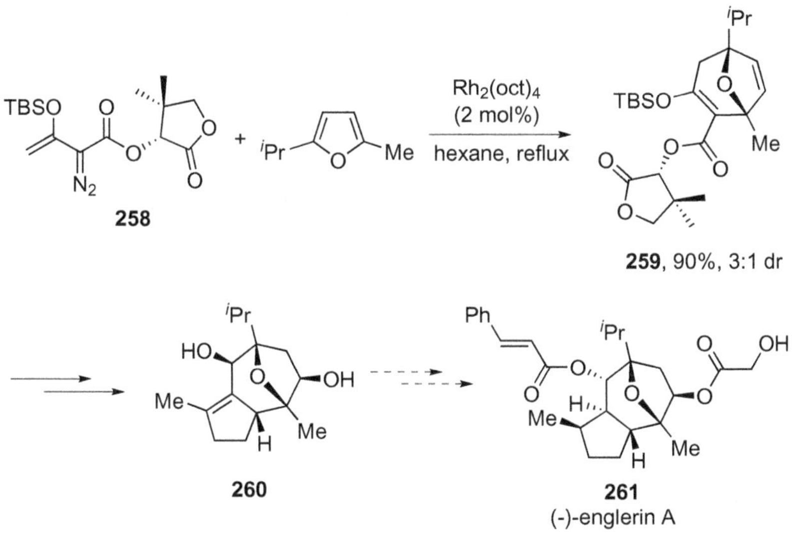

Scheme 15.62 Enantioselective formal synthesis of (−)-englerin A.

With chiral vinyl diazoester, the cyclopropanation/Cope rearrangement was also successfully applied in the enantioselective formal synthesis of (−)-englerin A **261**.[74] Notably, the initial Rh(II)-catalyzed formal [4+3] cycloaddition followed by intramolecular aldol condensation sequence provided a facile route to access tricyclic skeleton **260** of englerin (Scheme 15.62).

In conclusion, from the selected examples shown in this chapter, it is apparent that the reactions with diazo compounds are powerful tools in the synthesis of natural products. In particular, the reactions through catalytic metal carbene, such as C–H insertions, cyclopropanations and ylide reactions, have been extensively explored in synthesis of natural products. The corresponding asymmetric catalysis of these transformations, which have been well developed in the past decades, further enhances their usefulness in this area.

References

1. Pouwer, R. H.; Richard, J.-A.; Tseng, C.-C.; Chen, D. Y.-K. *Chem. Asian J.* **2012**, *7*, 22–35.
2. Lu, Y.; Yao, H.; Sun, B. *Chin. J. Org. Chem.* **2012**, *32*, 1–12.
3. Gao, Z.; Li, Y.; Cooksey, J. P.; Snaddon, T. N.; Schunk, S.; Viseux, E. M. E.; McAteer, S. M.; Kocienski, P. J. *Angew. Chem. Int. Ed.* **2009**, *48*, 5022–5025.
4. Xu, H.; Zhang, W.; Shu, D.; Werness, J. B.; Tang, W. *Angew. Chem. Int. Ed.* **2008**, *47*, 8933–8936.
5. Liu, R.; Zhang, M.; Wyche, T. P.; Winston-McPherson, G. N.; Bugni, T. S.; Tang, W. *Angew. Chem. Int. Ed.* **2012**, *51*, 7503–7506.
6. Linder, J.; Blake, A. J.; Moody, C. J. *Org. Biomol. Chem.* **2008**, *6*, 3908–3916.
7. Dakin, L. A.; Schaus, S. E.; Jacobsen, E. N.; Panek, J. S. *Tetrahedron Lett.* **1998**, *39*, 8947–8950.
8. Wu, J.; Panek, J. S. *J. Org. Chem.* **2011**, *76*, 9900–9918.
9. Gong, J.; Lin, G.; Sun, W.; Li, C.-C.; Yang, Z. *J. Am. Chem. Soc.* **2010**, *132*, 16745–16746.
10. Zhang, W.-B.; Lin, G.; Shao, W.-B.; Gong, J.-X.; Yang, Z. *Chem. Asian J.* **2015**, *10*, 903–909.
11. Pan, S.; Xuan, J.; Gao, B.; Zhu, A.; Ding, H. *Angew. Chem. Int. Ed.* **2015**, *54*, 6905–6908.
12. Wada, H.; Williams, H. E. L.; Moody, C. J. *Angew. Chem. Int. Ed.* **2015**, *54*, 15147–15151.
13. Taber, D. F.; Tian, W. *J. Org. Chem.* **2008**, *73*, 7560–7564.

14. Taber, D. F.; Nelson, C. G. *J. Org. Chem.* **2011**, *76*, 1874–1882.
15. Natori, Y.; Tsutsui, H.; Sato, N.; Nakamura, S.; Nambu, H.; Shiro, M.; Hashimoto, S. *J. Org. Chem.* **2009**, *74*, 4418–4421.
16. Koizumi, Y.; Kobayashi, H.; Wakimoto, T.; Furuta, T.; Fukuyama, T.; Kan, T. *J. Am. Chem. Soc.* **2008**, *130*, 16854–16855.
17. Wang, D.-H.; Yu, J.-Q. *J. Am. Chem. Soc.* **2011**, *133*, 5767–5769.
18. Lu, P.; Mailyan, A.; Gu, Z.; Guptill, D. M.; Wang, H.; Davies, H. M. L.; Zakarian, A. *J. Am. Chem. Soc.* **2014**, *136*, 17738–17749.
19. Wee, A. G. H.; Fan, G.-J.; Bayirinoba, H. M. *J. Org. Chem.* **2009**, *74*, 8261–8271.
20. Annadi, K.; Wee, A. G. H. *J. Org. Chem.* **2015**, *80*, 5236–5251.
21. Hirooka, Y.; Ikeuchi, K.; Kawamoto, Y.; Akao, Y.; Furuta, T.; Asakawa, T.; Inai, M.; Wakimoto, T.; Fukuyama, T.; Kan, T. *Org. Lett.* **2014**, *16*, 1646–1649.
22. Egger, J.; Bretscher, P.; Freigang, S.; Kopf, M.; Carreira, E. M. *Angew. Chem. Int. Ed.* **2013**, *52*, 5382–5385.
23. Villalobos, M. N.; Wood, J. L. *Tetrahedron Lett.* **2009**, *50*, 6450–6453.
24. Zhuo, X.; Xiang, K.; Zhang, F.-M.; Tu, Y.-Q. *J. Org. Chem.* **2011**, *76*, 6918–6924.
25. Higashi, T.; Isobe, Y.; Ouchi, H.; Suzuki, H.; Okazaki, Y.; Asakawa, T.; Furuta, T.; Wakimoto, T.; Kan, T. *Org. Lett.* **2011**, *13*, 1089–1091.
26. Mahoney, S. J.; Fillion, E. *Chem. Eur. J.* **2012**, *18*, 68–71.
27. Taber, D. F.; Sheth, R. B.; Tian, W. *J. Org. Chem.* **2009**, *74*, 2433–2437.
28. Taber, D. F.; Paquette, C. M. *J. Org. Chem.* **2014**, *79*, 3410–3413.
29. Sawada, T.; Nakada, M. *Org. Lett.* **2013**, *15*, 1004–1007.
30. Shen, L.; Zhang, M.; Wu, Y.; Qin, Y. *Angew. Chem. Int. Ed.* **2008**, *47*, 3618–3621.
31. Zhang, M.; Huang, X.; Shen, L.; Qin, Y. *J. Am. Chem. Soc.* **2009**, *131*, 6013–6020.
32. He, B.; Song, H.; Du, Y.; Qin, Y. *J. Org. Chem.* **2009**, *74*, 298–304.
33. Huang, H.-X.; Jin, S.-J.; Gong, J.; Zhang, D.; Song, H.; Qin, Y. *Chem. Eur. J.* **2015**, *21*, 13284–13290.
34. Levin, S.; Nani, R. R.; Reisman, S. E. *Org. Lett.* **2010**, *12*, 780–783.
35. Levin, S.; Nani, R. R.; Reisman, S. E. *J. Am. Chem. Soc.* **2011**, *133*, 774–776.
36. McDowell, P. A.; Foley, D. A.; O'Leary, P.; Ford, A.; Maguire, A. R. *J. Org. Chem.* **2012**, *77*, 2035–2040.
37. Hou, S.-H.; Tu, Y.-Q.; Liu, L.; Zhang, F.-M.; Wang, S.-H.; Zhang, X.-M. *Angew. Chem. Int. Ed.* **2013**, *52*, 11373–11376.
38. Stewart, C.; McDonald, R.; West, F. G. *Org. Lett.* **2011**, *13*, 720–723.
39. Rosset, I. G.; Dias, R. M. P.; Pinho, V. D.; Burtoloso, A. C. B. *J. Org. Chem.* **2014**, *79*, 6748–6753.
40. Lu, P.; Herrmann, A. T.; Zakarian, A. *J. Org. Chem.* **2015**, *80*, 7581–7589.

41. Hodgson, D. M.; Angrish, D.; Erickson, S. P.; Kloesges, J.; Lee, C. H. *Org. Lett.* **2008**, *10*, 5553–5556.
42. Hodgson, D. M.; Man, S. *Chem. Eur. J.* **2011**, *17*, 9731–9737.
43. Clark, J. S.; Baxter, C. A.; Dossetter, A. G.; Poigny, S.; Castro, J. L.; Whittingham, W. G. *J. Org. Chem.* **2008**, *73*, 1040–1055.
44. Clark, J. S.; Berger, R.; Hayes, S. T.; Thomas, L. H.; Morrison, A. J.; Gobbi, L. *Angew. Chem. Int. Ed.* **2010**, *49*, 9867–9870.
45. Clark, J. S.; Delion, L.; Farrugia, L. J. *Org. Lett.* **2014**, *16*, 4300–4303.
46. Clark, J. S.; Delion, L.; Farrugia, L. J. *Chem. Eur. J.* **2015**, *21*, 4772–4780.
47. Clark, J. S.; Romiti, F. *Angew. Chem. Int. Ed.* **2013**, *52*, 10072–10075.
48. Yakura, T.; Ozono, A.; Matsui, K.; Yamashita, M.; Fujiwara, T. *Synlett* **2013**, *24*, 65–68.
49. Padwa, A.; Chughtai, M. J.; Boonsombat, J.; Rashatasakhon, P. *Tetrahedron* **2008**, *64*, 4758–4767.
50. England, D. B.; Padwa, A. *J. Org. Chem.* **2008**, *73*, 2792–2802.
51. Mejía-Oneto, J. M.; Padwa, A. *Helv. Chim. Acta* **2008**, *91*, 285–302.
52. Li, H.; Bonderoff, S. A.; Cheng, B.; Padwa, A. *J. Org. Chem.* **2014**, *79*, 392–400.
53. Padwa, A.; Zou, Y.; Cheng, B.; Li, H.; Downer-Riley, N.; Straub, C. S. *J. Org. Chem.* **2014**, *79*, 3173–3184.
54. Shimada, N.; Hanari, T.; Kurosaki, Y.; Anada, M.; Nambu, H.; Hashimoto, S. *Tetrahedron Lett.* **2010**, *51*, 6572–6575.
55. Hanari, T.; Shimada, N.; Kurosaki, Y.; Thrimurtulu, N.; Nambu, H.; Anada, M.; Hashimoto, S. *Chem. Eur. J.* **2015**, *21*, 11671–11676.
56. Kim, C. H.; Jang, K. P.; Choi, S. Y.; Chung, Y. K.; Lee, E. *Angew. Chem. Int. Ed.* **2008**, *47*, 4009–4011.
57. Fang, C.; Shanahan, C. S.; Paull, D. H.; Martin, S. F. *Angew. Chem. Int. Ed.* **2012**, *51*, 10596–10599.
58. Du, C.; Li, L.; Li, Y.; Xie, Z. *Angew. Chem. Int. Ed.* **2009**, *48*, 7853–7856.
59. Lippert, A. R.; Naganawa, A.; Keleshian, V. L.; Bode, J. W. *J. Am. Chem. Soc.* **2010**, *132*, 15790–15799.
60. Mo, F.; Li, F.; Qiu, D.; Wang, J. *Tetrahedron* **2010**, *66*, 1274–1279.
61. Pinho, V. D.; Burtoloso, A. C. B. *Tetrahedron Lett.* **2012**, *53*, 876–878.
62. Bernardim, B.; Pinho, V. D.; Burtoloso, A. C. B. *J. Org. Chem.* **2012**, *77*, 9926–9931.
63. Tam, N. T.; Jung, E.-J.; Cho, C.-G. *Org. Lett.* **2010**, *12*, 2012–2014.
64. Gaugele, D.; Maier, M. E. *Synlett* **2013**, *24*, 955–958.
65. Braun, M.; Kergoët, G.; Kruska, F.; Frank, W. *Synthesis* **2010**, 2023–2026.
66. O'Connor, M.; Sun, C.; Lee, D. *Angew. Chem. Int. Ed.* **2015**, *54*, 9963–9966.
67. Lee, S.-S.; Kim, W.-Y.; Lee, H.-Y. *Chem. Asian J.* **2012**, *7*, 2450–2456.
68. Kim, Y.-J.; Yoon, Y.; Lee, H.-Y. *Tetrahedron* **2013**, *69*, 7810–7816.

69. Kang, T.; Song, S. B.; Kim, W.-Y.; Kim, B. G.; Lee, H.-Y. *J. Am. Chem. Soc.* **2014**, *136*, 10274–10276.
70. Geum, S.; Lee, H.-Y. *Org. Lett.* **2014**, *16*, 2466–2469.
71. Schwartz, B. D.; Denton, J. R.; Davies, H. M. L.; Williams, C. M. *Aust. J. Chem.* **2009**, *62*, 980–982.
72. Schwartz, B. D.; Denton, J. R.; Lian, Y.; Davies, H. M. L.; Williams, C. M. *J. Am. Chem. Soc.* **2009**, *131*, 8329–8332.
73. Jackson, K. L.; Henderson, J. A.; Motoyoshi, H.; Phillips, A. J. *Angew. Chem. Int. Ed.* **2009**, *48*, 2346–2350.
74. Xu, J.; Caro-Diaz, E. J. E.; Theodorakis, E. A. *Org. Lett.* **2010**, *12*, 3708–3711.

Chapter 16

Miscellaneous Reactions

In addition to the reactions summarized in the previous chapters, the rich chemistry of diazo compounds has also been demonstrated by a large number of transformations that cannot be categorized in the earlier chapters. Some of these reactions are unique and highly useful in organic synthesis. This chapter will summarize some of these miscellaneous reactions.

16.1 Reduction and oxidation

The diazo moiety may be partially reduced to hydrazine or fully reduced to methylene moiety. In 2009, a three-step sequence involving the deamination of α-aminoesters **1** was reported, which proceeded *via* diazotization, $NaBH_4$-medaited selective reduction to afford α-hydrazonoesters **3** and eventually DBU-promoted reduction along with N_2 extrusion (Scheme 16.1).[1] This sequential approach has wide substrate scope and has been utilized in the synthesis of a series of CysLT1 antagonists.

Nishida and co-workers reported a partial reduction of diazo moiety of **5** for the synthesis of heterocyclic compounds cinnoline **7** and 1-aminoindoline **8**. Thus, the diazo substrate **5**, which was prepared by a one-pot procedure utilizing phase-transfer catalysis, was partially reduced to afford hydrazone **6** (*E*) (Scheme 16.2).[2] The hydrazone **6** (*E*) could be further converted to cinnoline **7** and 1-aminoindoline **8**. Besides, as an alternative approach for deoxygenative reduction of ketones or aldehydes, Wang and co-workers reported an efficient Pd/C-catalyzed hydrogenative reduction of diazo compounds that were generated *in situ* from *N*-tosylhydrazones.[3]

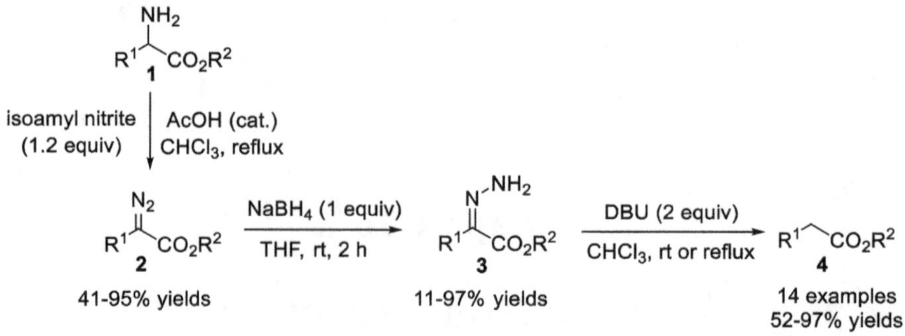

Scheme 16.1 Deamination of α-aminoesters through diazotization and reduction.

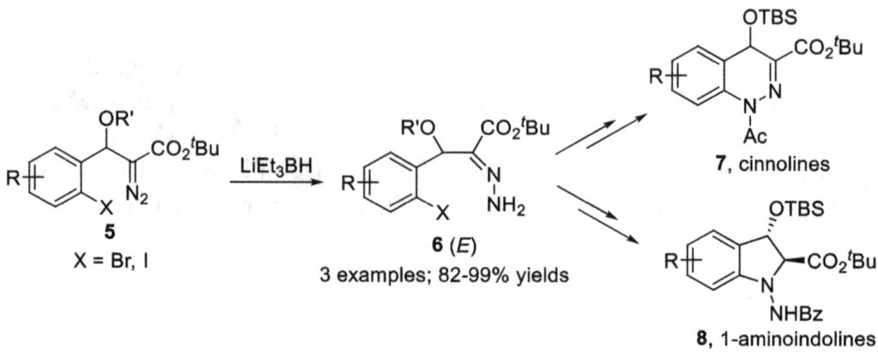

Scheme 16.2 Partial reduction of diazo moiety and its applications in the synthesis of cinnolines and 1-aminoindolines.

On the other hand, the diazo moiety can be readily oxidized to carbonyl group, which may also find applications in organic synthesis.[4,5] For example, Doyle and co-workers reported a novel Zn(II)-catalyzed Mukaiyama–Michael reaction/3,3-dimethyldioxirane (DMDO) oxidation cascade starting from silyl enoldiazoacetates **9** with α,β-unsaturated ketones.[4] The corresponding tetracarbonyl compounds **10** were obtained with high efficiency through the intermediates of functionalized ζ-keto-α-diazo-β-ketoesters. These 2,3,7-triketoesters **10** subsequently underwent highly stereoselective aldol reaction to afford either **11** or **12**, respectively, in the presence of $Yb(OTf)_3$ or Mes-SO_3H (Scheme 16.3).

Moreover, Doyle and co-workers developed asymmetric carbonyl-ene reactions of 2,3-diketoesters **14**, which were obtained by the oxidation of α-diazo-β-ketoesters **13** with *tert*-butyl hypochlorite as key intermediates

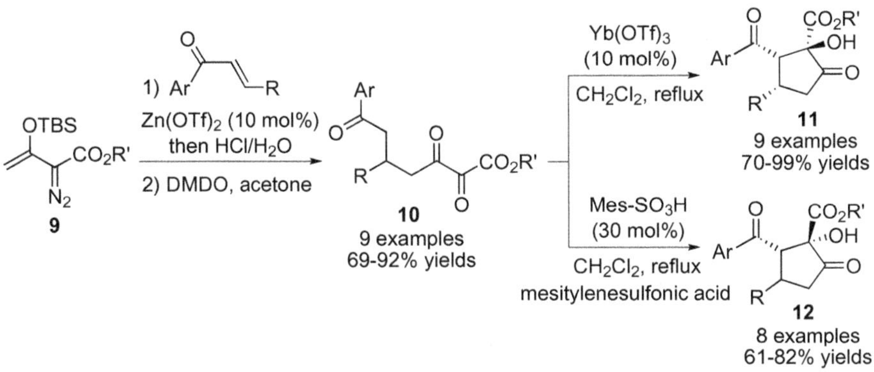

Scheme 16.3 Stereoselective synthesis of highly functionalized cyclopentanones through aldol reaction/diazo oxidation cascade.

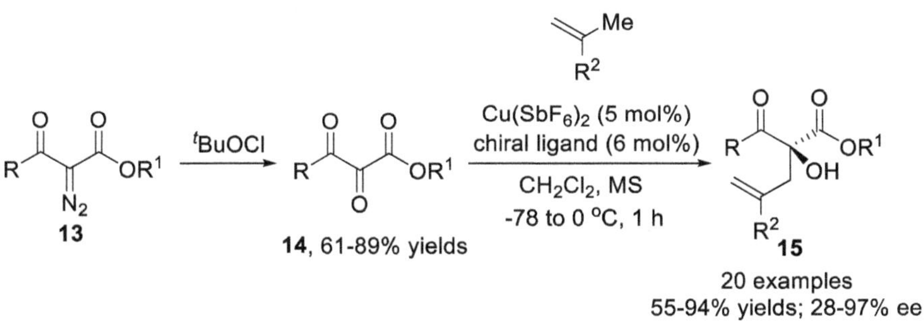

Scheme 16.4 Asymmetric carbonyl-ene reactions of 2,3-diketoesters.

(Scheme 16.4).[5] In the presence of Cu(II) catalysts and chiral bis(oxazoline) ligand, carbonyl-ene reaction afforded the chiral α-functionalized α-hydroxy-β-ketoesters with up to 94% yield and 97% *ee*.

16.2 Fragmentation

The fragmentation of the diazo moiety in a cyclic substrate provides a unique method for organic synthesis. Brewer and co-workers established a series of ring fragmentation transformations of cyclic α-diazocarbonyl compounds, which involved selective C–C bond cleavage process.[6–10] In 2008, they reported a Lewis acid-promoted fragmentation process of γ-silyloxy-β-hydroxy-α-diazoesters **16**, providing an

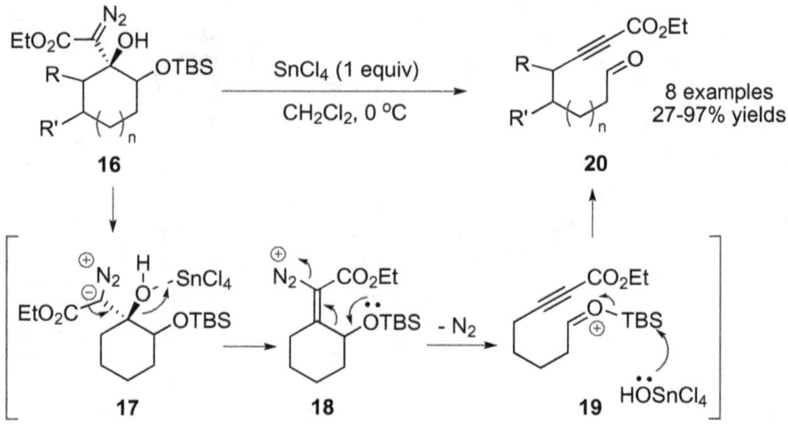

Scheme 16.5 Lewis acid-promoted fragmentation of γ-silyloxy-β-hydroxy-α-diazoesters.

efficient route to access tethered aldehyde ynoates **20** with varying tether lengths under mild conditions (Scheme 16.5).[6,7] This intriguing reaction proceeded *via* acid-mediated generation of vinyl diazonium species 17 followed by β-hydroxyl elimination. Subsequently, the C_β–C_γ bond cleavage and N_2 extrusion to form **19** were followed by the loss of silyl group to afford the tethered aldehyde ynoates **20**.

Moreover, the same group established a method to synthesize polycyclic 2,5-dihydropyrroles from α-silyloxyketones.[8] This three-step sequence involved the nucleophilic addition of EDA to generate γ-silyloxy-β-hydroxy-α-diazoesters **22**, followed by ring fragmentation to give tethered aldehyde ynoate **23** (Scheme 16.6). In the presence of amino acid silylesters, the sequential decarboxylation/intramolecular azomethine ylide 1,3-dipolar cycloaddition occurred, affording the 2,5-dihydropyrrole products **25** in good yields.

This fragmentation reaction was applied in the total synthesis of the steroidal alkaloid demissidine **29**, with epiandrosterone **26** as the starting material (Scheme 16.7).[9] The notable feature of this synthesis was the ring fragmentation that ruptured the D-ring of the diazo ester derivative **27**, generating an aldehyde tethered ynoate product **28** as the key intermediate.

Furthermore, they reported an efficient synthesis of medium-sized ynolides by the Lewis acid-mediated fragmentation of bicyclic

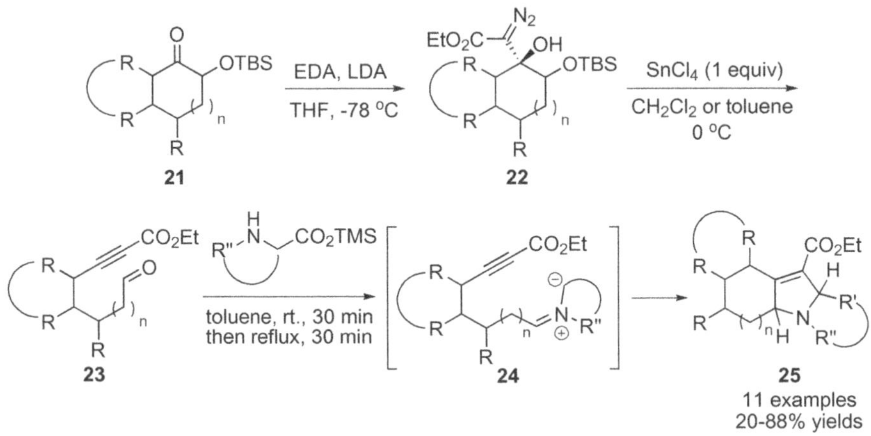

Scheme 16.6 Cascade synthetic route to polycyclic 2,5-dihydropyrroles from α-silyloxyketones.

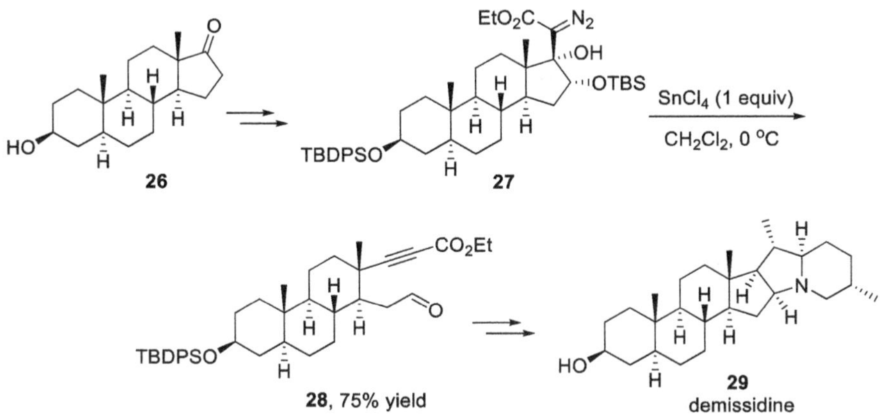

Scheme 16.7 Synthesis of demissidine *via* ring fragmentation and 1,3-dipolar cycloaddition cascade.

γ-silyloxy-β-hydroxy-α-diazolactones (Scheme 16.8).[10] The reaction afforded 11-membered ynolides **31** and **33** with good yields from bicyclo[5.4.0] and −[4.5.0] diazo substrates.

Moreover, the fragmentation of furan derivatives containing diazo moiety was reported by Tu and Cao.[11] Thus, an intriguing approach

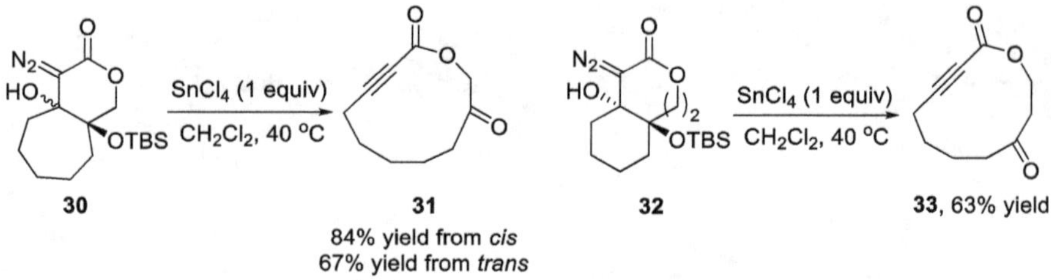

Scheme 16.8 Synthesis of ynolides *via* the fragmentation of bicyclic γ-silyloxy-β-hydroxy-α-diazolactones.

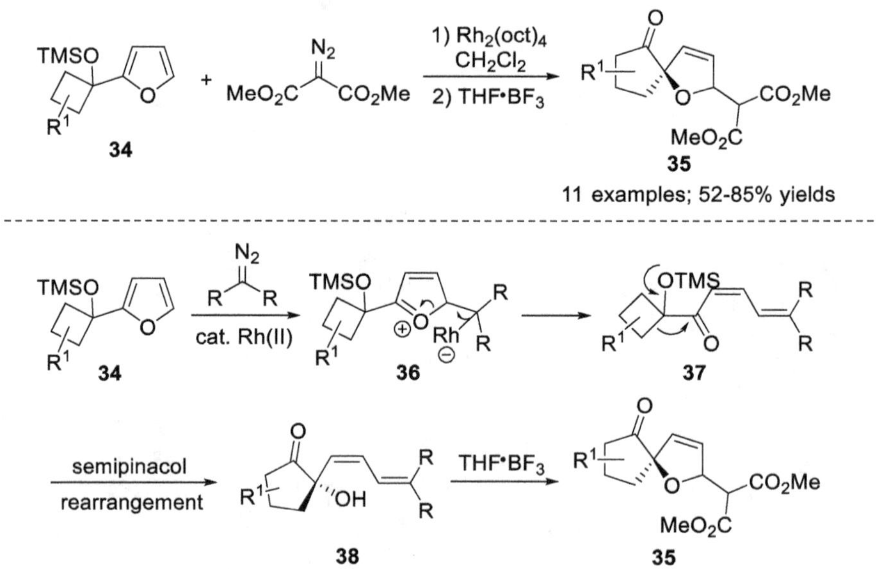

Scheme 16.9 Rh(II)-catalyzed synthesis of spirohydrofurans.

toward spirohydrofurans was realized by tunable tandem transformations from simple furan derivatives with diazo compounds in the presence of Rh(II) catalyst (Scheme 16.9). The proposed reaction mechanism of this tandem process involves the nucleophilic reaction of furan moiety with Rh(II) carbene to generate **36**, followed by ring-opening reaction to afford **37**. Subsequently, the semipinacol rearrangement occurred to give **38**, which was converted to the final spirohydrofuran **35** by the treatment of $Et_2O{\cdot}BF_3$.

16.3 Diazo compounds in chemical biology

Due to its rich reactivity, the diazo moiety was explored as efficient probes or other functional building blocks in chemical biology. For example, with the method of preparing diazo compounds from the corresponding azide by the deimidogenation process,[12,13] Raines and co-workers synthesized Ac_4ManDiaz **39** as a diazo derivative of *N*-acetylmannosamine, which was found to endure metabolism in cellulo and label the surface of a mammalian cell.[14] Another example was *C*-(α-D-glucopyranosyl)-4-cyanophenyldiazomethane **40**, which was stable in neutral or basic media (Figure 16.1).[15] Inhibitory activity evaluation indicated that this compound could serve as an effective irreversible inhibitor against α-glucosidase from *Saccharomyces cerevisiae*.

Moreover, Raines and co-workers developed a mild and efficient esterification method of various carboxylate acids with diazo compounds in an aqueous media near neutral pH.[16] The method was applied to proteins of bovine pancreatic ribonuclease. The esters in the ensuing protein turned out to be hydrolyzed by human cellular esterases to regenerate protein carboxyl groups. Thus, the *O*-alkylation strategy was expected to broaden the utility of antibodies, enzymes, hormones and other proteins in biochemical research.

On the other hand, the carbene transfer reactions have been explored as probes by Arnold and co-workers for testing the catalytic reactivity of engineered enzymes developed by directed evaluation. For example, they demonstrated that engineered variants of cytochrome $P450_{BM3}$ could serve as the catalyst for highly diastereo- and enantioselective cyclopropanations of functionalized styrenes with α-diazoesters (Eq. (16.1)).[17] In 2014, Arnold and co-workers further identified a highly active and

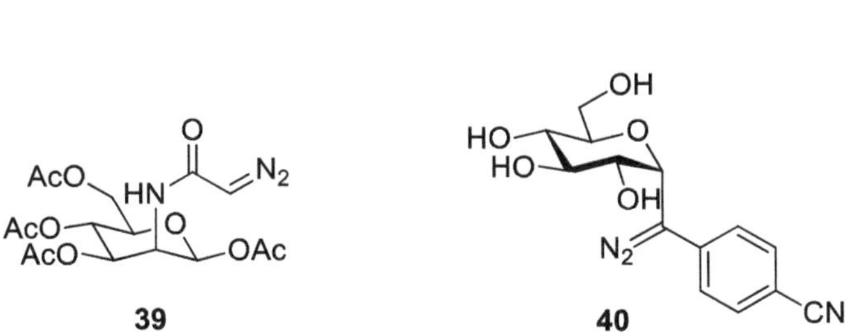

39 Ac4ManDiaz

40

Figure 16.1 Ac_4ManDiaz **39** and **40**.

enantioselective *in vivo* cyclopropanation catalyst through mutating the amino acid at the axial site of $P450_{BM3}$.[18]

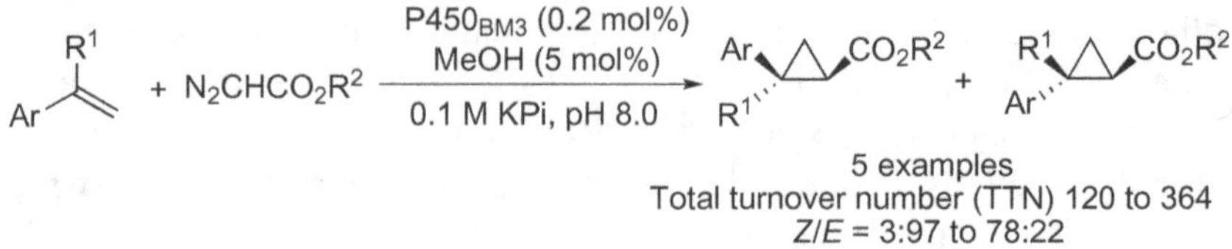

(16.1)

On the other hand, classic metal carbene reactions have been explored in the modification of peptides or proteins. For example, Francis and co-workers realized the significant improvement of tryptophan modification with Rh(II)-catalyzed carbene insertions.[19] With *tert*-BuNHOH as the promoter, this process proceeded under mild conditions in aqueous solution, allowing the selective modification of tryptophan residues in peptides and proteins at pH 6. The solvent accessibility of the indole side chains turned out to be a key factor in determining the outcome of labeling experiments with Rh(II) carbene. Besides, progress toward developing a general protein labeling strategy bearing solvent accessible tryptophan residues followed by chemoselective modification with Rh(II) carbene was also demonstrated.

In 2011, Ball and co-workers disclosed the remarkable side-chain modification on the basis of molecular shape, which was enabled by a novel coiled-coil assembly of substrate peptides with dirhodium metallopeptide catalysts ($K3_{a,e}Rh_2$) (Scheme 16.10).[20] This modification method was suitable for a wide range of substrates, including various amino acids and carboxamide derivatives. It also allowed the achievement of covalent

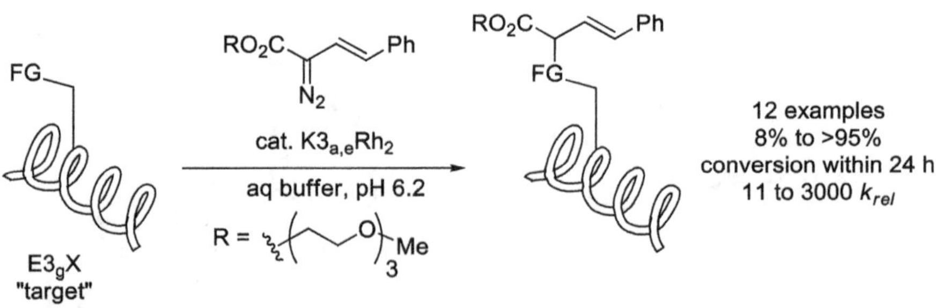

Scheme 16.10 Covalent modification of $E3_gX$ peptides with vinyl diazoacetates.

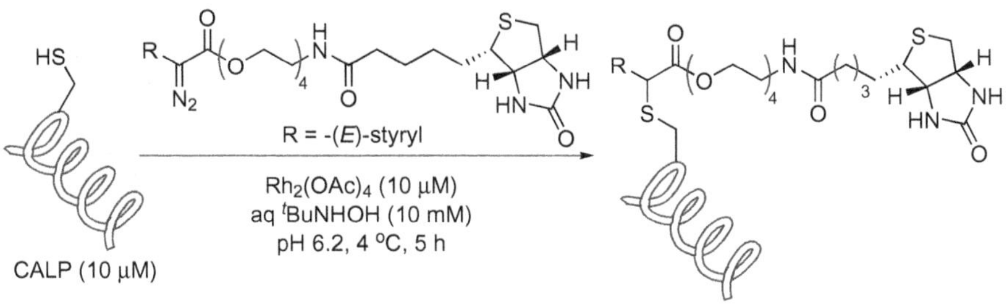

Scheme 16.11 Rh(II)-catalyzed cysteine modification of CALP.

structure-dependent modification of the c-FosbZip domain at different residues.

Moreover, the proximity-driven modification of peptide and protein substrates catalyzed by rhodium metallopeptides was also developed.[21] The newly designed metallopeptide catalysts were found to be inhibited in a sequence-specific fashion, mimicking the inhibition and regulation of natural enzymes.

In 2013, the cysteine modification with biotinylated diazo compound involving metallo-carbene species was realized with simple Rh(II) catalyst (Scheme 16.11).[22] The key feature of this process was the high cysteine-selectivity under mild conditions.

16.4 Diazo compounds in natural products

While the diazo moieties are relatively unstable, they also occur naturally in some cases. The notable examples are diazobenzoquinone and diazonaphthoquinone derivatives. Roughly half of these natural products belong to the diazofluorene (kinamycin) family.[23] Recently, the synthesis of various natural products containing diazo groups attracted significant attention. In this section, we will briefly cover the synthesis of natural products and relative derivatives bearing diazo groups.

In 2008, Moody and co-workers completed the first synthesis of naturally occurring Cremeomycin **43** (Scheme 16.12).[24] Starting with **41**, the synthesis involved nitration, hydrolysis, nitro group reduction, affording **42** as the key precursor. The final diazotization afforded cremeomycin **43** with identical spectroscopic properties to the natural product.

In 2010, Herzon and co-workers reported an enantioselective total synthesis of (−)-kinamycin F **48** in 12 linear steps with TIPS-protected

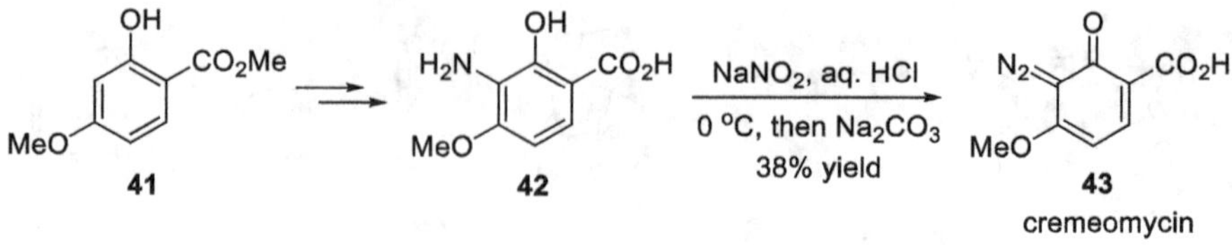

Scheme 16.12 Synthesis of cremeomycin.

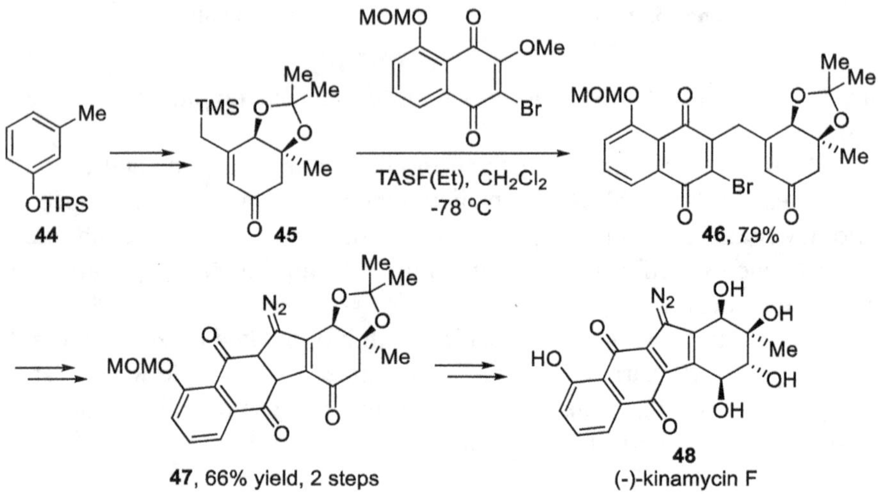

Scheme 16.13 Enantioselective total synthesis of kinamycin F.

m-cresol **44** (Scheme 16.13).[25] With **44** as the starting material, the synthesis involved Birch reduction, Sharpless asymmetric dihydroxylation and other transformations to give the β-(trimethyl-silylmethyl)-cyclohexenone **45** efficiently. Subsequently, γ-alkylation product **46** was obtained through 1,4-addition-elimination process. Intramolecular Heck and diazotization allowed the installation of the diazo component into benzofluorene ring scaffold of **47**, which was further elaborated to the target kinamycin F **48** *via* a three-step sequence.

The lomaiviticins, a family of cytotoxic marine natural products, have attracted great attention from both synthetic and biological chemists because of their intricate molecular scaffolds and potent biological activity. In 2011, Herzon and co-workers reported the first total synthesis of (−)-lomaiviticin aglycon **49**, which represented a complex natural product

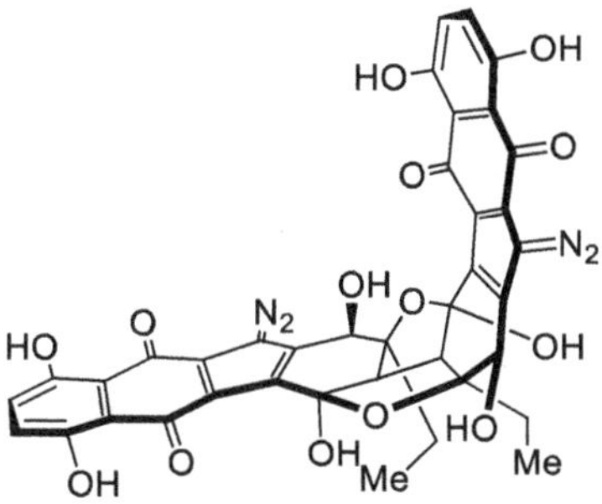

49
(-)-lomaiviticin aglycon

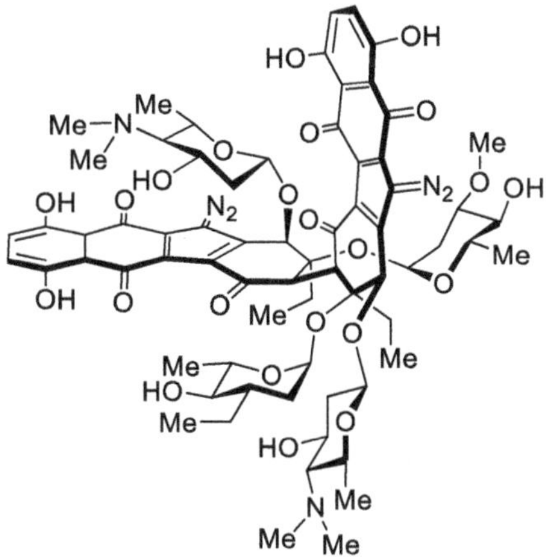

50
(-)-lomaiviticin A

Figure 16.2 (−)-Lomaiviticin aglycon and (−)-lomaiviticin A.

containing a diazotetrahydrobenzo[*b*]fluorene skeleton.[26,28] Moreover, the enantioselective synthesis of (−)-kinamycin F **49** and (−)-lomaiviticin aglycon **50** was completed (Figure 16.2).[27]

Furthermore, Herzon and co-workers investigated the DNA cleavage with (−)-lomaiviticin A. (−)-Lomaiviticin A inhibits the growth of cultured human cancer cells at nanomolar-picomolar concentrations.[29] As a result, the plasmid DNA cleavage pathway was independent of reactive oxygen species and iron, and the potent cytotoxicity of (−)-lomaiviticin A arose from the induction of DNA double-strand breaks. Mechanically, vinyl radical species was found to mediate this cleavage activity. Both diazofluorenes of (−)-lomaiviticin A contribute to strand cleavage.

Porco and co-workers achieved the asymmetric total synthesis of epoxykinamycin FL-120B′ **59** from readily available substrates (Scheme 16.14).[30] The synthesis involved Stille coupling reaction between functionalized arylstannanes **54** and chiral brominated epoxyketone fragment **52**, which was generated through Sharpless asymmetric epoxidation. Another key step of the synthesis was an intramolecular Friedel–Crafts acylation reaction of **56**, resulting in the construction of the desired five-membered ring ketone skeleton of the cyclization product.

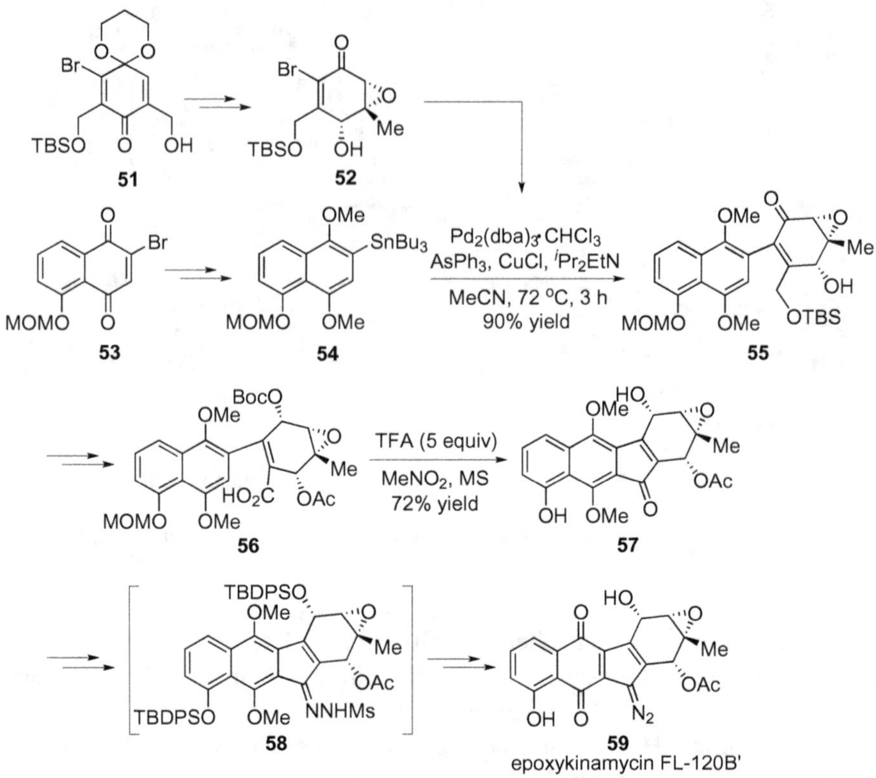

Scheme 16.14 Enantioselective total synthesis of epoxykinamycin FL-120B′.

16.5 Reactions of diazo compounds through radical process

Radical species may be generated from diazo compounds, and subsequently radical-type transformations may follow. In 2015, Wan and co-workers reported an intriguing Co-catalyzed multicomponent reaction of styrenes, diazocarbonyl compounds and tertiary amines in the presence of TBHP, affording highly functionalized β-ester-γ-amino ketones **60** (Scheme 16.15).[31] This tandem process incorporates carbene, radical and ionic transformations, constructing two C–C bonds and one C–O bond in one-pot procedure. Mechanistically, the reaction was assumed to take place *via* the interception of a cobalt-based carbene radical species **62** with the α-aminoalkyl radical **61** to generate a peroxide intermediate **63**, followed by the Kornblum–DeLaMare reaction to deliver the final product **60**.

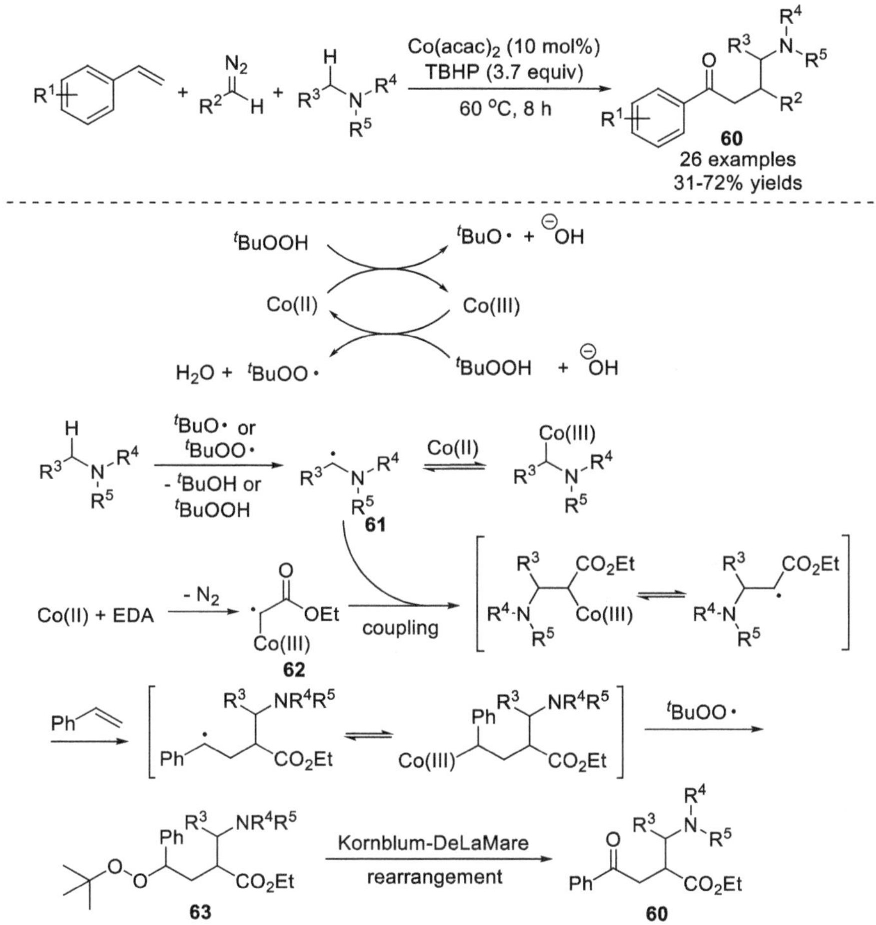

Scheme 16.15 Co-catalyzed carbene radical-type reaction to provide *β*-ester-*γ*-amino ketones.

16.6 Reaction with fullerene

The diazo compounds have been explored in the diverse modification of fullerenes. The formal carbene addition process proceeds *via* 1,3-dipolar [3+2] cycloaddition to generate pyrazoline intermediate **64**, followed by nitrogen extrusion to give a regioisomeric mixture of [6,6] closed methanofullerene **65** and [5,6] open fulleroids **66** and **67** (Scheme 16.16). Since this area has been summarized by some elegant reviews,[32–34] herein a very brief introduction and summary of the topic is presented.

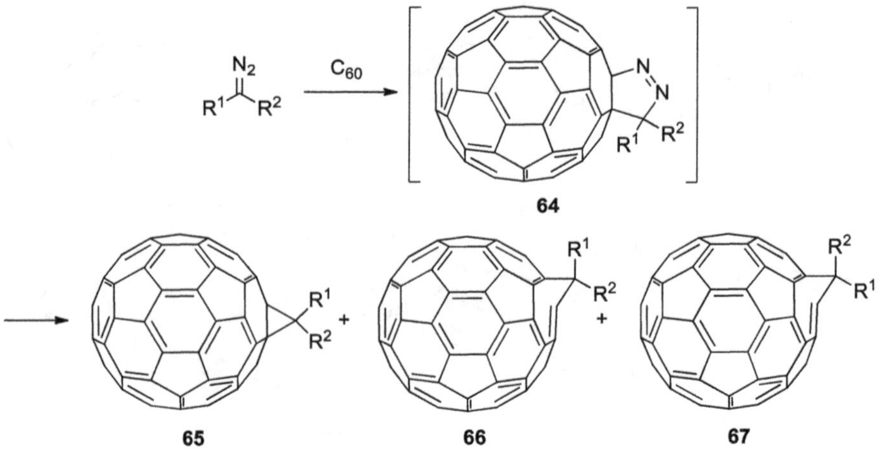

Scheme 16.16 General mode of the reaction of fullerene with diazo compounds.

A series of modified fullerenes have been prepared by thermal and catalytic reactions with various diazo compounds. Tuktarov and co-workers reported systemically their studies on [2+1] cycloaddition of fullerene with the employment of three-component catalytic system of $Pd(acac)_2$-PPh_3-Et_3Al or simple palladium catalysis.[35–50]

In addition, a series of covalently derivatized fullerenes with diazo compounds were subjected to the investigation of their electrochemical/ light absorption/photovoltaic properties, and other relevant properties. In particular, [6,6]-phenyl-C61-butyric acid methyl ester (PCBM) **68** and their analogues have emerged as important electron acceptors in organic thin-film photovoltaic cells (Figure 16.3). A wide range of diverse PCBM including bis-cycloadducts,[51] fluorene-containing PCBM,[52] 56 π-electron methano-$PC_{61}BM$ with 1,2-dihydromethano (CH_2) group,[53] and dumb-bell-shapeddimeric C_{60} derivatives[54] were synthesized from the corresponding *N*-tosylhydrazones in the presence of base or *via* one-pot procedure under aerobic conditions.[55] Recently, both the mono-adducts of $PC_{61}BM$/$PC_{71}BM$ and the bis-adduct analogues were successfully prepared by microwave-assisted one-step procedure.[56] The aqueous two-phase system was also developed for the facile synthesis of $PC_{61}BM$ or $PC_{71}BM$ under photoirradiation conditions.[57]

More recently, a Ag-mediated three-component annulation of [60] fullerene with *N*-tosylhydrazone and nitrile was developed, enabling the

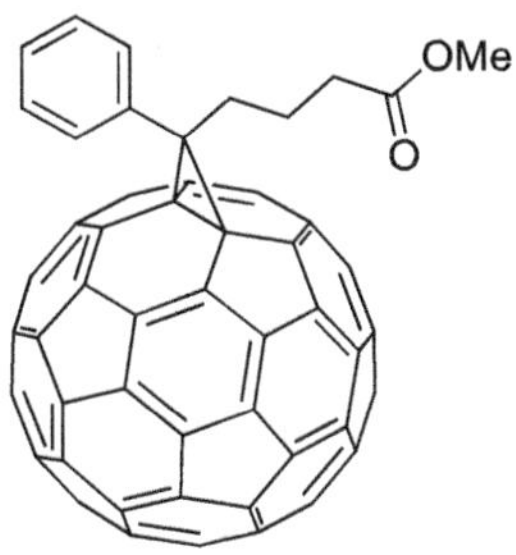

68, PCBM

Figure 16.3 [6,6]-phenyl-C61-butyric acid methyl ester (PCBM).

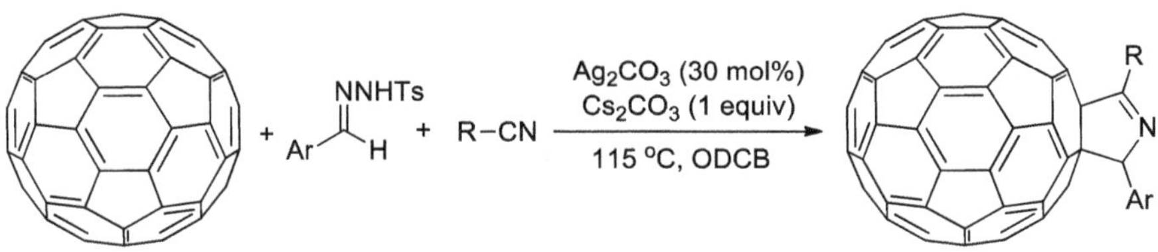

69
34 examples; 14-53% yields

Scheme 16.17 Ag(I)-mediated three-component annulation of fullerene, sulfonylhydrazones and nitriles.

synthesis of diverse disubstituted [60]fullerene-fused dihydropyrrole derivatives (Scheme 16.17).[58] Mechanistically, the reaction may follow a ylide-type [3+2] cycloaddition pathway.

Endofullerene derivatives, especially metallofullerenes, demonstrated superior activity in photo-induced charge transfer and other photovoltaic characteristics. Thus, chemists have utilized *N*-tosylhydrazones to prepare and characterize a wide variety of PCBM-like endohedral clusters, including $Ce_2@I_h\text{-}C_{80}$,[59] $La_2@I_h\text{-}C_{80}$-ZnPor and $Sc_3N@I_h\text{-}C_{80}$-ZnPor,[60] porphyrin linkedLa@C_{82} hybrids($M_{1\text{-}3}$Por)[61] and lithium ion-encapsulated fullerene of $[Li_+@PCBM]PF_6^-$.[62] Besides, [5,6] methano adduct of $Sc_3N@I_h\text{-}C_{80}$ was prepared directly from diaryl diazo compounds under mild reaction conditions.[63]

To summarize this section, the reaction of diazo compounds with C_{60} provides facile methods for accessing modified fullerenes, as well as endohedral and metallofullerenes. Owing to the potential importance in photovoltaics and materials research of these compounds, diazo compounds and other related carbene precursors will find further applications in the modification of fullerenes.

16.7 Reaction with CO and isocyanides

Carbonylation of metal carbene species, which is generated from transition-metal-mediated reaction of diazo compounds, provides a concise and efficient approach for ketenes.[64] The high reactivity of ketene intermediates allows the sequential cycloaddition or nucleophilic addition to occur, resulting in unique transformations in organic synthesis.

Ungváry and co-workers systemically studied the cobalt-mediated carbonylation of EDA with CO and the subsequent transformation.[65–70] In 2011, Wang and co-workers reported a Pd(0)-catalyzed carbonylation of α-diazocarbonyl compounds or *N*-tosylhydrazone salts under atmospheric pressure of CO.[71] The palladium-complexed ketene intermediate **70** could be trapped by nucleophiles (alcohols and amines), or imines. This reaction provided an efficient approach for the synthesis of esters and amides **71**, β-lactams **72** and 1,3-dioxin-4-one derivatives **73** (Scheme 16.18). Moreover, carbonylation of diazo compounds or *in situ* generated diazo species was also achieved in the presence of Ni(0)[72] or molybdenum hexacarbonyl[73] for amino-carbonylation.

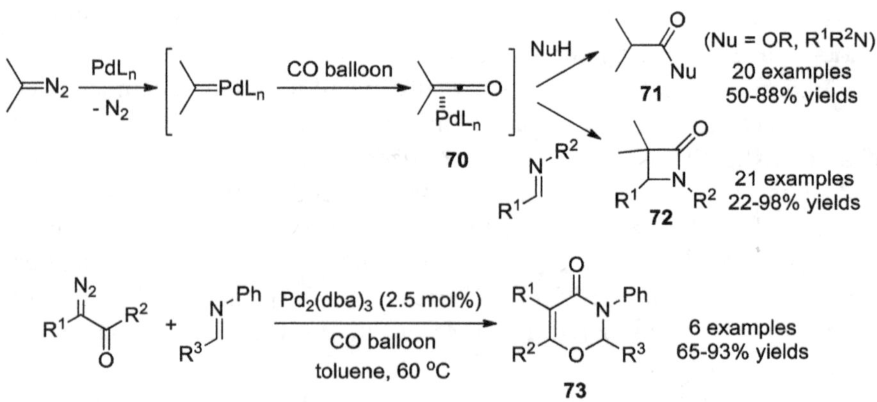

Scheme 16.18 Pd-catalyzed carbonylation of diazo compounds and the sequential nucleophilic addition, [2+2] or [4+2] cycloaddition.

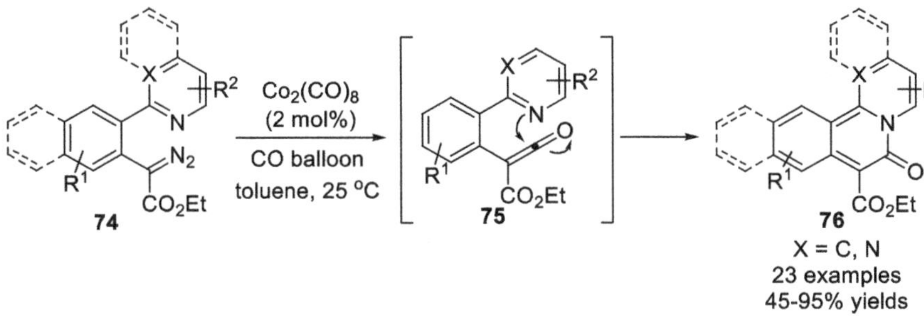

Scheme 16.19 Co-catalyzed carbonylative cyclization of pyridinyl diazoacetates.

Zhang and de Bruin reported a novel radical-type Co(II)Por-catalyzed carbene carbonylation with sequential ketene trapping process by diverse nucleophiles and imines, affording esters, amides or *trans*-β-lactams in high yields with one-pot cascade procedure.[74] The metalloradical [Co(II)TPP] complex has thus emerged as an efficient catalyst for the homologation of both α-diazocarbonyl compounds and *N*-tosylhydrazones.

Furthermore, de Bruin and co-workers reported the ketene and ketene imine synthesis through Rh(II)-catalyzed carbonylation of metal carbenes in the presence of PNN pincer complexes.[75] DFT calculations indicated that diazo activation to generate the rhodium-carbene species was the rate-determining step.

More recently, a carbonylative intramolecular cyclization of pyridinyl diazoacetates **74** was achieved with dicobalt octacarbonyl catalyst under CO atmosphere (Scheme 16.19).[76] A semi-one-pot reaction from pyridinyl aryl acetates was also devised to afford the functionalized pyridoisoquinolinone derivatives **76**. The reaction was initiated by Co-catalyzed carbonylation to generate ketene intermediate **75**, followed by sequential intramolecular cyclization with the pyridinyl moiety.

In addition to the catalytic carbonylation of diazo compounds, Pd-catalyzed homologation of diazo component with isocyanide **77**, affording ketenimine **78** as the key intermediate, was explored by Cai and co-workers (Scheme 16.20).[77,78] They achieved the Pd(0)-catalyzed aminocarbonylation of *N*-tosylhydrazones in the presence of base.[77] The reaction was initiated by the generation of Pd-carbene species, followed by reaction with isocyanide to form ketenimine **78** with sequential nucleophilic addition of water to finally afford the amides.

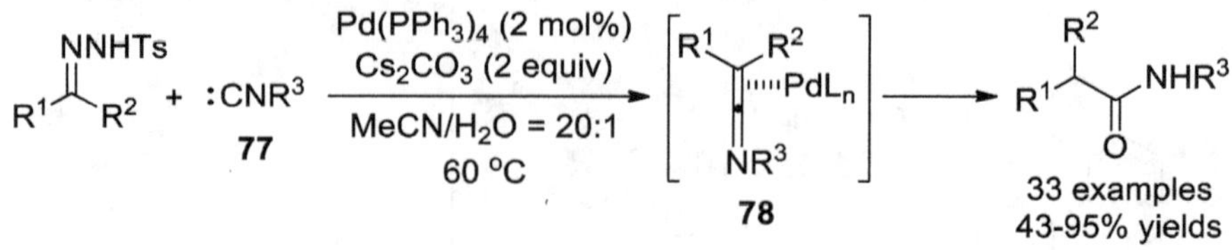

Scheme 16.20 Pd-catalyzed transformation of diazo compounds with isocyanides.

In summary, the selected examples of the transformations of diazo compounds that have not been categorized in the previous chapters further demonstrate the rich chemistry of diazo compounds. It can be expected that more intriguing applications of diazo compounds will emerge in the following years, particularly in the fields of materials science and chemical biology.

References

1. González, A.; Pérez, D.; Puig, C.; Ryder, H.; Sanahuja, J.; Solé, L.; Bach, J. *Tetrahedron Lett.* **2009**, *50*, 2750–2753.
2. Hasegawa, K.; Kimura, N.; Arai, S.; Nishida, A. *J. Org. Chem.* **2008**, *73*, 6363–6368.
3. Zhou, L.; Liu, Z.; Liu, Y.; Zhang, Y.; Wang, J. *Tetrahedron* **2013**, *69*, 6083–6087.
4. Truong, P.; Shanahan, C. S.; Doyle, M. P. *Org. Lett.* **2012**, *14*, 3608–3611.
5. Truong, P. M.; Zavalij, P. Y.; Doyle, M. P. *Angew. Chem. Int. Ed.* **2014**, *53*, 6468–6472.
6. Draghici, C.; Brewer, M. *J. Am. Chem. Soc.* **2008**, *130*, 3766–3767.
7. Jabre, N. D.; Brewer, M. *J. Org. Chem.* **2012**, *77*, 9910–9914.
8. Draghici, C.; Huang, Q.; Brewer, M. *J. Org. Chem.* **2009**, *74*, 8410–8413.
9. Zhang, Z.; Giampa, G. M.; Draghici, C.; Huang, Q.; Brewer, M. *Org. Lett.* **2013**, *15*, 2100–2103.
10. Bayir, A.; Brewer, M. *J. Org. Chem.* **2014**, *79*, 6037–6046.
11. Li, B.-S.; Liu, W.-X.; Zhang, Q.-W.; Wang, S.-H.; Zhang, F.-M.; Zhang, S.-Y.; Tu, Y.-Q.; Cao, X.-P. *Chem. Eur. J.* **2013**, *19*, 5246–5249.
12. Myers, E. L.; Raines, R. T. *Angew. Chem. Int. Ed.* **2009**, *48*, 2359–2363.
13. Chou, H.-H.; Raines, R. T. *J. Am. Chem. Soc.* **2013**, *135*, 14936–14939.
14. Andersen, K. A.; Aronoff, M. R.; McGrath, N. A.; Raines, R. T. *J. Am. Chem. Soc.* **2015**, *137*, 2412–2415.
15. Geng, Y.; Kumar, A.; Faidallah, H. M.; Albar, H. A.; Mhkalid, I. A.; Schmidt, R. R. *Bioorg. Med. Chem.* **2013**, *21*, 4793–4802.
16. McGrath, N. A.; Andersen, K. A.; Davis, A. K. F.; Lomax, J. E.; Raines, R. T. *Chem. Sci.* **2015**, *6*, 752–755.

17. Coelho, P. S.; Brustad, E. M.; Kannan, A.; Arnold, F. H. *Science* **2013**, *339*, 307–310.
18. Wang, Z. J.; Renata, H.; Peck, N. E.; Farwell, C. C.; Coelho, P. S.; Arnold, F. H. *Angew. Chem. Int. Ed.* **2014**, *53*, 6810–6813.
19. Antos, J. M.; McFarland, J. M.; Iavarone, A. T.; Francis, M. B. *J. Am. Chem. Soc.* **2009**, *131*, 6301–6308.
20. Popp, B. V.; Ball, Z. T. *Chem. Sci.* **2011**, *2*, 690–695.
21. Popp, B. V.; Chen, Z.; Ball, Z. T. *Chem. Commun.* **2012**, *48*, 7492–7494.
22. Kundu, R.; Ball, Z. T. *Chem. Commun.* **2013**, *49*, 4166–4168.
23. Nawrat, C. C.; Moody, C. J. *Nat. Prod. Rep.* **2011**, *28*, 1426–1444.
24. Varley, L. M.; Moody, C. J. *Synthesis* **2008**, 3601–3604.
25. Woo, C. M.; Lu, L.; Gholap, S. L.; Smith, D. R.; Herzon, S. B. *J. Am. Chem. Soc.* **2010**, *132*, 2540–2541.
26. Herzon, S. B.;Lu, L.; Woo, C. M.; Gholap, S. L. *J. Am. Chem. Soc.* **2011**, *133*, 7260–7263.
27. Woo, C. M.; Gholap, S. L.; Lu, L.; Kaneko, M.; Li, Z.; Ravikumar, P. C.; Herzon, S. B. *J. Am. Chem. Soc.* **2012**, *134*, 17262–17273.
28. Herzon, S. B. *Synlett* **2011**, 2105–2114.
29. Colis, L. C.; Woo, C. M.; Hegan, D. C.; Li, Z.; Glazer, P. M.; Herzon, S. B. *Nat. Chem.* **2014**, *6*, 504–510.
30. Scully, S. S.; Porco, J. A., Jr. *Angew. Chem. Int. Ed.* **2011**, *50*, 9722–9726.
31. Zhang, J.; Jiang, J.; Xu, D.; Luo, Q.; Wang, H.; Chen, J.; Li, H.; Wang, Y.; Wan, X. *Angew. Chem. Int. Ed.* **2015**, *54*, 1231–1235.
32. Yamada, M.; Akasaka, T.; Nagase, S. *Chem. Rev.* **2013**, *113*, 7209–7264.
33. Tuktarov, A. R.; Dzhemilev, U. M. *Russ. Chem. Rev.* **2010**, *79*, 585–610.
34. Lu, X.; Bao, L.; Akasaka, T.; Nagase, S. *Chem. Commun.* **2014**, *50*, 14701–14715.
35. Tuktarov, A. R.; Akhmetov, A. R.; Kamalov, R. F.; Khalilov, L. M.; Pudas, M.; Ibragimov, A. G.; Dzhemilev, U. M. *Russ. J. Org. Chem.* **2009**, *45*, 1168–1174.
36. Tuktarov, A. R.; Akhmetov, A. R.; Sabirov, D. Sh.; Khalilov, L. M.; Ibragimov, A. G.; Dzhemilev, U. M. *Russ. Chem. Bull. Int. Ed.* **2009**, *58*, 1724–1730.
37. Tuktarov, A. R.; Korolev, V. V.; Khalilov, L. M; Ibragimov, A. G.; Dzhemilev, Yu. M. *Russ. J. Org. Chem.* **2009**, *45*, 1594–1597.
38. Tuktarov, A. R.; Korolev, V. V.; Dzhemilev, U. M. *Russ. J. Org. Chem.* **2010**, *46*, 588–589.
39. Tuktarov, A. R.; Korolev, V. V.; Tulyabaev, A. R.; Yanybin, V. M.; Khalilov, L. M.; Dzhemilev, U. M. *Russ. Chem. Bull. Int. Ed.* **2010**, *59*, 977–983.
40. Dzhemilev, U. M.; Tuktarov, A. R.; Korolev, V. V.; Khalilov, L. M. *Petrol. Chem.* **2011**, *51*, 123–127.
41. Tuktarov, A. R.; Korolev, V. V.; Sabirov, D. Sh.; Dzhemilev, U. M. *Russ. J. Org. Chem.* **2011**, *47*, 41–47.

42. Tuktarov, A. R.; Khuzin, A. A.; Korolev, V. V.; Dzhemilev, U. M. *Russ. J. Org. Chem.* **2012**, *48*, 99–103.
43. Tuktarov, A. R.; Akhmetov, A. R.; Khalilov, L. M.; Dzhemilev, U. M. *Russ. Chem. Bull. Int. Ed.* **2010**, *59*, 611–614.
44. Tuktarov, A. R.; Khuzin, A. A.; Popod'ko, N. R.; Dzhemilev, U. M. *Tetrahedron Lett.* **2012**, *53*, 3123–3125.
45. Tuktarov, A. R.; Khuzina, L. L.; Popod'ko, N. R.; Dzhemilev, U. M. *Tetrahedron Lett.* **2013**, *54*, 2146–2148.
46. Tuktarov, A. R.; Korolev, V. V.; Tulyabaev, A. R.; Popod'ko, N. R.; Khalilov, L. M.; Dzhemilev, U. M. *Tetrahedron Lett.* **2011**, *52*, 834–836.
47. Khalilov, L. M.; Tulyabaev, A. R.; Tuktarov, A. R. *Magn. Reson. Chem.* **2011**, *49*, 768–774.
48. Tuktarov, A. R.; Akhmetov, A. R.; Kirichenko, G. N.; Glazunova, V. I.; Khalilov, L. M.; Dzhemilev, U. M. *Russ. J. Appl. Chem.* **2010**, *83*, 1238–1242.
49. Tuktarov, A. R.; Khuzina, L. L.; Dzhemilev, U. M. *Russ. Chem. Bull. Int. Ed.* **2011**, *60*, 662–666.
50. Bulgakov, R. G.; Gazeeva, D. R.; Tuktarov, A. R.; Khuzina, L. L.; Galimov, D. I.; Dzhemilev, U. M. *Russ. Chem. Bull. Int. Ed.* **2013**, *62*, 2389–2393.
51. Bouwer, R. K. M.; Hummelen, J. C. *Chem. Eur. J.* **2010**, *16*, 11250–11253.
52. Kuhlmann, J.-C.; de Bruyn, P.; Bouwer, R. K. M.; Meetsma, A.; Blom, P. W. M.; Hummelen, J. C. *Chem. Commun.* **2010**, *46*, 7232–7234.
53. Li, C.-Z.; Chien, S.-C.; Yip, H.-L.; Chueh, C.-C.; Chen, F.-C.; Matsuo, Y.; Nakamura, E.; Jen, A. K.-Y. *Chem. Commun.* **2011**, *47*, 10082–10084.
54. Morinaka, Y.; Nobori, M.; Murata, M.; Wakamiya, A.; Sagawa, T.; Yoshikawa, S.; Murata, Y. *Chem. Commun.* **2013**, *49*, 3670–3672.
55. Kumar, R.; Naqvi, S.; Gupta, N.; Chand, S. *RSC Adv.* **2014**, *4*, 15675–15677.
56. Hadad, C.; Syrgiannis, Z.; Bonasera, A.; Prato, M. *Eur. J. Org. Chem.* **2015**, 1423–1427.
57. Iwai, T.; Matsumoto, F.; Hida, K.; Moriwaki, K.; Takao, Y.; Ito, T.; Mizuno, T.; Ohno, T. *Synlett* **2015**, *26*, 960–964.
58. Chao, D.; Liu, T.-X.; Ma, N.; Zhang, P.; Fu, Z.; Ma, J.; Liu, Q.; Zhang, F.; Zhang, Z.; Zhang, G. *Chem. Commun.* **2016**, *52*, 982–985.
59. Guldi, D. M.; Feng, L.; Radhakrishnan, S. G.; Nikawa, H.; Yamada, M.; Mizorogi, N.; Tsuchiya, T.; Akasaka, T.; Nagase, S.; Herranz, M. Á.; Martín, N. *J. Am. Chem. Soc.* **2010**, *132*, 9078–9086.
60. Feng, L.; Radhakrishnan, S. G.; Mizorogi, N.; Slanina, Z.; Nikawa, H.; Tsuchiya, T.; Akasaka, T.; Nagase, S.; Martín, N.; Guldi, D. M. *J. Am. Chem. Soc.* **2011**, *133*, 7608–7618.
61. Feng, L.; Slanina, Z.; Sato, S.; Yoza, K.; Tsuchiya, T.; Mizorogi, N.; Akasaka, T.; Nagase, S.; Martín, N.; Guldi, D. M. *Angew. Chem. Int. Ed.* **2011**, *50*, 5909–5912.

62. Matsuo, Y.; Okada, H.; Maruyama, M.; Sato, H.; Tobita, H.; Ono, Y.; Omote, K.; Kawachi, K.; Kasama, Y. *Org. Lett.* **2012**, *14*, 3784–3787.
63. Izquierdo, M.; Cerón, M. R.; Olmstead, M. M.; Balch, A. L.; Echegoyen, L. *Angew. Chem. Int. Ed.* **2013**, *52*, 11826–11830.
64. Zhang, Z.; Zhang, Y.; Wang, J. *ACS Catal.* **2011**, *1*, 1621–1630.
65. Ungvári, N.; Fördős, E.; Kégl, T.; Ungváry, F. *Inorg. Chim. Acta* **2009**, *362*, 1333–1342.
66. Ungvári, N.; Fördős, E.; Kégl, T.; Ungváry, F. *Inorg. Chim. Acta* **2010**, *363*, 2016–2028.
67. Balogh, J.; Kégl, T.; Ungváry, F.; Skoda-Földes, R. *Tetrahedron Lett.* **2009**, *50*, 4727–4730.
68. Balogh, J.; Kégl, T.; Párkányi, L.; Kollár, L.; Ungváry, F.; Skoda-Földes, R. *J. Organomet. Chem.* **2011**, *696*, 1394–1403.
69. Balogh, J.; Csók, Z.; Párkányi, L.; Ungváry, F.; Kollár, L.; Skoda-Földes, R. *J. Organomet. Chem.* **2012**, *718*, 131–138.
70. Fördős, E.; Tuba, R.; Párkányi, L.; Kégl, T.; Ungváry, F. *Eur. J. Org. Chem.* **2009**, 1994–2002.
71. Zhang, Z.; Liu, Y.; Ling, L.; Li, Y.; Dong, Y.; Gong, M.; Zhao, X.; Zhang, Y.; Wang, J. *J. Am. Chem. Soc.* **2011**, *133*, 4330–4341.
72. Barcs, B.; Kollár, L.; Kégl, T. *Organometallics* **2012**, *31*, 8082–8097.
73. Rao, K. P.; Basak, A. K.; Raju, A.; Patil, V. S.; Reddy, K. *Tetrahedron Lett.* **2013**, *54*, 5510–5513.
74. Paul, N. D.; Chirila, A.; Lu, H.; Zhang, X. P.; de Bruin, B. *Chem. Eur. J.* **2013**, *19*, 12953–12958.
75. Tang, Z.; Mandal, S.; Paul, N. D.; Lutz, M.; Li, P.; van der Vlugt, J. I.; de Bruin, B. *Org. Chem. Front.* **2015**, *2*, 1561–1577.
76. Baek, Y.; Kim, S.; Jeon, B.; Lee, P. H. *Org. Lett.* **2016**, *18*, 104–107.
77. Zhou, F.; Ding, K.; Cai, Q. *Chem. Eur. J.* **2011**, *17*, 12268–12271.
78. Yan, X.; Liao, J.; Lu, Y.; Liu, J.; Zeng, Y.; Cai, Q. *Org. Lett.* **2013**, *15*, 2478–2481.

Index

www.ingramcontent.com/pod-product-compliance
Lightning Source LLC
LaVergne TN
LVHW010339220926
848484LV00004BB/934